Introduction to Electric Circuit Analysis

Merrill's International Series in Electrical and Electronics Technology

BATESON	*Introduction to Control System Technology*, 2nd Edition, 8255–2
BOYLESTAD	*Introductory Circuit Analysis*, 4th Edition, 9938–2
	Student Guide to Accompany Introductory Circuit Analysis, 4th Edition, 9856–4
BOYLESTAD/KOUSOUROU	*Experiments in Circuit Analysis*, 4th Edition, 9858–0
NASHELSKY/BOYLESTAD	*BASIC Applied to Circuit Analysis*, 20161–6
BREY	*Microprocessor/Hardware Interfacing and Applications*, 20158–6
FLOYD	*Digital Fundamentals*, 2nd Edition, 9876–9
	Electronic Devices, 20157–8
	Essentials of Electronic Devices, 20062–8
	Principles of Electric Circuits, 8081–9
	Electric Circuits, Electron Flow Version, 20037–7
STANLEY, B. H.	*Experiments in Electric Circuits*, 9805–X
GAONKAR	*Microprocessor Architecture, Progamming and Applications With the 8085/8080A*, 20159–4
ROSENBLATT/FRIEDMAN	*Direct and Alternating Current Machinery*, 2nd Edition, 20160–8
SCHWARTZ	*Survey of Electronics*, 2nd Edition, 8554–3
SEIDMAN/WAINTRAUB	*Electronics: Devices, Discrete and Integrated Circuits*, 8494–6
STANLEY, W. D.	*Operational Amplifiers With Linear Integrated Circuits*, 20090–3
TOCCI	*Fundamentals of Electronic Devices*, 3rd Edition, 9887–4
	Electronic Devices, 3rd Edition, Conventional Flow Version, 20063–6
	Fundamentals of Pulse and Digital Circuits, 3rd Edition, 20033–4
	Introduction to Electric Circuit Analysis, 2nd Edition, 20002–4
WARD	*Applied Digital Electronics*, 9925–0

Introduction to Electric Circuit Analysis
2nd Edition

RONALD J. TOCCI

Monroe Community College

```
621.319    T631I     87-01595
TOCCI RONALD J
INTRODUCTION TO ELECTRIC C
COPY    0001
```

Charles E. Merrill Publishing Company
A Bell & Howell Company
Columbus   Toronto   London   Sydney

Published by Charles E. Merrill Publishing Company
A Bell & Howell Company
Columbus, Ohio 43216

This book was set in Times Roman
Production Editor: Cynthia Brunk
Cover Designer: Tony Faiola
Cover Photograph: Larry Hamill

Copyright © 1983, 1974, by Bell & Howell Company. All rights reserved. No part of this book may be reproduced in any form, electronic or mechanical, including photocopy, recording, or any information storage and retrieval system, without permission in writing from the publisher.

Library of Congress Catalog Card Number: 82-082105
International Standard Book Number: 0-675-20002-4
Printed in the United States of America
 3 4 5 6 7 8—87 86 85

*This work is dedicated to those
students whose pursuit of knowledge
makes a teacher's profession so
challenging and rewarding.*

PREFACE

Introduction to Electric Circuit Analysis, Second Edition is intended for use in postsecondary, technician level programs. The level, however, would be appropriate for those secondary school curricula which include a coverage of basic trigonometry since it is required in the second half of the text.

The key word in the title of this text is *analysis*. Any basic circuits book written on the technician level should have as its major premise the fact that a technician must be able to *analyze*. Invariably, starting technicians are employed in positions involved with testing, troubleshooting, and modifying circuits and systems. As such, a basic circuits course should begin to develop the student's ability to reason in a logical manner. Unfortunately, most circuits texts attempt to achieve this by presenting numerous mathematical techniques for solving all kinds of complex circuits while the important physical concepts are often subordinated. Although mathematical techniques are useful, they should be presented judiciously and should complement a thorough coverage of fundamental concepts and principles; otherwise, the student tends to become a "formula seeker" who rarely attempts to solve *any* problem using logical reasoning based on these basic principles.

Typically, the analysis which an electronic technician is called on to perform involves: (1) making judgments on circuit operation based on measurements and observations and (2) modifying the circuit to produce more optimum operation. Very rarely does the technician perform a complete mathematical analysis to determine which modifications to make. Instead, based on understanding of the circuit (or system) and the effects of each component on the circuit operation, he or she can make intelligent judgments.

A circuits text should be designed to prepare the student for this latter type of analysis by presenting a thorough coverage of the fundamental concepts and repeatedly applying these fundamentals using an inductive approach. In the inductive approach, a circuit is analyzed using typical component values; then the effects of each component on the circuit's behavior are discussed qualitatively using basic principles. Very often a quantitative analysis is then performed to confirm the discussion. It should be stressed here that the analysis should rarely be wholly qualitative because the student must obtain a "feel" for the "quantities" involved and he must be allowed to develop the confidence in his knowledge that a quantitative approach can provide.

Care must be taken, however, to ensure that the mathematics is presented as a useful, practical tool, and does not create unnecessary stumbling blocks to the student's understanding of the concepts being studied. Artificially contrived, complex circuit analysis examples should be kept to a minimum. In this era of computer-aided circuit analysis, very few technicians (or engineers, for that matter) need to become involved in such tedious tasks. Besides, all circuits and systems, no matter how complex, can usually be subdivided into basic circuit types. It is these basic circuit types (voltage dividers, RC filters, resonant circuits, etc.) which a technician should be made to fully understand.

It is the author's contention that *Introduction to Electric Circuit Analysis*, to a considerable degree, is effective in fulfilling the objectives stated in the last three paragraphs. However, he harbors no illusions concerning this effectiveness since it is the instructor who most singularly controls the tone and philosophy of the course and the objectives to be met. Unless the instructor agrees with the philosophy of the text, any advantages to be gained from using it will be lost.

Retained Features

This new edition has retained the following features from the first edition:

(1) Clear, step-by-step explanations based on the fact that most students who will use this book have a minimal math background, and little, if any, science background.

(2) An unusually large number of detailed illustrative examples that help to reinforce much of the text material. A minimum number of these examples are of the "plug-in" variety; many of them are used to demonstrate reasoning processes utilizing the previously developed concepts.

(3) End-of-chapter questions and problems chosen to provide the student with the opportunity to exercise both quantitative and qualitative understanding of the material. Some of these problems are used to introduce new applications or extensions of the material discussed in the chapter (i.e., relays, amplifier coupling, impedance measurement). Many of the problems are *troubleshooting* problems designed to allow the student to use reasoning in attacking practical situations. A small number of exercises involving semiconductor devices are included throughout the text in order to accommodate those programs which offer a simultaneous coverage of electronics. These exercises allow the student to apply the circuit principles to electronic circuits.

(4) Following most chapters, a comprehensive "chapter summary" which provides a step-by-step review of the important concepts and is often used to compare (usually by tabulation) similar circuits or concepts from previous chapters.

New Features

This edition is substantially revised from the first edition as a result of many valuable suggestions and criticisms from users and nonusers. The major change has been a reorganization of topics so that dc and time-varying circuits are *not* taught concurrently but are treated separately. This will allow the student to use this textbook in a conventional dc/ac sequence.

Following are other significant changes:

(1) Many new end-of-chapter problems have been added, and all problems are now keyed to the pertinent chapter sections.

(2) There are new illustrative examples in several of the chapters.

(3) The *siemens* has replaced the *mho* as the SI unit for conductance.

(4) Chapter 2 has been retitled "Basic Electrical Quantities" and includes a basic introduction to voltage, current, charge, resistance, and power.

(5) The chapter on scientific notation and prefixed units now includes a section dealing with "estimating" answers, a topic which is sorely needed by calculator-dependent students.

(6) The old Chapter 4 entitled "Waveforms" has been moved to Chapter 12. All of the material in Chapters 5-11 that utilized waveform sources has now been deleted. Chapters 1-11 are now essentially devoted to dc circuits.

(7) The new Chapter 4 covers voltage and current conventions, sources, symbology, and so forth.

(8) Emphasis has been placed on determining voltages with respect to ground (i.e. finding V_x, the voltage at node x relative to ground).

(9) Chapter 10 has been replaced by the old Chapter 12, "Physical Characteristics of Resistance."

(10) Chapter 11 is a combination of the old Chapters 10 and 11 so that all of the dc circuit analysis techniques are in this chapter.

(11) Chapter 12 now covers the material on waveforms and signals which was previously in Chapter 4.

(12) The chapters on capacitance and inductance now include the mathematical technique for solving single time-constant circuits, in addition to the graphical technique.

Organization

The book is divided into four general areas. *Chapters 1–11* are concerned with the dc behavior of resistive circuits. *Chapters 12–15* deal with time-varying signals and the transient response of RC and RL circuits. *Chapters 16–19* concentrate on the analysis of simple ac circuits using phasor diagrams and simple trig. *Chapters 20–26* deal with advanced ac circuit concepts using the *j-operator* and complex numbers in the study of resonance, filters, power, and network analysis.

It should be pointed out that Chapter 22 is devoted to the study of filters, a very important topic that typically receives only superficial treatment in most texts. Chapter 25 provides a thorough and practical coverage of transformers, another topic that is often given only a cursory treatment.

The several appendices include an introduction to basic meter principles. The author chose to place this topic in an appendix to allow the instructor more flexibility as to where it will occur in a given course. It is the author's feeling that this material is best suited for coverage in the lab portion of the course.

ACKNOWLEDGMENTS

I am indebted to the many individuals who provided ideas for the improvement of the second edition, and especially to my editor, Chris Conty, who devoted so much time and energy to the task of soliciting these suggestions and discussing them with me at great length. I am confident that our joint efforts have produced an edition which is a great improvement upon the original, and which should be attractive to those of you who have used the first edition as well as many of you who did not.

Reviewers and users of the previous edition whose suggestions have been gratefully incorporated include Samuel L. Oppenheimer of Broward Community College, John W. Walstrum and John Schwartzman of Catonsville Community College, Frank T. Duda, Jr. of the Community College of Allegheny County, John T. Long of Hawkeye Technical Institute, John Ringsred of the University of Minnesota @ Duluth, Ted Rodriguez of Skagit Valley College, James V. Malone of University of Houston, Kenneth M. Dunn and Maurice G. Bevis of Pensacola Junior College, Ed Stevens of Mount Wachusett Community College, Jake Ebey of Oklahoma State University Technical Institute, Charles Reck of DeVry Institute of Technology-Dallas, James G. Brazee of Dutchess Community College, R. C. Jones of Gulf Coast Community College, Roger Thielking and Sam Smith of Onondaga Community College, Philip Knouse of Black Hawk Community College, Walter Hedges of Fox Valley Community College, Raymond Cook of Delgado Community College-West Bank Campus, William P. Fisher of Asheville-Buncombe Technical College, Gerald R. Schickman of Miami-Dade Community College, and, of course, last but not least, my colleagues and students at Monroe Community College.

Ronald J. Tocci

CONTENTS

1 Basic Concepts — 1

1.1 Introduction, 1
1.2 Mass, Force, Work, and Energy, 1
1.3 Units, 5
1.4 Electrical Charges, 6
1.5 The Atom, 8
1.6 Electrical Classification of Matter, 9
    Chapter Summary, 11
    Questions/Problems, 12

2 Basic Electrical Quantities — 15

2.1 Introduction, 15
2.2 Unit of Charge, 15
2.3 Coulomb's Law, 16
2.4 Electrostatic Potential Energy, 17
2.5 Potential Difference (Voltage), 19
2.6 Sources of Voltage, 20
2.7 Electric Current, 21
2.8 Current Direction, 23
2.9 Units of Current, 25
2.10 Voltage Causes Current, 27
2.11 Speed of Electricity, 28
2.12 Circuit Terminology, 29
2.13 Opposition in Electrical Circuits, 30
2.14 Electrical Power, 31
    Chapter Summary, 32
    Questions/Problems, 34

3 Scientific Notation and Prefixed Units — 39

3.1 Introduction, 39
3.2 Scientific Notation—Powers of Ten, 39
3.3 Scientific Notation—Converting from Ordinary Notation, 41
3.4 Converting from Scientific to Ordinary Notation, 43
3.5 Addition and Subtraction Using Scientific Notation, 43
3.6 Multiplication and Division Using Scientific Notation, 45
3.7 Finding Square Roots in Scientific Notation, 46
3.8 Calculations Using Scientific Notation, 46
3.9 Prefixed Units, 47
3.10 Converting Between Basic Units and Prefixed Units, 49
3.11 Converting Between Different Prefixed Units, 51
3.12 Calculations Using Prefixed Units, 52
3.13 Estimating Results, 53
Questions/Problems, 54

4 More on Voltage and Current — 57

4.1 Introduction, 57
4.2 Voltage Conventions, 57
4.3 Combining Voltages, 59
4.4 Combining Voltage Sources, 62
4.5 Ground or Common, 62
4.6 Assigning Positive Current Direction, 64
4.7 Time-Varying Voltage and Current, 65
4.8 Ideal Voltage and Current Sources, 66
Chapter Summary, 67
Questions/Problems, 68

5 Current-Voltage Graphs and Resistive Devices — 71

5.1 Introduction, 71
5.2 Current-Voltage Graph with Constant Slope, 72
5.3 Nonlinear I-V Graphs, 83
5.4 Dynamic Resistance and Conductance, 85
Chapter Summary, 89
Questions/Problems, 90

6 Resistive Devices in Series Circuits — 95

6.1 Introduction, 95
6.2 The Series Connection, 95
6.3 Inserting Elements in Series, 98

6.4 Linear Resistors in Series, 98
6.5 Kirchhoff's Voltage Law (KVL), 105
6.6 General Statement of KVL, 111
6.7 Algebraic Method for Using KVL, 114
6.8 Voltage Dividers (Attenuators), 117
6.9 Multiple Output Voltage Dividers, 123
6.10 Voltage Dividers with More Than One Voltage Source, 125
6.11 Internal Resistance of Voltage Sources, 127
     Chapter Summary, 132
     Questions/Problems, 134

7 Resistive Devices in Parallel Circuits 143

7.1 Introduction, 143
7.2 The Parallel Connection, 143
7.3 Inserting Elements in Parallel, 145
7.4 Linear Resistors in Parallel, 147
7.5 Kirchhoff's Current Law, 153
7.6 Current Divider Rule, 156
7.7 Effects of Open Circuits and Short Circuits, 158
7.8 Connecting Voltage Sources in Parallel, 159
7.9 Comparison of Series and Parallel Circuits, 160
     Chapter Summary, 160
     Questions/Problems, 161

8 Resistive Devices in Series-Parallel Circuits 169

8.1 Introduction, 169
8.2 Simple Series-Parallel Circuits, 169
8.3 Equivalent Resistance of Series-Parallel Circuits, 176
8.4 Linearity Principle in Series-Parallel Circuits, 178
8.5 Troubleshooting Series-Parallel Circuits, 180
8.6 Bridge Circuits, 182
8.7 Loaded Voltage Dividers, 183
     Chapter Summary, 186
     Questions/Problems, 186

9 Electrical Power in Resistive Devices 191

9.1 Introduction, 191
9.2 Definition of Power, 191
9.3 Electrical Power, 192
9.4 Power Dissipation in Resistive Devices, 194
9.5 Device Power Ratings, 197

xiv Contents

9.6 Power Conversion, 198
9.7 Maximum Power Transfer, 199
9.8 Kilowatthours, 201
Chapter Summary, 201
Questions/Problems, 202

10 Physical Characteristics of Resistance 205

10.1 Introduction, 205
10.2 Factors Affecting Resistance, 205
10.3 Basic Resistance Equation, 207
10.4 Wire Tables, 209
10.5 Other Units for ρ, 210
10.6 Effect of Temperature on Resistance, 212
Chapter Summary, 214
Questions/Problems, 215

11 Resistive Circuit Analysis Techniques 219

11.1 Introduction, 219
11.2 The Method of Superposition, 219
11.3 Practical Voltage Source, 223
11.4 Thevenin's Theorem, 223
11.5 Thevenizing (Finding a Thevenin Equivalent Circuit), 225
11.6 Maximum Power Transfer From a Network, 228
11.7 Current Sources, 228
11.8 Norton's Theorem, 232
11.9 Nortonizing (Finding a Norton Equivalent Circuit), 233
11.10 Conversions Between Current Sources and Voltage Sources, 235
11.11 The Delta-Wye Transformations, 236
11.12 The Branch Current Method, 241
11.13 The Node Voltage Method, 243
11.14 Complex Circuit Example, 245
Chapter Summary, 247
Questions/Problems, 248

12 Introduction to Waveforms 253

12.1 Introduction, 253
12.2 Functions, 253
12.3 Graphs, 254
12.4 Concept of Slope, 255
12.5 Waveforms—General Discussion, 257
12.6 Basic Waveforms, 259
12.7 Pulse Waveforms, 260

12.8   Periodic Waveforms, 266
12.9   Time-Varying Sources, 269
12.10 Response of a Linear Resistor to Voltage Waveforms, 273
12.11 Power in Time-Varying Circuits, 274
       Chapter Summary, 277
       Questions/Problems, 278

13 Capacitance 283

13.1   Introduction, 283
13.2   Electrostatic Force, 283
13.3   Capacitors, 285
13.4   Capacitance, 286
13.5   Charging and Discharging a Capacitor—Qualitative Discussion, 288
13.6   Charging a Capacitor—Quantitative Discussion, 291
13.7   Capacitor Charging Rate, $\Delta V/\Delta t$, 293
13.8   Effect of E_s, R, and C on Charging Time, 295
13.9   Time Constant, 297
13.10 Discharging a Capacitor—Quantitative Discussion, 301
13.11 Time-Constant Graph, 303
13.12 The Exponential Form, 308
13.13 Capacitors in Series and in Parallel, 313
13.14 Capacitor Connected to a Complex Network, 316
13.15 Physical Factors Determining Capacitance, 318
13.16 Capacitor Leakage, 320
13.17 Capacitor Energy Storage, 322
13.18 Applications of Capacitors, 323
13.19 Parasitic Capacitances, 326
13.20 Linearity of Capacitors, 326
13.21 Final Comments, 327
       Chapter Summary, 327
       Questions/Problems, 329

14 Magnetism 337

14.1   Introduction, 337
14.2   Some Basic Ideas, 337
14.3   Magnetic Field and Lines of Force, 340
14.4   Theory of Magnetism, 342
14.5   Preferred Paths for Flux Lines, 344
14.6   Magnetic Field Produced by Current Flow, 345
14.7   Magnetic Field Around a Solenoid (Coil), 347
14.8   Solenoids with Ferromagnetic Cores, 349
14.9   Magnetic Flux, Φ, 351
14.10 Flux Density, B, 352
14.11 Magnetomotive Force, $\mathcal{F}$, 354

14.12 Magnetic Field Intensity, H, 355
14.13 Magnetic Permeability, μ, 357
14.14 *B-H* Curves, 358
14.15 Hysteresis, 360
14.16 Ohm's Law for Magnetic Circuits, 363
       Chapter Summary, 367
       Questions/Problems, 369

15 Inductance 375

15.1   Introduction, 375
15.2   Electromagnetic Induction, 375
15.3   Faraday's Law, 377
15.4   Mutual Induction, 378
15.5   Mathematical Statement of Faraday's Law, 380
15.6   Lenz's Law, 381
15.7   Self-Inductance, 384
15.8   The Inductor, 385
15.9   Voltage Polarity Across an Inductance, 389
15.10 Factors Which Determine Inductance, 391
15.11 Rise of Current in an Inductor, 392
15.12 Time Constant of an *RL* Circuit, 395
15.12 Time-Constant Graph, 396
15.14 Exponential Form Applied to *RL* Circuits, 400
15.15 Comparison of *RL* and *RC* Circuits, 401
15.16 Discharging an Inductor, 403
15.17 Using the Inductive Kick, 407
15.18 Radiated Transients, 407
15.19 Energy Stored in an Inductor, 407
15.20 Comparison of *C* and *L* Energy Storage, 409
15.21 Steady-State Solution of Circuits Containing *R, L,* and *C*, 409
15.22 Combinations of Inductors, 409
       Chapter Summary, 410
       Questions/Problems, 412

16 The Sinusoidal Waveform (AC) 421

16.1   Introduction, 421
16.2   Rotating Vectors, Angles, and Quadrants, 422
16.3   Finding the Sine, Cosine, and Tangent of Any Angle, 426
16.4   Radians, 429
16.5   Plot of the Sine Function, 430
16.6   Sine Wave of Voltage (and Current), 433
16.7   Period and Frequency, 435
16.8   Sine Wave Expressed in Terms of t, 437
16.9   Amplitude of a Sine Wave, 440
16.10 Effective (rms) Value of a Sine Wave, 440

16.11 Power Calculations Using rms Values, 445
16.12 Phase Angles, 447
16.13 Determining Phase Angle from Waveforms, 449
16.14 Phasor Representation of Phase Angle, 450
16.15 Including φ in the Sine Wave Expression, 452
Chapter Summary, 453
Questions/Problems, 455

17 Opposition in AC Circuits 459

17.1 Introduction, 459
17.2 AC Voltage Applied to Resistor, 459
17.3 Inductance Opposes Change in Current, 460
17.4 Inductive Opposition to AC: Reactance, 460
17.5 Series and Parallel Inductive Reactances, 465
17.6 Inductive Phase Angle, 467
17.7 Complete Inductor Response to AC, 469
17.8 A Capacitor Blocks Direct Current, 472
17.9 Capacitive Opposition to AC, 472
17.10 Series and Parallel Capacitive Reactances, 477
17.11 Capacitive Phase Angle, 479
17.12 Complete Capacitor Response to AC, 481
Chapter Summary, 483
Questions/Problems, 484

18 RC Circuit Response to AC 489

18.1 Introduction, 489
18.2 Review of X_C and R, 489
18.3 X_C and R in Series, 490
18.4 Impedance of a Series RC Circuit, 495
18.5 Phasor Notation—Polar Form, 497
18.6 Finding R and C from Z, 501
18.7 Variation of Z and θ with Frequency, 502
18.8 Decibels, dB, 508
18.9 Filters—General, 510
18.10 RC Circuit as a High-Pass Filter, 511
18.11 RC Circuit as a Low-Pass Filter, 515
18.12 Low-Pass and High-Pass Comparison, 518
18.13 R and X_C in Parallel, 518
18.14 Impedance of Parallel RC Circuit, 522
18.15 Effect of Frequency on Parallel RC, 524
18.16 Bypass Capacitor, 526
18.17 Comparison of Series RC and Parallel RC, 527
Chapter Summary, 527
Questions/Problems, 530

19 RL Circuit Response to AC — 537

19.1 Introduction, 537
19.2 Review of X_L, 537
19.3 X_L and R in Series, 538
19.4 Impedance of a Series RL Circuit, 542
19.5 The Practical Inductor, 544
19.6 Frequency Effects on Series RL, 548
19.7 RL Circuit as a High-Pass Filter, 552
19.8 RL Circuit as a Low-Pass Filter, 555
19.9 Comparison of RC and RL Filters, 557
19.10 R and X_L in Parallel, 558
19.11 Impedance of Parallel RL Circuit, 560
19.12 Frequency Effects on Parallel RL, 562
19.13 Comparison of Series RL and Parallel RL, 564
19.14 Practical Inductor in a Parallel RL Circuit, 564
Chapter Summary, 565
Questions/Problems, 567

20 Vector Algebra for AC Circuits — 573

20.1 Introduction, 573
20.2 Vectors in Polar Form, 573
20.3 Vectors in Rectangular Form, 575
20.4 More About j, 577
20.5 Representing R, X, and Z in Rectangular Form, 579
20.6 Currents and Voltages in Rectangular Form, 583
20.7 Converting from Rectangular to Polar Form, 584
20.8 Vector Operations, 587
Chapter Summary, 591
Questions/Problems, 592

21 RLC Circuits and Resonance — 595

21.1 Introduction, 595
21.2 X_L and X in Series, 595
21.3 Series RLC, 599
21.4 Resonance in Series RLC Circuit, 602
21.5 Voltage Magnification at Resonance, 605
21.6 Series Resonance Curves, 608
21.7 Q of a Coil, 613
21.8 Measuring f_r of Series Resonant Circuits, 614
21.9 Measuring Q of a Series Resonant Circuit, 615
21.10 Tuning a Series Resonant Circuit, 616
21.11 R, X_C, and X_L in Parallel, 617
21.12 Ideal Parallel Resonant Circuit, 620
21.13 Practical Parallel Resonance, 623

21.14 Selectivity and Bandwidth, 627
21.15 Measurements on a Parallel Resonant Circuit, 630
21.16 Damping a Parallel Resonant Circuit, 630
Chapter Summary, 632
Questions/Problems, 633

22 Filters 641

22.1 Introduction, 641
22.2 Filter Types, 641
22.3 Sharpness of Filter Response, 644
22.4 Low-Pass Filter Circuits, 646
22.5 High-Pass Filter Circuits, 649
22.6 Band-Pass Filter Circuits, 651
22.7 Band-Reject Filter Circuits, 654
22.8 Phase-Shifting Networks, 654
Chapter Summary, 656
Questions/Problems, 657

23 AC Network Analysis 661

23.1 Introduction, 661
23.2 Impedances in Series, 661
23.3 Impedances in Parallel, 663
23.4 Conductance, Susceptance, and Admittance, 666
23.5 Series-Parallel Impedances, 671
23.6 The Superposition Method, 673
23.7 Thevenin's Theorem, 675
23.8 Norton Equivalent Circuits, 677
23.9 Delta (Δ)-Wye (Y) Conversion, 678
23.10 Bridge Circuits, 679
Questions/Problems, 685

24 Power in Sinusoidal Circuits 689

24.1 Introduction, 689
24.2 Power in a Pure Resistance, 689
24.3 Power in a Pure Inductance, 690
24.4 Power in a Pure Capacitance, 692
24.5 Apparent Power, 693
24.6 The Power Triangle, 695
24.7 Power Factor, 698
24.8 Determining Power Without Using Z_T, 700
24.9 Power Factor Correction, 704
24.10 Maximum Power Transfer Principle, 706
24.11 Wattmeters, 707
Chapter Summary, 709
Questions/Problems, 709

xx Contents

25 Transformers 713

25.1 Introduction, 713
25.2 Mutual Induction and Transformer Action, 713
25.3 AC Voltages Applied to the Primary, 716
25.4 Loaded Secondary, 719
25.5 Transformer Current Ratio, 724
25.6 Impedance Transformation, 727
25.7 Multiple Secondaries and Tapped Secondary, 731
25.8 Phase Inverting Transformer, 732
25.9 The Autotransformer, 734
25.10 Transformer as an Isolation Device, 735
25.11 Transformer Power Losses, 737
25.12 Transformer Leakage Flux, 740
25.13 Loading Effects in the Practical Transformer, 741
25.14 Transformer Frequency Response, 744
25.15 Air-Core Transformers, 746
25.16 Mutual Inductance, 746
25.17 Air-Core Transformer with AC Voltages, 748
Chapter Summary, 749
Questions/Problems, 750

26 Nonsinusoidal Waveforms and Harmonics 757

26.1 Introduction, 757
26.2 Harmonics, 757
26.3 Combining Harmonics, 758
26.4 Composition of Nonsinusoidal Waveforms, 759
26.5 Harmonic Analysis of a Square Wave, 761
26.6 Waveforms with Only Odd Harmonics, 763
26.7 Finding Average Value of a Waveform, 764
26.8 Nonsinusoidal Waveforms Applied to Linear Networks, 768
Chapter Summary, 771
Questions/Problems, 772

Appendices 775

A. Tables of $\log_{10} N$, 777
B. Trigonometric Functions: Decimal Degrees, 781
C. Values of ϵ^x and ϵ^{-x}, 787
D. Meters, 791
  1. The Moving-Coil (d'Arsonval) Meter Movement, 791
  2. Voltmeters, 794
  3. The Ohmmeter, 797
  4. Meter Loading, 797
  5. AC Meters, 799

Answers to Selected Problems 803
Index 807

CHAPTER 1

Basic Concepts

1.1 Introduction

Electricity is the study of the movement of electrons. *Electronics* deals with controlling the movement of electrons (or other charges) so as to perform some useful function. Before undertaking a study of either of these areas it is worthwhile to first examine the electron; and, since all electrons come from atoms, it is imperative that some time be spent discussing atomic structure. The study of atomic structure will also serve to provide some meaning to the difference in electrical characteristics among various materials.

Since the concept of *energy* is prevalent throughout the study of electricity and electronics as well as atomic structure, we will begin by considering this concept and the related concepts of *mass*, *force*, and *work*.

1.2 Mass, Force, Work, and Energy

Mass *Mass* is a fundamental physical quantity which is a basic property of all material things. Mass is the property of a material body that causes it to resist any change in its state of motion. If a body is at rest it will remain at rest unless it is acted on by some external agent. For example, a football placed on a kicking tee will remain there until the kicker hits it with his foot (unless a gust of wind blows it off the tee first). The kicker quickly becomes aware that the football has mass because he feels the resistance it presents against his foot.

Similarly, a body in motion will maintain its speed and direction of motion unless it is acted on by an external agent. A baseball player trying to hit a thrown baseball and change its direction of motion can feel the effect of the baseball's mass. Mass is the property of matter which gives rise to the sensations experienced by the kicker and batter in these two cases. In general, a body with a greater mass requires a greater effort to change its state of motion.

Force In order to change the state of motion of a mass an external influence must act on it. This external influence is the physical quantity called *force*. In the case of the football player the force was provided by his foot and leg muscles. The baseball player used a bat propelled by his arm muscles to exert the force on the baseball. Both of these forces are examples of *contact* forces because the object exerting the force had to make contact with the mass before it could cause a change in the motion of the mass.

In our everyday experience we observe the action of forces which are not contact forces. The earth exerts a force or pull on objects which are on its surface. This *gravitational* force is also exerted on objects which are not on the earth's surface such as airplanes, birds, thrown baseballs, etc. Another example is the force that a magnet exerts on certain magnetic materials without being in contact with them. These gravitational and magnetic forces are examples of *field* forces. A field force can act on a mass at a considerable distance from it without making contact.

The gravitational field force is the most common field force known to the layman, especially since the advent of televised moonshots. It is, however, sometimes a source of confusion because it produces the concept of *weight*, which is often confused with mass. *Weight is actually a force* and is essentially the force of gravity acting on an object. The gravitational force exerted by one object (earth) on another object depends on several factors: the mass of each object and the distance between the objects. For larger masses the gravitational force (weight) that one object exerts on the other is greater. The further apart the masses become the smaller will be the gravitational force. Thus, as a spaceship speeds away from the earth it eventually gets far enough away from the earth so that the earth's gravitational pull is essentially zero and the spaceship is rendered weightless.

Work If we push a heavy box across the floor we quickly become aware that we are expending a large amount of energy. The technical term for what we are doing is *work*. The technical meaning of work is not exactly the same as the common everyday meaning. For example, suppose a man is standing up and holding a large book in the palm of his hand for a considerable length of time. Most of us would agree that because his task is not particularly pleasurable and because it requires a certain degree of strength the man is doing work. However, the technical definition of work would indicate otherwise; that is,

Work takes place whenever motion is accomplished against the opposition of a force which is trying to prevent the motion.

Since the man holding the book is not causing the book to move he is *not* doing work. If he had picked the book up off the floor he would have performed work because he would have moved the book against the opposition of the force of gravity which tries to pull it back down to the floor.

The greater the opposing force is, the greater the amount of work required to move an object. In pushing a heavy box across the floor it takes more work if the floor is rough than it would for a smooth floor; here, the opposing force is a frictional force. If the opposing force is too great, no motion will take place and therefore no work will be done. For example, in trying to push a car out of a snowbank no work is actually performed if the car doesn't budge.

The mathematical definition for the amount of work, W, performed in moving an object a certain distance, d, by exerting a force, F, is given by

$$W = F \times d \tag{1-1}$$

The units of W, F, and d will be discussed later. For now this expression tells us that the amount of work increases proportionately with the force exerted and the distance through which the force is acting. It takes twice as much work to lift a 100-pound weight five feet off the floor as it would to lift a 50-pound weight the same distance; similarly, it takes twice as much work to lift the 100-pound weight ten feet off the floor as it would to lift it only five feet.

Energy From the foregoing discussion it should be clear that work is accomplished by exerting a force through a distance. Typical examples of working agents are an electric motor, a combustion engine, a man's arm, etc. Any working agent accomplishes work at the expense of another physical quantity called *energy*. Through the process of work, the energy of the working agent is converted from one form of energy to another. For example, in pushing the box across the floor the chemical energy in the man's body is converted to heat energy both in the man's body and at the surface of the floor due to friction. In order to perform work, a working agent must possess a storage of energy. Stated formally:

Energy is the capacity to do work.

In other words, a body that possesses any amount of energy is capable of doing work. A body can exert all of its energy to do work. In such a case the amount of work performed is numerically equal to the amount of energy which the body exerted. That is, energy and work are mathematically the same quantity and possess the same units (joules, as we shall see later). The chief difference between work and energy is in the cause-effect relationship. Energy must be present before work can be done. In the process of doing work the energy is converted from one form to another. Typical forms of energy include mechanical, chemical, electrical, magnetic, thermal, nuclear, sonic, and optical (light).

As energy is being transformed from one form to another, one basic principle is always adhered to: namely, the principle of the conservation of energy, which states that:

The total energy in a system remains constant.

Energy can be transformed from one form to another, but after the transformation is completed the total energy remaining must be the same as the total energy before the transformation.

Potential Energy The energy associated with a body is usually considered from two viewpoints. The first concerns the position of the body and is called *potential energy*. The second depends on the velocity at which the body is moving and is called *kinetic energy*. An example of potential energy occurs when a mass is lifted off the ground. To lift this mass a certain amount of work must be performed against the force of gravity. The amount of work can be calculated using Eq. (1-1) since we know the required force, F, will be equal to the weight of the mass, and the distance, d, is the distance the mass is lifted off the ground. In lifting the mass the lifter expends an amount of energy equal to the work ($F \times d$) performed. The energy expended by the lifter is converted into potential energy concentrated in the mass as it is held a distance d above the ground. We say that the mass has a certain potential energy due to its position above the ground. How do we know that the mass has potential energy? Because, if we let it go it can perform work (like driving a stake into the ground or breaking open walnuts). *It has the potential to do work.*

Potential energy depends on the position of a body since the higher we lift the mass the more work it takes and the more energy is converted into potential energy. This is intuitively obvious because we all would agree that a hard ball dropped on our heads from one foot above is much less painful than one dropped from ten feet above.

Kinetic Energy Before potential energy can be used to do work it is usually converted to *kinetic energy* or energy of motion. When the mass which has been lifted off the ground is released, it immediately begins moving back toward the ground picking up speed as it does. Because it has speed it possesses kinetic energy. The potential energy of the mass gradually decreases as it falls since it is getting closer to the ground; essentially, the potential energy is being converted to kinetic energy. Thus, the kinetic energy of the mass increases as it falls toward the ground and the speed increases.

The kinetic energy (KE) of a body is related to its mass (M) and its velocity (v). Mathematically,

$$\text{KE} = \frac{1}{2} M v^2 \qquad (1\text{-}2)$$

so that we can calculate the kinetic energy of a given mass if we know how fast it is moving.

In the study of electricity and electronics the concepts of force, work, and energy play an important role in understanding some of the basic electrical principles. As such, we must have some convenient means of expressing these concepts quantitatively. In the next section we will present the system of units to be used for these and other quantities in this text.

1.3 Units

In any study dealing with physical quantities it is necessary to establish a consistent system of units which can be used to express these quantities. Although many systems of units have been used in the past, the Institute of Electrical and Electronics Engineers has adopted the International System of Units (abbreviated SI) as the basic system of units. The SI units contain all of the basic units needed to derive any mechanical or electrical unit.

In the International System the basic unit for the measurement of *length* is the *meter*. In equations the letter symbol for length is l and the abbreviation for meter is m. For example, the following equation states that the length of a certain object is equal to 4.3 meters:

$$l = 4.3 \text{ m}$$

For those of us who are accustomed to using the *inch* as a basic unit for length, the meter represents a length of 39.37 inches. That is,

$$1 \text{ meter} = 39.37 \text{ inches}$$

The SI unit for *mass* is the *kilogram*. The letter symbol for mass is M and the abbreviation for kilogram is kg. Actually a kilogram is equal to 1000 grams. That is, 1 kg = 1000 g. The following equation states that the mass of a certain object is equal to 3 kilograms or 3000 grams:

$$M = 3 \text{ kg} = 3000 \text{ g}$$

The SI unit for *time* is the *second*. The letter symbol for time is t and the abbreviation for seconds is s. The following equation tells that the time duration (t) of a certain event is equal to 3.2 seconds:

$$t = 3.2 \text{ s}$$

The fourth basic SI unit is the *ampere*, which is the unit of electrical current. We will of course discuss current in much more detail later. However, for now let us just state that the symbol for current is the letter I and the abbreviation for ampere is A. The following equation states that the current I in a certain circuit is equal to 1.5 amperes:

$$I = 1.5 \text{ A}$$

These four units of measurement are called basic or fundamental units because the units of all other mechanical and electrical quantities can be derived from them. For example, consider the physical quantity of *force*, which was discussed in the last section. The unit of force in the International System is the *kilogram-meter per second per second;* that is,

$$\frac{\text{kg-m/s}}{\text{s}} = \text{kg-m/s}^2$$

This derived unit contains a combination of the basic units for mass, length, and time. The International System prefers not to use combinations of units

Chapter One

(too cumbersome) so it calls this derived unit of force the *newton*. In other words, a newton is equal to a kg-m/s². The abbreviation for the newton is N. Thus,

$$1 \text{ N} = 1 \text{ kg-m/s}^2$$

A force of 1 N is equivalent to a force of 0.22 lb in our everyday system.

Another example of a derived unit is the unit for work or energy. In the International System work and energy would have the units of kilogram-meter² per second per second; that is,

$$\frac{\text{kg-m}^2/\text{s}}{\text{s}} = \text{kg-m}^2/\text{s}^2$$

This derived unit for work and energy also combines the basic units of mass, length, and time. The SI prefers not to use this cumbersome combination so it calls this derived unit of work and energy the *joule*. That is, a joule is equal to a kg-m²/s². The abbreviation for joule is J. Thus,

$$1 \text{ J} = 1 \text{ kg-m}^2/\text{s}^2$$

It should be restated that the units for work and energy are the same since both represent the same physical quantity. It also should be noted that the unit for work and energy is the same as the unit for force except for an extra unit of length m. That is,

$$1 \text{ joule} = 1 \frac{\text{kg-m}^2}{\text{s}^2} = \left(\frac{1 \text{ kg-m}}{\text{s}^2}\right) \times \text{m}$$
$$= \text{N} \times \text{m}$$

This is consistent with our previous statement (sec. 1.2) that work is equal to force × distance.

As we encounter more mechanical and electrical quantities throughout the text we will present their SI units as part of their definitions rather than present them here. Table 1-1 summarizes the SI units given thus far.

1.4 Electrical Charges

If we take a glass rod and rub it with a silk cloth we can use the rod to attract and pick up small pieces of paper against the force of gravity. Since the rod is

TABLE 1-1

Physical Quantity	SI Unit
Length	meter, m
Mass	kilogram, kg
Time	second, s
Current	ampere, A
Work	joule, J
Energy	joule, J
Force	newton, N

now performing work, we must have supplied it with some form of energy when we rubbed it. Actually, what we did was to place an electric charge on the rod. The term *charge* is used to imply energy; that is, we charged the rod with a form of electric energy.

If we continue the experiment by touching the charged rod to a pair of suspended lightweight balls an interesting phenomenon occurs, as illustrated in Fig. 1-1(a). By touching the balls with the charged rod some of the charge is transferred to each ball. Once the balls become charged they begin to *repel* each other as shown.

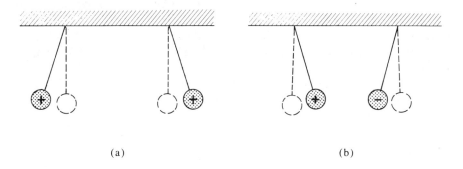

(a)   (b)

FIGURE 1-1 (a) *Balls with the same charge repel each other.* (b) *Balls with unlike charges attract each other.*

Now, if we take a hard-rubber rod and rub it with a piece of fur we will find that it is also *charged;* we can transfer the charge from the rod to the suspended balls with the result again being that the balls repel each other. From these last two observations we can conclude that:

Bodies possessing the same type of charge will repel each other.

Or, we can state it even more simply:

Like charges repel each other.

Apparently, then, like charges exert some kind of force on each other which tends to push them further apart.

Continuing even further, if we take the charged glass rod and touch it to one of the two suspended balls and touch the other ball with the charged rubber rod, the phenomenon of Fig. 1-1(b) occurs: the two balls are *attracted* toward each other. From this result we can reason that the charge on the glass rod and the charge on the rubber rod must be *different* since the balls did not repel each other but instead were attracted toward each other. Thus, we can say,

Unlike charges attract each other.

Apparently, then, the balls in Fig. 1-1(b) exert some kind of attractive force on one another.

It is now well known that there are only two different types of electric charge. We distinguish between the two types of electric charge by calling one a positive (+) charge and the other a negative (−) charge. The charge on the glass rod was a positive charge while that on the hard-rubber rod was a negative charge.

The repulsive and attractive forces observed in Fig. 1-1 must be *field* forces since the balls are not in contact but affect each other from a distance. A French physicist named Charles de Coulomb discovered that the magnitude of the force exerted between two bodies which are electrically charged *increases* if the amount of charge on either body is *increased*. He also found that the force *increases* if the charged bodies are moved closer together. The combination of these two principles is called *Coulomb's law of electrostatic force*. The forces between charged bodies are often referred to as *electrostatic*, or Coulombic, forces.

The concept of charge at this point is still somewhat superficial. To obtain a more meaningful understanding of charge we must go into the structure of matter on the atomic scale.

1.5 The Atom

The explanation of basic electrical effects must start within the structure of the *atom*. An atom is defined as the smallest part of an element* that can take part in ordinary chemical changes. The atom can be thought of as composed of two parts:

(1) The *nucleus*, which possesses most of the mass of the atom and is physically located at the center of the atom.
(2) *Electrons*, small particles which orbit around the nucleus in a manner similar to planets orbiting around the sun.

For our purposes, the atom can be visualized as a miniature solar system where the nucleus acts as the sun and the electrons correspond to orbiting planets. Figures 1-2(a) and (b) symbolize the structure of the hydrogen atom and lithium atom respectively. The hydrogen atom has a nucleus consisting of one *proton*, and it has one *electron* revolving around this nucleus. The proton is a particle which possesses a *positive* electrical charge. The electron is a particle which weighs much less than a proton but possesses a *negative* charge equal to the charge of the proton. The proton is approximately 1837 times heavier than the electron but they both contain the same amount of charge.

The lithium atom has a nucleus consisting of three protons and four *neutrons*. The neutrons are *uncharged* particles having about the same size and

---

*All matter consists of basic materials called *elements*. There is definite proof that at least 106 elements exist here on earth. Copper, iron, oxygen, and nitrogen are examples of elements. All elements are distinct (i.e., one element does not contain any other elements).

Basic Concepts 9

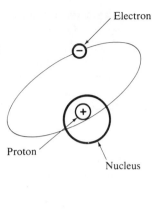
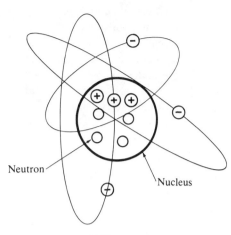

Hydrogen atom

Lithium atom

FIGURE 1–2 *Representation of hydrogen and lithium atoms.*

weight as protons. There are three electrons orbiting around the lithium nucleus.

After observing the structure of the hydrogen and lithium atoms, several general conclusions can be made:

(1) Almost all of the mass of an atom is contained in the nucleus because the electrons are so much lighter than the protons and neutrons.
(2) An atom is electrically neutral since the number of positive charges (protons in the nucleus) is the same as the number of orbiting electrons.
(3) The electrons of an atom have different orbital paths at different distances from the nucleus (see lithium atom).
(4) Protons and neutrons exist only in the nucleus, and electrons exist only outside the nucleus.

Besides hydrogen and lithium there are 104 other elements, each of which has a different atom and atomic structure. The atoms of the various elements differ in the number of particles in the nucleus and in the number of orbital electrons. All the atoms of a given element have the same structure. It is the difference in the atoms of the various elements that gives each element its own distinct properties.

1.6 Electrical Classification of Matter

In the atom the major force which keeps the electrons in their orbits is the electrostatic force exerted on them by the positively charged nucleus. The magnitude of this force is smallest for those electrons which are in orbital paths furthest from the nucleus. These remote electrons are called *valence* electrons and they play a principal role in determining the electrical properties of an atom (see Fig. 1–3).

When valence electrons are able to break free from the attraction of the nucleus, they become *free* electrons. Since *electric current* is defined as the movement of charges, it is logical to conclude that the ability of a material to conduct electricity depends on the number of free electrons available within the material.

There are three electrical classifications of matter based mainly on the ability of a material's valence electrons to break free from the nucleus. These three classifications are *conductors, insulators,* and *semiconductors*.

In a conductor the valence electrons associated with the individual atoms are very loosely bound to the nucleus; thus, it is easy for them to become free electrons available for conducting electric current. Most of the good conductors of electricity are metals. The most notable conductors are (in decreasing superiority) silver, copper, gold, and aluminum. In most circuit work copper wire is used as the conductor which connects the various circuit elements. Silver is a slightly better conductor than copper since it contains about 5 percent more free electrons per unit volume; however, copper is more abundant, and therefore less expensive.

In an insulator, the valence electrons are tightly bound to the nucleus. As such, very few valence electrons can become free electrons under normal conditions. It would take a strong external force (usually electrostatic) to pull these electrons away from the atom. Most of the commonly used insulating materials such as rubber, plastic, paper, oil, asbestos, and mica are polymers. Practically speaking, even these good insulators have some free electrons available to conduct current due to impurities or moisture. However, the number is relatively small so that these materials are used whenever a nonconducting material is required (such as for covering bare wire). Even air is an insulator, although its ability to conduct current increases when it becomes moist.

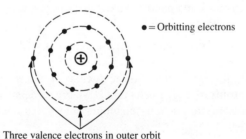

Three valence electrons in outer orbit

FIGURE 1-3 *Structure of aluminum atom showing three valence electrons.*

In a semiconductor there are many more free electrons available than in an insulator, but not nearly as many as in a conductor. Typical semiconductors include silicon, germanium, gallium arsenide, and carbon. The first three are used extensively in the manufacture of semiconductor devices such as diodes, transistors, and integrated circuits. Carbon is used in the manufacture of carbon composition resistors that are found in almost all electronic circuits.

Chapter Summary

1. Mass is the property of a material body that causes it to resist any change in its state of motion.
2. Force is an external influence acting on a mass so as to change its state of motion.
3. A contact force is one where the agent exerting the force makes contact with the mass.
4. A field force acts on a mass at a distance from the mass without making contact.
5. Weight is a force (gravitational force).
6. Mass is independent of any gravitational force.
7. Work is the accomplishment of motion against the opposition of a force trying to prevent the motion.
8. Amount of work increases directly with the force exerted and the distance through which it is exerted.
9. Energy is the capacity to do work. To perform work an agent must possess energy.
10. The total energy in a system remains constant.
11. Potential energy depends on the position of a body; kinetic energy depends on the movement of a body. Both types of energy are proportional to the mass of the body.
12. The basic SI unit of measurement for length (l) is the meter (m).
13. The basic SI unit for mass (M) is the kilogram (kg).
14. The basic SI unit for time (t) is the second (s).
15. The SI unit for electrical current (I) is the ampere (A).
16. The SI unit for force (f) is the newton (N).
17. The SI unit for work and energy is the joule (J).
18. Bodies possessing the same type of electric charge will repel each other.
19. Bodies possessing different types of charge will attract each other.
20. Forces between charged bodies are called electrostatic forces and these forces are field forces.
21. Electrostatic forces increase as the size of the charges is increased or as the distance between charges is decreased.
22. The atom is composed of a nucleus, which contains the relatively heavy protons and neutrons, and relatively light electrons which orbit around the nucleus.
23. Electrons possess a negative charge; protons an equal positive charge; and neutrons no charge.
24. Valence electrons are those orbiting furthest from the nucleus.

25. The electrical properties of a material depend on the ease with which the valence electrons can break free of the attraction of the nucleus.

26. The three electrical classes of matter are conductors, insulators, and semiconductors.

Questions/Problems

Section 1.2

1-1 Indicate which of the following are examples of contact forces, field forces, or neither:
   a. magnetism
   b. wind blowing leaves
   c. gravity
   d. a ball rolling on a frictionless surface
   e. a satellite orbiting the earth

1-2 Indicate which of the following constitutes *work* being done:
   a. running up the stairs
   b. holding a weight over your head
   c. wind blowing leaves
   d. lifting a bag of groceries
   e. carrying the bag of groceries on level ground

1-3 Which of the following requires the most *work* to accomplish:
   a. a 150-pound man climbing 10 feet on the earth's surface
   b. a 200-pound man climbing 6 feet on the earth's surface
   c. a 200-pound man climbing 6 feet on the moon.

1-4 What is meant by the term *energy?*

1-5 Give as many examples as possible of devices or systems in which energy is transformed from one form to another. For example, the internal combustion engine used in automobiles converts the chemical energy of the fuel into mechanical energy of motion.

1-6 Indicate which of the following are examples of potential energy and which are examples of kinetic energy:
   a. a car on the speedway
   b. a raised pile driver
   c. a jet airliner in flight
   d. a set mousetrap
   e. a falling star

1-7 Suppose there are two identical helicopters A and B. Helicopter A is moving at a speed of 100 mph at an altitude of 200 feet above the earth's surface. Helicopter B is also moving at the same speed and at an altitude of 200 feet above the moon's surface.
   a. Which of the helicopters has the greatest potential energy?
   b. Which of the helicopters has the greatest kinetic energy?

Section 1.3

1-8 Indicate which of the following are basic SI units and which are derived SI units:
   a. kilogram
   b. joule
   c. pound
   d. second
   e. newton

1-9 Give the SI units for the following quantities: length, time, force, mass, work, energy, current.

Section 1.4

1-10 What is your concept of electric charge?

1-11 Consider the charged balls in Fig. 1-1(a). If the balls were hung at a greater distance from each other, what would happen to the angle of deflection of each ball?

1-12 Consider the charged balls in Fig. 1-1(b). What would happen to the ball on the right if the other ball were suddenly yanked off the string?

1-13 Based on your answer to the preceding question, what kind of energy would you associate with the charged balls as they are situated in Fig. 1-1(b)?

Section 1.5

1-14 Describe the basic structure of an atom.

1-15 Explain the differences between electrons, protons, and neutrons.

Section 1.6

1-16 What are valence electrons?

1-17 Describe conductors, insulators, and semiconductors in terms of their valence electron properties.

CHAPTER 2

Basic Electrical Quantities

2.1 Introduction

An *electric circuit* is the physical means by which electrical energy is delivered to a user of the energy. All of us come in contact with electric circuits as part of everyday living; for example, every light in our homes is connected in a simple electric circuit. In almost all electric circuits the energy is conveyed by electric charges which are forced to move in a certain manner. In this chapter we will discuss the movement of charges, called *current*, and the stimulus which causes them to move, called *voltage*.

The quantities of current and voltage are extremely important for two reasons: (1) they are easily measurable using modern instruments such as voltmeters, ammeters, and oscilloscopes; and (2) the operation of most circuits, no matter how complex, can be expressed in terms of the circuit voltages and currents. As such, these quantities will be considered carefully in this chapter so as to make the following chapters more easily understood.

We will also introduce the concepts of electrical opposition (resistance) and electrical power, two other important electrical quantities that will be dealt with in more detail in later chapters.

2.2 Unit of Charge

The concept of electric charges was introduced in chap. 1 and it was seen that all materials contain charged particles (electrons and protons) as part of their basic structure. As mentioned in the introduction, it is these charges (most often electrons) that are used to convey electric energy.

Chapter Two

At this point the SI unit for electric charge should be introduced. At first, it might seem reasonable that since the charge on an electron is the smallest unit of charge, it should be made the basic unit of charge. However, the charge of a single electron is too small for practical purposes. The SI unit of charge is chosen instead to be the *coulomb*. It has been experimentally determined that *a coulomb of charge represents the combined charge of* 6.24×10^{18} *(6,240,000,000,000,000,000) electrons (or protons)*.

The symbol for coulomb is C. When we talk about quantity of charges from now on, the symbol q will be used to represent the quantity of charge. For example, the statement

$$q = 1.5 \text{ C}$$

means that the amount of charge (q) is 1.5 coulombs (C).

2.3 Coulomb's Law

Coulomb's law of electrostatic force, which was discussed in chap. 1, can now be stated more formally as:

The electrostatic force between two electric charges is directly proportional to the product of the two charges and inversely proportional to the square of the distance between the charges.

In equation form,

$$F = \frac{kq_1q_2}{d^2} \qquad (2\text{-}1)$$

where F represents the force in newtons, q_1 and q_2 represent the respective magnitudes of the two charges in coulombs, and d is the distance between the charges in meters. The constant of proportionality k has a value of approximately 9,000,000,000 (9×10^9) newton-meter² per coulomb². That is,

$$k = 9 \times 10^9 \frac{\text{N-m}^2}{\text{C}^2}$$

Consider Fig. 2–1(a). Two charges, labeled q_1 and q_2, are shown a certain distance, d, apart. From a previous discussion we know that if q_1 and q_2 are both of the *same polarity* (*both positive or both negative*), they will exert a repelling force on each other. Conversely, if they are of different polarities, they will exert an attracting force on each other. In either case, the magnitude of the

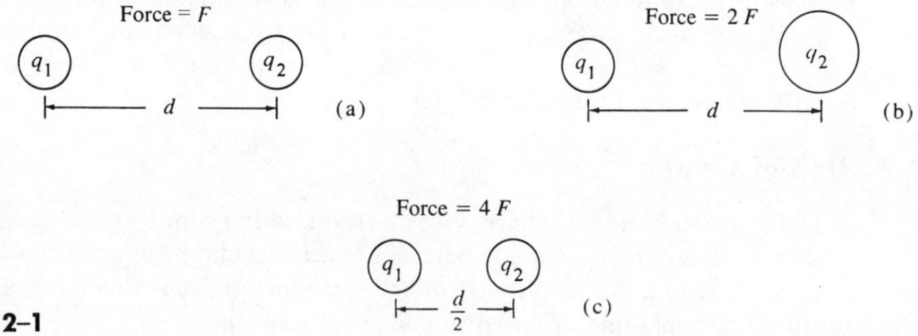

FIGURE 2–1

force can be calculated using Eq. (2-1). It is important to understand that this force is a *mutual* force that is equally felt by both charges. That is, q_1 exerts a force on q_2, while q_2 exerts the same force on q_1.

The magnitude of this force depends on the size or quantity of each charge. For example, if either charge q_1 or q_2 is *doubled*, Eq. (2-1) indicates that the resultant force will also *double* [see part (b) of Fig. 2-1]. Similarly, if q_1 or q_2 is made smaller, the force will proportionately decrease.

The magnitude of the force also depends on the distance between the charges. For example, if the distance between the charges is reduced by *half* [part (c) of Fig. 2-1], then the amount of the force will increase. However, the force will increase by *four* times because of the d^2 term in Eq. (2-1). Similarly, as the charges are moved further apart the force between them becomes smaller.

It should again be emphasized that the nature of these electrostatic forces is very similar to the gravitational forces we experience daily. In fact, the equation which describes the gravitational force is similar to Eq. (2-1) with q_1 and q_2 replaced by m_1 and m_2, the masses of two objects. In our own case, m_1 would represent the earth's mass and m_2 would represent the mass of any object. The gravitational attraction force between these two masses is also a mutual force. However, because the earth is so massive ($m_1 \gg m_2$) the force is not enough to cause it to move, but the other mass will be moved toward the earth instead.

2.4 Electrostatic Potential Energy

We know that any object which is a distance d above the earth's surface has *potential energy*. The fact that it has this potential energy can be reasoned as follows. Consider Fig. 2-2, which shows a ball being supported by a string a certain distance above the ground. The string is exerting a force on the ball in an upward direction which counteracts the gravitational force acting to pull the ball downward.

We say that the ball has potential energy because we know that if the string is broken, the ball will drop toward the ground and, in doing so, will pick up speed. By virtue of this speed it will have *kinetic energy*. This kinetic energy can be made to perform work such as driving a stake into the ground. Thus, when an object has potential energy it means that it has the *potential* to do work.

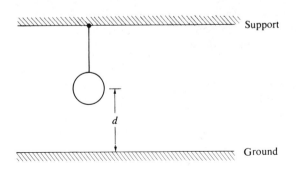

FIGURE 2-2

A completely analogous situation exists in an electrostatic system such as the one illustrated in Figure 2–3. Here, a negative charge q_1 is shown a distance d from a metal plate which is filled with positive charges. Let's imagine that some restraining force is being used to keep q_1 from moving despite the obvious electrostatic attraction of the positively charged plate. If this restraining force is removed, it is apparent that q_1 will immediately begin moving toward the metal plate. Thus, we can conclude that q_1 has potential energy. To be exact, q_1 has *electrostatic potential energy* by virtue of the force exerted on it by the positive plate.

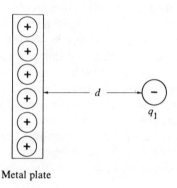

Metal plate

FIGURE 2–3

EXAMPLE 2.1 If the amount of charge on q_1 is doubled, what would happen to its potential energy?

Solution: The increase in q_1's charge will cause a proportionate increase in the electrostatic force of attraction which the plate has on q_1 (recall Coulomb's law). Thus, q_1 will have its potential energy increased by a factor of two; the larger force will give q_1 the potential to move at a greater speed (kinetic energy) once it is allowed to move.

This last example illustrates an important point. Although the potential energy of q_1 depends on the amount of q_1's charge, the amount of potential energy *per unit charge* is constant at a given distance from the plate. To illustrate, suppose $q_1 = 2$ coulombs and this resulted in q_1 having a potential energy of 2 joules. If q_1 had its charge increased to 4 coulombs, its potential energy would increase to 4 joules. In both cases, q_1 has 1 joule of potential energy for each coulomb of its charge. In other words, for the situation in Fig. 2–3, charge q_1 has

$$\text{joules/coulomb} = 1$$

This quantity (joules/coulomb) is given a special name—**electrostatic potential** (or simply, *potential*). Potential is not to be confused with *potential energy,* although they are related. Potential is the amount of potential energy per unit charge. In Fig. 2–3, at a given distance from the plate, q_1 will have potential energy in proportion to its charge, but its *potential* will be the same regardless of the size of q_1's charge.

2.5 Potential Difference (Voltage)

Consider the situation shown in Fig. 2–4(a). Metal plate A, which is charged positively, is separated by a distance d from plate B, which is charged negatively. Charge q_1, which is negative, is shown next to plate B. In this position, it is apparent that q_1 has a certain amount of potential energy because of the combined effects of the repulsive force of plate B and the attractive force of plate A, which will both cause q_1 to move toward A. Thus, q_1 has a certain potential (potential energy per coulomb) in its position at plate B. This is analogous to the gravitational situation in Fig. 2–4(b) where the ball has the maximum amount of potential energy at the top of the stairs.

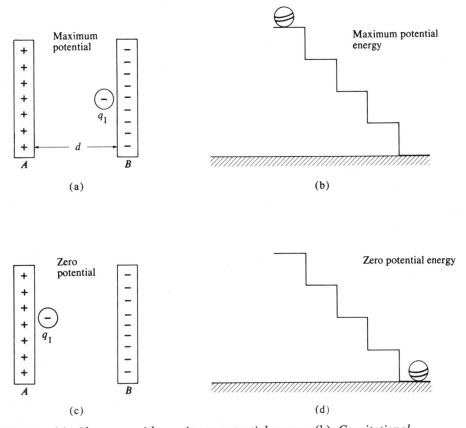

FIGURE 2–4 (a) *Charge q_1 with maximum potential energy.* (b) *Gravitational analogy.* (c) *q_1 with zero potential energy.* (d) *Gravitational analogy.*

Now consider the situation shown in Fig. 2–4(c), where q_1 is shown next to plate A. In this position, q_1 has *no* potential energy because it will not move from this position. This is because it is attracted by positive plate A and no other force is present which will pull it away from A. Thus, when q_1 is next to A it has a potential of zero. This is analogous to the gravitational situation in Fig. 2–4(d) where the ball is resting at the bottom of the stairs and therefore has zero potential energy.

We can say, then, that there is a *difference in potential* between plates A and B just as there is a difference of gravitational potential energy between the top of the stairs and the bottom of the stairs. This difference in potential could be any amount depending on the amount of charge on the plates. The difference in potential does not depend upon q_1's charge because, recall, potential means potential energy per coulomb of q_1.

Voltage Let's assume that the difference in potential between A and B is 2 joules-per-coulomb. Another way to say this is that the difference in potential is 2 **volts**. In other words, a **volt** is defined as

$$1 \text{ volt} = 1 \text{ joule/coulomb}$$

In most electrical work the volt is used instead of the joule-per-coulomb to represent potential. Also, the term potential difference is often referred to as *voltage*. Thus, there are several ways to express the situation in Fig. 2–4. The two ways which will be used henceforth in the text are (1) potential difference between A and B equals 2 volts, and (2) voltage between A and B equals 2 volts. The unit symbol for volts is V.

A potential difference or voltage will exist wherever an imbalance in electrical charges is present. In the situation of Fig. 2–4 the imbalance was obvious because plate A was positively charged and plate B negatively charged. If both plates were charged with the same polarity, a voltage could exist between them if one plate has more charge than the other. Such a case is seldom incurred in electronics, but it emphasizes the point that *voltage is a difference in potential between two points*. As we shall see later, differences in potential in an electric circuit are purposely produced by *voltage sources* which are designed to produce a particular charge imbalance resulting in a particular voltage between two points.

2.6 Sources of Voltage

A voltage is produced whenever an imbalance of electric charges is present. In all electric circuits a source of voltage is needed to produce current. The various types of voltage sources in use all employ some method of producing a charge imbalance (and, thus, a voltage) between two points. In general, this text will not be overly concerned with the mechanisms by which voltage sources operate. Our main concern will be using these voltage sources and analyzing the electric circuits which contain them.

The four most common voltage sources are:

(1) *Dry cells and storage batteries.* These sources of voltage utilize chemical reaction principles to develop a constant (unchanging) potential difference between two metal rods called electrodes. The chemical action in the battery causes one electrode to have an excess of electrons and thus have a negative charge; the other electrode has a deficiency of electrons and is thus charged positively. This type of voltage source will be used often. A typical battery is shown in Fig. 2–5(a). The circuit symbol for a battery is given in Fig. 2–5(b). The battery in this figure is a 3-V battery, which means that the voltage between electrodes (terminals) A and B is a constant 3 V.

Basic Electrical Quantities   21

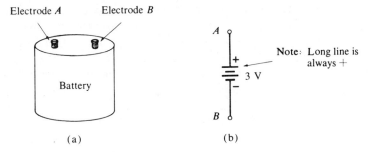

FIGURE 2-5 (a) *Typical battery.* (b) *Symbol for battery.*

(2) *Electronic power supplies.* These are actually voltage converters, which are special electronic circuits (employing diodes, transistors, etc.) that convert a changing voltage to a constant voltage. The changing voltage is usually provided by a power line source.

(3) *Power line source.* This is the type which is familiar to all of us since it is used to provide electricity in most of our homes and buildings. The voltage from these sources is produced by converting mechanical energy (such as falling water) to electrical energy (by using turbines, for example). Power line voltage is a changing voltage. That is, the voltage between the two sides of an electrical wall socket varies in a predetermined manner. This variation, called *sinusoidal*, will be discussed in a later chapter.

(4) *Signal generators.* These are special electronic circuits designed to produce changing voltages (signals). Some of the many common types of signals will be discussed in a later chapter.

2.7 Electric Current

In all electric circuits the electrical energy is conveyed by electric charges (usually electrons). As we discussed in chap. 1, most materials contain some number of *free conduction electrons*. These free electrons are used to convey electrical energy. In order to make use of these free electrons and make them move in a predetermined manner, a source of voltage is used. For example, refer to Fig. 2-6.

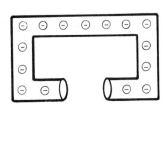
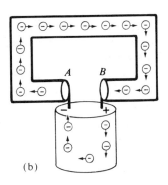

FIGURE 2-6

In part (a) of the figure a copper wire is shown unconnected at both ends. We know that copper contains many free electrons that are able to move through the copper material with relative ease. In the situation of part (a), the free electrons have no preference in their movements because there is no force acting on them. Thus, their movement will be *random*, meaning that over a period of time there will be just as many electrons moving in one direction as in any other.

In part (b) of the figure the same copper wire is shown with its ends connected to opposite terminals of a battery. The battery has a terminal A charged negative and terminal B charged positive so that a potential difference exists between the ends of the copper wire. The free electrons in the copper material will move under the influence of this potential difference. They will be repelled by negative terminal A and attracted by positive terminal B so that movement will take place in a clockwise direction as denoted by the arrows in Fig. 2–6(b).

The flow of electrons in Fig. 2–6(b) is a continuous one; that is, electrons flow through the wire from A to B and then through the battery from terminal B to A and back through the wire etc., etc. This continuous flow can be explained as follows.

(1) Because terminal B of the battery is positive, it will attract electrons out of the end of the wire that is connected to B.

(2) These electrons leaving the wire at B cause that part of the wire to become positively charged since there are now some protons there that are not being neutralized by corresponding free electrons.

(3) Other electrons in the wire now move toward the part of the wire that has become positive. As this happens, the positions vacated by these electrons become positive so that still other electrons will move toward these positions and so on and so forth.

(4) As the electron movement occurs in the wire, eventually the end of the wire at terminal A will contain protons that are not neutralized. However, these protons will attract electrons out of the negative electrode of the battery so that a continuous supply of electrons is present.

From this description it might appear that eventually the positive charge on electrode B would be neutralized as electrons are drawn out of the wire at B and that the negative charge on electrode A would be neutralized as electrons are drawn off this electrode. However, this does not happen because the chemical action internal to the battery is continuously removing electrons from electrode B and depositing them on electrode A. As a result, a constant potential difference is present across the battery terminals and thus across the ends of the wire. This constant potential difference causes the continuous flow of electrons to be maintained.

The path through which the electrons are forced to move is called an *electric circuit* and the movement of electrons under the influence of a potential difference is called *electric current*. It is important to realize that in order for a *continuous* electric current to exist there has to be a complete circuit path such as the one in Fig. 2–6(b). If any part of the path is broken the flow of electrons will stop because the path will now have an insulating material (usually air) between the broken ends of the wire and this material will not conduct current.

Electrons are not the only electric charges that can be made to move although they are, by far, the most prevalent in electric circuits. There are some cases where positive ions or negative ions, which are charged particles, are set in motion by a potential difference. For example, in certain gaseous electron tubes the flow of positive gas ions occurs during normal operation. Another example occurs in nature's brilliant display of large electric currents, called a lightning stroke, where both positive and negative ions in the atmosphere are set in motion. With this in mind, a more precise definition of electric current can be stated.

Electric current is the directed movement of charged particles through any material.

This means that electric current need not necessarily exist only in a wire conductor such as copper. In fact, current will exist in almost any material subjected to a potential difference because there will usually be some free electrons available. As we already know, certain materials possess a greater abundance of free electrons than do other materials. More will be said on this later.

To summarize the last few paragraphs, we can say that the presence of a voltage (potential difference) within a material which possesses free electrons will cause these electrons to move toward the more positive potential. This movement of electrons through a continuous path constitutes the electric current which provides the means by which electrical energy is conveyed.

2.8 Current Direction

Figure 2–7(a) shows a diagram of the simple circuit discussed in the last section. The circuit symbol for the battery should be recognized from Fig. 2–5(b). The conducting wire is symbolized by a heavy black line. From the preceding discussion we know that the voltage produced by the battery will cause electrons to move around the circuit in the direction shown in Fig. 2–7(a) toward the *positive* terminal of the battery. It is apparent, then, that the direction of current in this circuit is clockwise as shown.

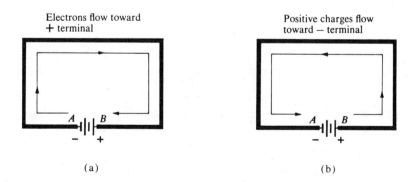

FIGURE 2–7 (a) *Direction of negative charge movement in a conductor.* (b) *Direction of positive charge movement.*

Unfortunately, there exists today in the field of electronics a difference of opinion concerning the direction of current primarily due to the traditional concept of positively charged current carriers. In the early days of electricity starting with Ben Franklin it was mistakenly thought that current in a wire was caused by the movement of *positive* charges. Refer to Fig. 2–7(b) and imagine that the wire is filled with many free positive charges (rather than electrons). In such a case, the movement of these charges would be in the direction shown, toward the *negative* terminal of the voltage source.

It is apparent, then, that the direction of charge movement depends on the polarity (+ or −) of the free charges in the wire. What is not so apparent is that both cases are essentially identical. This can be understood if we realize that when an electron moves *away* from a point it leaves that point more positive (less negative charge). The same result would occur when a positive charge moved *toward* that same point; that is, the point would become more positive. Similarly, when an electron moves toward a point it makes that point more negative, which is the same result that would occur if a positive charge were to move away from that same point. In other words,

Electrons moving in one direction have the same effect as positive charges moving in the opposite direction.

Many textbooks, especially in the areas of physics and higher level electronics, use the concept of positive charge movement when discussing current. Those textbooks would state that the direction of current in Fig. 2–8 is clockwise. When this choice is used for the direction of current, the current is called *conventional* or *traditional current*. Though it has been proven that electrons are the actual moving charges, conventional current is widely used in the field of electronics.

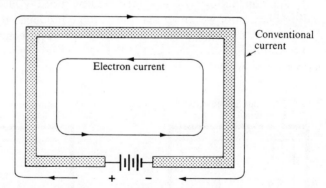

FIGURE 2–8 *Electron current is always in opposite direction to conventional current.*

Throughout most of this text we will assume that the direction of current is the direction in which electrons move. This is called *electron current*. Figure 2–8 emphasizes the fact that electron current is always in the opposite direction to conventional (positive-charge) current. Students of electronics should eventually gain experience in using both current conventions because of the lack of standardization in the field. Unless otherwise indicated, whenever we talk about current we will be referring to *electron current*.

2.9 Units of Current

Electric current is one of the electrical quantities which it is often necessary to measure in a circuit. In order to do this we must establish the basic unit for current. Current, as we know, is the movement of charge. To measure current we must measure the *rate* at which this charge is moving. In other words, we must measure the amount of charge that passes a given point in a given period of time.

Consider, for example, Fig. 2–9, where a length of copper wire is shown. If a potential difference is applied to the ends of the wire electrons will begin to move toward the positive potential. If we could count the number of electrons which pass through a cross-section of the wire, say at point A, in one second, then we could determine how many coulombs per second are flowing through the wire. If we were to count exactly 6.24×10^{18} electrons, this would represent exactly one coulomb per second. If we counted only half that many electrons passing through point A this would represent only one-half coulomb per second of charge flow.

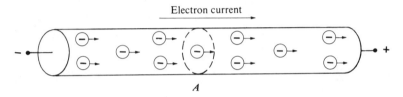

FIGURE 2–9 *Current is measured as coulombs per second passing through a given point.*

To summarize the preceding paragraphs we can say that *the quantity of electric current is defined as the rate of flow of electric charge and the quantity of current is measured in coulombs per second.*

The units of current could be expressed as coulombs per second (C/s). However, in honor of André Marie Ampère, a pioneer in electrical physics, the International System of Units perfers to use the *ampere* as the basic unit of current. Thus,

An ampere is defined as the amount of current produced when one coulomb of charge passes a certain point in a circuit in one second.

The unit symbol for the ampere is A. Thus, we have

$$1 \text{ A} = 1 \text{ C/s}$$

The letter symbol for current is I. For example, the expression

$$I_x = 2.3 \text{ A}$$

means that the current past point x in a certain circuit is equal to 2.3 A (coulombs per second).

We can express the equivalence of current and coulombs per second in equation form as

$$I = \frac{q}{t} \tag{2-2}$$

where I is current in amperes, q is charge in coulombs, any t is time in seconds.

EXAMPLE 2.2 In Fig. 2–9 what would be the value of the current when 6.0 coulombs of charge pass through point A in 2 minutes?

Solution: Using Eq. (2-2)

$$I = \frac{6.0 \text{ C}}{120 \text{ s}} = 0.05 \text{ C/s} = \mathbf{0.05 \text{ A}}$$

Notice that t had to be expressed in seconds in order for I to come out in amperes.

EXAMPLE 2.3 If 3 A is the current in a particular circuit, how long will it take for 6 coulombs to pass through a point in the circuit?

Solution: Equation (2-2) can be rearranged to solve for time t as follows:

$$t = \frac{q}{I}$$

Thus, to find t in our problem, we use the above expression

$$t = \frac{6 \text{ C}}{3 \text{ A}} = \mathbf{2 \text{ s}}$$

Note that for t to come out in seconds, q has to be expressed in coulombs, and I has to be expressed in amperes.

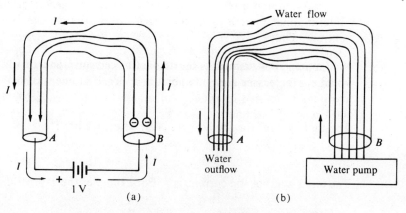

FIGURE 2–10 *Rate of electron flow is constant throughout the circuit.*

Refer to Fig. 2–10(a), which shows a 1-V battery connected across the ends of an odd-shaped wire. In this circuit there will be a steady flow of electrons in a ccw direction. It is important to understand that the current in this circuit is the same at any point in the circuit regardless of the cross-sectional area. In other words, the rate of charge flow is the same at point A as it is at point B even though the wire is wider at point B.

This concept may be understood more easily if we consider the analogy of water being pumped through an odd-shaped pipe as in Fig. 2–10(b). The water pump at B is pumping water into the pipe at a certain rate, say 100 gallons per second. We know that the water will be coming out of the pipe at A at the rate of 100 gallons per second, the same as the input rate. As the water travels through the pipe it will move at a slower speed through the wider portion of the pipe and at a higher speed through the narrow portion. It has to move faster through the narrow portion in order to get the same amount of water through in the same amount of time as in the wider pipe.

In the electrical circuit the electrons act like the water and the battery acts like the pump so that the rate of charge flow (C/s) is constant throughout the circuit. The shape of the wire *does* determine the amount of this current but the amount does not change throughout the current path. Later on we will investigate the effects of the wire shape on the amount of current flow.

2.10 Voltage Causes Current

The amount of current that will flow through a material depends on several factors. The voltage applied to the material is the actual impetus or push which causes the electrons to move. Normally, if the amount of voltage is increased, the current through the material will also increase. The additional voltage causes the electrons to move at a faster rate and, therefore, the amount of charge per second passing through the circuit will be increased.

It was mentioned in chap. 1 that certain materials called *insulators* were greatly deficient in the number of free electrons available to conduct current. If a potential difference is applied across an insulator there will normally be only a minute current flow. This small current is due to the fact that even in an insulator there will always be a relatively small number of valence electrons which can acquire the energy needed to become free conduction electrons.

Under certain conditions, an insulator, or any material for that matter, will conduct large amounts of current. This will happen whenever a large enough voltage is impressed across the insulator. This is illustrated in Fig. 2–11, where a voltage is applied to a length of rubber (a good insulator). For a voltage of 500 V a negligible amount of current is produced through the rubber. However, when 500,000 V is applied to the piece of rubber a large current will be produced because of a phenomenon called *insulator breakdown*. Insulator breakdown occurs when the applied voltage is so large that it actually pulls bound valence electrons out of their orbits and causes them to conduct current. When this happens the insulator behaves like a conductor because of the abundance of free electrons. Usually, the high current that is produced during breakdown destroys the structure of the material, making it useless as an insulator.

28 Chapter Two

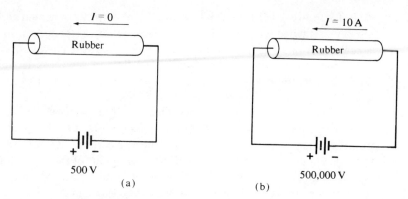

FIGURE 2-11 *Insulator breakdown.*

Another example of insulator breakdown is the *arcing* over that occurs when a switch is open in a circuit. The arcing occurs because the potential difference across the contacts of the switch ionizes the air molecules (pulls electrons out of the atoms) causing current to flow through the air.

To summarize:

(1) Voltage applied across a material produces current.
(2) As voltage is increased, the current usually increases.
(3) The amount of current for a given voltage depends on the electrical properties of the material.
(4) Any material will conduct current readily if a large enough voltage is applied to it.

2.11 Speed of Electricity

Consider the situation in Fig. 2–12. In the diagram the circuit is shown disconnected because the switch is open. No current will flow in this condition. Suppose the switch is closed to complete the circuit path. We know that the electrons in the wire at A will be attracted by the + side of the battery and the electrons in the wire at B will be repelled by the − side of the battery. As a result, electron flow will take place in a ccw direction.

The question at this point is this: How long does it take for the current to start flowing after the switch is closed? We know from experience that it doesn't take very long. Typically, it might take *one-millionth* of a second. This means that the current is established in the wire almost instantaneously. However, the speed of any one electron moving through the wire might only be 0.001 inches per second (1 inch in 16 minutes). In other words, even though the electrons themselves are moving slowly, the effect of the changes in position of the electrons ripples through the wire almost instantaneously, producing the resultant current. This is analogous to the situation where a long tube is completely filled with marbles. If one marble is pushed into the one end of the tube, then one marble must be forced out of the other end of the tube. Thus, although it may take a long time for a given marble to travel completely through the tube, the effect of inserting a marble at one end is almost instantly felt at the other end.

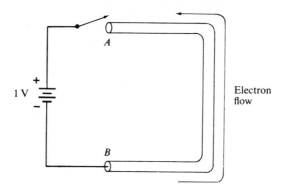

FIGURE 2-12 *Electron flow begins almost instantaneously upon closing the switch.*

2.12 Circuit Terminology

In working with electrical circuits one must become familiar with the terminology commonly associated with electrical circuits. The terms explained here will be used throughout the text. Many more terms will be introduced in the chapters to follow.

Every electrical circuit has prescribed paths through which current will flow. These paths are often called circuit *branches*. These branches contain electrical or electronic devices such as *resistors, coils, transistors*, etc., that are interconnected by conducting wire (usually copper) to one or more sources of voltage. For example, consider Fig. 2-13, where three devices, A, B, C, are connected in a certain arrangement to voltage source E.

This circuit has *three* branches; that is, there are three separate current paths. These paths are: (1) from Z to Y and through the voltage source to X; (2) from X to Z through device B; and (3) from X to Z through device C. There is no limit to the number of branches in a circuit, but in practice most of the circuits encountered will have from one to five branches.

The points in the circuit where two or more branches are joined together are called *nodes* or *junctions*. In Fig. 2-13 points Z and X are examples of nodes. Point Y is not really a node because it is in the middle of a branch. However, if desired, it can be considered as a node by breaking up branch 1 into two smaller branches.

The symbol E is often used to represent voltage sources. This E stands for *electromotive force* (emf) which comes from the idea that the voltage source

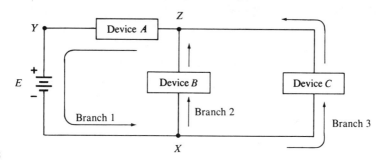

FIGURE 2-13

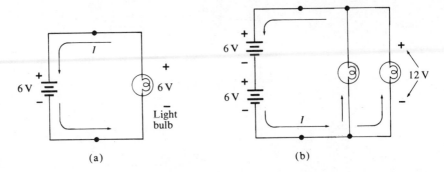

FIGURE 2-14

produces a *force* which causes *electron motion*. In fact, voltage sources are also called *sources of emf*. We will use both terms interchangeably.

In our circuit work, a distinction must be made between a *source* of voltage and a *user* of voltage. The source of voltage produces the emf which causes current to flow. The user of voltage, which is called a *load*, is connected to the voltage source, and the current which flows *through* the load provides energy *to* the load. Figure 2-14(a) shows an example of this. Here a 6-V battery is connected across a small light bulb which serves as the *load*. The 6-V battery provides a 6-V emf to the light bulb. The resulting current through the bulb heats up the bulb's filament and causes the bulb to glow.

In some circuits there may be more than one source of emf and more than one load. Such arrangements will be encountered at a later time, but an example of one is shown in Fig. 2-14(b).

Usually we refer to the voltage sources in a circuit as the *input* since they supply the input energy for the electrons. The voltage across the load or the current through the load is usually called the *output* because the load uses up the input energy. Circuits can have more than one input or more than one output as we shall see.

2.13 Opposition in Electrical Circuits

When a given voltage is applied to a wire or to an electrical device, the amount of current which flows will depend on the physical properties of the material. For example, if 6 V is applied to a copper rod, the resultant current will be much different than if that 6 V were applied to an identical carbon rod, or to a copper rod with different physical dimensions.

In other words, every specimen of material and every electrical device has its own specific effect on the flow of current. We say that every material or device presents a certain degree of *opposition* to the flow of current. This opposition is another important electrical quantity; it is given the special name *resistance* and the special symbol R.

Ohm's Law This is a basic electrical law, which we study thoroughly in chap. 5, that relates current, voltage, and resistance. It says that for many materials the amount of current, I, which will flow for an applied voltage, V, can be determined from the following formula

$$I = V/R \qquad (2\text{--}3)$$

where R is the material's resistance.

The basic unit for resistance is the *ohm*, given the symbol Ω, which looks like an inverted horseshoe. The following examples will illustrate the use of this relationship. A much more detailed study of Ohm's law and its various forms and implications will be undertaken in chap. 5.

EXAMPLE 2.4 How much current will flow through a light bulb that has a resistance of 30 Ω if a 1.5-V battery is connected across the bulb?

Solution: $I = \dfrac{V}{R} = \dfrac{1.5\ \text{V}}{30\ \Omega} = \mathbf{0.05\ A}$

EXAMPLE 2.5 The device in Fig. 2–15 has a resistance of 100 Ω. Determine the direction and magnitude of current in the circuit.

Solution: The current will flow clockwise out of the negative terminal of the battery and around to the positive terminal. The size of the current is

$$I = \dfrac{V}{R} = \dfrac{30\ \text{V}}{100\ \Omega} = \mathbf{0.3\ A}$$

EXAMPLE 2.6 Repeat the previous example for an applied voltage of 60 V.

Solution: $I = \dfrac{V}{R} = \dfrac{60\ \text{V}}{100\ \Omega} = \mathbf{0.6\ A}$

This assumes that the device's resistance, R, has not changed from the value given when the voltage was 30 V.

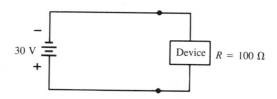

FIGURE 2–15

2.14 Electrical Power

We have said that electrical energy is present whenever a voltage produces current in a circuit. This energy is supplied by the voltage source to the various devices in the circuit. In fact, we can say that the voltage source is actually an energy source. The *rate* at which this energy is supplied is called *electrical power*. We will study power in great detail in later chapters, but for now we will simply state the basic relationship between power, voltage, and current.

$$\text{Power} = \text{Voltage} \times \text{Current}$$
$$P = V \times I \quad (2\text{--}4)$$

The basic unit for power is the *watt* which is given the symbol W. A single watt of power represents energy being supplied at the rate of one joule per second. The relationship above can be used to calculate the power supplied to a circuit by a voltage source and the power consumed by each electrical device in the circuit. We will demonstrate this in chap. 9, but for now the following examples will illustrate how this basic formula is used.

EXAMPLE 2.7 When 120 V is applied to a certain light bulb, the current through the bulb is 0.833 A. How much power is the voltage source delivering to the light bulb?

Solution: $P = V \times I$
$= 120 \text{ V} \times 0.833 \text{ A}$
$= \mathbf{100 \text{ W}}$

EXAMPLE 2.8 How much power is being delivered to the device in Fig. 2.15?

Solution: $I = \dfrac{V}{R} = \dfrac{30 \text{ V}}{100 \text{ }\Omega} = 0.3 \text{ A}$
$P = V \times I = 30 \text{ V} \times 0.3 \text{ A}$
$= \mathbf{9 \text{ W}}$

Chapter Summary

1. The coulomb (C) is the basic unit of charge.
2. The electrostatic force between two electric charges is directly proportional to the product of the two charges and inversely proportional to the square of the distance between the charges.
3. Electrostatic force between charges is a mutual force felt equally by both charges.
4. An object which is above the earth's surface has gravitational potential energy.
5. An object with potential energy has the potential to do work.
6. A charge has electrostatic potential energy if a force is being exerted on it by another charge (or charges).
7. The amount of electrostatic potential energy per coulomb possessed by a charge is called electrostatic potential (joules/coulomb).
8. A volt is equal to a joule/coulomb.

Basic Electrical Quantities

9. When a charge has a different potential at one point than it has at another point, then the two points are said to have a difference in potential.
10. A voltage (*V*) is a difference in potential between two points. The basic unit for voltage is the volt (V).
11. A voltage source employs some method of producing a charge imbalance between two points.
12. A voltage source produces continuous electron movement in a wire in a direction toward the positive terminal of the source.
13. A complete circuit path is required in order for a continuous current to flow.
14. Electric current is the movement of any charged particles through any material.
15. Electrons moving in one direction have the same effect as positive charges moving in the opposite direction.
16. Unless otherwise indicated, whenever we talk about current we will be referring to electron current.
17. Electric current is measured as the rate of flow of electric charge (coulombs per second).
18. The basic unit of current is the ampere (A).
19. An ampere of current is one coulomb of charge passing a certain point in a circuit in one second.
20. The letter symbol for current is *I*.
21. In a continuous circuit path the current has the same value at all points along the path, regardless of the shape of the path.
22. An insulator will conduct large currents if the voltage impressed across it causes insulator breakdown.
23. Current is established in a circuit almost instantaneously upon the application of a voltage to the current.
24. Branches are current paths in a circuit.
25. Nodes (or junctions) are points in a circuit where two or more branches are joined together.
26. Voltage sources are often called sources of electromotive force (emf).
27. A voltage source is a supplier of input electrical energy.
28. A load is a user of electrical energy.
29. The voltage source is usually called the input and the load voltage or current is usually called the output.
30. Resistance is opposition to current flow and is measured in ohms (Ω).
31. Ohm's law: $I = V/R$
32. Electrical power is the rate at which energy is being supplied or used: $P = V \times I$. The basic unit for power is the watt (W).

Questions/Problems

Sections 2.2–2.3

2–1 What is the basic unit of charge, and what is its unit symbol?

2–2 State Coulomb's law in words and then in equation form.

2–3 Consider Fig. 2–16(a). Charges q_1 and q_2 are both *positive* charges, and are a distance d apart. The repelling force between them happens to be 2 newtons. (a) What would be the magnitude of the force if the distance between the charges is doubled? (b) What would the force be if q_1 is doubled and q_2 is halved? (c) What would the force be if q_1 is made a *negative* charge but with the same amount of charge?

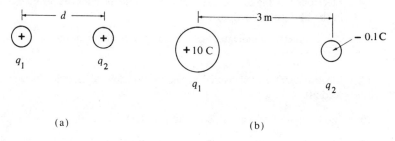

(a)   (b)

FIGURE 2–16

Sections 2.4–2.5

2–4 In the situation of Fig. 2–16(b), which charge would you expect to move faster as a result of electrostatic attraction? What will eventually happen if the two charges are not restrained from moving?

2–5 How do we know that an object held above the earth's surface has potential energy?

2–6 How do we know that a charge which is held a certain distance from a charged plate has potential energy?

2–7 Refer to Fig. 2–3. Suppose that when $q_1 = 0.5$ C it had a potential energy of 4 joules. If q_1 were increased to 2 C, what would be its potential energy?

2–8 What is the *potential* of q_1 in both cases of question 2–7?

2–9 Refer to Fig. 2–4. Assume q_1 has a potential of 10 volts at B and zero volts at A. (a) Use the gravitational analog (staircase) to determine the potential of q_1 if it were halfway between the plates. (b) Repeat part (a) if q_1's charge is doubled.

2–10 Express the volt in terms of its basic units.

Sections 2.6–2.8

2–11 Describe some common voltage sources.

2–12 Explain how a battery will produce a continuous current in a wire.

2–13 Define electric current.

2–14 Which of the following constitute electric current?
  a. electrons flowing through a copper wire
  b. electrons flowing through a solution of salt water
  c. nitrogen ions moving through the atmosphere
  d. neutrons moving at high speed in a cyclotron
  e. charges resting on a charge belt

2–15 What is the difference between *electron current* and *conventional current*?

2–16 Determine the current directions in Fig. 2–17.

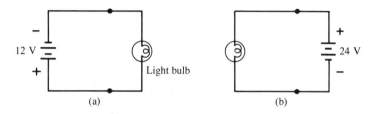

FIGURE 2–17

Section 2.9

2–17 What is the basic unit of current?

2–18 Determine the value of current in each of the following cases:
  a. 0.1 coulomb passing a given point in 0.01 second
  b. 100 coulombs passing a given point in 1 minute
  c. 5 coulombs passing a given point in 1 second

2–19 Which of the following represents the largest amount of current:
  a. 100 electrons moving at a speed of 1 m/s
  b. 1 electron moving at a speed of 1000 m/s

2–20 If a circuit has a current of 10 A how long will it take for 0.1 C of charge to pass through a point in the circuit?

2–21 In Fig. 2–18 both wires have charge flowing through them at the rate of 30 coulombs per minute. Which wire has the largest current?

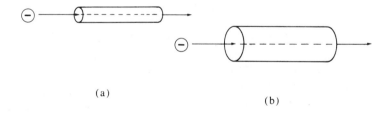

FIGURE 2–18

2–22 If a current of 5 A exists in a circuit, how much charge will pass a certain point in the circuit in 0.2 second?

2–23 Consider the situation in Fig. 2–19. Current produced by the battery enters point X. This current I_X is 0.3 A. What is the value of current I_Y at point Y?

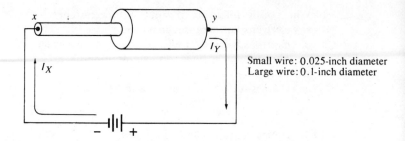

FIGURE 2-19

Section 2.10

2-24 If the battery in Fig. 2-19 has its voltage increased what would happen to I_X and I_Y?

2-25 Under what condition will an insulator conduct large currents?

2-26 Indicate which of the following factors affect the amount of current through a certain object:
 a. voltage across the object
 b. material of the object
 c. shape and size of the object

2-27 Consider the circuit in Fig. 2-20.
 a. Label the major nodes in the circuit.
 b. Determine how many branches the circuit contains.

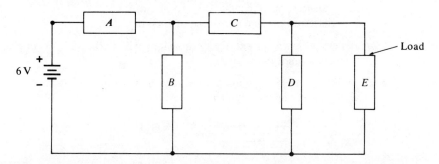

FIGURE 2-20

2-28 If the emf in Fig. 2-20 is increased, what do you intuitively expect to happen to the current through the load?

Section 2.13

2-29 120 V is applied to a motor's armature coil which has a resistance of 5 Ω. How much current flows?

2-30 A small light bulb draws 0.03 A from a 9-V battery. What is the bulb's resistance? (Hint: transpose the Ohm's law equation and solve for R.)

2-31 A technician applies a voltage from a variable power supply to a light bulb which has $R = 75 \, \Omega$. He increases the voltage until the current is 0.2 A. What is the size of the voltage at that point? (Hint: transpose the Ohm's law formula and solve for V.)

Section 2.14

2-32 Calculate the power in problem 2-29.

2-33 Calculate the power in problem 2-30.

2-34 How much current will a 75-W light bulb draw from a 120-V source?

CHAPTER 3

Scientific Notation and Prefixed Units

3.1 Introduction

In the study of electricity and electronics it is often necessary to make mathematical calculations in order to estimate how a circuit will operate or to help determine whether it is operating satisfactorily. To help in these calculations, which usually involve numbers ranging from the very small (0.000000000001) to the very large (1,000,000,000,000), several techniques are available. In this chapter we will explain and illustrate these techniques, and they will be used throughout the text to facilitate problem solving.

Although modern electronic calculators have made it unnecessary to do complex calculations by hand, mastery of the material in this chapter will help you to get quick, approximate answers to simple problems without a calculator. While the calculator is indeed a valuable tool, it is important for anyone working in technology to develop a "feel" for numbers and not become totally dependent on a little box with buttons and flashing lights.

3.2 Scientific Notation — Powers of Ten

Numbers expressed in arabic numerals such as 47, 119.5, and 30,000 are said to be expressed in *ordinary* notation. This is the type of notation we are most familiar with. This ordinary notation is satisfactory for certain calculations.

Note: Most of this chapter may be skipped over if the student is familiar with powers of ten and scientific notation. However, secs. 9 and 10 on *prefixed units* should be covered before going on to chap. 4.

However, in the sciences and technologies the size of the numbers involved covers a wide range. For example, the mass of an electron is equal to 0.000000000000000000000000000911 kg, while the number of electrons in one coulomb of charge is 6,240,000,000,000,000,000.

Both of these quantities are expressed in ordinary notation. It would be very cumbersome to use these numbers to make a calculation. To illustrate, suppose we wanted to calculate the total mass of 1 coulomb of negative charge. To do this we have to multiply the mass of one electron by the number of electrons that make up one coulomb of charge. We would have

$$0.000000000000000000000000000911 \text{ kg}$$
$$\times \quad 6,240,000,000,000,000,000 \text{ electrons/coulomb}$$

There is no need to proceed any further with this calculation. It is obvious that the procedure is cumbersome and quite inconvenient.

Fortunately, there is an alternative to representing numbers in ordinary notation. It is called *scientific notation*. In scientific notation we overcome the inconvenience of ordinary notation by using *powers of ten* to represent large or small values. Before explaining scientific notation, let's review powers of ten and their relationship to ordinary notation. Table 3–1 lists various powers of ten and their equivalents in ordinary notation (decimal equivalents).

TABLE 3–1

Power of Ten	Decimal Equivalent
10^{-12}	0.000000000001
10^{-9}	0.000000001
10^{-6}	0.000001
10^{-3}	0.001
10^{0}	1.0
10^{+1}	10.0
10^{+2}	100.0
10^{+3}	1,000.0
10^{+6}	1,000,000.0
10^{+9}	1,000,000,000.0

Any number raised to the zero power is equal to one. Thus, we can say that

$$10^0 = 1.0$$

This expression is the basis for converting powers of ten to their decimal equivalent or vice versa. For example, to determine the decimal equivalent of 10^3 we would look at the *exponent* which is $+\underline{3}$. This $+3$ tells us to take the basic expression 1.0 and move the decimal point $\underline{3}$ places to the *right*. This results in 1,000. Thus, $10^3 = 1000$. Thus, we can generally state that:

To convert a positive power of ten to its decimal equivalent, take the basic expression 1.0 and move the decimal point to the right a number of places equal to the exponent.

This procedure can be verified by checking Table 3–1.

EXAMPLE 3.1 Convert 10^{+5} to decimal form.

Solution: $10^{+5} = 1.00000 = 100{,}000$ (5 places to right)

Now if we wanted to determine the decimal equivalent of 10^{-3} we would look at the exponent which is -3. This negative exponent tells us to take the basic expression 1.0 and move the decimal point 3 places to the *left*. This results in 0.0010 or simply 0.001, since the last zero is not needed. In general, we can state that:

To convert a negative power of ten to its decimal equivalent, take the basic expression 1.0 and move the decimal point to the left a number of places equal to the exponent.

This procedure can be verified by checking Table 3–1.

EXAMPLE 3.2 Convert 10^{-5} to decimal form.

Solution: $10^{-5} = 00001.0 = 0.00001$ (5 places to left)

The procedure for converting a decimal equivalent to a power of ten is the complete reverse of the process just described. In going from a power of ten to decimal, we started with 1.0 and moved the decimal point so many places to the right (positive exponent) or to the left (negative exponent). In going from decimal to powers of ten, we take the decimal point and move it to the left or right until we are left with 1.0, counting the number of places as we go along. If the movement of the decimal point was to the left, then the power of ten is positive. If it was to the right, then the power of ten is negative. In either case, the magnitude of the exponent is equal to the number of places the decimal point was moved. To illustrate:

$$0.00001 \rightarrow 0.00001 \quad \text{(5 places to } right\text{)}$$

Therefore,

$$0.00001 = 10^{-5}$$

Another illustration:

$$10{,}000 \rightarrow 1\,0{,}000. \quad \text{(4 places to } left\text{)}$$

Therefore,

$$10{,}000 = 10^{+4}$$

This procedure can be verified by checking Table 3–1.

It should be pointed out that the conversion processes carried out here are only valid for *integral* (whole number) powers of ten. However, this is not a drawback because in scientific notation, as we shall see, only integral powers of ten are used.

3.3 Scientific Notation — Converting from Ordinary Notation

Following are some numbers written in both ordinary notation (left-hand column) and scientific notation (right-hand column). The key to the scientific notation may be apparent from these examples. That is, each number is expressed as a product of *two* numbers. The first is *always* a number between 1.000 . . . and 9.999 . . . and the second is a power of ten. The power of ten is used to "keep track" of the decimal point.

$$53.4 = 5.34 \times 10^{+1}$$
$$3782.57 = 3.78257 \times 10^{+3}$$
$$0.037 = 3.7 \times 10^{-2}$$
$$0.825 = 8.25 \times 10^{-1}$$

The procedure for converting from ordinary notation to scientific notation is as follows:

Step (1). If the number in ordinary notation is greater than 1.0, then shift the decimal point to the *left* until there is only *one* digit to the left of the decimal point. The number that remains is called the *coefficient*.

Step (2). Count the number of places the decimal point was moved. This number becomes the *positive* exponent for the power of ten.

Step (3). The number expressed in scientific notation is the product of the coefficient and the power of ten.

Step (4). If the number in ordinary notation is less than 1.0 (a decimal fraction like 0.257), then shift the decimal point to the *right* until there is one digit to the left of the decimal point. The number that remains is the coefficient.

Step (5). Count the number of places the decimal point was moved. This number becomes the *negative* exponent for the power of ten.

Step (6). The number expressed in scientific notation is the product of the coefficient and the power of ten.

Several examples follow which employ this procedure.

I. $3275 = 3275. \rightarrow 3.275.$ (3 places to left, step 1)
    coefficient = 3.275
    exponent = +3   (step 2)
    $3275 = 3.275 \times 10^{+3}$ (step 3)

II. $637.615 \rightarrow 6.37.615$ (2 places to left, step 1)
    coefficient = 6.37615
    exponent = +2   (step 2)
    $637.615 = 6.37615 \times 10^{+2}$ (step 3)

III. $0.0754 \rightarrow 0.07.54$ (2 places to right, step 4)
    coefficient = 7.54
    exponent = -2 (step 5)
    $0.0754 = 7.54 \times 10^{-2}$ (step 6)

IV. $0.00043 \rightarrow 0.0004.3$ (4 places to right, step 4)
    coefficient = 4.3
    exponent = -4 (step 5)
    $0.00043 = 4.3 \times 10^{-4}$ (step 6)

V. 3.761   (no shift of decimal point)
    coefficient = 3.761
    exponent = 0
    $3.761 = 3.761 \times 10^{0}$

This last illustration is the special case where the number falls between 1.0 and 9.999 .... In this case no decimal point shift is needed. Notice in all the illustrations that the coefficient is always a number between 1.0 ... and 9.999 ..., and the exponent of the power of ten is equal to the number of places that the decimal point had to be moved in order to produce such a coefficient. Also notice that the exponent of ten is negative only when the original number is a decimal fraction (less than 1.0).

For a more dramatic illustration of the advantage of scientific notation, consider the mass of an electron, which we can convert to scientific notation as follows:

$0.\underline{0000000000000000000000000000009}11$ kg (31 places to right)
coefficient = 9.11; exponent = -31
scientific notation = 9.11×10^{-31} kg

It is obvious that it is much more efficient to express the mass of an electron in scientific notation. This is one advantage of scientific notation. An even more significant benefit will become apparent when we perform later calculations.

3.4 Converting from Scientific to Ordinary Notation

As it might be expected, to convert a number represented in scientific notation to ordinary notation we simply reverse the process applied in the last section. Consider the following examples:

(a) $3.75 \times 10^{+2} = 375$
(b) $2.43 \times 10^{-5} = 0.0000243$
(c) $8.6 \times 10^{+3} = 8600$
(d) $1.1 \times 10^{-2} = 0.011$

The process of conversion is fairly straightforward. The decimal point in the coefficient is moved to the right or to the left depending on whether the exponent in the power of ten is positive or negative respectively. The number of places the decimal point is moved is equal to the magnitude of the exponent.

In example (a), the power of ten is $+2$. Thus, the decimal point is moved 2 places to the right so that $3.75 \times 10^{+2}$ becomes 375. In example (b) the exponent is -5 which requires that the decimal point be moved 5 places to the left. Thus, 2.43×10^{-5} becomes 0.0000243.

In doing electrical circuit problems one must often convert back and forth between ordinary notation and scientific notation. As one gains experience in performing such conversions it should be possible to visualize the actual value of a number whether the number is in ordinary notation or scientific notation. Also, with practice the steps in the conversion processes usually can be condensed into one step. This facility will develop on its own.

3.5 Addition and Subtraction Using Scientific Notation

In order to add or subtract quantities expressed in scientific notation, the powers of ten must be identical. Consider the following examples:

(a) $(3.5 \times 10^2) + (1.8 \times 10^2) = (3.5 + 1.8) \times 10^2 = 5.3 \times 10^2$
(b) $(6.7 \times 10^{-4}) - (2.5 \times 10^{-4}) = (6.7 - 2.5) \times 10^{-4} = 4.2 \times 10^{-4}$

(c) $(2.8 \times 10^{-3}) - (7.1 \times 10^{-3}) = (2.8 - 7.1) \times 10^{-3} = -4.3 \times 10^{-3}$
(d) $(3.1 \times 10^{2}) - (3.0 \times 10^{2}) \quad = (3.1 - 3.0) \times 10^{2} = 0.1 \times 10^{2}$
(e) $(8.4 \times 10^{-1}) + (6.6 \times 10^{-1}) = (8.4 + 6.6) \times 10^{-1} = 15.0 \times 10^{-1}$

Several important points are brought out in these examples. First of all, notice that the + sign has been left out when positive powers of ten are represented (10^2 instead of 10^{+2}). This is done because the + sign is understood as it is in all algebraic operations. The student may want to keep writing the + sign in the exponent if it helps to make things clearer.

Secondly, notice that in examples (d) and (e) the results do not come out in standard scientific notation form. That is, the coefficients of the results do not lie between 1.0 and 9.999 . . . . These results can be converted to standard form by shifting the decimal point and changing the power of ten accordingly. To illustrate, the result of case (d) was 0.1×10^2. Thus, we have

$$0.1 \times 10^{2} = 1.0 \times 10^{1} \quad \text{(one place to right, subtract one from exponent)}$$

By moving the decimal point *one* place to the right we essentially increased the coefficient by a factor of ten. In order to keep the value of the number unchanged, then, the power of ten has to be *decreased* by *one*. To further illustrate, the result of case (e) was 15.0×10^{-1}. Thus, we have

$$15.0 \times 10^{-1} = 1.5 \times 10^{0} \quad \text{(one place to left, add one to exponent)}$$

Since the decimal point has to be shifted *one* place to the left (decreasing the coefficient by factor of ten), the power of ten has to be increased by *one*.

Frequently, when numbers in scientific notation are to be added or subtracted, the exponents of the various quantities are not the same. For example, in order to add 3.75×10^5 to 4.1×10^4 the powers of ten must be made identical. There are two possibilities:

$$(3.75 \times 10^{5}) + (4.1 \times 10^{4}) = (37.5 \times 10^{4}) + (4.1 \times 10^{4})$$
$$= 41.6 \times 10^{4} = 4.16 \times 10^{5}$$

or

$$(3.75 \times 10^{5}) + (4.1 \times 10^{4}) = (3.75 \times 10^{5}) + (0.41 \times 10^{5})$$
$$= 4.16 \times 10^{5}$$

Either number may be changed to nonstandard scientific notation in order to obtain identical powers of ten before the addition is performed.

Similar procedures are followed in subtraction. To illustrate:

$$(4.5 \times 10^{-2}) - (1.7 \times 10^{-1}) = (0.45 \times 10^{-1}) - (1.7 \times 10^{-1})$$
$$= -1.25 \times 10^{-1}$$

or

$$(4.5 \times 10^{-2}) - (1.7 \times 10^{-1}) = (4.5 \times 10^{-2}) - (17 \times 10^{-2})$$
$$= -12.5 \times 10^{-2} = -1.25 \times 10^{-1}$$

The key principle in all the operations performed thus far can be summarized as: *in order to keep the value of a number unchanged, each time we increase the power of ten by one we must also move the decimal point one place to the left; and each time we decrease the power of ten by one we must also move the decimal point one place to the right; and vice versa.* After a while, it will become second nature as more and more problems are performed.

3.6 Multiplication and Division Using Scientific Notation

Multiplication of quantities written in scientific notation *does not* require identical powers of ten. Consider the following example:

$$\begin{aligned}(7.5 \times 10^2) \times (4.0 \times 10^3) &= (7.5 \times 4.0) \times (10^2 \times 10^3) \\ &= 30.0 \times 10^{2+3} \\ &= 30.0 \times 10^5 \\ &= 3.0 \times 10^6\end{aligned}$$

In this example, we first multiplied the coefficients (7.5 and 4.0) together, resulting in 30.0. Then, we multiplied the powers of ten (10^2 and 10^3), which is performed by *adding* the exponents (2 and 3), resulting in 10^5. The total result is 30.0×10^5, which can be put in standard form by shifting the decimal point one place to the *left* and *adding one* to the power of ten (3.0×10^6).

The exponents of the powers of ten are always added in multiplication. The addition is an *algebraic* one, which means that the addition process takes account of the signs of the exponents. For example:

$$\begin{aligned}(8.1 \times 10^{-2}) \times (5 \times 10^1) &= (8.1 \times 5) \times (10^{-2} \times 10^1) \\ &= 40.5 \times 10^{-2+1} \\ &= 40.5 \times 10^{-1} \\ &= 4.05 \times 10^0\end{aligned}$$

In this example, the exponents -2 and 1 were added to give a resultant exponent of -1.

Consider the following example of division of two numbers in scientific notation:

$$\frac{3.2 \times 10^4}{1.6 \times 10^2} = \left(\frac{3.2}{1.6}\right) \times \left(\frac{10^4}{10^2}\right) = 2 \times 10^{4-2} = 2 \times 10^2$$

In this example we first divided the coefficient of the numerator (3.2) by the coefficient of the denominator (1.6), which results in 2. Then, we divided the powers of ten (10^4 by 10^2), which is performed by subtracting the exponent of the denominator from the exponent of the numerator, resulting in 10^2. The total result is 2×10^2.

The exponents of the powers of ten are always subtracted in division. The subtraction is an *algebraic* one so that signs must be taken into account. For example:

$$\begin{aligned}\frac{8.1 \times 10^3}{9.0 \times 10^{-2}} &= \left(\frac{8.1}{9.0}\right) \times \left(\frac{10^3}{10^{-2}}\right) = 0.9 \times 10^{3-(-2)} \\ &= 0.9 \times 10^5 = 9 \times 10^4\end{aligned}$$

Chapter Three

In this example, the exponent -2 is subtracted from 3 to give a resultant of $3 - (-2) = 3 + 2 = 5$. Notice, also, that the division of the coefficients results in a number less than one so that the decimal point has to be shifted to the right to bring it back to standard scientific notation.

3.7 Finding Square Roots in Scientific Notation

Consider these examples:

$$\sqrt{6.25 \times 10^2} = \sqrt{6.25} \times \sqrt{10^2}$$
$$= 2.5 \times 10^1$$
$$\sqrt{3.24 \times 10^{-6}} = \sqrt{3.24} \times \sqrt{10^{-6}}$$
$$= 1.8 \times 10^{-3}$$

The procedure in finding the square root is to find the square root of the coefficient and then the square root of the power of ten. The square root of the power of ten is simply obtained by dividing the exponent by two.

Consider these examples:

$$\sqrt{6.25 \times 10^3} = \sqrt{62.5 \times 10^2} = \sqrt{62.5} \times \sqrt{10^2}$$
$$= 7.9 \quad \times 10^1$$
$$\sqrt{1.6 \times 10^{-5}} = \sqrt{16 \times 10^{-6}} = \sqrt{16} \times \sqrt{10^{-6}}$$
$$= 4 \quad \times 10^{-3}$$

In both of these examples it was necessary to change the number to nonstandard form in order to have an *even* power of ten. This is necessary if we are to end up with an integral exponent in the final result. The easiest approach is to increase the coefficient by a factor of ten (shift decimal point one place to the right) and subtract one from the exponent. In this way, the final result always comes out in standard scientific notation.

Very often, the square root of a number is written as the one-half power of that number. That is,

$$\sqrt{3.5 \times 10^4} \quad \text{and} \quad [3.5 \times 10^4]^{1/2}$$

are equivalent ways of representing the square root of 3.5×10^4. Both representations will be used in the text.

3.8 Calculations Using Scientific Notation

The following examples will illustrate the use of scientific notation in making calculations as part of solving a problem.

EXAMPLE 3.3 Determine the mass of one coulomb of negative charge (one coulomb's worth of electrons).

Solution: In sec. 3.2, this same problem was set up to be performed using ordinary notation and was found to be extremely cumbersome. The mass of an electron can be expressed in scientific notation as

$$0.000000000000000000000000000000911 \text{ kg} = 9.11 \times 10^{-31} \text{ kg}$$

The number of electrons that make up one coulomb of electric charge can be written as

$$6{,}240{,}000{,}000{,}000{,}000{,}000 = 6.24 \times 10^{18} \text{ electrons per coulomb}$$

The mass of one coulomb of electrons is thus obtained as

$$(9.11 \times 10^{-31} \text{ kg}) \times (6.24 \times 10^{18}) = (9.11 \times 6.24) \times (10^{-31} \times 10^{18}) \text{ kg}$$
$$= 56.85 \times 10^{-13} \text{ kg} = \mathbf{5.685 \times 10^{-12} \text{ kg}}$$

EXAMPLE 3.4 A certain length of material has a voltage applied to it. The voltage causes charge to flow through the material at the rate of 0.000125 amperes. In one-thousandth of a second, how much charge will pass through the material?

Solution: Using Eq. (2–2) and rearranging to solve for Q, we have

$$Q = It$$

In this problem, $I = 0.000125$ A, which can be written as 1.25×10^{-4} A. The time t is 0.001 s, which can be written as 10^{-3} s. Substituting, we have

$$Q = (1.25 \times 10^{-4} \text{ A}) \times (10^{-3} \text{ s})$$
$$= (1.25 \times 10^{-4} \times 10^{-3}) \text{ A} \times \text{s}$$
$$= (1.25 \times 10^{-7}) \frac{C}{\cancel{s}} \times \cancel{s} = \mathbf{1.25 \times 10^{-7} \text{ C}}$$

Notice in the last step that the unit A (amperes) was changed to its equivalent C/s (coulombs per second). More work with units will be presented in a later section.

EXAMPLE 3.5 How much current will flow through a device which has a resistance of 2500 Ω when 5 V is applied to it?

Solution: $I = \dfrac{V}{R} = \dfrac{5 \text{ V}}{2500 \text{ Ω}} = \dfrac{5 \text{ V}}{2.5 \times 10^{3} \text{ Ω}}$
$= 2 \times 10^{-3} \text{ A} = \mathbf{0.002 \text{ A}}$

3.9 Prefixed Units

We have considered the basic units for all of the physical quantities discussed thus far. The unit of voltage is the volt, of current is the ampere, of charge is the coulomb, of time is the second, of length is the meter, and so on. In practice, when these physical quantities are encountered they are often much smaller or much larger than one basic unit. For example, in most transistor circuits the amount of current is usually in the *thousandths* of ampere range (10^{-3} amperes). In integrated circuits, physical dimensions are usually in the range of *millionths* of a meter (10^{-6} meter). In power transmission, voltages are typically in the range of *thousands* of volts (10^{3} volts).

As such, we often make use of *prefixed units*, which are derived from the basic unit by the addition of a prefix to the name of the basic unit. Consider the examples shown in Table 3-2.

TABLE 3-2

Basic Unit	Prefixed Units
ampere	milliampere, microampere, nanoampere
coulomb	microcoulomb, picocoulomb
volt	millivolt, microvolt, kilovolt, megavolt
meter	millimeter, kilometer
second	millisecond, microsecond, nanosecond, picosecond

Each of the prefixed units, such as *milliampere*, consists of a prefix (*milli*) and the basic unit (ampere). The prefix in each case is a *multiplier* which is used to modify the basic unit by a certain power of ten. For instance, the prefix "milli" means 10^{-3} or one-thousandth. Thus, we can say that

$$1 \text{ milliampere} = 1 \times 10^{-3} \text{ amperes} = 1/1000 \text{ amperes}$$

and

$$1 \text{ millimeter} = 1 \times 10^{-3} \text{ meters} = 1/1000 \text{ meters}$$

The prefix "milli," then, *multiplies* the basic unit by 10^{-3}.

Another example is the prefix "kilo," which stands for 10^3 or one thousand. Thus,

$$1 \text{ kilovolt} = 1 \times 10^3 \text{ volts} = 1000 \text{ volts}$$

and

$$1 \text{ kilometer} = 1 \times 10^3 \text{ meters} = 1000 \text{ meters}$$

The prefix "kilo" multiplies the basic unit by 10^3.

The prefixed units milliampere, millimeter, millisecond, etc., are examples of *submultiple* units because they are *smaller* (by a factor of 10^{-3}) than the basic unit. The prefixed units kilometer, kilovolt, etc., are examples of multiple units because they are *greater* (by a factor of 10^3) than the basic unit.

TABLE 3-3

Prefix	Value by Which Basic Unit Is Multiplied	Symbol
giga	$1,000,000,000 = 10^9$	G
mega	$1,000,000 = 10^6$	M
kilo	$1,000 = 10^3$	K*
none	$1 = 10^0$	(none)
milli	$0.001 = 10^{-3}$	m
micro	$0.000001 = 10^{-6}$	μ
nano	$0.000000001 = 10^{-9}$	n
pico	$0.000000000001 = 10^{-12}$	p

*Although lowercase k is the accepted symbol for "kilo," we will use capital K so that all prefixes greater than one will be symbolized with capital letters (G, M, and K). Later, when these prefixes become second nature, we will use lowercase k for kilo.

Table 3-3 lists the most commonly used prefixes encountered in electronics, their values as multipliers, and their symbols. The contents of this table should eventually be memorized. Repeated use of these prefixes will help to commit them to memory.

In looking over Table 3-3, several points should be emphasized. Notice that the three prefixes *giga*, *mega*, and *kilo*, are all *greater than one* so that they will always produce multiples of the basic unit which they modify. A megavolt is 1,000,000 volts, a kilometer is 1,000 meters, and so forth. The letter symbols for these prefixes are all uppercase letters, G, M, and K.

On the other hand, the prefixes *milli*, *micro*, *nano*, and *pico* are all *less than one;* they will always produce submultiples of the basic unit. A microampere is 0.000001 amperes, a millivolt is 0.001 volts, and so forth. The letter symbols for these prefixes are all lowercase letters m, μ, n, and p. The symbol for micro is the greek letter μ (mu).

EXAMPLE 3.6 Use symbols to express the following: (a) 75 millivolts, (b) 0.07 microamperes, (c) 10 kilovolts, (d) 22 nanoseconds, (e) 820 pico-coulombs, (f) 3.3 megohms

Solution: (a) 75 (milli)(volts) =
   75   m   V   = 75 mV

(b) 0.07 (micro)(amperes) =
   0.07 μ A   = 0.07 μA

(c) 10 kilovolts = 10 KV (d) 22 nanoseconds = 22 ns
(e) 820 picocoulombs = 820 pC (f) 3.3 megohms = 3.3 MΩ

Table 3-3 indicates that the prefix values are all powers of ten in which the powers are all multiples of three (10^3, 10^6, 10^9, etc.). With this in mind, it is apparent that each prefix is 10^3 (1000) times bigger than the next smallest prefix. For example, micro is 10^3 times greater than nano, and milli is 10^3 times greater than micro. Similarly, giga is 10^3 times greater than mega, and mega is 10^3 times greater than kilo. Note that kilo is *not* 10^3 times greater than milli because in between kilo (10^3) and milli (10^{-3}) there is 10^0. 10^0 is *one*, which means a multiplier of *one*, which of course means the basic unit is left unchanged. No prefix is needed for 10^0.

3.10 Converting Between Basic Units and Prefixed Units

Suppose we wanted to convert the quantity 0.37 A to its equivalent expression in units of mA. From Table 3-3 we know that a mA is 10^{-3} A. That is, a mA is smaller than an A by a factor of a thousandth. Another way to say this is that an A is one-thousand times larger than a mA. Thus, 1 A = 1000 mA = 10^3 mA. Since this is true, we can convert 0.37 A to mA by replacing the unit A by its equivalent in terms of mA as follows:

$$0.37(A) = 0.37(10^3 \text{ mA}) = 0.37 \times 10^3 \text{ mA}$$
$$= 3.7 \times 10^2 \text{ mA}$$
$$\text{or} = 370 \text{ mA}$$

Suppose we wanted to express the quantity 3725 V in terms of megavolts (MV). From Table 3–3 it is seen that a MV equals 10^6 V. Thus,

$$1 \text{ MV} = 10^6 \text{ V}$$

or equivalently (by multiplying both sides by 10^{-6})

$$10^{-6} \text{ MV} = 10^{-6} \times 10^6 \text{ V} = 10^0 \text{ V} = 1 \text{ V}$$

Thus, we can convert 3725 V to MV as follows:

$$3725(\text{V}) = 3725(10^{-6} \text{ MV}) \quad \text{(replace V by } 10^{-6} \text{ MV)}$$
$$= 3725 \times 10^{-6} \text{ MV}$$
$$= 3.725 \times 10^{-3} \text{ MV}$$

These procedures can be summarized as follows:

(1) To convert from a basic unit to a prefixed unit, replace the basic unit (for example, V) by the desired prefixed unit (MV) multiplied by the appropriate power of ten.

(2) The appropriate power of ten will always be the opposite of the power of ten of the prefix (for example, 10^{-6} because M = 10^6). This has to be true, because then the two powers of ten (10^{-6} M = $10^{-6} 10^6$) will result in a net of 10^0 or 1 so that the *value* of the quantity remains unchanged.

EXAMPLE 3.7 Convert 5×10^{-10} C to pC.

Solution: Replace C by pC, and since p = 10^{-12} multiply by 10^{+12}. Thus,

$$5 \times 10^{-10}(\text{C}) = 5 \times 10^{-10}(10^{12} \text{ pC})$$
$$= 5 \times 10^2 \text{ pC}$$

EXAMPLE 3.8 Convert 7×10^4 ohms to kilohms.

Solution: $7 \times 10^4 (\Omega) = 7 \times 10^4 (10^{-3} \text{ K}\Omega)$
$$= 7 \times 10^1 \text{ K}\Omega$$

EXAMPLE 3.9 Express 0.0073 A in μA.

Solution: $0.0073(\text{A}) = 0.0073(10^6 \text{ }\mu\text{A})$
$$= 0.0073 \times 10^6 \text{ }\mu\text{A}$$
$$= 7.3 \times 10^3 \text{ }\mu\text{A} = 7300 \text{ }\mu\text{A}$$

The procedure for converting from a prefixed unit to a basic unit is somewhat similar. Suppose we wanted to express the quantity 375 mV in units of V. From Table 3–3 we know that a mV is 10^{-3} V. Thus, we can write

$$375(\text{m})\text{V} = 375 \times (10^{-3})\text{V}$$
$$= 3.75 \times 10^{-1} \text{ V}$$

Using the same procedure, the quantity 4.5 KV can be expressed in volts as

$$4.5 \text{ (K)V} = 4.5 \times (10^3)\text{V} = 4500 \text{ V}$$

Scientific Notation and Prefixed Units 51

To convert from a prefixed unit to a basic unit, then, involves simply replacing the prefix (for example, m) by its power of ten equivalent (10^{-3}).

EXAMPLE 3.10 Convert 3000 pC to C.

Solution: $3000(p)C = 3000 (10^{-12})$ C
$= 3 \times 10^3 \times 10^{-12}$ C
$= \mathbf{3 \times 10^{-9}}$ **C**

EXAMPLE 3.11 Convert 720 KΩ to Ω.

Solution: $720 (K)\Omega = 720 (10^{+3})\Omega$
$= 7.2 \times 10^2 \times 10^3$ Ω
$= \mathbf{7.2 \times 10^5}$ **Ω**

EXAMPLE 3.12 Three identically shaped segments of different material have a voltage of 120 mV applied between their ends. As a result of the voltage, segment A has a current of 2500 μA, segment B has a current of 10^5 mA, and segment C has a current of 0.000002 A. Rank each of the three materials according to its ability to conduct current.

Solution: Segment A; $I = 2500$ μA $= 2.5 \times 10^3$ μA
$= 2.5 \times 10^3 \times 10^{-6}$ A
$= 2.5 \times 10^{-3}$ A

Segment B; $I = 10^5$ mA $= 10^5 \times 10^{-3}$ A
$= 10^2$ A

Segment C; $I = 0.000002$ A
$= 2 \times 10^{-6}$ A

Thus, segment B has the greatest current, segment A the next highest current, and segment C has the least current. Segment B is probably a conductor, segment A a semiconductor, and segment C an insulator.

3.11 Converting Between Different Prefixed Units

Occasionally, it is necessary to convert a quantity which is expressed in one prefixed unit to another prefixed unit. For example, suppose we wanted to express the quantity 2000 μA in terms of mA. Since μ is 10^{-6} and m is 10^{-3}, then a μA is smaller than a mA by a factor of 10^{-3}. Thus, 1 μA $= 10^{-3}$ mA. Using this fact,

$$2000 (\mu A) = 2000 (10^{-3} \text{ mA})$$
$$= 2000 \times 10^{-3} \text{ mA}$$
$$= 2 \text{ mA}$$

EXAMPLE 3.13 Convert 3 MV to KV.

Solution: Since $MV = 10^6$ V and $KV = 10^3$ V, then $MV = 10^3$ KV. Therefore,

$$3(MV) = 3(10^3 \text{ KV})$$
$$= 3 \times 10^3 \text{ KV} = \mathbf{3000 \text{ KV}}$$

EXAMPLE 3.14 Convert 3200 pA to nA.

Solution: Since $pA = 10^{-12}$ A and $nA = 10^{-9}$ A, then $pA = 10^{-3}$ nA. Therefore,

$$3200(pA) = 3200(10^{-3} \text{ nA})$$
$$= \mathbf{3.2 \text{ nA}}$$

3.12 Calculations Using Prefixed Units

Consider the following example.

EXAMPLE 3.15 33 mV is applied to a device with a resistance of 2.2 MΩ. Calculate the current and express it in nA and in μA.

Solution: $I = \dfrac{V}{R} = \dfrac{33 \text{ mV}}{2.2 \text{ M}\Omega} = \dfrac{33 \times 10^{-3} \text{ V}}{2.2 \times 10^6 \text{ }\Omega}$
$$= 15 \times 10^{-9} \text{ A}$$
$$= \mathbf{15 \text{ nA} = 0.015 \text{ }\mu\text{A}}$$

The preceding example illustrates the general procedure used in solving formulas which contain prefixed units. *All quantities in the formula are converted to basic units regardless of what units they are given in. The formula is then evaluated and the result is always in basic units.* The result can then be converted to a prefixed unit if desired.

EXAMPLE 3.16 Calculate the power delivered to a load which draws 75 A of current from a 4.8 KV source. Express P in kilowatts and in megawatts.

Solution: $P = I \times V$
$$= 75 \text{ A} \times 4.8 \text{ KV}$$
$$= 75 \text{ A} \times 4800 \text{ V}$$
$$= 360,000 \text{ W}$$
$$= \mathbf{360 \text{ KW} = 0.36 \text{ MW}}$$

To summarize: When dealing with quantities in a formula, if the quantities are expressed in basic units before the formula is evaluated, then the result will always be in its basic unit. This procedure is a simple way to avoid complicated unit determinations, especially in some of the more complex formulas. If this procedure is followed, there is never any doubt about the units of the quantity being calculated from the formula.

3.13 Estimating Results

When one is testing or troubleshooting a circuit, it is not always necessary to know the exact theoretical values for the circuit quantities being measured. Often a technician makes a simple estimate of the *theoretical* value of a quantity so that he can check to see that the *measured* value is reasonably close to what is expected. This allows the technician to catch any gross errors caused by component failures, bad connections, and so forth.

The key to making an estimate is this: *make the numbers easy.* In other words, round off the values used in the calculation so that only easy numbers are involved (e.g., multiples of 2, 5, and 10). The following examples will illustrate:

EXAMPLE 3.17 Estimate the current flowing through a 4.7-kΩ resistance that has 16 V applied to it.

Solution: $I = \dfrac{V}{R} = \dfrac{16 \text{ V}}{4.7 \text{ k}\Omega}$

The first step is to round off the numbers in the numerator and denominator. Thus, $16 \approx 15$, and $4.7 \approx 5$. The next step is to perform the calculation using the rounded-off values.

$$I \approx \dfrac{15 \text{ V}}{5 \text{ k}\Omega} = \dfrac{3 \text{ V}}{\text{k}\Omega}$$

$$\approx 3 \dfrac{\text{V}}{10^3 \Omega} = 3 \times 10^{-3} \dfrac{\text{V}}{\Omega} = 3 \text{ mA}$$

This last step can be simplified by remembering that 1 V/1 kΩ is always 1 mA.

The resultant estimate of 3 mA should, of course, be treated as an approximation and not as a solution to a problem (the exact value for I is 3.4 mA). The usefulness of the estimated value occurs when the technician measures the actual current. If, for example, it measures at 3.25 mA, the circuit can be assumed to be working properly. If it measures, say, at 1.3 mA or 5.2 mA, the circuit can be assumed to be malfunctioning.

EXAMPLE 3.18 Estimate the current through a 2.2-MΩ resistance as a result of 38 V applied to it.

Solution: $I = \dfrac{V}{R} = \dfrac{38 \text{ V}}{2.2 \text{ M}\Omega}$

$$\approx \dfrac{40 \text{ V}}{2 \text{ M}\Omega} = 20\left(\dfrac{\text{V}}{\text{M}\Omega}\right) = 20\left(\dfrac{\text{V}}{10^6 \Omega}\right)$$

$$\approx 20\left(10^{-6} \dfrac{\text{V}}{\Omega}\right) = 20 \text{ }\mu\text{A}$$

This last step can be simplified by remembering that 1 V/1 M is always 1 μA.

EXAMPLE 3.19 Estimate the power being supplied to a 2.7-kΩ load from a 12-V source.

Solution: $I = \dfrac{V}{R} = \dfrac{12 \text{ V}}{2.7 \text{ k}\Omega} \approx \dfrac{12 \text{ V}}{3 \text{ k}\Omega} = 4 \text{ mA}$

$\therefore I \approx 4 \text{ mA}$
$P = V \times I$
$\approx 12 \text{ V} \times 4 \text{ mA} = 48(\text{V} \times 10^{-3} \text{ A}) = 48 \times 10^{-3}(\text{V} \times \text{A})$
$\approx \mathbf{48 \text{ mW}}$

Note that $1 \text{ V} \times 1 \text{ mA} = 1 \text{ mW}$.

Questions/Problems

Section 3.2

3-1 Convert the following powers of ten to their decimal equivalent:
   a. 10^{-10} b. 10^{+5}
   c. 10^{-7} d. 10^{+4}

3-2 Convert the following numbers to powers of ten:
   a. 0.00001   b. 100,000
   c. 10,000,000   d. 0.01

3-3 For each of the following numbers, indicate whether it is written in ordinary notation, scientific notation, or neither:
   a. 376.57   b. 91.53×10^5
   c. 9.153×10^6 d. 0.42×10^{-3}
   e. 0.0047   f. 4.7×10^{-7}
   g. 3.7×10^0

Sections 3.3–3.4

3-4 Convert the following numbers to scientific notation:
   a. 0.0000563   b. 12,940,000
   c. 37.77   d. 125×10^{-3}
   e. 8360.9   f. 6.4

3-5 Convert the following numbers to ordinary notation:
   a. $7.35 \times 10^{+2}$ b. 8.15×10^{-5}
   c. $6.653 \times 10^{+7}$ d. 1.5×10^{-2}
   e. 6×10^0 f. 35.7×10^{-3}

Section 3.5

3-6 Perform the following arithmetic operations. Express all results in scientific notation.
   a. $7.2 \times 10^3 + 4.8 \times 10^3$
   b. $7.2 \times 10^{-3} - 4.8 \times 10^{-3}$
   c. $3.75 \times 10^5 + 6.2 \times 10^4$

Scientific Notation and Prefixed Units   55

    d. $4.1 \times 10^{-3} - 9.3 \times 10^{-2}$
    e. $4.7 \times 10^{2} - 4.6 \times 10^{2}$
    f. $7.9 \times 10^{7} + 4.1 \times 10^{9}$

Section 3.6

3-7 Perform the following arithmetic operations. Express each result in scientific notation.
    a. $(3.9 \times 10^{3}) \times (7 \times 10^{-5})$
    b. $(8.7 \times 10^{-6}) \times (3 \times 10^{-3})$
    c. $(4.5 \times 10^{-5}) \times (6 \times 10^{4})$
    d. $(2.5 \times 10^{4}) \times (3.2 \times 10^{-4})$
    e. $(3.2 \times 10^{7}) \div (4 \times 10^{5})$
    f. $(5.5 \times 10^{-3}) \div (-1.1 \times 10^{3})$
    g. $(1.5 \times 10^{2}) \div (1.5 \times 10^{-3})$
    h. $(4.5 \times 10^{3}) \div (6 \times 10^{3})$
    i. $(42.42 \times 10^{-7}) \div (7 \times 10^{-3})$
    j. $(64) \times (-2.57 \times 10^{-3})$

Section 3.7

3-8 Find the square roots of the following numbers:
    a. 3.24×10^{-6} b. 4.55×10^{5}
    c. 957×10^{-3} d. 4.9×10^{7}

Section 3.8

3-9 A certain wire has 0.000005 C of charge flowing past a given point in 0.00025 s. Determine the value of current I in the circuit. Use scientific notation.

3-10 In the formulas below, $X = 0.0047$, $Y = 33{,}000$, and $Z = 377$. Calculate P in each case using scientific notation.
    a. $P = \dfrac{XY}{Z}$ b. $P = \sqrt{Z^{2} + XY}$

Sections 3.9–3.11

3-11 Match each of the items in column I below with the item in column II which means the same thing.

I	II
mega	(a) 10^{-12}
nano	(b) 10^{-3}
milli	(c) M
kilo	(d) micro (μ)
1,000,000	(e) K
p	(f) n
10^{-6}	(g) G
10^{9}	
0.001	
0.000000001	

3-12 For each of the following groups of expressions, indicate the largest quantity and smallest quantity in each group.
   a. 1 mA, 0.01 A, 10^{-6} A
   b. 1 μV, 0.001 V, 10^{-2} V
   c. 1 pC, 0.0000000001 C, 10^{-9} C

3-13 Perform the following conversions:
   a. 0.257 A = ____ mA
   b. 5247 V = ____ KV
   c. 3.7×10^{-8} C = ____ μC
   d. 2.5×10^{-8} A = ____ nA = ____ μA
   e. 4,753,000 Ω = ____ KΩ
   f. 950×10^{-9} A = ____ pA
   g. 9,275 mv = ____ V
   h. 1.5×10^7 μA = ____ A
   i. 4900 nC = ____ C
   j. 25 MΩ = ____ KΩ
   k. 2500 pC = ____ μC
   l. 95 GV = ____ MV

Sections 3.12–3.13

3-14 A certain wire has a current of 35 nA. How much charge will pass a given point in the wire in 5 ms?

3-15 Determine the amount of voltage needed to produce a 2-mA current through a 47-KΩ resistance.

3-16 How much current flows when a 25-V source supplies 100 μW to a load?

3-17 Estimate the value of current resulting from each of the following calculations:
   a. 20 V/3.3 KΩ
   b. 100 V/4.7 MΩ
   c. 60 mV/82 Ω
   d. 6 V/68 KΩ

CHAPTER 4

More on Voltage and Current

4.1 Introduction

Now that we have learned the basic ideas behind voltage and current, we are ready to find out more about these important electrical quantities. In this chapter we will see how to represent voltages symbolically and how to combine several voltages in a circuit. We will introduce the idea of a "ground" reference and will describe the difference between *dc* and *ac* circuit operation. Many of these ideas will be introduced in basic terms here and expanded on in later chapters.

4.2 Voltage Conventions

Differences in potential are common to all electric circuits. In fact, it is differences in potential (voltages) which produce electric current in these circuits, as we shall see. Because voltages are probably the most commonly used and measured electrical quantities, it is necessary to establish certain conventions and notation that will enable us to express voltages quantitatively.

Refer to Fig. 4–1(a). From our previous discussion we know that a voltage (difference in potential) will exist between A and B. If this voltage is, say, 5 volts, then we can express this as follows,

$$V_{AB} = +5 \text{ V}$$

which states that the *voltage at A relative to B* is $+5$ volts. In representing voltages this way, two subscripts are used to indicate the two points between

Chapter Four

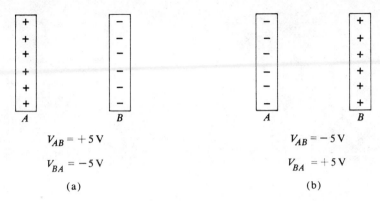

$V_{AB} = +5\,\text{V}$ $\qquad\qquad\qquad\qquad V_{AB} = -5\,\text{V}$

$V_{BA} = -5\,\text{V}$ $\qquad\qquad\qquad\qquad V_{BA} = +5\,\text{V}$

(a) $\qquad\qquad\qquad\qquad\qquad$ (b)

FIGURE 4-1 *Various voltage situations.*

which the voltage exists. Since voltages are differences in potential between *two* points, it is *always* necessary to specify both points when indicating a voltage.

The polarity of the voltage depends on which point is more positive. V_{AB} indicates the voltage at point A relative to point B. If A is more positive than B [as in Fig. 4–1(a)] then V_{AB} will be a positive quantity. If A is less positive (or more negative) than B, then V_{AB} will be a negative quantity. This is illustrated in Fig. 4–1(b) where A is now negative relative to B. Thus, we have

$$V_{AB} = -5\,\text{V}$$

where the minus sign indicates that point A (the first subscript) is negative with respect to point B (the second subscript).

It is usually arbitrary as to which point is chosen as the first subscript and which point is chosen as the second subscript. However, depending on the choice, the voltage will be either positive or negative with the magnitude remaining the same. For example, in the situation of Fig. 4–1(a) if we wanted to express the voltage at point B relative to point A, then we would have

$$V_{BA} = -5\,\text{V}$$

because B is more negative than A. Similarly, in Fig. 4–1(b) we can write

$$V_{BA} = +5\,\text{V}$$

because B is more positive than A. Whether we express the voltage as V_{AB} or V_{BA} is a matter of choice, although often the choice is dictated by other considerations which will become apparent later.

EXAMPLE 4.1 Refer to Fig. 4–2. In each case shown there is a difference of potential of 2 V. Express the values of V_{xy} and V_{yx} for each case.

Solution: (a) $V_{xy} = +2\,\text{V}$; $V_{yx} = -2\,\text{V}$
(b) $V_{xy} = -2\,\text{V}$; $V_{yx} = +2\,\text{V}$ (x is more negative than y)
(c) $V_{xy} = -2\,\text{V}$; $V_{yx} = +2\,\text{V}$

In all cases, notice that by reversing the subscripts of a voltage expression the polarity is reversed. That is, $V_{xy} = -V_{yx}$ or, equivalently, $V_{yx} = -V_{xy}$.

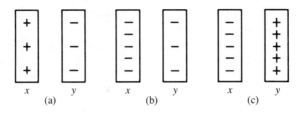

FIGURE 4–2

Circuit Representation Of Voltages Figure 4–3 shows how we represent the voltage across a device in a circuit. The block represents any two-terminal device, and points A and B are the terminals to which connections are made to the device. These terminals are labelled either $+$ or $-$ depending on which point is more positive.

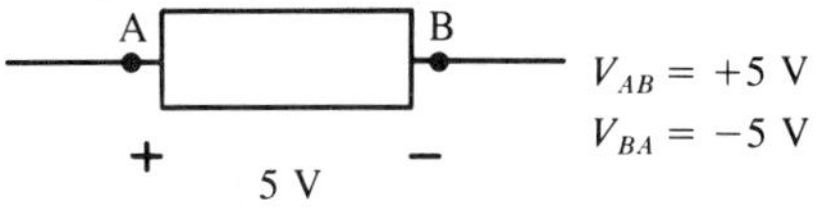

$$V_{AB} = +5 \text{ V}$$
$$V_{BA} = -5 \text{ V}$$

FIGURE 4–3 *Circuit representation of device voltage.*

4.3 Combining Voltages

Using the voltage notation and conventions described in the previous section we can now address ourselves to the situation where more than one voltage is present. Such a situation is most often the case in electric circuits. Refer to Fig. 4–4(a) which shows a common two-voltage situation. From this figure it can be readily seen that point C is 3 volts more positive than point D. That is, $V_{CD} = +3$ V. Also, point A is 6 volts more positive than point B so that $V_{AB} = +6$ V.

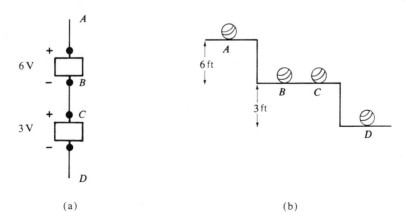

FIGURE 4–4 (a) *Two-voltage situation.* (b) *Gravitational analogy.*

Suppose, now, that we wanted to know the difference in potential between A and D (that is, V_{AD}). In order to do this something must be said concerning points B and C. Notice that these two points are connected by a wire. This wire is almost always made from a highly conductive metal (usually copper) and, as we will see later, we can normally assume that there will be practically no potential difference between the ends of such a wire. In other words, we can assume $V_{BC} = 0$. Such a connecting wire is often called a *short circuit* wire, and it can be said that points B and C are *shorted* together by this wire.

Since B and C have no potential difference between them, then it should be apparent that point B must also be 3 V more positive than point D. Points B and C are at the same voltage relative to D. Thus, $V_{BD} = +3$ V also. Now, because A is 6 V more positive than B, it follows that A must be 9 V more positive than D. Thus, $V_{AD} = +9$ V.

Figure 4–4(b) shows the gravitational analog of the electrical situation just considered. The ball resting at the bottom of the staircase at point D is at a lower potential energy than the other balls shown. The balls at B and C are on the same level at a somewhat higher potential energy than the ball at D. The ball at A is at a still higher level and has a potential energy greater than the balls at B and C.

Consider another example shown in Fig. 4–5(a). By inspection, we can see that C is 3 V more positive than $D (V_{CD} = +3$ V) and A is 4 V negative relative to $C (V_{AC} = -4$ V). Thus, it is clear that A must be 1 V negative relative to D so that $V_{AD} = -1$ V. Another way to say this is that A is at a voltage lower than D by 1 V. The gravitational analog for this case is shown in Fig. 4–5(b).

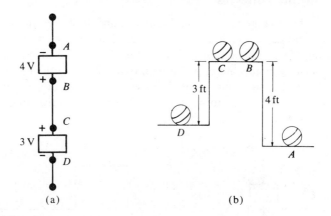

FIGURE 4–5 (a) *Another two-voltage situation.* (b) *Gravitational analogy.*

The reasoning followed in the preceding examples is not necessarily required if a more rapid method of determining voltages is used. A quicker procedure for determining V_{AD} in Fig. 4–5 proceeds as follows. Start at point D and move through the circuit to point A. In going from D to A add up all the rises in potential that are encountered and add up all the drops in potential that are encountered. The net rise or drop in voltage can then be found.

In Fig. 4–5(a), in going from point D to point C we encounter a *voltage rise* of 3 V because C is $+3$ V above D. Going from C on up to B produces no rise or drop in voltage. Going from B on up to A produces a *voltage drop* of 4 V. Thus, we experienced a rise of 3 V and a drop of 4 V resulting in a net 1-V drop in going from D to A. This means that A is 1 V lower than D or $V_{AD} = -1$ V. The following examples will further illustrate this technique for more complicated situations.

EXAMPLE 4.2 Determine V_{AD} in Fig. 4–6(a).

Solution: From D to C: 2.5 V rise
          From C to B: 3.2 V drop
          From B to A: 6.0 V rise

                   Total 8.5 V rise   3.2 V drop
                   Net   5.3 V rise

Therefore, A is 5.3 V higher than D so that $V_{AD} = +5.3$ V.

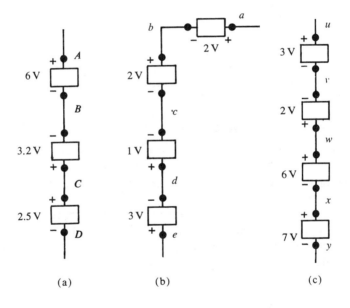

(a)   (b)   (c)

FIGURE 4–6

EXAMPLE 4.3 Determine V_{ae} in Fig. 4–6(b).

Solution: From e to d: 3 V drop
          From d to c: 1 V drop
          From c to b: 2 V rise
          From b to a: 2 V rise

                   Total 4 V rise   4 V drop
                   Net   0 V rise

Therefore point a is at the same voltage as point e so that $V_{ae} = 0$.

In going through these examples several important points should be noticed: (1) when finding V_{AD} we always start at D and work toward A; (2) in going from one point to another a *voltage rise* is experienced if the point we're going to is more positive; and (3) a *voltage drop* is experienced if the point we're going to is more negative; (4) points at the same potential need not be labeled with separate letters [for example, the letter C represents two points in Fig. 4–6 (a)]; (5) a point might be positive relative to one point but negative relative to another [see point d in Fig. 4–6(b)].

The procedure followed in Exs. 4.2 and 4.3 can be streamlined somewhat after doing many such problems. The student can use the procedure followed in these examples until he is confident enough to use shortcuts to save time. These shortcuts will develop naturally.

EXAMPLE 4.4 Find V_{XU} in Fig. 4–6(c).

Solution:

From U to V:			3 V drop
From V to W:	2 V rise		
From W to X:			6 V drop
Total	2 V rise		9 V drop
Net			7 V drop

Thus, $V_{XU} = -7$ V (X is 7 V lower than U).

4.4 Combining Voltage Sources

Sometimes there will be two or more voltage sources connected in the same circuit. A common situation is illustrated in Fig. 4–7 where, in each case, two voltage sources are connected in the same current path. The sources are said to be connected in *series*.

Whenever two or more voltage sources are connected in series, their effect on the circuit can be determined by combining them into one *equivalent* voltage source. For example, in Fig. 4–7(a) the two 3-V sources are both acting to produce current flow in a ccw direction around the circuit. The two sources are said to be *aiding* each other. Their net effect is the same as would be produced by a single voltage source with a voltage equal to the sum of their voltages.

In Fig. 4–7(b) the sources are connected in *opposition* to each other. That is, the 6-V source is trying to push current in a ccw direction, and the 4-V source is trying to push current in a cw direction. The net effect is the same as would be obtained by a single 2-V source connected to push current ccw.

To summarize, two voltage sources connected in series *aiding* can be replaced by a single voltage source equal to the *sum* of their voltages; two voltage sources connected in series *opposition* can be replaced by a single voltage source equal to the *difference* in their voltages.

4.5 Ground or Common

In most circuits there is one point that is chosen as a reference point from which all circuit voltages are to be measured. This point is called the *common*

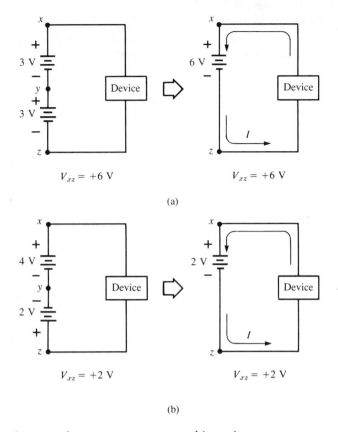

FIGURE 4-7 *Combining voltage sources connected in series.*

or *ground* point and is given the special symbol ⏚, as shown in Fig. 4–8. Here point z is the ground point.

In many cases, the ground point of a circuit is tied to the metal *chassis* on which the circuit is built and sometimes it is even tied to actual earth ground, which is a conducting metal rod driven into the earth. One side of the 120-V ac power line used in residential wiring is actually connected to earth ground usually by connection to a cold-water pipe. This is done for safety reasons to reduce the possibility of shock. However, in electronic equipment the chassis ground usually just serves as a common reference point for the source voltages and output voltages. Most of the time, the power sources used in electronic equipment have one of their terminals (either + or −) tied to chassis ground. In many of the circuits which we use throughout the text, one of the terminals will serve as the common or ground terminal.

Voltages Relative to Ground When specifying a voltage from one point relative to ground, the second subscript is usually omitted. For example, in Fig. 4–8, V_{xz} represents the voltage at point x relative to ground, and V_{yz} represents the voltage at y relative to ground. These voltages are often written as V_x and V_y respectively. In fact, we rarely even assign a letter symbol to the ground node.

Whenever you see a voltage specified with only one subscript, such as V_x, it is understood that this voltage is measured at that point relative to ground. We will use this convention in our later work.

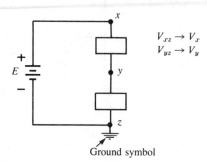

FIGURE 4–8 *Point z is the chosen ground point.*

4.6 Assigning Positive Current Direction

We have seen that the voltages between two points can be either positive or negative depending on which point is used as the reference. Until now we have not assigned any polarity (+ or − sign) to current. A current can be designated as either positive or negative depending on which direction it is flowing. To do this we must assign one particular current direction as the *positive* direction. Sometimes this assignment is obvious, but sometimes it is arbitrary.

Figure 4–9(a) is a simple circuit containing a constant 6-V source. Here the choice is easy. We simply choose the ccw direction as the positive direction for current because we know that the current will always flow ccw in this simple circuit due to the polarity of the 6-V source. This is called a *direct current (dc)* circuit and the 6-V source is a *dc source* since the current will flow in only one direction.

There will be some complex circuits, encountered later, where we cannot tell in which direction the current will be flowing until we do a detailed analysis of the circuits. In these circuits, we can arbitrarily assign one current direction as the positive direction. This is illustrated in Fig. 4–9(b) where the load current arrow, I_L, is drawn *upward;* this upward direction is assigned positive polarity. If, after making calculations or measurements to determine the size and direction of I_L, we find that the load current actually flows *downward*, we can simply reverse the current arrow on the diagram. Alternatively, we can leave the arrow as it is and indicate that I_L is *negative*. For example, if we state that $I_L = -2$ mA, this indicates that I_L has a magnitude of 2 mA and is flowing downward (opposite to the assigned positive current direction).

The direction of current can alternate from one direction to the other in circuits that are driven by voltage sources whose polarity changes periodically. These sources are called *alternating current (ac) sources*. Since the current direction will be periodically changing in an ac circuit, we again assign one of the directions on the circuit diagram to be the positive direction, and the other direction automatically becomes the negative direction. We will say more about this a little later.

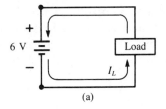
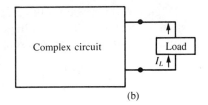

FIGURE 4-9

In summary, just remember that the current arrow drawn on the diagram is *always* the assumed *positive* direction for current. When the current actually flows in that direction, it is a *positive* current; when it actually flows in the opposite direction, it is a *negative* current. The positive and negative notations simply serve as a means for indicating the actual direction of current relative to the arrow drawn on the diagram.

EXAMPLE 4.5 (a) In what direction is the actual current flow in Fig. 4-10 if $I_L = +4$ mA? (b) Repeat for $I_L = -4$ mA.

Solution: (a) Since I_L is positive, it must be flowing in the same direction (downward) as the indicated positive direction on the diagram.

(b) Since I_L is negative it must be flowing in the opposite direction (upward) to that indicated on the diagram.

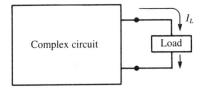

FIGURE 4-10

4.7 Time-Varying Voltage and Current

Only a relatively small number of circuits operate with voltage and current values that are constant; that is, they do not change with time. In general, the voltages and currents in an electrical circuit are *time-varying,* meaning that their values change with time. Time-varying voltage (or current) can be represented by a graph showing how the voltage (or current) varies with time. This graph is called a *waveform*.

Figure 4-11(a) shows one of the most common voltage waveforms. Note that the vertical axis is voltage and the horizontal axis is time, t. The graph shows that the voltage is positive at certain times and negative at other times. This is an *ac* voltage similar to what the power company supplies to its customers.

Using this waveform graph we can determine the *instantaneous* value of voltage at any point in time. For example, at $t = 1$ ms, the instantaneous value of voltage is $+4$ V (point x); at $t = 3$ ms, the value is -4 V (point y).

Figure 4–11(b) shows the graph of a *pure dc* voltage. It is a horizontal line showing that the value of voltage is constant at 6 V at all instants of time. This is the kind of voltage that is supplied by a battery or dc power supply.

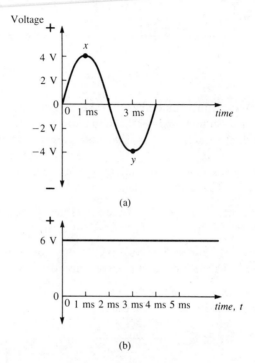

FIGURE 4–11 *Waveforms of ac voltage and dc voltage.*

We are not going to do too much with waveforms until later in the text. For most of our work on resistive circuits we will use dc voltages and currents in order to simplify the learning of the important circuit principles. It should be understood, however, that although we will be using dc sources, many of the principles and laws that we study can be applied to circuits which use time-varying sources. We will keep reminding ourselves of this as we go along.

4.8 Ideal Voltage and Current Sources

A voltage source produces a specific voltage difference between its terminals. Ideally, this voltage will be the same no matter what we connect to the source terminals. As we shall see later, this is not possible in practice. For now, however, we will treat all our voltage sources as *ideal* voltage sources. This

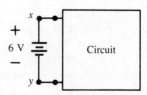

FIGURE 4–12 *An ideal voltage source produces a voltage across its terminals which is independent of what is connected to these terminals.*

means, for example. that the voltage across terminals x-y in Fig. 4–12 will be kept at +6 V by the dc source regardless of what circuit we connect to these terminals.

Ideal Current Sources The energy sources with which we are most familiar are voltage sources such as batteries, dc power supplies, and the ac voltage from our electrical outlets. Most electronic circuits utilize one or more of these voltage sources. There are, however, sources that are designed to produce a specific terminal current. These *current sources* ideally will produce the same flow of current no matter what circuit is connected to their terminals.

Figure 4–13 shows a current source symbolized by an arrow inside a circle. The arrow indicates the direction of current through the current source and around the circuit. The value of this current is shown as 4 mA. If this were an *ideal* current source, the 4 mA would flow through whatever circuit we connected to terminals x-y. As we shall see later, this is not possible in practice. For now, however, we will treat all current sources as ideal sources.

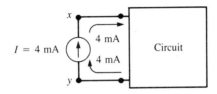

FIGURE 4–13 *An ideal current source produces a current through its terminals that is independent of the circuit connected to its terminals.*

Chapter Summary

1. The symbol V_{AB} represents the voltage at point A relative to point B.
2. If A is more positive than B, then V_{AB} will be a positive quantity. If A is more negative than B, then V_{AB} will be a negative quantity.
3. It is always true that $V_{BA} = -V_{AB}$.
4. If, in going from one point in a circuit to another, there is an increase in potential (going from − to +), then we call this a *voltage rise*.
5. If there is a decrease in potential (going from + to −), this is called a *voltage drop*.
6. Two or more voltage sources connected in series can be replaced by a single voltage source producing the same net voltage.
7. Ground (or common) is a reference point or node in a circuit from which all circuit voltages are measured.
8. When a voltage is specified with only one subscript (e.g., V_X), it is understood that it represents the voltage at that point relative to ground.

Chapter Four

9. A current is said to be positive if it flows in the same direction as the assumed positive direction on the circuit diagram; it is negative if it flows in the opposite direction.

10. A time-varying voltage or current is one whose value changes with time.

11. A dc voltage or current is one whose value is constant with time.

12. An ac voltage or current is one whose polarity or direction alternates periodically.

13. An ideal voltage source produces a voltage across its terminals which is independent of what is connected to these terminals.

14. An ideal current source produces a current flow through its terminals which is independent of the circuitry connected to these terminals.

Questions/Problems

Sections 4.2–4.3

4–1 In Fig. 4–14 the plates have a difference of potential of 3 V. Express V_{XY} and V_{YX} for each case shown in the figure.

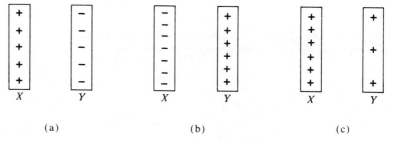

FIGURE 4–14

4–2 Determine the value of V_{AD} for each of the cases in Fig. 4–15.

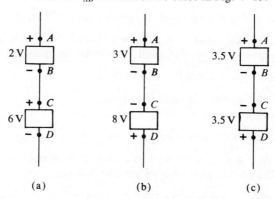

FIGURE 4–15

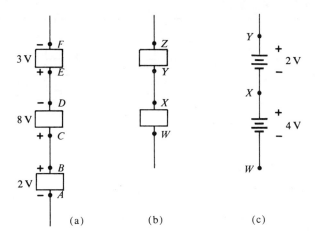

FIGURE 4–16

4–3 For the situation in Fig. 4–16(a), determine the values of V_{FA}, V_{AF}, V_{AD}, and V_{CF}.

4–4 In Fig. 4–16(b), the value of $V_{ZY} = +2$ V and $V_{WZ} = -3$ V. Determine V_{XW}.

4–5 Determine V_{YW} in Fig. 4–16(c).

Section 4.4

4–6 Replace each pair of sources in Fig. 4–17 with an equivalent single source, and determine the direction of current flow.

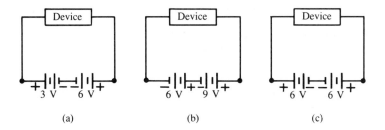

FIGURE 4–17

4–7 Calculate the current which flows in the circuit of Fig. 4–17(b) if the device has a resistance of 1.5 kΩ.

Section 4.5

4–8 What is the significance of a ground or common point of a circuit? Draw the ground symbol.

4–9 Refer to Fig. 4–18. If point z is chosen as the ground terminal, what are the values of V_x and V_y?

4–10 If point x is chosen as ground in Fig. 4–18, what are the values of V_y and V_z?

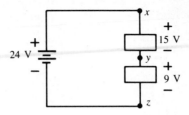

FIGURE 4–18

Sections 4.6–4.8

4–11 Refer to Fig. 4–9(b). Suppose you determine, by calculation or measurement, that the load current is 12 mA and flows downward through the load. Would you specify the current as $I_L = +12$ mA or as -12 mA?

4–12 What is the difference between a dc source and an ac source?

4–13 What is a time-varying voltage or current?

4–14 Sketch the waveform of a time-varying current whose values at various instants of time are given in the table below.

Time (ms)	Current (mA)
0	0
1	6.3
2	8.6
3	9.5
4	9.8
5	10.0
6	10.0

4–15 Describe the difference between an ideal voltage source and an ideal current source. Draw the circuit symbols for both.

CHAPTER 5

Current-Voltage Graphs and Resistive Devices

5.1 Introduction

Refer to Fig. 5–1 where a pure dc voltage source E_s is connected to a certain electrical device. We know that, in general, there will be a flow of current through this device as a result of E_s and the electrical characteristics of the device. The device electrical characteristic which is the most important is its *current-voltage graph* (abbreviated, *I-V* graph or *I-V* curve).

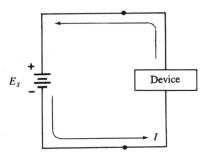

FIGURE 5–1

The *I-V* curve of the device is simply a graph of how the current through the device will vary as the voltage across the device varies. In this chapter we will study the *I-V* curves of a class of devices called *resistive devices*. A resistive device is characterized by the fact that as soon as a constant voltage is applied across its terminals (as in Fig. 5–1), a constant flow of current will be produced

71

through the device. A common example of a resistive device is a flashlight bulb. As soon as the switch is closed connecting the battery to the bulb, a constant current is supplied through the bulb.

In later chapters we will investigate devices which are classified as *inductive* or *capacitive*. Such devices do not exhibit the same behavior as resistive devices because the current through these devices will not reach a steady value instantly when a pure dc voltage is applied to them. Instead, the current will be changing for some time until a steady value is reached. In resistive devices there is no time factor involved; the current reaches its steady value instantaneously.

5.2 Current-Voltage Graph With Constant Slope

The most common resistive device is one which has a current-voltage graph that is *linear*. Figure 5-2 shows a typical linear I-V graph for such a device. The graph shows how the current I will vary as the voltage across the device, V_{xy}, is varied. V_{xy} can be varied by using a variable voltage source as illustrated in Fig. 5-2. The arrow through the dc source indicates that its voltage can be varied.

Note that both V_{xy} and I can be of either polarity. When V_{xy} is made positive by the applied source, I will flow in the direction shown on the circuit diagram; this is the *positive* direction for I. When V_{xy} is made negative by reversing the voltage source, I will flow in the direction opposite to that shown on the diagram; thus, I will be *negative*.

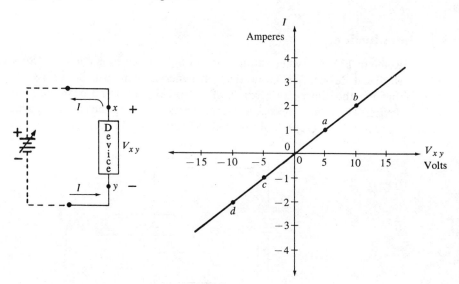

FIGURE 5-2

The linear *I-V* graph shows that for such a device the current is directly proportional to the applied voltage. For example, when $V_{xy} = 5$ V the current is $I = 1$ A (point *a* on the graph). If V_{xy} is doubled to 10 V the current also doubles to 2 A (point *b*). The same holds true for negative values of V_{xy} (points *c* and *d*). It seems that for this device the current will increase by 1 ampere for every 5 volts across the device. Note that the graph passes through the origin ($I = 0$, $V_{xy} = 0$) because there can be no current without a potential difference across the device.

We learned in basic algebra that a straight line passing through the origin is expressed by the equation:

$$y = mx$$

where m is the "constant of proportionality" called the slope. For the *I-V* graph the variable I is on the *y*-axis and the variable V is on the *x*-axis. Thus, we can say that I will be equal to some constant of proportionality times V_{xy}. This constant of proportionality is given the symbol G, which results in

$$I = G \cdot V_{xy} \qquad (5\text{-}1)$$

To determine the value of G, we take any point on the line and substitute into Eq. (5-1) the values for I and V_{xy} at that point. For example, using point *a* we have

$$1 \text{ A} = G \times 5 \text{ V}$$

or

$$G = \frac{1 \text{ A}}{5 \text{ V}} = 0.2 \text{ A/V}$$

Similarly, using point *d*

$$-2 \text{ A} = G \times (-10 \text{ V})$$
$$G = \frac{-2 \text{ A}}{-10 \text{ V}} = +0.2 \text{ A/V}$$

For any point chosen the value of G will be the same. G is the slope of the line in amps per volt and is constant for any value of voltage across the device.

Conductance The slope G has an important physical significance. G is a measure of the device's ability to *conduct* current. It tells us how much current will flow through the device per volt across the device. For this reason G is called the *conductance* of the device. The conductance of the device depends on many factors, as we shall see later.

If one device has a *greater* value of conductance than a second device, then the first device will conduct a *greater* current than the second device for the same voltage applied to each. For example, a device which has $G = 2 \text{ A/V}$ will be twice as good a conductor as a device which has $G = 1 \text{ A/V}$.

Resistance For the straight-line graph in Fig. 5-2, we could just as easily say that voltage V_{xy} is directly proportional to I. In other words, V_{xy} is equal to some constant of proportionality times I. That is,

$$V_{xy} = R \cdot I \qquad (5\text{-}2)$$

where R is the constant of proportionality.

The value of R can be easily obtained from the graph by taking any point on the graph and substituting into Eq. (5–2) the values of I and V_{xy} at that point. For example, taking point b gives us

$$10\text{ V} = R \times 2\text{ A}$$

or

$$R = 5\text{ V/A}.$$

Similarly, taking point c gives us

$$-5\text{ V} = R \times (-1\text{ A})$$

or

$$R = \frac{-5\text{ V}}{-1\text{ A}} = 5\text{ V/A}$$

For any point chosen, the value of R will be the same. R tells us how many volts are needed per ampere of current. Since $R = 5$ V/A in our example device, then this device needs 5 V for each ampere of current through the device.

The constant R has an important physical significance. R is a measure of the device's ability to *resist* (or oppose) current. A larger value of R indicates that a greater voltage is needed to produce one ampere of current. Saying it another way, a device with a *large* value of R will conduct *less* current than a device with a small value of R. Since a larger value of R indicates a greater opposition to current, R is called the *resistance* of the device. The resistance of the device depends on many factors which will be pointed out in a subsequent chapter.

At this point the student might be wondering if there is a relationship between a device's conductance G and its resistance R. Intuitively, we would expect that a device with a *high* value of *conductance* (ability to *conduct* current) will have a correspondingly *low* value of *resistance* (ability to *resist* current). Mathematically, we can show that this is true. First, let's rewrite Eq. (5–2) and rearrange it

$$V_{xy} = R \cdot I \qquad (5\text{–}2)$$

or

$$I = \frac{V_{xy}}{R} = \left(\frac{1}{R}\right)V_{xy} \qquad (5\text{–}3)$$

But, Eq. (5–1) tells us that

$$I = G \cdot V_{xy} \qquad (5\text{–}1)$$

Comparing (5–2) and (5–3) leads us to conclude that

$$G = \frac{1}{R} \qquad (5\text{–}4a)$$

or equivalently

$$R = \frac{1}{G} \qquad (5\text{–}4b)$$

In other words, resistance and conductance are the *reciprocals* of one another. We can verify this for our example device of Fig. 5-2 as follows:

$$G = \frac{0.2 \text{ A}}{\text{V}} = \frac{1}{R} = \frac{1}{5 \text{ V/A}} = \frac{1 \text{ A}}{5 \text{ V}} = \frac{0.2 \text{ A}}{\text{V}}$$

Thus, if a device has a low value of R it will have a large value of G and vice versa.

Ohm's Law Equations (5-1) through (5-3) are equivalent mathematical statements of *Ohm's law*. Ohm was the scientist who first discovered that certain materials have a linear electrical characteristic. In other words, Ohm's law states that:

In certain materials the amount of current through the material is directly proportional to the applied voltage.

We will have occasion to use Ohm's law many times throughout the study of electricity and electronics.

Before proceeding further, the more commonly used units for R and G should be introduced. The elctrical unit which is used for R is called the "ohm" and has the symbol Ω which was introduced in chap. 2.

An ohm (Ω) is equal to one volt per ampere. A device with a resistance of one ohm will conduct a current of one ampere when one volt is applied to it.

Thus, in our example we had

$$R = 5 \frac{\text{V}}{\text{A}} = 5 \text{ ohms} = 5 \: \Omega$$

The SI electrical unit which is used for G is called the *siemens*, and has the symbol S.*

A *siemens* is equal to one ampere per volt. A device with a conductance of one *siemens* will conduct a current of one ampere when one volt is applied to it.

Thus, in our example we had

$$G = 0.2 \frac{\text{A}}{\text{V}} = 0.2 \text{ S}$$

From the reciprocal relationship established in Eqs. (5-4) we can conclude that

$$1 \text{ ohm} = \frac{1}{\text{S}}$$

Equations (5-1) to (5-3) are all equivalent expressions of Ohm's law. In most of our work we will be using Eqs. (5-2) and (5-3) since *resistance* is more com-

---

* The siemens replaces the previously used unit for conductance, the *mho*, which has the symbol ℧. Some people still use the mho, and you will find it in most literature published prior to 1975.

monly used than *conductance* in expressing the electrical characteristics of a device. Table 5–1 summarizes the most important Ohm's law relationships for a device that has a linear *I-V* curve. This table should eventually be committed to memory. Actually, one need only remember the first relationship because the others are derived from it by rearranging the quantities.

TABLE 5–1

Ohm's Law Summary	
$I = \dfrac{V}{R}$	1 ampere = 1 volt per ohm
$V = IR$	1 volt = 1 amp × 1 ohm
$R = \dfrac{V}{I}$	1 ohm = 1 volt per ampere

Circuit Symbol for Constant Resistance In much of our work we will be dealing with devices that have a linear *I-V* curve. Such devices have a constant resistance to the flow of current. Whenever we use such a device in a circuit the symbol will be that shown in Fig. 5–3. The device is called a *resistor* or *resistance* and we usually speak of a resistor as having a certain amount of resistance.

FIGURE 5–3 *Circuit symbol for a constant resistance (resistor).*

EXAMPLE 5.1 Figure 5–4 shows the *I-V* graph for a certain resistor R. Determine the following:

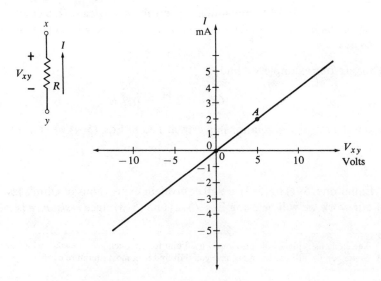

FIGURE 5–4

a. the resistance of the resistor
b. the conductance of the resistor
c. the amount of current that will flow through the resistor when 20 volts are applied to it.

Solution: (a) The resistance can be found by taking the ratio of V_{xy} over I for any point on the line. Using point A on the line we have

$$R = \frac{V_{xy}}{I} = \frac{5 \text{ V}}{2 \text{ mA}} = \frac{5 \text{ V}}{2 \times 10^{-3} \text{ A}} = 2.5 \times 10^3 \frac{V}{A}$$

or

$$R = 2500 \text{ } \Omega$$

This value of resistance can be expressed in the prefixed unit kilohm (kΩ),* which is equal to 10^3 ohms. That is,

$$R = 2500 \text{ } \Omega = 2.5 \times (10^3 \text{ } \Omega) = \mathbf{2.5 \text{ k}\Omega}$$

(b) The conductance is simply the reciprocal of R. Therefore, we have

$$G = \frac{1}{R} = \frac{1}{2500 \text{ } \Omega} = 0.0004 \text{ S}$$
$$= 4 \times 10^{-4} \text{ S}$$

The value can be expressed in the prefixed unit millisiemens (mS) which is equal to 10^{-3} S. That is,

$$G = 4 \times 10^{-4} \text{ S} = 0.4 \times (10^{-3} \text{ S}) = \mathbf{0.4 \text{ mS}}$$

Notice in these calculations we always express the quantities in their basic units in the formulas. In this way the result always comes out in its basic unit. Then, if desired, the result can be expressed in a prefixed unit.

(c) There are several ways to determine I for $V_{xy} = 20$ V. We could extend the graph somewhat to include the 20-V point. An easier method takes advantage of the Ohm's law formula $I = V/R$ [Eq. (5–3)]. Since we know R from part (a), we can proceed to find I.

$$I = \frac{V}{R} = \frac{20 \text{ V}}{2500 \text{ } \Omega} = 0.008 \frac{V}{\Omega} = 0.008 \text{ A} = \mathbf{8 \times 10^{-3} \text{ A}}$$

This can be expressed in milliamperes as

$$I = 8 \times (10^{-3} \text{ A}) = \mathbf{8 \text{ mA}}$$

We could have also used Eq. (5–1) since G is also known. However, in most calculations requiring the use of Ohm's law we will only use the equations which contain R (Table 5–1) as is the common practice in the electrical and electronics fields.

---

* From now on we will use a lowercase *k* for *kilo*, as is the standard practice in the electronics industry.

EXAMPLE 5.2 Draw the *I-V* graph for a resistor with $R = 10$ kilohms.

Solution: Since the *I-V* graph will be a straight line, all we have to do is to find two points and draw a line through them. One point which is always on an *I-V* graph is the $I = 0$, $V = 0$ point. To find a second point we can use one of the Ohm's law equations in Table 5-1. Let's use $V = I \times R$. We know $R = 10$ kΩ. We can now pick any value for *I* and solve for the value of *V* needed to produce that *I*. Let's use $I = 5$ mA. This gives us

$$V = 5 \text{ mA} \times 10 \text{ k}\Omega$$
$$= 5 \times 10^{-3} \text{ A} \times 10 \times 10^{3} \text{ }\Omega$$
$$= 50 \text{ V}$$

Thus, another point on our graph will be $I = 5$ mA, $V = 50$ V. The graph is shown in Fig. 5-5.

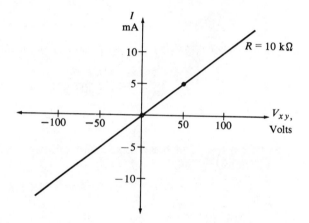

FIGURE 5-5

To check our results we can take any point on the line and determine the value of *R* to see if it checks out to be 10 kΩ. You should verify this for yourself.

EXAMPLE 5.3 Figure 5-6 shows three *I-V* graphs for three different resistors (*R*1, *R*2, and *R*3) plotted on the same set of *I-V* axes. Determine which resistor has the largest resistance and which resistor has the smallest resistance.

Solution: There are several ways to attack this problem. We could calculate the conductance or resistance from each graph and compare the three values. Before doing this, we can reason out the answer without making any calculations.

Since *R*1 has the steepest slope, then it follows that *R*1 experiences the greatest current change for a given voltage change. Similarly, *R*3 has the least steep slope and it experiences the smallest current change for a given voltage change. Thus, *R*1 has the least resistance and *R*3 has the largest resistance of the three.

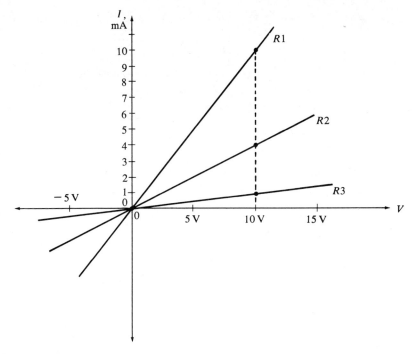

FIGURE 5–6

To verify this reasoning we can calculate the resistance values of each resistor. We will use the $V = 10\text{-V}$ point for each line.

$$R1 = \frac{V}{I} = \frac{10 \text{ V}}{10 \text{ mA}} = \frac{10 \text{ V}}{0.010 \text{ A}} = 1000 \text{ ohms} = 1 \text{ kilohm}$$

$$R2 = \frac{10 \text{ V}}{4 \text{ mA}} = \frac{10 \text{ V}}{0.004 \text{ A}} = 2500 \text{ ohms} = 2.5 \text{ kilohms}$$

$$R3 = \frac{10 \text{ V}}{1 \text{ mA}} = \frac{10 \text{ V}}{0.001 \text{ A}} = 10{,}000 \text{ ohms} = 10 \text{ kilohms}$$

To summarize, an I-V graph with a steeper slope indicates less resistance to current flow, while a flatter slope indicates greater resistance.

EXAMPLE 5.4 Consider the circuits shown in Fig. 5–7(a)–(c). Find the current through each resistor.

Solution: (a) In Fig. 5–7(a) the 25-ohm resistor has a 6-volt dc source connected across its terminals. Thus, the voltage across R is $V_{xy} = 6$ V. As a result of this voltage, current will flow through the circuit in a ccw direction (the battery pushes electrons out of its negative terminal). Using Ohm's law we have

$$I = \frac{V_{xy}}{R}$$

$$= \frac{6 \text{ V}}{25 \text{ ohms}} = 0.24 \text{ A} \qquad (5\text{–}3)$$

(b) In Fig. 5-7(b) the 100-kΩ resistor has 50 V applied to it. Thus, we have

$$I = \frac{50 \text{ V}}{100 \text{ k}\Omega} = \frac{50 \text{ V}}{100 \times 10^3 \text{ }\Omega}$$
$$= 0.5 \times 10^{-3} \text{ A}$$
$$= \mathbf{0.5 \text{ mA}}$$

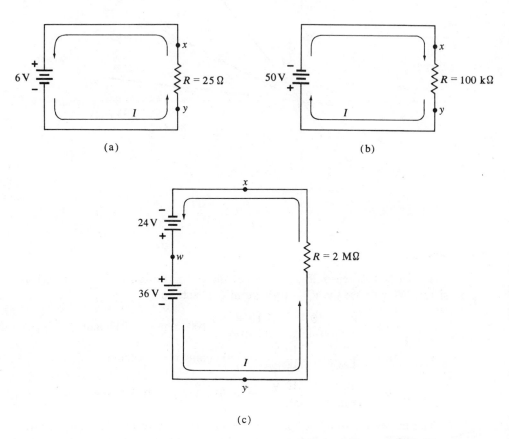

FIGURE 5-7

(c) Figure 5-7(c) shows a different situation. There are two dc batteries connected in series. In order to find the current in this circuit we must find the total voltage that appears across R. To find V_{xy} we will use the procedure outlined in chap. 4.

Starting at point y and moving to point w, we go through a voltage rise of 36 V. Then, going from w to x we encounter a voltage *drop* of 24 V. The net change in going from y to x is a voltage *rise* of 12 V. Thus, point x is 12 V greater than point y so that $V_{xy} = +12$ V. Since the 36-V battery is the larger battery, it will dominate in determining the flow of current. Current will thus flow out of its negative terminal in a ccw direction.

The value of I can now be calculated as

$$I = \frac{V_{xy}}{R} = \frac{12 \text{ V}}{2 \text{ M}\Omega}$$

Since 2 MΩ = 2 megohms = 2×10^6 ohms

$$I = \frac{12 \text{ V}}{2 \times 10^6 \text{ ohms}} = 6 \times 10^{-6} \text{ A}$$

This can be expressed in the prefixed unit microamperes (μA) since a μA is 10^{-6} A.

$$I = 6 \times (10^{-6} \text{ A}) = 6 \text{ μA}$$

Notice that in this last example we did not have to use any *I-V* graphs in order to determine the circuit current. This was because the value of *R* was known in each case. The *I-V* graph would give us no new information. In other words, *for a linear resistor (one with a linear I-V graph) the value of R is all the information we will ever need to determine its behavior in an electric circuit.*

In using Ohm's law to make circuit calculations we will usually be using an equation that contains *three* quantities (*I*, *V*, and *R*) such as those in Table 5-1. As such, we can always determine any one of the three quantities if the other two are known. The following examples will illustrate.

EXAMPLE 5.5 Figure 5-8(a) shows a portion of a circuit containing a 500-ohm resistor which has a current of 1.6 μA flowing through it. Determine the resistor voltage V_{xy}.

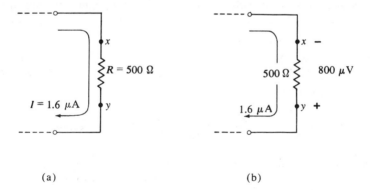

(a)   (b)

FIGURE 5-8

Solution: First, we must determine the polarity of the resistor voltage. *I* is shown flowing from *x* to *y* through the resistor. This is the direction it would flow if a battery were connected to *R* with its negative terminal at *x*. Thus, point *x* is negative relative to *y*.

The magnitude of the resistor voltage is equal to the current times the resistance ($V = IR$). That is,

$$1.6 \text{ μA} \times 500 \text{ ohms} = 1.6 \times 10^{-6} \text{ A} \times 500 \text{ ohms}$$
$$= 800 \times 10^{-6} \text{ V}$$
$$= 800 \text{ μV (microvolts)}$$

Thus,

$$V_{xy} = -800 \text{ μV}$$

82   Chapter Five

or, equivalently,

$$V_{yx} = +800 \ \mu V$$

This result is shown in Fig. 5-8(b).

This last example points out that *in a resistor the electron current always flows from the negative terminal to the positive terminal through the resistor* [see Fig. 5-8(b)]. This is easy to remember by realizing that the terminal of the resistor where the electrons are entering will be made more negative because of the negative electrons, and the terminal where the electrons are leaving will be made more positive. This is just the opposite to what happens in a voltage source. *In a voltage source, the electron current flows out of the negative terminal.*

EXAMPLE 5.6 A certain length of copper conducting wire has a voltage of 10 mV across its ends. The resultant current through the copper wire is 1 ampere. What is the resistance of the wire?

Solution: $R = \dfrac{V}{I} = \dfrac{10 \text{ mV}}{1 \text{ A}} = \dfrac{10 \times 10^{-3} \text{ V}}{1 \text{ A}} = 0.01 \text{ ohm}$

This low value of resistance indicates why copper wire is used extensively in connecting elements in a circuit. Its low resistance allows current to flow through it very easily without the need for any significant voltage across it.

EXAMPLE 5.7 A certain segment of rubber insulator has a voltage of 200 V across its ends. The resultant current through the rubber is 0.1 nA. What is the resistance of the rubber segment?

Solution: $R = \dfrac{V}{I} = \dfrac{200 \text{ V}}{0.1 \text{ nA}} = \dfrac{200 \text{ V}}{10^{-10} \text{ A}} = 2 \times 10^{12} \text{ ohms}$

This extremely large value of resistance indicates why rubber is a good insulator. It can be used to cover copper wires that are touching each other and thus prevent current from flowing from one wire to the other.

Short Circuits and Open Circuits These last two examples represent the extremes of resistance values. Actually, in theory, a resistance can be as low as zero ohms. If such a resistance exists, it is called a *short circuit*. In practice, zero ohms is impossible to obtain under normal conditions. However, in many cases the resistance of connecting wires is sufficiently small, as in Ex. 5.6, to allow us to refer to these wires as short circuits. In other words, if a certain resistance in a circuit is very small compared to the other resistances in the circuit, we can treat it as a short circuit with $R = 0$. This is why in our analysis of circuits we will not usually concern ourselves with the resistance of the connecting wires. A piece of connecting conductor is treated as a short circuit, as illustrated in Fig. 5-9(a). *A short circuit will have no voltage across it* because $V = I \times R = I \times 0 = 0$.

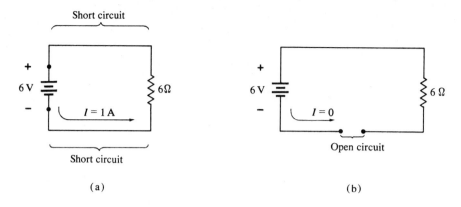

FIGURE 5–9 (a) *Example of short circuits between two points.* (b) *Example of open circuit.*

On the other hand, a resistance can theoretically be infinitely large. If such a resistance exists it is called an *open circuit*. In practice, an infinite* resistance is not possible. However, in many cases the resistance of insulating materials is very large, as in Ex. 5.7, and can be considered an open circuit. Because of its infinite resistance *an open circuit will not conduct current*, since $I = V/R = V/\infty = 0$.

When a circuit has a broken connecting wire or an open switch, such a discontinuity in the current path constitutes an open circuit through which electrons cannot flow. Figure 5–9(b) illustrates.

5.3 Nonlinear I-V Graphs†

Many of the important electrical devices have *I-V* curves which are not straight lines. Instead of having perfectly linear *I-V* curves, these devices have nonlinear *I-V* curves. An example of a nonlinear *I-V* curve is shown in Fig. 5–10. This *I-V* curve happens to be the characteristic of a device called a *diode*. The electronic symbol for a diode is shown in the figure.

Examination of this *I-V* curve reveals several important points. First, it is apparent that for negative values of V_{xy} there is essentially no current flow through the device. In other words, when terminal *x* is made negative relative to *y*, no current will flow through the device. This is a characteristic of most diodes.

On the other hand, when V_{xy} is positive there will be a current *I* produced through the device in the direction shown. However, unlike a linear resistor, the current is not directly proportional to the voltage. The *I-V* curve for positive values of V_{xy} is not a straight line. The slope of the curve is not constant. In fact, near $V_{xy} = 0$ the slope is very small since the curve is essentially flat. As V_{xy} increases in the positive direction, the curve becomes steeper and the slope increases.

---

*We represent the value of infinity by the symbol ∞. An open circuit has $R = \infty$.
†Sections 5.3–5.4 may be omitted if desired without affecting the continuity of the material.

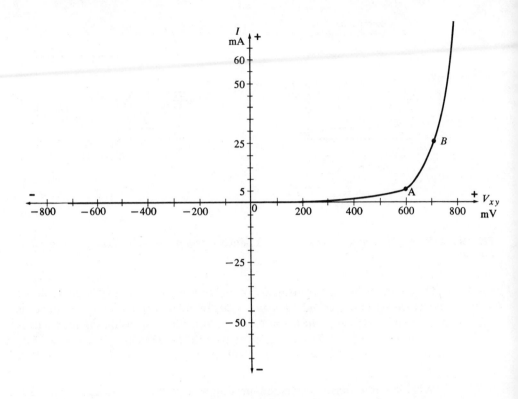

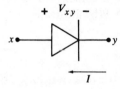

FIGURE 5-10 *Typical nonlinear I-V curve for a diode.*

For any nonlinear device such as the diode we *cannot* express the current as $I = GV$ or $I = V/R$. These expressions, recall, were only valid for a linear *I-V* curve passing through the origin. The values of conductance G and resistance R were constants for a linear *I-V* curve. For nonlinear *I-V* curves the values of G and R will, in general, change from point to point. For example, at point A on the curve in Fig. 5–10, the voltage is 600 mV and the current is 5 mA. Thus, the conductance at point A is

$$G_A = \frac{5 \text{ mA}}{600 \text{ mV}} = \frac{5 \times 10^{-3} \text{ A}}{600 \times 10^{-3} \text{ V}} = 0.0083 \text{ S}$$
$$= 8.3 \text{ mS}$$

and the resistance at point A is

$$R_A = \frac{600 \text{ mV}}{5 \text{ mA}} = \frac{600 \times 10^{-3} \text{ V}}{5 \times 10^{-3} \text{ A}} = 120 \text{ }\Omega$$

At point B the voltage is 700 mV, resulting in $I = 25$ mA. Thus,

$$G_B = \frac{25 \text{ mA}}{700 \text{ mV}} = \frac{25 \times 10^{-3} \text{ A}}{700 \times 10^{-3} \text{ V}} = 0.0357 \text{ S}$$
$$= 35.7 \text{ mS}$$

and

$$R_B = \frac{700 \text{ mV}}{25 \text{ mA}} = \frac{700 \times 10^{-3} \text{ V}}{25 \times 10^{-3} \text{ A}} = 28 \text{ }\Omega$$

As these calculations show, both G and R have different values at different points on the curve.

Thus, when we are indicating the values of G or R for a nonlinear device we must always stipulate the point on the I-V curve which these values correspond to. For example, according to the previous calculations we have

$$R[\text{at } I = 5 \text{ mA}] = 120 \text{ }\Omega$$

and

$$R[\text{at } I = 25 \text{ mA}] = 28 \text{ }\Omega$$

Static Resistance and Conductance Accurately speaking, the quantity R, which is obtained by taking the ratio of V to I at a particular point on the curve, should be called the *static resistance* at that point. The term *static* resistance* is used because this is the resistance the device would exhibit if a constant (static) voltage were applied to it. Similarly, G is called *static conductance*. The terms *dc resistance* and *dc conductance* are often used to mean the same thing as static resistance and static conductance.

For a linear resistor, the value of its static resistance, R, will be the same at all points on its I-V graph. As we just saw, for a nonlinear device the value of R will generally be different at different points on the device's I-V curve.

5.4 Dynamic Resistance and Conductance

Whereas static resistance R indicates the resistance of a device at a given constant voltage and current, *dynamic resistance* of a device indicates its resistance at a changing voltage and current. To illustrate dynamic resistance, let's first consider the case of a linear I-V graph shown in Fig. 5-11. It is apparent that this graph corresponds to a static resistance R of 10 ohms (since $I = 1$ A when $V = 10$ volts).

We can show on this graph that any change in voltage across the resistor will produce a proportional change in current through the resistor. Refer to points a and b on the graph. Point a corresponds to $V = 10$ V, $I = 1$ A, and point b corresponds to $V = 20$ V, $I = 2$ A. Thus, if the voltage applied to the resistor is changed from 10 V to 20 V the current will change from 1 A to 2 A. This increase in voltage of 10 V results in an increase in current of 1 A.

---

*"*Static*" means at rest or in a steady state of no movement.

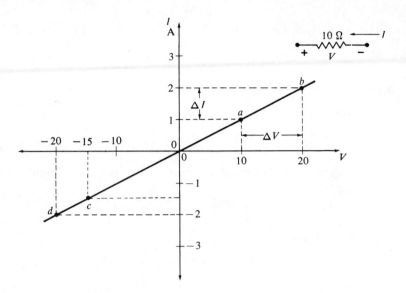

FIGURE 5-11 *In a linear resistor ΔI is proportional to ΔV.*

The ratio of the change in voltage to the resultant change in current is called *dynamic resistance*. That is,

$$r = \frac{\Delta V}{\Delta I} \qquad (5\text{-}5)$$

In this expression Δ (Greek letter "delta") stands for "change in," so that ΔV means "change in V" and ΔI means "change in I." The symbol for dynamic resistance is lowercase *r* to distinguish it from static resistance, which is uppercase *R*.

The quantity ΔV will be *positive* if we are taking an *increase* in V. In other words, going from point *a* to point *b* in Fig. 5-11 entails an *increase* in V from 10 V to 20 V. Thus, ΔV = 20 V − 10 V = +10 V. Similarly, ΔV will be *negative* if we are taking a *decrease* in V. In other words, going from point *c* to point *d* in Fig. 5-11 entails a *decrease* in V from −15 V to −20 V because −20 V is algebraically less (more negative) than −15 V. Thus, ΔV would be −20 V − (−15 V) = −5 V. The same statements hold true for ΔI.

The dynamic resistance going between points *a* and *b* in Fig. 5-11 can now be calculated using Eq. (5-5).

$$r = \frac{\Delta V}{\Delta I} = \frac{20\text{ V} - 10\text{ V}}{2\text{ A} - 1\text{ A}} = \frac{10\text{ V}}{1\text{ A}} = 10 \text{ ohms}$$

Similarly, the dynamic resistance going between points *c* and *d* can be determined as

$$r = \frac{\Delta V}{\Delta I} = \frac{-20\text{ V} - (-15\text{ V})}{-2\text{ A} - (-1.5\text{ A})} = \frac{-20\text{ V} + 15\text{ V}}{-2\text{ A} + 1.5\text{ A}} = \frac{-5\text{ V}}{-0.5\text{ A}} = 10 \text{ ohms}$$

And, in fact, going between any two points on this linear graph will always result in the same value of dynamic resistance and this value is always equal to the static resistance. In other words,

A linear resistor has a dynamic resistance r which is equal to its static resistance R at all points.

Thus, in a linear resistor, not only is current proportional to voltage, but changes in current are proportional to changes in voltage.

Dynamic conductance, g, can be similarly defined as

$$g = \frac{\Delta I}{\Delta V} \tag{5-6}$$

Obviously, g and r are related such that

$$g = \frac{1}{r} \text{ or } r = \frac{1}{g} \tag{5-7}$$

For a linear resistor, then, dynamic conductance g will be equal to its static conductance G.

For a nonlinear device, in general, the dynamic resistance will *not* be a constant value any place on the *I-V* curve of the device. This can be shown for the diode *I-V* curve which is redrawn in Fig. 5-12. The values of voltage and current at points A, B, and C are tabulated:

Point	V	I
A	600 mV	5 mA
B	700 mV	25 mA
C	500 mV	2.5 mA

Consider first the dynamic resistance in going between points A and B. Using Eq. (5-5):

$$r_{A\text{-}B} = \frac{\Delta V}{\Delta I} = \frac{700 \text{ mV} - 600 \text{ mV}}{25 \text{ mA} - 5 \text{ mA}} = \frac{100 \text{ mV}}{20 \text{ mA}}$$

$$= \frac{100 \times 10^{-3} \text{ V}}{20 \times 10^{-3} \text{ A}}$$

$$= 5 \text{ ohms}$$

Next, consider the dynamic resistance between points C and A.

$$r_{C\text{-}A} = \frac{\Delta V}{\Delta I} = \frac{600 \text{ mV} - 500 \text{ mV}}{5 \text{ mA} - 2.5 \text{ mA}}$$

$$= \frac{100 \text{ mV}}{2.5 \text{ mA}} = \frac{100 \times 10^{-3} \text{ V}}{2.5 \times 10^{-3} \text{ A}}$$

$$= 40 \text{ ohms}$$

88   Chapter Five

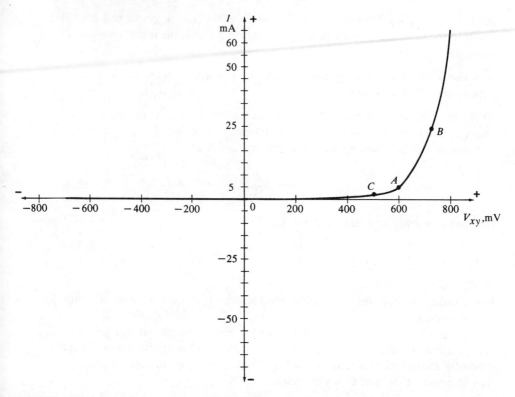

FIGURE 5–12 *Dynamic resistance has different values at different points on a nonlinear I-V curve.*

Thus, we can see that on different portions of the curve the dynamic resistance can have different values.

We could have predicted that r_{C-A} would be greater than r_{A-B} because in the region between A and C the curve is flatter (has less slope) than in the region between A and B. This means that for the same ΔV there will be less of a current change ΔI, so that $r = \Delta V/\Delta I$ will be greater. In general, then, we can say that

Dynamic resistance is greater for those portions of an I-V curve with a smaller (flatter) slope.

When indicating a value of r for a nonlinear device, we also have to indicate where on the *I-V* curve the calculation or measurement of r is made. The usual method is to stipulate the midpoint of the region where r was calculated. For example, in Fig. 5–12 the value of r determined between points A and C would be expressed as

$$r[\text{at } I = 3.75 \text{ mA}] = 40 \text{ ohms}$$

because 3.75 mA is halfway between points A and C.

Similarly, the value of r between points A and B is expressed as

$$r[\text{at } I = 15.0 \text{ mA}] = 5 \text{ ohms}$$

Current-Voltage Graphs and Resistive Devices 89

The concepts of dynamic resistance, r, and conductance, g, are very important for the study of semiconductor devices such as diodes and transistors. These devices all have nonlinear I-V curves and so cannot be characterized by a single value of resistance as can a linear resistor. We will not be concerned with r and g in the remainder of this text; hence, whenever we talk about resistance and conductance, we will be talking about static values (R and G) of a linear resistor.

Chapter Summary

1. An I-V curve for a device is a graph of current through the device as a function of voltage across the device.
2. In resistive devices the current responds instantaneously to a change in voltage.
3. In a device which has a linear I-V curve, current is directly proportional to voltage.
4. The static conductance G of a linear device is the ratio of current to voltage (I/V) and is constant any place on the I-V curve. In fact, G is the slope of the linear I-V curve.
5. The static resistance R of a linear device is the ratio of voltage to current (V/I) and is constant any place on the I-V curve.
6. G is a measure of a device's ability to conduct current while R is a measure of a device's ability to resist the flow of current.
7. A device with a high G has a low R and vice versa since $G = 1/R$.
8. Ohm's law states that in certain materials the amount of current is directly proportional to the applied voltage ($I = V/R$ or $I = GV$).
9. The unit for resistance is the ohm (Ω), which equals 1 volt per ampere. The unit for conductance is the siemens (S) which equals 1 ampere per volt.
10. An I-V graph with a steeper slope indicates less resistance to current flow and a flatter slope indicates greater resistance.
11. For a linear resistor the value of R is all we need to determine its behavior in an electric circuit.
12. In a linear resistor current always flows through the resistor from the negative potential terminal to the positive potential terminal.
13. A short circuit has a resistance of zero ohms and can have no voltage across it.
14. An open circuit has an infinitely large resistance and can have no current through it.
15. A nonlinear I-V curve does not have a constant slope.
16. In a nonlinear device I is not directly proportional to V.
17. For a nonlinear I-V curve the values of G and R change for different points on the curve.

18. Dynamic resistance r is a measure of a device's resistance to a changing current flow, and equals the ratio of change in voltage (ΔV) to change in current (ΔI).

19. Dynamic conductance g is a measure of a device's ability to conduct a changing current and is equal to the ratio of change in current (ΔI) to change in voltage (ΔV).

20. A linear resistor has a constant value of r at all points on its *I-V* curve and, in fact, $r = R$, the static resistance. Similarly, $g = G$.

21. For a nonlinear device, r and g will, in general, vary for different points on the *I-V* curve.

22. Dynamic resistance is greater for those portions of an *I-V* curve that have a flatter slope and vice versa for dynamic conductance.

Questions/Problems

Sections 5.1–5.2

5-1 What is the principal characteristic of resistive devices that distinguishes them from inductive or capacitive devices?

5-2 A certain resistive device has a linear *I-V* curve. What will happen to the current through the device if the voltage across the device is cut in half?

5-3 Consider the *I-V* graph in Fig. 5–13. Determine the value of conductance from this graph. Do it for at least two points on the line.

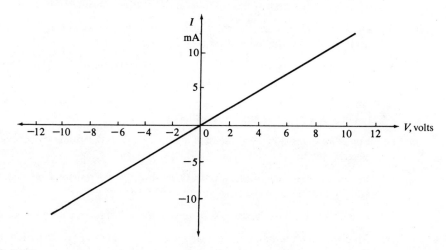

FIGURE 5–13

5-4 Determine which of the following cases corresponds to the greatest value of conductance.
  a. $I = 5$ mA when $V = 0.2$ V
  b. $I = 5$ A when $V = 20$ V
  c. $I = 5$ μA when $V = 200$ μV

Current-Voltage Graphs and Resistive Devices

5-5 Determine the value of resistance for the *I-V* graph in Fig. 5-13. Use at least two points on the line.

5-6 Determine which of the cases in prob. 5-4 corresponds to the greatest value of resistance.

5-7 Refer to Fig. 5-13. On the same set of axes sketch an *I-V* graph that has *twice* the resistance of the graph shown.

5-8 Repeat 5-7 for *half* the resistance of the original *I-V* graph.

5-9 Indicate which of the following statements are *not* valid:
  a. A device with a high resistance to current flow always has a low value of conductance.
  b. The greater the slope of the *I-V* curve the greater will be the value of resistance.
  c. If the voltage across a resistor is increased, the current through the resistor increases. Thus, the resistance to current flow decreases.
  d. In an open circuit the current is zero and for a short circuit the voltage is zero.

5-10 Determine the current (magnitude and direction) through each of the resistors in Fig. 5-14.

5-11 In the circuit in Fig. 5-14(c), by how much should the voltage source *be increased* in order to produce a current of 10 mA?

5-12 In the circuit of Fig. 5-14(a), by how much should the resistance *be decreased* in order to produce a current of 10 μA?

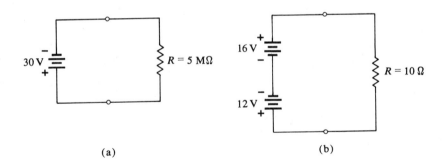

(a)   (b)

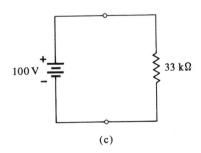

(c)

FIGURE 5-14

92 Chapter Five

5-13 Consider the circuit of Fig. 5-14(b). Indicate whether the current will increase, decrease, or remain the same as the following changes are made in the circuit values:
  a. R is decreased.
  b. 16-V battery is increased.
  c. 12-V battery is increased.
  d. Both batteries are doubled and R is also doubled.

5-14 Consider the circuit in Fig. 5-15. Calculate I for the following two cases:
  a. The box contains a short circuit between x and y.
  b. The box contains an open circuit between x and y.

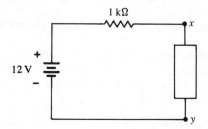

FIGURE 5-15

5-15 Refer to Fig. 5-10. Calculate G and R at the following values of voltage:
  a. $V = 650$ mV
  b. $V = 750$ mV
  c. $V = -400$ mV

5-16 Indicate which of the following statements pertain to linear devices and which to nonlinear devices:
  a. The static resistance is constant.
  b. The dynamic resistance is constant.
  c. Current always flows equally well for both polarities of voltage.
  d. Current could more than double if the applied voltage is doubled.
  e. The device obeys Ohm's law.

5-17 Calculate r and g for the I-V curve in Fig. 5-13.

5-18 Calculate r and g for the I-V curve in Fig. 5-10 between $V = 650$ mV and $V = 750$ mV.

5-19 Refer to the I-V curve in Fig. 5-16. Consider the four regions labeled A, B, C, and D.
  a. Which of the four regions exhibits the largest dynamic resistance? Smallest dynamic resistance?
  b. Which region exhibits the largest dynamic conductance? Smallest dynamic conductance?
  c. In which region will the current change the most for a small ΔV?
  d. In which region will the current change the least for a small ΔV?

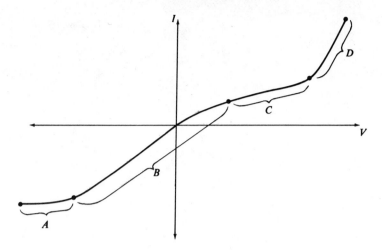

FIGURE 5-16

CHAPTER 6

Resistive Devices in Series Circuits

6.1 Introduction

Most useful circuits contain more than one electrical device. Usually, several devices (linear and nonlinear types) are connected together in a circuit to produce a desired result. There are an innumerable number of possible circuits. However, there are only a few basic ways of connecting devices in a circuit. In this chapter we will study the *series* connection and some of its properties. In the process we will be introduced to one of the two most important electrical circuit laws, *Kirchhoff's Voltage Law*.

6.2 The Series Connection

A series connection is defined as two or more devices connected end to end to form only one path for current. In other words, the same current flows through each device in the circuit since there is only one path for current. To illustrate, consider Fig. 6-1(a) where a constant voltage source E_s is connected to a circuit containing two resistors $R1$ and $R2$ connected *in series*. The constant voltage source will produce a constant current I in a ccw direction around the circuit. The electrons which are pushed out of the negative terminal of E_s travel this ccw path and pass through both $R1$ and $R2$ on their way back to the positive terminal of E_s. The important thing here is that the same exact movement of charge (current) is taking place everywhere along the path shown. Thus, we can state that $R1$ and $R2$ and E_s are connected in *series*. Stated formally,

Two or more electrical devices are said to be **in series** if the same current passing through one device passes through each of the other devices.

96   Chapter Six

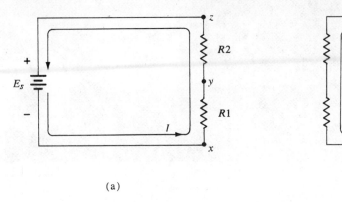

FIGURE 6–1 *Simple series circuits.*

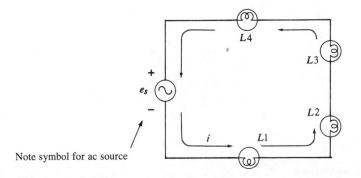

Note symbol for ac source

FIGURE 6–2 *Christmas tree lamps connected in series.*

Any number of devices can be connected in series. Figure 6–1(b) shows a series circuit with five resistors.

An example of a series circuit with which we are all familiar is the notorious string of Christmas tree lamps connected in series as shown in Fig. 6–2. The current produced by the source e_s (usually the ac voltage from your home outlet) passes through each lamp L_1, L_2, L_3, and L_4. As long as each lamp filament is unbroken, current will flow and each lamp will radiate light. Unfortunately, if any one of the lamp filaments burns out (breaks apart), the current path will be broken and the current will cease to flow. In other words, the burned-out filament is an *open circuit* and since there are no alternate paths for current, the flow of electrons cannot take place and all the lamps will go out. The burned-out lamp has to be replaced to complete the circuit and provide a current path.

Figure 6–3 shows a circuit which is not a simple series circuit. It is a more complex circuit which will be discussed in detail in a later chapter. It is presented here to help illustrate the series connection. In the figure the current which is leaving the negative terminal of E_s is labeled I_S. This current passes through resistors $R1$ and $R2$. Thus, $R1$ and $R2$ are in series since the same current flows through each.

Now, looking at $R3$ we might be tempted to say that $R3$ is in series with $R2$ and $R1$. However, going back to our basic definition of a series connection we

Resistive Devices in Series Circuits

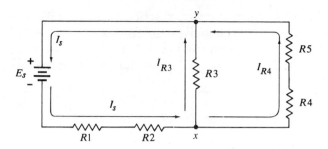

FIGURE 6-3 *Complex circuit illustrating series connection of* R1 *with* R2, R4 *with* R5.

must check to see whether $R3$ has the same current flowing through it that flows through $R1$ and $R2$, namely I_S. Looking at I_S, we can see that when this current reaches point x it is presented with two possible *parallel** paths that it can take up to point y. As a result, a portion of I_S will take the path through $R3$, and the other portion of I_S will take the path through $R4$ and $R5$. This situation is much like that occurring in a system of water pipes. I_S is like the water flowing through the main pipeline. When the water reaches point x it can flow through either of the two *parallel* pipes to get to y.

What this all means is that the current through $R3$ (labeled I_{R3}) is *not* the same current as I_S. Also, the current through $R4$ and $R5$ (labeled I_{R4}) is *not* the same current as I_S. I_{R3} and I_{R4} are both portions of I_S. In other words, electrons reaching point x will split up into two paths so that some electrons will flow through each parallel path producing a current through each branch. Therefore, $R3$ *cannot* be in series with $R1$ and $R2$. Similarly, $R4$ and $R5$ *are not* in series with $R1$ and $R2$.

Looking at $R4$ and $R5$ we can easily see that they are in series because the same current I_{R4} flows through each resistor. Before leaving this example, note that the current shown flowing back into the $+$ terminal of E_S is also labeled I_S. This is because we know that the number of electrons leaving the negative terminal of E_S will always be exactly balanced out by the same number of electrons entering the positive terminal. Thus, *the current into the $+$ terminal equals the current out of the $-$ terminal.*

EXAMPLE 6.1 Which resistors would be in series in the circuit in Fig. 6-3 if we were to remove $R3$ from the circuit?

Solution: With $R3$ removed, there would be only one path for I_S. This path would be through $R1$, $R2$, $R4$, and $R5$. Thus, these remaining four resistors would now be in series.

In summary, it is necessary that two devices have the same current passing through them if they are connected in series. In a complex circuit, such as in Fig. 6-3, it is helpful to draw all the circuit current arrows in order to determine

---

*Parallel connections are discussed in chap. 7.

which elements are, in fact, in series. In drawing these current arrows, we always start from the negative terminal of the source. When a node or junction is reached (such as point x in Fig. 6–3), the current splits up and takes more than one path. Later on we will find out how to determine how much current flows through each path.

6.3 Inserting Elements in Series

There are many times during the course of troubleshooting a circuit or performing an experiment that it is necessary to insert an electrical element or device into a circuit such that it is in series with certain other elements. For example, consider Fig. 6–4(a) where we have a simple series circuit with resistors $R1$ and $R2$ connected to a source E_S. The current I_S flows through the series circuit. Suppose we wanted to insert a third resistor $R3$ in series with both $R1$ and $R2$. How could we accomplish this? Actually, there are several possibilities because the only thing we have to be concerned with is that the current flowing through $R3$ is the same as the current flowing through $R1$ and $R2$. Three possibilities are shown in Figs. 6–4(b), (c), and (d).

In each of these three circuits $R3$ has been inserted so that it lies in the current path for I_S. In all three cases $R3$ is in series with $R1$ and $R2$ since I_S flows through all three resistors. The value of I_S, however, will be changed from its value in the original circuit before $R3$ was added. As we shall soon see, adding resistance in a series circuit always causes a decrease in current.

An element which is often inserted in series with other circuit elements during experimentation or testing of a circuit is an *ammeter*. An ammeter is an electrical measuring device used to measure and indicate the amount of current flowing through a circuit. The symbol for an ammeter is a circle with an A inside of it and two terminals. Figure 6–5 shows a circuit containing a battery and several resistors. It also contains two ammeters A_1 and A_2. Ammeter A_1 is connected in series with $R1$ and ammeter A_2 is connected in series with $R4$ and $R5$. If this circuit were constructed in a laboratory, ammeter A_1 would have a calibrated scale with a pointer or needle that moves along the scale to indicate the amount of current which is flowing through A_1. In other words, A_1 would indicate the value of the current I_{R1} passing $R1$. Similarly, A_2 would indicate the value of I_{R4}, the current through $R4$ and $R5$.

Usually ammeters are constructed with a very low resistance to current so that they do not cause an appreciable drop in current when they are inserted in the circuit. *To measure the current through any path in a circuit, an ammeter must be inserted in the circuit such that this same current passes through the ammeter.* You will undoubtedly learn more about the use of ammeters and some of the principles behind them in the laboratory portions of your electricity courses.

6.4 Linear Resistors in Series

From our work in chap. 5 we know that a voltage applied across a resistor will produce a current, according to Ohm's law, of $I = V/R$. Suppose we wanted to know how much current would flow if the same voltage were applied to a series circuit of two or more resistors. To arrive at a method for solving this problem

Resistive Devices in Series Circuits   99

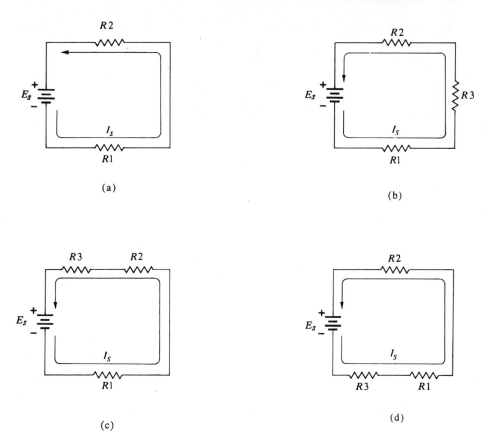

FIGURE 6–4 *Equivalent ways to connect an element in series with other elements.*

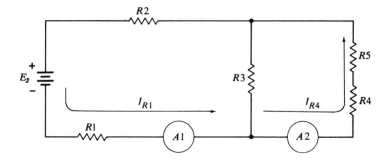

FIGURE 6–5 *An ammeter is inserted in series so that the current to be measured passes through it.*

consider Fig. 6–6. In part (a) of the figure a voltage source E_S is connected in series with two resistors, $R1 = 25$ ohms and $R2 = 50$ ohms. In part (b) of the figure the *I-V* graphs for $R1$ and $R2$ are plotted. For convenience, only positive values of current and voltage are shown.

Looking at the circuit, suppose we wanted to know how much voltage to apply in order to produce a current I of 0.1 A through the series circuit. To

100   Chapter Six

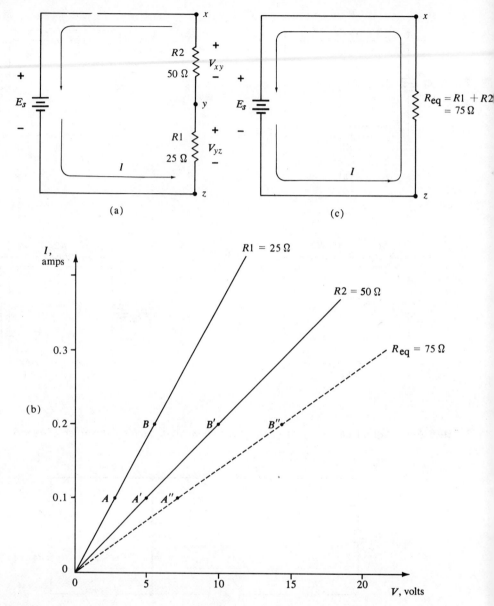

FIGURE 6-6 (a) $R1$ and $R2$ in series. (b) I-V graphs for $R1$, $R2$, and R_{eq}.
(c) $R_{eq} = R1 + R2$.

answer this we can reason as follows. According to $R1$'s I-V curve (point A), the voltage across $R1$ has to be equal to 2.5 V in order for a current of 0.1 A to flow through it. Similarly, $R2$ requires 5 V for a current of 0.1 A (see point A' on $R2$'s I-V curve). Thus, in order to produce a current $I = 0.1$ A through the series combination of $R1$ and $R2$ we must have $V_{yz} = 2.5$ V and $V_{xy} = 5$ V.

What, then, will the value of E_S have to be in order to produce $I = 0.1$ A? Looking at the circuit diagram, it is clear that the voltage between points x and z has to be equal to the applied voltage E_S. That is, $E_S = V_{xz}$. According to our reasoning in the last paragraph, the voltage between x and z will have to equal a total of 2.5 V + 5 V = 7.5 V if a current of 0.1 A is to be produced. Thus, E_S has to have a value of 7.5 V in order to produce a current of 0.1 A through this circuit.

We can follow this same reasoning for a current of 0.2 A. From point B on $R1$'s *I-V* curve we read 5 V. From point B' on $R2$'s curve we read 10 V. Thus, to produce a current of 0.2 A through the series combination of $R1$ and $R2$ we need E_S = 5 V + 10 V = 15 V.

This procedure can be followed for any desired value of I. Table 6–1 presents some of the results for various values of I. If these results are plotted on the same *I-V* axis, the result is a straight line shown dotted in Fig. 6–6(b). This dotted line is essentially a representation of the current-voltage relationship for the *series combination* of $R1$ and $R2$. If we look at it closely, we see that this dotted line corresponds to a resistance of 75 ohms since 7.5 V produces 0.1 A, and 7.5 V/0.1 A = 75 ohms.

TABLE 6–1

Value of I	Required Value of E_S
0 A	0 V
0.1 A	7.5 V (point A" on graph)
0.2 A	15 V (point B" on graph)
0.25 A	18.75 V
0.3 A	22.5 V

What all this means is that $R1$ = 25 ohms and $R2$ = 50 ohms in series have the same electrical characteristics as a single resistor with a resistance of 75 ohms. In other words, as far as the flow of current is concerned, the circuit in Fig. 6–6(a) acts exactly the same as the circuit in Fig. 6–6(c). In Fig. 6–6(c) the two resistors $R1$ and $R2$ have been replaced by one resistor of 75 ohms. This one resistor is said to be the *equivalent resistance* or *equivalent resistor* for the series combination of $R1$ and $R2$, and is given the symbol R_{eq}.

From the results of our example, then, we can conclude that:

Two resistances, $R1$ and $R2$, connected in series present the same resistance to current flow as does a single resistor, R_{eq}, whose resistance is equal to the sum of $R1$ and $R2$.

In general

$$R_{eq} = R1 + R2 \qquad (6\text{–}1)$$

for two resistors in series. The resistance R_{eq} is the resistance the series circuit presents to the flow of current. This is the resistance "seen" by the voltage source.

The value of R_{eq} is always greater than either $R1$ or $R2$. This means that $R1$ and $R2$ connected in series always present more resistance to the flow of current than either one taken by itself. This should be fairly obvious since the current must travel a path which contains both resistors. As a result, less current will flow for a given applied voltage if $R1$ and $R2$ are in series.

EXAMPLE 6.2 Two resistors of 2 kilohms each are connected in a series circuit. What is their equivalent resistance? How much current will flow through these resistors if 20 V are applied to the circuit?

Solution: (a) $R_{eq} = R1 + R2$
$= 2 \text{ k}\Omega + 2 \text{ k}\Omega$
$= 4 \text{ k}\Omega$

(b) The total circuit resistance is 4 kilohms. This is the resistance to which the 20 V are applied. Using Ohm's law we have

$$I = \frac{20 \text{ V}}{4 \text{ k}\Omega} = \frac{20 \text{ V}}{4 \times 10^3 \text{ }\Omega} = 5 \times 10^{-3} \text{ A} = 5 \text{ mA}$$

This last example introduces an important concept. It shows that Ohm's law can be applied to a series circuit if we use the total applied voltage and equivalent series resistance. That is,

$$I = \frac{E_S}{R_{eq}} = \frac{E_S}{R1 + R2} \qquad (6\text{--}2)$$

EXAMPLE 6.3 Two resistors are connected in series in a certain circuit. With an applied source voltage of 12 V an ammeter connected in series with the resistors indicates that $I = 4$ mA in the circuit. What is the equivalent series resistance? What are the values of each of the series resistors?

Solution: (a) Using Eq. (6–2) we can write

$$I = \frac{E_S}{R_{eq}}$$

$$4 \text{ mA} = \frac{12 \text{ V}}{R_{eq}}$$

and solving for R_{eq}

$$R_{eq} = \frac{12 \text{ V}}{4 \text{ mA}} = \frac{12 \text{ V}}{4 \times 10^{-3} \text{ A}} = 3 \times 10^3 \text{ }\Omega = 3 \text{ k}\Omega$$

(b) Simply knowing the value of R_{eq} will not tell us the values of the two resistors in series. They could be 1.5 kΩ and 1.5 kΩ or they could be 1 kΩ and 2 kΩ, or, in fact, any combination that adds up to 3 kΩ. As long as they total 3 kΩ they will have the same effect on the circuit current as any combination that totals up to 3 kΩ.

EXAMPLE 6.4 Calculate I for each circuit in Fig. 6–7.

Solution: (a) In circuit *a* the 6-V battery sees a total series resistance of 2 ohms + 4 ohms. Thus,

$$I = \frac{6 \text{ V}}{6 \text{ }\Omega} = 1 \text{ A}$$

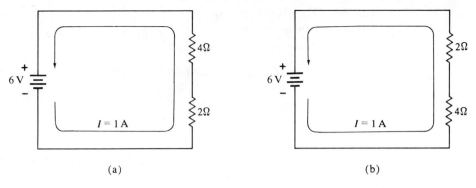

FIGURE 6–7 *The order in which resistors are connected in series has no effect on the value of current.*

(b) In circuit *b* the total series resistance is still 6 ohms so that *I* is still 1 A. Even though the 2-ohm resistor and 4-ohm resistor have alternated positions, the total resistance to current in circuit *b* is the same as in circuit *a*. In other words, it doesn't matter in which order resistors are placed in series as far as the circuit current is concerned.

More Than Two Resistors in Series The discussion concerning two series-connected resistors can be readily extended to cases where more than two resistors are in series. Consider Fig. 6–8(a) where three resistors are connected in series. These three resistors can be combined into one equivalent resistance by simply adding the resistances together. That is, $R_{eq} = R1 + R2 + R3$ as shown in Fig. 6–8(b). In fact, we can state more generally that

> The equivalent resistance of any number of resistors in series is equal to the sum of the resistor values.

As more resistors are added in series, the total resistance to current flow increases so that it requires a greater voltage to produce a given current.

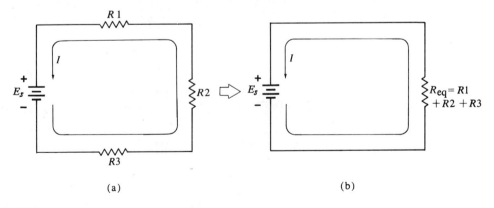

FIGURE 6–8

EXAMPLE 6.5 For the circuit in Fig. 6–8(a) determine what value of E_S is required to produce $I = 2$ mA for: (a) $R1 = R2 = 5$ kΩ, $R3 = 0$ Ω (short circuit); (b) $R1 = R2 = R3 = 5$ kΩ.

Solution: (a) The total resistance is $R_{eq} = 5\ \text{k}\Omega + 5\ \text{k}\Omega + 0\ \text{k}\Omega = 10\ \text{k}\Omega$. The amount of voltage needed across 10 kΩ to produce 2 mA of current is obtained using Ohm's law:

$$V = 2\ \text{mA} \times 10\ \text{k}\Omega$$
$$= 2\ \text{mA} \times 10\ \text{k}\Omega$$
$$= 2 \times 10^{-3}\ \text{A} \times 10 \times 10^{3}\ \Omega$$
$$= 20\ \text{V}$$

Therefore, E_S has to be 20 V to produce 2 mA of current.

(b) The total resistance is now 15 kΩ because of the addition of $R3 = 5\ \text{k}\Omega$. Therefore, to produce 2 mA we need

$$E_S = 2\ \text{mA} \times 15\ \text{k}\Omega = \mathbf{30\ V}$$

EXAMPLE 6.6 If $E_S = 30\ \text{V}$, $R1 = 30\ \Omega$, $R2 = 20\ \Omega$, and $R3 = 10\ \Omega$, determine the voltage across each resistor.

Solution: The circuit is redrawn in Fig. 6-9(a) with the values given. To determine the voltage across each resistor we have to first determine how much current is flowing through the circuit. Knowing this current, we can use Ohm's law to calculate the amount of voltage each resistor requires across its terminals in order to produce that value of current through the resistor.

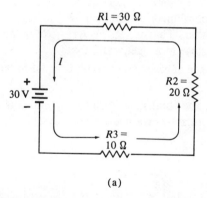
(a)

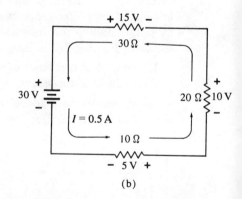
(b)

FIGURE 6-9

The equivalent series resistance to which the 30-V source is supplying current can be calculated as

$$R_{eq} = 30\ \Omega + 20\ \Omega + 10\ \Omega = 60\ \Omega$$

Therefore,

$$I = \frac{E_S}{R_{eq}} = \frac{30\ \text{V}}{60\ \Omega} = 0.5\ \text{A}$$

This 0.5 A flows through each resistor. If we consider first the 30-ohm resistor we have to ask the question, "how much voltage is needed across this resistor to cause 0.5 A to flow through it?" Using $V = I \times R$ we can determine that

$$V_1 = 0.5 \text{ A} \times 30 \text{ }\Omega = \mathbf{15 \text{ V}} \text{ (across } R1)$$

is required across this resistor. In a similar manner we can determine that $R2$ requires

$$V_2 = 0.5 \text{ A} \times 20 \text{ }\Omega = \mathbf{10 \text{ V}} \text{ (across } R2)$$

and $R3$ requires

$$V_3 = 0.5 \text{ A} \times 10 \text{ }\Omega = \mathbf{5 \text{ V}} \text{ (across } R3)$$

These voltages are shown in Fig. 6–9(b). Notice the polarity of the resistor voltages. The electron current flows into the negative terminal through the resistor and out the positive terminal. Also note that $V_1 + V_2 + V_3 = E_S$.

EXAMPLE 6.7 Determine I and the voltage across each resistor in Fig. 6–10.

Solution: $R_{eq} = 1.8 \text{ M}\Omega + 1.5 \text{ M}\Omega + 1 \text{ M}\Omega = 4.3 \text{ M}\Omega$

$$\therefore I = \frac{24 \text{ V}}{4.3 \text{ M}\Omega} = \mathbf{5.58 \text{ }\mu\text{A}}$$

and

$$V_3 = 5.58 \text{ }\mu\text{A} \times 1.8 \text{ M}\Omega = \mathbf{10.04 \text{ V}}$$
$$V_2 = 5.58 \text{ }\mu\text{A} \times 1.5 \text{ M}\Omega = \mathbf{8.37 \text{ V}}$$
$$V_1 = 5.58 \text{ }\mu\text{A} \times 1 \text{ M}\Omega = \mathbf{5.58 \text{ V}}$$

Note that $V_3 + V_2 + V_1 = 24 \text{ V} = E_s$.

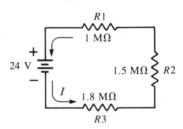

FIGURE 6–10

Summary We can summarize the discussion and examples of this section by saying that *a series combination of linear resistors can always be combined into one equivalent linear resistor for the purpose of determining the value of the current passing through the series circuit.* Of course, once the circuit current is determined, then the voltage across each series resistor can be determined as was done in Ex. 6.6 using Ohm's law.

6.5 Kirchhoff's Voltage Law (KVL)

In all of the discussion and examples in the preceding sections an implicit assumption was made concerning the distribution of voltages around the series circuit. It turned out that the voltage applied to the circuit by the voltage source was always equal to the sum of the voltages across the resistive elements in series. For example, in Fig. 6–9 (Ex. 6.6) the source voltage is 30 V. The voltages across $R1$, $R2$, and $R3$ are 15 V, 10 V, and 5 V respectively, and they sum up to 30 V. Mathematically, we can see why this has to be true. Since the applied voltage is E_s, we have

$$I = \frac{E_S}{R_{eq}}$$

or

$$E_S = I \times R_{eq}$$
$$= I \times (R1 + R2 + R3)$$
$$= I\,R1 + I\,R2 + I\,R3$$
$$E_S = V_{R1} + V_{R2} + V_{R3}$$

In other words, E_S equals the sum of the voltages across each resistor. This same result is true for any number of elements in series, and the elements do not have to be linear resistors.

In a series circuit the sum of the voltages across each series element must equal the total voltage supplied by the voltage source (or sources) at any instant of time.

This statement is a particular case of a more general circuit law, namely Kirchhoff's Voltage Law (hereafter abbreviated KVL), which will be discussed more generally later. Notice that the statement includes the possibility of more than one voltage source in the circuit. The following examples will illustrate the validity of this particular case of KVL.

EXAMPLE 6.8 For the circuit of Fig. 6–11(a) calculate I and then verify KVL.

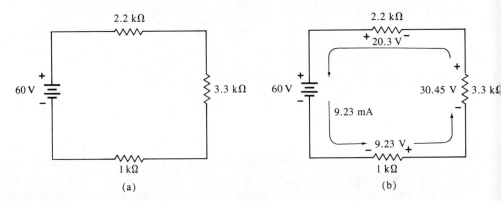

FIGURE 6–11

Solution: The series resistors have a total equivalent resistance of $R_{eq} = 6$ kΩ. Thus, I is given by

$$I = \frac{E_S}{R_{eq}} = \frac{60\text{ V}}{6.5\text{ k}\Omega} = 9.23\text{ mA}$$

(Notice that a volt divided by a kilohm, V/kΩ, is always equal to a milliampere, mA. This is a good shortcut to remember.)

The current through each resistor is equal to 9.23 mA. The voltage across the 2.2-kΩ resistor must therefore be

$$2.2\text{ k}\Omega \times 9.23\text{ mA} = \mathbf{20.30\text{ V}}$$

Similarly, the voltages across the other resistors are

$$3.3 \text{ k}\Omega \times 9.23 \text{ mA} = \textbf{30.45 V}$$

and

$$1 \text{ k}\Omega \times 9.23 \text{ mA} = \textbf{9.23 V}$$

respectively. Summing these voltages produces a total of 59.98 ≈ 60 which equals the source voltage. The results are shown in Fig. 6–11(b). (Notice that a kilohm times a milliamp, kΩ × mA, always equals a volt. This is another good shortcut to remember.)

EXAMPLE 6.9 In the circuit of Fig. 6–12(a) determine the circuit current and verify KVL.

Solution: In this series circuit there are two dc voltage sources connected in series with the resistors. In chap. 4 we saw that two voltage sources connected in series could be connected so as to *aid* each other or to *oppose* each other. Looking at Fig. 6–12(a) we can tell that the two sources are *aiding* each other because both of them will attempt to push current in a ccw direction around the circuit. Thus, the circuit will act as if it has a total source voltage of 10 V + 15 V = 25 V.

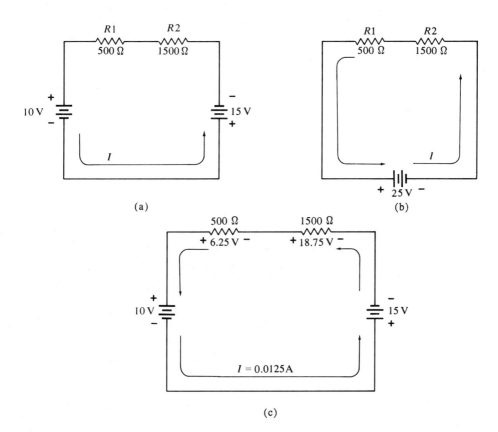

FIGURE 6–12

In other words, the two sources can be represented by one equivalent source of 25 V as shown in Fig. 6–12(b). The equivalent series resistance is R_{eq} = 500 ohms + 1500 ohms = 2000 ohms. This gives a circuit current of

$$I = \frac{25 \text{ V}}{2000 \text{ }\Omega} = 0.0125 \text{ A} = \textbf{12.5 mA}$$

Notice that we use the *total* source voltage of 25 V divided by the *total* series resistance of 2000 ohms to determine I.

The voltage across each resistor can now be calculated

$$V_{R1} = 500 \text{ }\Omega \times 0.0125 \text{ A}$$
$$= \textbf{6.25 V}$$

$$V_{R2} = 1500 \text{ }\Omega \times 0.0125 \text{ A}$$
$$= \textbf{18.75 V}$$

Since $V_{R1} + V_{R2}$ = 25 V we can see that KVL is verified. The original circuit with all the current and voltage values is shown in part (c) of Fig. 6–12.

EXAMPLE 6.10 Reverse the 15-V battery and repeat Ex. 6.9.

Solution: The revised circuit is shown in Fig. 6–13(a). The 15-V battery is now attempting to push current in the cw direction in opposition to the 10-V battery. The net effect, then, is that of a single 5-V source (15 V − 10 V) pushing current in a cw direction. The circuit is redrawn in part (b) of the figure with the two sources combined into one equivalent source.

$$I = \frac{5 \text{ V}}{2000 \text{ }\Omega} = 0.0025 \text{ A} = \textbf{2.5 mA}$$

so that the voltages across the resistors are given by

$$V_{R1} = 500 \text{ }\Omega \times 0.0025 \text{ A} = \textbf{1.25 V}$$

and

$$V_{R2} = 1500 \text{ }\Omega \times 0.0025 \text{ A} = \textbf{3.75 V}$$

respectively. Figure 6–13(c) shows the final result for the original circuit.

KVL can be easily verified by adding these resistor voltages and comparing them to the total source voltage (5 V). The result is 1.25 V + 3.75 V = 5 V, which does verify KVL for this circuit. By checking to see that KVL is satisfied we are provided with a means of catching any mistakes we might have made in calculating the circuit current or the resistor voltages. If the sum of the resistive voltages did not come out equal to the total source voltage, then this would indicate that we had made a mistake somewhere in our calculations.

Note that the polarity of the voltages across $R1$ and $R2$ is different from Ex. 6.9, Fig. 6–12(b). This is because the current is actually flowing in the cw direction so that the left-hand terminal of each resistor will be negative.

Resistive Devices in Series Circuits 109

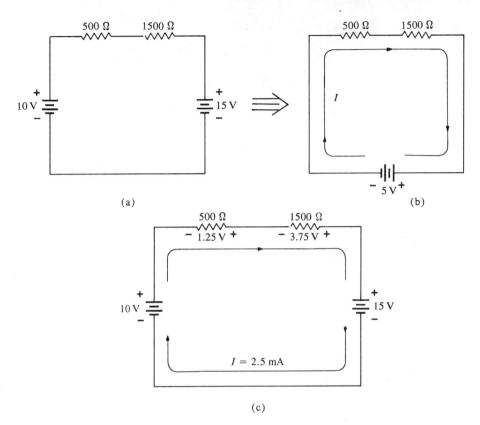

FIGURE 6–13

EXAMPLE 6.11 Figure 6–14 shows a series circuit containing a dc source of 12 V connected to three devices in series. Two of the devices are resistors of 20 kilohms and 50 kilohms respectively. The third device is labeled "device X." An ammeter placed in series with the circuit indicates that $I = 0.15$ mA. Determine the voltage across device X and also determine its resistance.

Solution: Since the circuit current is 0.15 mA, we can calculate the voltage across each of the resistors.

$$V_{R1} = 0.15 \text{ mA} \times 20 \text{ k}\Omega$$
$$= 3 \text{ volts}$$

$$V_{R2} = 0.15 \text{ mA} \times 50 \text{ k}\Omega$$
$$= 7.5 \text{ volts}$$

Since we don't know the electrical characteristic of device X it may appear that we cannot determine its voltage. However, we can utilize KVL to determine the unknown voltage. KVL tells us that the voltages across the series elements must add up to the applied voltage of 12 V. Thus, we have

$$V_{R1} + V_{R2} + V_X = 12 \text{ V}$$

But we know V_{R1} and V_{R2} so that

$$3\text{ V} + 7.5\text{ V} + V_X = 12\text{ V}$$

or

$$V_X = 12\text{ V} - 10.5\text{ V}$$
$$= \mathbf{1.5\text{ V}}$$

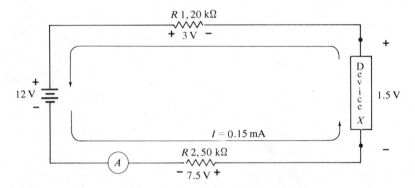

FIGURE 6–14

Since device X has 1.5 V across its terminals and a current of 0.15 mA flowing through it, we can calculate its resistance as

$$R_X = \frac{1.5\text{ V}}{0.15\text{ mA}} = \frac{1.5\text{ V}}{0.15 \times 10^{-3}\text{ A}} = 10 \times 10^3\text{ }\Omega$$
$$= \mathbf{10\text{ k}\Omega}$$

This value of R_X is the resistance of device X at the specific value of current $I = 0.15$ mA. If device X is a linear resistor then its resistance must always be 10 kΩ. If device X is a nonlinear device, then its resistance will, in general, be different at a different value of current.

The Linearity Principle What will happen to the current in a series circuit if the source voltage is doubled? What will happen to the voltages across each series resistor? We can answer this by noting that since $I = E_S/R_{eq}$, doubling E_S will cause I to double. This two-fold increase in I will, in turn, cause the voltage across each resistor to double since $V = I \times R$.

This result is an example of the *linearity principle* which applies to *all* linear circuits.

In any circuit which contains only linear resistors, any change in the total applied voltage will produce a proportionate change in all of the circuit currents and voltages.

This principle is useful in determining the new values of current and voltage when the applied voltage is changed in a linear circuit.

EXAMPLE 6.12 Determine the current in the circuit of Fig. 6–11 if the source is decreased to 50 V. Also determine the voltage across the 3.3-kΩ resistor.

Solution: We can use the linearity principle by noting that the ratio of the new value of I to the old value (9.23 mA) will be the same as the ratio of the new value of E_S (50 V) to the old value of E_S (60 V).

$$\frac{I}{9.23 \text{ mA}} = \frac{50 \text{ V}}{60 \text{ V}} = \frac{5}{6}$$

Thus,

$$I = \frac{5}{6} \times 9.23 \text{ mA} = \mathbf{7.69 \text{ mA}}$$

Similarly, the voltage across the 3.3-kΩ resistor can be found using this same proportion

$$\frac{V_{3.3 \text{ k}\Omega}}{30.45 \text{ V}} = \frac{5}{6}$$

or $V_{3.3 \text{ k}\Omega} = \mathbf{25.38 \text{ V}}$. This same value could be obtained using $V = IR$ with $I = 7.69$ mA.

6.6 General Statement of KVL

KVL is not restricted to series circuits. It is applicable to *any* electrical circuit. A more general statement of Kirchhoff's Voltage Law is as follows:

> **If we start at any point in a circuit at any instant of time and take any path through the circuit that returns to the starting point, the net change in voltage which we encounter will be zero volts.**

In other words, suppose we make one complete *round*-trip excursion from any point in a circuit and we sum up the drops in voltage and the rises in voltage which we encounter along the way. According to KVL, the total of the voltage drops will equal the total of the voltage rises at all times, resulting in a net voltage change of zero.

To help illustrate, consider the circuit in Fig. 6–15(a). The voltages across each resistor have already been calculated and are shown on the diagram. We can verify the preceding statement of KVL by picking any starting point. Let's start at point x and proceed in a cw direction to point y, then point z, and then point w, and back to point x. As we do so, let's keep track of the voltage rises and drops we encounter by using a table as in part (b) of the figure.

Starting at x and proceeding to y we go through the 3-kilohm resistor from $+$ to $-$ so that we encountered a voltage *drop* of 30 V. From y to z we go through the 1-kilohm resistor and encounter a voltage *drop* of 10 V. Going from z to w, we go through the 60-V source from $-$ to $+$, which is a voltage *rise* of 60 V. Finally, going from w to x we encounter a voltage *drop* of 20 V. If we look at our table, it is easy to see that in the complete trip from point x

back to point x the total of the voltage drops encountered is 60 V. The total voltage rises encountered is also 60 V. Thus, the net change in voltage in going around the circuit is *zero* volts.

According to KVL we can also start at point x and proceed in a ccw direction around the circuit with the same result. The table in part (c) of Fig. 6–15 shows the values encountered for this case. Again, the net change in voltage is *zero*.

The same results will be obtained using any starting point in the circuit. If we use z as the starting point and proceed in a cw direction the values encountered are those shown in Fig. 6–15(d). These values again verify that the net change in voltage is *zero* if we start at one point in the circuit and end up at the same point.

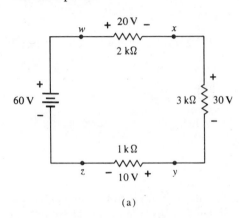

Going from	Voltage rise	Voltage drop
point x to y	—	30 V
point y to z	—	10 V
point z to w	60 V	—
point w to x	—	20 V
Totals	60 V	60 V

(b)

Going from	Voltage rise	Voltage drop
point x to w	20 V	—
point w to z	—	60 V
point z to y	10 V	—
point y to x	30 V	—
Totals	60 V	60 V

(c)

Going from	Voltage rise	Voltage drop
point z to w	60 V	—
point w to x	—	20 V
point x to y	—	30 V
point y to z	—	10 V
Totals	60 V	60 V

(d)

FIGURE 6–15 *Verifying KVL:* (a) *circuit diagram,* (b) *starting at* x *and going cw,* (c) *starting at* x *and going ccw, and* (d) *starting at* z *and going cw.*

The general statement of KVL given here is applicable to any electrical circuit. The statement of KVL given at the beginning of this section is a special case of KVL for series circuits. We will now apply the general statement of KVL in several example problems.

EXAMPLE 6.13 Consider the circuit in Fig. 6–16(a). A current of 100 mA flows through the circuit. Determine V_{xz}.

Resistive Devices in Series Circuits

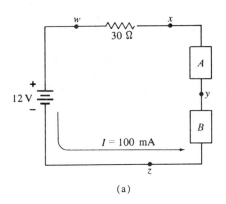 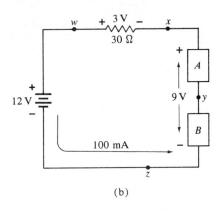

(a) (b)

FIGURE 6-16

Solution: We can first determine V_{wx}, the voltage across the 30-ohm resistors, as

$$V_{wx} = 100 \text{ mA} \times 30 \text{ ohms}$$
$$= 0.1 \text{ A} \times 30 \text{ ohms}$$
$$= +3 \text{ V}$$

We know that V_{wx} is positive because current is flowing through the resistor from x to w, so that w is more positive than x [see part (b) of the figure].

We can now use KVL to determine V_{xz}. If we start at point x and proceed in a ccw direction, we first encounter a voltage *rise* of 3 V in going from x to w. Going from w to z we encountered a *drop* of 12 V. We are now left with the problem of going from z back to x. So far, we have encountered along the path a *rise* of 3 V and a *drop* of 12 V or a *net voltage drop* of 9 V. In order to satisfy KVL we must therefore encounter a 9-V *rise* in going from z to x because the net change has to be 0 V when we return to the starting point x. This means that in the circuit the potential at x must be 9 V more positive than the potential at z. In other words,

$$V_{xz} = +9 \text{ V}$$

Part (b) of Fig. 6-16 shows the circuit with the voltage values calculated here.

Notice that it is not possible to calculate the individual voltages across device A or device B because we do not know the resistance of either device.

EXAMPLE 6.14 Figure 6-17 shows a complex circuit. The voltages across some of the resistors are shown. Determine V_{bd}.

Solution: This problem is a little more difficult than those encountered in a series circuit. In a series circuit there was only one path which could be used. Here there are several paths. Since our objective is to determine V_{bd}, we have to pick a path that contains V_{bd}. There are two such paths, namely: *b-a-e-f-d-b* and *b-c-f-d-b*.

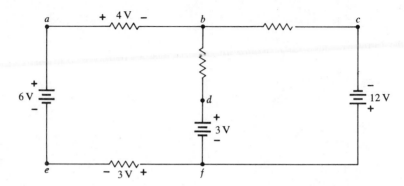

FIGURE 6-17

Path b-a-e-f-d-b has all the voltages given except for the unknown V_{bd}, whereas path b-c-f-d-b has two unknown voltages, V_{bd} and V_{bc}. Therefore, we cannot use this second path because it won't help us find V_{bd}. The first path, however, can be used.

Starting at point b and proceeding through the path from b to a to e to f to d we encounter a net voltage *rise* of 4 V. Therefore, the last step in the path, d back to b, must contain a voltage *drop* of 4 V in order to satisfy KVL. Thus, point b is more negative than d by 4 V so that

$$V_{bd} = -4 \text{ V}.$$

Notice in this example that we could also have used point d as the starting point and obtained the same result. *When determining the voltage between two points using KVL, one of the two points should be used as the starting point.*

Usefulness of KVL The usefulness of KVL will become increasingly more apparent as you gain more experience in electric circuits. As we have seen, it can be used to determine unknown voltages in a circuit. It is also an invaluable means of checking the results of a circuit analysis in which all the currents and voltage in a circuit have been calculated. Checking to see that KVL is satisfied around every path in the analyzed circuit provides one means for determining whether any errors were made in the calculations.

In much of the work to follow in the text we will call upon KVL as a means for helping us analyze circuit operation. In the next chapter an equally useful circuit law, Kirchhoff's Current Law, will be added to our "circuit analysis toolkit." Gradually, we are developing many useful techniques that will help us become better equipped to understand and analyze electric circuits.

6.7 Algebraic Method for Using KVL

The mathematics of algebra is often a useful tool that can be applied to circuit analysis. Although algebra is not always necessary to help solve a problem, it provides a more systematic, and often more efficient, means to a solution. In addition, algebra serves as a means for formulating an analysis so that others can easily follow what is being done. The algebraic process can be applied to circuit problems involving KVL.

In algebraic terms KVL can be stated as

$$\sum_{\circlearrowleft} V = 0 \tag{6-3}$$

In this expression, the symbol "Σ" (greek letter sigma) stands for "the sum of." The symbol "$\circlearrowleft$" stands for a "round-trip path." In words, the expression says that *the sum of (Σ) the voltages (V) encountered in making a complete round trip ($\circlearrowleft$) anywhere in a circuit is equal to zero.*

Refer to Fig. 6–18 where we will write the algebraic expression of KVL for the path b-a-e-f-d-b. The process is actually quite easy to perform. As we follow the desired path, we simply write down in symbolic form the voltage across each element encountered. To begin, in going from point b to a the voltage encountered is V_{ab}. In going from a to e the voltage encountered is V_{ea} and so on. The sum of the voltages around the complete path thus results in

$$V_{ab} + V_{ea} + V_{fe} + V_{df} + V_{bd} = 0$$

according to KVL.

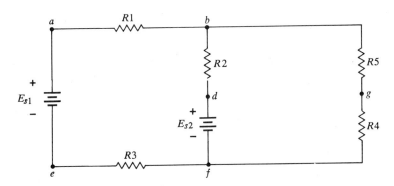

FIGURE 6–18

Notice that in going around the path, each time we went through a circuit element or a source, we wrote down the voltage symbol with the *second* subscript being the terminal which is encountered *first*. For example, in going through R3 from point e to point f, we wrote down the voltage as V_{fe}. This has to be done each time to be consistent. The resultant expression can be put in another form, if desired, by substituting the source voltages E_{S1} and E_{S2}. Looking at the circuit diagram it should be clear that

$$E_{S1} = V_{ae}$$

or equivalently,

$$-E_{S1} = -V_{ae} = V_{ea}$$

Similarly,

$$E_{S2} = V_{df}$$

Substituting for V_{ea} and V_{df} in the total expression results in

$$V_{ab} - E_{S1} + V_{fe} + E_{S2} + V_{bd} = 0$$

At this point there is not much else we can do with this expression. For now, we can regard this expression as simply a mathematical statement of KVL around the path b-a-e-f-d-b.

EXAMPLE 6.15 For the circuit of Fig. 6–17 (redrawn in Fig. 6–19) write the algebraic expression of KVL for the path b-d-f-e-a-b. Then use this expression to determine V_{bd}.

Solution: Following the procedure outlined, the KVL expression for the desired path is

$$V_{db} + V_{fd} + V_{ef} + V_{ae} + V_{ba} = 0$$

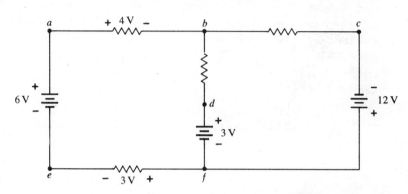

FIGURE 6–19

All the voltages in this expression, except V_{db}, are known and are shown on the diagram. That is,

$$V_{fd} = -3 \text{ V}, V_{ef} = -3 \text{ V}$$
$$V_{ae} = +6 \text{ V}, V_{ba} = -4 \text{ V}$$

Substituting these values in the KVL equation results in

$$V_{db} - 3 \text{ V} - 3 \text{ V} + 6 \text{ V} - 4 \text{ V} = 0$$

or combining terms,

$$V_{db} - 4 \text{ V} = 0$$

This simple equation can be solved for V_{db} as

$$V_{db} = 4 \text{ V}$$

Therefore,

$$V_{bd} = -V_{db} = -4 \text{ V}$$

which is the same result obtained in Ex. 6.14.

The alert student might be wondering about the efficiency of the method employed in the last example. It seems as though a certain amount of time and effort could have been saved by writing the KVL expression immediately, including the *known* voltages rather than substituting them in later. Indeed, this can be and should be done once one understands what is being done. In doing

Resistive Devices in Series Circuits 117

it this quicker way, we write down the known voltage as it is encountered in our trip around the path. If we encounter the known voltage as a voltage *drop* (from + to −) we write it as a *negative* quantity in the KVL expression. If it is a voltage *rise* we write it as a positive quantity.

To illustrate, for the circuit path considered in the last example (Fig. 6–19) we would first write down V_{db} as we went from b to d. Then, in going from d to f we would write down -3 V since we are going through a voltage *drop*. Similarly, from f to e a voltage *drop* occurs. Continuing from e to a we encounter a rise of 6 V so this is written as $+6$ V. Finally, from a back to b we encounter a drop of 4 V, which is therefore written as -4 V. The whole expression is then set equal to 0, as follows:

$$V_{db} - 3\text{ V} - 3\text{ V} + 6\text{ V} - 4\text{ V} = 0$$

which is exactly the expression obtained in Ex. 6.15.

To summarize, when some voltages around the path are known, they can be written directly into the KVL expression. To further illustrate, the KVL expressions are given here for two other possible paths in Fig. 6–19. The student should verify these to see that they are consistent with the preceding discussion:

Path *e-f-d-b-a-e:*

$$3\text{ V} + 3\text{ V} + V_{bd} + 4\text{ V} - 6\text{ V} = 0$$

Path *b-c-f-e-a-b:*

$$V_{cb} + 12\text{ V} - 3\text{ V} + 6\text{ V} - 4\text{ V} = 0$$

At this point, the student might be a little confused by all of the various ways of stating KVL and all of the various techniques for using it in circuit problems. This confusion may be somewhat cleared up by stating that the various statements and techniques concerning KVL are all equivalent. This might have already been apparent by the several examples which were worked out.

If the student is wondering which method should be used for certain problems, the answer at this point should be "use whichever method you are most comfortable with and understand best." As we go on, it will become clear that in certain situations some methods are better than others. Whenever this is true, it will be pointed out. For now, simply try to understand as much as possible about KVL because its importance cannot be overemphasized.

6.8 Voltage Dividers (Attenuators)

We are now prepared to discuss one of the most useful types of circuits. It is actually not a new circuit because we have discussed it thoroughly in the preceding sections. We're just going to give it a new name. The circuit is nothing other than a series circuit which is used to produce an output voltage that is a certain fraction of the input voltage. We saw earlier that in a resistive series circuit the input voltage is distributed among all the series elements. Each series element has a portion of the input voltage across its terminals. If we take as the output of the circuit the voltage across one of the series elements, then this output voltage will always be some fraction of the input voltage. A series circuit used in this manner is called a *voltage divider* circuit because the input voltage is divided up among the series elements.

Consider Fig. 6–20(a), where a simple voltage divider circuit is shown with an input source E_{in} and two resistors $R1$ and $R2$. The equivalent series resistance is $R1 + R2$ so that the circuit current i will be given by

$$I = \frac{E_{in}}{R1 + R2}$$

Thus, the voltage V_{xy} across resistor $R1$ will be given by

$$V_{xy} = I \times R1$$
$$= \frac{E_{in}}{R1 + R2} \times R1$$
$$= \frac{R1}{R1 + R2} \times E_{in} \quad (6\text{--}4)$$

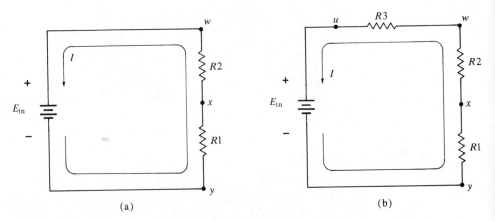

FIGURE 6–20 *Simple series circuits as voltage dividers.*

Obviously, the factor $R1/R1 + R2$ is a fraction which will always be less than one since the numerator can never be greater than the denominator. Thus, V_{xy} will be some fraction of E_{in}.

In a similar manner, the voltage V_{wx} across resistor $R2$ will be given by

$$V_{wx} = I \times R2$$
$$= \frac{E_{in}}{R1 + R2} \times R2$$
$$= \frac{R2}{R1 + R2} \times E_{in} \quad (6\text{--}5)$$

Again, the factor $R2/R1 + R2$ is a fraction less than one so that V_{wx} is always some fraction of E_{in}.

Voltage Divider Equation Looking at the expressions for the resistor voltages given by (6–4) and (6–5), we see that the voltage across a given resistor is equal to its resistance divided by the total series resistance, all multiplied times the input voltage E_{in}. This result can be extended to a series circuit with any number of resistors. For example, in Fig. 6–20(b) the individual resistor voltages will be given by

Resistive Devices in Series Circuits

$$V_{xy} = \frac{R1}{R1 + R2 + R3} \times E_{in}$$

$$V_{wx} = \frac{R2}{R1 + R2 + R3} \times E_{in}$$

$$V_{uw} = \frac{R3}{R1 + R2 + R3} \times E_{in}$$

This process can be extended to any number of series resistors. We can write a general expression to cover any number of series resistors. If we want the voltage across any of the resistors we can use

$$V_{out} = \frac{R_{out}}{R_{total}} \times E_{in} \qquad (6\text{--}6)$$

where R_{out} is the resistor across which the output V_{out} is being taken, and R_{total} is the total series resistance. This equation is called the *voltage divider equation*.

EXAMPLE 6.16 For the voltage divider in Fig. 6–21(a) determine the voltage across each resistor using the voltage divider equation.

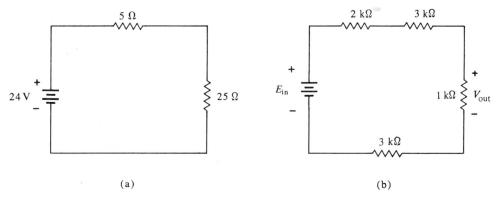

FIGURE 6–21

Solution: The total series resistance is 30 ohms. To find the voltage across the 5-ohm resistor we will use Eq. (6–6) with $R_{out} = 5$ ohms. Thus,

$$V_{out} = \frac{5 \, \Omega}{30 \, \Omega} \times E_{in}$$

$$= \frac{1}{6} \times 24 \text{ V} = \textbf{4 V}$$

Similarly, the voltage across the 25-Ω resistor is found using (6–6) with $R_{out} = 25 \, \Omega$. Thus,

$$V_{out} = \frac{25 \, \Omega}{30 \, \Omega} \times 24 \text{ V} = \textbf{20 V}$$

EXAMPLE 6.17 For the voltage divider in Fig. 6–21(b) the output is taken across the 1-kilohm resistor. Determine the value of V_{out} if $E_{in} = 18$ V.

Solution:
$$R_{total} = 2\text{ k}\Omega + 3\text{ k}\Omega + 1\text{ k}\Omega + 3\text{ k}\Omega$$
$$= 9\text{ k}\Omega$$
$$R_{out} = 1\text{ k}\Omega$$

Thus,
$$V_{out} = \frac{1\text{ k}\Omega}{9\text{ k}\Omega} \times 18\text{ V} = 2\text{ V}$$

In these last two examples the resistor voltages could have been calculated by first calculating the circuit current and then multiplying by the resistance in question. However, using the voltage divider equation, (6–6), is somewhat of a shortcut because it is not necessary to calculate the circuit current. In addition, the voltage divider equation is more useful in cases where the input voltage is not constant but is changing with time. Its usefulness comes from the fact that the ratio (R_{out}/R_{total}) is constant no matter what the value of E_{in}. To illustrate, in Fig. 6–21(b) the value of V_{out} can be written in terms of E_{in} as

$$V_{out} = \frac{R_{out}}{R_{total}} \times E_{in}$$
$$= \frac{1\text{ k}\Omega}{9\text{ k}\Omega} \times E_{in}$$
$$= \frac{1}{9} E_{in}$$

This tells us that V_{out} will be 1/9 of E_{in} no matter what the value of E_{in}. Even if E_{in} were a time-varying waveform, the V_{out} waveform would be exactly 1/9 the size of the E_{in} waveform.

This (R_{out}/R_{total}) ratio is often called the *voltage divider ratio*. It is fixed by the resistor values in the circuit. It is also called the *attenuation factor*. In fact, a voltage divider circuit is often called an *attenuator* or *attenuation circuit*. An input voltage to a voltage divider is said to be attenuated by a factor equal to the attenuation factor resulting in an output which is smaller than the input.

EXAMPLE 6.18 (a) Determine the voltage divider ratio (attenuation factor) for the circuit in Fig. 6–21(b) if the output is to be taken across the 2-kilohm resistor.

(b) Determine the output voltage when $E_{in} = 90$ V.

Solution: (a) $\dfrac{R_{out}}{R_{total}} = \dfrac{2\text{ k}\Omega}{9\text{ k}\Omega} = \mathbf{0.222}$

(b) $V_{out} = 0.222\, E_{in} = \mathbf{20\text{ V}}$

EXAMPLE 6.19 What will happen to the output voltage of Ex. 6–18 if (a) the 1-kilohm resistor is increased in value; (b) the 2-kilohm resistor is increased in value?

Solution: (a) By increasing the 1-kilohm resistor the total series resistance R_{total} will increase. This will cause a decrease in the voltage divider ratio R_{out}/R_{total}. Thus, V_{out} will be a smaller fraction of E_{in}.

(b) By increasing the 2-kilohm resistor, R_{out} will increase. R_{total} will also increase, but the ratio R_{out}/R_{total} will have an overall increase. For instance, if the 2-kilohm is changed to 3-kilohm, then R_{out} = 3 kilohms, R_{total} = 10 kilohms, and the voltage divider ratio is 3 kilohms/10 kilohms = 0.3, which is an increase over the original ratio of 0.222. Thus, V_{out} will be a greater portion of E_{in}.

This last example brings us to an important characteristic of series circuits and voltage dividers.

In a series circuit the input voltage is distributed among the series elements in proportion to their resistance values. The larger resistances will have a greater portion of the voltage and vice versa.

In other words, since the current is the same for all the series elements it is obvious that the larger resistors will have a greater $I \times R$ and thus a greater voltage. Furthermore, the voltages across the resistors are proportional to their resistance. If $R1$ is twice as great as $R2$, then the voltage across $R1$ will be twice the voltage across $R2$, and so on. In fact, we can set up a proportion between the resistor voltages and their resistances as follows:

$$\frac{V_{R1}}{V_{R2}} = \frac{R1}{R2} \qquad (6-7)$$

where $R1$ and $R2$ are any two series resistors. For example, in Fig. 6–21(a) we determined that there were 20 V across the 25-ohm resistor and 4 V across the 5-ohm resistor. We can use these values to check Eq. (6–7). Let $R1$ = 5 ohms, $R2$ = 25 ohms. Therefore

$$\frac{4 \text{ V}}{20 \text{ V}} = \frac{5 \text{ ohms}}{25 \text{ ohms}}$$

$$\frac{1}{5} = \frac{1}{5} \quad \text{(checks)}$$

This proportionality relationship can be useful in circuit calculations.

EXAMPLE 6.20 In the circuit of Fig. 6–21(b) determine the voltage across the 2-kilohm resistor if there is a voltage of 7.9 V across the 3-kilohm resistor.

Solution: We can use the proportionality relationship of Eq. (6–7).

$$\frac{V_{(across\ 2k\Omega)}}{7.9 \text{ V}} = \frac{2 \text{ k}\Omega}{3 \text{ k}\Omega} = \frac{2}{3}$$

$$V(\text{across 2 kilohms}) = \frac{2}{3} \times 7.9 \text{ V}$$

$$= 5.27 \text{ V}$$

Effect of an Open Circuit In a series circuit voltage divider, if there is one resistor that is much greater than the others, then most of the input voltage will appear across this large resistor. For example, in Fig. 6–22(a) the voltage across the 1-MΩ resistor will be given by

$$V_{out} = \frac{1 \text{ M}\Omega}{1 \text{ M}\Omega + 1 \text{ k}\Omega} \times E_{in}$$

$$= \frac{10^6 \ \Omega}{10^6 \ \Omega + 10^3 \ \Omega} \times E_{in}$$

$$V_{out} = \frac{1,000,000}{1,001,000} \times E_{in} = 0.999 \ E_{in} = 9.99 \text{ V}$$

Almost all (9.99 V) of the input voltage will be across the 1-MΩ resistor and conversely very little of the input will be across the 1-kΩ resistor (0.01 V).

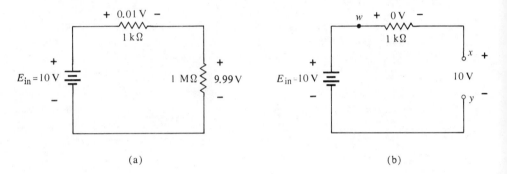

FIGURE 6–22

If the 1-MΩ resistor is made even larger there will be very little change in its voltage since the largest it can get is 10 V. In fact, if it is made an open circuit (essentially infinite), then *all* of the input will be across it. This situation is shown in Fig. 6–22(b) where the 1 MΩ has been replaced by an open circuit. The voltage V_{xy} across the open circuit *has* to equal E_{in}. Since there can be no current through the open circuit, then there can be no $I \times R$ voltage across the 1 kΩ. Thus, $V_{wx} = 0$. If we write the KVL equation for the series circuit going through the path y-x-w-y, the result is

$$V_{xy} + V_{wx} - E_{in} = 0$$

Since $V_{wx} = 0$, then

$$V_{xy} - E_{in} = 0$$

or

$$V_{xy} = E_{in} = 10 \text{ V}$$

which verifies that all of the input voltage appears across the open circuit.

Open circuits occasionally occur in a circuit as a result of a component failure due to overcurrent, aging, or possibly a bad solder connection. A good indication that such a failure has occurred is the presence of an unexpectedly large voltage across the component.

In this analysis of Fig. 6–22(b), note that we applied KVL to a circuit that was not a *complete* circuit. In other words, the circuit did not contain a complete path for current because of the open circuit. It is important to understand that *KVL is valid around any path that begins and ends with the same point even if that path contains an open circuit.*

6.9 Multiple Output Voltage Dividers

In some applications it is necessary to take an input voltage and divide it down into two or more output voltages each referenced to the same point. Figure 6–23 illustrates a two-output voltage divider circuit. Output voltage V_{yz} is taken between points y and z (across $R1$). Output voltage V_{xz} is taken between points x and z (across $R1 + R2$). Notice that point z is common to both output voltages and to the input source. Point z thus serves as a common point to which the input voltage and output voltages are referenced. That is, E_{in} is the voltage at w relative to z; V_{yz} is the voltage at y relative to z; and V_{xz} is the voltage at x relative to z. A common reference point like z is usually shown tied to a ⏚. This is the symbol for a *common* or *ground* terminal to which all the circuit voltages are referenced. Point z would be called the common or ground point of the circuit. As we stated in chap. 4, V_{xz} and V_{yz} can be written V_x and V_y since z is ground.

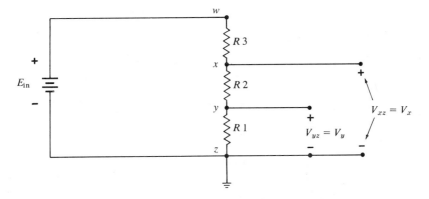

FIGURE 6–23 *Two-output voltage divider with z the common reference point.*

It is a relatively simple matter to calculate V_y using the voltage divider equation (6–6). The result is

$$V_y = \frac{R1}{R1 + R2 + R3} \times E_{in}$$

The output V_x is taken across the series combination of $R1$ and $R2$. We can still use the voltage divider equation if we let $R_{out} = R1 + R2$. Thus, we have

$$V_x = \frac{R_{out}}{R_{total}} \times E_{in}$$

$$= \frac{R1 + R2}{R1 + R2 + R3} \times E_{in}$$

R_{out} need not be a single resistor as this illustration shows, but it can be a series combination of more than one resistor.

EXAMPLE 6.21 Calculate V_x and V_y for the circuit in Fig. 6–24.

Solution: $V_y = \dfrac{24 \text{ k}\Omega}{10 \text{ k}\Omega + 24 \text{ k}\Omega + 100 \text{ k}\Omega} \times 12 \text{ V}$

$= \dfrac{24}{134} \times 12 \text{ V} = \mathbf{2.15 \text{ V}}$

$V_x = \dfrac{10 \text{ k}\Omega + 24 \text{ k}\Omega}{134 \text{ k}\Omega} \times 12 \text{ V} = \dfrac{34}{134} \times 12 \text{ V} = \mathbf{3.04 \text{ V}}$

EXAMPLE 6.22 Calculate V_{xy} in Fig. 6–24.

Solution (method 1): $V_{xy} = \dfrac{10 \text{ k}\Omega}{134 \text{ k}\Omega} \times 12 \text{ V} = \mathbf{0.89 \text{ V}}$

Solution (method 2): Since $V_x = 3.04$ V, point x must be 3.04 V more positive than ground. Likewise, $V_y = 2.15$ V means that point y is 2.15 V more positive than ground. Clearly then, x must be $3.04 - 2.15 = 0.89$ V more positive than y. Thus, $V_{xy} = \mathbf{+0.89 \text{ V}}$.

The reasoning used in this example can be generally stated as follows: *the voltage between two points (x and y) is equal to the difference in their respective voltages relative to ground.* That is,

$$V_{xy} = V_x - V_y \tag{6-8}$$

This relationship can be useful in certain circuit calculations, and we will employ it wherever it is applicable.

EXAMPLE 6.23 Determine V_{wx} in Fig. 6–23 if $E_s = 29$ V and $V_x = 4$ V.

Solution: $V_{wx} = V_w - V_x$
$= 29 \text{ V} - 4 \text{ V} = \mathbf{25 \text{ V}}$

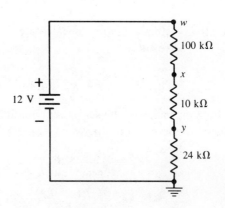

FIGURE 6–24

6.10 Voltage Dividers With More Than One Voltage Source*

Frequently a voltage divider circuit may contain more than one voltage source. In such cases, the voltage divider equation (6–6) can be used to find the voltage across any resistor if the series sources are combined into *one* equivalent

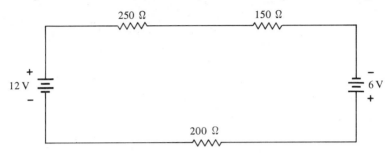

FIGURE 6–25

source. Consider the circuit in Fig. 6–25. We can combine the two sources into one. Since both the 12-V source and the 6-V source are acting to push current in a ccw direction they are equivalent to a single source of 18 V. Thus, we can calculate the voltage across the 150-ohm resistor using the voltage divider equation.

$$v_{out} = \frac{R_{out}}{R_{total}} \times E_{in}$$

$$= \frac{150 \ \Omega}{600 \ \Omega} \times 18 \ V$$

$$= 4.5 \ V \tag{6-6}$$

EXAMPLE 6.24 In the circuit of Fig. 6–26(a) both sources are connected to the ground terminal. The output of the circuit is taken from point y to ground, V_y. Determine the value of V_y.

Solution: Earlier in this chapter we stated that to find an unknown voltage using KVL, we must find a circuit path that contains the desired voltage and no other unknown voltages. Looking at the circuit in question there are two circuit paths that contain V_y. They are the paths w-y-z-w and w-y-x-w. However, both of these paths contain other unknown voltages across resistors.

In order to use either of these paths to calculate V_y, we have to first calculate the voltage across the resistor in the path. Let's use path w-y-z-w. Thus, we have to determine the voltage across the 180-ohm resistor. The two voltage sources are opposing each other; thus, the *net* source voltage is 24 V − 10 V = 14 V, and it pushes current in a ccw direction, making point y positive relative to z. The value of V_{yz} is obtained using the voltage divider equation.

$$V_{yz} = \frac{180 \ \Omega}{300 \ \Omega} \times 14 \ V = 8.4 \ V$$

---

* This section may be omitted without any loss of continuity.

This value is shown on the circuit diagram in Fig. 6–26(b). To determine the value of V_y we can write the KVL equation around the w-y-z-w path.

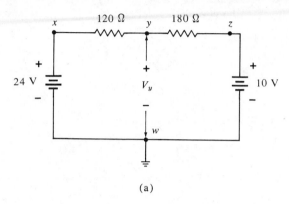

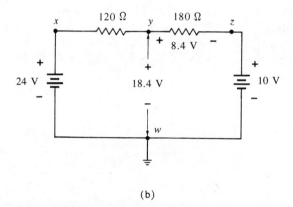

FIGURE 6–26

$$V_y - 8.4 \text{ V} - 10 \text{ V} = 0$$

so that

$$V_y = \textbf{18.4 V}$$

Alternatively, we can use Eq. 6–8

$$V_{yz} = V_y - V_z$$
$$\therefore V_y = V_{yz} + V_z$$
$$= 8.4 \text{ V} + 10 \text{ V} = \textbf{18.4 V} \tag{6–8}$$

EXAMPLE 6.25 Find V_y in Fig. 6–26 by first calculating the voltage across the 120-Ω resistor.

$$V_{xy} = \frac{120 \text{ }\Omega}{300 \text{ }\Omega} \times 14 \text{ V} = 5.6 \text{ V}$$
$$V_{xy} = V_x - V_y$$
$$\therefore V_y = V_x - V_{xy} = 24 \text{ V} - 5.6 \text{ V}$$
$$= \textbf{18.4 V}$$

Simplified Representation of Voltage Sources In many practical circuits, the same voltage sources are connected to several different points. For this reason most industrial schematics use a simplified method for representing voltage sources. This is illustrated in Fig. 6–27 where the actual circuit is shown in (a) and its simplified representation in (b).

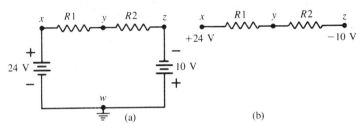

FIGURE 6–27

In the simplified schematic the voltage source symbols have been deleted. Instead, point x is labelled as $+24$ V to indicate that it is actually connected to the *positive* end of the 24-V source. Similarly, point z is labelled -10 V to indicate that it is actually connected to the negative end of the 10-V source. The negative terminal of the 24-V source and the positive end of the 10-V source are not shown; but, it is understood that they are both connected to the system ground point which is also not shown in the simplified schematic. In other words, the ground point is assumed to be connected to the source terminals that are not shown on the schematic.

6.11 Internal Resistance of Voltage Sources

Up to this point, all of the voltage sources which we have considered have been *ideal* voltage sources. They were ideal because the voltage produced at their terminals did not depend on the amount of current which they were producing through a circuit. Unfortunately, there are no ideal voltage sources in the real world of electronics, although many times a source can behave as an ideal source under certain conditions.

In practical voltage sources, such as batteries, electronic power supplies, and generators, the voltage at the terminals of the source will *decrease* as more current is drawn through the source. An example of this effect occurs when your automobile headlights dim as the starter motor draws current from the battery. Another example is the way your home lights dim whenever the air conditioning unit or furnace turns on. In each of these cases, the voltage produced at the terminals of the source is decreased because of the effect of the current being drawn from it.

We can conveniently take this effect into account by representing a *practical* voltage source by an *ideal* voltage source *in series with* a resistance. This resistance is called the *internal resistance of the source* or, as it is more commonly called, the *source resistance*. Figure 6–28 shows the representation of a practical voltage source. The terminals x and y are the actual physically accessible terminals of the source. The source resistance R_s is physically part of the voltage source and is inside the terminals x-y.

Effects of Source Resistance Let's examine the effects of the source resistance R_s on the voltage at the terminals of a practical voltage source. In Fig. 6–29(a) a 6-V constant dc battery with a source resistance of 10 ohms is shown. The terminals of the battery are x and y. The voltage at these terminals is V_{xy}. If we wanted to measure this voltage we would use an instrument called a *voltmeter*. The voltmeter is the voltage counterpart of the *ammeter* (sec. 6.3), which measures current.

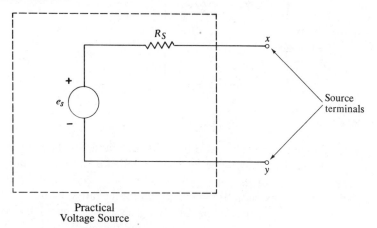

FIGURE 6–28 *Representation of a practical voltage source.*

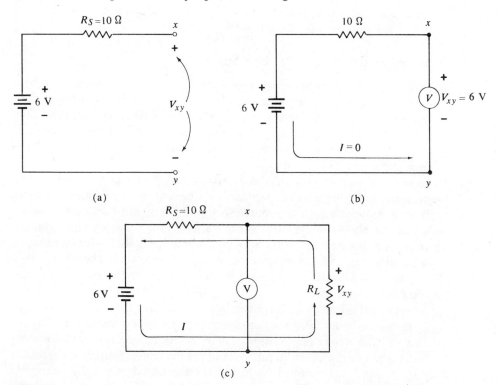

FIGURE 6–29 *Effect of R_s on terminal voltage of a practical voltage source.*

Resistive Devices in Series Circuits 129

A voltmeter has two terminals and is represented by a circle with a V in it. To measure a voltage such as V_{xy} a voltmeter is connected to terminals x and y as shown in part (b) of the figure. The voltmeter would then indicate the value of V_{xy}. Ideally, a voltmeter has a very high resistance so that it draws very little current. For now we will assume that a voltmeter has infinite resistance. In other words, we will treat it as an open circuit. Thus, in Fig. 6–29(b) there will be no current flow ($I = 0$). With $I = 0$ there can be no voltage across R_s since $V = I \times R = 0$.

As such, the 6 V supplied by the source must be across the voltmeter in order to satisfy KVL (recall that all of the input voltage appears across an open circuit in a series circuit). Thus, the voltmeter indicates that the voltage at the battery terminals is 6 V.

During normal operation a battery will be used to supply current to some *load* such as a lamp, a motor, or a heater. We will represent the load by a resistor. In Fig. 6–29(c) a load resistor R_L is shown connected across the battery terminals x-y. The voltmeter is still there to measure the terminal voltage V_{xy}. There is now a complete path for current as shown in the diagram. It should be obvious that the circuit will now function as a *voltage divider* between R_S and R_L. The 6 V from the ideal source will divide between R_S and R_L according to their respective values. As such, there will never be exactly 6 V across R_L. Using the voltage divider equation the voltage across R_L will be

$$V_{xy} = \frac{R_L}{R_L + R_S} \times 6 \text{ V}$$

Since V_{xy} is the actual voltage at the battery's terminals, we can see that V_{xy} will always be less than 6 V with a load connected to it. We can use the preceding expression to calculate V_{xy} for various values of R_L. However, it might be more helpful to calculate I for various values of R_L and then use it to calculate V_{xy}. The results are shown in Table 6–2.

TABLE 6–2 $R_S = 10 \ \Omega$

R_L	I	$I \cdot R_s$	Terminal voltage, $V_{xy} = 6 \text{ V} - I \cdot R_s$
10,000 Ω	~0.6 mA	0.006 V	5.994 V
1,000	~6	0.060	5.94
100	54.5	0.545	5.45
50	100	1	5
20	200	2	4
10	300	3	3
1	545	5.45	0.545
0.1	~600	5.94	0.06
0	600	6	0

The results in the table show that as R_L is decreased in value, the terminal voltage V_{xy} decreases. As R_L decreases it causes a greater current to flow which, in turn, means that there will be a larger voltage across R_S ($V = I \times R_S$). Since R_S has a larger portion of the available 6 V then it follows that there will be less voltage across R_L and the battery terminals.

The table also shows that when R_L is much greater than R_S (i.e., $R_L = 1000 \, \Omega$ and $R_S = 10 \, \Omega$), there is only a slight drop in the battery terminal voltage. On the other hand, when R_L is much smaller than R_S (i.e., $R_L = 1 \, \Omega$ or $0.1 \, \Omega$) the voltage at the battery terminals is a very small portion of the total 6 V. Of course, for the extreme case where $R_L = 0$ (i.e., a short circuit) there will be no voltage at the battery terminals. This condition is also the condition of maximum current.

The voltage measured at the source terminals when there was no load connected [Fig. 6–29(b)] is referred to as the *no-load voltage* (E_{NL}) or the *open-circuit voltage* (E_{OC}) of the source. This is the voltage one measures when a high-resistance voltmeter is placed across the source terminals before the source is placed in a circuit. As we have seen, the voltage divider action between R_L and R_S will, in general, cause the voltage at the source terminals to be *less than* the no-load voltage when the source is connected in a circuit. For this reason, one cannot adjust a voltage source to produce a certain voltage until the source is connected to the circuit in which it is to operate. In other words, the voltage should be adjusted under loaded conditions.

EXAMPLE 6.26 A certain dc voltage source has $E_{NL} = 50$ V and $R_S = 1 \, k\Omega$. Determine the maximum current which this source can produce before its terminal voltage drops by 10 percent.

Solution: Since $E_{NL} = 50$ V and $R_S = 1 \, k\Omega$ the source can be represented as shown in Fig. 6–30. A load R_L is connected to its terminals. A 10 percent drop in the terminal voltage would be a drop of 5 V. Thus, there would be only 45 V delivered to the load and 5 V would be across R_S. Since $R_S = 1 \, k\Omega$, this 5 V would produce a current equal to

$$I = \frac{5 \text{ V}}{1 \text{ k}\Omega} = 5 \text{ mA}$$

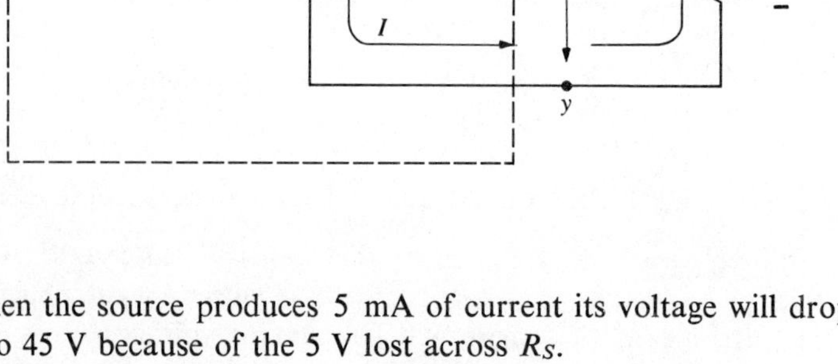

FIGURE 6–30

Thus, when the source produces 5 mA of current its voltage will drop by 10 percent to 45 V because of the 5 V lost across R_S.

Another way to solve this problem takes advantage of the proportionality relationship between series resistor voltages [Eq. (6–7)]. We can solve for the value of R_L which results in a voltage of 45 V across R_L:

$$\frac{R_L}{R_S} = \frac{R_L}{1 \text{ k}\Omega} = \frac{45 \text{ V}}{5 \text{ V}}$$

or, rearranging

$$\frac{R_L}{1 \text{ k}\Omega} = 9$$

Thus,

$$R_L = 9 \text{ k}\Omega$$

With this value of R_L the current produced by the battery will equal

$$\frac{50 \text{ V}}{R_S + R_L} = \frac{50 \text{ V}}{10 \text{ k}\Omega} = 5 \text{ mA}$$

which agrees with our previous result.

Measurement of R_S Often it becomes necessary to measure the source resistance of a practical voltage source. The value of R_S can be determined in two steps: (1) measure the no-load voltage of the source; (2) connect a load resistor to the source and measure the source terminal voltage and the current supplied to the load. Using these measurements the value of R_S can be calculated.

To illustrate, consider the situation shown in Fig. 6–31. The no-load voltage of the source was previously measured to be 24 V. With the 100-ohm resistor connected to the source, the voltmeter indicates a voltage of 20 V across the source terminals and the load. The ammeter indicates a current flow of 0.2 A through the load. Since only 20 V appears across the load, then the difference between 24 V and 20 V must be across R_S. With 4 V across R_S and 0.2 A flowing through it, the value of R_S must be

$$R_S = \frac{V}{I} = \frac{4 \text{ V}}{0.2 \text{ A}} = 20 \text{ ohms}$$

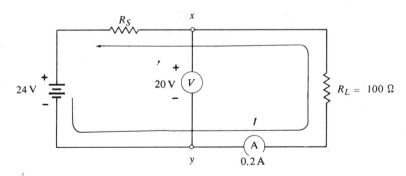

FIGURE 6–31 *One method for taking measurements to determine R_S.*

Notice that we cannot measure the voltage across R_S directly because R_S is internal to the source. Also, note that it is not necessary to know the value of R_L to use this method. If R_L is accurately known then it is not necessary to measure both the current and load voltage. The following example illustrates this point.

EXAMPLE 6.27 A certain practical dc voltage source has a terminal voltage of 32 V under no-load conditions. This voltage drops to 25 V when a load resistor of 30 ohms is placed across its terminals. Determine the value of the source resistance.

Solution: We can use the proportionality relationship of Eq. (6–7) to solve for R_S. Since there are 25 V across $R_L = 30$ ohms, there must be 32 V − 25 V = 7 V across R_S. Thus,

$$\frac{7 \text{ V}}{25 \text{ V}} = \frac{R_S}{30 \text{ }\Omega}$$

Rearranging,

$$\frac{R_S}{30 \text{ }\Omega} = \frac{7 \text{ V}}{25 \text{ V}} = 0.28$$

Thus,

$$R_S = 0.28 \times 30 \text{ }\Omega$$
$$= 8.4 \text{ }\Omega$$

Some Comments on Source Resistance From the preceding discussion it may be concluded that since the internal resistance of a voltage source causes its output voltage to drop when the source is connected to a load, it is desirable to have as small a value of R_S as possible. It is impossible to have $R_S = 0$, but there are many modern electronic power supplies that have source resistances of much less than 1 ohm. On the other hand, it is not always necessary to have a very small R_S because it all depends on the size of the load being driven by the source. For example, suppose a certain source is used to supply current to loads on the order of 10 kilohms to 100 kilohms. In such a case even an R_S of 100 ohms would not result in an appreciable drop in source voltage under loaded conditions.

In the remainder of the text whenever we use a voltage source we will assume it is a practical voltage source unless otherwise indicated. In many cases, however, we will not include the source resistance in the circuit diagrams; in those instances we will assume R_S is small enough so that we can neglect its effects on the source's terminal voltage.

Chapter Summary

1. A series connection is one which contains two or more devices connected end to end to form only one path for current.

2. Two or more devices are in series if the same current passes through each device.

3. When the current flowing through a branch in a circuit reaches a junction where two or more branches come together, then this current splits up among the various branches.

4. To measure the current through any path in a circuit, an ammeter must be inserted in the circuit such that this same current passes through the ammeter.

5. Any number of resistors connected in series has an equivalent resistance which is equal to the sum of the series resistors. $R_{eq} = R_1 + R_2 + R_3 \ldots$

6. This equivalent resistance can be used to determine the amount of current which will flow through the series circuit for a given applied voltage by using Ohm's law.

7. As more resistors are added in series the total resistance to current increases so that less current will flow for a given applied voltage.

8. Current always flows into the negative terminal and out of the positive terminal of a resistor.

9. In a series circuit the sum of the voltages across each series resistive element must equal the voltage applied by the voltage source (or sources) at any instant of time (Kirchhoff's Voltage Law).

10. If two voltage sources are connected in series so that they try to push current in the same direction, then they are aiding each other and their voltages will add to produce a total source voltage.

11. If two series voltage sources are connected such that they are trying to push current in opposite directions, then they are opposing each other and their voltages will subtract to produce a total source voltage.

12. Kirchhoff's Voltage Law (KVL) has to be satisfied at every instant of time in any electric circuit.

13. In a circuit containing only linear resistors, any change in the total applied voltage will produce a proportionate change in all the currents and voltages in the circuit.

14. If we start at any point in a circuit at any instant of time and take any path through the circuit that returns to the starting point, the total of all the voltage drops encountered will equal the total of all the voltage rises encountered, resulting in a net voltage change of zero volts (KVL).

15. When determining the voltage between two points using KVL, one of the two points should be used as the starting point, and the path chosen must contain no unknown voltages except for the one being determined.

16. When writing the algebraic expression for KVL around a path in a circuit, voltage rises are given a + sign while voltage drops are given a − sign.

17. A series circuit in which the output is the voltage across one of the series elements is called a voltage divider.

18. In a voltage divider the voltage across any given resistor is equal to its resistance, divided by the total series resistance, and multiplied times the input voltage.

19. The ratio R_{out}/R_{total} is called the voltage divider ratio or attenuation factor.

20. In a series circuit the input voltage is distributed among the series elements in proportion to their resistance values. The larger resistances will have a greater portion of the voltage and vice versa.

21. In a series circuit all of the input voltage will appear across an open circuit in the circuit path.

134   Chapter Six

22. A point in a circuit which is common to the input and output(s) voltages is called the common or ground point.

23. In a practical voltage source the voltage at the terminals of the source will decrease as current is drawn from the source.

24. A voltmeter is a measuring instrument used to measure the voltage between two points in a circuit. The two terminals of the voltmeter are connected to the two points of interest.

25. For values of load resistance which are much greater than the source resistance, the voltage output of a practical source will remain unchanged.

26. As R_L decreases, the current drawn from a voltage source increases and more voltage is lost across R_S, resulting in a lower voltage at the source terminals.

27. The voltage measured at the source terminals with no load connected is called the no-load voltage or open-circuit voltage.

28. A practical voltage source can act as an ideal source if R_S is much smaller than any R_L which the source is expected to supply current to.

Questions/Problems

Sections 6.2–6.3

6–1 For each of the circuits in Fig. 6–32 indicate which devices are connected in series.

6–2 Replace resistor $R4$ by a short circuit (conducting wire) in Fig. 6–3. Which resistors are now in series in this modified circuit?

6–3 Place a resistor between point y and the $+$ terminal of E_S in Fig. 6–3 and repeat question 6–2.

6–4 Where would you insert an ammeter in Fig. 6–32(e) to measure the current flowing through $R5$? Through $R1$? Through the voltage source?

6–5 At how many different places could you insert an ammeter to measure the current through $R4$ in Fig. 6–32(d)?

Section 6.4

6–6 A 100-ohm resistor and a 50-ohm resistor are connected in series. Sketch the I-V graph of each resistor and the I-V graph of their equivalent resistor.

6–7 A 5-kilohm resistor and 3-kilohm resistor are connected in series. How much current will flow through these resistors if 12 V is applied to the series circuit?

6–8 In problem 6–7 what voltage would be required to produce a current of 250 μA through the series circuit?

6–9 Assume that a certain current is flowing in the circuit of Fig. 6–1. Indicate what will happen to the current I if:
 a. Resistor $R1$ is increased in value.
 b. Resistor $R2$ is decreased in value.

c. E_S is increased in value.
d. $R1$ and $R2$ are doubled and E_S is doubled.
e. $R1$ and $R2$ are interchanged.

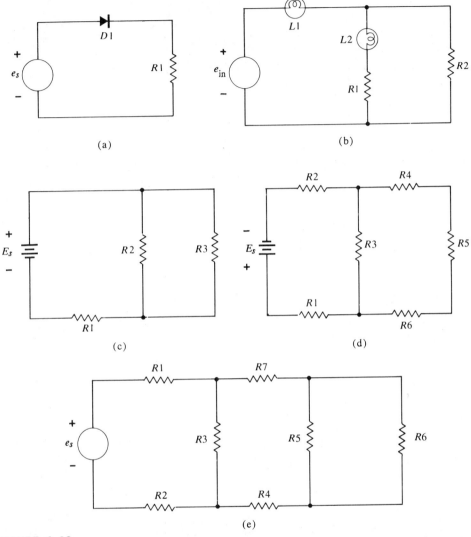

FIGURE 6–32

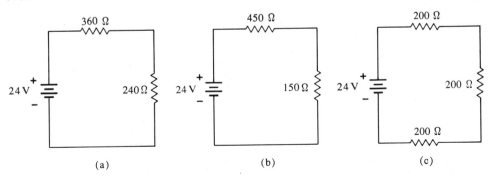

FIGURE 6–33

6-10 Calculate the current for each circuit in Fig. 6-33.

6-11 Calculate R_{eq} for each circuit in Fig. 6-34. Then determine the value of I for each case.

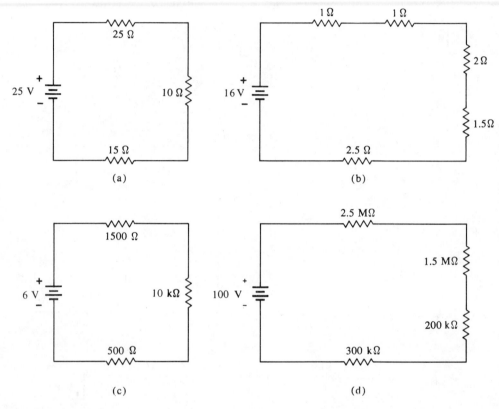

FIGURE 6-34

6-12 For each of the circuits of Fig. 6-34 calculate the amount of voltage needed to produce a current of 1 mA (see Ex. 6.5).

6-13 For the circuits in Fig. 6-34(a) and (d) determine the voltage across each resistor (see Ex. 6.6).

6-14 In the circuit of Fig. 6-34(a) what will happen to the voltage across the 15-ohm resistor if the source voltage is increased? What will happen to this voltage if the 15-ohm resistor is decreased in value?

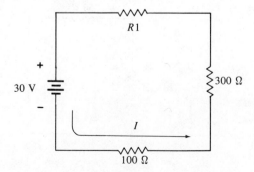

FIGURE 6-35

6–15 Calculate the voltage across $R1$ in Fig. 6–35 for the following cases: (a) $R1 = 200\ \Omega$; (b) $R1 = 20\ \Omega$; (c) $R1 = 2000\ \Omega$.

6–16 Suppose a string of Christmas tree lamps are connected in series. Can you explain why the lamps get dimmer if we add more lamps to the string?

6–17 Consider the circuit in Fig. 6–36.
   a. If $R1$ is increased in value what will happen to the brightness of the lamp?
   b. If E_S is decreased in value what will happen to the brightness of the lamp?
   c. For what value of $R1$ will the lamp be the brightest if E_S is not changed?
   d. For what value of $R1$ will the lamp be the dimmest?

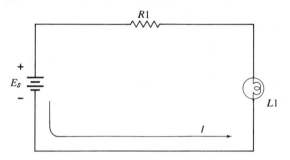

FIGURE 6–36

Section 6.5

6–18 For each of the circuits of Fig. 6–37 determine the circuit current and verify KVL.

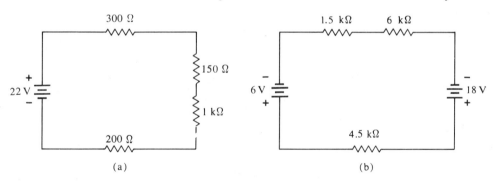

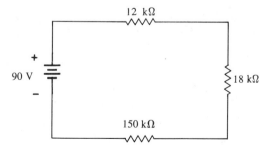

FIGURE 6–37

6-19 Use the linearity principle to determine the current and voltages in Fig. 6-37(c) if the source voltage is decreased to 45 V. Verify KVL.

Section 6.6

6-20 Indicate which of the following statements concerning KVL are valid:
   a. KVL holds true for series circuits that contain only linear devices.
   b. KVL holds true for circuits which are connected to any type of voltage source.
   c. KVL holds true only for circuits that contain resistive devices.
   d. If KVL is not satisfied by the voltages around a series circuit, then an error must have been made in determining these voltages.

6-21 Consider the circuit in Fig. 6-38. When E_S is set to 10 V the ammeter indicates a current of 2 mA. Determine the voltage across "device X" and its resistance.

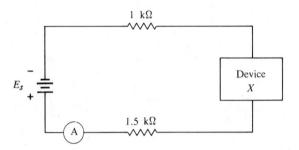

FIGURE 6-38

6-22 In the same circuit the current increases to 2.5 mA when E_S is set to 15 V. Determine whether device X is a linear or nonlinear device.

6-23 If the voltage source in Fig. 6-38 is doubled, will the current necessarily double?

6-24 If the voltage source in Fig. 6-37(a) is doubled will the current double?

6-25 In your own words give a general statement of KVL.

6-26 For the circuit of Fig. 6-37(b) verify KVL by summing up the voltage rises and voltage drops.

6-27 Redo Ex. 6.13, Fig. 6-16(a), by using point z as the starting point.

6-28 Determine V_{CB} in the circuit of Fig. 6-17. Use two different paths to arrive at the same result.

6-29 What is the equivalent resistance "seen" by the source in Fig. 6-16(a)? What is the equivalent resistance of the series combination of device A and device B?

Section 6.7

6-30 Write the algebraic expression for KVL around the path *d-f-g-b-d* in the circuit of Fig. 6-18. Repeat for the path *b-g-f-e-a-b*.

6-31 Consider the circuit in Fig. 6-39.
   a. Write the algebraic expression for KVL around the path *a-b-c-d-e-a*. Then, solve for the unknown V_{cd}.
   b. Write the algebraic expression for KVL around the path *c-d-e-f-c*. Then, solve for the unknown V_{cf}.
   c. Determine V_{ce} and V_{be}.

6-32 Can the voltage V_{cf} be found in Fig. 6-39 without first finding V_{cd}? If so, show how.

Section 6.8

6-33 Use the voltage divider equation to find the voltage across the 1-kilohm resistor in Fig. 6-37(a).

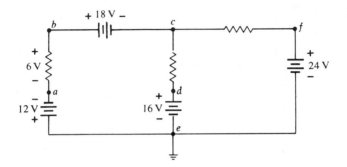

FIGURE 6-39

6-34 Use the voltage divider equation to find the voltage across the 18-kilohm resistor in Fig. 6-37(c).

6-35 Give at least four ways in which the size of the voltage in the last problem may be decreased by changing one of the circuit values.

6-36 Design as many different voltage divider circuits as you can which have a voltage divider ratio of 0.25. Use only resistors with the following values: 1 kilohm, 2.5 kilohms, 10 kilohms, and 15 kilohms. You can use more than one of each value.

6-37 Indicate which of the following statements are not valid:
  a. In a voltage divider circuit, the output is always some fraction of the input.
  b. If all the resistors in a voltage divider circuit are doubled, the output voltage will double.
  c. As the value of a resistor in a series circuit is decreased, the circuit current increases and so does the resistor's voltage.
  d. To increase the output of a voltage divider circuit, the voltage divider ratio has to be increased.
  e. Since there is no current through an open circuit, there can be no voltage across an open circuit.
  f. If the output resistor in a voltage divider is doubled in value, the output voltage will *not* double in value.
  g. If any one resistor in a series circuit is short-circuited, the voltage across each of the other series resistors will increase.

6-38 A potentiometer is a linear resistor with *three* terminals as shown in Fig. 6-40(a). Between the two end terminals, the resistance is a constant value R. The other terminal is called the *wiper-arm* terminal because it is connected to a movable conductor (wiper-arm) that is allowed to move up and down the extremities of the resistor R. Depending on the position of the wiper-arm, there will be a certain resistance between the wiper-arm and the upper end of R, and there will be a certain resistance between the wiper-arm and the lower end of R. Obviously, the sum of these two resistances has to equal R. In other words, the wiper-arm essentially divides R into two series resistors, R_{upper} and R_{lower}.

The potentiometer can be used as a *variable* voltage divider by taking the output voltage between the wiper-arm terminal and one of the outer terminals as shown in Fig. 6–40(b).

Determine the voltage divider ratio and output voltage for the circuit in Fig. 6–40(b) for the following potentiometer settings:

a. $R_{upper} = R_{lower} = 5\ k\Omega$
b. $R_{upper} = 2\ k\Omega$; $R_{lower} = 8\ k\Omega$
c. wiper-arm three-fourths of the way toward upper terminal
d. wiper-arm all the way down to bottom terminal

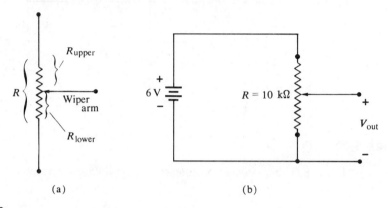

FIGURE 6–40

6–39 A potentiometer can be used as a variable resistor as shown in Fig. 6–41 by using only one end terminal. If $R = 100$ kilohms, determine the range of output voltages as the wiper-arm is varied over the extremities of the potentiometer.

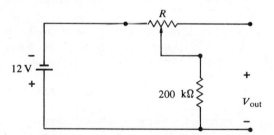

FIGURE 6–41

6–40 Refer to Fig. 6–21(b). The voltage across the 2-kΩ resistor is 13 V. Without calculating the current, determine the voltages across each resistor, and then determine the value of E_{in}.

Section 6.9

6–41 Determine the values for V_w, V_x, and V_y for the circuit in Fig. 6–42.

6–42 Use the values calculated in the previous problem to determine V_{xy}, V_{wy}, V_{wt} and V_{yt}. (Hint: use Eq. 6–8)

Section 6.10

6–43 (a) Change the 10-V source in Fig. 6–26 to 36 V and find V_y. (b) Reverse the polarity of the 36-V source and repeat.

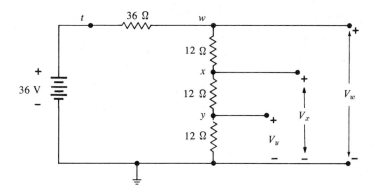

FIGURE 6-42

6-44 Calculate V_y for the circuit of Fig. 6-43. (Hint: redraw the circuit showing the sources connected to common ground.)

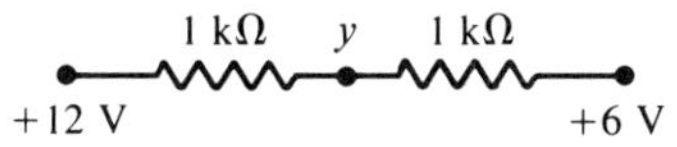

FIGURE 6-43

Section 6.11

6-45 What is the principal difference between an ideal and a practical voltage source?

6-46 A certain dc voltage has $E_{NL} = 60$ V and $R_S = 22$ ohms. Determine the voltage at the source terminals for the following values of load resistor connected to the source terminals: 0.2 ohm, 2 ohms, 22 ohms, 200 ohms, 2000 ohms.

6-47 For the source in the last problem determine the minimum value of R_L which could be used if it is required that the source terminal voltage never drops below 50 V.

6-48 How would the answer to problem 6-47 be affected (increases, decreases, remains the same) if
  a. R_S is made larger
  b. E_{NL} is increased

6-49 A certain voltage source is measured to have $E_{NL} = 12.6$ V. When a 200-ohm load is placed across its terminals the source voltage drops to 11.7 V. Calculate R_S.

6-50 Under what conditions does a practical source behave similarly to an ideal source?

CHAPTER 7

Resistive Devices in Parallel Circuits

7.1 Introduction

In chap. 6 we studied one of the basic circuit connections — the series connection. This chapter will concentrate on the second basic circuit connection, called the *parallel* connection. The properties of parallel circuits will be examined, and the second of the important circuit laws, *Kirchhoff's Current Law*, will be introduced.

7.2 The Parallel Connection

A parallel connection is defined as two or more devices connected such that one end of each device is tied to a common point and the other end of each device is tied to a second common point. To illustrate, consider Fig. 7-1(a), where resistors $R1$ and $R2$ are connected *in parallel* because their upper ends are connected to point x and their lower ends are connected to point y. Figure 7-1(b) shows an equivalent way of drawing this same situation. The arrangement in (b) is more common.

From these first two examples it might appear that devices connected in parallel *electrically* must always be parallel *geometrically*. This is not true. Figure 7-1(c) shows $R1$ and $R2$ connected in parallel but not drawn parallel geometrically. As long as two devices have their ends connected to common points, they are connected in parallel electrically.

Figure 7-1(d) shows three resistors connected in parallel. The resistors have points x and y in common. An equivalent way of drawing three parallel resistors is shown in part (e) of the figure. In this latter representation $R1$ is tied directly to x, $R2$ is connected to x by a connecting wire, and $R3$ is connected to the upper end of $R2$. Thus, either directly or indirectly, the three resistors each have their upper ends tied to point x. The same is true for the lower ends of the resistors, which are tied to y.

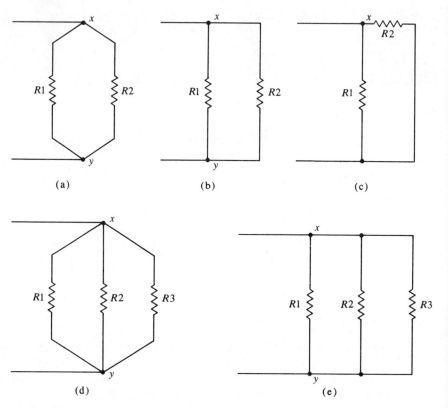

FIGURE 7–1 *Various examples of parallel connections.*

In Fig. 7–2(a) a constant dc voltage source is connected across the parallel combination of two resistors. In this circuit the 12-V source and resistors $R1$ and $R2$ are all in parallel with one another since they have two points in common. Because of the voltage source, point x will be 12-V positive relative to y. That is, $V_{xy} = +12$ V. Since both resistors are connected between x and y, it is apparent that *each* resistor has 12 V across it. In fact, no matter how many resistors or other devices are connected in parallel, the 12 V will appear across each device [see part (b) of the figure]. We can now generalize and say that:

When two or more devices are connected in parallel, each device will have the same identical voltage impressed across its terminals.

Of course, there will be current through each of the parallel devices in the circuits of Fig. 7–2 as a result of the applied 12 V. The current through each device will in general be different and depends on its electrical characteristics. As we shall see momentarily, the sum of all the currents through the various devices equals the current supplied through the battery.

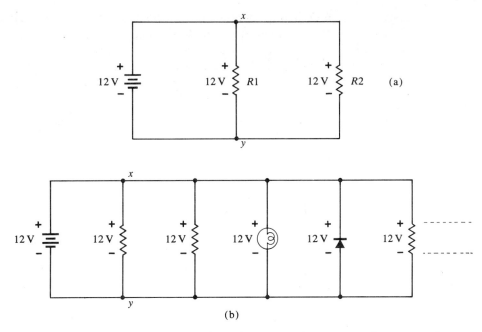

FIGURE 7-2 *Devices in parallel always have the same voltage across their terminals.*

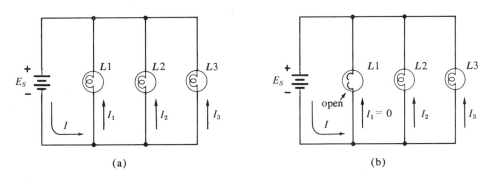

FIGURE 7-3 *Christmas tree lamps connected in parallel.*

A common example of a parallel circuit is a string of Christmas tree lamps connected in parallel as shown in Fig. 7–3(a). The source E_S is connected across each lamp. As a result, each lamp has current flowing through it. If one of the lamp filaments burns out, the current path through that lamp will be broken and no current will flow through that lamp. However, this will not affect the other lamps since the current paths through these lamps back through the source are still intact. This is illustrated in Fig. 7–3(b) where we see that even though $L1$ is open-circuited, current still flows through the other lamps and they will remain lit.

7.3 Inserting Elements in Parallel

It is a relatively simple matter to connect an electrical element or device in parallel with any portion of a circuit. Consider the circuit in Fig. 7–4(a) and suppose that it is desired to connect a resistor R4 in parallel (or in shunt)* with R2. It is only necessary that the terminals of R4 be connected to the terminals of R2. There are two possibilities as shown in (b) and (c) of Fig. 7–4. Both of these possibilities are equivalent electrically as long as R4 and R2 have two points in common.

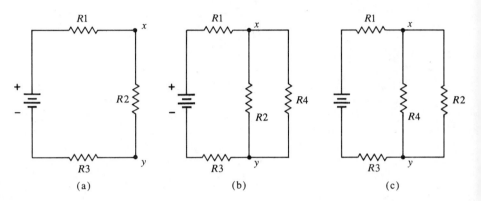

FIGURE 7–4

When connecting a circuit element in parallel it is never necessary to break the circuit before doing so. This is only true when inserting an element in series as we saw in chap. 6.

In sec. 6.10 the voltmeter was introduced as an instrument for measuring voltages in a circuit. The voltmeter is *always* connected in parallel (shunt) with that portion of the circuit where the voltage is being measured. The voltmeter has to be subjected to the same voltage that it is trying to measure, and to accomplish this the voltmeter terminals have to be connected to the points in the circuit between which the voltage is to be measured. Figure 7–5(a) shows a voltmeter connected to measure the voltage across R1. Figure 7–5(b) shows it connected to measure the voltage across the R1 and R2 series combination.

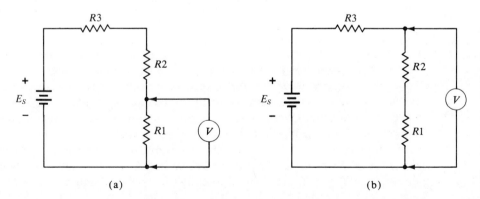

FIGURE 7–5 *Connecting voltmeters in parallel with that portion of the circuit across which the voltage is to be measured.*

---

*The term "shunt" is often used interchangeably with the term "parallel."

7.4 Linear Resistors in Parallel

Figure 7–6(a) shows a simple parallel circuit. The 6-V source is in parallel with a 1-ohm resistor and a 2-ohm resistor. Consider the problem of determining the currents I_1 and I_2 through the respective resistors and the current I_s through the source. We know that the source will produce a flow of electrons *out* of its negative terminal. These electrons travel from the negative terminal toward point x. When they reach point x these electrons can take either of the two parallel paths to point y. A certain number of the electrons will travel up through the 1-ohm resistor. This flow of electrons constitutes I_1. The remaining electrons will travel up through the 2-ohm resistor. This constitutes I_2. Since the currents I_1 and I_2 are produced by dividing up the source current I_s, then it should be apparent that

$$I_s = I_1 + I_2 \qquad (7\text{–}1)$$

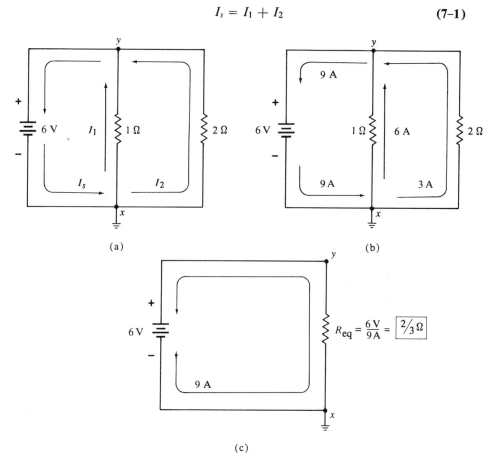

FIGURE 7–6 *Two parallel resistors are equivalent to one resistor which draws an equivalent amount of current from the source.*

In other words, the sum of the currents through the parallel branches equals the current produced by the source. The currents I_1 and I_2 flow up through their respective resistors toward point y. At y they recombine to form I_s, which flows back to the + terminal of the source. (Recall the water pipe analogy in sec. 6.2.)

Now, the values of I_1 and I_2 can be easily calculated using Ohm's law since each resistor has 6 V across it. Thus, we have

$$I_1 = \frac{6 \text{ V}}{1 \text{ ohm}} = 6 \text{ A}$$

and

$$I_2 = \frac{6 \text{ V}}{2 \text{ ohms}} = 3 \text{ A}$$

The source current is therefore

$$I_s = I_1 + I_2 = 9 \text{ A}$$

These currents are shown in Fig. 7-6(b).

The parallel combination of the 1-ohm and 2-ohm resistors results in a 9-A current produced by the 6-V source. As far as the source is concerned the total resistance it sees is

$$\frac{6 \text{ V}}{9 \text{ A}} = \frac{2}{3} \Omega$$

In other words, a 2/3-ohm resistor connected by itself across the 6-V source would also draw 9 A from the source. Thus, we can say that the 1 ohm in parallel with 2 ohms is equivalent to 2/3 ohms [see part (c) of Figure 7-6].

From the results of this example several statements can be made concerning the parallel combination of two resistors.

The equivalent resistance of two parallel resistors is that value of resistance which draws the same current from the source as the parallel combination does.

In other words, the parallel combination of two resistors presents a certain total resistance to the flow of current. A single resistor with a value equal to this total resistance is equivalent to the parallel combination.

The equivalent resistance of two parallel resistors always has a smaller value than either parallel resistor.

This last statement is easily reasoned because the two resistors in parallel draw more total current than either resistor alone.

The method used in this circuit to find R_{eq} can be used to find the equivalent resistance of any combination of two or more parallel resistors. However, it would be valuable if we could develop a simple formula which can be used to calculate R_{eq}. To help us develop such a formula, let's consider the same circuit of Fig. 7-6(a) redrawn using conductance rather than resistance values as in Fig. 7-7(a). The 1-ohm resistor has a conductance of 1/1 ohm = 1 S. The 2-ohm resistor has a conductance of 1/2 ohm = 0.5 S.

The equivalent circuit of Fig. 7-6(a) is redrawn in Fig. 7-7(b) with the $R_{eq} = 2/3$ ohm replaced by its conductance which is

$$G_{eq} = \frac{1}{R_{eq}} = \frac{1}{2/3 \; \Omega} = 1.5 \text{ S}$$

As can be seen in Fig. 7-7, the value of G_{eq} is equal to the sum of the con-

Resistive Devices in Parallel Circuits

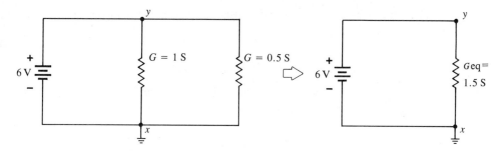

FIGURE 7-7 *Conductances in parallel add to give equivalent conductance.*

ductances of the parallel resistors. This principle is true for any number of parallel resistors. In other words,

The equivalent conductance of any number of parallel resistors is equal to the sum of the conductances of the individual resistors.

EXAMPLE 7.1 Find the equivalent resistance of two 1-kΩ resistors in parallel.

Solution: Each 1-kΩ resistor has a conductance of

$$1/1 \text{ k}\Omega = 10^{-3} \text{ S}$$
$$= 1 \text{ mS}$$

Thus, the equivalent conductance is 2 mS. The equivalent resistance is therefore

$$R_{eq} = \frac{1}{G_{eq}} = \frac{1}{2 \times 10^{-3} \text{ S}} = 500 \text{ ohms}$$

Note that R_{eq} is less than 1 kilohm.

For the conductances in parallel, then, we have

$$G_{eq} = G_1 + G_2 \qquad (7\text{--}2)$$

In general, for any number of conductances in parallel we have

$$G_{eq} = G_1 + G_2 + G_3 \ldots \qquad (7\text{--}3)$$

It was mentioned earlier that resistance values rather than conductance values are usually used on circuit diagrams and for analyzing circuits. For this reason it would be more valuable to convert the equivalent conductance formulas to resistance formulas. This can be done easily using $G = 1/R$. Thus, for two resistors in parallel, Eq. (7–2) becomes:

$$\frac{1}{R_{eq}} = \frac{1}{R_1} + \frac{1}{R_2} \qquad (7\text{--}4)$$

For any number of resistors in parallel, Eq. (7–3) becomes

$$\frac{1}{R_{eq}} = \frac{1}{R_1} + \frac{1}{R_2} + \frac{1}{R_3} + \cdots \qquad (7\text{--}5)$$

The special case of two parallel resistors [Eq. (7–4)] can be rearranged to solve for R_{eq} directly as

$$R_{eq} = \frac{R_1 \times R_2}{R_1 + R_2} \tag{7-6}$$

This is often known as the *product-over-sum* formula since R_{eq} equals the product of the resistances over the sum of the resistances.

EXAMPLE 7.2 Determine the equivalent resistance of a 600-Ω and a 300-Ω resistor connected in parallel.

Solution: Using "product over sum"

$$R_{eq} = \frac{600\ \Omega \times 300\ \Omega}{600\ \Omega + 300\ \Omega} = \frac{18 \times 10^4}{9 \times 10^2}$$
$$= 200\ \Omega$$

EXAMPLE 7.3 Determine the equivalent resistance of the parallel combination of Fig. 7–8.

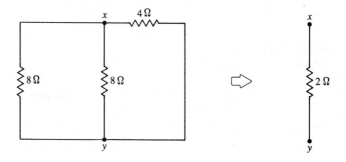

FIGURE 7–8

Solution: Using Eq. (7–5) we have

$$\frac{1}{R_{eq}} = \frac{1}{8} + \frac{1}{8} + \frac{1}{4}$$
$$= \frac{1}{4} + \frac{1}{4}$$
$$\frac{1}{R_{eq}} = \frac{1}{2}$$

Inverting both sides,

$$R_{eq} = 2\ \text{ohms}$$

EXAMPLE 7.4 Two resistors are connected in parallel across a 24-V source. An ammeter indicates that the current through the source is 12 mA. If one of the resistors, R_1, is 6 kΩ, what is the value of the other resistor?

Solution : The current through R_1 will be

$$I_1 = \frac{24 \text{ V}}{6 \text{ k}\Omega} = 4 \text{ mA}$$

The current through R_2 has to be the difference between the source current and the current through R_1. That is,

$$I_2 = I_S - I_1 = 12 \text{ mA} - 4 \text{ mA} = 8 \text{ mA}$$

Since R_2 has 8 mA flowing through it when its voltage is 24 V, its resistance must be

$$R_2 = \frac{24 \text{ V}}{8 \text{ mA}} = 3 \text{ k}\Omega$$

EXAMPLE 7.5 Consider the circuit in Fig. 7-9. Determine the values of I_1, I_2, and I_S.

Solution: $I_1 = \dfrac{120 \text{ V}}{10 \text{ k}\Omega} = 12 \text{ mA}$

$I_2 = \dfrac{120 \text{ V}}{15 \text{ k}\Omega} = 8 \text{ mA}$

$I_S = I_1 + I_2 = 20 \text{ mA}$

Alternatively, I_S can be found by first determining the equivalent resistance seen by the source.

$$R_{eq} = \frac{10 \text{ k}\Omega \times 15 \text{ k}\Omega}{10 \text{ k}\Omega + 15 \text{ k}\Omega} = 6 \text{ k}\Omega$$

$$I_S = \frac{E_S}{R_{eq}} = \frac{120 \text{ V}}{6 \text{ k}\Omega} = 20 \text{ mA}$$

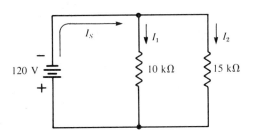

FIGURE 7-9

EXAMPLE 7.6 Consider the circuit in Fig. 7-10(a). The source current and branch current values are given. Indicate what effect the following changes would have on the values of I_s, I_1, I_2, and R_{eq}.
    (a) an increase in the value of R_2
    (b) a decrease in the source voltage
    (c) adding a third resistor in parallel with R_1 and R_2

Solution: (a) If the value of R_2 is increased the current I_2 through R_2 will decrease. Since R_1 is unchanged its current I_1 will stay at 4 A. Because of the decrease in I_2, the source current I_s will decrease. Thus, the total resistance seen by the source must *increase* since less current is being drawn from the source.

(b) A decrease in the source voltage will cause a proportional decrease in the circuit currents. Since R_1 and R_2 are unchanged the value of R_{eq} will be unchanged.

(c) The situation where a third resistor is added in parallel is shown in Fig. 7–10(b). The addition of R_3 will result in a current I_3. This current I_3 must be supplied from the source along with I_1 and I_2 (which are unchanged) so that the source current I_S must be larger. The larger I_S means that the source sees a *smaller* equivalent resistance.

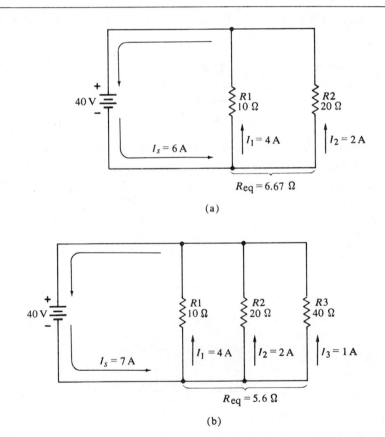

FIGURE 7–10

This last example can allow us to draw several conclusions about parallel linear resistors.

(1) Increasing (decreasing) any one resistor of a parallel combination will increase (decrease) the equivalent resistance.

(2) Adding one or more resistors in parallel with another resistor (or resistors) will decrease the equivalent resistance of the combination.

(3) The value of the source voltage has no effect on the circuit resistance.

Special Case of Equal Parallel Resistors If a number of *equal*-valued resistors are connected in parallel the equivalent resistance of the parallel combination can be found using the simple relationship

$$R_{eq} = \frac{R}{n} \qquad (7\text{–}7)$$

where R is the common resistance value and n is the number of resistors.

For example, using Eq. (7–7) the equivalent resistance of *two* parallel resistors of 12 ohms will be 12 Ω/2 = 6 Ω; *three* resistors of 12 ohms will be 12 Ω/3 = 4 Ω; *four* resistors of 12 ohms will be 12 Ω/4 = 3 Ω; and so on.

7.5 Kirchhoff's Current Law

In the last section we saw that in a simple parallel circuit the source current was equal to the *sum* of the currents through the individual parallel branches [see Figs. 7–6(a) and 7–10 and Exs. 7.4 and 7.5]. This case is a special case of a general circuit law called Kirchhoff's Current Law (abbreviated KCL) which states that

At any junction point or node in an electric circuit the sum of the currents flowing into the junction must equal the sum of the currents flowing away from the junction at any instant of time.

This important circuit law along with its voltage counterpart (KVL) comprise the basis for many of the methods of circuit analysis.

To help illustrate KCL consider Fig. 7–11. The ammeter indicates that I_s equals 50 mA. The current I_1 is easily determined as 45 V/1 kΩ = 45 mA since there are 45 V across the 1-kΩ resistor. To find I_2 we must use KCL since the value of R_2 is not given. Considering the junction point x we can apply KCL. The only current flowing *into* the junction is I_s. The currents flowing away from x are I_1 and I_2. Thus, according to KCL

$$I_s = I_1 + I_2$$

Thus,

$$50 \text{ mA} = 45 \text{ mA} + I_2$$

or

$$I_2 = 5 \text{ mA}$$

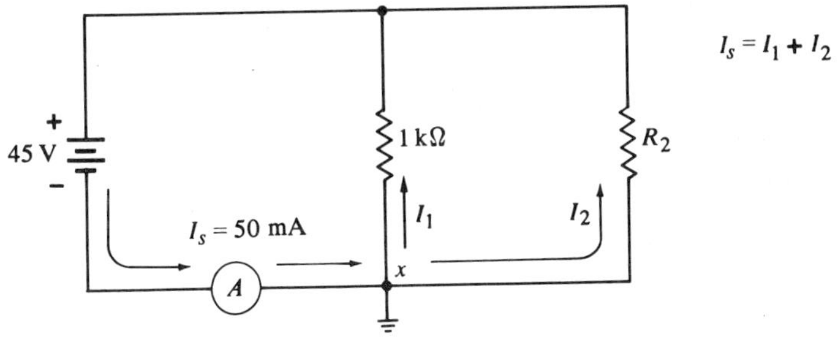

FIGURE 7–11 *Illustration of KCL.*

We can apply KCL like KVL to any circuit configuration. We shall do so whenever it is necessary to find an unknown current entering or leaving a point and the other currents into or out of the junction are known. The following examples will further illustrate KCL.

EXAMPLE 7.7 Figure 7–12(a) shows a portion of a certain circuit. Determine I_x.

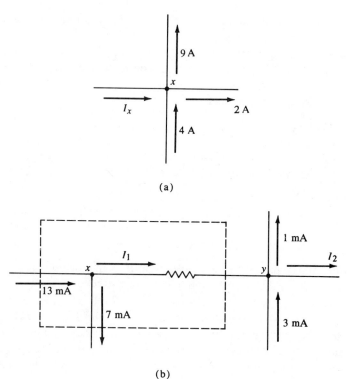

FIGURE 7–12

Solution: The currents flowing into junction point x are I_x and 4 A. The currents flowing away from x are 9 A and 2 A. Thus, we have

$$\text{currents in} = \text{currents out}$$
$$I_x + 4\text{ A} = 9\text{ A} + 2\text{ A}$$
$$= 11\text{ A}$$

or

$$I_x = 11\text{ A} - 4\text{ A} = 7\text{ A}$$

EXAMPLE 7.8 Determine I_2 in Fig. 7–12(b).

Solution: In order to determine I_2 we must know all of the other currents into or out of junction y. Since I_1 flows into point y, we must first determine I_1 in order to determine I_2. Thus, consider junction x first (dotted lines in figure). Using KCL we can write

$$13 \text{ mA} = 7 \text{ mA} + I_1$$

or

$$I_1 = 6 \text{ mA}$$

Now consider junction y. We can write

$$I_1 + 3 \text{ mA} = I_2 + 1 \text{ mA}$$

Since $I_1 = 6$ mA, this gives $I_2 = 8$ mA.

$$6 \text{ mA} + 3 \text{ mA} = I_2 + 1 \text{ mA}$$

or

$$I_2 = 8 \text{ mA}$$

EXAMPLE 7.9 Figure 7–13 shows a complex circuit which occurs quite often in electronic circuits. It will be analyzed in detail later in the text. The circuit has two voltage sources. The E_{S1} source is attempting to push a current I_1 in the direction shown. The E_{S2} source is attempting to produce a current I_2 in the direction shown.

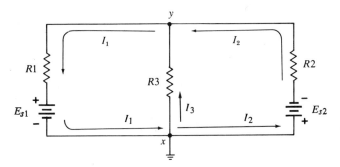

FIGURE 7–13

The current I_3 through $R3$ will depend on the values of I_1 and I_2. If I_1 is larger than I_2 then current will flow *up* through $R3$. If I_1 is smaller than I_2 the current will flow *down* through $R3$.

The values of E_{S1}, E_{S2}, $R1$, $R2$, and $R3$ will determine which of the currents I_1 or I_2 is the larger. Even when the source and R values are known, it is not easy to predict which current is larger until a complete analysis is performed. Thus, it is not always possible to predict beforehand which way current will flow through $R3$. For this reason, when we label the currents in a circuit diagram such as Fig. 7–13 it is purely arbitrary as to which direction we show I_3 flowing. When the circuit is eventually analyzed the actual direction of I_3 will manifest itself.

To illustrate, assume I_3 is in the direction shown on the diagram and calculate I_3 for the following cases: (a) $I_1 = 5$ mA, $I_2 = 3$ mA; (b) $I_1 = 5$ mA, $I_2 = 10$ mA; (c) $I_1 = 7$ μA, $I_2 = 7$ μA.

Solution: (a) Using KCL we can write for junction x

$$I_1 = I_2 + I_3$$

Thus, for the values given

$$5 \text{ mA} = 3 \text{ mA} + I_3$$

or

$$I_3 = 2 \text{ mA}$$

(b) For the values given

$$5 \text{ mA} = 10 \text{ mA} + I_3$$

or

$$I_3 = -5 \text{ mA}$$

The *negative* sign of this result indicates that I_3 is flowing *opposite* to the assumed direction. The *magnitude* of the current is correct at 5 mA; however, the current through $R3$ actually flows *downward*. Whenever a minus sign appears in the result for a current, it simply indicates that the current direction is opposite to the direction assumed.

(c) For the values given

$$7 \text{ μA} = 7 \text{ μA} + I_3$$

or

$$I_3 = 0$$

which indicates that the current through $R3$ is zero.

7.6 Current Divider Rule

Consider the parallel circuit in Fig. 7–14. Using Ohm's law we can write

$$I_1 = \frac{E_s}{R1} \quad \text{and} \quad I_2 = \frac{E_s}{R2}$$

We also know that

$$I_s = \frac{E_s}{R_{eq}}$$

or

$$E_s = I_s R_{eq}$$

Thus, we have

$$I_1 = \frac{E_s}{R1} = I_s \frac{R_{eq}}{R1} \tag{7-8}$$

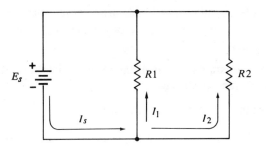

FIGURE 7-14

Similarly,

$$I_2 = \frac{I_s R_{eq}}{R2} \tag{7-9}$$

Substituting $R_{eq} = R1 \cdot R2/(R1 + R2)$ into (7-8) and (7-9) finally yields

$$I_1 = \left(\frac{R2}{R1 + R2}\right) \times I_s \tag{7-10}$$

and

$$I_2 = \left(\frac{R1}{R1 + R2}\right) \times I_s \tag{7-11}$$

These last two expressions are called the *current divider formulas*. They show how the source current I_s divides up between I_1 and I_2. Note that I_1 is obtained by multiplying I_s by the ratio of $R2$ (not $R1$) to the sum $R1 + R2$. This ratio is of course always less than one so that $I_1 < I_s$. Similarly, I_2 is found by multiplying I_s by the ratio of $R1$ (not $R2$) to the sum $R1 + R2$.

EXAMPLE 7.10 In Fig. 7-15 find I_1 and I_2 if $I_s = 12$ mA.

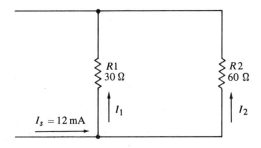

FIGURE 7-15

Solution: $I_1 = \dfrac{60}{30 + 60} \times 12 \text{ mA} = 2/3 \times 12 \text{ mA} = 8 \text{ mA}$

$I_2 = \dfrac{30}{30 + 60} \times 12 \text{ mA} = 4 \text{ mA}$

Note that $I_1 + I_2 = 12 \text{ mA} = I_s$

Notice from the results of the last example that twice as much current flows through the 30-ohm resistor as through the 60-ohm resistor. In general, it can be stated that

For two resistors in parallel, the ratio of their currents is equal to the inverse ratio of their resistances.

In equation form this statement becomes

$$\frac{I_1}{I_2} = \frac{R_2}{R_1} \qquad \text{(note reversal of subscripts in resistor ratio)} \qquad (7\text{--}12)$$

This ratio relationship can be useful in cases where more than two resistors are in parallel. The following example illustrates.

EXAMPLE 7.11 Find I_1, I_3, and I_S in Fig. 7–16.

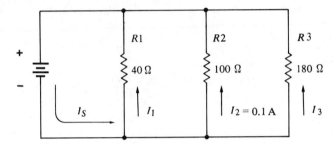

FIGURE 7–16

Solution: $\quad \dfrac{I_1}{I_2} = \dfrac{R_2}{R_1} = \dfrac{100\ \Omega}{40\ \Omega} = 2.5$

Thus, $I_1 = 2.5\, I_2 = 2.5 \times 0.1\text{ A} = \mathbf{0.25\text{ A}}$
Similarly,

$$\frac{I_3}{I_2} = \frac{R_2}{R_3} = \frac{100\ \Omega}{180\ \Omega} = 0.55$$

so that

$$I_3 = 0.55\, I_2 = \mathbf{0.055\text{ A}}$$

The total source current is therefore

$$I_S = I_1 + I_2 + I_3 = 0.25 + 0.1 + 0.055 = \mathbf{0.405\text{ A}}$$

7.7 Effects of Open Circuits and Short Circuits

In a parallel circuit if one of the parallel elements becomes open-circuited, the equivalent resistance of the parallel combination increases since one of the available current paths is removed. This usually manifests itself as a *decrease* in total circuit current unless a constant current source is being used.

If one of the parallel elements becomes short-circuited the equivalent resistance becomes *zero*. This is easy to see from the formula for the equivalent resistance of two parallel resistors

Resistive Devices in Parallel Circuits   159

$$R_{eq} = \frac{R1 \times R2}{R1 + R2} \tag{7-6}$$

If either $R1$ or $R2$ becomes *zero* then it is obvious that R_{eq} will be zero. This usually manifests itself as an *increase* in total circuit current. Also, as a result of the short circuit all of the current flows through the short circuit and none through the other resistors. This is illustrated in Fig. 7–17.

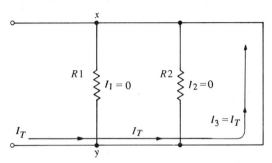

FIGURE 7–17   *All the circuit current flows through the short circuit.*

Because of the short circuit the voltage V_{xy} will be zero, which means that there is no voltage across $R1$ or $R2$. Thus, no current will flow through either resistor. As a result, all of I_T flows through the short circuit.

7.8   Connecting Voltage Sources in Parallel

Suppose we had a 6-V battery and a 12-V battery and we connected them in parallel as shown in Fig. 7–18(a). At first, it might seem that such a situation is an impossible one because one battery is trying to make $V_{xy} = +6$ V while the other is trying to make $V_{xy} = +12$ V. The dilemma is not as impossible as it seems if we remember to include the internal source resistance of each battery as in Fig. 7–18(b). Now, the circuit is actually a series circuit with two sources and two resistors similar to the type discussed in chap. 6. The difference in the source voltages will appear across R_{s1} and R_{s2}.

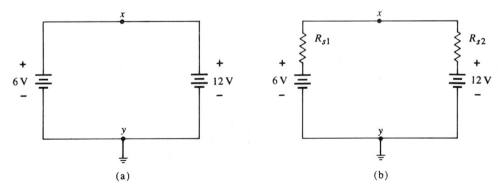

FIGURE 7–18   *Connecting voltage sources in parallel.*

A practical application for connecting sources in parallel occurs when one battery is used to charge another. The battery with the greater voltage is used to supply current to the battery with the lower voltage. As current flows to the low-voltage battery it recharges and its voltage gradually builds up until it equals the voltage of the high-voltage battery. At that point, there is no net voltage and the charging current stops flowing.

7.9 Comparison of Series and Parallel Circuits

Now that series circuits and parallel circuits have been thoroughly discussed, we can make a comparison between the two types. Table 7–1 summarizes the major characteristics of each.

TABLE 7–1

Simple Series Circuits	Simple Parallel Circuit
1. *Current* is the same through each series element.	1. *Voltage* is the same across each parallel element.
2. *Voltage* across each series element is less than input source voltage.	2. *Current* through each parallel element is less than the input source current.
3. Total series resistance is *larger* than any of the individual resistances.	3. Total parallel resistance is *less* than any of the individual resistances.
4. Sum of the *voltages* across each series element must equal the input source voltage.	4. Sum of the *currents* through each parallel element must equal the input source current.
5. Changing the resistance of any of the series elements changes the *voltages* across each element.	5. Changing the resistance of parallel elements *does not* change the voltages of other elements.
6. The input source voltage is distributed among the series elements in direct proportion to their *resistance* values.	6. The input source current is distributed among the parallel elements in *inverse* proportion to *resistance* values.
7. Open circuiting any series element causes all circuit current to stop flowing.	7. Open circuiting any parallel element does not affect the current through the other elements if the voltage is unchanged.
8. Adding resistances in series *increases* the total resistance.	8. Adding resistors in parallel *decreases* the total resistance.

Chapter Summary

1. A parallel connection is defined as two or more devices connected such that they have two points in common.

2. Devices connected in parallel always have the same identical voltage across their terminals.

3. Devices connected in parallel will in general have different currents flowing through them.

4. When connecting a circuit element in parallel with another it doesn't matter in which order the elements are connected.

Resistive Devices in Parallel Circuits   161

5. In a circuit with parallel resistors the sum of the currents through the parallel branches equals the current produced by the source.

6. The equivalent resistance of two or more parallel resistors has a smaller resistance than any of the parallel resistors.

7. The equivalent conductance of two or more parallel resistors is equal to the sum of the conductances of the individual resistors.

8. When a time-varying voltage is applied to a parallel combination of resistors the sum of the currents through the individual resistors equals the source current at every instant of time.

9. Increasing (decreasing) any one resistor of a parallel combination will increase (decrease) the equivalent resistance.

10. Adding more resistors in parallel with another resistor (or resistors) will decrease the equivalent resistance of the combination.

11. The equivalent resistance of n parallel resistors with the same resistance R is equal to R/n.

12. At any junction point in a circuit the sum of the currents flowing into the junction must equal the sum of the currents flowing away from the junction at any instant of time (KCL).

13. For two resistors in parallel the ratio of their currents is equal to the inverse ratio of their resistances.

14. In a parallel circuit a short circuit in one of the parallel elements will reduce R_{eq} to zero and all of the circuit current will flow through the short.

15. When an element of a parallel circuit becomes open-circuited the value of R_{eq} will increase.

Questions/Problems

Sections 7.1–7.3

7-1 For each of the circuits in Fig. 7-19 indicate which devices are connected in parallel.

7-2 Show how to connect a voltmeter to measure:
   a. the voltage across $R4$ in Fig. 7-19a
   b. the voltage across $R2$ in Fig. 7-19d
   c. the voltage across the parallel combination of $R3$ and $R5$ in Fig. 7-19d.

Section 7.4

7-3 Consider Fig. 7-20. Determine the equivalent resistance seen by the source by calculating I_1, I_2, I_3, and I_s.

7-4 Determine the equivalent conductance in Fig. 7-20 by using Eq. (7-3). Then calculate R_{eq} from G_{eq}.

162   Chapter Seven

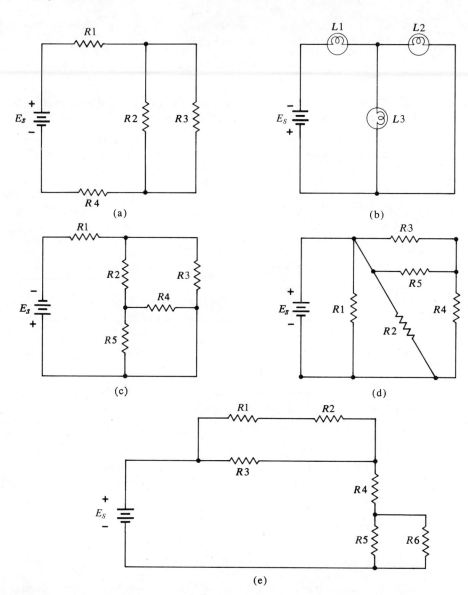

FIGURE 7–19

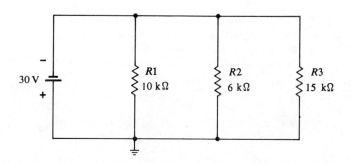

FIGURE 7–20

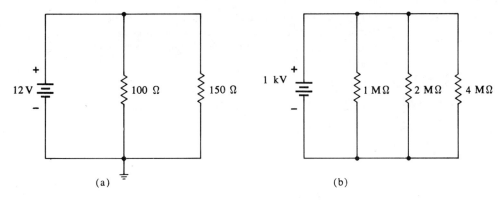

FIGURE 7-21

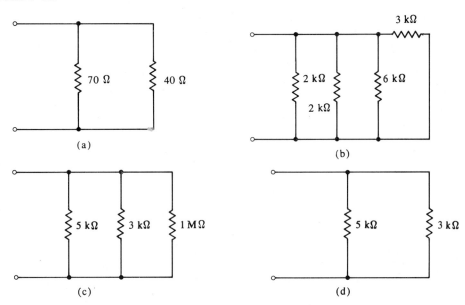

FIGURE 7-22

7-5 For the circuit of Fig. 7-21(a) determine I_s by first determining R_{eq} using the product-over-sum formula.

7-6 Find I_s in Fig. 7-21(b) by first finding R_{eq} using Eq. (7-5).

7-7 Calculate R_{eq} for each case in Fig. 7-22.

7-8 Compare the values of R_{eq} for Fig. 7-22(c) and (d). What conclusion can be made concerning the effect of the 1-MΩ resistor?

7-9 The ammeter in Fig. 7-23 indicates a current of 0.3 A. Determine the value of R_3.

7-10 Determine the values of I_1, I_2, and I_S for the circuit in Fig. 7-24.

7-11 Indicate which of the following statements are always true:
  a. Adding resistors in parallel decreases the current drawn from the source.
  b. Decreasing one of the resistors in Fig. 7-24 will increase the current through the other resistor.

c. Doubling each resistor in Fig. 7–24 will cause i_s to decrease by half.
d. Removing one resistor from a parallel combination will *increase* R_{eq}.
e. The equivalent resistance of a parallel combination is less than any of the parallel resistors.

7–12 Find the R_{eq} of two 32-kΩ resistors in parallel.

7–13 Find the R_{eq} of *five* 390-ohm resistors in parallel.

7–14 Calculate I_s in Fig. 7–25.

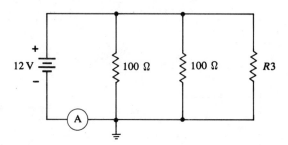

FIGURE 7–23

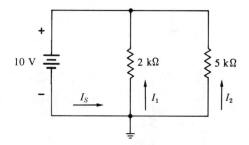

FIGURE 7–24

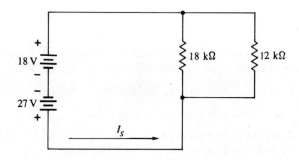

FIGURE 7–25

Section 7.5

7–15 Determine I_3 in Fig. 7–26.

7–16 Determine R_3 in Fig. 7–26.

7–17 Determine the value of I_x in each of the cases in Fig. 7–27.

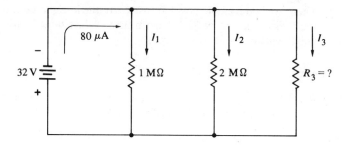

FIGURE 7–26

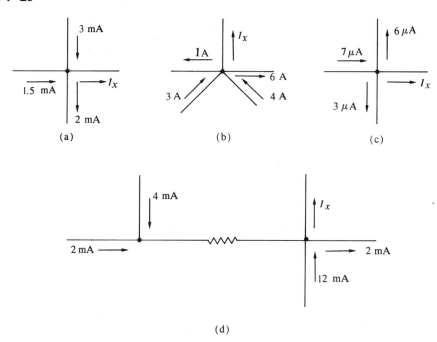

FIGURE 7–27

Section 7.6

7–18 Indicate which of the following statements refer to a *parallel* circuit and which refer to a *series* circuit:
  a. The current is the same through each element.
  b. The current through any one element is less than the source current.
  c. The equivalent resistance is greater than any one of the resistors.
  d. The waveforms of the circuit currents have the same shape as the input waveform.
  e. The voltage across each element is the same.

7–19 Determine the unknown currents in Fig. 7–28.

7–20 A constant current of 100 μA is being supplied to a parallel combination of two resistors. Choose values for the two resistors such that one resistor gets three times as much current as the other.

7–21 There are innumerable possible answers to the last question. If the voltage across the resistors has to be 75 V what resistor values must be used to satisfy the requirements of the last question?

Sections 7.7–7.8

7–22 Recalculate I_1, I_2, and I_3 in Fig. 7–28(a) if the 5-ohm resistor is open-circuited. Compare to the values obtained in problem 7–19. (Assume I_s stays at 14 A.)

7–23 Recalculate I_1, I_2, and I_3 in Fig. 7–28(a) if the 5-ohm resistor is short-circuited.

7–24 A good 12-V battery is being used to recharge a used battery whose terminal voltage has dropped to 11 V. The good battery has a source resistance of 0.05 ohm and the poor battery has a source resistance of 0.07 ohm. The batteries are connected in parallel (+ to +, − to −). Calculate the charging current.

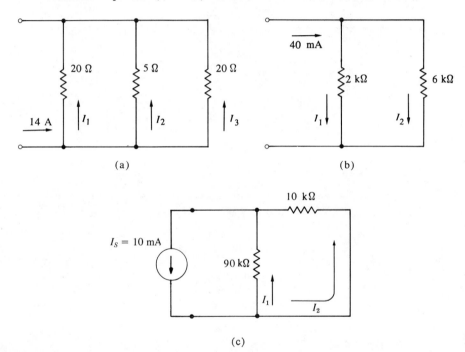

FIGURE 7–28

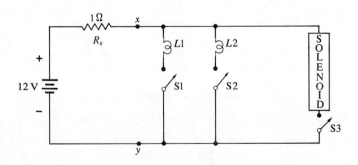

FIGURE 7–29

Section 7.9

7-25 Indicate which of the following statements pertain to series circuits and which pertain to parallel circuits:
  a. Voltage across each resistor is less than input voltage.
  b. Current through each resistor is less than input current.
  c. R_{eq} is larger than any individual resistor.
  d. R_{eq} is smaller than any individual resistor.
  e. Sum of the individual resistor voltages must equal input voltage.
  f. Increasing any of the resistors increases the total resistance.
  g. Doubling the input voltage doubles the current through each resistor.
  h. Open-circuiting one of the resistors causes the total source voltage to appear across the open circuit.
  i. Doubling each resistor doubles the total resistance.
  j. The currents through the various resistors are in proportion to their *conductance* values.

7-26 Consider the circuit drawn in Fig. 7-29. A 12-V storage battery is used to power two lamps, $L1$ and $L2$, and a solenoid which controls a valve. The two lamps are rated at 1 A @ 12 V and the solenoid is rated at 2 A @ 12 V. The storage battery has a source resistance of 1 Ω as shown. The solenoid will "kick in" whenever its voltage exceeds 9.5 V. The three switches, $S1$, $S2$, $S3$, are used to switch the various elements into the circuit.

A technician tests the circuit and observes that the solenoid will kick in if neither lamp switch is closed or if only one lamp switch is closed, but will fail to energize if *both* lamps are switched on.

Try to explain why the circuit does not work properly. Discuss how it may be modified. (Hint: calculate the resistance of each element.)

CHAPTER 8

Resistive Devices in Series-Parallel Circuits

8.1 Introduction

In the last two chapters considerable time was devoted to simple series and simple parallel circuits. In practice, most electric circuits are more complex and consist of a combination of these two simple connections. Circuits which contain both series and parallel connections are called *series-parallel circuits*. In this chapter we will use many of the principles developed in the previous two chapters to help us analyze various types of series-parallel circuits. Consideration will be focused only on those types of series-parallel circuits that occur frequently in practice rather than wasting time on circuits which are presented purely for mathematical exercise.

8.2 Simple Series-Parallel Circuits

Figure 8-1(a) shows an example of a simple series-parallel circuit. In this circuit we can see that $R2$ and $R3$ are connected in parallel since they have points x and y in common. The resistor $R1$ is not in series with $R2$ or $R3$ because the current i_1 flows through $R1$ but splits up between $R2$ and $R3$. Rather, we can say that $R1$ is in series with *the parallel combination of $R2$ and $R3$*. This can be more readily seen if we replace the $R2$-$R3$ combination by its equivalent resistance, as shown in Fig. 8-1(b).

As far as the source is concerned, both circuits in Fig. 8-1 are equivalent and the current I_S drawn from the source is the same in both circuits. As can be seen in Fig. 8-1(b), the original circuit has been simplified to a series circuit where $R1$ is in series with R_{eq}. This series circuit can be analyzed using the methods discussed in chap. 6. The following example illustrates.

$R1$ is in series with R_{eq}. This series circuit can be analyzed using the methods discussed in chap. 6. The following example illustrates.

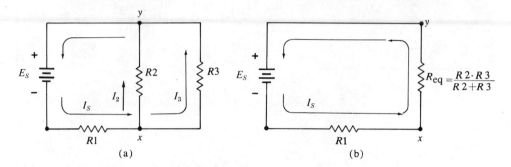

FIGURE 8–1 (a) *Simple series-parallel circuit.* (b) *Simplified version.*

EXAMPLE 8.1 For the circuit in Fig. 8–2(a) determine the currents and voltages for each resistor.

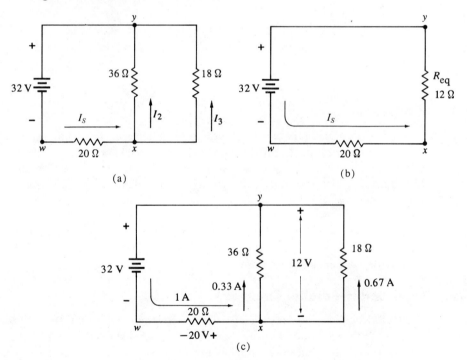

FIGURE 8–2 (a) *Circuit of Ex. 8.1.* (b) *Simplified version.* (c) *Original circuit with I and V values.*

Solution: The 36-ohm and 18-ohm resistors are connected in parallel between points x and y. Thus, we can replace them by their equivalent resistance.

$$R_{eq} = 36 \, \Omega \parallel 18 \, \Omega$$
$$= \frac{36 \times 18}{36 + 18} = 12 \, \Omega$$

(Note: the symbol "∥" stands for *the parallel combination of*.)

In Fig. 8-2(b) the parallel resistors are replaced by R_{eq} = 12 ohms. As can be seen, the revised circuit is now a series circuit. The total series resistance is 20 ohms + 12 ohms = 32 ohms. Thus, the current I_S produced by the source will be

$$I_S = \frac{32 \text{ V}}{32 \text{ }\Omega} = 1 \text{ A}$$

Since $I_S = 1$ A the voltage across the 20-ohm resistor can easily be found as

$$V_{xw} = 1 \text{ A} \times 20 \text{ ohms} = 20 \text{ V}$$

The voltage across R_{eq} is also found as

$$V_{yx} = 1 \text{ A} \times 12 \text{ ohms} = 12 \text{ V}$$

Notice that these two voltages add up to 32 V, which equals the source voltage and satisfies KVL.

We can now return to the original circuit to determine I_2 and I_3. The voltage across R_{eq} was calculated as 12 V in the revised circuit; this means that there are 12 V across the parallel combination in the original circuit [see part (c) of the figure]. In other words, both the 36-ohm and 18-ohm resistors have 12 V across their terminals. Thus, we can calculate I_2 and I_3 as

$$I_2 = \frac{12 \text{ V}}{36 \text{ }\Omega} = 0.33 \text{ A} \quad \text{and} \quad I_3 = \frac{12 \text{ V}}{18 \text{ }\Omega} = 0.67 \text{ A}$$

Note that $I_2 + I_3 = 1$ A $= I_S$, which satisfies KCL.

In the series-parallel circuit of Fig. 8-2 we essentially have the source voltage being divided between the series resistance (20 ohms) and the parallel combination. This situation occurs quite often in voltage divider circuits where the parallel combination is present as the result of a load being connected across the output of the voltage divider. This situation will be discussed later.

The following example is intended to help us see how the various elements in the circuit of Fig. 8-2 affect the overall circuit current and voltage values. The example will be performed without making any calculations but by using the principles developed previously.

EXAMPLE 8.2 Indicate the effect each of the following changes will have on the circuit currents and voltages in Fig. 8-2: (a) a decrease in the 18-ohm resistor; and (b) a decrease in the 20-ohm resistor.

Solution: (a) A decrease in the 18-ohm resistor will reduce R_{eq}. Therefore, the total resistance seen by the source will be less and the current I_S will correspondingly increase.

We might now suspect that both I_2 and I_3 should increase since I_S has been made greater. However, this is not the case. With a larger I_S the voltage across the 20-ohm series resistor must be larger. As a result, there has to be less voltage across the parallel combination if KVL is to be satisfied. For this reason the current I_2 will be smaller since there is less voltage across the

172   Chapter Eight

36-ohm resistor. On the other hand, the current I_3 has to be larger in order to satisfy KCL: $I_2 + I_3 = I_S$.

In summary, the circuit values will change as follows, where the arrows indicate increase (↑), decrease (↓), or remain the same (↔).

$$I_S \uparrow \; ; I_2 \downarrow \; ; I_3 \uparrow$$
$$V_{xw} \uparrow \; ; V_{yx} \downarrow$$

(b) A decrease in the 20-ohm resistor will reduce the total resistance seen by the source. This will cause an increase in I_S. The equivalent resistance of the parallel combination is unchanged. Since the current I_S passes through R_{eq}, then it follows that a larger I_S means that there is a larger voltage across the parallel combination; that is, V_{yx} must be greater. In order to satisfy KVL, then, the voltage across the series resistor, V_{xw}, must be less.

This last conclusion at first may not appear to be consistent with the fact that I_S is larger. However, realize that the 20-ohm resistor has been decreased in value so that a smaller value of voltage V_{xw} can result in a larger current through this resistor.

The larger value of V_{xy} across the parallel resistors indicates an increase in both I_2 and I_3. The results are summarized as follows:

$$I_1 \uparrow \; ; I_2 \uparrow \; ; I_3 \uparrow$$
$$V_{xw} \downarrow \; ; V_{yx} \uparrow$$

Another type of series-parallel circuit is shown in Fig. 8–3(a). It can be seen that R2 and R3 are connected in series. Resistor R1 is *not* in parallel with R2 or R3 because R1 does not have two points in common with either R2 or R3. Rather, we can say that R1 is in parallel with the *series combination of R2 and R3*. This can be readily seen if we replace the R2-R3 combination by its equivalent resistance as shown in Fig. 8–3(b).

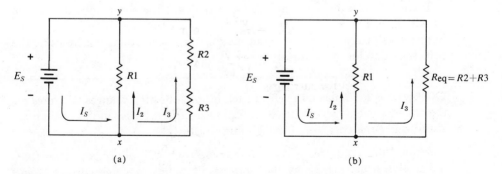

FIGURE 8–3 (a) *Another series-parallel circuit.* (b) *Simplified version.*

In the revised circuit it is apparent that the total source voltage appears across R1 and also across R_{eq}. The current through each branch will depend on the resistance in each branch. The total current produced by the source I_S will equal the sum of the branch currents.

EXAMPLE 8.3 Determine the currents and voltages in the circuit of Fig. 8–4(a). Also determine the total circuit resistance.

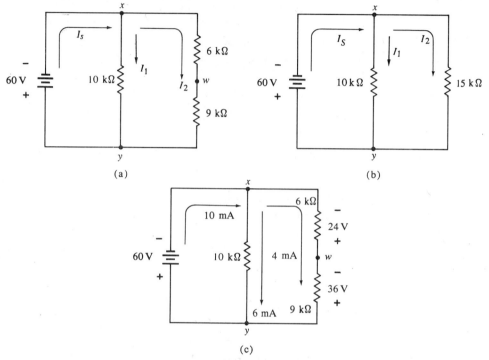

FIGURE 8–4

Solution: The 6-kΩ and 9-kΩ resistors are connected in series between points x and y. Thus, we can replace them by their equivalent resistance of 15 kΩ as in part (b) of the figure. Since there are 60 V across both branches, the currents I_1 and I_2 can be calculated as

$$I_1 = \frac{60 \text{ V}}{10 \text{ k}\Omega} = 6 \text{ mA}$$

$$I_2 = \frac{60 \text{ V}}{15 \text{ k}\Omega} = 4 \text{ mA}$$

Thus, the total source current is

$$I_s = I_1 + I_2 = 10 \text{ mA}$$

In order to find the voltages across the 6-kΩ and 9-kΩ resistors we return to the original circuit in Fig. 8–4(c). Since $I_2 = 4$ mA flows through both resistors, we have

$$V_{wx} = 4 \text{ mA} \times 6 \text{ k}\Omega = 24 \text{ V} \quad \text{and} \quad V_{yw} = 4 \text{ mA} \times 9 \text{ k}\Omega = 36 \text{ V}$$

Notice that the voltages across these two series resistors add up to 60 V. This must be so in order to satisfy KVL around the *x-w-y-x* path.

It was *not* necessary to determine the total resistance seen by the source in order to solve for the currents and voltages in this circuit. The total circuit resistance can be found by using the source voltage 60 V and the source current 10 mA so that $R_T = 60\text{ V}/10\text{ mA} = 6\text{ k}\Omega$. Alternatively, the same result is obtained by finding the equivalent resistance of 10 kΩ and 15 kΩ in parallel.

$$R_T = 10\text{ k}\Omega \parallel 15\text{ k}\Omega$$
$$= \frac{10 \times 15}{10 + 15} = 6\text{ k}\Omega$$

EXAMPLE 8.4 Four lamps are to be wired to a 240-V source. The lamps are all rated at 120 V/1 A for normal brilliance. How can the lamps be wired to the 240-V source so that each lamp is operating at normal brilliance?

Solution: If the four lamps were connected in parallel across the source, each would have 240 V. This is twice the lamps' rating and would probably result in burned-out filaments.

If the lamps were connected in series across the 240-V source, each lamp would only get 60 V (one-fourth of the source voltage). With such a low voltage the lamps would be operating at much less than normal brilliance.

By connecting *two* lamps in series across the 240-V source each lamp will be provided with 120 V, its normal operating voltage. Therefore, the four bulbs can be connected as shown in Fig. 8–5. The two parallel branches each contain two lamps in series. Each parallel branch has 240 V across it which divides equally between the two series lamps.

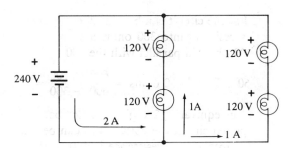

FIGURE 8–5

The total current supplied by the source will be 2 A, since each parallel branch draws 1 A of current (which is the rated lamp current at 120 V). Thus, the four lamps present an equivalent resistance of 240 V/2 A = 120 ohms to the source.

There are many variations of the simple series-parallel circuits discussed so far. The following example illustrates a more complex series-parallel circuit.

EXAMPLE 8.5 Determine the voltage V_{xy} in Fig. 8–6(a).

Solution: V_{xy} is across the 200-ohm resistor. If V_{wy} can be determined, then V_{xy} can be easily found by using the voltage divider rule or by finding I_3.

Resistive Devices in Series-Parallel Circuits   175

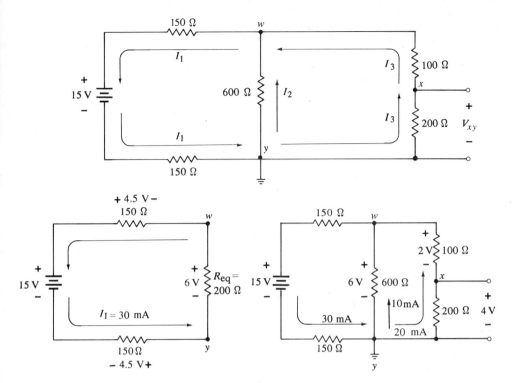

FIGURE 8-6

To help us find V_{wy} the circuit can be simplified by combining the 100-ohm and 200-ohm series resistors into 300 ohms and then finding the equivalent resistance of this 300 ohms in parallel with the 600-ohm resistor. The result is

$$R_{eq} = 600 \text{ ohms} \| 300 \text{ ohms} = \frac{600 \times 300}{600 + 300} = 200 \text{ ohms}$$

Thus, we can place the equivalent 200-ohm resistor between w and y as in Fig. 8-6(b). In this simplified circuit the voltage V_{wy} can be calculated by first calculating I_1. Since the total circuit resistance is $150 + 150 + 200 = 500$ ohms, the current I_1 is given by

$$I_1 = \frac{15 \text{ V}}{500 \text{ }\Omega} = 30 \text{ mA}$$

Therefore, V_{wy} is calculated as

$$V_{wy} = 30 \text{ mA} \times 200 \text{ ohms} = 6 \text{ V}$$

V_{wy} could also have been found using the voltage divider rule

$$V_{wy} = \frac{200 \text{ }\Omega}{500 \text{ }\Omega} \times 15 \text{ V}$$
$$= 0.4 \times 15 \text{ V} = 6 \text{ V}$$

We can now return to the original circuit [part (c) of the figure] to find V_{xy}. The voltage $V_{wy} = 6$ V is across the two parallel branches. Thus, there are 6 V across the branch with the series combination of 100 ohms and 200 ohms. The current through this branch is therefore $I_3 = 6\,\text{V}/300$ ohms $= 20\,\text{mA}$. V_{xy} can now be calculated as

$$V_{xy} = 20\,\text{mA} \times 200\,\text{ohms} = 4\,\text{V}$$

V_{xy} could also be found using the voltage divider rule

$$V_{xy} = \frac{200\,\Omega}{300\,\Omega} \times 6\,\text{V} = 4\,\text{V}$$

Notice that in using the voltage divider formula the source voltage of 15 V is *not* used. Instead, the voltage used was 6 V because 6 V is the total voltage across the series combination. The source voltage of 15 V does *not* appear across the series combination of the 100-ohm and 200-ohm resistors.

8.3 Equivalent Resistance of Series-Parallel Circuits

The equivalent resistance of a series-parallel combination of resistors can be found by successive applications of the rules for combining series resistors and parallel resistors. This technique was used in some of the examples of the last section. The technique is formally described as a step-by-step process by the flow diagram in Fig. 8–7.

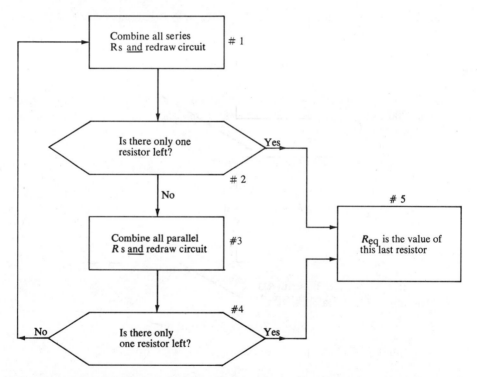

FIGURE 8–7 *Flow diagram of the step-by-step procedure for finding R_{eq} of a series-parallel network.*

This flow diagram will be explained by way of an example. We will use it to find the equivalent resistance between x and y for the series-parallel network of Fig. 8–8(a). The first step (box #1 in Fig. 8–7) is to combine all series resistors. Looking at the circuit in Fig. 8–8(a) we can combine $R5$ and $R6$ into one re-

Resistive Devices in Series-Parallel Circuits 177

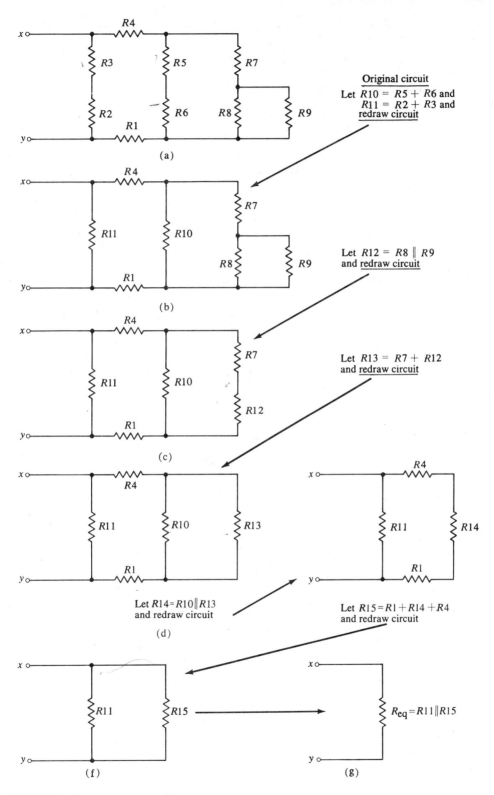

FIGURE 8-8

sistor, $R10 = R5 + R6$. Similarly, $R3$ and $R2$ can be combined into $R11 = R3 + R2$. Then, the circuit is redrawn in Fig. 8-8(b). The next step (box #2) is to ask the question, "Is there only one resistor left?" The answer at this point is obviously "no." Thus, we proceed to the next step (box #3), which indicates that all parallel resistors should be combined.

Looking at Fig. 8-8(b) we can now combine $R8$ and $R9$ into one resistor $R12 = R8 \parallel R9$ and redraw the circuit as in Fig. 8-8(c). The next step (box #4) again asks if there is only one resistor left. Since the answer is still "no," the flow diagram directs us back to the original first step (box #1) from which point we go through the same process again.

Eventually as we go through the steps of combining series resistors and then parallel resistors we will reach a point where only one resistor is left. This resistor will then be the required R_{eq} for the circuit. The remaining steps in the process for our example network are given in Fig. 8-8(d)-(g). The student can verify that the correct procedure was followed.

8.4 Linearity Principle in Series-Parallel Circuits

The series-parallel circuits discussed in the preceding sections contained one input source and only linear resistors. This type of circuit has two significant properties: (1) The voltage across any resistor in the circuit is always equal to some fraction of the input voltage (unless the resistor is in parallel with the source); and (2) the currents in the circuit are always some fraction of the total source input current. Stated in another way, *the currents and voltages can never be greater than the source current and voltage.*

A third even more important property is due to the *linearity principle* discussed in chap. 6. The linearity principle states that

In a circuit containing a single voltage source and only linear devices, all the voltages and currents in the circuit are proportional to the input voltage.

As a result of this property, any change in the input voltage results in a proportional change in all the circuit currents and voltages. For example, if the input voltage increases by 10 percent, then every current and voltage in the circuit will increase by 10 percent.

The linearity principle is useful in circuits where the source voltage is not constant but varies with time or in situations where circuit currents and voltages have to be calculated for various values of source voltage. The following examples illustrate its application.

EXAMPLE 8.6 Determine values of I_S and V_{xy} in Fig. 8-9 for $E_S = 9$ V and $E_S = 12$ V.

Solution: The circuit will be first analyzed for the condition when $E_S = 9$ V. To find the total source current we must determine the total resistance seen by the source. We begin by combining the 9-kΩ and 6-kΩ resistors in series to give 15 kΩ. This 15 kΩ is in parallel with the 10-kΩ resistor; the parallel combination of 15 kΩ and 10 kΩ is

Resistive Devices in Series-Parallel Circuits 179

$$R_{eq} = 15 \text{ k}\Omega \parallel 10 \text{ k}\Omega$$
$$= \frac{15 \times 10}{15 + 10} = 6 \text{ k}\Omega$$

Thus, we can place a 6-kΩ resistor between points w and y as shown in Fig. 8–9(b). Using this equivalent circuit we can calculate the source current when $E_S = 9$ V.

$$I_S = \frac{9 \text{ V}}{9 \text{ k}\Omega} = 1 \text{ mA}$$

The voltage V_{wy} is therefore

$$V_{wy} = 1 \text{ mA} \times 6 \text{ k}\Omega = 6 \text{ V}$$

Returning to the original circuit [part (c) of the figure] we can utilize the fact that $V_{wy} = 6$ V. This 6 V is present across the 9-kΩ–6-kΩ series combination. The desired voltage V_{xy} across the 6-kΩ resistor can be found using the voltage divider rule

$$V_{xy} = \frac{6 \text{ k}\Omega}{15 \text{ k}\Omega} \times 6 \text{ V} = \mathbf{2.4 \text{ V}}$$

Thus, we have determined that when the input is at 9 V the source current I_S is 1 mA and the output V_{xy} is 2.4 V. We can now repeat the same analysis for when E_S is at 12 V. However, it is much simpler to take advantage of the

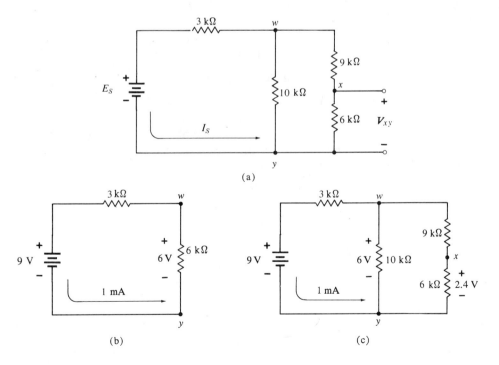

FIGURE 8–9

linearity principle. An increase in E_S from 9 V to 12 V is an increase of 3 V which is 3/9 = 1/3 of its original value. Therefore, each of the circuit currents and voltages will increase by 1/3 of the value they had when E_S = 9 V. Their new values at E_S = 12 V are calculated as follows:

$$I_S = 1 \text{ mA} + \frac{1}{3} \times 1 \text{ mA} = \mathbf{1.33 \text{ mA}}$$

$$V_{xy} = 2.4 \text{ V} + \frac{1}{3} \times 2.4 \text{ V} = \mathbf{3.2 \text{ V}}$$

EXAMPLE 8.7 What value of E_S will make V_{xy} = 32 V?

Solution: The ratio of V_{xy} to E_S can be calculated from the results of the previous example for either E_S = 9 V or E_S = 12 V.

$$\frac{V_{xy}}{E_S} = \frac{2.4 \text{ V}}{9 \text{ V}} = \frac{3.2 \text{ V}}{12 \text{ V}} = 0.267$$

This ratio (often called the *attenuation ratio*) is fixed by the circuit resistance values and will be the same for any value of E_S. Thus, $V_{xy} = 0.267 E_S$ can be used to find the value of E_S needed to produce V_{xy} = 32 V.

$$V_{xy} = 0.267 E_S = 32 \text{ V}$$
$$\therefore E_S = \frac{32 \text{ V}}{0.267} = \mathbf{120 \text{ V}}$$

8.5 Troubleshooting Series-Parallel Circuits

In every circuit there are things that can go wrong, thus causing the circuit to function improperly. In circuits containing resistors a common mishap is the "burning out" of a resistor due to excess current flow. As we shall see later, every resistor has a limit on how much current it can safely pass before it becomes overheated and burns out. If a resistor burns out, its resistance can become almost infinite and it will essentially act like an open circuit.

Occasionally a resistor will be operating with a current through it that is below the burn-out level but is still high enough to cause the resistor to overheat. When overheated the resistance value of the resistor will change from its normal value. It could either increase or decrease depending on the resistor material.

In circuits containing semiconductor devices such as diodes and transistors, both short circuits and open circuits can occur if semiconductors are operated above their rated current or voltage levels.

Thus, it is important for a good technician to be able to recognize the effects of some of these possible component failures on the overall operation of a circuit. In general, when one component in a circuit is faulty, it can affect the currents and voltages throughout the entire circuit. The following examples will provide an introduction to the kind of reasoning used when troubleshooting a circuit. Notice in these examples that *only voltages* will be measured in the circuits discussed. In practice, the troubleshooter finds it much easier and quicker to measure the voltages in a circuit than to check the currents. This is because current measurements require breaking the circuit open to insert a current meter, which is especially prohibitive when checking printed circuit boards or soldered-in circuits.

EXAMPLE 8.8 Consider the circuit in Fig. 8–10. The voltage values shown are the nominal values which should be present when the circuit is operating properly. A technician measures each of the voltages with a voltmeter and obtains the following readings:

$V_{wx} = 30$ V across $R1$; $V_{xy} = 0$ V across $R3$; $V_{wz} = 60$ V source
$V_{xz} = 30$ V across $R2$; $V_{yz} = 30$ V across $R4$

Determine a probable cause for the circuit's not operating properly.

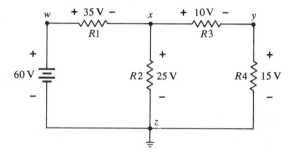

FIGURE 8–10

Solution: The reading which stands out most significantly is the zero voltage across $R3$. There are two possible causes for this reading: $R3$ could be shorted-out ($R3 = 0$), in which case $V = IR$ would be zero; or there is no current through $R3$ because $R4$ is open-circuited so that $V = IR$ would be zero.

Let's look at the first possibility. If $R3$ were shorted-out, then the series combination $R3 + R4$ would have less resistance. Thus, the equivalent resistance of the parallel combination between points x and z would be smaller than normal. This would mean that the voltage division ratio between $R1$ and the parallel combination would be changed, with $R1$ getting a *greater* portion of the supply voltage, and the parallel combination getting *less*. However, according to the voltage readings the voltage V_{wx} across $R1$ is *lower* (30 V) than it is supposed to be (35 V) and, correspondingly, the voltage v_{xz} across the parallel combination is greater (30 V) than it is supposed to be (25 V). Thus, we can rule out a shorted $R3$ as the cause of the circuit malfunctions.

Let's check out the second possibility. If $R4$ were open-circuited, then the series combination $R3 + R4$ would have an infinite resistance. Thus, the parallel combination between x and z would have a larger resistance than normal. This would mean that $R1$ would get a *smaller* portion of the source voltage and the parallel combination would get a *greater* portion. As the readings show, this is actually the case. Therefore, an open-circuited $R4$ is probably the cause of the circuit's not operating properly.

EXAMPLE 8.9 Refer again to the same circuit in Fig. 8–10. A technician observes the following voltage readings:

$V_{wx} = 10$ V across $R1$; $V_{xy} = 20$ V across $R3$; $V_{wz} = 60$ V source
$V_{xz} = 50$ V across $R2$; $V_{yz} = 30$ V across $R4$

Determine a probable cause for these unexpected readings.

Solution: The first thing to notice about these measured readings is that the $R3$ and $R4$ voltages are in the correct proportion (see Fig. 8–10) although the values are too high. This should indicate that the $R3$ and $R4$ resistance values are probably correct, so we can rule out these resistors as possible faulty components.

This leaves $R1$ and $R2$ as possible causes of the trouble. The voltage across $R1$ was measured lower (10 V) than its expected value (35 V) and the voltage across $R2$ was measured higher (50 V) than its expected value (25 V). Thus, there are two possibilities: $R1$'s resistance is too *low* or $R2$'s resistance is too *high*. Either of these possibilities could cause the observed voltage readings. There is no easy way to determine which of these two resistors is the bad one without removing them from the circuit and measuring their resistance values. An educated guess would be that $R2$'s resistance is too high because *faulty* resistors usually show up with a greater resistance than normal, especially if the change in resistance is due to overheating because of too much current. Occasionally, a low-valued resistor shows up but such cases are relatively uncommon.

8.6 Bridge Circuits

A type of series-parallel circuit that finds extensive use in certain measurement circuits and in control systems is the *bridge circuit*. In its simplest form a bridge circuit is a circuit with two parallel branches each of which has two series resistors as shown in Fig. 8–11.

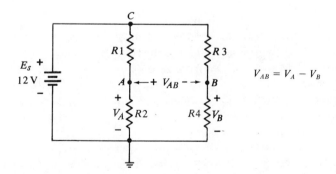

FIGURE 8–11 *Bridge circuit.*

Each of the parallel branches has a voltage across it equal to the source voltage. For each branch this voltage divides between the two series resistors. The output of the circuit is usually taken between points A and B. That is,

Resistive Devices in Series-Parallel Circuits

V_{AB} is the output voltage. Depending on the resistance values in each parallel branch, the output can be positive, negative, or zero. Recall from chap. 6 that $V_{AB} = V_A - V_B$. Thus, V_{AB} will be *positive* when $V_A > V_B$, or *negative* when $V_A < V_B$.

The Balanced Condition If $V_A = V_B$, then points A and B will be at the same potential relative to ground. With no difference in potential between A and B, V_{AB} must be *zero*. The bridge circuit is said to be *balanced* when $V_{AB} = 0$.

In order for the bridge to be balanced the resistor values have to be chosen so that $V_A = V_B$. If $R2$ and $R1$ are chosen in the same proportion as $R4$ and $R3$, then the voltage division ratio in each series combination will be the same so that V_A will equal V_B. In other words,

$$\frac{R2}{R1} = \frac{R4}{R3}$$

for the balanced condition. For example, if $R2 = 10\text{ k}\Omega$, $R1 = 5\text{ k}\Omega$, $R4 = 2\text{ k}\Omega$, and $R3 = 1\text{ k}\Omega$, the above equality is satisfied. With these resistor values we can calculate V_A and V_B.

$$V_A = \frac{R2}{R1 + R2} \times E_s = \frac{10}{5 + 10} \times 12 \text{ V}$$
$$= 8 \text{ V}$$

$$V_B = \frac{R4}{R3 + R4} \times E_s = \frac{2}{1 + 2} \times 12 \text{ V}$$
$$= 8 \text{ V}$$

As predicted, $V_A = V_B$ so that $V_{AB} = 0$.

If the resistor ratios in each of the parallel branches are not equal, then $V_A \neq V_B$ and V_{AB} will not be zero. For example, if $R2$ is changed from 10 kΩ to 15 kΩ the value of V_A becomes

$$V_A = \frac{15}{5 + 15} \times 12 \text{ V}$$
$$= 9 \text{ V}$$

Since V_B is still 8 V, then point A must be 1 V positive relative to B. That is, $V_{AB} = +1$ V. Changing *any* of the resistor values from the balanced condition will cause V_{AB} to move from zero.

8.7 Loaded Voltage Dividers

In chap. 6 we saw how series circuits act as voltage dividers. The output of a voltage divider is taken across one of the series resistors, and this output voltage is a certain fraction of the input. When a voltage divider is designed, the resistor values are chosen to provide a certain voltage divider ratio. For example, in Fig. 8-12(a) the resistor values produce a voltage divider ratio of 0.25. That is,

$$V_{\text{out}} = \frac{1 \text{ k}\Omega}{1 \text{ k}\Omega + 3 \text{ k}\Omega} \times E_{\text{in}}$$
$$V_{\text{out}} = 0.25 \, E_{\text{in}}$$

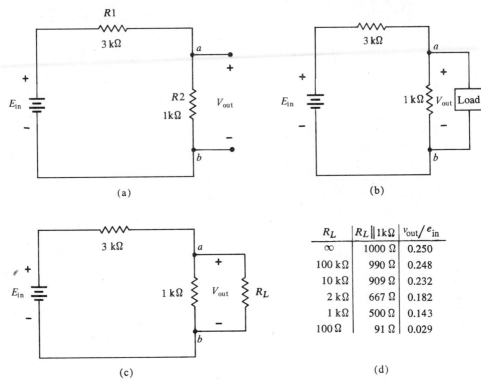

FIGURE 8-12 *Effect of a load on the output of a voltage divider.*

so that this voltage divider *cuts down* (attenuates) the input voltage by a factor of 0.25.

Usually, when a voltage divider is used it is desired that the voltage divider ratio remain relatively constant. This would present no problem if the output voltage were not used for any purpose. However, V_{out} is normally used to supply voltage to some other circuit or load as illustrated in Fig. 8–12(b). In such cases the load is in parallel with the output resistor of the voltage divider. Let's examine the effect of the load on the output voltage V_{out}.

In Fig. 8–12(c) the load is represented as a load resistor R_L. The value of R_L is the equivalent resistance of the load. The load might be a circuit element such as a lamp or motor, or it could be the input resistance of an amplifier. The effect that R_L will have on the output voltage depends on the value of R_L. In Fig. 8–12(c), R_L is in parallel with the 1-kΩ resistor. Thus, the actual equivalent output resistance between points a and b will be $R_L \parallel 1$ kΩ. The table in Fig. 8–12(d) gives the values of this equivalent output resistance for various values of R_L.

As the table shows, the equivalent output resistance between points a and b can vary greatly depending on the value of R_L. What this means is that the voltage divider ratio will also vary. For example, when $R_L = 100$ kΩ the resistance between a and b is 990 ohms. This results in

$$V_{out} = \frac{990 \text{ ohms}}{3990 \text{ ohms}} \times E_{in}$$
$$= 0.248 \, E_{in}$$

Similarly, when $R_L = 10 \text{ k}\Omega$,

$$V_{out} = \frac{910 \, \Omega}{3910 \, \Omega} \times E_{in}$$
$$= 0.232 \, E_{in}$$

The voltage divider ratio values for the various values of R_L are also tabulated in Fig. 8–12(d). As can be seen, the **voltage divider ratio decreases as R_L decreases.**

For larger values of R_L the effect is not as great. When R_L is much larger than the voltage divider output resistor (1 kΩ, in our example), it has very little effect on the output resistance and therefore on the voltage divider ratio. Thus, for a given voltage divider there is a limit to the minimum size of R_L which can be used without causing the voltage divider ratio to decrease significantly. A useful rule-of-thumb is to insure that R_L is always at least 10 times greater than the voltage divider output resistor. The following example illustrates.

EXAMPLE 8.10 A voltage divider is to be designed with a voltage divider ratio of 0.4. The output of the divider has to supply voltage to loads ranging from 5 kΩ to 25 kΩ. Choose resistor values for the divider.

Solution: The circuit is shown in Fig. 8–13. Since the minimum R_L which is to be used is 5 kΩ, R2 should be made at least 10 times smaller than 5 kΩ. Let's use R2 = 500 ohms.

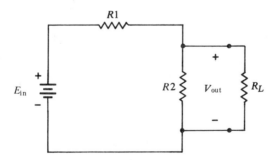

FIGURE 8–13

The value of R1 can now be chosen to provide a 0.4 ratio. Thus,

$$\frac{R2}{R1 + R2} = 0.4$$

$$\frac{500}{R1 + 500} = 0.4$$

Solving this for R1 yields the result R1 = **750 ohms**.

We could have chosen an even smaller value for R2 which would have resulted in a smaller value for R1. The smaller R2 would insure that R_L would have even less effect on v_{out}. However, care must be taken that R2 and R1 do not draw excessive current from the source.

Sometimes the loading on a voltage divider is specified in terms of the amount of current it will draw. The load on the voltage divider might be a motor or lamp or some other device whose I-V characteristic is nonlinear so that its resistance is not constant. For example, suppose we needed a voltage divider which would take a constant 16-V source and produce a 6-V output. Further suppose that we are given that the load on the divider will draw a *maximum* current of 10 mA. At 6 V and 10 mA the load has an effective resistance of 6 V/10 mA = 600 ohms. Using this value for R_L the values for R1 and R2 can be chosen to provide the desired voltage divider ratio as in Ex. 8.10.

Note that the *maximum* load current must be considered because this corresponds to the *minimum* load resistance, and therefore the greatest effect on V_{out}.

Chapter Summary

1. In series-parallel circuits containing linear resistors the voltage across any resistor is always less than the input voltage (unless the resistor is connected in parallel with the input source).
2. The currents in these series-parallel circuits are always less than the total source input current.
3. In a circuit containing a single voltage source and linear elements, all the voltages and currents are directly proportional to the input voltage.
4. If a voltage is represented with only one subscript (such as V_x) then the reference point is assumed to be ground. Thus, V_x is the voltage at point x relative to ground.
5. A bridge circuit is said to be balanced when its output voltage is zero.
6. Bridge balance occurs when the resistor ratios in each branch are the same.
7. If a load resistor R_L is connected to the output of a voltage divider, the voltage divider ratio will decrease as R_L decreases.
8. The equivalent resistance of a series-parallel circuit can be found using the method illustrated in Figs. 8–7 and 8–8.

Questions/Problems

Section 8.2

8-1 For each circuit in Fig. 8–14 determine which resistors are in series and which resistors are in parallel.

Resistive Devices in Series-Parallel Circuits   187

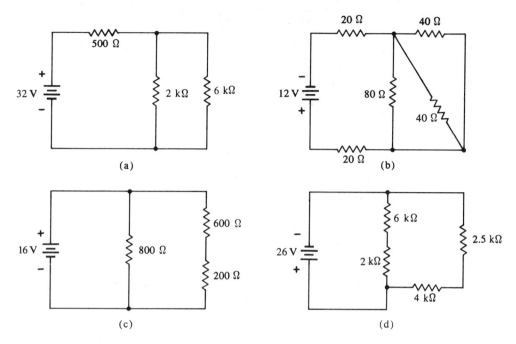

(a)   (b)   (c)   (d)

FIGURE 8-14

8-2 Determine the currents and voltages for each resistor in Fig. 8-14(a). *Use KVL and KCL to check your results.*

8-3 Repeat problem 8-2 for Fig. 8-14(b).

8-4 Repeat problem 8-2 for Fig. 8-14(c).

8-5 Repeat problem 8-2 for Fig. 8-14(d).

8-6 Determine the value of V_{xy} in Fig. 8-15.

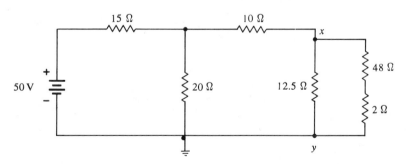

FIGURE 8-15

8-7 Indicate the effect each of the following changes will have on the value of V_{xy} in Fig. 8-15:
   a. Decrease the 48-ohm resistor.
   b. Double the input voltage.
   c. Decrease the 15-ohm resistor.
   d. Increase the 10-ohm resistor.
   e. Open-circuit the 12.5-ohm resistor.

188   Chapter Eight

8-8   Five identical 60 V/1 A lamps and six identical 30 V/1 A lamps are to be connected to a 120-V source so that each lamp is operating at its normal voltage. Show how to connect the lamps to produce the desired results.

Section 8.3

8-9   Find the equivalent resistance seen by the source in each of the networks in Fig. 8–16. (Hint: it may help to redraw the circuit in some cases.)

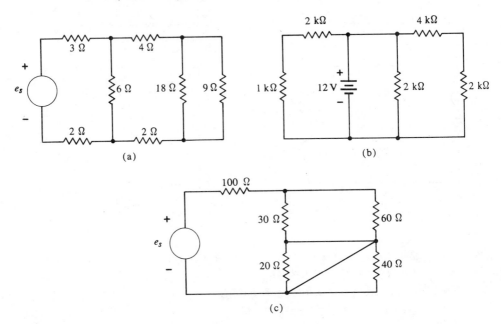

FIGURE 8–16

Section 8.4

8-10   Use the results of problem 8–2 and the linearity principle to determine all the currents and voltages in Fig. 8–14(a) for a battery voltage of (a) 48 V; and (b) −16 V.

8-11   Use the results of problem 8–6 and find what value of E_S will produce $V_{xy} = 24$ V.

8-12   If the input voltage to a linear circuit is doubled, what has to be done to the circuit resistance values in order to have the circuit currents remain at the same values? What will happen to the circuit voltages in this case?

Section 8.5

8-13   In illustrative Ex. 8.8 why couldn't the cause of the circuit malfunction be an open-circuited R3? Why couldn't it be an open circuit in the path connecting R3 and R4?

8-14   Consider the circuit in Fig. 8–17. The voltage values shown are the correct ones if the circuit is operating properly. A technician measures the circuit voltages and obtains the following readings:

$V_{wx} = 3$ V $\qquad V_{xy} = 4$ V
$V_{xz} = 8$ V $\qquad V_{yz} = 4$ V
$V_{zu} = 19$ V

Determine, if possible, the probable cause of the incorrect circuit operation.

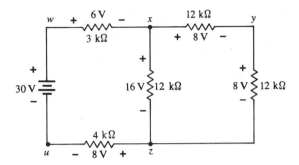

FIGURE 8–17

Section 8.6

8–15 Consider the bridge circuit in Fig. 8–19. What value of $R4$ must be used for the bridge to be balanced?

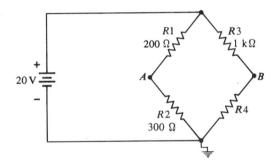

FIGURE 8–18

8–16 Determine V_{AB} for the bridge circuit in Fig. 8–18 for $R4 = 1$ kilohm and then for $R4 = 2$ kilohms.

Section 8.7

8–17 A certain voltage divider has as its input a sine wave of 120 volts amplitude. It is desired that the output have a 24-volt amplitude and be able to drive loads ranging from 150 ohms to 3000 ohms. Choose appropriate resistors for the circuit.

8–18 A 32-V dc source is to be divided down to 6 V to supply to various loads. Design a voltage divider which will produce the desired output. Assume that the load on the output is a 6-V lamp that draws 12 mA of current.

8–19 Repeat 8–18 if the voltage divider output has to supply 6 V to *three* 6 V/12 mA lamps and one 6 V/24 mA motor, all connected in parallel.

8–20 Draw a circuit that contains a voltage source driving a circuit that contains two resistors $R1$ and $R2$ which are in series with a parallel combination of $R3$, $R4$, and $R5$.

CHAPTER 9

Electrical Power in Resistive Devices

9.1 Introduction

In this chapter we will begin a more in-depth study of power and its presence in resistive circuits. As we shall see, power is the rate at which energy is transferred in a circuit. For this reason, it is important that we understand the concept of power and be able to calculate the various amounts of power in a circuit. We will concentrate on power in dc circuits for now and leave the study of ac power for later chapters.

9.2 Definition of Power

Power is the rate at which we do work or consume energy. For example, when we mow our lawn, we call this *work* because we have to expend a certain amount of *energy*. If we mow the lawn more rapidly, we use more *power* than if we mow it slowly because we are expending energy at a more rapid rate.

As another example, suppose our automobile is climbing a steep hill. This requires a certain amount of work in overcoming the force of gravity. Therefore a certain amount of energy is expended (chemical energy of the fuel) in climbing the hill. If we climb at a faster rate, then the energy will be expended at a faster rate. This means that more *power* is required.

For still a further example, consider the application of a voltage to an electric circuit. The applied voltage, you may recall, represents potential energy that is imparted to the free charges (electrons) in the circuit. As a result of this imparted energy, the electrons move through the circuit at a certain rate. If the applied voltage is increased, the energy imparted to the electrons increases and they move at a faster rate. Since energy is being imparted to the electrons at a faster rate, this requires more *power* from the voltage source.

As these examples show, the faster energy is used up, the more power is required. Power is the rate at which energy is being consumed. In chap. 1 it was stated that the SI approved unit for energy is the *joule* (J). If we expend one joule of energy in a time interval of one second, the rate at which we expend energy is 1 J/s. If we expend five joules in two seconds, this represents a rate of 2.5 J/s. Thus, we can say that

$$\text{Power} = \frac{\text{Energy}}{\text{Time}} \tag{9-1}$$

The letter symbol for power is P. The units for P are joules/seconds (J/s). However, the SI prefers to use the watt (W) as the basic unit for power. That is,

$$1 \text{ watt} = 1 \text{ joule/second}$$

Thus, if energy is being expended at the rate of 1 joule per second, this requires a power P equal to 1 watt.

EXAMPLE 9.1 When applied to a certain circuit, a certain voltage source supplies 240 joules of energy over a period of 2 minutes. How much power does this represent?

Solution: $\text{Power} = \dfrac{\text{Energy}}{\text{Time}}$

$$P = \frac{240 \text{ J}}{120 \text{ s}} = 2 \text{ J/s} = 2 \text{ watts} = \mathbf{2 \text{ W}}$$

EXAMPLE 9.2 The same voltage source supplies 720 joules of energy to the circuit over a period of 6 minutes. How much power does this represent?

Solution: $P = \dfrac{720 \text{ J}}{360 \text{ s}} = 2 \text{ J/s} = \mathbf{2 \text{ W}}$

Note that the power in this case is the same as in the last example because the *rate* at which energy is being supplied is the same for both cases.

9.3 Electrical Power

There are as many forms of power as there are forms of energy. Mechanical energy per unit time represents mechanical power; heat (thermal) energy per unit time represents heat (thermal) power; electrical energy per unit time represents electrical power; and so on. Since our main interest here is in electrical power, we will examine it more closely.

When a voltage source is applied to a circuit, it causes charge to flow through the circuit. It may be recalled (chap. 2) that *voltage* is energy per unit charge (joules per coulomb). Since *current* is the rate of charge flow (coulombs per second), we have

Electrical Power in Resistive Devices

$$\text{(current)} \times \text{(voltage)} = \text{(amperes)} \times \text{(volts)}$$
$$= \frac{\text{cou}\cancel{\text{lombs}}}{\text{second}} \times \frac{\text{joules}}{\cancel{\text{coulomb}}}$$
$$= \frac{\text{joules}}{\text{second}}$$
$$= \text{watts}$$
$$= \text{power}$$

Stated mathematically,

$$P = I \times E \tag{9-2}$$

where P represents the power supplied by the source, E represents source voltage (emf), and I is the current produced by the source. This expression is the basic relationship for electrical power. It states that *the power supplied by a source equals the product of its voltage and current.* If the current is expressed in its basic unit (A) and the voltage in its basic unit (V), then the resultant power is in its basic unit (W).

EXAMPLE 9.3 A certain light bulb draws 2.5 A when connected to a 120-V source. How much power is being used?

Solution: $P = I \times E$
$\quad\quad\quad\quad = 2.5 \text{ A} \times 120 \text{ V}$
$\quad\quad\quad\quad = \mathbf{300 \text{ W}}$

EXAMPLE 9.4 A 12-V source produces 15 mA of current in a certain circuit. How much power is the source supplying?

Solution: $P = I \times E$
$\quad\quad\quad\quad = 15 \text{ mA} \times 12 \text{ V}$
$\quad\quad\quad\quad = 0.015 \text{ A} \times 12 \text{ V}$
$\quad\quad\quad\quad = \mathbf{0.18 \text{ W}}$

This result can also be expressed in the prefixed unit *milliwatts*, mW. A mW equals 10^{-3} W. Thus,

$$P = 0.18 \text{ W} = 180 \times 10^{-3} \text{ W} = \mathbf{180 \text{ mW}}$$

EXAMPLE 9.5 A 100-V source is connected to a 2-ohm resistor. How much power is supplied by the source?

Solution: The current produced by the source is

$$I = \frac{100 \text{ V}}{2 \text{ } \Omega} = 50 \text{ A}$$

Thus the power delivered is

$$P = 50 \text{ A} \times 100 \text{ V}$$
$$= \mathbf{5000 \text{ W}}$$

This can be expressed in the prefixed unit *kilowatts*, kW. A kW equals 10^3 W. Thus,

$$P = 5 \times 10^3 \text{ W} = 5 \text{ kW}$$

EXAMPLE 9.6 Which draws less current from the 120-V power source, a 50-W light bulb or a 75-W light bulb?

Solution: A 50-W light bulb will be supplied with 50 W of power from the 120-V source. Thus,

$$P = I \times E$$
$$50 \text{ W} = I \times 120 \text{ V}$$

or

$$I = \frac{50 \text{ W}}{120 \text{ V}} = 0.417 \text{ A}$$

This is the current produced through the 50-W bulb.

The 75-W light bulb will be supplied with 75 W of power from the 120-V source. Thus,

$$75 \text{ W} = I \times 120 \text{ V}$$

or

$$I = \frac{75 \text{ W}}{120 \text{ V}} = 0.625 \text{ A}$$

which is the current through the 75-W bulb.

The larger wattage bulb draws the larger current from the source.

9.4 Power Dissipation in Resistive Devices

To produce a current flow through a resistive device, a voltage source must supply a certain amount of power. As current flows through the resistive material, *heat* is produced because the free electrons are continually making collisions with the material's atoms as they move through the material. We can think of it as an "atomic friction" that produces this heat. The heat is a form of energy and its presence in the resistive device indicates that power is being supplied to the device.

Any device which possesses resistance to the flow of current will *dissipate* energy in the form of heat. The heat energy is said to be *dissipated* because it is radiated into the surrounding medium and cannot be returned to the circuit as electrical energy. The rate at which this energy is dissipated is the *power dissipation* of the device.

In a circuit, the various resistances will dissipate power in the form of heat. The total power dissipated in the circuit must be supplied by the input source (often called a "power supply") since it is the only source of energy in the circuit. The source supplies *electrical power* to the circuit, and the resistances in the circuit dissipate this power as *heat*. At any instant of time, the power

Electrical Power in Resistive Devices

being supplied by the source must equal the total power being dissipated by the circuit.

To calculate the power dissipation in a resistive device we can again use the current-voltage product. That is,

$$P = I \times V \tag{9-3}$$

This expression is actually the same as Eq. (9–2) except the device voltage, V, is used instead of the source voltage, E.

EXAMPLE 9.7 A certain diode has a current of 35 mA flowing through it when its voltage is 0.8 V. How much power is the diode dissipating?

Solution: $P = I \times V$
$= 35 \text{ mA} \times 0.8 \text{ V}$
$= 0.035 \text{ A} \times 0.8 \text{ V} =$ **0.028 W**   or   **28 mW**

EXAMPLE 9.8 A 12-kΩ resistor has 24 V applied to it. How much power does the resistor dissipate? At what rate is heat energy being lost to the surroundings?

Solution: $I = \dfrac{V}{R} = \dfrac{24 \text{ V}}{12 \text{ k}\Omega} = 2 \text{ mA}$

$P = I \times V = 2 \text{ mA} \times 24 \text{ V} =$ **48 mW**

This power dissipation of 48 mW represents heat energy being dissipated at the rate of 48 millijoules/second.

EXAMPLE 9.9 A 30-V source is applied to a series circuit containing a 25-ohm and a 50-ohm resistor. Determine the power dissipated in each resistor and the power supplied by the source.

Solution: The circuit current is

$$I = \frac{30 \text{ V}}{75 \text{ }\Omega} = 0.4 \text{ A}$$

Thus, the voltage across the 25-ohm resistor is

$$V_1 = 0.4 \text{ A} \times 25 \text{ ohms} = 10 \text{ V}$$

so that the power dissipated in this resistor is

$$P_1 = 0.4 \text{ A} \times 10 \text{ V} = 4 \text{ W}$$

Similarly, the voltage across the 50-ohm resistor is 20 V so that its power dissipation is

$$P_2 = 0.4 \text{ A} \times 20 \text{ V} = 8 \text{ W}$$

The power supplied by the source is

$$P_s = I \times E$$
$$= 0.4 \text{ A} \times 30 \text{ V} = \textbf{12 W}$$

Note that the power supplied by the source equals the sum of the resistor power dissipations.

Power in Linear Resistors When considering the power dissipation in a *linear resistor*, we can take advantage of Ohm's law to develop expressions for power in terms of the resistance value, R. For example, we can substitute $V = I \times R$ into the basic power formula [Eq. (9-3)] as follows:

$$P = I \times V$$
$$= I \times (I \times R)$$
$$P = I^2 R \tag{9-4}$$

This last expression relates the resistor's power dissipation to its current and resistance.

EXAMPLE 9.10 How much power is being dissipated by a 3-kΩ resistor which is passing 4 mA of current?

Solution: $P = I^2 R$
$= (4 \text{ mA})^2 \times 3 \text{ k}\Omega$
$= (4 \times 10^{-3})^2 \times (3 \times 10^3)$
$= 16 \times 10^{-6} \times 3 \times 10^3$
$= 48 \times 10^{-3}$ W
$=$ **48 mW**

An alternative expression can be obtained by substituting $I = V/R$ into the basic power formula as follows:

$$P = I \times V$$
$$= \frac{V}{R} \times V$$
$$P = \frac{V^2}{R} \tag{9-5}$$

If the voltage across a resistor is known, this expression can be used to calculate P.

EXAMPLE 9.11 How much power will a 250-ohm resistor dissipate if 10 V are applied to it?

Solution: $P = V^2/R$
$= \frac{(10^2)}{250} = \frac{100}{250} =$ **0.4 W**

The three relationships [in Eqs. (9-3), (9-4), and (9-5)] are all equivalent expressions of the resistor power. The one to be used in a given situation depends on which factors are known. However, Eq. (9-3) is valid for any device while (9-4) and (9-5) are used only for linear resistors.

Useful Power Dissipation As stated earlier, whenever current flows through a resistor it causes power to be dissipated as heat. The amount of heat generated depends on the amount of power dissipation, which, in turn, depends on the amount of current flow. Thus, whenever resistive devices are present in a circuit, heat will be present as a result of the necessity for providing current through these devices.

In certain applications, the heat dissipation is used to good advantage. For example, the heat dissipation is put to use in an electric stove, toaster, or soldering iron. In incandescent light bulbs the power dissipation is used to heat up the filament so that it becomes white-hot and glows. In such devices the amount of heat is usually controlled by controlling the current through the device.

9.5 Device Power Ratings

Every resistive device will become heated as it dissipates power. If a device becomes overheated it will probably burn up and cease to function. For this reason every resistive device is given a *power dissipation rating*. A device's power rating gives the amount of wattage that the device can *safely* dissipate in the form of heat without becoming damaged. For example, a 1-watt resistor can safely dissipate up to 1 W of power.

Many types of devices are used to meet the requirements of electronic circuits. Because of the many different power requirements for these circuits, the power ratings of the devices can vary over a considerable range. Resistors, for example, are available in power ratings from 0.1 W to 250 W. The resistors with the larger power ratings are physically larger so that they can radiate a greater amount of heat.

It is important to know the value of power that a resistor will be called on to dissipate in a given circuit in order to insure that the device used has a sufficient power rating. Otherwise the resistor may overheat and possibly burn out. Even if it didn't burn out, the large amount of heat it radiates could affect other components in the circuit. To minimize the possibility of overheating and burning out, a safety factor of 100 percent is often used when selecting resistors for a circuit. In other words, if the calculated power dissipation for a resistor in a circuit is 0.5 W, a resistor with a 1-W rating should be used.

EXAMPLE 9.12 Consider the circuit in Fig. 9–1(a). Determine the necessary power rating for each resistor.

Solution: The current in this series circuit is found to be

$$I = \frac{20 \text{ V}}{200 \text{ }\Omega} = 0.1 \text{ A}$$

To find the power in each resistor we can use $P = I^2R$. Thus,

$$P_{R1} = (0.1 \text{ A})^2 \times 100 \text{ ohms}$$
$$= 1 \text{ W}$$

Similarly,

$$P_{R2} = 0.25 \text{ W}$$

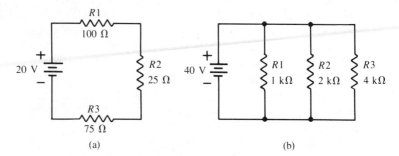

FIGURE 9-1

and

$$P_{R3} = 0.75 \text{ W}$$

Thus, for $R1$ we should use a resistor with a power rating of at least $2 \times 1 \text{ W} = 2 \text{ W}$; for $R2$, a power rating of at least $2 \times 0.25 \text{ W} = 0.5 \text{ W}$; for $R3$, a power rating of at least $2 \times 0.75 \text{ W} = 1.5 \text{ W}$.

Note that *in a series circuit the resistor with the most resistance dissipates the most power* since $P = I^2 R$ and I^2 is the same for each resistor. Thus, P is directly proportional to R.

EXAMPLE 9.13 Determine the necessary power ratings for each resistor in Fig. 9-1(b).

Solution: Each resistor has 40 V across its terminals. We can calculate the power dissipation of each resistor by using $P = E^2/R$. Thus,

$$P_{R1} = \frac{(40 \text{ V})^2}{1000} = \frac{1600}{1000} = 1.6 \text{ W}$$

$$P_{R2} = \frac{(40)^2}{2000} = 0.8 \text{ W}$$

$$P_{R3} = \frac{(40)^2}{4000} = 0.4 \text{ W}$$

The power rating of each resistor should be chosen to be at least twice its calculated power dissipation.

Note that *in a parallel circuit the smallest resistor dissipates the most power* since $P = E^2/R$ and E^2 is the same for each resistor. Thus, P is inversely proportional to R.

9.6 Power Conversion

In many applications a source of electrical power is used to supply power to a device that converts the electrical power to another useful form of power. A light bulb or heater, as we have seen, is an example of this. Another example of *power conversion* is an electric motor. An electric motor is driven by a voltage source. The motor converts the electrical power from the source to *mechanical power*, which can be used to rotate a shaft.

Mechanical power is often measured in units called *horsepower*, abbreviated, hp. The horsepower is a larger unit than the watt. In fact,

$$1 \text{ hp} = 746 \text{ W} \approx 0.75 \text{ kW}$$

Thus, a 10-hp motor can produce $10 \times 746 = 7460$ W of power. Of course, to do so it must be supplied with electrical power.

In general, the electrical power the motor requires will be greater than the mechanical output power. This is because some of the input electrical power is lost as heat in the motor. For example, the 10-hp motor might require an electrical power input of 9 kW or 9000 W. Since the motor puts out only 7460 W of mechanical power, then $9000 - 7460 = 1540$ W of power are lost as heat. The *efficiency* of the power conversion process is usually stated as the ratio of percentage of useful output power to input power. That is,

$$\text{Percent efficiency} = \frac{P_{\text{out}}}{P_{\text{in}}} \times 100\% \tag{9-6}$$

In our 10-hp motor, then, P_{out} would be 7460 W and P_{in} would be 9000 W so that

$$\text{Percent efficiency} = \frac{7460}{9000} \times 100\% = 82.9\%$$

The efficiency of a motor or other power conversion device would ideally be 100 percent. In practice, 100 percent efficiency is impossible to achieve.

EXAMPLE 9.14 A dc motor delivers 2 hp of mechanical power to a load. The 50-V source connected to the motor windings is supplying 41.7 A to the motor. What is the efficiency of the motor's conversion from electrical to mechanical power?

Solution: Electrical power input $= 50 \text{ V} \times 41.7 \text{ A} = 2085 \text{ W}$

Mechanical power output $= 2 \text{ hp} \times \dfrac{746 \text{ W}}{\text{hp}} = 1492 \text{ W}$

Percentage efficiency $= P_{\text{out}}/P_{\text{in}}$
$= \dfrac{1492 \text{ W}}{2085 \text{ W}} = \mathbf{71.5\%}$

9.7 Maximum Power Transfer

In chap. 6 we discussed practical voltage sources and the effects of internal source resistance. We saw that the terminal voltage of a battery or source will drop as current is drawn by a load. Another important effect of source resistance is concerned with the maximum power that a source can deliver to a load.

In Fig. 9-2(a) a practical 10-V dc source is shown with a source resistance of 100 ohms. A load resistor, R_L, is connected across the source terminals x-y. If the value of R_L is varied, the circuit current will vary, the terminal voltage will vary, and the power delivered to the load will vary. In Table 9-1 the various circuit quantities are tabulated for various values of R_L. Let's examine the table closely.

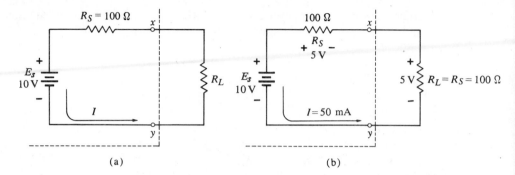

FIGURE 9–2 (a) *Practical source with a load.* (b) $R_L = R_S$ *for maximum load power.*

Starting with maximum R_L (open circuit) and decreasing toward minimum R_L (short circuit) the table shows that the current I increases as expected. Since the source is supplying more current, then the input source power increases as shown. However, the output load power does something interesting. It increases from zero (for open-circuit condition) up to a maximum value and then decreases back to zero (for short-circuit condition). The load power reaches a maximum of 250 mW when $R_L = 100$ ohms. Any other value of R_L results in less than 250 mW of power in the load. The significant fact here is that this

TABLE 9–1

R_L	Current, I	Output Load Power $I^2 R_L$	Source Input Power $I \times E_s$	$P_{out}/P_{in}(\%)$
(open)	0.0	0.0	0.0	0
900 Ω	10.0 mA	90.0 mW	100.0 mW	90
400 Ω	20.0 mA	160.0 mW	200.0 mW	80
150 Ω	40.0 mA	240.0 mW	400.0 mW	60
100 Ω	50.0 mA	250.0 mW	500.0 mW	50
50 Ω	66.7 mA	222.2 mW	666.7 mW	33
10 Ω	90.9 mA	82.7 mW	909.1 mW	11
0 Ω	100.0 mA	0.0 mW	1000.0 mW	0

value of R_L is the same as the source resistance value. This can be stated as the *maximum power principle:*

A practical source with a given R_S will deliver maximum power to a load when the load resistance equals the source's internal resistance ($R_L = R_S$).

In other words, a load will consume maximum power when its resistance is the same as the source resistance. Note that the total source input power is *not* maximum when $R_L = R_S$ (500 mW from Table 9–1); it is maximum when $R_L = 0$ since this represents the condition for maximum current drawn from the source.

Also note that the ratio of output load power (P_{out}) to source input power (P_{in}) is only 50 percent for the $R_L = R_S$ condition. That is, the *efficiency* is only 50 percent because half of the input power reaches R_L while the other half is dissipated in the source resistance R_S.

For the maximum load power condition ($R_L = R_S$) we can develop a general formula for load power. Since $R_L = R_S$, then the load voltage will be half of E_S. Therefore, the load power when $R_L = R_S$ becomes

$$P_{\text{load}}(\text{max}) = \frac{(E_S/2)^2}{R_L} = \frac{E_S^2}{4\,R_L} = \frac{E_S^2}{4\,R_S} \qquad (9\text{--}7)$$

EXAMPLE 9.15 Which source can deliver more power to a load: a 20-V source with $R_S = 10$ ohms or a 20-V source with $R_S = 50$ ohms?

Solution: Looking at Eq. (9–9), we can see that a smaller R_S will yield a larger value of maximum power for the same E_S. To check for $R_S = 10$ ohms

$$P_{\text{load}}(\text{max}) = \frac{(20)^2}{4 \times 10} = 10 \text{ W}$$

For $R_S = 50$ ohms

$$P_{\text{load}}(\text{max}) = \frac{(20)^2}{4 \times 50} = 2 \text{ W}$$

The maximum power principle plays an important part in electronics. It is particularly significant in certain amplifier circuits. When an amplifier with a certain internal output resistance supplies power to a load (such as a loudspeaker), it is usually desirable to make the load resistance equal to the internal resistance of the amplifier in order to obtain maximum load power. This is called *matching* the load to the amplifier.

9.8 Kilowatthours

A unit of electrical energy which is in common use especially by utility companies is the *kilowatthour* (kWh). We receive electrical energy from our utility company and we pay for this energy at the rate of about 8 cents per kWh. The amount of energy consumed can be calculated in units of kWh by multiplying the power consumed in kilowatts (kW) by the time during which the power is being consumed, in hours (h). That is,

$$\text{Energy(kWh)} = \text{power(kW)} \times \text{time(h)}$$

As an example, if a 100-W bulb is burning for 16 hours, the total energy consumed is $0.1 \text{ kW} \times 16 \text{ h} = 1.6 \text{ kWh}$. At 8 cents per kWh, this would cost only 13 cents! A typical household consumes about 1000 kWh of energy per month.

Chapter Summary

1. Power is the rate at which we do work or consume energy.
2. The more rapidly we expend energy, the more power is required.
3. The letter symbol for power is P and the SI unit for power is the watt (W).

4. One watt equals one joule per second.
5. A voltage source which produces current in a circuit supplies an amount of power equal to the product of the current and voltage ($P = I \times E$).
6. When a source supplies power to a resistive device, this power is dissipated in the device in the form of heat.
7. In a circuit the total of the power dissipations must equal the power supplied by the source.
8. The power dissipated in a resistive device is equal to $I \times V$.
9. The power dissipation in a linear resistor can also be expressed as $P = I^2 R$ or $P = E^2/R$.
10. A device's power rating is the maximum wattage it can safely dissipate.
11. In a series circuit a resistor's power dissipation is directly proportional to its resistance.
12. In a parallel circuit a resistor's power dissipation is inversely proportional to its resistance.
13. Efficiency of power conversion is $(P_{out}/P_{in}) \times 100\%$.
14. A practical source will deliver maximum power to a load when the load resistance equals the source's internal resistance.
15. A kilowatthour is a unit of electrical energy. It is the amount of energy consumed when 1 kW of power is consumed for 1 hour.

Questions/Problems

Sections 9.2–9.3

9–1 Define "power."

9–2 Which of the following requires more power:
   a. climbing a 500-ft mountain in 2 hours
   b. climbing a 1000-ft mountain in 3 hours.

9–3 A certain voltage source can supply 300 J of energy to a circuit in 6 seconds. How much power is this?

9–4 How many joules of energy can the voltage source in problem 9–3 supply in one hour?

9–5 How much electrical power is being used by a motor which draws 30 mA from a 120-V power source?

9–6 If a 100-W light bulb and a 300-W light bulb are both connected across the 120-V source, how much power is the source supplying?

9–7 A small light bulb is supplied with 7 W of power from a 120-V source. How much current does the bulb draw?

9-8 For the light bulb in the last problem, how much energy is being used up every minute?

9-9 A transistor has a current of 2.5 mA flowing through it when its voltage is 4.2 V. How much power is the transistor dissipating?

Section 9.4

9-10 A 470-ohm resistor has 10 mA of current flowing through it. How much power is the resistor dissipating?

9-11 A 6-V source is applied to a series circuit containing resistors $R1$ and $R2$. The total power supplied by the source is 24 mW. If $R1$ dissipates 14 mW determine the voltage across resistor $R2$.

9-12 Calculate the power dissipation in a 10-ohm resistor which is passing 12 mA of current.

9-13 Calculate the power dissipation in a 12-kΩ resistor when 24 V are applied to it.

9-14 How much voltage must be applied to a 40-ohm resistor if it is to dissipate 36 mW?

9-15 A certain heater element for an electric toaster has a resistance of 12 ohms. How much voltage must be applied to it if it is to convert 1.2 kW of electrical power into heat?

9-16 Two resistors $R1$ and $R2$ are immersed in separate glasses of water. $R1 = 100$ ohms and has 10 V applied to it for 5 seconds. $R2 = 200$ ohms and has 16 V applied to it for 3 seconds. In which glass will the water be warmer? Explain.

9-17 A certain heating element has $R = 2$ ohms. How much resistance must be placed in series with the heater if the heater must draw 180 W of power? Assume $E_S = 90$ V.

Section 9.5

9-18 A series circuit contains a 120-ohm, 220-ohm, and 360-ohm resistor. Determine the necessary power rating for each resistor if $E = 35$ V is the source voltage.

9-19 A parallel circuit contains a 5-kΩ, 10-kΩ, and 20-kΩ resistor. If $E = 120$ V, determine the desired power rating for each resistor.

9-20 Consider the series circuit in Fig. 9-3. Indicate what will happen to the power dissipated in $R3$ and the total power supplied by the source for each of the following changes:
 a. $R1$ is increased in value.
 b. $R2$ is increased in value.
 c. $R4$ is short-circuited.
 d. E is increased.
 e. A resistor $R5$ is added to the circuit.

9-21 For the parallel circuit in Fig. 9-4 indicate what will happen to the power dissipation of $R3$ and the total power supplied by the source for each of the following changes:
 a. $R1$ is decreased in value.
 b. $R2$ is open-circuited.
 c. E is increased.
 d. A resistor, $R5$, is added in parallel.

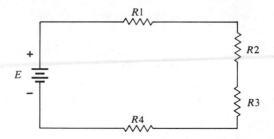

FIGURE 9-3

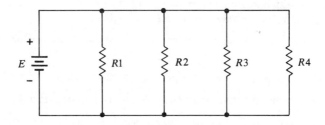

FIGURE 9-4

Section 9.6

9-22 A 200-V dc source supplies 8 A of current to a motor. Calculate the motor's efficiency if the motor produces 0.5 hp of mechanical power.

9-23 How much current does a 120-V source have to supply to a 1.5-hp motor operating at 70% efficiency?

Section 9.7

9-24 A 50-V source has $R_S = 25 \, \Omega$. What is the maximum power which it can deliver to a load?

9-25 A practical 24-V source delivers 400 mW of power to a 1-kΩ load. What is the maximum power this source can ever deliver to a load?

9-26 A load R_L is connected to a source with a source resistance R_S.
    a. For what R_L will the most current be drawn from the source?
    b. For what R_L will the load voltage be the greatest?
    c. For what R_L will the load power be the greatest?
    d. For what R_L will the source's internal power dissipation be greatest?

Section 9.8

9-27 Determine how many joules are contained in a kWh.

9-28 How many 300-W light bulbs can be operated continuously for 50 dollars a week?

CHAPTER 10

Physical Characteristics of Resistance

10.1 Introduction

In this chapter we will look at some of the physical properties which are factors in determining the electrical resistance of a material. In doing so, the concepts of *resistivity* and *temperature coefficient* will be introduced.

10.2 Factors Affecting Resistance

The flow of charge through any material experiences a hindering force similar to mechanical friction. This opposition to charge flow (current) is called *electrical resistance* and is a sort of an "atomic friction" that is characterized by conduction electrons continuously making collisions with the material's atoms. As we already know, the unit for resistance is the ohm (Ω).

The electrical resistance of a piece of material depends on several factors. The four principal factors are: (a) kind of material (copper, carbon, etc.); (b) length of the material along the current path; (c) cross-sectional area through which current is flowing; and (d) temperature of the material.

Let's consider factors (b) and (c) first. In Fig. 10–1 two wires made of the same material and with the same cross-sectional area are shown. Wire A is twice as long as wire B. If a voltage is applied across the ends of each wire electrons will begin to flow toward the positive end. The amount of resistance that electrons will meet in flowing from one end of the wire to the other will be twice as great in wire A because the electrons will have twice as many opportunities for colliding with atoms of the material. Thus, we can say that

Resistance is directly proportional to length.

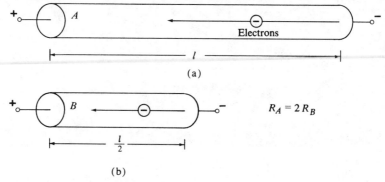

FIGURE 10-1 *Resistance is directly proportional to length.*

Another way to reason the situation in Fig. 10–1 is to view wire A as being the same as *two* wire Bs connected in *series* so that $R_A = R_B + R_B = 2R_B$.

In Fig. 10–2 wires A and B are of the same material and same length, but A has a cross-sectional area twice as great as B. If a voltage is applied to each wire, the larger wire, A, will conduct twice as much current as wire B. This can

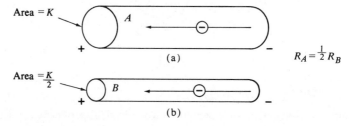

FIGURE 10-2 *Resistance is inversely proportional to cross-sectional area.*

be reasoned by viewing wire A as being the same as two wire Bs connected in *parallel*. The two parallel wires will pass twice as much current as the single wire. Since wire A passes twice as much current as B for the same applied voltage, then it follows that A's resistance is half of B's resistance. Thus, we can conclude that

Resistance is inversely proportional to cross-sectional area.

EXAMPLE 10.1 A certain wire with a cross-sectional area of 10^{-8} meter² is 1 meter long and has a resistance of 2 ohms. What will be the resistance of a wire 20 meters long of the same material with a cross-sectional area of 3×10^{-8} meter²?

Solution: If the 1-meter-long wire is lengthened to 20 meters, its resistance will *increase* by 20 times to $20 \times 2\,\Omega = 40\,\Omega$. If its cross-sectional area is then increased from 10^{-8} meter² to 3×10^{-8} meter², its resistance will *decrease* by *three* times to $40\,\Omega/3 = 13.3\,\Omega$. Thus, the overall effect of increasing the length by a factor of 20 and increasing the area by a factor of 3 is an increase in resistance by a factor of $20/3 = 6.67$ times, so that $R = \mathbf{13.34\,\Omega}$.

We noted in chap. 1 that some materials contain more free electrons than others, and those materials with more free electrons per unit volume are better conductors of electric current. For this reason, we can state that

Resistance depends on the atomic structure of the material.

Thus, a wire made out of copper has a lower resistance than an aluminum wire *with identical dimensions*.

Temperature also plays an important part in determining the resistance of a piece of material. As a material is heated the heat energy is absorbed by the atoms and electrons. Depending on the material, the absorbed energy can result in an increase or decrease in resistance. More will be said about this later.

In the following sections we will develop quantitative relationships for these factors to help give us a better feel for how each contributes to a material's resistance.

10.3 Basic Resistance Equation

Any wire with a length l and a cross-sectional area A has a resistance R given by the following formula:

$$R = \rho \frac{l}{A} \qquad (10\text{--}1)$$

The symbol ρ (Greek letter rho) is a constant which has a value that depends on the material. ρ represents the *resistivity* of the material. *Resistivity* is also called *specific resistance*, but to avoid confusion between specific resistance and resistance, we will use the term resistivity for ρ.

The formula in Eq. (10–1) shows that the resistance of the wire is directly proportional to its length and inversely proportional to its cross-sectional area as we previously reasoned. It also shows that R is directly proportional to ρ. For two identically shaped wires (same l and A) made of different materials, the one with the greater value of resistivity will have the greater resistance. Thus, resistivity is used as a means of comparing the electrical current opposition of different materials.

Resistivity is a property of each material and does not depend on the shape or dimensions of the specimen. In other words, resistivity is an innate characteristic of a material, the same as color and density are innate characteristics. On the other hand, the resistance of a piece of material depends on its resistivity and also on its dimensions as expressed in Eq. (10–1). Thus, it is possible for a piece of low resistivity material to have more *resistance* than a piece of higher resistivity material.

When Eq. (10–1) is used to determine the resistance of a wire, the length l is expressed in units of feet (ft) and the area A is expressed in units called *circular mils* (cmil). *A wire that has a diameter equal to 1 mil (0.001 inch) has a cross-sectional area of 1 cmil*. In other words, a cmil is the area of a circle that has a diameter of 1 mil (0.001 inch).

In Fig. 10–3(a) the wire shown has a diameter of 1 mil. Thus, its area is equal to 1 cmil. In part (b) of the figure the wire shown has a diameter of 2 mils. As we know, the area of a circle increases as the *square* of the diameter

($A = \pi d^2/4$). Thus, the wire in (b) must have *four* times the area of the circle in (a) so that its area is 4 cmil. Similarly, a wire with a diameter of 3 mils will have an area of $3^2 = 9$ cmil, a wire with a diameter of 4 mils has an area of $4^2 = 16$ cmil, and so on. In general, then, we can say that the area in cmils is equal to the square of the diameter in mils. That is,

$$A(\text{in cmils}) = [D (\text{in mils})]^2 \qquad (10\text{--}2)$$

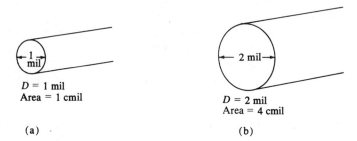

D = 1 mil
Area = 1 cmil

(a)

D = 2 mil
Area = 4 cmil

(b)

FIGURE 10–3 *Doubling the wire diameter increases the area by a factor of four.*

We can now use Eq. (10–1) to determine the proper units for resistivity ρ. If we solve (10–1) for ρ, we have

$$\rho = \frac{A \times R}{l}$$

$$= \frac{\text{cmil} \times \text{ohms}}{\text{ft}} = \frac{\text{cmil} - \text{ohms}}{\text{ft}}$$

Thus, the units for ρ are cmil-Ω per ft.

EXAMPLE 10.2 Calculate the resistance of 300 ft of aluminum wire which has a diameter of 10 mils. The resistivity of aluminum is 17 cmil-Ω/ft.

Solution: First, the area A is calculated as $D^2 = 10^2 = 100$ cmil. Using the basic resistance formula (12–1)

$$R = \rho \frac{l}{A} = 17 \left(\frac{\text{cmil-}\Omega}{\text{ft}}\right) \times \frac{300 \text{ ft}}{100 \text{ cmil}}$$

$$= 51 \text{ ohms}$$

Note that all the units cancel out except for the ohms.

EXAMPLE 10.3 Copper has a resistivity of 10.4 cmil-Ω/ft. Calculate the resistance of a copper wire which has the same dimensions as the aluminum wire in the last example.

Solution: $R = \rho \dfrac{l}{A}$

$$= 10.4 \times \frac{300}{100}$$

$$= 31.2 \text{ ohms}$$

Thus, the copper wire has a lower resistance than the aluminum wire of the same dimensions. Although both copper and aluminum are good conductors, copper is a better conductor because it has a *lower* resistivity.

Table 10–1 gives the resistivity values for some of the common conducting materials. The smaller values of ρ are the better conductors of electricity. Note that the table indicates the values of resistivities at a specific temperature of 20°C.* A specific temperature has to be indicated because the resistivities of most materials will change with temperature. This point will be discussed later.

TABLE 10–1

Material	Resistivity (ρ) in cmil-Ω/ft at 20°C	
Silver	9.9	←best conductor
Copper	10.4	
Gold	14.0	
Aluminum	17.0	
Tungsten	33.8	
Nickel	52.0	
Iron	58.0	
Steel	100.0	
Nichrome	676.0	

10.4 Wire Tables

Table 10–2 lists the standard wire sizes used by North American wire manufacturers which conform to the American Wire Gauge (AWG) system. Each gauge number has associated with it a certain diameter (in mils) and cross-sectional circular area (in cmils).

Examination of this table reveals several points: (1) As the AWG numbers increase from 1 to 40, the wire diameter decreases and the area decreases. Higher gauge numbers indicate thinner wires. (2) The circular area *decreases* by one half for every three gauge sizes. (Example: AWG #28 has an area of 159.8 cmil while AWG #31 has an area of 79.7 cmil, approximately a 50 percent decrease). (3) As the AWG number increases, the wires get thinner and thus will have a greater resistance for any given length.

EXAMPLE 10.4 Calculate the resistance of 100 ft of AWG #22 gold wire at 20°C.

Solution: From Table 10–2 an AWG #22 wire has an area of 642.4 cmil. Since the wire is made of gold we need to know the resistivity of gold given in Table 10–1 as 14 cmil-Ω/ft.

---

*20°C is used as the standard reference temperature at which values of ρ are given. 20°C = 68°F, approximate room temperature.

$$R = \rho \frac{l}{A}$$

$$= 14 \times \frac{100}{642.4}$$

$$= 2.18 \ \Omega$$

In general, lower gauge wires can safely handle more current because they have a greater cross-sectional area. In applications where currents in the milli-ampere range are flowing (such as in radio receiver circuits), the hookup wire used is generally AWG #22. In residential wiring where currents up to 15 A are being conducted, AWG #12 can be used. Safety standards set up by fire under-writers specify the minimum sizes for residential wiring.

10.5 Other Units for ρ

To find the resistance of a piece of material that is not in the form of a wire, we can still use Eq. (10–1). However, the units for l, A, and ρ are usually centimeters (cm), square centimeters (cm²), and ohm-centimeters (Ω-cm) respectively. For example, pure silicon (a semiconductor material) has a resistivity $\rho = 55 \times 10^3$ Ω-cm. Thus, a silicon bar which is 10-cm long and has a cross-sectional area of 2 cm² will have

$$R = 55 \times 10^3 \ \Omega\text{-cm} \times \frac{10 \text{ cm}}{2 \text{ cm}^2} = 275{,}000 \ \Omega$$

Resistivity is usually given in Ω-cm for materials that are not used in making electrical wires. All the materials listed in Table 10–1 are used to make wire and their values of ρ are usually given in cmil-Ω/ft. For those who are interested, the conversion factor between these two different units for ρ is given by

$$1 \ \Omega\text{-cm} = 6 \times 10^6 \ \frac{\text{cmil-}\Omega}{\text{ft}}$$

Table 10–3 gives the values of ρ for several materials at 20°C.

EXAMPLE 10.5 12 volts are applied across the ends of the germanium bar in Fig. 10–4. How much current will flow through the bar?

Solution: To determine the current we must first calculate the resistance of the bar. The bar has a length $l = 2$ cm and a cross-sectional area $A = (0.1 \text{ cm}) \times (0.1 \text{ cm}) = 0.01$ cm². Thus, the bar's resistance is

$$R = \rho \frac{l}{A}$$

$$= 65 \ \Omega\text{-cm} \times \frac{2 \text{ cm}}{0.01 \text{ cm}^2}$$

$$= 13{,}000 \ \Omega = 13 \text{ k}\Omega$$

Using Ohm's law the current will be

$$I = \frac{12 \text{ V}}{13 \text{ k}\Omega} = 0.92 \text{ mA}$$

Physical Characteristics of Resistance

TABLE 10-2 *American Wire Gauge (AWG) Wire Sizes.*

Gage No.	Diameter, Mils	Circular-mil Area	Gage No.	Diameter, Mils	Circular-mil Area
1	289.3	83,690	21	28.46	810.1
2	257.6	66,370	22	25.35	642.4
3	229.4	52,640	23	22.57	509.5
4	204.3	41,740	24	20.10	404.0
5	181.9	33,100	25	17.90	320.4
6	162.0	26,250	26	15.94	254.1
7	144.3	20,820	27	14.20	201.5
8	128.5	16,510	28	12.64	159.8
9	114.4	13,090	29	11.26	126.7
10	101.9	10,380	30	10.03	100.5
11	90.74	8,234	31	8.928	79.70
12	80.81	6,530	32	7.950	63.21
13	71.96	5,178	33	7.080	50.13
14	64.08	4,107	34	6.305	39.75
15	57.07	3,257	35	5.615	31.52
16	50.82	2,583	36	5.000	25.00
17	45.26	2,048	37	4.453	19.83
18	40.30	1,624	38	3.965	15.72
19	35.89	1,288	39	3.531	12.47
20	31.96	1,022	40	3.145	9.88

TABLE 10-3

Material	Resistivity in Ω-cm at 20°C	
Silver	1.6×10^{-6}	(conductor)
Copper	1.7×10^{-6}	(conductor)
Carbon	4×10^{-3}	(semiconductor)
Germanium	65	(semiconductor)
Silicon	55×10^{3}	(semiconductor)
Glass	17×10^{12}	(insulator)
Hard rubber	10^{18}	(insulator)

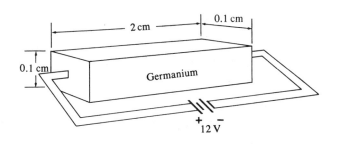

FIGURE 10-4

10.6 Effect of Temperature on Resistance

Temperature will have some effect on the resistivity of most materials. As a result, the resistance of most materials will change with temperature. Certain materials will have their resistance *increase* with an *increase* in their temperature. Most metal conductors such as copper, silver, nichrome, and tungsten behave in this manner. Certain other materials will have their resistance *decrease* with an *increase* in temperature. Most semiconductor materials (carbon, silicon, germanium, etc.) are in this category. A relatively small number of materials are essentially unaffected by changes in temperature. The metal alloys *constantan* and *manganin* are such materials.

Every material has associated with it a characteristic called its *temperature coefficient of resistance*, which is given the symbol α (Greek letter *alpha*). The temperature coefficient α states how much the material's resistance will change for a change in temperature. If the value of α is known, the following formula can be used to determine the value of resistance at any temperature.

$$R_X = R_{20} + \alpha R_{20} (T_X - 20°) \qquad (10\text{--}3)$$

In this formula R_{20} represents the resistance at 20°C (room temperature), T_X is the temperature at which we are calculating R_X (the new value of resistance).

As an example, consider a copper wire that has a resistance of 1.25 Ω at 20°C. Thus, $R_{20} = 1.25$ Ω. Copper has a temperature coefficient $\alpha = 0.004$. To determine this wire's resistance at 50°C we can use Eq. (10–3) with $T_X = 50°C$:

$$\begin{aligned} R_X &= 1.25 + 0.004 \times 1.25 \ (50 - 20) \\ &= 1.25 + 0.15 \\ &= 1.4 \ \Omega \ (\text{at } 50°C) \end{aligned}$$

The resistance of the copper wire increases with the increase in temperature.

Table 10–4 lists the values of α for various materials. Note that the metals tungsten, copper, silver, and nichrome have positive values of α while carbon has a negative value of α. A *negative* temperature coefficient indicates that resistance will decrease with an increase in temperature as the following example shows.

TABLE 10–4

Material	Temperature Coefficient α
Copper	+0.004
Silver	+0.004
Nichrome	+0.0002
Tungsten	+0.005
Carbon	−0.0003
Manganin	≈ 0
Constantan	≈ 0

EXAMPLE 10.6 A certain specimen of carbon has a resistance of 325 Ω at 20°C. What will its resistance be at 70°C?

Solution: Using (10–3) with $T_X = 70°C$, $R_{20} = 325\ \Omega$, and $\alpha = -0.0003$ we have

$$R_X = 325 + (-0.0003) \times 325 \times (70 - 20)$$
$$= 325 - 4.88$$
$$= \mathbf{320.12\ \Omega}$$

Thus, the resistance has decreased with a 50°C increase in temperature.

The table also shows that manganin and constantan have values of α which are approximately zero, indicating that these metals will not exhibit a significant change in resistance with temperature [formula (10–3) becomes $R_X = R_{20}$ if $\alpha = 0$]. As such, these two materials are often used to make precision wire-wound resistors whose resistance is unaffected by temperature.

Temperature Increase Due to Power Dissipation If a wire conductor is conducting a large amount of current, then we can expect that the power dissipated in the wire's resistance will produce heat. The heat which is produced will cause the wire's resistance to increase (assuming α is positive). For example, a 100-W incandescent light bulb will have a current of 0.83 A when it is operated at 120 V ($I = P/E$). Thus, we can calculate the resistance of the tungsten filament as

$$R = \frac{120\ \text{V}}{0.83\ \text{A}} = 144\ \text{ohms}$$

when the filament is hot. If the resistance of the filament is measured with an ohmmeter when the bulb is not hot, it would measure around 10 ohms.

Thus, the tungsten filament has a "cold" resistance of 10 ohms, but when it conducts current and becomes heated its "hot" resistance is 144 ohms. This means that when voltage is initially applied to the bulb, the filament current will be extremely high (12 A) since R (cold) is 10 ohms. However, as the filament heats up, its resistance increases and the current decreases to 0.83 A. The large "surge" of current when the bulb is turned on could be a problem except for the fact that the thin filament heats up very rapidly (in less than 1 ms) and the current quickly drops to its low value. Otherwise, the surge current would probably blow a fuse or trip a circuit breaker.

Because of the characteristic described here, tungsten is not used in heater elements for electrical appliances (such as stoves) which take a few seconds to heat up. These heater elements are usually made from some alloy metal such as nichrome which has a very low temperature coefficient. The lower temperature coefficient means that there won't be such a large difference in "cold" and "hot" resistances.

Fuses Many circuits use *fuses* as protection against excessive current flow. Fuses are simply very thin wires usually made of aluminum, nickel, or tin-coated copper. If the current through the fuse increases to an excessive amount, the heat generated by the power dissipation (I^2R) in the fuse will cause the thin wire to melt and break open. The "blown" fuse will then open the circuit so that current cannot flow. The purpose is to let the fuse "blow" before the excessive current can damage any of the other elements in the circuit.

A fuse is rated according to the amount of current it takes to cause it to open. Fuses are available in current ratings from 2 mA up to several hundred amperes. A fuse also has a voltage rating. This is because an *open* fuse will often have the full applied voltage across its terminals, as illustrated in Fig. 10–5. As such, an excessive amount of voltage across the open fuse could cause it to *arc* and conduct current. A typical fuse rating might be 5 A, 125 V.

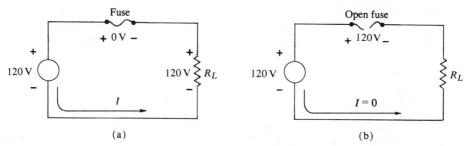

FIGURE 10–5 (a) *A good fuse acts as a very low resistance (short circuit) with almost 0 V across its terminals.* (b) *A blown fuse is an open circuit and has the full 120 V across its terminals.*

Thermal Circuit Breakers As we know, a blown fuse has to be replaced before the circuit will operate again. Often, to avoid this inconvenience circuit breakers are used to interrupt the flow of current. Circuit breakers generally consist of a thermal element in the form of a spring which expands with heat. Excessive current causes the spring to expand and snap open. The circuit breaker can be manually reset after the cause of the excessive current is removed.

Chapter Summary

1. The four major factors which determine the resistance of a piece of material are: (a) type of material; (b) length; (c) cross-sectional area; and (d) temperature.
2. Resistance is directly proportional to length (l) but inversely proportional to cross-sectional area (A).
3. The basic formula for finding the resistance of a piece of material is $R = \rho l / A$ where ρ represents the resistivity or specific resistance of the material.
4. Resistivity is an innate characteristic of each type of material and is also temperature-dependent.
5. The circular mil (cmil) is a unit of area equal to the area of a circle that has a diameter of 1 mil (10^{-3} inches).
6. The circular cross-sectional area (A) of a wire is given by $A = D^2$ where A is in cmils and D is in mils.

Physical Characteristics of Resistance 215

7. The units of resistivity ρ are cmil-Ω/ft. These units are mostly used for calculation of wire resistance.
8. North American wire manufacturers conform to the AWG system of wire sizes listed in Table 10–2.
9. The Ω-cm is another unit which is used for resistivity ρ. This unit is usually used when determining the resistance of a piece of material which is not in the shape of a wire.
10. The temperature coefficient (α) for a given material is a measure of how much the material's resistance will change with a change in temperature.
11. A positive value for α indicates that R will increase as temperature increases, while a negative α indicates that R will decrease as temperature increases.
12. Usually, a wire conductor has a greater resistance when it is conducting current ("hot" resistance) than when it is not conducting ("cold" resistance).

Questions/Problems

10–1 List the four major factors that determine the resistance of a piece of material.

10–2 A certain length of copper wire, AWG #22, has a room-temperature resistance of 3.5 Ω. Determine the effect each of the following changes will have on its resistance:
  a. doubling its length
  b. doubling its cross-sectional diameter
  c. heating it to 50°C
  d. cutting it in half and putting the two halves in parallel
  e. changing to an AWG #25 wire of the same length

10–3 Using the wire table (Table 10–2) calculate the resistance of 250 ft of AWG #20 nickel wire at 20°C.

10–4 Repeat 10–3 for 70°C ($\alpha = 0.006$ for nickel).

10–5 Indicate which of the following statements are *true:*
  a. The resistance of all metals increases with temperature.
  b. Any AWG #22 copper wire has more resistance than any AWG #21 copper wire.
  c. A negative temperature coefficient indicates that resistance will increase with a decrease in temperature.
  d. Cutting a piece of material in half will reduce its resistivity and its resistance by half.
  e. A sheet of silicon which is 1 meter on each side and is 1-cm thick will have the same resistance as a cube of silicon which is 1 cm $\times$ 1 cm $\times$ 1 cm.
  f. A higher gauge wire can handle more current than a lower gauge wire.

10–6 The circuit in Fig. 10–6 is to be protected by a fuse as shown. The fuse has to be chosen so that it will blow if the load is short-circuited. Which of these fuse ratings would be satisfactory for this application: (a) 5 A, 125 V, (b) 5 A, 250 V, (c) 10 A, 250 V?

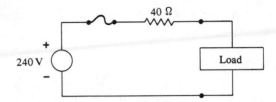

FIGURE 10-6

10-7 A *thermistor* is a *resistor* whose resistance changes with temperature. It is used in certain circuit applications including temperature compensation and temperature sensing. Figure 10-7 shows a thermistor being used in a circuit which produces an output voltage which varies proportionally with temperature. The 100-V source and series 100-kΩ resistor act as an approximate 1-mA constant current source since the thermistor R_T has a resistance of only around 1 kΩ. If the thermistor has a resistance of 1000 ohms at 20°C and its temperature coefficient $\alpha = 0.008$, determine the output voltage at the following temperatures: 0°C, 20°C, 30°C, 40°C, 50°C, and 100°C. Plot V_{out} vs temperature (°C).

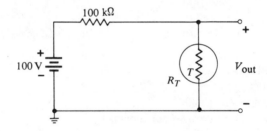

FIGURE 10-7

10-8 When large amounts of current are delivered to a load which is a long distance away, the wires which connect the source to the load must be chosen so that the voltage dropped across the wire resistance is not significant. For example, consider Fig. 10-8 where a 120-V source is supplying voltage and current to a 12-ohm load over a distance of 200 feet.
   a. Determine the current and voltage delivered to the load if AWG #16 copper wire is used. AWG #16 copper wire has 0.004-ohm resistance per foot.
   b. Repeat for AWG #19 copper wire.

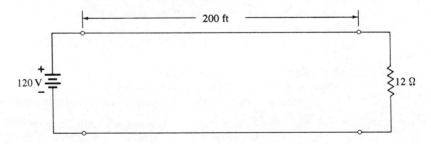

FIGURE 10-8

10-9 The tungsten bulb in the circuit of Fig. 10-9 is a 120-V, 300-W bulb. Its "cold" resistance is 8 ohms (measured with an ohmmeter). Calculate the surge current when the switch is initially closed. Calculate the bulb's "hot" resistance.

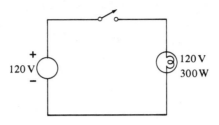

FIGURE 10-9

CHAPTER 11

Resistive Circuit Analysis Techniques

11.1 Introduction

In the preceding chapters we have seen how to analyze simple series, parallel, and series-parallel circuits using the principles of Ohm's law, KVL, and KCL. Many circuits that occur in electronics can be analyzed using these techniques. However, there are many circuits that are complex or contain more than one source and so require different analysis methods.

In this chapter we will study circuit analysis techniques that can be applied to almost all linear resistive circuits. Some of these techniques will involve some form of circuit simplification to facilitate reaching a solution, while others will use a systematic application of KVL and KCL.

11.2 The Method of Superposition

Many electronic circuits contain more than one voltage source. Except for the simple case of two sources in series we have not yet considered such *multisource* circuits. An example of such a circuit is shown in Fig. 11-1. Such circuits cannot be analyzed using any of the techniques of the previous chapters.

A principle called the *superposition principle* is the basis of a method by which linear, multisource networks can be analyzed. The *superposition principle* can be stated as follows:

> In a linear circuit containing more than one source, the current (or voltage) for any element in the circuit can be found by superimposing the individual

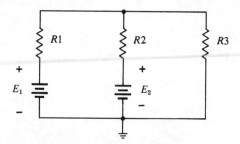

FIGURE 11-1 *Example of a multisource circuit.*

currents (or voltages) produced through the element by each source acting alone with the other sources de-energized.

In other words, a multisource circuit may be analyzed by considering only *one* source at a time and then superimposing the results due to each source. When using one source at a time, all the other sources must be de-energized. A *voltage source* is de-energized by not allowing it to generate any voltage. Thus, its voltage is reduced to 0 V, which means it is actually replaced by a *short circuit**. A *current source* (discussed briefly in chap. 4) is de-energized by not allowing it to produce any current. Thus, it is replaced by an *open circuit*.

Applying the Superposition Method Consider the circuit in Fig. 11-2(a). This circuit is merely a series circuit with two sources. We have previously analyzed such a circuit by algebraically combining the two sources. We can do the same thing here. Since the sources are opposing each other, the net source voltage is 15 V and is pushing electron current in a cw direction. Since the total R in the circuit is 5 kΩ, the current I is 15 V/5 kΩ = 3 mA. This method is simple enough and we would normally not solve this circuit using the superposition principle. However, this simple circuit allows us to more clearly illustrate the method of superposition. The step-by-step procedure is as follows:

(1) Consider the circuit with only the 30-V source. Replace the 45-V source by a short circuit as in Fig. 11-2(b). The 30-V source produces a ccw current I' as shown in the figure. The magnitude of I' is easily calculated as 30 V/5 kΩ = 6 mA.

(2) Consider the circuit with only the 45-V source and replace the 30-V source by a short circuit as in part (c) of the figure. The 45-V source produces a cw current I''. The magnitude of I'' is calculated as 45 V/5 kΩ = 9 mA.

(3) Redraw the original circuit and combine the results of steps (1) and (2) as in part (d) of the figure. Since the currents I' and I'' are flowing in opposite directions, the net result is $I = I'' - I' = 3$ mA in a cw direction. This agrees with our previous result.

Consider the circuit of Fig. 11-3(a) which *cannot* be solved directly using Ohm's law. Let's assume that it is desired that we determine the current through the load resistor $R_L = 3$ kΩ. Using the superposition method, the step-by-step procedure is as follows:

(1) Consider the circuit with only the 16-V source present and the 8-V source replaced by a short circuit as in part (b) of the figure. The current I'_L

---

*We assume that any internal source resistance is already included as part of the circuit.

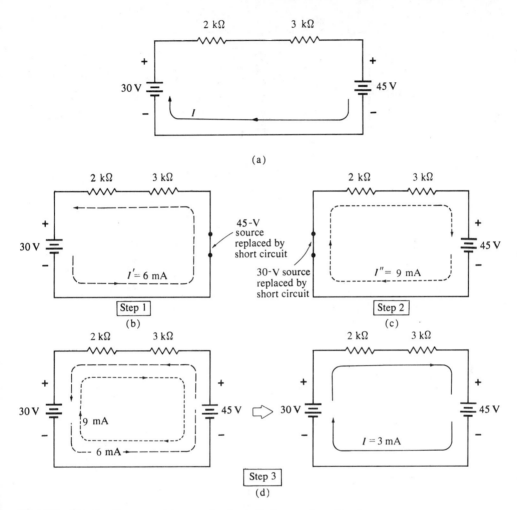

FIGURE 11-2 *Superposition method applied to series circuit.*

through the load produced by the 16-V source can be determined using the techniques for series-parallel circuits discussed in chap. 8. The result is $I'_L = 4/3$ mA flowing *up* through R_L as shown. (This result should be verified by the student.)

(2) Consider the circuit with the 8-V source present and the 16-V source replaced by a short circuit as shown in part (c) of the figure. The current I''_L through the load produced by the 8-V source can be found to be 2/3 mA flowing *down* through R_L as shown.

(3) Redraw the original circuit and combine I'_L and I''_L to find the total resultant load current I_L. This is done in Fig. 11–3(d). Since I'_L and I''_L flow in opposite directions the resultant I_L is calculated as $I'_L - I''_L = 2/3$ mA flowing upward. The voltage across the load can now be calculated as 2/3 mA $\times$ 3 kΩ = 2 V.

Limitations of the Superposition Method If the superposition method is to be used, one must be aware of the limitations of the method so that it can be used

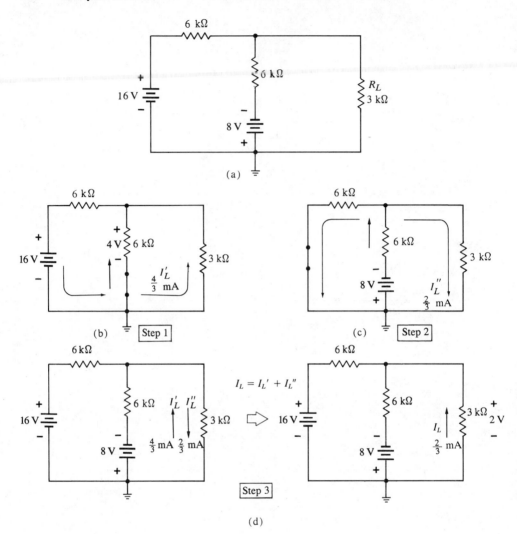

FIGURE 11-3 *Another example of applying the superposition method to a multisource network.*

intelligently. The superposition principle is valid only in circuits where all the circuit elements are *linear, bilateral* devices. In other words, the devices must have a linear *I-V* relationship and they must conduct current equally well in both directions. Thus, a circuit containing diodes, for instance, generally cannot be analyzed using the superposition method.

The superposition method *cannot* be used to find the total resultant power in a load by finding the load power produced by each source, one at a time. For example, in the circuit of Fig. 11–3, the power in the load when considering only the 16-V source is

$$P'_L = (I'_L)^2 \times R_L = (4/3 \text{ mA})^2 \times 3 \text{ k}\Omega$$
$$= 5.33 \text{ mW}$$

The power in the load when considering only the 8-V source is

$$P''_L = (I''_L)^2 \times R_L = (2/3 \text{ mA})^2 \times 3 \text{ k}\Omega = 1.33 \text{ mW}$$

The actual load power is produced by the resultant value of load current I_L. Thus,

$$P_L = I_L^2 \times R_L$$
$$= (2/3 \text{ mA})^2 \times 3 \text{ k}\Omega$$
$$= 1.33 \text{ mW}$$

Note that $P_L \neq P_L' + P_L''$. In other words, we can't superimpose P_L' and P_L'' to find P_L. *The superposition method holds only for currents and voltages.*

11.3 Practical Voltage Source

It may be recalled that a practical voltage source can be represented by an ideal voltage source in series with a resistance. Because of this series resistance the terminal voltage of the practical source will never equal the ideal source voltage except in the open-circuit (no-load) condition. When a load is connected to the source terminals, the terminal voltage decreases.

This same characteristic can be said to be true of *any* linear network containing one or more sources. In other words, the output voltage of such a network will be greatest when there is no load connected to the output terminals. When a load is connected, the output voltage generally decreases. Thus, any linear network that contains sources behaves like a single practical voltage source as far as the output is concerned. In fact, in the next section we will introduce a technique for representing such a network by an equivalent practical voltage source.

11.4 Thevenin's Theorem

Named after a French engineer, M. L. Thevenin, this theorem forms the basis of one of the most useful methods of simplifying complex networks. Thevenin's theorem states that:

Any network containing linear resistances and voltage and/or current sources connected between terminals x-y may be replaced by a single practical voltage source connected between these same terminals.

This is illustrated in Fig. 11–4.

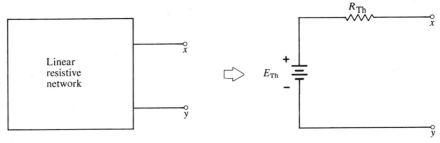

FIGURE 11–4 *Thevenin's theorem allows us to replace the complex network at the left by the practical voltage source at the right.*

In this figure the box at the left contains a linear, resistive network. In other words, the box can contain resistors and sources, any number of each connected in any manner. Terminals x and y represent any two points in the network. Frequently, x and y are the output terminals of the network. Thevenin's theorem states that the entire network in the box connected to terminals x and y can be replaced by a voltage source E_{Th} in series with a resistor R_{Th} connected to x and y as shown in the righthand part of the figure. This series combination is called the *Thevenin equivalent circuit* for the original network between terminals x and y. The source E_{Th} is the Thevenin equivalent source and the resistor R_{Th} is the Thevenin equivalent resistance.

Usefulness of Thevenin's Theorem Before discussing the method for determining E_{Th} and R_{Th}, let's consider the implications of Thevenin's theorem by referring to Fig. 11–5. The circuit in part (a) of the figure is a complex series-parallel circuit. Suppose we connect a load to the x-y terminals (shown dotted in the figure), and we want to know the load voltage for several different values of the load. We could use the methods of chap. 8 since it is a series-parallel circuit. However, we would have to repeat the analysis for each different value for the load. This could be a laborious process.

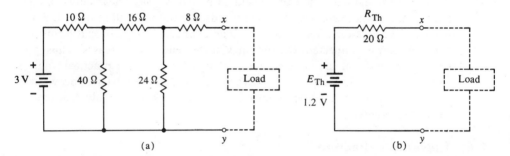

FIGURE 11–5 *Thevenin's equivalent used to determine output voltage for varying load.*

An alternative approach would be to replace the complex network by its Thevenin equivalent circuit [part (b) of the figure]. This equivalent circuit will produce the same voltage at terminals x and y as the original network. Thus, we can use the simple equivalent circuit to find the load voltage for any load by connecting the load between x and y of the equivalent circuit. The resultant circuit is nothing but a simple series circuit which can easily be analyzed for various values of the load.

Finding E_{Th} The value of the voltage source E_{Th} to be used in the Thevenin equivalent circuit is found by calculating or measuring the voltage across the x-y terminals with no external circuit element or load connected to x-y. In other words, E_{Th} is the open-circuit voltage across the x-y terminals. That is, $E_{Th} = V_{xy}$ (no load).

Finding R_{Th} The value of the resistance, R_{Th}, is found by calculating or measuring the equivalent resistance seen looking *back into* the network at terminals x-y with no load and with all sources de-energized. In other words, if all the sources in the network are de-energized then only resistors will remain. The equivalent resistance between x-y is R_{Th}. It can be measured by putting an ohmmeter across x and y with all the network sources de-energized and with no load connected to x-y.

11.5 Thevenizing (Finding a Thevenin Equivalent Circuit)

Consider the network in Fig. 11–6(a). A voltage divider is formed by $R1$ and $R2$. The voltage-divider output terminals are x and y. It is desired to find the output voltage for various sized loads connected to x-y. The procedure will be to take the voltage-divider network (without a load) and find its Thevenin equivalent circuit between x and y. The equivalent circuit will then be used to drive the load. This process is called *Thevenizing*.

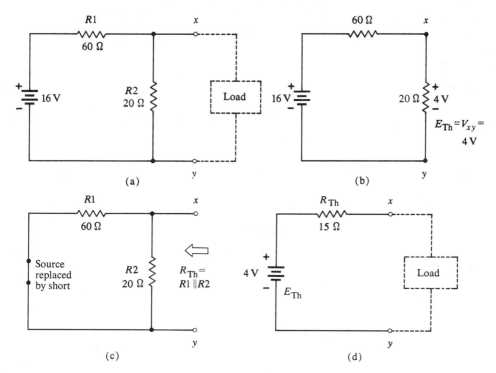

FIGURE 11–6 *Thevenizing a circuit.*

We will begin the process by determining E_{Th}. This is done by taking the load off the circuit and calculating V_{xy} as shown in part (b) of the figure. Using the voltage-divider rule, V_{xy} is calculated as

$$V_{xy} = \left(\frac{20}{60 + 20}\right) \times 16 \text{ V} = 4 \text{ V}$$

Thus, the Thevenin equivalent source is $E_{Th} = 4$ V.

The next step is to determine R_{Th}, which is the equivalent resistance seen between x and y with the 16-V source replaced by a short circuit as in Fig. 11–6(c). Looking in from the x-y terminals the 60-ohm and 20-ohm resistors are in parallel. In other words, if an ohmmeter were placed across the x-y terminals, it would read a resistance value of

$$60 \text{ } \Omega \parallel 20 \text{ } \Omega = \frac{60 \times 20}{60 + 20} = 15 \text{ } \Omega$$

Thus, $R_{Th} = 15$ Ω.

We can now construct the complete Thevenin equivalent circuit using the calculated values of E_{Th} and R_{Th} as shown in Fig. 11–6(d). Note that the terminals x and y correspond to the same terminals in the original circuit. This equivalent circuit can now be used to calculate the voltage or current for any load connected to the x-y terminals. The results so obtained will be exactly the same as they would be if the load were connected to the original circuit. To verify this, let us calculate V_{xy} with a load $R_L = 60$ ohms for both the original circuit [Fig. 11–6(a)] and the equivalent circuit.

Consider first the 60-ohm load on the original circuit. The 60-ohm load is in parallel with $R2$. This parallel combination is equivalent to 15 ohms between x and y. Using the voltage divider rule between this 15 ohms and $R1$ results in

$$V_{xy} = \frac{15}{15 + 60} \times 16 \text{ V}$$
$$= 3.2 \text{ V}$$

This same 60-ohm load applied to the Thevenin equivalent circuit will produce a V_{xy} of (using divider rule)

$$V_{xy} = \frac{60}{15 + 60} \times 4 \text{ V}$$
$$= 3.2 \text{ V}$$

Thus, the two results are the same, as expected.

Some Comments on R_{Th} When we found R_{Th} for the circuit of Fig. 11–6 it may have been disconcerting that $R1$ and $R2$ were said to be in parallel [Fig. 11–6(c)]. $R1$ and $R2$ are indeed in parallel if we are looking at the circuit from terminals x-y. In other words, if we placed a source between x-y in Fig. 11–6(c), it would see an equivalent resistance of $R1 \parallel R2$.

As far as the 16-V source is concerned, $R1$ and $R2$ are in series as seen from the source terminals. What this means is that a network's equivalent resistance will generally be different depending on the terminals at which it is being measured or calculated.

Another Thevenizing Example We will now apply Thevenin's theorem to the circuit which was previously analyzed using the superposition method (Fig. 11–3). The circuit diagram is repeated in Fig. 11–7(a). Let's suppose that we want to determine the output voltage V_{xy} for various values of R_L. To do this we can Thevenize the circuit between x-y and use the Thevenin equivalent circuit to drive the load.

The first step is to remove the load and calculate the voltage V_{xy} which will be Thevenin equivalent voltage E_{Th}. This is shown in Fig. 11–7(b). With the load removed the circuit is just a series circuit. The two sources are aiding each other and pushing electron current in a ccw direction. The magnitude of this current is $I = (16 \text{ V} + 8 \text{ V})/12 \text{ k}\Omega = 2$ mA as shown. This 2-mA current indicates that the voltage across each resistor is 2 mA $\times$ 6 kΩ = 12 V. We can now find $V_{xy} = +4$ V by taking either path from y to x. Thus, $E_{Th} = +4$ V.

The next step is to find R_{Th} by replacing *both* sources by short circuits and calculating the resistance looking in from terminals x-y. This is shown in Fig. 11–7(c). As can be seen, the value of R_{Th} is equal to 6 k$\Omega \parallel$ 6 kΩ, which

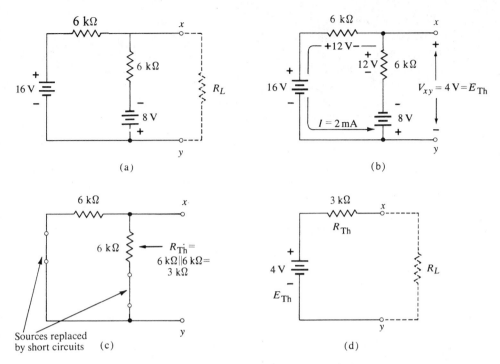

FIGURE 11-7 *Another example of applying Thevenin's theorem.*

equals 3 kΩ. We can now construct the Thevenin equivalent circuit shown in part (d) of the figure. This equivalent circuit can be used to calculate the voltage for any value of R_L.

For example, with $R_L = 3$ kΩ the equivalent circuit produces

$$V_{xy} = \frac{3 \text{ k}\Omega}{6 \text{ k}\Omega} \times 4 \text{ V} = 2 \text{ V}$$

which agrees with the result obtained previously using the method of superposition (Fig. 11-3).

Limitations of Thevenin's Theorem Like the superposition method, the application of Thevenin's theorem is limited to networks containing only *linear* elements. In other words, the portion of the network which is being Thevenized must contain only linear elements. This *doesn't* mean that the load on the circuit has to be a linear device. For example, in Fig. 11-7 the load could be a nonlinear device such as a diode or transistor. This is all right because the load is not included in that portion of the network which is being Thevenized.

Although Thevenin's theorem is restricted to linear networks, it is very often used in circuits containing nonlinear devices if these devices are operating on the linear portions of their *I-V* curves. As such, you will find Thevenin's theorem often applied to complicated amplifier circuits as a means of characterizing the amplifier by its simple Thevenin equivalent circuit.

When a network is replaced by its Thevenin equivalent circuit, the equivalent circuit can only be used to determine voltages, currents, and powers in the *external circuit* connected to its terminals. In other words, any load or circuit connected to the Thevenin equivalent circuit terminals will have the same I, V, and P that it would have if connected to the original un-Thevenized network. However, any I, V, or P which is internal to the original network will be lost or obscured when the network is replaced by its Thevenin equivalent. For example, using Fig. 11-7(d) we cannot determine the current supplied by the 16-V source in the original network of 11-7(a).

If one is interested in the currents and voltages internal to the network as well as in the load, the superposition method is probably more useful since it can be used to solve for all circuit currents and voltages.

11.6 Maximum Power Transfer From a Network

In chap. 9 it was shown that a *practical* voltage source will deliver maximum power to a load when the load resistance equals the internal source resistance. Now that Thevenin's theorem has been introduced, we can extend this maximum power principle to the situation where a *network* is delivering power to a load.

A network will deliver maximum power to a load when the load resistance is equal to R_{Th}, the Thevenin equivalent resistance of the network.

For example, for the circuit of Fig. 11-7 the network containing the two sources will deliver maximum power to a load resistor of 3 kilohms. This isn't hard to understand because the network can be replaced by its Thevenin equivalent circuit, which is essentially a practical voltage source with an internal source resistance equal to R_{Th}.

11.7 Current Sources

In chap. 4 we were briefly introduced to the *ideal current source*. An ideal current source is a source of electrical energy that allows a fixed amount of current to flow through it. The symbol for an ideal *dc* current source is shown in Fig. 11-8(a) where the current source terminals are connected to a load resistor R_L.

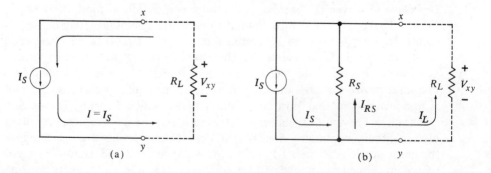

FIGURE 11-8 (a) *Ideal dc current source, and* (b) *practical dc current source driving a load* R_L.

The ideal current source will produce a constant current equal to I_S through the load no matter what the value of R_L is. The voltage V_{xy} will equal the product $I_S \times R_L$ and therefore depends on R_L. For example, suppose $I_S = 0.1$ A. Then, if $R_L = 100\ \Omega$ the value of V_{xy} will be $0.1 \times 100 = 10$ V. If $R_L = 1000\ \Omega$ the value of V_{xy} will be $0.1 \times 1000 = 100$ V. Since V_{xy} is also the voltage across the current source, then it can be stated that:

An ideal constant current source I_S will produce a constant current through any load connected to its terminals; but its terminal voltage will change according to the value of $I_S \times R_L$.

In other words, the terminal voltage of the current source will vary but its current will always be the same. This is in contrast to an ideal constant voltage source, which will have a constant terminal voltage but its current will vary with the load.

Practical Current Source An ideal current source cannot be realized in practice because it would have to be able to generate impossibly high voltages for large values of load resistance. To illustrate, suppose $I_S = 0.1$ A and R_L was 10 MΩ. The 0.1 A flowing through the 10-MΩ resistance would result in $V_{xy} = 0.1 \times 10,000,000 = 1,000,000$ volts! Most practical current sources would not be able to generate such a high voltage. As such, we would find that with such a *large* value of R_L across its terminals, the *output current* of a *practical* current source would *drop*. This is in contrast to a practical voltage source whose *output voltage* will *drop* when a *small* R_L is connected to its terminals.

How do we represent this characteristic of a practical current source? We represent it by including an *internal source resistance* R_S in *parallel* with the ideal current source as shown in Fig. 11–8(b). This parallel R_S is the counterpart of the series R_S included with a practical voltage source. Let us analyze the effect of R_S when a load is connected to the practical current source.

The load R_L and the source resistance are in parallel. Thus, the current I_S will divide up between the two. When R_L is much smaller than R_S, most of I_S will flow through the load. When R_L is much larger than R_S, most of I_S flows through resistor R_S. When R_L is much larger than R_S, most of I_S will flow through R_S and very little through the load. For example, let's suppose $I_S = 100$ mA, and $R_S = 1000$ ohms. Table 11–1 shows the results for the various values of R_L. Examine in the table especially the values of load current, I_L. For any value of R_L we can use the rules for parallel circuits to determine the load current. As the table shows, the load current is maximum when the load is a short circuit ($R_L = 0$). In this instance all of $I_S = 100$ mA flows through the short circuit, and none flows through R_S. As R_L is increased, the load current decreases and the current through R_S increases. When $R_L = R_S = 1000$ ohms, the 100 mA divides evenly between R_L and R_S. Further increases in R_L find the value of I_L dropping further. The extreme case of $R_L = \infty$ (open circuit) results in zero load current.

From these results it can be concluded that for a practical current source the load current will be maximum when R_L is very small compared to R_S. Thus, a current source with a *large* value of R_S is desirable since it means that the load current will be constant at its maximum value over a wide range of R_L values. This is in contrast to a voltage source where a *low* value of R_S is desirable. In

TABLE 11-1

R_L	I_L Current Through R_L	I_{RS} Current Through R_S	V_{xy}, Terminal Voltage
0 Ω	100 mA	0 mA	0 V
10 Ω	99 mA	1 mA	1 V
100 Ω	91 mA	9 mA	9 V
1000 Ω	50 mA	50 mA	50 V
10000 Ω	9 mA	91 mA	91 V
100000 Ω	1 mA	99 mA	99 V
∞ Ω	0 mA	100 mA	100 V

fact, if $R_S = \infty$ (open circuit) the current source becomes an ideal current source.

Thevenin's Equivalent of a Practical I Source Figure 11-9(a) shows a practical current source with source current I_S and source resistance R_S. We can find the Thevenin equivalent of this current source by Thevenizing between terminal x-y. The first step, shown in part (b) of the figure, is to calculate E_{Th}, which is the voltage V_{xy} with no load across the output. As can be seen, $V_{xy} = I_S \times R_S$ since all of I_S flows through R_S. Thus, $E_{Th} = I_S R_S$.

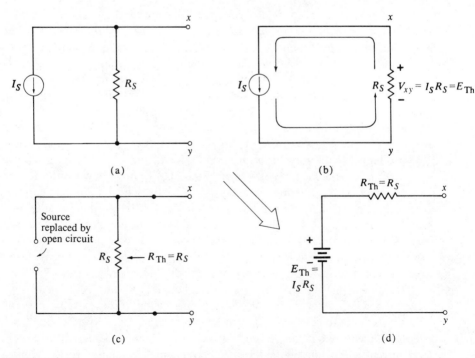

FIGURE 11-9 Converting a practical current source (a) to its Thevenin equivalent (d).

To find the Thevenin equivalent resistance R_{Th}, the current source is de-energized. This is shown in part (c) where the current source has been replaced by an open circuit ($I_S = 0$). Thus, R_{Th} is easily seen to be equal to R_S. Part (d) of the figure shows the complete Thevenin equivalent.

What this shows us is that a practical current source I_S with source resistance R_S is equivalent to a practical voltage source with $E_S = I_S R_S$ and a source resistance also equal to R_S. For example, a current source with $I_S = 0.1$ A and $R_S = 1000$ ohms is equivalent to a practical voltage source with $E_S = 0.1$ A $\times$ 1000 ohms = 100 V and $R_S = 1000$ ohms.

EXAMPLE 11.1 Find the equivalent practical voltage source for a current source with $I_S = 20$ mA and $R_S = 10$ kΩ. Then verify that they are equivalent by determining the load current for each when $R_L = 15$ kΩ.

Solution:
$$E_S = I_S \times R_S = 20 \text{ mA} \times 10 \text{ k}\Omega$$
$$= 200 \text{ V}$$
$$R_S = 10 \text{ k}\Omega$$

The current source and its equivalent voltage source are both shown in Fig. 11–10 connected to a 15-kΩ load.

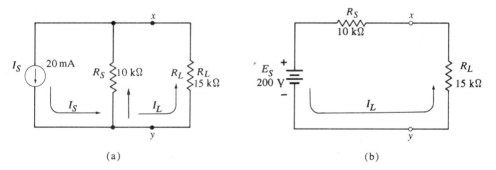

FIGURE 11–10

Considering the current source first, we can calculate the current in the load by first calculating V_{xy}. The equivalent resistance of $R_S \parallel R_L$ is 6 kΩ. Thus, the 20-mA current flows through an equivalent of 6 kΩ and produces $V_{xy} = 20$ mA $\times$ 6 kΩ = 120 V. Since this 120 V is across R_L, the value of I_L is

$$I_L = \frac{120 \text{ V}}{15 \text{ k}\Omega} = 8 \text{ mA}$$

For the voltage source, the R_S and R_L are in series. Thus, I_L can be found as

$$I_L = \frac{200 \text{ V}}{25 \text{ k}\Omega} = 8 \text{ mA}$$

As expected, the two results agree. Both circuits are equivalent as far as the load is concerned.

Current Source Applications Current sources are not as common in electronic circuits as are voltage sources. Most electronic laboratories contain many sources of voltage, including batteries and electronic power supplies. If a constant current source is needed for a certain application, it is often constructed by adding a large resistor in series with a voltage source. This is reasonable since we have seen that a practical current source is equivalent to a voltage source with a series resistor. One of the problems at the end of this chapter will illustrate this.

The concept of a current source is often used in the analysis of certain types of amplifier circuits. In such applications, a current source is used as part of the equivalent circuit for amplifying devices such as vacuum tubes, transistors, and FETs. In the next section we will briefly discuss the current source counterpart of Thevenin's theorem.

11.8 Norton's Theorem

In secs. 11.4 and 11.5 we saw how a linear resistive network could be replaced by its Thevenin equivalent practical voltage source. There is a related technique which allows us to replace the same linear resistive network by a *practical current source*. This should not be surprising since we saw in the last section that a practical current source and a practical voltage source are related.

Named after E.L. Norton, a Bell Telephone Laboratory scientist, Norton's theorem states that:

Any network containing linear resistances and voltage and/or current sources connected to terminals *x-y* may be replaced by a single practical current source connected between these same terminals.

This is illustrated in Fig. 11–11 where I_N is the Norton equivalent current source, and R_N is the Norton equivalent resistance.

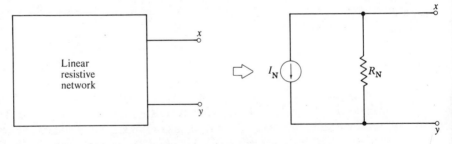

FIGURE 11–11 *Norton's theorem allows us to replace the network at the left by the practical current source at the right.*

Finding I_N The value of the Norton equivalent current source I_N is found by putting a *short circuit* across the *x-y* terminals and finding the value of the current that flows through this short. The value of this short-circuit current is I_N.

Finding R_N The Norton equivalent resistance R_N is found in the same manner as R_{Th}. The network's sources are de-energized and the resistance between terminals *x-y* is measured or calculated. This value is R_N. Obviously, for any network it can be seen that

$$R_N = R_{Th} \tag{11–1}$$

11.9 Nortonizing (Finding a Norton Equivalent Circuit)

To illustrate the application of Norton's theorem we will use the network which we previously Thevenized in Fig. 11–6. It is redrawn in Fig. 11–12(a). To find the Norton equivalent current source I_N we put a short circuit between x-y as shown in Fig. 11–12(b) and calculate the current through the short circuit.

The 20-ohm resistor is in parallel with the short circuit. Thus, its effect can be ignored. The total resistance seen by the 16-V source is 60 ohms. As such, the current through the source is 16 V/60 ohms = 0.267 A. All this current will flow through the short; none will flow through the 20-ohm resistor. The value of I_N is therefore 0.267 A.

To find R_N we follow the procedure illustrated in Fig. 11–12(c). The 16-V source is replaced by a short circuit, and the equivalent resistance between x-y is found. As can be seen, R_N equals the parallel combination of 60 ohms and 20 ohms. Thus,

$$R_N = \frac{60 \times 20}{60 + 20} = 15 \text{ ohms}$$

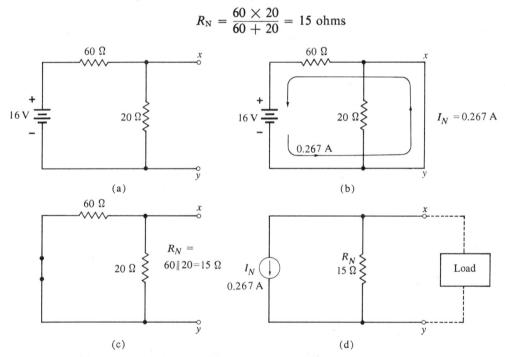

FIGURE 11–12 *Nortonizing a circuit.*

The Norton equivalent circuit can now be constructed as in Fig. 11–12(d). Note the direction of the current source arrow. It is shown as downward. This is because when a short is placed across x-y, current must flow upward through the short from y to x, the same as for the original network [part (b) of the figure]. Another way to look at this is that the current source has to push electron current in a ccw direction, the same direction that the 16-V source in the original circuit would push current.

We can now use this equivalent circuit to calculate the current and voltage for any load connected to the x-y terminals. For example, let's calculate V_{xy} for a load R_L = 60 ohms. The load of 60 ohms is in parallel with the R_N of 15 ohms. Thus, the total resistance between x and y seen by the source is

15 Ω ∥ 60 Ω = 12 Ω. The current of 0.267 A flows through this 12-ohm resistance, thereby resulting in

$$V_{ry} = 0.267 \text{ A} \times 12 \text{ ohms}$$
$$= 3.2 \text{ V}$$

This result agrees with the result obtained using the Thevenin equivalent circuit of Fig. 11-6(d). Thus, we can conclude that both the Norton and Thevenin equivalent circuits can replace a given network for purposes of determining the current and voltage for *any* load on the network.

Comparing Thevenin and Norton Equivalents   If we compare the Thevenin equivalent circuit of Fig. 11-6(b) with the Norton equivalent circuit of Fig. 11-12(d) for the same original network, we can see that $R_N = R_{Th}$ as expected. Also, it appears that E_{Th} and I_N are related according to

$$E_{Th} = I_N \times R_N \qquad (11\text{-}2)$$

To verify,

$$4 \text{ V} = 0.267 \text{ A} \times 15 \text{ ohms}$$
$$= 4 \text{ V}$$

This relationship can also be written as

$$I_N = \frac{E_{Th}}{R_N} = \frac{E_{Th}}{R_{Th}} \qquad (11\text{-}3)$$

Thus, if the Thevenin equivalent circuit for a network is known, we can easily find the Norton equivalent and vice versa.

EXAMPLE 11.2   Find the Norton equivalent circuit for the network of Fig. 11-7(a).

Solution: The Thevenin equivalent for this network was already found, as is shown in Fig. 11-7(d). Using Eq. (11-3) we can find the value of the Norton current source I_N.

$$I_N = \frac{E_{Th}}{R_{Th}} = \frac{4 \text{ V}}{3 \text{ k}\Omega} = 1.33 \text{ mA}$$

Since $R_N = R_{Th} = 3 \text{ k}\Omega$, we can conclude that the Norton equivalent circuit is a 1.33-mA current source in parallel with 3 kΩ.

From the relationship in Eq. (11-2) we can see that

$$R_N = R_{Th} = \frac{E_{Th}}{I_N} \qquad (11\text{-}4)$$

This tells us that the network's Norton or Thevenin resistance can be found by finding the open-circuit voltage (E_{Th}) and dividing by the short-circuit current (I_N). This gives us an alternate method for calculating R_{Th} or R_N. It also provides a method for measuring R_{Th} and R_N experimentally without the necessity for de-energizing the network's sources.

11.10 Conversions Between Current Sources and Voltage Sources

The Thevenin equivalent for a practical current source is a practical voltage source. The Norton equivalent for a practical voltage source is a practical current source. What this means is that any practical voltage source can be converted to a practical current source and vice versa without altering the results at the source terminals. This is illustrated in Fig. 11–13.
varying sources.

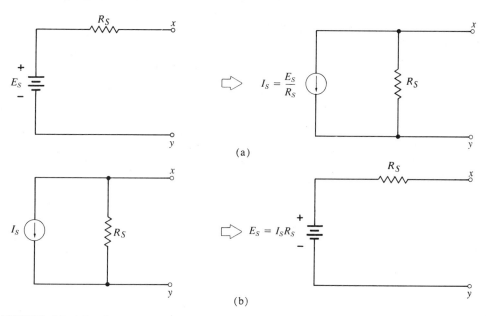

FIGURE 11–13 *Source conversions.*

In part (a) of the figure a practical voltage source is shown being converted to a practical current source. Note that $I_S = E_S/R_S$ and that R_S is the same for both. In part (b) of the figure a practical current source is shown being converted to a practical voltage source. Note that $E_S = I_S \times R_S$ and that R_S is the same for both.

These source conversions can be useful in analyzing certain types of networks. To illustrate, consider the circuit in Fig. 11–14(a). This network contains three voltage sources. Suppose we wanted to find the Thevenin equivalent circuit for this network so we could use it to determine the current through various loads at *x-y*.

Since each branch in the circuit contains a voltage source with a series resistor, we can treat each branch as a practical voltage source. We can take each of the sources and convert it to its equivalent practical current source. This is shown in part (b) of the figure. For example, the 8-V source in series with 100 ohms is converted to an 8-V/100-Ω = 80-mA current source in parallel with 100 ohms. The 12-V and 4-V source branches are treated likewise.

The resultant equivalent circuit in Fig. 11–14(b) can now be considered. It consists of three different current sources in parallel with three different resistors. The three parallel resistors can be combined into one equivalent

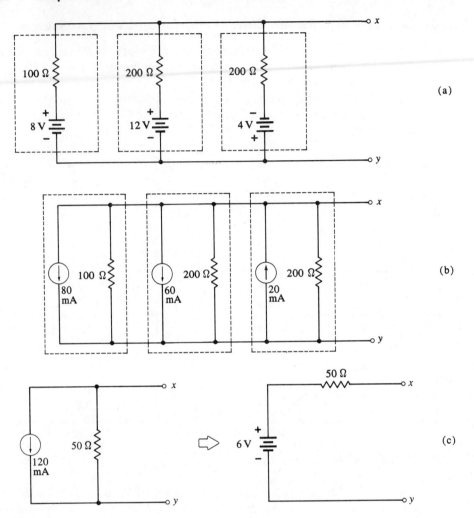

FIGURE 11-14 *Finding the Norton and Thevenin equivalents by using source conversions.*

resistance. That is, 100 Ω || 200 Ω || 200 Ω || = 50 Ω. The 80-mA and 60-mA current sources are both pushing current downward while the 20-mA source is pushing current upward. These three current sources have the same effect as *one* current source pushing 120 mA downward (80 mA + 60 mA − 20 mA). In Fig. 11-14(c) the three current sources have been replaced by one 120-mA source in parallel with the 50-ohm equivalent resistance.

The circuit in Fig. 11-14(c) is in the form of a practical current source which can be easily converted to a practical voltage source. The final result is shown in part (d) of the figure. Note that E_S = 120 mA × 50 Ω = 6 V. This final circuit is the Thevenin equivalent of the original network. It should be apparent that the use of source conversions made the solution of this problem fairly easy.

11.11 The Delta-Wye Transformations

Consider the networks in Fig. 11-15. Suppose we wanted to determine the equivalent resistance as seen by the source for each of these networks. After

Resistive Circuit Analysis Techniques   237

looking at these networks for a while, it becomes apparent that we cannot simplify either of them by combining resistors because neither network contains any series combinations or parallel combinations. In this section we will discuss a new method of equivalent resistances that will help us to solve problems like these.

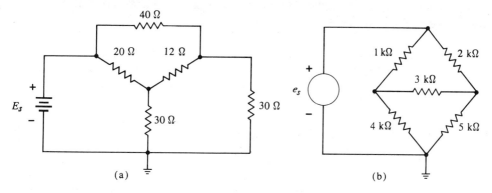

FIGURE 11–15 *Networks where resistors cannot be combined to find* R_{eq}.

The network in Fig. 11–16(a) is called a Y (wye) network, as suggested by its shape. The Y network is characterized by having three resistors joined together at one common point. The network in Fig. 11–16(b) is called a T (tee) network because of its shape. It should be clear that the Y network and the T network are essentially the same network except that R_A and R_B are at an angle in the Y network.

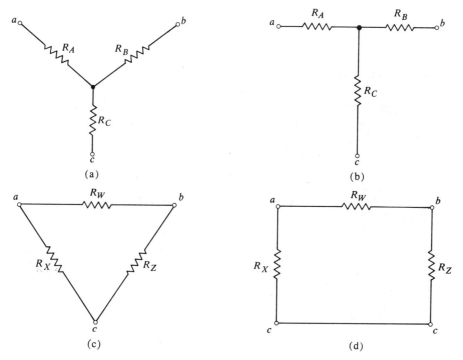

FIGURE 11–16 (a) Y, (b) T, (c)Δ, *and* (d) π *networks.*

The network in Fig. 11–16(c) is called a Δ (delta) network because it is shaped like a Δ. The Δ network is characterized by three resistors connected together end-to-end in a triangular arrangement. The network in Fig. 11–16(d) is called a π (pi) network because the resistors are arranged like the Greek letter π. It should be apparent that the π network and the Δ network are essentially the same network, the only difference being that R_X and R_Z are connected by a short circuit in the π network.

Y-to-Δ Conversion Formulas As we will see later, it is often helpful to be able to convert a Y network (or T) to a Δ network (or π) and vice versa. Let us first consider the conversion from a Y network to a Δ network. In other words, we want to take a Y network like that in Fig. 11–16(a) and convert it to a Δ like that in Fig. 11–16(c). For the conversion to be valid, both networks must have the same equivalent resistance seen between any pair of terminals a, b, c. To help us see how the conversion takes place, refer to Fig. 11–17.

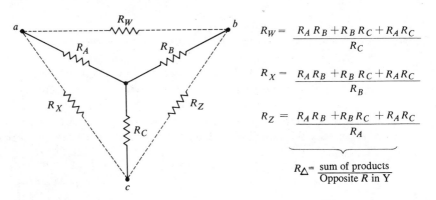

$$R_W = \frac{R_A R_B + R_B R_C + R_A R_C}{R_C}$$

$$R_X = \frac{R_A R_B + R_B R_C + R_A R_C}{R_B}$$

$$R_Z = \frac{R_A R_B + R_B R_C + R_A R_C}{R_A}$$

$$R_\Delta = \frac{\text{sum of products}}{\text{Opposite } R \text{ in Y}}$$

FIGURE 11–17 *Conversion formulas for* Y-*to-*Δ, *or* T-*to-*π *conversion.*

Here we have the Y network, with terminals a, b, and c, shown in solid lines. It is to be converted to an equivalent Δ network connected between the same terminals and shown in dotted lines. The conversion formulas relating the Δ resistors (R_W, R_X, and R_Z) to the Y resistors (R_A, R_B, and R_C) are shown below the diagram in the figure. These formulas can be used to convert a Y or T network to an equivalent Δ or π network.

Note that the three formulas have the same general form, as indicated by the basic rule on the right. This basic rule is probably easier to remember and use than the three individual formulas. The basic rule states that any of the Δ resistors (R_Δ) can be found by first taking all possible products of the Y resistors, using two at a time. There are three such products. The *sum* of these *products* is then divided by the value of the Y resistor, which is opposite the Δ resistor being calculated. For example, when calculating R_W, the opposite Y resistor is R_c (see figure); for R_X the opposite Y resistor is R_B; and for R_Z the opposite Y resistor is R_A.

EXAMPLE 11.3 Convert the Y network in Fig. 11–18(a) to its equivalent Δ network.

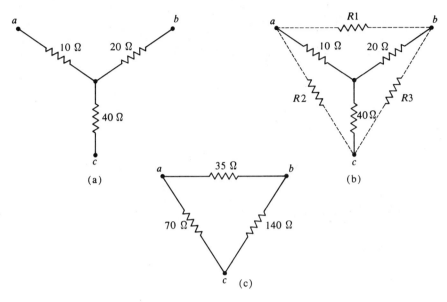

FIGURE 11–18

Solution: The Y network is redrawn in part (b) of the figure along with the desired Δ network shown in dotted lines. To find the values of the Δ resistors we will use the basic rule given in Fig. 11–17.

$$R1 = \frac{\text{sum of products}}{\text{opposite } R \text{ in Y}}$$
$$= \frac{10 \times 20 + 10 \times 40 + 20 \times 40}{40}$$
$$= \frac{1400}{40} = 35 \text{ ohms}$$
$$R2 = \frac{1400}{20} = 70 \text{ ohms}$$
$$R3 = \frac{1400}{10} = 140 \text{ ohms}$$

The equivalent Δ network is shown in Fig. 11–18(c).

Δ-to-Y Conversion Formulas Let us now consider the conversion from a Δ network to a Y network. Refer to Fig. 11–19. Here we have the Δ network, with terminals *a*, *b*, and *c* shown in solid lines. It is to be connected to an equivalent Y network connected between the same terminals and shown in dotted lines. The conversion formulas relating the Y resistors (R_A, R_B, and R_C) to the Δ resistors (R_W, R_X, R_Z) are shown below the diagram in the figure. These formulas can be used to convert a T or Y network to a Δ or π network.

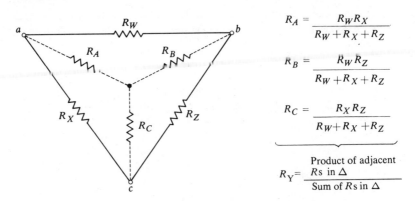

FIGURE 11-19 *Conversion formulas for Δ-to-Y or π-to-T conversions.*

Again, note that the three formulas have the same general form as indicated by the basic rule on the right. The basic rule states that any of the Y resistors (R_Y) can be found by taking the *product* of the two adjacent Δ resistors and dividing by the *sum* of the three Δ resistors. For example, when calculating R_A, the two adjacent Δ resistors are R_W and R_X; for R_B the two adjacent Δ resistors are R_W and R_Z and for R_C the two adjacent Δ resistors are R_X and R_Z.

EXAMPLE 11.4 In Ex. 11.3 the Y network of Fig. 11–18(a) was converted to the Δ network of Fig. 11–18(c). Convert this Δ network back to a Y network using the Δ-to-Y conversion formulas.

Solution: The Δ network is redrawn in Fig. 11–20(a) along with the desired Y network shown in dotted lines. The values of Y resistors are found as follows:

$$R1 = \frac{\text{product of adjacent } R\text{'s in } \Delta}{\text{sum of } R\text{'s in } \Delta}$$

$$= \frac{70 \times 35}{70 + 35 + 140} = \textbf{10 ohms}$$

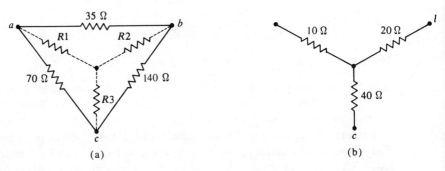

FIGURE 11-20

$$R2 = \frac{35 \times 140}{245} = 20 \text{ ohms}$$

$$R3 = \frac{70 \times 140}{245} = 40 \text{ ohms}$$

The equivalent Y network is shown in part (b) of Fig. 11–20. As expected, it is the same as the original Y network in Fig. 11–18(a).

11.12 The Branch Current Method

The network analysis techniques that have been presented thus far are sufficient for solving any linear, resistive circuit. However, for extremely complex circuits there are other, more general, techniques that can be used. These other techniques do not utilize any circuit simplifications or equivalent circuits; rather, they are based on the application of KVL and KCL.

The circuit in Fig. 11–21(a) will be analyzed by using the method of *branch currents*. The procedure begins by drawing current arrows through each branch in the circuit as shown. As we shall see, it doesn't matter what direction we assume for the individual branch currents. Once the current arrows are drawn, the voltage polarities across each resistor can be assigned consistent with the current directions. For example, the 20-ohm resistor is shown with a polarity − to + going from left to right since I_1 (electron flow) flows from left to right.

The three unknown branch currents I_1, I_2, and I_3 must now be found. Since we have *three* unknowns, then we need *three* independent equations relating

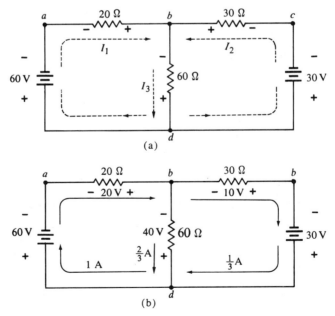

FIGURE 11–21 *Application of branch current method.*

these unknowns in order to solve for them. We can get these equations by applying KCL and KVL.

First, we can apply KCL at the junction point b. Since I_1 and I_2 are shown flowing toward b and I_3 is flowing away from b, we have

$$I_1 + I_2 = I_3 \tag{11-5}$$

We can also write the KVL equations around any of the complete paths in the circuit. Consider the path a-b-d-a. If we traverse this complete path the net voltage change in doing so should be zero according to KVL. Starting at point a and proceeding around this path, the KVL equation is

$$I_1 \times 20 + I_3 \times 60 - 60 \text{ V} = 0$$

or rearranging

$$20 I_1 + 60 I_3 = 60 \tag{11-6}$$

Consider now the path b-c-d-b. Starting at b and proceeding around this path, the KVL equation is

$$-I_2 \times 30 + 30 \text{ V} - I_3 \times 60 = 0$$

or rearranging

$$30 I_2 + 60 I_3 = 30 \tag{11-7}$$

Equations (11–5) to (11–7) are the required three relationships necessary to solve for I_1, I_2, and I_3.

The procedure now is to solve this set of three *simultaneous* equations. We will use the method of elimination as follows.*

(1) Using Eq. (11–5) substitute for I_3 in Eqs. (11–6) and (11–7).

$$20 I_1 + 60(I_1 + I_2) = 60 \tag{11-6a}$$

$$30 I_2 + 60(I_1 + I_2) = 30 \tag{11-7a}$$

(2) Rearrange Eqs. (11–6a) and (11–7a).

$$80 I_1 + 60 I_2 = 60 \tag{11-6b}$$

$$60 I_1 + 90 I_2 = 30 \tag{11-7b}$$

(3) Multiply (11–6b) by 3 and (11–7b) by -2 and then add the equations:

$$240 I_1 + 180 I_2 = 180 \tag{11-6c}$$

$$-120 I_1 - 180 I_2 = -60 \tag{11-7c}$$

$$120 I_1 \qquad\quad\; = 120 \tag{11-8}$$

(4) Solve for $I_1 = 1$ A.

---

* The method of determinants could also be used.

(5) Use (11–6b) to solve for I_2 when $I_1 = 1$ A.

$$80 + 60\,I_2 = 60$$

$$I_2 = -1/3 \text{ A}$$

(6) Use (11–5) to find I_3.

$$I_3 = I_1 + I_2 = 1 - 1/3 = 2/3 \text{ A}$$

The results are summarized:

$$I_1 = 1 \text{ A}$$
$$I_2 = -1/3 \text{ A}$$
$$I_3 = 2/3 \text{ A}$$

Looking at these results we see that the result for I_2 came out *negative*. The negative sign in the result for I_2 simply means that our assumed *direction* for I_2 in Fig. 11–21(a) was *wrong*. In other words, the current through the 30-ohm resistor actually flows from left to right. Since I_1 and I_3 came out positive, their assumed directions are correct. Figure 11–21(b) shows the circuit with correct value and direction for each current.

Now that the branch currents have been determined they can be used to calculate the voltages across the various resistors. These voltages are shown in the diagram. Note that the polarity of the voltage across the 30-ohm resistor is opposite to the assumed polarity in Fig. 11–21(a) because I_2's direction was reversed.

In conclusion, it can be seen that this circuit has been solved using only the basic circuit laws (Ohm's law, KVL, and KCL) without simplifying the circuit in any way. Any linear resistive network can be solved in a similar manner.

11.13 The Node Voltage Method

In the branch current method the unknowns were branch currents that were determined by writing the KVL equations around the circuit loops. Another analysis method uses *node voltages* as the unknowns and solves for them using KCL.

A *node* is simply a junction point where two or more circuit elements are connected. A *principal node* is a junction point where *three* or more branches are connected. In Fig. 11–22, points x and y are principal nodes. However, we will specify node y as the reference node or ground node. In the diagram it is shown connected to ground. We will be interested in determining the value of voltage at node x relative to node y. In other words, V_{xy} is to be determined. Since y is ground, then we can write V_{xy} as V_x.

If V_x can be found, then all the resistor currents and voltages can easily be calculated. The circuit in Fig. 11–22 is the same circuit which was solved in Fig. 11–21 by the branch current method. We will now solve it using the node voltage method.

Chapter Eleven

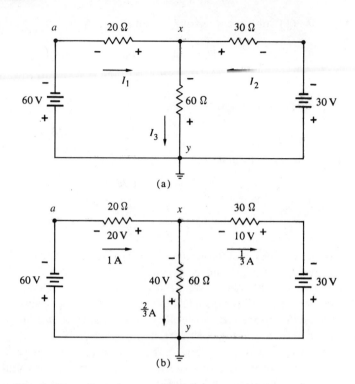

FIGURE 11-22 *Circuit for node voltage analysis.*

Writing the Node Equations The currents through each branch are labeled I_1, I_2, and I_3 respectively. We will express these currents in terms of the unknown node voltage V_x as follows:

(1) The current I_1 is produced by the potential difference (V_{xa}) across the 20-ohm resistor. The right side of this resistor is at V_x volts relative to ground. The other side of the resistor is at -60 V relative to ground because of the 60-V source ($V_a = -60$ V). Thus, we have

$$V_{xa} = V_x - V_a$$
$$= V_x - (-60)$$
$$= V_x + 60$$

Therefore:

$$I_1 = \frac{V_{xa}}{20\ \Omega} = \frac{V_x + 60}{20\ \Omega} \qquad (11\text{-}9a)$$

(2) Similarly, we can write for I_2:

$$I_2 = \frac{V_x + 30}{30\ \Omega} \qquad (11\text{-}9b)$$

(3) Finally, for I_3 we can write

$$I_3 = \frac{-V_x}{60\ \Omega} \qquad (11\text{-}9c)$$

Resistive Circuit Analysis Techniques   245

The negative sign is necessary because if V_x is positive, I_3 will flow opposite to its assumed positive direction.

(4) Using KCL at node x, $I_3 = I_1 + I_2$ so that

$$\frac{-V_x}{60} = \frac{V_x + 60}{20} + \frac{V_x + 30}{30} \qquad (11\text{-}10)$$

This last equation contains only the one unknown, V_x. It can be solved for V_x with the result

$$V_x = -40 \text{ V}$$

(5) With V_x determined, the resistor voltages and currents can be easily determined using Eqs. 11–9(a)–(c) with $V_x = -40$ V.

$$I_1 = \frac{V_x + 60}{20 \, \Omega} = 1 \text{ A}$$

$$I_2 = \frac{V_x + 30}{30 \, \Omega} = -1/3 \text{ A}$$

$$I_3 = I_1 + I_2 = 2/3 \text{ A}$$

The resistor voltages are then determined using Ohm's law. The results are shown on the diagram in Fig. 11–22(b). The results agree with the solution of Fig. 11–21.

This node voltage method seemed to provide a simpler solution for this particular circuit than did the branch current method. However, when a more complex circuit is solved there will be more than one unknown node voltage to solve for.

In general, the node voltage method consists of taking each principal node (except ground) and using KCL at each node with the currents expressed in terms of unknown node voltages. In the next section we will look at a more complex circuit solved by both the branch current and node voltage methods.

11.14 Complex Circuit Example

In this section a more complex circuit will be used to demonstrate the branch current and node voltage methods. The equations will be set up for each method without going through the actual numerical solution.

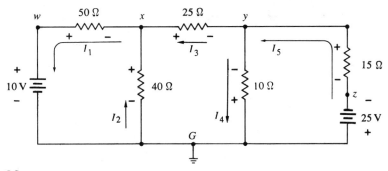

FIGURE 11–23

Chapter Eleven

In Fig. 11-23 the circuit is shown, followed by necessary equations for solving for the various branch currents using the branch current method. These equations should be examined closely and verified by the student. The 5 equations for the 5 unknown branch currents can be solved by the process of elimination or by using determinants.

Branch Current Equations

I. KCL at node x:

$$I_1 = I_2 + I_3$$

II. KCL at node y:

$$I_5 = I_3 + I_4$$

III. KVL around path w-x-g-w:

$$-50\, I_1 - 40\, I_2 + 10 = 0$$

IV. KVL around path x-y-g-x:

$$-25\, I_3 + 10\, I_4 + 40\, I_2 = 0$$

V. KVL around path y-z-g-y:

$$-15\, I_5 + 25 - 10\, I_4 = 0$$

Note there are 5 equations in 5 unknowns.

Node Voltage Equations Following are the steps and equations for solving for the various node voltages. Note that node G is the ground reference node so that V_x and V_y are the two unknown node voltages.

I. Write currents in terms of V_x and V_y:

$$I_1 = \frac{10 - V_x}{50}; \quad I_2 = \frac{V_x}{40}; \quad I_3 = \frac{V_x - V_y}{25}$$

$$I_4 = \frac{-V_y}{10}; \quad I_5 = \frac{V_y + 25}{15}$$

II. KCL at node x:

$$I_1 = I_2 + I_3$$

$$\boxed{\frac{10 - V_x}{50} = \frac{V_x}{40} + \frac{V_x - V_y}{25}}$$

III. KCL at node y:

$$I_5 = I_3 + I_4$$

$$\boxed{\frac{V_y + 25}{15} = \frac{V_x - V_y}{25} - \frac{V_y}{10}}$$

Substitute expressions for I_1, I_2, I_3, I_4, I_5

These last two equations can now be used to solve for V_x and V_y.

For those interested in working out the solutions for either of these methods, the approximate results are:

$$V_x = -1.6 \text{ V}; V_y = -8.4 \text{ V}; I_1 = 0.23 \text{ A}; I_2 = -0.04 \text{ A};$$
$$I_3 = 0.27 \text{ A}; I_4 = 0.85 \text{ A}; I_5 = 1.1 \text{ A}$$

Chapter Summary

1. *Superposition Principle.* In a linear circuit containing more than one source, the current (or voltage) for any element in the circuit can be found by superimposing the individual currents (or voltages) produced in the element by each source acting alone with the other sources de-energized.

2. To de-energize a voltage source, replace it by a short circuit; to de-energize a current source, replace it by an open circuit.

3. The superposition method is valid only for linear, bilateral circuits and for superimposing voltages and currents (not powers).

4. Any linear resistive network with two terminals x and y can be replaced by a practical voltage source (E_{Th} in series with R_{Th}) connected to these same terminals x-y (Thevenin's theorem).

5. The value of E_{Th} is the voltage produced at the terminals x-y with no external load connected to these terminals. The value of R_{Th} is the equivalent resistance seen across the x-y terminals with all sources de-energized.

6. A network will deliver maximum power to a load when the load resistance equals the network's R_{Th}.

7. An ideal constant current source I_S will produce a constant current through any load connected to its terminals, but its terminal voltage will change according to the value $I_S \times R_L$.

8. A practical current source is represented by an ideal current source with an internal source resistance R_S in parallel.

9. The Thevenin equivalent of a practical current source has $E_{Th} = I_S \times R_S$ and $R_{Th} = R_S$.

10. A practical current source I_S which has a source resistance R_S is equivalent to a practical voltage source with $E_S = I_S R_S$ and with a source resistance R_S.

11. Any linear resistive network with two terminals x and y can be replaced by a practical current source (I_N in parallel with R_N) connected to these same terminals x-y (Norton's theorem).

12. The value of I_N is the current through a short circuit which is connected between terminals x-y. The value of R_N is the equivalent resistance seen across the x-y terminals with all sources de-energized.

13. For a given network: $E_{Th} = I_N \times R_N$, $I_N = E_{Th}/R_{Th}$, and $R_N = R_{Th}$.

14. Resistance networks that cannot be simplified using ordinary series-parallel techniques can often be simplified using Δ-Y transformations.

Questions/Problems

Section 11.2

11-1 Use the superposition method to determine the current and voltage for the load in Fig. 11-24.

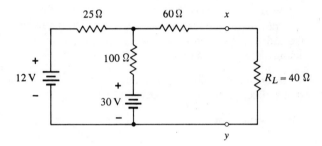

FIGURE 11-24

11-2 Determine the load current in the circuit of Fig. 11-24 if the 30-V source is increased to 45 V. (Hint: use the linearity principle.)

11-3 Determine the voltage V_x in Fig. 11-25 using the superposition method.

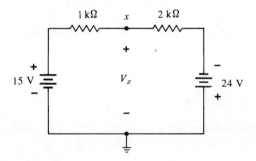

FIGURE 11-25

11-4 Consider the circuit which was analyzed in problem 11-1. What would the 12-V source have to be changed to in order to have a resultant load current of 210 mA? (Hint: use the results of 11-1.)

Sections 11.4–11.6

11-5 Use Thevenin's theorem to find the load current and voltage in the circuit of Fig. 11-24. Compare with the results of problem 11-1.

11-6 Find the Thevenin equivalent circuit for the T network connected to terminals x-y in Fig. 11-26. Then, determine what load to use for maximum load power and calculate P_{max}.

11-7 Use Thevenin's theorem to determine what value of R_L is needed in the circuit of Fig. 11-27 to produce a load current of 10 mA.

11-8 Repeat 11-7 with the 50-volt source changed to 200 V. What happens to E_{Th}?

11-9 In the circuit of Fig. 11-27 replace R_L by a silicon diode with its cathode at y and its anode at x. Determine the diode current.

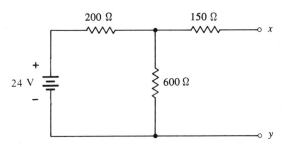

FIGURE 11-26

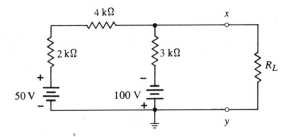

FIGURE 11-27

Section 11.7

11-10 Indicate which of the following statements is *always* true:
   a. For an ideal current source the voltage across its terminals never varies, while for a practical current source it does vary.
   b. A practical current source delivers maximum current when $R_L = 0$.
   c. A practical voltage source delivers maximum current when $R_L = 0$.
   d. A practical current source should have a small R_S if it is to be close to ideal.
   e. A practical current source can always be converted to a practical voltage source.

11-11 A certain practical current source has $I_S = 3$ mA and $R_S = 50$ kΩ. Determine the output current for the following R_L values: 0, 1 kΩ, 10 kΩ, 50 kΩ, and 200 kΩ.

11-12 Find the Thevenin equivalent for the current source in question 11-11. Then use it to find the load current for the various R_L values given in question 11-11 and compare results.

11-13 Figure 11-28 shows a practical current source. Determine the load voltage for $R_L = 200$ ohms.

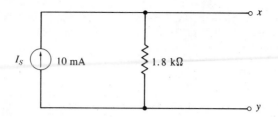

FIGURE 11-28

11-14 Find the Thevenin equivalent for the source in Fig. 11-28. Use it to determine the load voltage for $R_L = 200$ ohms. Compare to 11-13 solution.

11-15 A pretty good practical constant-current source can be made using a voltage source and a large series resistor. Figure 11-29 shows such a circuit. Determine the *maximum output* current. Then determine the range of R_L values over which the current remains within 10 percent of this maximum value.

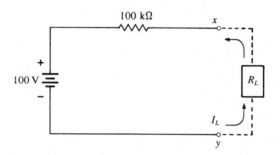

FIGURE 11-29

Sections 11.8–11.9

11-16 Find the Norton equivalent circuit for the circuit in Fig. 11-27 at terminals x-y.

11-17 Use the results of the last problem to determine what value of R_L will draw 10 mA. Compare to the results of problem 11-7.

11-18 A certain complex network has $E_{Th} = 12$ V and $R_{Th} = 200\ \Omega$ at terminals x-y. What is the Norton equivalent circuit at x-y?

11-19 A certain complex network has output terminals x and y. The following measurements are made on the network output: $V_{xy} = 2.7$ V with no load; $I_{xy} = 17$ mA with x-y short-circuited. Find the Thevenin and Norton equivalents for this network.

11-20 A technician is asked to take a certain complex network and determine its Thevenin equivalent in the laboratory by measuring E_{Th} and R_{Th}. He takes the network and measures the open-circuit voltage across the output terminals with a voltmeter. He reads 10 volts and therefore concludes that $E_{Th} = 10$ V. He then de-energizes the sources in the network and measures $R_{Th} = 100$ kΩ using an ohmmeter. Using this Thevenin equivalent with a 50-kΩ load, he measures a load voltage of 3.33 V. However, using the same load on the original network he measures a load voltage of 4 V. From the following list, choose one or more as possible reasons for the discrepancy (explain each choice):

a. Original network contains nonlinear devices.
b. Original network contains a burnt-out component.
c. Voltmeter used to measure E_{Th} caused loading.
d. Technician should have used Norton's equivalent circuit.
e. Network contains nonideal sources.

Section 11.10

11–21 Use source conversions to change the complex circuit of Fig. 11–30 to a single practical voltage source between terminals x-y.

Section 11.11

11–22 Convert the Y network in Fig. 11–31(a) to a Δ network.

11–23 Convert the Δ network in Fig. 11–31(b) to a Y network.

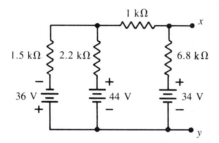

FIGURE 11–30

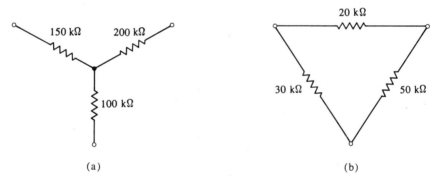

(a)   (b)

FIGURE 11–31

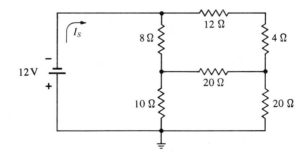

FIGURE 11–32

11-24 Use a Δ-to-Y conversion to find the equivalent resistance seen by the source in Fig. 11-32. Determine I_S.

11-25 Repeat Problem 11-24 using a Y-to-Δ conversion.

Sections 11.12–11.14

11-26 For the circuit in Fig. 11-21 the solution shows an electron current flowing into the negative terminal of the 30-V source. How can this happen?

11-27 Use the branch current method to solve for all currents and voltages in the circuit of Fig. 11-33.

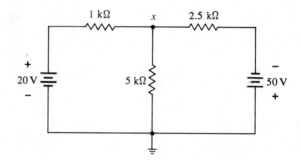

FIGURE 11-33

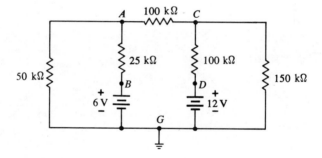

FIGURE 11-34

11-28 Assign current directions to each branch in the circuit of Fig. 11-34. Write the complete set of equations relating these currents. Do not solve for the current values.

11-29 Solve the circuit of Fig. 11-33 using the node voltage method.

11-30 Write the complete set of equations to be used for solving for the node voltages in Fig. 11-34. Solve the equations for V_A and V_C and then determine all the circuit currents and voltages.

CHAPTER 12

Introduction to Waveforms

12.1 Introduction

In electronics we are always concerned with how the voltage and currents in a circuit are behaving. In general, the voltages and currents will change as time goes on. In other words, we say that these voltages and currents are *time varying*, meaning that they vary with time. In this chapter we will introduce some of the basic voltage and current *waveforms*. Waveforms are essentially graphical representations of how the voltage (or current) is changing as time goes on.

We will start by discussing the concept of a *function*, because the voltages and currents are functions of time. Then we will review some of the basics of *graphing*, since waveforms are actually graphs of voltage (or current) versus time.

12.2 Functions

Whenever some quantity X depends on one or more other quantities Y, then we say that X is a function of Y. For example, X might represent the quantity temperature. Temperature depends on several factors, including location, time of the day, and so forth. Thus, we can say that the temperature is a function of all these factors.

The area of a rectangle is given by

$$A = l \times w$$

where l is the length and w is the width. The value of A depends on both l and w. Thus, we say that A is a function of l and also a function of w.

In science and technology, many of the quantities encountered will be *functions of time*. That is, the value of the quantity will be different at each instant of time. To illustrate, suppose we were up in a helicopter at 1000 feet and we dropped a heavy package out the window. If we wanted to know how far the package has fallen a certain amount of time later we can use the equation

$$d = 16\, t^2$$

In this equation d is the distance the package has fallen in feet, and t represents the time, in seconds, after the package was dropped.

When one quantity has a value which is a function of another quantity, we call the first quantity a *dependent quantity* or *dependent variable*. In the functional relationship $d = 16t^2$ the distance d is the dependent variable because its value is *dependent* on the value of t. The quantity which the dependent variable depends on is called the *independent variable* because its value is free to vary in the particular situation being discussed. In the preceding expression the quantity t is the independent variable.

In the relationship $A = l \times w$, the area A is the dependent variable and l and w are both independent variables. In general, a dependent quantity will be a function of several independent variables.

12.3 Graphs

In many cases one quantity will be a function of another quantity and there may be no easy way of representing it in an equation. For example, the temperature at a given location, on a certain day, will be different at different times of the day. We cannot normally express this variation of temperature with time in an equation. Instead we usually make a table of values that lists the value of temperature at various times in the day. Another way to represent the relationship is by using a graph (see Fig. 12–1). The graph is essentially a continuous record of temperature for the 24-hour period. In the graph, the vertical scale represents the possible values of temperature and the horizontal scale represents the various values of time. Almost always, the *dependent quantity* is represented by the *vertical scale,* while the *independent quantity* is represented by the *horizontal scale*.

The graph in Fig. 12–1 is an irregularly shaped curve which cannot be easily expressed mathematically. In order to determine what the value of temperature was at any given time during the day, it is necessary to find the value of time on the horizontal scale (also called horizontal axis) and draw a vertical line up to the curve; the point where the vertical line intersects the curve is then used as a starting point for drawing a horizontal line over to the vertical scale (or axis). This horizontal line intersects the vertical scale at the desired value of temperature. For example, if we wanted to know the temperature at 9:00 A.M., we would do it as shown in Fig. 12–1 in *dotted* lines. The result is approximately 67°F.

Conversely, if we wanted to know at what time the temperature was 72°F we would reverse the above process. This is illustrated in Fig. 12–1 by the *dashed* lines. The result is approximately 10:00 A.M.

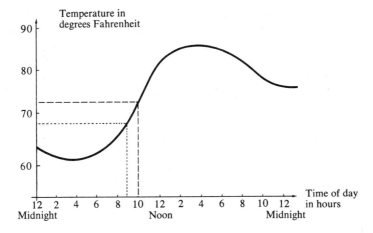

FIGURE 12-1

12.4 Concept of Slope

If we examine Fig. 12–1, it is clear that the temperature varies quite nonuniformly with time. Between 12:00 midnight and 5:00 A.M. the temperature is slowly decreasing. From 5:00 A.M. to 2:00 P.M. the temperature rises relatively sharply. Between 2:00 P.M. and 4:00 P.M. the temperature remains fairly constant. After 4:00 P.M. it begins to decrease substantially. In other words, the curve in some places slopes or slants downward while in other places it slopes upward or stays level (no slope).

In speaking about the shape of a curve we will often be interested in its *slope* at various points or regions of the curve. In the next chapter we will discuss a method for measuring the slope of a curve. For the present discussion let us simply consider some of the important points concerning slope:

(1) When a portion of a curve slopes upward from left to right, we say that the curve has a *positive slope* on that portion of the curve. In Fig. 12–1 the curve has a positive slope between 5:00 A.M. and 2:00 P.M. When a curve has a positive slope it means that the dependent variable *increases* as the independent variable *increases*.

(2) Conversely, a curve has a *negative slope* when it slopes downward from left to right. In Fig. 12–1 the curve has a negative slope between 12:00 midnight and 5:00 A.M. When a curve has a negative slope it means that the dependent variable *decreases* as the independent variable *increases*.

(3) When a portion of a curve is flat we say that the curve has a *zero slope*. In Fig. 12–1 the curve has a zero slope between 2:00 P.M. and 4:00 P.M. A zero slope indicates that the dependent variable is not changing as the independent variable increases.

(4) When a curve is changing from a negative (downward) slope to a positive (upward) slope, or vice versa, the slope at the point on the curve where the change takes place also has a zero slope. In other words, the slope is zero at the top of the peaks and the bottom of the valleys that a curve might contain. In Fig. 12–1 a valley occurs at approximately 5:00 A.M. so the slope at that point is zero.

(5) When a portion of a curve has a slope that is uniform (a straight line) we say that the slope is constant. In Fig. 12–1 the curve between 9:00 A.M. and 11:00 A.M. slopes upward and is very close to being a straight line (constant slope). A constant slope indicates that the dependent variable is changing in direct proportion to the changes in the independent variable. The term *linear slope* or simply *linear* is used to describe a constant slope.

(6) The magnitude of the slope of a curve is essentially the steepness of the slope. A steeper slope is a greater slope. In Fig. 12–1 the slope is greater (more steep) between 6:00 P.M. and 10:00 P.M. than it is between 12:00 midnight and 4:00 A.M. although they are both negative (downward) slopes.

EXAMPLE 12.1 Figure 12–2 shows a graph of voltage as a function of time. That is, voltage (the dependent variable) is represented by the vertical axis and time (independent variable) by the horizontal axis. This particular curve is called a *sine curve* or *sinusoid* and is very important in electronics work. For this graph determine the following:
   a. portions of the curve with a positive slope
   b. portions of the curve with a negative slope
   c. portions of the curve with a zero slope
   d. portions of the curve with a linear slope
   e. portions of the curve with the greatest slope

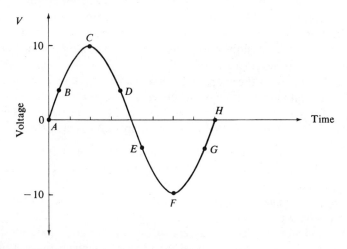

FIGURE 12–2 *Sine curve.*

Solution: (a) Between points A and C and between points F and H the curve slopes upward (positive slope).
(b) Between points C and F the curve has a negative slope.
(c) There are no regions where the curve is *perfectly* flat. However, at the peak points C and F the slope is zero.
(d) A sine curve theoretically has no portion where its slope is constant. However, practically speaking, we can see from Fig. 12–2 that between points A and B, between points D and E, and between points G and H the slope is approximately constant.

(e) For the sine curve the slope is steepest in the same regions that the slope is almost linear [same answer as (d)].

12.5 Waveforms—General Discussion

If a voltage (or current) in a circuit is time varying (changing with time) so that its value is dependent on time, then a graph can be constructed with voltage plotted as a function of time. This graph is called a *waveform*. Figure 12–2 is a typical voltage waveform. Before discussing the specific types of waveforms, a few general comments should be made.

Circuit Representation of Time-varying Voltage and Current Figure 12–3(a) shows a device which has a time-varying voltage, v_{xy}, across its terminals, and a time-varying current, i, flowing through it. Note that *lowercase* letters are used for v_{xy} and i. This indicates that they are time-varying; uppercase I and V are used only for values which are not changing with time.

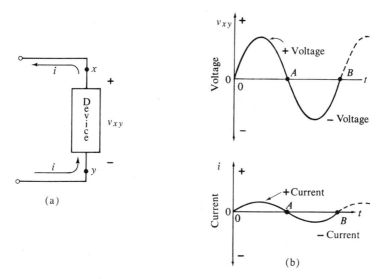

FIGURE 12–3 *Waveforms of voltage and current can, in general, go positive and negative.*

The waveform for v_{xy} might appear as shown in Fig. 12–3(b). It starts out at zero volts at time $t = 0$ and goes positive for a certain time interval (point A) and then it goes negative for a certain time interval (from A to B). During the interval when v_{xy} is *positive*, the x terminal is *positive* relative to the y terminal. During the interval when v_{xy} is *negative*, the x terminal is *negative* relative to y. So during the normal operation the polarity of x relative to y will be changing. However, to represent v_{xy} on the circuit diagram we *always* put a + sign next to the x terminal and a − sign next to the y terminal. In other words, we label the circuit diagram as if v_{xy} were positive. During the time intervals when v_{xy} is negative (e.g., between A and B), the actual polarity is the reverse of that shown on the diagram.

258 Chapter Twelve

The waveform for i might appear as shown in Fig. 12–3(b). It starts out at zero at $t = 0$ and goes positive until point A, then goes negative until point B. During the time when the i waveform is positive, the actual circuit current flows in the assumed positive direction shown on the diagram. For those portions of the i waveform where i is negative, the actual current direction is opposite to that shown on the diagram.

Thus, the waveforms represent the exact behavior of the voltage and current at all instants of time. In some cases we will want to use the waveforms to determine the *instantaneous* values of current and voltage at some particular instant of time. In other cases we will only be interested in the amplitude (size) and shape of the waveforms.

dc and ac Currents and Voltages The current and voltage waveforms in Fig. 12–3 are examples of *alternating current (ac)* waveforms because they change from one polarity to another at various times. Waveforms which do not change polarity are called *direct current (dc)* waveforms. Some examples of dc waveforms are shown in Fig. 12–4.

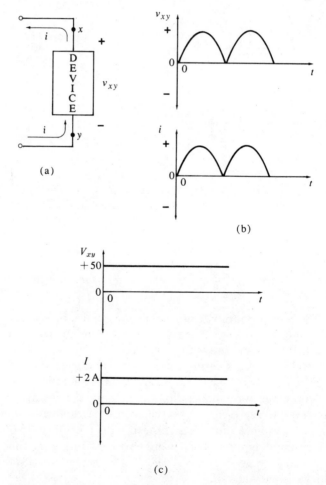

FIGURE 12–4 (a) *Circuit.* (b) *dc waveforms of* v_{xy} *and* i. (c) *Pure (constant) dc waveforms for* V_{xy} *and* I.

Introduction to Waveforms 259

In Fig. 12-4(b) a typical dc waveform is shown for v_{xy}. The waveform indicates that v_{xy} is changing with time but that it always has the same polarity (positive in this case). The resultant current i is changing with time (see waveform) but it also always remains positive. These waveforms are classified as dc because they never reverse polarity.

In Fig. 12-4(c) a special case of dc is shown. Here the V_{xy} waveform is simply a straight line indicating a constant voltage (in this case 50 V). The current waveform is also a constant value (+2 A). When the voltages and currents are *constant* we call them *constant dc* or *pure dc*. Pure dc is the type of voltage and current produced by a battery or dc power supply. Uppercase V_{xy} and I are used since the values do not change with time.

In the next few sections we will examine some of the basic waveforms that occur in electronics. Some of these waveforms will be ac waveforms while others will be dc in nature. The difference between ac, dc, and pure dc is summarized in Table 12-1.

TABLE 12-1

Type	Magnitude	Polarity
Alternating current — ac	Changes with time	Alternates from one to the other
Direct current — dc	Changes with time	Always has the same polarity
Pure dc	Constant with time	Always has the same polarity

12.6 Basic Waveforms

Although there are an unlimited number of possible waveforms that can occur in electronics, all waveforms are comprised of one or more of only a few basic types of waveforms. There are essentially five basic waveforms and they are illustrated in Figs. 12-5 through 12-9. In the illustrations we will use only voltages, but it should be understood that they could just as well be current waveforms.

Figure 12-5 shows a constant voltage or pure dc. Part (a) of the figure shows a constant positive voltage, while part (b) shows a constant negative voltage.

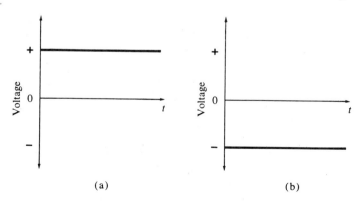

FIGURE 12-5 *Pure dc (constant voltage).*

Figure 12–6 illustrates the *step voltage waveform*. Its major characteristic is the fact that it changes from one constant voltage to another instantaneously. In part (a) of the figure the voltage is at zero volts from time $t = 0$ until time $t = 1$ s. At $t = 1$ s the voltage jumps or steps instantaneously up to $+5$ V where it remains indefinitely. This is an example of a *positive-going step* of 5 V since it jumped in the positive direction.

Part (b) of Fig. 12–6 shows a *negative-going step* occurring at $t = 1$ s. Part (c) shows a positive-going step of 5 V again, but this time it is jumping from a $+1$ V level to a $+6$ V level. The size of the step is the actual change in voltage at the instant the step occurs.

In practice, a perfect step waveform is impossible to produce. This is because changes in voltage and current cannot take place instantaneously. In other words, it will take a certain amount of time for the voltage to jump from one value to another. It is not unusual for this amount of time to be less than 1 μs or even 1 ns in some circuits.

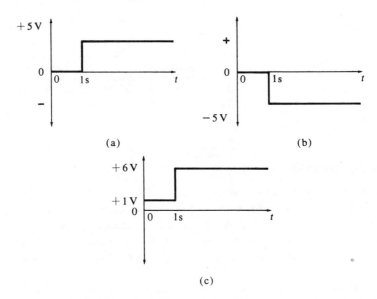

FIGURE 12–6 *Examples of step voltage waveforms.*

Figure 12–7 illustrates the *exponential waveforms*. The exponential waveform is like a step voltage except that the step is a gradual one. In part (a) of the figure the voltage is at 0 V at $t = 0$ and begins increasing positively. The slope of the waveform is relatively large to begin with but it gradually becomes less and less until the waveform eventually levels off at $t = 5$ s. This is an example of a *positive-going exponential*. A *negative-going exponential* is illustrated in Fig. 12–7(b).

Exponential waveforms do not always start at zero volts or at $t = 0$. Figure 12–7(c) illustrates a negative-going exponential that starts at $+6$ V (at $t = 5$ s) and ends up at $+1$ V (at $t = 10$ s). The exponential waveforms are very important in the analysis of certain types of circuits. For this reason, more will be said about exponential waveforms in the following chapters.

Figure 12–8 illustrates a *sinusoidal waveform* or *sine wave* as it is often called. This waveform is an ac waveform. In fact, it is often called *pure ac* because it is

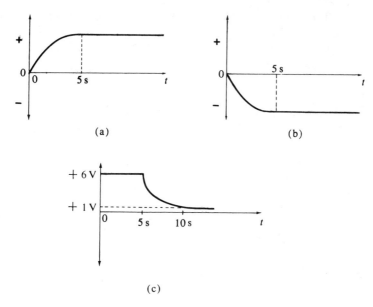

FIGURE 12-7 *Examples of exponential waveforms.*

completely symmetrical above and below the horizontal axis. In other words, its shape and size are the same during the positive portion as during the negative portion of the waveform.

The waveform in Fig. 12-8 starts as 0 V at $t = 0$ and increases positively until it reaches 10 V at $t = 1$ ms (millisecond). Then it begins decreasing. At $t = 2$ ms the voltage is again at 0 V but continues to decrease until it reaches -10 V at $t = 3$ ms. At that point it begins to increase again until at $t = 4$ ms it is back to 0 V. This complete waveform ($t = 0$ to $t = 4$ ms) is called a *cycle*. Thus, Fig. 12-8 shows one cycle of a sine wave. The length of the cycle is seen to be 4 ms. That is, the cycle takes 4 ms to complete. During the first *half-cycle* ($t = 0$ to $t = 2$ ms) the sine wave is positive. During the second *half-cycle* ($t = 2$ ms to $t = 4$ ms) the sine wave is negative.

The sine wave is seen to have two *peaks* or points where it reaches a limit and then begins going the other way. The first peak is at $+10$ V and the second peak

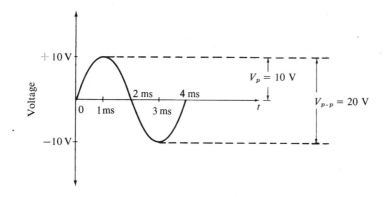

FIGURE 12-8 *Sine wave.*

is at -10 V. The two peaks are always the same number of volts above or below zero. We can indicate the size of a sine wave by expressing the *magnitude* of the voltage at one of its peaks. In our example, we would say that the sine wave has a *peak voltage* of 10 V. That is,

$$V_p = 10 \text{ V}$$

where V_p represents the voltage at the peak of the waveform.

Another way to indicate the magnitude of the sine wave is to use the amount of voltage between the two peaks. In our example, the positive peak is at $+10$ V and the negative peak is at -10 V so that there is a difference of 20 V between the peaks. Thus, we could say that the sine wave has a *peak-to-peak voltage* of 20 V. That is,

$$V_{p\text{-}p} = 20 \text{ V}$$

where $V_{p\text{-}p}$ represents the voltage between peaks.

It should be obvious that for a sine wave the value of $V_{p\text{-}p}$ is always twice the value of V_p. There is no preference as to which means of indicating the size of the sine wave we use at this point. However, when expressing the magnitude of a sine wave we must clearly indicate whether we are using peak voltage or peak-to-peak voltage.

From now on we will begin using the word *amplitude* to mean *magnitude* of a waveform. *Amplitude* is used throughout electronics as an indication of the size of a waveform. Thus, in Fig. 12–8 the sine wave has an amplitude of 10 V (peak) or 20 V (peak-to-peak).

The sine wave is the most important waveform in electricity and electronics:

(1) Voltage generated by power companies and distributed to our homes and factories, etc., is a pure ac sine wave.

(2) Radio transmission consists of sine waves.

(3) Sine waves are useful to help analyze the operation of many types of circuits.

For these reasons and others the sine wave will appear throughout the text. Each time more will be said about it so that eventually a thorough knowledge of sine waves will be developed.

EXAMPLE 12.2 The current i in a certain circuit starts out at 5 mA at $t = 0$ and remains there until $t = 10$ ms at which time a negative-going step of 10 mA occurs. Sketch the waveform for i.

Solution: The waveform is shown in Fig. 12–9(a). Notice that the negative-going step takes the current from a $+5$ mA to a -5 mA.

EXAMPLE 12.3 For the sine wave of current in Fig. 12–9(b) determine its amplitude, and the length of its cycle.

Solution: Amplitude = 3 μA (peak) or 6 μA (peak-to-peak); cycle = 10 μs

---

EXAMPLE 12.4 For the same waveform determine the times at which the slope is zero.

Solution: The slope is zero at the peaks. The peaks of a sine wave always occur at one-quarter of a cycle and three-quarters of a cycle. From the figure we can see that these peaks occur at $t = 2.5$ μs and $t = 7.5$ μs.

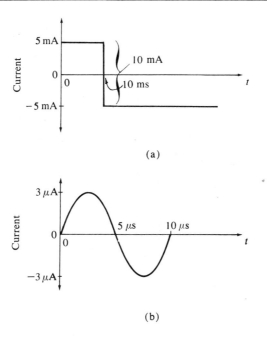

FIGURE 12-9

12.7 Pulse Waveforms

Most of the waveforms encountered in electronic circuits are derived from various combinations of the basic waveforms discussed in the last section. Several examples of derived waveforms will be discussed in this section.

Figure 12-10 illustrates the *rectangular pulse waveform*. Part (a) of the figure is a positive-going pulse and part (b) is a negative-going pulse. The pulse in Fig. 12-10(a) is simply a combination of a positive-going step occurring at $t = t_1$ and a negative-going step occurring at $t = t_2$. Between t_1 and t_2 the waveform remains a constant. The negative-going pulse in Fig. 12-10(b) is simply a combination of a negative-going step at $t = t_1$ and a positive-going step at $t = t_2$. Between t_1 and t_2 the voltage remains constant.

The amplitude of the rectangular pulse is shown in the figures. It is measured as the difference between the constant portions of the pulse. Another important characteristic of pulses is *pulse width*. The width of the pulses in Fig. 12-10 is labeled t_p and can be seen to be the duration of time between the leading edge (first step) and trailing edge (second step) of the waveform. In other words $t_p = t_2 - t_1$.

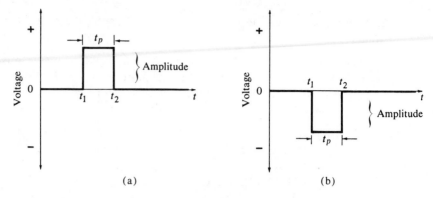

FIGURE 12-10 *Rectangular pulse waveforms.*

EXAMPLE 12.5 A certain negative-going pulse has an amplitude of 2 V and a pulse width of 10 ms. Sketch the waveform for the pulse if it occurs at $t = 10$ ms.

Solution: (See Fig. 12-11).

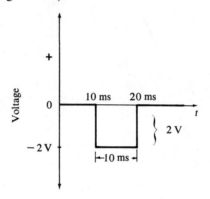

FIGURE 12-11

The rectangular pulses of Fig. 12-10 are highly ideal because, as we will find out, a voltage or current cannot be made to change instantaneously from one value to another. In other words, it is impossible to have a perfectly rectangular pulse with vertical edges and square corners. In practice, a rectangular pulse might appear as shown in Fig. 12-12. Here the pulse has edges which are very much like exponential waveforms. The transitions between the top and bottom of the pulse are shown taking place gradually. This is a *practical* or *non-ideal* rectangular pulse. In practice, however, the transitions sometimes can be made to take place in 1 μs or even 1 ns. In such cases the pulse can often be considered ideal.

In Fig. 12-13 a type of pulse waveform called a *spike* is illustrated. Part (a) of the figure shows a positive-going spike, while part (b) shows a negative-going spike. The positive-going spike is a combination of a positive-going step occurring at $t = t_1$ and a negative-going exponential that begins at the peak of the step. The negative-going spike is a combination of a negative-going step at $t = t_1$ and a positive-going exponential that begins at the peak of the step.

Introduction to Waveforms 265

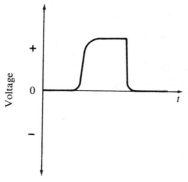

FIGURE 12-12 *Practical rectangular pulse.*

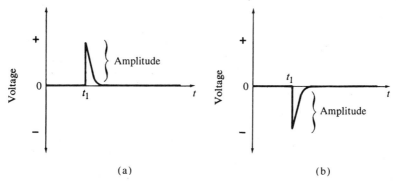

FIGURE 12-13 *Examples of spike waveforms.*

The spike waveform is common to many types of electronic circuits. It will occur later in our work concerning waveform-producing circuits.

EXAMPLE 12.6 Sketch a positive-going spike voltage that occurs at $t = 5$ ms, has an amplitude of 5 V, and starts out at $+1$ V before $t = 5$ ms.

Solution: Figure 12-14 shows the desired waveform. The spike is said to be resting on a 1-V level because the waveform has a value of 1 V before the spike occurs and after the spike is completed.

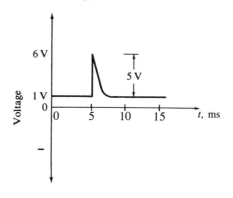

FIGURE 12-14

12.8 Periodic Waveforms

In electronic circuits the waveforms discussed in the last two sections are often found to repeat themselves at uniform intervals of time. When a waveform repeats itself regularly, then it is called a *periodic* waveform. Figure 12–15 shows an example of a periodic rectangular pulse waveform.

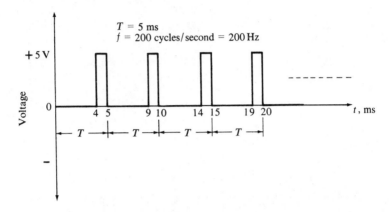

FIGURE 12–15 *Periodic rectangular pulse waveform.*

Between $t = 0$ and $t = 5$ ms the waveform shows a rectangular pulse with a pulse width of 1 ms occurring at $t = 4$ ms. Looking at the time interval between $t = 5$ ms and $t = 10$ ms, we see that the exact same waveform occurs during this interval as during the first interval. Between $t = 10$ ms and $t = 15$ ms the same behavior occurs. In fact, we can see that during every successive 5-ms time interval the same identical waveform is repeated. Each 5-ms interval constitutes one *period* of the waveform. The letter symbol for period is T. Thus, in this example, the *period* is

$$T = 5 \text{ ms}$$

That portion of the waveform that occurs during each period is called a *cycle*. The waveform goes through a complete cycle (beginning to end) during one period and then repeats the same cycle during the next period.

For the waveform of Fig. 12–15 one complete cycle consists of zero volts for 4 ms and 5 V for 1 ms. Thus, one cycle occurs in a time interval equal to 5 ms (one period); two cycles would take 10 ms (two periods); and so on. An important characteristic of a periodic waveform is its *rate of repetition* or *frequency*, which is essentially the number of cycles that occur per second. For the pulse waveform in Fig. 12–15, the number of cycles that occur per second can be calculated as

$$\frac{1 \text{ second}}{5 \text{ milliseconds}} = \frac{1 \text{ s}}{5 \times 10^{-3} \text{ s}}$$

$$= \frac{1}{5} \times 10^3$$

$$= 0.2 \times 10^3$$

$$= 200$$

Therefore, we can say that

$$f = 200 \text{ cycles per second}$$
$$= 200 \text{ cps}$$

for the waveform in Fig. 12–15. The symbol f is used for frequency. The symbol *cps* is used for cycles per second.

Up until recently the units cps were always used for frequency. In the past few years, however, another unit has been accepted as the standard unit for frequency. The *hertz*, abbreviated Hz, is the modern unit for frequency. It is the same as the cps. That is, 1 Hz = 1 cps. Thus, the frequency in Fig. 12–15 is 200 Hz.

In general, the frequency of a periodic waveform is equal to $1/T$ where T is the period. Stated formally,

$$f = \frac{1}{T} \tag{12–1}$$

where T is in seconds and f is in Hz.

EXAMPLE 12.7 Determine the period and frequency of the waveforms in Fig. 12–16.

Solution: (a) Figure 12–16(a) is a periodic sine wave. One period of the sine wave contains one complete cycle. In this case one cycle occurs in 16.6 ms. Thus:

$$T = \mathbf{16.6 \text{ ms}}$$

The frequency, then, is obtained using Eq. 12–1

$$f = \frac{1}{16.6 \text{ ms}} = \frac{1}{16.6 \times 10^{-3} \text{ s}}$$
$$= 0.06 \times 10^3 = 60 \text{ cps}$$
$$= \mathbf{60 \text{ Hz}}$$

This frequency is the frequency of the ac power which is supplied by power companies in the United States.

(b) Figure 12–16(b) shows another rectangular pulse waveform. The period is easily seen to be

$$T = 2 \text{ }\mu\text{s}$$

This gives a frequency of

$$f = \frac{1}{2 \times 10^{-6} \text{ s}} = 0.5 \times 10^6 \text{ Hz}$$
$$= \mathbf{500{,}000 \text{ Hz}}$$

This frequency can be expressed in the prefixed unit *kilohertz*, abbreviated kHz. A kilohertz is equal to 10^3 Hz. Thus,

$$f = 500{,}000 \text{ Hz} = 500 \times 10^3 \text{ Hz} = 500 \text{ kHz}$$

268 Chapter Twelve

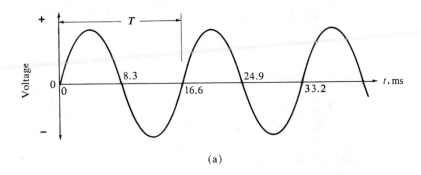

(a)

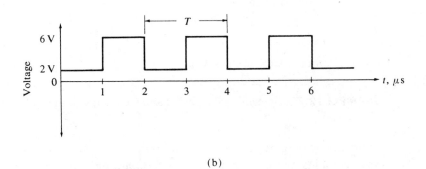

(b)

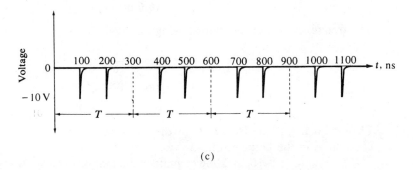

(c)

FIGURE 12–16

Another important characteristic of this particular pulse waveform is the fact that its pulse width t_p is exactly one-half of the period. This type of rectangular pulse waveform is called a *square wave*. A square wave has two voltage levels and it switches back and forth between these levels at equal time intervals. In Fig. 12–16(b) the square wave stays at 2 V for 1 μs and at 6 V for 1 μs during each 2-μs period.

(c) The waveform in Fig. 12–16(c) presents a somewhat more difficult problem. At first glance it might appear that the first cycle of this waveform extends from $t = 0$ to $t = 100$ ns and includes one pulse. On closer examination it can be seen that after two such cycles there is a gap where no pulse occurs. Since a cycle must be repeated each period, it should be clear that the above 100-ns cycle is not the correct one.

To determine the cycle we must look for a pattern that repeats itself periodically. In the figure, dotted lines are used to show the various cycles. The length of one cycle is seen to be

$$T = 300 \text{ ns}$$

This gives a frequency of

$$f = \frac{1}{300 \times 10^{-9} \text{ s}} = 0.0033 \times 10^9 \text{ Hz}$$

$$= 3{,}300{,}000 \text{ Hz} = \mathbf{3.3 \times 10^6 \text{ Hz}}$$

This frequency can be expressed in the prefixed unit *megahertz* (MHz). A megahertz is equal to 10^6 Hz. Thus,

$$f = 3.3 \times 10^6 \text{ Hz}$$
$$= \mathbf{3.3 \text{ MHz}}$$

12.9 Time-varying Sources

There are many kinds of time-varying sources which produce voltage and current waveforms. Many of them fall into the general category called *waveform generators* (also referred to as *signal generators* and *function generators*). These are pieces of electronic equipment which contain the necessary circuitry for generating the desired output waveform. Waveform generators are primarily used as laboratory sources for testing circuits and systems. The most common waveform generators produce sine waves and square waves.

Voltage Waveform Sources Waveform sources often appear as inputs to electronic circuits. As such, symbols are needed to represent these sources. Figure 12–17(a) shows a commonly used circuit symbol for a voltage source. The letter symbol for this voltage source can be either e_s (emf) or v_s (voltage). If the source happens to be the *input* to a circuit the symbol can be e_{in} or v_{in}. In any case, lowercase letters are used for e and v to denote the fact that the voltage produced at the terminals of the source is a time-varying voltage. In the special case where the voltage source produces a constant voltage (pure dc), the circuit symbol for a dc battery is used along with uppercase E_s or V_s.

When the circuit symbol for a time-varying voltage source appears in a circuit, it is meaningless unless it is accompanied by a waveform diagram or some type of description of how the voltage changes with time. Figure 12–18 shows two different presentations of the same information concerning a voltage source e_{in}, which is the input to a certain circuit.

In Fig. 12–18 the voltage source e_{in} is shown connected to input terminals x and y of the circuit. The variation of e_{in} is shown in the accompanying waveform diagram. Clearly, e_{in} is a sine wave as the diagram shows. As the sine wave alternates between positive and negative voltages, the voltage at the terminals of e_{in} will also change polarity.

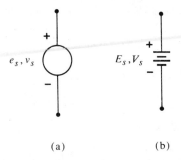

FIGURE 12–17 (a) *Symbol for time-varying voltage source.* (b) *Symbol for constant (dc) voltage source.*

Between $t = 0$ and $t = 0.5$ ms the waveform diagram shows that e_{in} is a positive voltage. During this time interval, then, the voltage source in the circuit diagram is *positive* at its upper terminal and negative at its lower terminal.

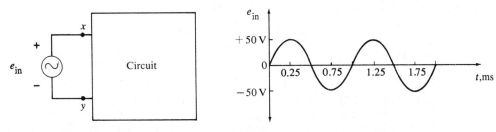

$e_{in} = 100$ V p-p. Sine-wave frequency = 1 kHz

FIGURE 12–18 *Two methods of defining the behavior of a time-varying source, e_{in}.*

For example, at $t = 0.25$ ms the value of e_{in} is seen to be $+50$ V. At this point in time there will be 50 V between the terminals of e_{in} with point x positive relative to y ($v_{xy} = +50$ V).

Between $t = 0.5$ ms and $t = 1$ ms the waveform shows that e_{in} will be a negative voltage. During this interval the voltage source will have a polarity opposite to that shown on the diagram. That is, the upper terminal will be *negative* and the lower terminal will be *positive*. For example, at $t = 0.75$ ms the value of e_{in} is seen to be -50 V. At this point there will be 50 V between the terminals of e_{in} with x negative relative to y ($v_{xy} = -50$ V).

In summary, for those times when the waveform diagram shows e_{in} as positive, the polarity of the voltage source is the polarity shown in the circuit diagram. When e_{in} is negative the polarity of the voltage source is opposite to the polarity shown in the circuit diagram.

The waveform diagram tells us the magnitude and polarity of the voltage source at any instant of time. In much of the work to follow we will refer to the *instantaneous voltage* between two points in a circuit. The instantaneous voltage is the voltage between the two points at a certain instant of time. For example, in Fig. 12–18 we can see that the instantaneous voltage at $t = 0.25$ ms is $+50$ V. Furthermore, the instantaneous voltage from the source varies sinusoidally (sine wave) with time.

Figure 12–18 illustrates another way to indicate the nature of e_{in}. In this case, stating that e_{in} is a sine wave with peak-to-peak amplitude of 100 V at a frequency of 1kHz is equivalent to presenting the waveform. From this information, the waveform of e_{in} may be constructed. It is not always possible to describe the variation of the source voltage in words and numbers. This is especially true for complex waveforms that are not easily described in terms of the basic waveforms. In such cases a waveform diagram is always used.

EXAMPLE 12.8 Consider Fig. 12–19. Determine the value of v_{ab} at $t = 0.5$ s and at $t = 3.5$ s.

Solution: The waveform diagram indicates that $e_{in} = -5$ V at $t = 0.5$ s. Thus, the voltage across the source terminals will be 5 V with the upper terminal *negative*. This will make $v_{ab} = -5$ V.

At $t = 3.5$ s the value of e_{in} is $+5$ V. Thus, the source will put out 5 V with the upper terminal *positive* so that $v_{ab} = +5$ V.

EXAMPLE 12.9 Describe the waveform in Fig. 12–19.

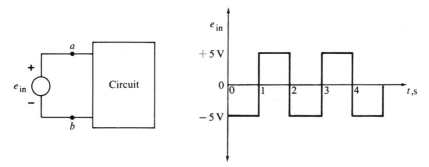

FIGURE 12–19

Solution: The waveform is a square wave with peak-to-peak amplitude of 10 V at a frequency of

$$f = \frac{1}{T} = \frac{1}{2 \text{ s}} = 0.5 \text{ Hz}$$

To describe the waveform more completely we should indicate that it is a *bipolar square wave*, which means that it goes both $+$ and $-$ rather than staying at one polarity. The bipolar square wave is an ac waveform.

Current Waveform Sources Voltage sources provide a voltage that has a certain relationship with time. As we have seen, voltage sources can supply dc, sine waves, square waves, etc., of voltage. Occasionally in electronic circuits we will encounter *current waveform sources*.

A current source supplies a current to a circuit and that current has a certain relationship with time. Figure 12–20(a) shows the circuit symbol for the time-varying current source. The letter symbol for this current is i_s. If the source happens to be the input to a circuit the symbol can be i_{in}. In either case, lowercase letters are used to denote the fact that the current supplied by the source is a time-varying current. In the special case where the current source produces a constant current (pure dc), the circuit symbol shown in Fig. 12–20(b) can be used. Here the uppercase symbol I_s is used since the current is constant.

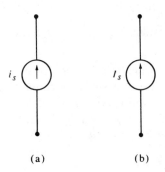

(a) (b)

FIGURE 12–20 (a) *Symbol for time-varying current source.* (b) *Symbol for constant current source.*

The arrows in the current source symbols represent the *positive* direction of electron current flow. The meaning of the arrows can best be illustrated by considering an example. Figure 12–21 shows a current source as the input to a certain circuit. The current arrow is shown pointing upward. The waveform for i_s is an ac square wave switching between +5 mA and −5 mA. Between $t = 0$ and $t = 5$ ms the waveform diagram shows that i_s is equal to +5 mA. This means that during this interval current flows from the source in the direction of the arrow.

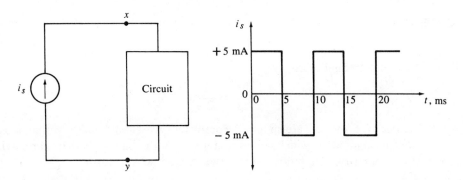

FIGURE 12–21

Between $t = 5$ ms and $t = 10$ ms the current is equal to −5 mA. This negative sign means that the current supplied by the source during this interval is flowing *opposite* to the direction of the arrow. In other words, for those instantaneous values of i_s which are negative, the current from the source is actually being supplied in a direction opposite to the arrow. The arrow indicates the current direction when i_s is positive.

12.10 Response of a Linear Resistor to Voltage Waveforms

As we saw in the previous chapters, a linear resistor has the characteristic that its current is directly proportional to the voltage across it. As the instantaneous voltage across a resistor changes, the instantaneous current through the resistor will change proportionately according to Ohm's law. In other words, Ohm's law is valid for a resistor even if the applied voltage is a time-varying voltage waveform. This is expressed as

$$i = \frac{v}{R} \qquad (12\text{–}2a)$$

or

$$v = iR \qquad (12\text{–}2b)$$

These are the Ohm's law relationships written for time-varying voltage and current.

What this means is that we can determine the waveform of current through a resistor for a given voltage waveform across the resistor. To illustrate, consider Fig. 12–22(a) where a square wave of voltage is applied to a 1-kΩ resistor. The input voltage switches between $+10$ V and -10 V at a frequency of 1 kHz. According to Eq. 12-2a we can determine the value of resistor current at any instant of time by dividing the voltage present across the resistor by the resistance R.

For the situation in Fig. 12–22(a), the resistor voltage v_{xy} will be either $+10$ V or -10 V. For those intervals of time when $v_{xy} = +10$ V we have

$$i = \frac{v_{xy}}{R} = \frac{10 \text{ V}}{1 \text{ k}\Omega} = \frac{10 \text{ V}}{1000 \text{ }\Omega} = 0.01 \text{ A}$$
$$= 0.01 \times 10^3 \text{ mA}$$
$$= 10 \text{ mA}$$

For those intervals of time when $v_{xy} = -10$ V we have

$$i = \frac{-10 \text{ V}}{1 \text{ k}\Omega} = -10 \text{ mA}$$

Thus, we can now construct the waveform for i which occurs in response to the input voltage waveform. The result is shown in Fig. 12–22(b). The current waveform is also a square wave.

Another illustration is shown in Fig. 12–23, where a sine wave of voltage is applied to a 2-megohm resistor. Following the procedure used in the last illustration we can calculate i for any point in time. At $t = 0$, $t = 4$ ms, and $t = 8$ ms the input voltage is at 0 V. Thus, $i = v/R = 0$ at these points in time. At $t = 2$ ms, $e_{in} = 100$ V so that $v_{xy} = 100$ V and we have

$$i = \frac{100 \text{ V}}{2 \text{ M}\Omega} = \frac{100 \text{ V}}{2 \times 10^6 \text{ }\Omega} = 50 \times (10^{-6} \text{ A})$$
$$= 50 \text{ }\mu\text{A}$$

Similarly, at $t = 6$ ms we have $e_{in} = -100$ V so that $v_{xy} = -100$ V and

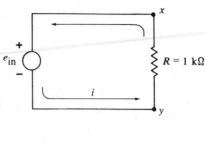

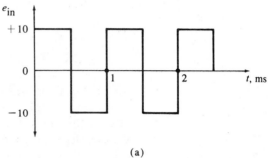

(a)

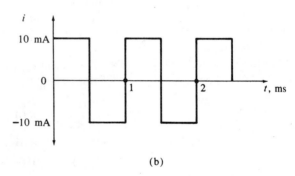

(b)

FIGURE 12-22 *Current response of a resistor for an input voltage square wave.*

$$i = \frac{-100 \text{ V}}{2 \text{ M}\Omega} = -50 \text{ }\mu\text{A}$$

We could continue this procedure for any point in time. The result would be the waveform for i shown in Fig. 12-23(b). The current waveform is also a sine wave.

Based on these two examples an important conclusion can be drawn:

The current through a resistor has a waveform which has exactly the same shape as the applied voltage waveform.

In other words, the current waveform in a resistor will be exactly the same as the voltage waveform, with each point on the current waveform being related to its corresponding point on the voltage waveform by Ohm's law ($i = v/R$).

12.11 Power in Time-Varying Circuits

Power equals the product of voltage and current. This equality will hold true for *all* conditions in *all* circuits. If the circuit contains a time-varying source, the

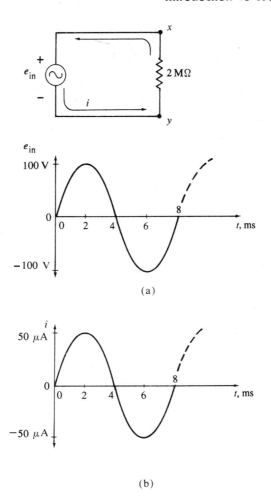

FIGURE 12-23 *Current response of a resistor for an input voltage sine wave.*

circuit currents and voltages will, in general, be time-varying quantities. In such cases it becomes convenient to talk about the *instantaneous power, p.* Thus,

$$p = i \times e \qquad (12\text{-}3)$$

or

$$p = i \times v \qquad (12\text{-}4)$$

where power, p, is now a time-varying quantity.

Consider the situation in Fig. 12-24 where a sinusoidal source of voltage is applied to a 10-ohm resistor. The source voltage has a peak amplitude of 10 V. The resultant circuit current is therefore a sine wave of current with a peak amplitude of 10 V/10 ohms = 1 A. The voltage and current waveforms are shown in part (b) of the figure.

Since the voltage and current are time-varying, the power dissipation, p, will be time-varying. The value for p at any instant of time is determined by the

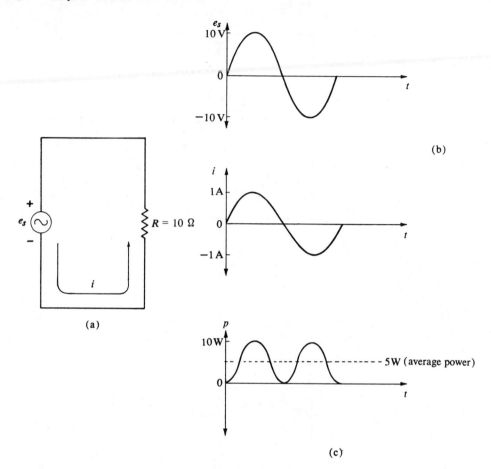

FIGURE 12-24 *Determining power in a sinusoidal circuit.*

product of the instantaneous voltage and current that occurs at the corresponding instant of time. If we take the e_s and i graphs and calculate $p = i \times e_s$ at each instant of time, the graph in Fig. 12–24(c) will result. It is the waveform of power, p. Several significant facts can be noted about this power waveform: waveform:

(1) The power curve is alway positive since i is $+$ when e_s is $+$, and i is $-$ when e_s is $-$.

(2) The power waveform has a frequency that is twice that of the i and e_s waveforms; that is, it goes through two complete cycles for each cycle of e_s.

(3) The peak value of the power curve (10 W) is equal to the peak voltage (10 V) multiplied by the peak current (1 A).

(4) The power waveform is symmetrical about an *average* value (5 W) that is equal to one half of the peak value of the power (0.5 × 10 W).

(5) The power varies from a minimum of *zero* when e_s and i_s are both *zero*, to a peak value of 10 W when e_s and i_s are at their peak values.

Average Power, P_{av} In time-varying circuits, the *average power* is an important quantity because it is the average power which is considered when determining the power rating for resistive devices. For example, in Fig. 12–24 the average power supplied to the 1-ohm resistor is seen to be 5 W. Thus, the 1-ohm resistor is chosen to have a safe power rating of $2 \times 5 \text{ W} = 10 \text{ W}$. For the *sinusoidal* case we can state that

$$\text{average power } P_{av} = 1/2 \ V_{\text{peak}} \times I_{\text{peak}} \qquad (12\text{–}5)$$

This is *not* a general formula for all waveforms, only for sine waves.

We will return to the concept of power in time-varying circuits later. For now the preceding example is a sufficient introduction to this important concept. Several problems at the end of the chapter will serve as further examples.

Chapter Summary

1. When one quantity X depends on one or more other quantities, then we say that X is a function of these other quantities.
2. If X is a function of Y, then X is the dependent variable and Y is the independent variable.
3. A graph has a positive slope where its curve slopes upward; it has a negative slope where its curve slopes downward.
4. A graph has a zero slope where its curve is flat or at the top of the peaks or bottom of the valleys.
5. A graph has a constant slope in regions where its curve is a straight line. A linear slope is another term for constant slope.
6. A slope is greater as it gets steeper.
7. A waveform of voltage (or current) is a graph showing how the voltage (or current) changes with time.
8. A waveform that changes polarity is an ac waveform.
9. A waveform that does *not* change polarity is a dc waveform.
10. A sine wave is a *pure ac* waveform because its shape and size are the same during the positive portion as during the negative portion of the waveform.
11. One complete sinusoidal waveform is called a cycle.
12. The amplitude of a sine wave can be expressed as peak voltage (voltage at the peaks) or peak-to-peak voltage (voltage difference between peaks).
13. When a waveform repeats itself at uniform intervals of time it is said to be a periodic waveform.
14. One complete cycle occurs during each period, T, of a periodic waveform.
15. The frequency of a periodic waveform is equal to the number of cycles that occur in one second. That is, $f = 1/T$ where the units are hertz (Hz).

16. When the waveform for a voltage source shows its voltage is positive, the polarity of the voltage source is the polarity shown in the circuit diagram. When the waveform shows the voltage is negative, the polarity of the voltage source is opposite to the polarity shown on the circuit diagram.

17. When the waveform for a current source shows the current is positive, the direction of the current produced by the source is the direction of the arrow on the current source symbol. When the current is negative the current produced is in the direction opposite to the arrow.

18. Ohm's law is valid for time-varying voltages applied to a linear resistor.

19. The waveform of current through a linear resistor has exactly the same shape as the applied voltage waveform.

20. The *instantaneous* power, p, in a time-varying circuit is given by $p = i \times v$ where all quantities are time-varying.

21. The *average* power, P_{av}, in a time-varying circuit is determined graphically from the waveform of instantaneous power. For a *sine wave* circuit, $P_{av} = 0.5 \times V_{peak} \times I_{peak}$.

Questions/Problems

Sections 12.2–12.4

12–1 Which of the following are functions of time?
  a. area of this page
  b. barometric pressure in New York City
  c. position of the moon
  d. volume of a cylinder

12–2 For each of the following choose the dependent variable and the independent variable(s).
  a. area of a circle $= \pi r^2$
  b. $F = k \dfrac{q_1 q_2}{d^2}$
  c. barometric pressure at different longitudes along the equator at 12:00 noon on a given day
  d. length of a day equals 24 hours

12–3 Refer to the graph in Fig. 12–1.
  a. Determine the temperature at 5:00 P.M.
  b. Determine at what time (or times) during the day the temperature was 85°F.

12–4 Consider the graph in Fig. 12–25 which shows the relationship between the current I through a particular electronic device and the voltage V applied across the device. Indicate the areas of the curve where the slope is
  a. positive
  b. negative
  c. zero
  d. linear (or almost linear)

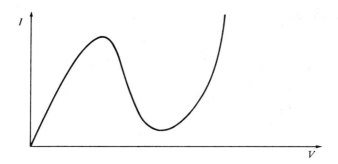

FIGURE 12-25

12-5 Which region of Fig. 12-25 has the greatest slope? Which has the smallest slope?

Section 12.5

12-6 Refer to the sinusoidal waveform of Fig. 12-2. Assume that the time scale increments are exactly 1 ms. Determine the instantaneous values of voltage at $t = 1$ ms; 4 ms; and 6 ms.

12-7 For the same conditions as the preceding problem, determine the approximate instant(s) of time when the instantaneous voltage will be $+5$ V. Repeat for -5 V.

12-8 Refer to Fig. 12-26. Indicate which waveforms are ac, which ones are dc, and which ones are *pure* dc.

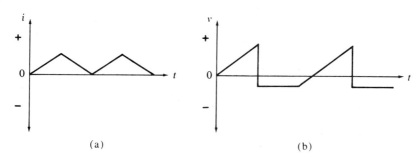

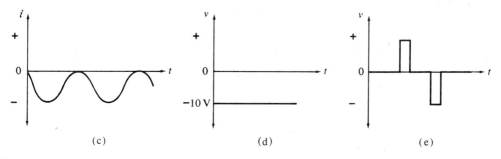

FIGURE 12-26

Section 12.6

12-9 Sketch the following waveforms:
   a. A constant current of -6 mA
   b. A positive-going 3-V step occurring at $t = 5$ ms.
   c. A constant $+2$ V until $t = 2$ ms; then a negative-going 5-V step.
   d. A constant -2 V until $t = 1$ ms; then an exponential increase to $+2$ V at $t = 5$ ms.
   e. A sine wave with a cycle of 100 μs and an amplitude of 60 mA peak-to-peak.

12-10 Determine V_p and $V_{p\text{-}p}$ for the waveform in Fig. 12-28(b).

12-11 In addition to peak amplitude and peak-to-peak amplitude there is a third way to express the amplitude of a sine wave. It is called rms amplitude and it will be discussed later because of its importance. For now, we will simply state that a sine wave has an rms amplitude equal to 0.707 of the peak amplitude. Thus, a sine wave with a 10-V peak amplitude will have an amplitude of 7.07 V-rms.
   a. Determine the rms amplitude of a sine wave that has a peak-to-peak amplitude of 340 V.
   b. Determine peak voltage of a sine wave that has an amplitude of 24 V-rms.

Section 12.7

12-12 In Fig. 12-27 a rectangular current pulse is shown. The pulse is resting on a 2-mA constant level. Determine the pulse amplitude and pulse width.

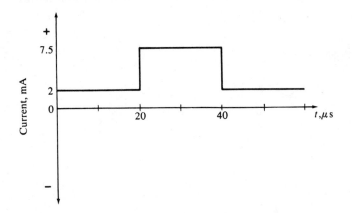

FIGURE 12-27

12-13 Sketch the waveform for a rectangular pulse which is sitting on a 2-V level, has a negative-going amplitude of 4 V and a pulse width of 50 μs.

Section 12.8

12-14 Determine the period and frequency of the waveforms in Fig. 12-28.

12-15 Draw a periodic pulse waveform with the following characteristics:
   a. $T = 20$ ms   b. $t_p = 5$ ms
   c. positive-going 5-V pulse resting on -1-V constant level.

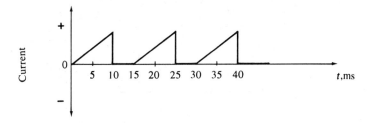

(a)

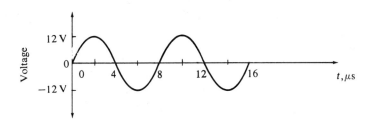

(b)

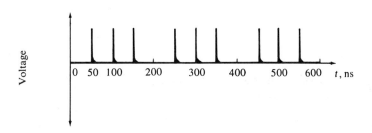

(c)

FIGURE 12-28

12-16 What is the frequency of the waveform of the previous problem?

Section 12.9

12-17 In the situation shown in Fig. 12-29 determine the values of v_{xy} at the following times: $t = 0.5$ ms, $t = 1.5$ ms, and $t = 2.5$ ms.

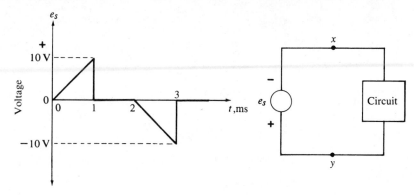

FIGURE 12–29

12–18 In the situation in Fig. 12–30 determine the value of current i at $t = 5$ ms and at $t = 15$ ms. Reverse the current source arrow and repeat.

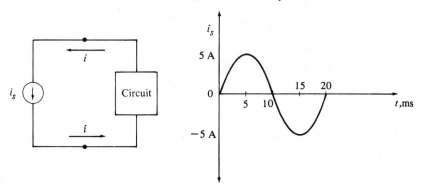

FIGURE 12–30

Section 12.10

12–19 The waveform of Fig. 12–28(b) is applied to a 150-Ω resistor. Sketch the resulting current waveform.

12–20 The waveform of Fig. 12–29 is applied to the 150-Ω resistor. Sketch the resulting current waveform.

12–21 A source e_S produces a sine wave with $V_p = 160$ V. Determine the peak-to-peak amplitude of the current waveform when this source is applied to a 6.8-kΩ resistor.

12–22 In chap. 5 we learned that a nonlinear device does not have a constant resistance at all points on its *I-V* curve. How do you think this might affect the current waveform when a time-varying voltage is applied across such a device?

Section 12.11

12–23 Calculate P_{av} for the circuit of Fig. 12–23.

12–24 Calculate P_{av} for the values of problem 12–21.

12–25 A 60-Hz sine wave with $V_{p\text{-}p} = 320$ V is applied to a light bulb rated at 100 W. Determine the peak amplitude of the current flowing through the bulb. The light bulb power rating is its P_{av} rating.

CHAPTER 13

Capacitance

13.1 Introduction

The previous chapters dealt with the electrical property of *resistance* and its effect in circuits. As such these circuits were called resistive circuits. In addition to resistance, two other electrical circuit properties are important. These properties are called *capacitance* and *inductance*. In this chapter the characteristics of capacitance and *capacitors* (devices which have capacitance) will be studied.

As we shall see, capacitance is much different from resistance in its effect on the voltages and currents in a circuit. The effects of capacitance on circuit operation will be examined thoroughly in this chapter because capacitance, like resistance, is present to some degree in every circuit.

13.2 Electrostatic Force

The nature of capacitance will be better understood if the concept of electrostatic force is reviewed. We observed in chap. 1 that an electrostatic force exists between any two electric charges. If the charges have the same polarity the force is one of repulsion, and if the charges have opposite polarities it is a force of attraction.

It was also observed in chap. 1 that electric force requires no physical contact and is therefore a *field* force. In the same way that we talk about the gravitational field of the earth, we can talk about an *electric field* associated with any charge. To illustrate, let us examine what happens when a small positive charge (such as one proton) is placed in the vicinity of a fixed *positive* charge. We know that the proton will be repelled away from the fixed charge and we can say that the proton is acted upon by the *electric field* of the fixed charge. This electric field can be represented as shown in Fig. 13–1(a) by drawing lines extending radially out from the + charge. These *lines of force* are drawn in a direction along which any other + charge will be repelled away. In contrast, the electric field surrounding a fixed *negative* charge will *attract* a small positive charge, resulting in the lines of force shown in Fig. 13–1(b).

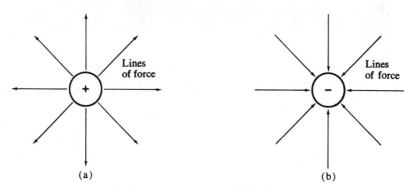

FIGURE 13–1 *Electric field surrounding electric charges.*

We can now say that an electric field exists in any region in which an electric charge is acted upon by an electric force. The electric lines of force associated with an electric field always point in the direction in which a *positive* charge will be forced to move by the electric force (see Fig. 13–1).

In studying capacitance we will not be particularly interested in the electric fields associated with isolated electric charges. Rather our interest will be on the electric field that exists between two metal plates that are charged with opposite polarities as shown in Fig. 13–2. In this situation, electric lines of force will exist between the plates and will extend from the positive plate to the negative plate as shown.

As we would expect, the negative charges on the right-hand plate will exert an attractive force on the positive charge on the left-hand plate and vice versa. This mutual force depends on the *intensity* of the electric field between the plates. We can represent the *intensity* or *strength* of the electric field symbolically by $\mathcal{E}$ (a script E).

The electric field intensity $\mathcal{E}$ can be expressed numerically as

$$\mathcal{E} = \frac{V}{d} \tag{13–1}$$

where V represents the potential difference (voltage) between the two charged plates and d is the distance between the plates in meters. Thus, $\mathcal{E}$ is expressed in units of volt/meter.

It should be obvious that the electric field intensity can be increased by moving the plates closer together (smaller d) or by increasing the number of charges on the plates (larger V). The field intensity is a direct measure of the electric force exerted by the electric field.

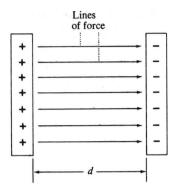

FIGURE 13-2 *Electric field between two charged plates.*

It can be shown that an electric field with $\mathcal{E} = 1$ V/m will exert 1 newton of force on a charge of 1 coulomb. In other words, we can write

$$F = Q \times \mathcal{E} \tag{13-2}$$

where F is the force in newtons exerted on a charge Q, in coulombs, by an electric field with an intensity $\mathcal{E}$ in volts/m.

13.3 Capacitors

A capacitor is an electrical device which consists of two conducting plates separated by an insulator. The insulator is often called the capacitor *dielectric*. To begin to understand the operation of a capacitor consider Fig. 13-3(a) where a square metal plate is connected to a negatively charged rod. When the rod is brought into contact with the plate, electrons from the rod will spread out to all parts of the plate (due to mutual repelling forces of electrons). Thus, the plate takes on a negative charge as shown.

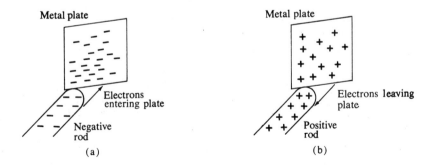

FIGURE 13-3 *Charging metal plates with* (a) *negative charges and* (b) *positive charges.*

Similarly, a metal plate brought into contact with a positively charged rod will take on a positive charge as electrons leave the plate because of the attraction of the positive rod. This is illustrated in Fig. 13-3(b).

If we now remove the charged rods from the plates and place the plates near each other, the situation is that shown in Fig. 13–4(a). *The plates will remain charged* because there is no way for the charges to move off the plates. There will be a force of attraction between the charges on the two plates but they cannot reach each other to intermingle because the air separating them is an insulator (or dielectric).

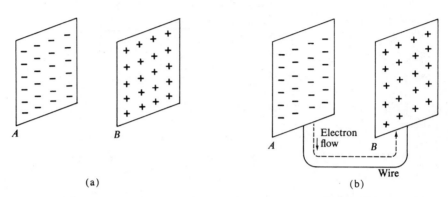

FIGURE 13–4 (a) *Plates storing charges and* (b) *discharging through connecting wire.*

These two metal plates separated by an insulator constitute one form of a *capacitor*. This capacitor has the ability to hold or store a charge that is placed on its plates.

If we now provide a conducting path between the two charged plates as in Fig. 13–4(b), the charges on the plates will be neutralized. This occurs because the electrons on plate A can now flow through the connecting wire to plate B. This movement of electrons constitutes a current which flows momentarily until the two plates are *discharged* (net charge = 0).

13.4 Capacitance

When the charges are placed on the two metal plates of the capacitor in Fig. 13–4, an electric field is present in the region between the plates. This occurs because the difference in the charges on the two plates produces a potential difference (voltage) between the plates. In other words, plate B is at a positive potential relative to plate A. The capacitor which is storing charges thus acts as a source of voltage similar to a battery. A battery produces a voltage by virtue of the different charges on its electrodes, while the capacitor's voltage is produced by the charges on its plates.

This is why when the wire is connected to the capacitor plates [Fig. 13–4(b)] a current will flow as shown. The current persists until all the charges have been removed. A similar situation occurs if a wire is connected between the terminals of a battery. A battery, however, will take longer to completely discharge because it has a more plentiful supply of charges produced by its internal chemical reactions. A capacitor only has the charges on its plates that were placed there by some means (e.g., a charged rod).

The voltage between the capacitor's plates depends on the amount of charges on the plates. In fact, a capacitor's voltage is directly proportional to this charge. To illustrate, refer to Fig. 13-5(a). Here a capacitor is represented by its standard circuit symbol. It is shown with positive charges on one plate and an equal number of negative charges on the other plate. Let's say there are Q coulombs on each plate. The result of these charges is a voltage V between the two plates.

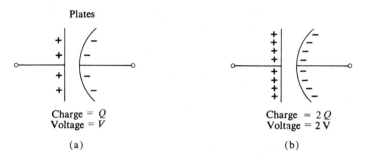

FIGURE 13-5 *Voltage between capacitor plates is proportional to the amount of charge on the plates.*

If we now double the amount of charges on each plate as shown in Fig. 13-5(b), the voltage between the plates will also double. Similarly, a decrease in the amount of charges will produce a proportional decrease in voltage. Thus, we can say that:

In a capacitor, the voltage V between its plates is proportional to the amount of charge Q on the plates.

Equivalently, we can say that Q is proportional to V. Stated this way, we can mathematically express the relationship between Q and V as

$$Q = CV \qquad (13\text{-}3\text{a})$$

where C is the proportionality constant and is called *capacitance*. This expression can be solved for V to give

$$V = \frac{1}{C} \times Q \qquad (13\text{-}3\text{b})$$

which shows V proportional to Q with the proportionality constant being $1/C$.

These two expressions are the basic capacitor relationships the same as $I = V/R$ and $V = IR$ are the basic relationships for a resistor. Note that in a capacitor Q and V are proportional whereas in a resistor I and V are proportional. This is an important difference as we shall see.

The units for *capacitance* can be determined from Eq. (13-3a) by solving for $C = Q/V$ = coulombs/volt. Thus, C can be expressed in coulombs/volt. However, capacitance is more usually expressed in units called *farads* (named after scientist Michael Faraday). A farad (F) is equal to a coulomb/volt. That is

$$1 \text{ farad} = 1 \frac{\text{coulomb}}{\text{volt}}$$

Thus, in using expressions (13-3a) or (b), C is expressed in farads, V in volts, and Q in coulombs.

EXAMPLE 13.1 A certain capacitor has a capacitance of 5 F. How much voltage will be developed across its plates if each plate has a charge of 20 coulombs?

Solution: Using (13–3b)

$$V = \frac{Q}{C} = \frac{20 \text{ C}}{5 \text{ F}} = 4 \text{ V}$$

Note that the value of Q used was 20 coulombs, which is the charge on *one* plate.

In practice, the farad is too large a unit to represent the capacitance of most practical capacitors. The prefixed units microfarad (μF) and picofarad (pF) are more commonly used. Capacitance values from about 1 pF (10^{-12} F) to about 10,000 μF are encountered in electronic circuits.

EXAMPLE 13.2 How much charge is on the plates of a 50-μF capacitor that has a voltage of 15 volts across its plates?

Solution: 50 μF = 50 $\times$ 10^{-6} F. Using (13–3a) we have

$$Q = 50 \times 10^{-6} \times 15$$
$$= 750 \times 10^{-6} \text{ coulombs}$$
$$= 750 \text{ μC}$$

Capacitance is a measure of how much charge is required to produce a given voltage between the plates. As C is increased it takes more charge to produce a given voltage [Eq. (13–3a)]. On the other hand a smaller value of C requires less charge to produce a given voltage. For a capacitor, the value of capacitance depends on certain physical characteristics. We will be concerned with these factors later. For now we will concentrate on the behavior of capacitors in circuits.

13.5 Charging and Discharging a Capacitor — Qualitative Discussion

In most circuit applications a capacitor is charged by connecting it to some source of voltage. Consider Fig. 13–6(a). With the switch in position 1 the capacitor is completely discharged and therefore has no voltage between its plates ($V_{xy} = 0$).

Charging Process In part (b) of the figure the switch has been moved to position 3. As we can see, the 10-V source is now connected to the capacitor plates. At the instant that the switch is moved to position 3, we might expect *two* things to happen: (1) the capacitor voltage would immediately become 10 V ($V_{xy} = 10$ V) and (2) no current would flow in the circuit because the insulator between the plates acts as open circuit.

In actuality, these two conditions do occur, but they do not occur *instantaneously* when the switch is closed. The reason is that in order to have a 10-V potential difference across the capacitor's plates, there must be a charge on

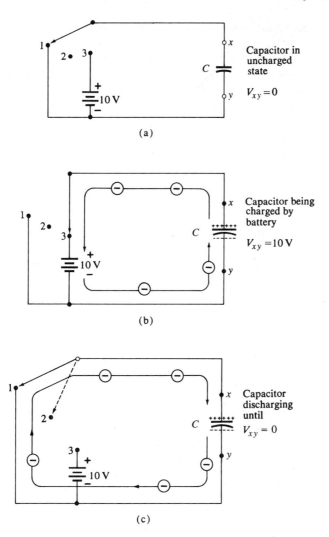

FIGURE 13-6 *Illustration of capacitor charging and discharging.*

these plates ($Q = CV$). Before the switch was moved to position 3, the capacitor had no charge. In position 3, the 10-V source is in the circuit and it will begin charging the capacitor as follows:

(1) The negative terminal of the source will produce a repelling force on the free electrons in the wire connecting it to the capacitor plate.

(2) As such, electrons in this wire will begin flowing toward the bottom plate of the capacitor. Once they get to this plate they can go no further because of the dielectric (insulator) and, therefore, a buildup of negative charges takes place on this plate.

(3) Each electron which is deposited on the lower plate will repel one electron off of the upper plate and these upper-plate electrons will be attracted toward the plus terminal of the source. This causes the upper plate to become positively charged (it is losing electrons).

(4) This action will continue until the buildup of negative charges on the bottom plate becomes great enough to counteract the force exerted by the negative terminal of the source. When this occurs, electrons in the bottom wire will no longer move toward the capacitor plate and the charging of the capacitor will be completed. This occurs when the capacitor charges have built up to the point where the voltage between the plates equals the source voltage of 10 V.

The important thing about this charging process is that it takes *time*. In other words, the capacitor voltage V_{xy} does not instantaneously jump to 10 V. It gradually increases to 10 V as electrons pile up on its bottom plate and electrons leave the upper plate. We will discuss the factors which determine *how long* this process takes in a later section.

During the charging process something happens which might at first be disturbing. We see that electrons are flowing out of the negative terminal of the source [Fig. 13-6(b)] toward the bottom capacitor plate. At the same time, electrons are flowing into the positive source terminal. This flow of electrons represents a *current* which could be detected by an ammeter placed in the circuit. Since for every electron deposited on the bottom plate an electron is driven off the upper plate, the current entering the source is equal to the current leaving the source. This current, as we shall see, persists only for a short interval of time until the capacitor is fully charged. This *charging* current is only temporary or *transient*. The part about this which is disturbing is that current is flowing even though the capacitor has essentially an open circuit between its plates. We learned in previous chapters that *no* current will flow if an open circuit is present in the current path. This is still true, but there is always a temporary or transient current that flows until the open-circuited points (in this case the capacitor plates) are charged up to a voltage equal to the source voltage. The duration of this transient current is usually very short (nanoseconds) when the open circuit is simply a disconnected wire or an open switch. However, with a capacitor the charging current could persist for longer periods of time before the capacitor plates become fully charged.

To summarize this discussion we can say that:

When a capacitor is being charged from a constant voltage source, a transient current will flow until the capacitor is charged up to the source voltage, after which no further current will flow and the capacitor will act like an open circuit.

Storing the Charge If the switch in Fig. 13-6(c) is now put in position 2, the source will be disconnected from the circuit. The capacitor, having been previously charged up to 10 V, will remain in this charged state because there is no path through which the electrons on the bottom plate can flow to reach the positive upper plate. There is a strong force of attraction between the positive charges on the upper plate and the negative charges on the bottom plate. However, the electrons cannot pass through the insulating dielectric to reach the positive charges. Thus, in position 2 the capacitor will remain fully charged indefinitely with $V_{xy} = 10$ V. (In practice, because no insulator is perfect, electrons will *slowly leak* through to the positive plate and eventually the charges will be completely neutralized. This *leakage* concept will be discussed in more detail later.)

Discharging If the switch in Fig. 13-6(c) is now placed in position 1, a path is provided through which electrons can flow from the bottom plate to neutralize the positive charges on the upper plate. Again, this discharge current will only be temporary until the capacitor is completely discharged. When completely discharged the situation reverts back to that in Fig. 13-6(a) with $V_{xy} = 0$ and no current flowing. As we stated earlier, during its discharge *the capacitor acts as the source of voltage* that produces the discharge current.

13.6 Charging a Capacitor — Quantitative Discussion

Now that we have a feel for how a capacitor charges, we can investigate it in more detail and develop the quantitative relationships that are important in analyzing capacitor circuits. Consider the circuit in Fig. 13-7(a). Here again we will investigate the capacitor charging when the switch is closed. The resistor R is in series with the source and the capacitor, and could represent the source resistance, the resistance of the connecting wires, and/or a resistor purposely put in the circuit to limit current. The effects of this resistance were not included in the discussion of the last section but must always be considered in practice.

For illustrative purposes, values for E_S, R, and C were chosen as shown in the figure. With the switch open in the circuit, the capacitor will not charge because electrons from the source cannot reach its plate. In part (b) of the figure, the switch is closed. The instant the switch is closed will be referred to as time $t = 0$. The situation at $t = 0$ is shown in Fig. 13-7(b).

At $t = 0$, the capacitor has not had a chance to acquire any charge yet. As such, the capacitor voltage v_{xy} is still at zero volts. *It cannot change instantaneously.* In order to satisfy KVL, then, it follows that the 10 V from the source must appear across the resistor. This 10 V across the 100-ohm resistor indicates that the charging current i_c* has a value of $10/100 = 0.1$ A at $t = 0$. In other words, at the instant the switch closes, the source produces a current equal to 0.1 A. This movement of charges begins to charge the capacitor.

Since the initial flow of current is 0.1 A, then charges are being supplied to the capacitor at the rate of 0.1 A = 0.1 coulomb/second. This movement of charges will cause the capacitor voltage to gradually increase. However, as the capacitor voltage v_{xy} increases from zero, the voltage across the resistor must decrease in order to satisfy KVL (resistor voltage + capacitor voltage = source voltage of 10 V). With a smaller voltage across R, the charging current i_c must also decrease according to Ohm's law ($i = v/R$). Stated another way, as the capacitor charges its voltage builds up and *opposes* the source voltage; thus, there is a smaller net voltage available to produce current through the resistor and therefore through the entire circuit.

For example, some time after the switch is closed the situation will be that shown in Fig. 13-7(c). The capacitor voltage has increased to 6 V so that only 4 V appear across the resistor, resulting in a current $i_c = 4/100 = 0.04$ A.

---

*Lowercase symbols are used for v_{xy} and i_c because these quantities are varying with time.

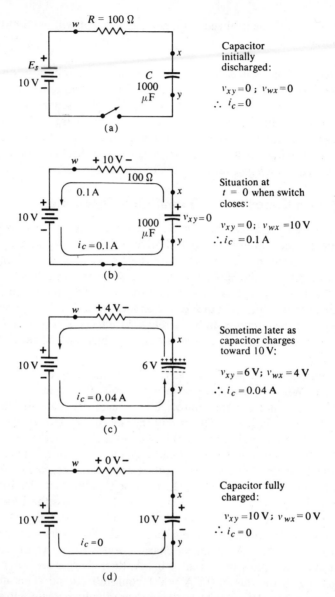

FIGURE 13-7 *Capacitor charging circuit (a) before closing switch; (b) at instant of closing switch, t = 0; (c) sometime later as capacitor charges toward 10 V; and (d) capacitor fully charged.*

According to the preceding discussion, then, we can state that *as the capacitor charges, the magnitude of the charging current decreases.* This gradually decreasing current means that the rate of charge flow will be decreasing. This, in turn, means that the rate at which the capacitor voltage is building up will be gradually decreasing. Thus, as the capacitor voltage builds up due to the charging current, the rate at which it builds up becomes less and less because the charging current becomes less and less. This process continues until the capacitor voltage reaches 10 V, at which time there is no net voltage left in the circuit to produce any charging current. After this point no further capacitor charging takes place [Fig. 13–7(d)].

The above process can be better visualized by referring to the waveform graphs shown in Fig. 13–8. Several important points can be observed concerning these waveforms:

(1) At the instant the switch is closed ($t = 0$), the total source voltage (10 V) appears across the resistor ($v_{wx} = 10$ V) because the capacitor voltage *cannot change instantaneously* ($v_{xy} = 0$).

(2) Thus, the charging current i_c at $t = 0$ is at its maximum value (0.1 A).

(3) The capacitor voltage initially increases at a relatively rapid rate (slope of v_{xy} curve is steepest at $t = 0$) but the rate of increase gradually becomes less and less because the charging current is gradually decreasing.

(4) The resistor voltage (and therefore i_c) begins at its maximum value and gradually decreases toward zero as the capacitor charges.

(5) After a certain time interval (0.5 s in this case) the capacitor reaches 10 V (equals E_s) and the resistor voltage and charging current have dropped to zero. With no charging current the capacitor is no longer being supplied with charges so v_{xy} will remain at 10 V indefinitely.

13.7 Capacitor Charging Rate, $\Delta V/\Delta t$

The rate at which the capacitor voltage increases depends directly on the rate at which charges are supplied to its plates. For example, suppose that a current of 1 A is charging a 0.1-farad capacitor. This means that 1 coulomb of charge per second will be accumulating on the capacitor's plates. Using the basic capacitor relationship

$$V = \frac{1}{C} \times Q \qquad (13\text{–}3\text{b})$$

each coulomb of charge will produce

$$V = \frac{1}{0.1 \text{ F}} \times 1 \text{ C} = 10 \text{ V}$$

Thus, in one second the 1-A current will cause the capacitor voltage to increase by 10 V. Stated mathematically, in a time interval Δt of 1 s the capacitor voltage will increase by an amount ΔV equal to 10 V. Thus, the *rate of change of voltage* is

$$\frac{\Delta V}{\Delta t} = \frac{10 \text{ V}}{1 \text{ s}} = 10 \text{ V/s}$$

where $\Delta V/\Delta t$ represents the rate of change of voltage, measured in volts per second.

If the charging current is doubled to 2 A it should be obvious that the rate of change of voltage will also double to 20 V/s since charges are being supplied at twice the rate. In fact, $\Delta V/\Delta t$ is directly proportional to the charging current in the same way that V is proportional to Q. That is,

$$V = \frac{1}{C} \times Q$$

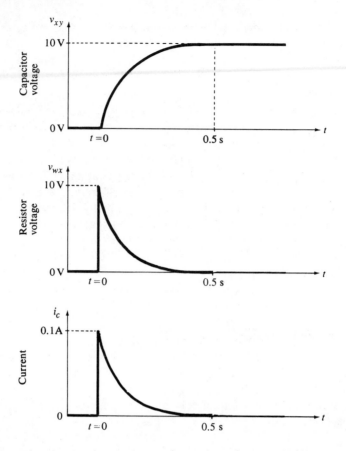

FIGURE 13-8 *Waveforms of capacitor and resistor voltage and charging current.*

so that

$$\Delta V = \frac{1}{C} \times \Delta Q$$

and

$$\frac{\Delta V}{\Delta t} = \frac{1}{C} \times \frac{\Delta Q}{\Delta t}$$

which becomes

$$\frac{\Delta V}{\Delta t} = \frac{1}{C} \times i_c \qquad (13\text{--}4)$$

since $\Delta Q/\Delta t$ (rate of charge flow) is the same as the charging current i_C.

The relationship in Eq. (13–4) relates the capacitor charging rate $\Delta V/\Delta t$ to the charging current and the capacitance. As expected, a larger i_c produces a greater rate of charge while a larger C causes the rate of charge to be less (slower). This relationship is valid at any instant of time and can be used to determine the rate of capacitor charging if the value of i_c is known.

Referring again to the example circuit considered in Figs. 13–7 and 13–8, we saw that the initial charging current was $i_c = 0.1$ A. Thus, using (13–4) we can calculate the initial rate of increase of capacitor voltage v_{xy} as

$$\frac{\Delta v_{xy}}{\Delta t} = \frac{1}{1000 \ \mu F} \times 0.1 \text{ A} = \frac{10^{-1}}{10^{-3}} = 100 \text{ V/s}$$

The capacitor voltage begins increasing at this rate. This 100 V/s represents the initial slope of the capacitor charging curve which is redrawn in Fig. 13–9. The dotted line drawn tangent to the curve at $t = 0$ has a slope equal to 100 V/s. As the charging current decreases, the capacitor charging rate will decrease. This is indicated by a smaller slope for times greater than $t = 0$ as illustrated in Fig. 13–9.

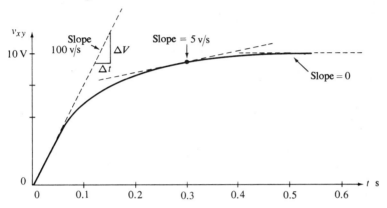

FIGURE 13–9 *Capacitor charging curve redrawn from Fig. 13–8.*

EXAMPLE 13.3 The waveforms in Fig. 13–8 show that at $t = 0.3$ s after the switch is closed the charging current has decreased to about 0.005 A. At what rate is the capacitor voltage increasing at $t = 0.3$ s?

Solution: $\dfrac{\Delta v_{xy}}{\Delta t} = \dfrac{i_c}{C} = \dfrac{0.005}{1000 \times 10^{-6}} = 5 \text{ V/s}$

This is the slope of the capacitor charging curve at time $t = 0.3$ s as shown in Fig. 13–9.

13.8 Effect of E_s, R, and C on Charging Time

In the example considered in Figs. 13–7 through 13–9, the total amount of time it took the capacitor to complete its charging was about 0.5 s. Let's examine the effect of the various circuit values on the total charging time. The circuit diagram is redrawn in Fig. 13–10(a).

Consider the effect of the resistor R. If the value of R is *decreased*, then the initial charging current will *increase*. In fact if R is reduced by half to 50 ohms, the initial charging current will *double* from 0.1 A to 10 V/50 ohms = 0.2 A. This greater current will supply charges to C twice as fast. Thus, the capacitor will charge to the full 10 V in only half the time (0.25 s). This is shown in Fig. 13–10(b) by comparing charge curve II (R = 50 ohms) with charge curve I of the original circuit.

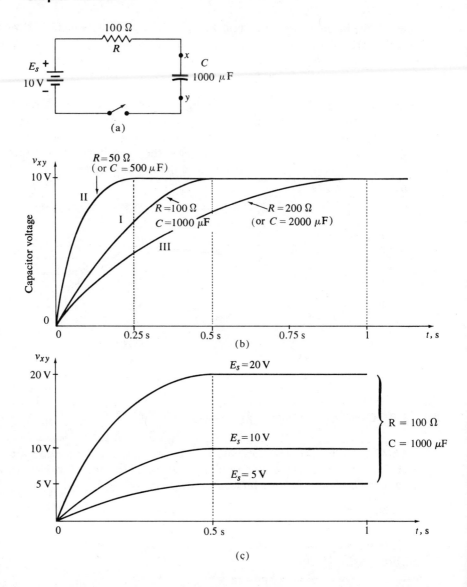

FIGURE 13-10 *Effects of circuit values on capacitor charging time.*

Conversely, an *increase* in R causes a smaller charging current and therefore a longer time to charge the capacitor to the full 10 V. See curve III for $R = 200$ ohms.

Consider now the effect of decreasing the capacitance C. The initial charging current depends only on E_s and R, and so it will be the same for any value of C. According to Eq. (13-4), a smaller capacitance will produce a greater $\Delta V/\Delta t$ for a given current, so that the smaller capacitor will charge at a faster rate. In fact, if C is reduced by half to 500 μF, the initial $\Delta V/\Delta t$ will *double* to 200 V/s. Thus, the capacitor charges twice as fast and the result is also curve II in Fig. 13-10(b).

As C is decreased further, the charging time will also decrease. But, as long as there is some capacitance (no matter how small), *the voltage across the* capacitor will not rise instantaneously but will take a certain amount of time to charge. In other words, *the capacitor opposes any quick change in voltage.*

An increase in C has the opposite effect: it will charge at a slower rate. If C is doubled the charge curve becomes curve III showing that the total charge time doubles to 1 s. Thus, C and R have the same effect on charge time as shown in the curves of Fig. 13–10(b).

Lastly, consider the effects of a change in the source voltage, E_s. If E_s is doubled to 20 V, this will produce twice the amount of charging current so that the capacitor will be charging at twice the rate. However, for the capacitor to be fully charged, it must now reach 20 V. Therefore, the net result is that the total charging time is unchanged because the capacitor is charging at twice the rate but has to charge twice as far. Using the same reasoning, the total charging time will not be affected by a decrease in the source voltage. The curves in Fig. 13–10(c) show that for various values of E_s the total charge time is still 0.5 s for the circuit in question, although the capacitor charges to different voltages.

13.9 Time Constant

For the circuit of Fig. 13–10(a) with the values shown, the capacitor becomes fully charged 0.5 s after the switch is closed. We say that the *transient response* of this circuit lasts for 0.5 s. The transient response occurs while the capacitor voltage, resistor voltage, and charging current are all in a transient (changing) state. After 0.5 s the circuit is in *steady state* because the voltage and current magnitudes are no longer changing.

To help us determine the amount of time it takes this RC circuit to reach steady state, we will utilize a quantity called the circuit's *time constant*. The time constant has the symbol τ (tau) and is given by the relationship

$$\tau = R \times C \qquad (13\text{–}5)$$

The basic unit for τ is *seconds*. In other words, τ is a quantity of time. To verify this we can use Eq. (13–5) and replace R and C by their basic units. That is,

$$\begin{aligned}
\tau &= R \times C \\
&= \text{ohms} \times \text{farads} \\
&= \frac{\text{volts}}{\text{amps}} \times \frac{\text{coulombs}}{\text{volts}} \\
&= \frac{\text{coulombs}}{\text{amps}} = \frac{\text{coulombs}}{\text{coulombs/s}} \\
&= \text{s}
\end{aligned}$$

Thus, the time constant, τ, is a measure of time. τ is in units of seconds if R is in ohms and C is in farads.

EXAMPLE 13.4 Calculate the value of τ for the circuit of Fig. 13–10(a).

Solution:
$$\begin{aligned}
\tau &= RC \\
&= 100 \text{ ohms} \times 1000 \ \mu\text{F} \\
&= 100 \times 1000 \times 10^{-6} \\
&= 10^{-1} \text{ s} = \mathbf{0.1 \text{ s}}
\end{aligned}$$

The time constant τ is a convenient interval of time because it can be shown that an RC circuit will essentially reach steady state in an amount of time equal to $5\,\tau$. Stated formally:

In a simple* RC circuit, all the currents and voltages have essentially reached their steady-state values after a time interval equal to $5\,\tau$.

For the circuit in Fig. 13–10(a), then, we can calculate $5\tau = 5 \times 0.1\text{ s} = 0.5\text{ s}$ as the time to reach steady state. After $5\,\tau$ has elapsed, the capacitor will be fully charged, the current will be at zero, and the resistor voltage will be at zero.

EXAMPLE 13.5 Figure 13–10 shows that when R is reduced to 50 Ω the time to reach steady state decreases to 0.25 s. Verify this using τ.

Solution: $\tau = RC = 50 \times 1000 \times 10^{-6}$
$\qquad\qquad\; = 0.05\text{ s}$
$\qquad\therefore\; 5\tau = 0.25\text{ s}$

which is the time needed to fully charge the capacitor.

Percentage Change in $1\,\tau$ Since $\tau = RC$, it is obvious that an increase in R or C will produce a greater time constant and therefore a longer time to reach steady state, and vice versa. Refer now to Fig. 13–11, where the RC charging circuit has been redrawn. The waveforms showing the gradual rise of capacitor voltage and the gradual fall of the resistor voltage are also redrawn in the figure. Since this circuit has $\tau = 0.1$ s, the waveforms reach steady state in $5\,\tau = 0.5$ s as shown.

Let's examine the two waveforms at $t = 0.1$ s. The capacitor voltage has increased to 6.3 V while the resistor voltage has dropped to 3.7 V. The capacitor voltage is ultimately going to increase by 10 V (from 0 to 10 V). Thus, in a time interval equal to one time constant (0.1 s) the capacitor voltage has undergone 63 percent of the total possible change. Similarly, the resistor voltage is ultimately going to decrease by 10 V (from 10 V to 0). Thus, the resistor voltage has undergone 63 percent of its total possible change in a time equal to $1\,\tau$.

A similar statement can be made about the charging current since its magnitude follows the resistor voltage exactly. In fact, it can be stated that in general:

In an RC circuit all currents and voltages will make 63 percent of their total change from initial value to steady-state value in a time interval of one time constant.

In other words, for any values of R and C, the circuit variables (currents and voltages) will always go 63 percent of the *remaining* distance toward their final steady state values in a time interval equal to $1\,\tau$.

This characteristic is true during any portion of the transient response. For example, the capacitor-charging waveform in Fig. 13–11 shows the voltage at 6.3 V after one time constant. At this point the capacitor will be 3.7 V *away* from its final value. Thus, after one more time constant it will complete 63%

---

*A simple RC circuit contains only *one* capacitor or equivalent capacitor, but can have *more* than one resistor as we shall see.

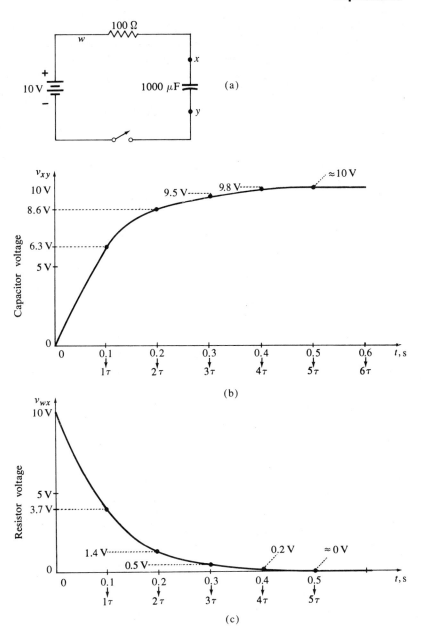

FIGURE 13-11 A closer look at capacitor and resistor voltage waveforms.

of this 3.7 V remaining distance. Since 63% × 3.7 V = 2.3 V, the capacitor voltage will increase by 2.3 V during the second time-constant interval (between 0.1 s and 0.2 s). This is verified by the curve where v_{xy} = 8.6 V at $t = 2\tau$ = 0.2 s.

The same procedure can be used to calculate the capacitor voltage at $t = 3\tau$, $t = 4\tau$, etc. In a similar fashion, the resistor voltage and the charging current values at $t = 2\tau$, 3τ, etc., can be found. The student should calculate some of these values and check them against the curves in Fig. 13-11.

Chapter Thirteen

Percent Changes in 2 τ, 3 τ, etc. A circuit variable will move 63% of the distance toward its final value in 1 τ. It can also be shown that it will move 86% of the distance toward its final value in 2 τ, 95% in 3 τ, 98% in 4 τ, and 99.3% in 5 τ. These values are shown in Table 13–1.

As the table shows, even after 10 τ a 100% change is not completed. For practical purposes, however, we will always consider the total change to be completed in 5 τ. This will not cause any significant error in any of our work.

TABLE 13–1

Time Interval	Percentage of Total Ultimate Change
1 τ	63%
2 τ	86
3 τ	95
4 τ	98
5 τ	99.3
6 τ	99.8
10 τ	99.9955

EXAMPLE 13.6 A 0.01-μF capacitor is being charged by a 30-V dc source through a 5-kΩ series resistor. Determine the capacitor voltage after it has been charging for 50 μs; 100 μs; 150 μs.

Solution: The charging time constant is

$$\tau = 5 \times 10^3 \times 0.01 \times 10^{-6} \text{ s}$$

$$\tau = 50 \times 10^{-6} \text{ s} = 50 \text{ } \mu\text{s}$$

Thus, after 50 μs the capacitor voltage will have made 63% of the total change from 0 V to 30 V. That is, $v_{cap} = 18.9$ V after 50 μs.

Since 100 μs represents 2 τ, then the capacitor voltage will have made 86% of the total change from 0 to 30 V. That is, $v_{cap} = 0.86 \times 30 = 25.8$ V after 100 μs. In the same manner, $v_{cap} = 0.95 \times 30 = 28.5$ V after 150 μs. The results are tabulated:

	t	v_{cap}
(τ)	50 μs	18.9 V
(2 τ)	100 μs	25.8 V
(3 τ)	150 μs	28.5 V

EXAMPLE 13.7 For the same circuit of the last example, determine the resistor voltage at $t = 50$ μs, 100 μs, and 150 μs.

Solution: There are two possible approaches that can be used here. The first method utilizes KVL, which states that the resistor voltage plus the capacitor voltage must add up to 30 V (the source voltage) *at all times*. Thus, we can use the results for v_{cap} determined in the last example to find $v_R = 30$ V $- v_{cap}$. The results are tabulated:

t	v_{cap}	v_R
50 μs	18.9 V	30 − 18.9 = 11.1 V
100 μs	25.8 V	30 − 25.8 = 4.2 V
150 μs	28.5 V	30 − 28.5 = 1.5 V

The second approach uses the time-constant method. The resistor voltage will change from 30 V to 0 V during the transient response. In 1 τ it will complete 63% of this change so that it will have decreased from 30 V by an amount equal to $0.63 \times 30 = 18.9$ V. Thus, $v_R = 30 - 18.9 = 11.1$ V after 50 μs. Similarly, in 2 τ it will complete 86% of its total change or $0.86 \times 30 = 25.8$ V, so that $v_R = 30 - 25.8 = 4.2$ V after 100 μs. The student should verify that after 3 τ the resistor voltage will be 1.5 V.

13.10 Discharging a Capacitor — Quantitative Discussion

For the circuit in Fig. 13–12(a), we know the capacitor will fully charge to 10 V if the switch is in position 3 for longer than 0.5 s (5 τ). Let's assume that the capacitor has been allowed to charge to 10 V. If the switch is now put in position 2, the capacitor will store the charge it previously acquired when the switch was in position 3.

To discharge the capacitor, we will throw the switch to position 1 at time $t = 0$ as shown in part (b) of the figure. In this situation the voltage source is out of the circuit so it is not shown. The charged capacitor is now the only energy source in the circuit. At the instant $t = 0$, the capacitor has 10 V across its plates. As can be seen in the figure, this 10 V is now impressed across the 100-ohm resistor with the right-hand terminal of the resistor being positive so that $v_{wx} = -10$ V.

The 10 V across the 100-ohm resistor result in a flow of current equal to 0.1 A in the direction shown with electrons flowing off the negatively charged capacitor plate. As this current flows, the charge on the capacitor is reduced. The decreased charge causes the capacitor voltage to decrease ($V = Q/C$). Since the capacitor is actually the battery in this circuit, as its voltage decreases the discharge current will also decrease. For example, some time after the discharge begins the capacitor voltage will be at 5 V so that the resistor voltage is also 5 V. At this time the discharge current will be only $5/100 = 0.05$ A.

What this means is that as the capacitor discharges, the *rate* at which it discharges gets gradually slower and slower. Figure 13–12(c) shows the capacitor discharged completely with $v_{xy} = 0$ and $i_c = 0$. A clearer picture of the discharge can be seen by viewing the waveforms for the capacitor voltage and discharge current shown in Fig. 13–13. These waveforms clarify several important points.

(1) At $t = 0$, the capacitor begins to discharge rapidly since the discharge current is maximum.

(2) As the capacitor discharges, the rate of discharge decreases (curve is less steep).

(3) Eventually the capacitor is completely discharged ($v_{xy} = 0$) and the discharge current is zero.

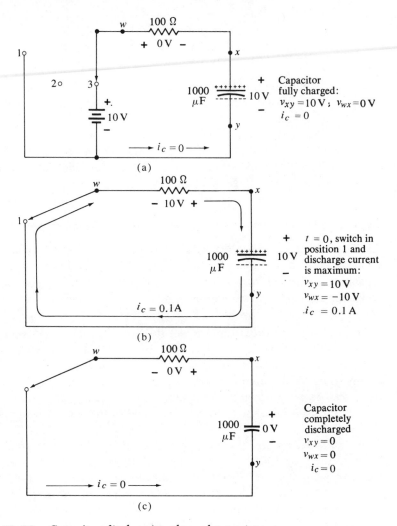

FIGURE 13–12 *Capacitor discharging through a resistor.*

Discharge Rate, ΔV/Δt In order to find the rate at which the capacitor voltage is decreasing, we can use the relationship developed in sec. 13.7.

$$\frac{\Delta V}{\Delta t} = \frac{1}{C} \times i_c \tag{13–4}$$

Using this expression we can determine the rate at which the capacitor voltage v_{xy} begins to decrease at $t = 0$ when $i_c = 0.1$ A. Thus,

$$\left.\frac{\Delta v_{xy}}{\Delta t}\right|_{(\text{at } t\,=\,0)} = \frac{1}{1000 \times 10^{-6}\,\text{F}} \times 0.1\,\text{A} = 100\,\text{V/s}$$

Actually, since v_{xy} is decreasing this rate will be *negative* so that $\Delta v_{xy}/\Delta t = -100$ V/s. As the capacitor continues its discharge, the value of i_c decreases and the discharge rate decreases as indicated by Eq. (13–4).

Discharge Time Constant The time-constant approach introduced in sec. 13.10 is equally applicable to capacitor discharge. The value of τ is again given by the

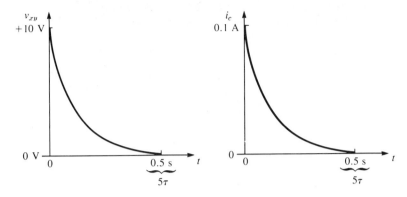

FIGURE 13-13

product of C and the total resistance in its discharge path. For the circuit in Fig. 13-12,

$$\tau = 100 \text{ ohms} \times 1000 \text{ }\mu\text{F} = 0.1 \text{ s}$$

Thus, the discharge will be complete after a time equal to $5\tau = 0.5$ s. This is seen to agree with the waveforms in Fig. 13-13.

We can also use τ to determine the capacitor voltage and discharge current values at times equal to τ, 2τ, etc. For example, at $t = \tau = 0.1$ s the capacitor voltage will complete 63% of its discharge. Thus, it will discharge by an amount equal to 6.3 V so that $v_{xy} = 10 - 6.3 = 3.7$ V at $t = 0.1$ s. This approach can be continued for other values of t using Table 13-1.

EXAMPLE 13.8 For the circuit in Fig. 13-12 determine the value of the discharge current and the value of $\Delta v_{xy}/\Delta t$ at $t = 0.3$ s.

Solution: According to Table 13-1, the current will complete 95% of its total change in an interval equal to $3\tau = 0.3$ s. Therefore, since i_c is initially 100 mA, it will drop by 95 mA to 5 mA at $t = 0.3$ s.
  Using $i_c = 0.005$ A in Eq. (13-4) yields

$$\frac{\Delta v_{xy}}{\Delta t} = \frac{0.005}{1000 \times 10^{-6}} = 5 \text{ V/s}$$

as the rate of discharge at $t = 0.3$ s. As expected it is much lower than the 100 V/s initial discharge rate.

13.11 Time-Constant Graph

Table 13-1 gives the percentage of the total ultimate change that a current or voltage in a simple RC circuit will make in time intervals that are integer multiples of τ. Very often we need to know how much change will occur in time intervals that are not integer multiples of τ, such as 0.2τ or 1.65τ, etc. To aid us in such calculations, we can make use of a more complete table or, better still, a graph which relates percentage change for any number of time constants. Such a graph is plotted in Fig. 13-14. For any number of time constants on the horizontal axis, the curve will tell us the percentage change which occurs during that time interval. The graph has a shape exactly like the shape of the capacitor-charging waveform. This graph will be used in the examples to follow so that its usefulness can be illustrated.

EXAMPLE 13.9 In the circuit of Fig. 13–15 the capacitor is initially completely discharged. At $t = 0$ the switch is closed and the 50-V source begins charging the capacitor.
(a) Determine the capacitor voltage at $t = 1.2$ ms after the switch is closed.
(b) Determine the amount of time it will take the capacitor to reach 25 V.
(c) Determine the charging current at $t = 0.8$ ms.

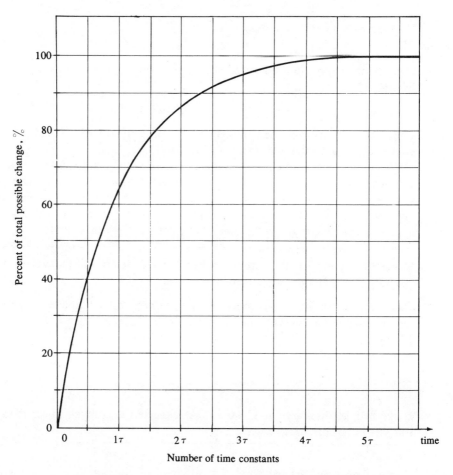

FIGURE 13–14 *Universal time–constant graph to be used for any simple RC circuit.*

Solution: (a) The charging time constant is given by

$$\tau = 2 \text{ k}\Omega \times 0.5 \text{ }\mu\text{F}$$
$$= 2000 \times 0.5 \times 10^{-6} \text{ s}$$
$$= 1 \text{ ms}$$

Thus, $t = 1.2$ ms represents 1.2 time constants. That is, $t = 1.2\ \tau$. Referring to the graph in Fig. 13–14 we see that in $1.2\ \tau$ a change of approximately 70% will take place.

Since the capacitor voltage is making a total possible change of 50 V, it will make $70\% \times 50\text{ V} = 35\text{ V}$ of this change in a time equal to $1.2\ \tau$. Thus, $v_{\text{cap}} = \mathbf{35\text{ V}}$ at $t = 1.2$ ms.

(b) We wish to find how long it takes the capacitor voltage to reach 25 V. Since the capacitor is charging from 0 V to 50 V, then it will have completed 50% of its total change when it reaches 25 V. According to the time-constant graph of Fig. 13–14, a 50% change occurs in a time interval equal to approximately $0.7\ \tau$. Consequently, the capacitor will reach 25 V at a time

$$t = 0.7\ \tau = 0.7 \times 1\text{ ms} = \mathbf{0.7\text{ ms}}$$

(c) The charging current is initially $50\text{ V}/2\text{ k}\Omega = 25$ mA at $t = 0$. Since $\tau = 1$ ms, 0.8 ms represents $0.8\ \tau$. From the graph in Fig. 13–14 we see that a change of 55% takes place in a time equal to $0.8\ \tau$.

The charging current will ultimately (in $5\ \tau$) drop from 25 mA to 0 mA, a total possible decrease of 25 mA. Therefore, in $0.8\ \tau$ it will *decrease* by $55\% \times 25\text{ mA} = 13.75$ mA so that at $t = 0.8$ ms we have

$$i_c = 25 - 13.75 = \mathbf{11.25\text{ mA}}$$

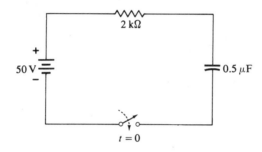

FIGURE 13–15

EXAMPLE 13.10 Consider again the circuit in Fig. 13–15. Change the resistor to 100 kΩ. If the capacitor were initially charged to 10 V *before* the switch was closed, determine the capacitor voltage 125 ms *after* the switch is closed. Sketch the capacitor voltage waveform.

Solution: In this situation, the capacitor voltage is 10 V before the switch is closed. At $t = 0$ when the switch closes, the capacitor will begin at 10 V and start charging toward 50 V, a *total change* of 40 V. It will reach 50 V after $5\ \tau$. In this case $\tau = 100\text{ k}\Omega \times 0.5\ \mu\text{F} = 50$ ms. The capacitor waveform is sketched in Fig. 13–16.

With $\tau = 50$ ms, $t = 125$ ms represents $125/50 = 2.5\ \tau$. From the time-constant graph of Fig. 13–14, we see that approximately 92% change occurs in $2.5\ \tau$. Thus, the capacitor voltage will *increase* by $92\% \times 40\text{ V} = 36.8$ V in

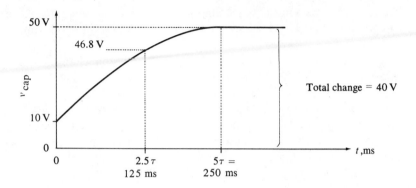

FIGURE 13-16

125 ms. Consequently, the capacitor voltage will be 10 V + 36.8 V = **46.8 V** at $t = 125$ ms.

EXAMPLE 13.11 Consider the circuit in Fig. 13-17(a). The switch is initially in position 1. At $t = 0$, the switch is moved to position 2 where it remains for 15 μs, after which it is moved to position 3. Determine the capacitor voltage 3 μs after the switch is moved to position 3. Sketch the capacitor waveform.

Solution: With the switch in position 2, the capacitor will charge toward 20 V with $\tau = 20$ kΩ × 100 pF = $2 \times 10^4 \times 100 \times 10^{-12} = 2 \times 10^{-6}$ s = 2 μs. The switch remains in this position for 15 μs, which is more than 5 τ. Thus, the capacitor voltage will be at 20 V at $t = 15$ μs when the switch is thrown to position 3.

At this instant the situation will be that shown in Fig. 13-17(b). The 12-V battery is now in the circuit. The capacitor acts initially as a 20-V source in opposition to this 12-V battery. Thus, current will begin to flow in the direction shown so as to *discharge* the capacitor. This current will flow until the capacitor voltage reaches 12 V. At that point there will be no net source voltage to produce current and the capacitor will discharge no further. This discharge will take 5 τ = 5 × 2 μs = 10 μs. The capacitor voltage waveform is shown in part (c) of the figure.

In position 3, the capacitor will discharge from 20 V to 12 V for a total *change* of 8 V. With $\tau = 2$ μs, 3 μs represents 1.5 τ. Using Fig. 13-14 we see that in 1.5 τ the change will be about 77% complete. Thus, in 3 μs the capacitor voltage will drop by 77% × 8 V = 6.2 V so that it will be at 20 V − 6.25 = 13.8 V. This occurs at $t = 18$ μs on the waveform plot.

EXAMPLE 13.12 Sketch the resistor voltage waveform for the situation in Fig. 13-17.

Solution: Refer to Fig. 13-18. With the switch in position 2 the resistor waveform (v_{wx}) jumps to 20 V and then decreases to 0 V as the capacitor charges to 20 V in 5 τ. At $t = 15$ μs when the switch is moved to position 3,

Capacitance 307

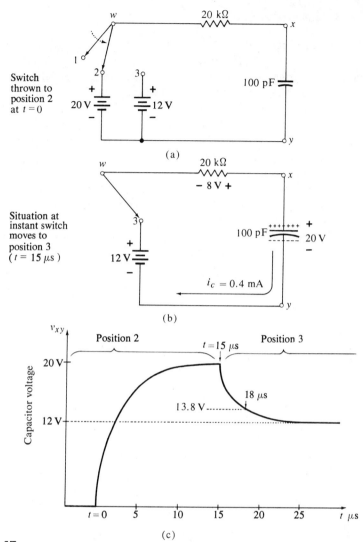

FIGURE 13–17

the voltage across the resistor reverses polarity ($v_{wx} = -8$ V) as shown in Fig. 13–17(b). This change from 0 V to -8 V occurs essentially instantaneously, after which v_{wx} heads back toward 0 V as C discharges from 20 V to 12 V.

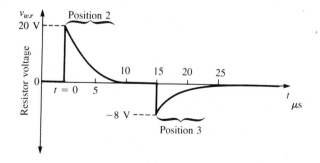

FIGURE 13–18

13.12 The Exponential Form

When solving circuits such as those in the previous sections, it is convenient to be able to mathematically express the variation of currents and voltages with time. Although graphical representations are more descriptive, mathematical expressions are easier to manipulate and are usually more accurate and require less space. In addition, the expressions for any current or voltage in a single time constant circuit (one that contains either one capacitor *or* one inductor) have the same general form. This general form is called the *exponential form*.

The key to the exponential form is the *exponential function* $\epsilon^{-t/\tau}$ in which ϵ (Greek letter epsilon) stands for the Naperian (natural) logarithm base and is approximately equal to 2.718. The t stands for the time variable, which can assume any value from 0 to ∞, and the constant τ is the time constant. Both t and τ have to be expressed in the same units.

The value of $\epsilon^{-t/\tau}$ for a given value of t can be determined by raising 2.718 to the appropriate power. For example, to determine the value of this function when $t = \tau$,

$$\epsilon^{-t/\tau} = \epsilon^{-1} = \frac{1}{\epsilon} = \frac{1}{2.718} = 0.368$$

For $t = 2\tau$,

$$\epsilon^{-t/\tau} = \epsilon^{-2} = \frac{1}{\epsilon^2} = \frac{1}{2.718^2} = 0.135$$

Notice from these examples that as t increases, the value of the exponential function decreases. These examples were relatively simple because the exponents were whole numbers. For fractional values of t/τ, either mathematical tables or a calculator must be used. A table of values of the exponential function is given in Appendix C if a calculator is not available.

The results of the tabulation of some values of $\epsilon^{-t/\tau}$ for various integer values of t are shown here.

t	$\epsilon^{-t/\tau}$
0	1.000
τ	0.368
2τ	0.135
3τ	0.050
4τ	0.018
5τ	0.007 ≈ 0
10τ	0.000045 ≈ 0

Again notice that increasing values of t result in rapidly decreasing values of $\epsilon^{-t/\tau}$. Even though the function will actually reach zero only when $t = \infty$, for the purposes of this text the function will be assumed to be zero when $t \geq 5\tau$. A plot of the exponential function versus t is shown in Fig. 13–19(a). Note that it has the same shape as a capacitor discharge waveform. Figure 13–19(b) shows a plot of $(1 - \epsilon^{-t/\tau})$; it has the shape of a capacitor charging waveform.

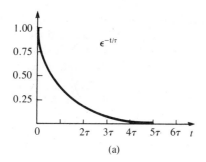
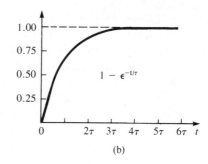

FIGURE 13–19 *Graphs of $\epsilon^{-t/\tau}$ and $1 - \epsilon^{-t/\tau}$.*

Exponential form for v The exponential function can now be used to express the behavior of currents and voltages in single time constant circuits. It can be shown using the methods of calculus and differential equations that in a linear circuit containing only one capacitance (or inductance), the voltage v across *any* element in the circuit in response to any *sudden change* can be expressed as follows:

$$v = V_F + (V_0 - V_F)\epsilon^{-t/\tau} \tag{13–6}$$

This expression shows how the voltage (v) varies with time, where V_0 represents the *initial* value of v at the instant the *change* is made (at $t = 0$), and V_F represents the *final* value of v attained in steady state (at $t \geq 5\tau$).

The definitions of V_0 and V_F can be verified by using equation (13–6) to calculate the initial and steady-state values of v. The initial value of v is found by setting $t = 0$ in Eq. 13–6

$$v(t = 0) = V_F + (V_0 - V_F)\epsilon^{-0}$$
$$= V_F + (V_0 - V_F) \times 1 = V_0$$

The final steady-state value of v is found at $t \geq 5\tau$, where $\epsilon^{-t/\tau}$ is essentially zero.

$$v(t \geq 5\tau) = V_F + (V_0 - V_F) \times 0$$
$$= V_F$$

This general voltage expression can be applied to the capacitor charging circuit shown in Fig. 13–20. Consider first the capacitor voltage v_{xy}. Assume the capacitor is initially discharged to zero volts. At $t = 0$ the switch is closed instantaneously (this is the sudden change), and the value of v_{xy} is still zero. Therefore, $V_0 = 0$.

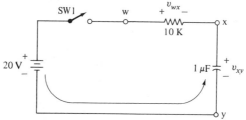

FIGURE 13–20 *Capacitor charging circuit*

To determine V_F, the steady-state value of v_{xy}, it is necessary to determine the voltage to which the capacitor will charge. This can be done by replacing the capacitor with an open circuit, since in steady state the capacitor draws no current (for a constant source voltage). In this condition, the value of v_{xy} will be 20 V, which is V_F.

The circuit time constant (τ) is determined by the capacitance value and the equivalent resistance seen by the capacitor. In this circuit the resistance is 10 kΩ. Thus, τ is (10 kΩ)(1 μF) = 10 ms.

These values of V_0, V_F, and τ are used in Eq. (13–6) to determine the mathematical expression for the variation of v_{xy} with time:

$$v_{xy} = V_F + (V_0 - V_F)\epsilon^{-t/\tau} \quad (13\text{–}6)$$
$$= 20 + (0 - 20)\epsilon^{-t/10 \text{ ms}}$$
$$= 20 - 20\epsilon^{-t/10 \text{ ms}}$$
$$v_{xy} = 20(1 - \epsilon^{-t/10 \text{ ms}}) \text{ V}$$

This last expression, if plotted versus t, will result in the same curve as that in Fig. 13–8. The student should verify this by taking several values of t and evaluating the expression. The expression reflects the fact that v_{xy} *increases* with time; the exponential term $\epsilon^{-t/10 \text{ ms}}$ is *subtracted* from one so that as the exponential decreases, the total term in parentheses increases.

EXAMPLE 13.13 Using the expression for v_{xy}, determine the voltage across the capacitor 5 ms after SW1 is closed. Repeat for $t = 100$ ms.

Solution: (a) To evaluate v_{xy} at $t = 5$ ms, set $t = 5$ ms in the expression for v_{xy}.

$$v_{xy} = 20(1 - \epsilon^{-t/10 \text{ ms}}) \text{ V}$$
$$v_{xy}(@ \ t = 5 \text{ ms}) = 20(1 - \epsilon^{-5/10}) \text{ V}$$
$$= 20(1 - \epsilon^{-0.5}) \text{ V}$$
$$= 20(1 - 0.606) \text{ V} = \mathbf{7.88 \text{ V}}$$

(b) Set $t = 100$ ms in the expression for v_{xy}.

$$v_{xy}(@ \ t = 100 \text{ ms}) = 20(1 - \epsilon^{-10}) \text{ V}$$
$$= 20(1 - 0.000045) \text{ V}$$
$$\approx \mathbf{20 \text{ V}}$$

This last result could have been predicted since 100 ms is greater than 5 τ.

EXAMPLE 13.14 (a) Determine how long it takes the capacitor in Fig. 13–20 to reach 10 V. (b) Repeat for 15 V.

Solution: (a) Set the expression for v_{xy} equal to 10 V and solve for t.

$$20(1 - \epsilon^{-t/10 \text{ ms}}) = 10$$
$$1 - \epsilon^{-t/10 \text{ ms}} = 0.5$$
$$\epsilon^{-t/10 \text{ ms}} = 0.5$$

Take the *natural log* (i.e., $\log_\epsilon = \ln$) of both sides to eliminate the exponential.

$$\frac{-t}{10 \text{ ms}} = \ln(0.5)$$

The ln(0.5) can be evaluated on a scientific calculator as -0.693. Thus,

$$\frac{-t}{10 \text{ ms}} = -0.693$$

$$\therefore t = \mathbf{6.93 \text{ ms}}$$

(b) Set $v_{xy} = 15$ V and solve for t.

$$20(1 - \epsilon^{-t/10 \text{ ms}}) = 15$$
$$1 - \epsilon^{-t/10 \text{ ms}} = 0.75$$
$$\epsilon^{-t/10 \text{ ms}} = 0.25$$
$$\frac{-t}{10 \text{ ms}} = \ln(0.25) = -1.39$$
$$\therefore t = \mathbf{13.9 \text{ ms}}$$

EXAMPLE 13.15 Referring again to Fig. 13–20, determine the mathematical expression for the resistor voltage v_{wx} in response to the closing of the switch, then find v_{wx} at $t = 13$ ms.

Solution: The values of V_0, V_F, and τ must be determined and substituted into the general expression for v_{wx}.

$$v_{wx} = V_F + (V_0 - V_F)\epsilon^{-t/\tau}$$

The initial voltage across the resistor, as determined previously, is 20 V; $V_0 = 20$ V. The steady-state resistor voltage will be zero, since the capacitor acts as an open circuit in steady state allowing no current to flow. Thus, $V_F = 0$ V.

The time constant τ is 10 ms, as determined previously. *For a given circuit the time constant is unique no matter what voltage or current we happen to be looking at.* The complete expression for the resistor voltage is therefore

$$v_{wx} = 0 + (20 - 0)\epsilon^{-t/10 \text{ ms}}$$
$$= 20\epsilon^{-t/10 \text{ ms}}$$

If this expression were plotted versus time, the result would be the same as previously determined in Fig. 13–8. The student should verify this.

$$v_{wx}(\text{at } t = 13 \text{ ms}) = 20\epsilon^{-13/10}$$
$$= 20\epsilon^{-1.3}$$
$$= 20 \times 0.273$$
$$= \mathbf{5.46 \text{ V}}$$

Expression for i The technique used in the preceding discussion can also be used to find the mathematical expression for *current* in a circuit. The general expression is

$$i = I_F + (I_0 - I_F)\epsilon^{-t/\tau} \quad (13\text{–}7)$$

I_0 is the initial value of current, and I_F is the steady-state value. The following example will illustrate the use of this expression:

EXAMPLE 13.16 Referring again to Fig. 13–20 determine the equation for the current that flows in response to the switch closure. Find i at $t = 5$ ms.

Solution: The circuit time constant is known (10 ms), and I_0 and I_F must be determined. The initial current flow is determined by the initial resistor voltage drop divided by the resistance. Thus, $I_0 = 20$ V/10 kΩ = 2 mA. The value of I_F is zero because in steady state the capacitor is open. Substituting the values for I_F, I_0, and τ into Eq. (13–7),

$$i = 0 + (2 - 0)\epsilon^{-t/10 \text{ ms}}$$
$$= 2\epsilon^{-t/10 \text{ ms}}$$

where the units are mA. As discussed earlier, this current decreases exponentially with a time constant of 10 ms. The same result could have been obtained using Ohm's law and the expression for resistor voltage obtained in Ex. 13.15. That is,

$$i = \frac{v_{wx}}{R} = \frac{20\epsilon^{-t/10 \text{ ms}}}{10 \text{ k}\Omega} = 2\epsilon^{-t/10 \text{ ms}} \text{ (mA)}$$

To find i at $t = 5$ ms, set $t = 5$ ms in the expression for i.

$$i = 2\epsilon^{-t/10 \text{ ms}} \text{ (mA)}$$
$$i(@\ t = 5 \text{ ms}) = 2\epsilon^{-5/10} \text{ mA}$$
$$= 2\epsilon^{-0.5} \text{ mA}$$
$$= 2 \times (0.607) \text{ mA}$$
$$= \mathbf{1.214 \text{ mA}}$$

EXAMPLE 13.17 Consider the circuit in Fig. 13–21(a). The switch has been closed for some time, so that the capacitor is charged to -12 V ($v_{xy} = -12$ V). If the switch is opened at $t = 0$, determine the expression for v_{xy} and plot it versus time. Then determine how long before v_{xy} reaches -2 V.

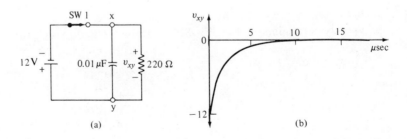

FIGURE 13–21

Solution: The instant the switch is opened the capacitor will hold at -12 V, so $V_0 = -12$ V. In steady state the capacitor voltage will be zero, since with the switch open there is no source in the circuit. Thus, $V_F = 0$. The circuit time constant is

$$\tau = RC$$
$$= (220 \, \Omega)(0.01 \, \mu F)$$
$$= (2.2 \times 10^2)(10^{-8}) \text{ s}$$
$$\tau = 2.2 \times 10^{-6} \text{ s} = 2.2 \, \mu s$$

Substituting these values into the general voltage expression (13–6) gives

$$v_{xy} = 0 + (-12 \text{ V} - 0)\epsilon^{-t/2.2 \, \mu s}$$
$$= -12\epsilon^{-t/2.2 \, \mu s} \text{ V}$$

The plot of this expression is shown in Fig. 13–21(b). Notice that v_{xy} is actually *algebraically increasing*, since it is going from a negative value to a less negative (more positive) value. However, the *magnitude* of capacitor voltage is *decreasing*.

To find at what time v_{xy} equals -2 V, set $v_{xy} = -2$ V and solve for t.

$$v_{xy} = -12\epsilon^{-t/2.2 \, \mu s} = -2 \text{ V}$$
$$\epsilon^{-t/2.2 \, \mu s} = +0.166$$
$$\frac{-t}{2.2 \, \mu s} = \ln(0.166) = -1.79$$
$$\therefore t = 3.94 \, \mu s$$

13.13 Capacitors in Series and in Parallel

Capacitors, like resistors, can be connected in series with each other or in parallel with each other. Although these combinations of capacitors do not occur as often as resistors in electrical circuits, it is necessary that we be able to determine the equivalent capacitance of capacitor combinations.

Parallel Capacitors In Fig. 13–22 the capacitors $C1$ and $C2$ are connected in parallel across the voltage source E_s. As such, both capacitors will be charged to a voltage equal to E_s. Thus, $C1$ must have an amount of charge on it equal to

$$Q1 = C1 \times E_s$$

Similarly,

$$Q2 = C2 \times E_s$$

These charges were originally supplied by the source when it was connected to the capacitors. The total charge Q_T supplied by the source is $Q1 + Q2$. Thus,

$$Q_T = Q_1 + Q_2 = C_1 E_s + C_2 E_s$$
$$= E_s(C_1 + C_2)$$

This last expression indicates that the two parallel capacitors draw as much charge from the source as one capacitor with a capacitance equal to $C1 + C2$. In fact, we can state that *for any number of capacitors connected in parallel, the equivalent capacitance is the sum of all the parallel capacitances*.

$$C_{eq} = C1 + C2 + C3 \ldots \text{etc.} \tag{13–8}$$

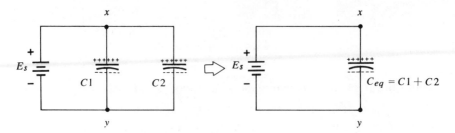

FIGURE 13-22 *Capacitors in parallel can be added to give an equivalent single capacitor.*

It is interesting to note that *capacitors in parallel* are added together the same as *resistors in series*. As we shall see, *capacitors in series* are combined the same as *resistors in parallel*.

Capacitors in Series Two capacitors connected in series can be replaced by an equivalent capacitor which is *smaller* than either series capacitor. The same is true for any number of series capacitors. The formula for equivalent series capacitance is

$$\frac{1}{C_{eq}} = \frac{1}{C1} + \frac{1}{C2} \ldots \text{etc.} \tag{13-9a}$$

This formula is similar to the formula for equivalent parallel resistance.

For two capacitors in series, this formula reduces to

$$C_{eq} = \frac{C1 \times C2}{C1 + C2} \quad \text{(product over sum)} \tag{13-9b}$$

For example, a 6-μF capacitor in series with a 3-μF capacitor will have an equivalent capacitance of

$$\frac{6\ \mu\text{F} \times 3\ \mu\text{F}}{6\ \mu\text{F} + 3\ \mu\text{F}} = 2\ \mu\text{F}$$

which is less than either capacitor. Also, note that two equal capacitors C in series will have an equivalent capacitance equal to $C/2$.

Figure 13-23 shows the 6-μF and 3-μF capacitors connected in series to a 60-V source. Because the capacitors are in series, it is obvious that the *sum* of the capacitor voltages must equal 60 V to satisfy KVL. It is also true that since the capacitors are in series, the charging current produced by the source will be the same for each. This means that the same amount of charge is supplied to both capacitors.

With both capacitors supplied with the *same amount of charge*, it is apparent that the capacitor voltages will vary inversely with capacitance ($V = Q/C$). That is, the smaller capacitance will have the larger voltage and vice versa. Thus, the 3 μF which has *half* the capacitance of the 6 μF will have *twice* the voltage as shown in the figure.

In general, then, it can be stated that *for capacitors in series the applied voltage divides in inverse proportion to the capacitance values.*

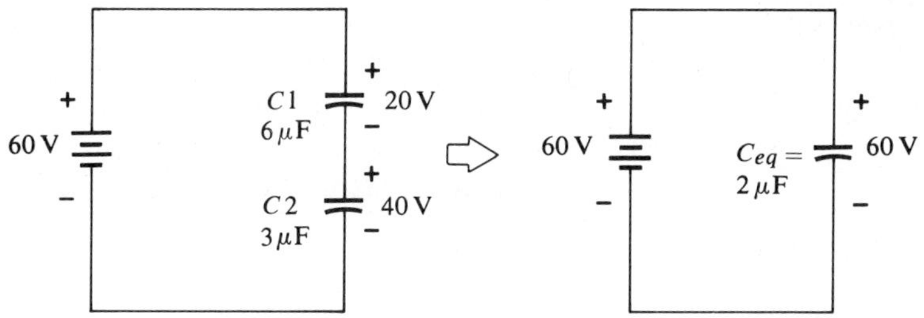

Capacitance

[Figure 13-23 diagram: Left circuit shows 60 V source with C1 = 6 μF (20 V) in series with C2 = 3 μF (40 V). Right circuit shows 60 V source with Ceq = 2 μF.]

FIGURE 13–23 *For capacitors in series, the voltage divides in inverse proportion to the capacitors.*

EXAMPLE 13.18 Three capacitors are connected in series in Fig. 13–24. Determine the equivalent capacitance and the voltage across each capacitor.

Solution: (a) $\dfrac{1}{C_{eq}} = \dfrac{1}{100\ \mu F} + \dfrac{1}{25\ \mu F} + \dfrac{1}{20\ \mu F}$

$\dfrac{1}{C_{eq}} = 10^4 + 4 \times 10^4 + 5 \times 10^4 = 10 \times 10^4 = 10^5$

so that

$$C_{eq} = \dfrac{1}{10^5} = 10^{-5}\ F = 10\ \mu F$$

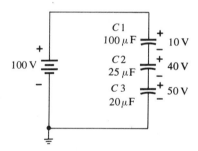

FIGURE 13–24

(b) The total charge produced by the source is the same as if it were charging a 10-μF capacitor. Thus,

$$Q = 10\ \mu F \times 100\ V = 1000\ \mu C$$

This amount of charge is supplied to *each* series capacitor. Using this amount of charge we can calculate each capacitor's voltage ($V = Q/C$).

$$V1 = \dfrac{Q}{C1} = \dfrac{1000\ \mu C}{100\ \mu F} = 10\ V$$

$$V2 = \dfrac{Q}{C2} = \dfrac{1000\ \mu C}{25\ \mu F} = 40\ V$$

$$V3 = \dfrac{Q}{C3} = \dfrac{1000\ \mu C}{20\ \mu F} = 50\ V$$

316 Chapter Thirteen

As expected, the largest capacitor has the smallest voltage and the smallest capacitor has the largest voltage. The capacitor voltages add up to 100 V, satisfying KVL.

To summarize the difference between parallel and series capacitors.

(1) Connecting capacitors in parallel *increases* the total capacitance; connecting capacitors in series *decreases* total capacitance.

(2) Capacitors in parallel have the same *voltage;* capacitors in series have the same *charge* but different voltages.

13.14 Capacitor Connected to a Complex Network

In many applications a capacitor is connected across two terminals of a complex network. Several examples of such circuits are shown in Fig. 13–25. In each of the circuits the capacitor is connected between terminals *x-y*. Note that the circuits in *a* and *b* are linear resistive networks, while *c* and *d* are *not* linear because of the diodes. Also note that these networks contain only constant dc sources.

In cases such as these we often need to know the voltage to which the capacitor will charge. These are not series *RC* circuits, so we cannot say that the capacitor voltage will charge to the source voltage. In fact, in Fig. 13–25(b) there are two voltage sources. To solve this type of problem, we have to remember that *a capacitor will continue to charge until the charging current drops to zero*. In a simple series *RC* circuit, this will happen when the capacitor volt-

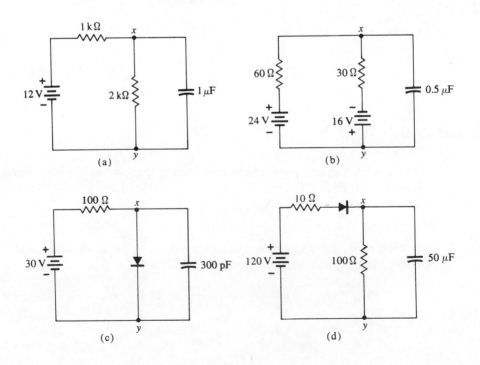

FIGURE 13–25 *Complex capacitor-charging networks.*

age equals and opposes the source voltage so that no net voltage is present to produce current.

In steady state, then, no current flows to the capacitor and it acts like an *open circuit*. Thus, for the cases in Fig. 13–25 we can determine the voltage which the capacitor charges to, by treating the capacitor as an open circuit and determining V_{xy}. This value of V_{xy} will be the capacitor voltage in steady state since the capacitor is connected between x and y. For example, in Fig. 13–25(a) if we treat the capacitor as an open circuit, we can calculate V_{xy} to be 8 V using the voltage-divider method. The capacitor will thus be charged to 8 V in steady state.

This procedure is, of course, valid only for networks with constant dc sources. If the sources are not constant the capacitor voltage will not reach a constant value, but will be continually changing as the input voltage changes.

To summarize, we can state that:

To find the steady-state capacitor voltage in any dc circuit, the capacitor can be replaced by an open circuit and its terminal voltage calculated.

This statement is valid even for circuits with more than one capacitor as long as the circuit is a dc circuit.

EXAMPLE 13.19 Determine the steady-state voltage across the capacitor in Fig. 13–25(c). Assume the diode is a silicon diode.

Solution: Treat the capacitor as an open circuit in steady state. This leaves only a series circuit with the diode forward-biased. Thus, the diode will have about 0.7 V across its terminals so that $V_{xy} = 0.7$ V. The capacitor, then, will be charged to 0.7 V.

Thevenizing the Complex Network If the network to which the capacitor is connected is a linear, resistive network, we can use Thevenin's theorem to replace the network by an equivalent source and series resistor. The Thevenin equivalent circuit can then be used to determine the *complete* waveform of capacitor voltage. The following example illustrates.

EXAMPLE 13.20 The switch in the circuit of Fig. 13–26(a) is closed at time $t = 0$. If the capacitor is initially discharged so that $v_{xy} = 0$, determine the waveform of capacitor voltage.

Solution: After the switch is closed, we can Thevenize the resistive network connected to terminals x-y. To do so, we temporarily remove the capacitor and calculate E_{Th} and R_{Th}.

E_{Th} is the voltage V_{xy} and is equal to $3/9 \times 90$ V = 30 V. R_{Th} is equal to 6 k$\Omega \parallel$ 3 kΩ = 2 kΩ. The Thevenin equivalent circuit is now connected to terminals x-y along with the 5-μF capacitor. This circuit [part (b) of the figure] is now a simple series RC circuit which can be analyzed easily.

The equivalent source which is charging the capacitor is 30 V. Therefore, the capacitor will charge from zero toward 30 V. The charging-time constant is equal to $R_{Th} \times C$ = 2-k$\Omega \times 5$ μF = 10 ms.

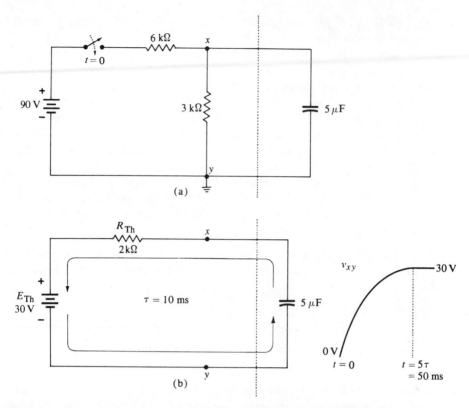

FIGURE 13–26 *Using Thevenin's theorem to determine the capacitor voltage waveform.*

Thus, the capacitor will complete its charging in $5\tau = 50$ ms. The capacitor voltage waveform is shown in Fig. 13–26(b).

A procedure identical to that in the preceding example can be used whenever a capacitor is connected to a *linear, resistive* network with dc sources. It can be generally stated that:

If a capacitor C is connected to a linear, resistive network with only dc sources, the capacitor will charge toward a voltage equal to the E_{Th} of the network. Furthermore, the charging τ will be determined by the R_{Th} of the network multiplied by C.

13.15 Physical Factors Determining Capacitance

Now that we have seen how to analyze certain capacitor circuits, we will pause to consider some of the physical characteristics of a capacitor. Figure 13–27 shows a capacitor consisting of two identical parallel plates each with a surface area A. The plates are separated by a distance d and in between the plates there is a dielectric material.

As we saw earlier, a capacitor has the ability to store charge depending on its capacitance. A larger capacitance will store more charge for a given voltage

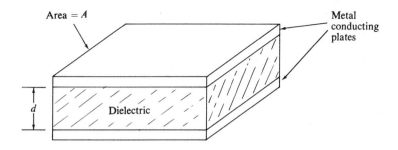

FIGURE 13-27 *Parallel plate capacitor.*

($Q = CV$). It is desirable to be able to produce capacitors with different values of capacitance for different applications. The capacitance of the capacitor in Fig. 13–27 can be varied by varying the area of the plates, the plate separation, and changing the type of dielectric. The formula relating capacitance to these three factors is:

$$C = \frac{8.9 \times 10^{-12} \times K_\epsilon \times A}{d} \text{ (farads)} \qquad (13\text{--}10)$$

In this formula, A is given in square meters and d in meters. The constant K_ϵ is called the *dielectric constant* of the dielectric material.

According to Eq. (13–10), capacitance is proportional to the area of the plates and inversely proportional to the plate separation. Doubling the area of the plates will double the capacitance. Doubling the plate separation will reduce the capacitance by half.

Capacitance is also proportional to the dielectric constant of the material between the plates. For the same plate area and separation, a material with a greater K_ϵ will produce a greater value of capacitance. Table 13–2 lists the typical values of K_ϵ for various dielectric materials.

As the table indicates, the ceramic barium-strontium titanate has an extremely high value of K_ϵ compared to the other materials. Development of these ceramics has made it possible to make capacitors which are much more compact for a given amount of capacitance.

TABLE 13-2

Dielectric	Average Dielectric Constant K_ϵ
Air	1
Porcelain (ceramic)	6
Glass	6
Mica	5
Oil	4
Paper	3
Barium-strontium titanate (ceramic)	7500

Dielectric Breakdown The dielectric material in a capacitor is always a good insulator. However, every insulator will begin to conduct current if a strong enough voltage is applied to it. A high voltage could cause *dielectric breakdown*. In the phenomenon of dielectric breakdown the valence electrons of the insulator are actually yanked out of their nuclear orbits by the strong electric field. These electrons become free electrons just like in a conductor, and the insulator is able to conduct current easily.

Dielectric materials have a rating called *dielectric strength*. Dielectric strength is the ability of an insulator to withstand a voltage without going into dielectric breakdown. The dielectric strength of a material is given in volts per mil (recall that a *mil* is 0.001 inch). For example, paper has a dielectric strength of around 1200 V/mil. This means that it takes 1200 V across a 1-mil thickness of paper to cause breakdown. For a 2-mil thickness it would take 2400 V to cause breakdown, and so on. This means that a thicker dielectric layer will withstand higher voltages. However, a thicker dielectric will reduce the capacitance, according to Eq. (13–10).

Because of dielectric breakdown, any capacitor is limited as to the maximum voltage that can be placed across its plates. In other words, every capacitor has a *maximum voltage rating* or *working voltage rating* that should not be exceeded. If exceeded, the dielectric will break down and a conducting path will be provided between the plates. Then the capacitor cannot store charges.

Voltage ratings for capacitors range from 6 V up to 20,000 V depending on the size and the type of dielectric. A capacitor should never be used in a circuit where its voltage could exceed its voltage rating.

By connecting capacitors in series, the total voltage will divide among the capacitors. Thus, a series combination of capacitors will have a greater overall voltage rating than a single capacitor. For example, two 100-V, 1-μF capacitors in series can handle a total of 200 V (100 V across each capacitor). Often, a series combination is used to obtain a greater voltage rating than is available in a single capacitor.

13.16 Capacitor Leakage

If we charge a capacitor from a dc voltage source and then disconnect the source, we expect the capacitor to hold its charge (voltage) indefinitely. This would only happen if the dielectric between the capacitor plates was a *perfect* insulator. Because there is no perfect insulator, the charge on the capacitor will eventually be neutralized as electrons on the negative plate *leak* through the insulator to the positive plate. The amount of time for this process to take place depends on the actual resistance of the dielectric. This resistance is called the *leakage resistance*, R_l, of the capacitor. For some dielectric materials like paper, ceramic, and mica the leakage resistance is very high (100 MΩ or more). Capacitors with an *electrolytic* dielectric have relatively low leakage resistance (typically 1-5 MΩ).

As shown in Fig. 13–28(a), the effect of this leakage resistance is taken into account by including R_l in parallel with the *ideal* capacitor C. Obviously R_l is not physically visible because it is internal to the capacitor similar to the internal resistance of a source.

Measuring R_l To measure the leakage resistance of a capacitor the procedure illustrated in Fig. 13-28(b) is used. The dc voltage source is used to charge the capacitor. When the capacitor is fully charged, there will still be a small current in the circuit flowing through R_l (not through C). The ammeter is used to

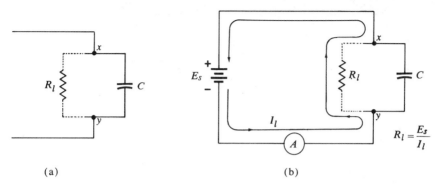

FIGURE 13-28 (a) *Representing R_1 in parallel with C. (b) Method for determining R_1.*

measure this leakage current, I_l. R_l is then calculated as E_S/I_l. For example, if $E_S = 10$ V and I_l is measured to be 0.2 μA, then R_l will equal 50 MΩ.

The value of R_l for a given capacitor usually decreases with temperature because the resistivity of the dielectric decreases. R_l also tends to decrease with age for certain type of capacitors, especially paper and electrolytic capacitors.

Effects of Capacitor Leakage The leakage resistance of a capacitor often has no significant effect on circuit operation. Sometimes, however, R_l can alter the charging characteristics of a capacitor. Consider Fig. 13-29(a) where a practical capacitor with a leakage resistance of 2 MΩ is being charged from a 10-V source in series with a 1-kΩ resistor. In this circuit the 1-kΩ resistor and the 2-MΩ leakage resistor form a voltage divider. The capacitor will thus charge to a voltage equal to

$$\frac{2 \text{ M}\Omega}{2 \text{ M}\Omega + 1 \text{ k}\Omega} \times 10 \text{ V} = \frac{2.000 \times 10^6}{2.001 \times 10^6} \times 10 \text{ V} = 9.995 \text{ V}$$

which for all practical purposes is 10 V. The leakage resistance in this case has a negligible effect.

If the series resistor is of a much higher value the results will be different. In Fig. 13-29(b) the series resistor is 1 MΩ. Now the capacitor will only charge to a voltage of

$$\frac{2 \text{ M}\Omega}{2 \text{ M}\Omega + 1 \text{ M}\Omega} \times 10 \text{ V} = \frac{2}{3} \times 10 \text{ V} = 6.67 \text{ V}$$

because of the voltage divider action. Thus, in this case the leakage resistance prevents the capacitor from charging to the full 10 V. This will occur whenever R_l is not a great deal larger than the series resistor.

When charging a capacitor, then, the effects of leakage resistance can usually be neglected except when the series-charging resistor is of the same order of magnitude as R_l.

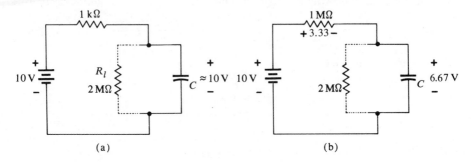

FIGURE 13–29 *Capacitor leakage resistance will cause the capacitor to charge to a lower voltage if R_1 is not $\gg$ the series resistor.*

EXAMPLE 13.21 A 50-μF capacitor is charged to 20 V by a dc voltage source. After the source is disconnected, how long will it take the capacitor's charge to leak off if $R_l = 2.5$ MΩ?

Solution: Essentially, the 50-μF capacitor will be discharging through its internal leakage resistance. The discharge time constant is

$$\tau = 2.5 \text{ M}\Omega \times 50 \text{ }\mu\text{F} = 125 \text{ s}$$

Thus, total discharge takes place in $5\tau = $ **625 s.**

13.17 Capacitor Energy Storage

We have seen that a capacitor can store charges for long periods of time. We also have seen that a charged capacitor can act as an energy source as it discharges through a resistor. In other words, we can say that *the capacitor stores energy*. This energy is originally provided by the voltage source which charges the capacitor. The capacitor stores this energy until it is made to discharge and produce a discharge current.

Intuitively, it seems reasonable that the higher the voltage to which a capacitor is charged, the greater will be the energy stored in the capacitor. We might also reason that for the same voltage a larger capacitance will store more energy because the greater capacitance will produce a discharge current for a longer period of time (slower discharge rate). These intuitive ideas are correct as verified by this formula for the energy stored in a capacitor.

$$\text{Energy} = \tfrac{1}{2} CV^2 \text{ (joules)} \tag{13–11}$$

In this formula C is in farads, V is the capacitor voltage in volts, and the resultant energy is in joules. For example, a 5-μF capacitor which is charged to 200 V will have a stored energy equal to

$$\tfrac{1}{2} \times 5 \times 10^{-6} \times (200)^2 = 0.1 \text{ joules}$$

This 0.1 J of energy was supplied by the source that charged the capacitor. If the capacitor is discharged, this stored energy will be dissipated as heat in the resistance through which the discharge current flows. The energy stored by a capacitor can serve many useful purposes as we shall see.

It should be noted at this point that *an ideal capacitor will not dissipate energy* (power) as a resistance does. It *stores* energy. A resistance cannot store energy. A practical capacitor will dissipate minute amounts of power because of its leakage resistance. This effect can usually be neglected.

EXAMPLE 13.22 Suppose a 2-μF capacitor is charged to 100 V through a 1-kΩ resistor. The energy stored by the capacitor is therefore 0.01 J [using Eq. (13–11)]. If the series resistor had been equal to 100 Ω, would the stored energy be different?

Solution: Intuitively, it might seem that with a smaller resistor, the energy supplied from the source to the capacitor would be greater because the charging current would be greater. However, with a smaller R the charging time will be shorter so that the larger charging current is flowing for a shorter period of time. The net result is that the value of R does not affect the amount of energy. It depends only on C and the capacitor voltage V as given by Eq. (13–11).

13.18 Applications of Capacitors

Capacitors find many applications in electricity and electronics. It would be impossible to enumerate them all. Instead, we will take a brief look at some typical uses for capacitors. Some of these applications will be discussed later in the text.

Energy Storage The ability of a capacitor to store energy is very useful in cases where we need a very high current for a short duration of time and we cannot obtain this current from our voltage source. For example, suppose we needed to supply a current *pulse* of 2 A to a 50-ohm load, and the current pulse has to last for 0.1 s. Furthermore, suppose we had a 100-V source with a source resistance of 100 ohms. We could not draw 2 A from this source. In fact, the most current we would draw would be 100 V/100 Ω = 1 A if its output terminals were short-circuited. Figure 13–30 shows a method for solving this problem.

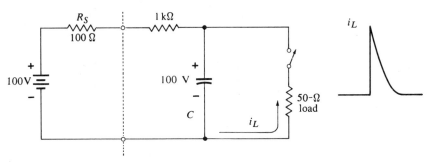

FIGURE 13–30 *Using a capacitor to supply a large pulse (spike) of current to a load.*

In this circuit the 100-V source is used to charge the capacitor C to 100 V through a total series resistance of 1.1 kΩ. During charging the current from the source is limited to a maximum of 100 V/1.1 k$\Omega \simeq$ 90 mA. After the capacitor is fully charged to 100 V, the switch can be closed connecting the 50-Ω load across the capacitor. The capacitor will then rapidly discharge through the load with the discharge current starting at a maximum of 100 V/50 Ω = 2 A. The discharge current rapidly drops to zero (see waveform in figure) in 5 τ where τ is equal to 50 $\Omega \times C$. The value of C is chosen to produce the desired duration of the current pulse. In our example, we want 5 τ = 0.1 s or τ = 20 ms This requires C = 20 ms/50 Ω = 400 μF.

Circuits similar to the one just described are used to obtain microsecond pulses for radar transmitters. They are also used in electric spot welding where a short-duration pulse of very high current is needed to make the weld.

Timing Circuits As a capacitor charges in an RC circuit, its voltage will always follow the exact same curve. For example, in Fig. 13–31(a) with the switch open the capacitor will charge toward 20 V with $\tau = RC$ = 10 ms. Thus, in 10 ms its voltage will reach 12.6 V; in 20 ms it will reach 17.2 V; and so on. If the switch is momentarily closed to rapidly discharge the capacitor, the capacitor voltage will follow the exact same charging curve when the switch is reopened. Thus, we can use an RC charging circuit as the basis for very accurate timing circuits.

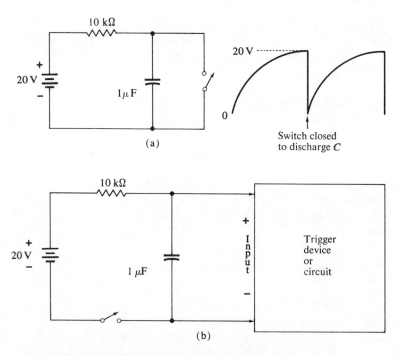

FIGURE 13–31 *Capacitor used in timing circuit applications.*

In such applications a capacitor is allowed to charge toward the source voltage [Fig. 13–31(b)] and the capacitor voltage is used as the input to some trigger device or circuit. The trigger device or circuit is simply one that requires its input voltage to be above a certain *threshold* value before it is triggered into operation. Examples of such devices are relays, unijunction transistors, zener diodes, etc. As the capacitor charges toward the source voltage, after a certain time interval its voltage will reach the value needed by the trigger device. Thus, we have a means for adjusting the amount of time between the closing of the switch and the triggering of the trigger device by adjusting the *RC* time constant.

EXAMPLE 13.23 The trigger device in Fig. 13–31(b) requires a voltage of 9 V in order to be triggered. (a) Determine the amount of time delay between the closing of the switch and the triggering of the device. (b) What are *three* ways in which this time delay can be increased?

Solution: (a) The capacitor is charging toward 20 V. When it reaches 9 V, the trigger device will trigger. This 9 V is 45% of the way toward 20 V. Using the graph in Fig. 13–14 it is seen that a 45% change occurs in $t \approx 0.6\,\tau$. Thus,

$$t = 0.6 \times R \times C$$
$$= 0.6 \times 10^4 \times 10^{-6} = \mathbf{6\ ms^*}$$

(b) The time delay can be decreased by slowing down the capacitor charging rate. This can be done by increasing R or C (which increases τ). It can also be done by *decreasing* the source voltage. A smaller E_s produces less charging current so that the capacitor voltage will take longer to reach 9 V. The student can verify that the time delay will be 9.2 ms if $E_s = 15$ V.

Waveshaping The ability of a capacitor to oppose any rapid changes in the voltage across its terminals makes it useful for certain types of *waveshaping* circuits. A waveshaping circuit is a circuit which takes an input voltage waveform and in some way changes the shape of the waveform. A simple series *RC* circuit can act as a waveshaping circuit with the output taken either across *R* or across *C*. Figure 13–32 shows some typical situations where the circuit input is a dc square wave at a frequency of 1 kHz (1000 cycles per second). The various possible output waveforms are obtained by varying the *RC* time constant. Waveshaping circuits are discussed in detail in a subsequent volume on pulse and switching circuits.

Filters Another important area where capacitors are employed is the area of *filters*. Filters are networks which are designed to respond differently to input voltages at different frequencies. As we know, capacitors are devices which oppose rapid *changes* in voltage. As such, a capacitor will respond differently to a voltage that is changing rapidly (a high-frequency sine wave for example) than it would to a voltage that is changing slowly (a low-frequency sine wave for example). This characteristic makes capacitors useful in different filter circuits. We shall see later that there are many types of filter circuits using various combinations of resistors, capacitors, and inductors.

---

* Another way to find t is to set up the exponential expression for v_{cap}, set it equal to 9 V and solve for t. That is, $v_{\text{cap}} = 20(1 - \epsilon^{-t/10\ \text{ms}}) = 9$ V.

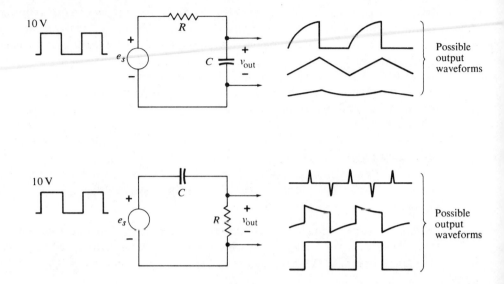

FIGURE 13-32 *Various types of waveshaping accomplished by a series RC circuit.*

The most commonly employed filters used every day are in our radios and TV sets. Some are tuning filters which are used to select the frequency of the desired radio station or TV channel. Other types of filters are employed in our dc power supplies.

13.19 Parasitic Capacitances

Capacitance is present whenever two electrical conductors are separated by an insulating material. This means that any *two* wire conductors have the property of capacitance between them. This *stray, wiring* capacitance is typically very small, on the order of a few pF. However, even these small capacitances can have a significant effect in systems where the voltages are changing at a rapid rate, such as in a digital computer or communications system.

Capacitance is also present across the PN junctions of semiconductor devices like diodes, transistors, and FETs. These *junction* capacitances can also affect the circuit operation at high frequencies. Stray wiring capacitances, junction capacitances, and other types of parasitic capacitances are extremely hard to measure and are very hard to eliminate in cases where they cause trouble.

13.20 Linearity of Capacitors

Capacitors are linear devices. We know this is true because if we double the source voltage in an *RC* circuit, the capacitor voltage will charge to two times the original value. Another way to look at it is that for a capacitor *voltage is directly proportional to charge and vice versa* as given by

$$V = \frac{Q}{C} \quad \text{and} \quad Q = CV$$

Capacitors and resistors being linear devices indicates that any *RC* circuit is a linear circuit and therefore obeys all the principles of linearity. This means, for example, that we can use the principle of superposition on any *RC* circuit with more than one source. Further implications of the linearity of capacitors will be encountered when we study the capacitor in ac circuits.

13.21 Final Comments

Our study of capacitance does not end here. We have just begun to get a feeling for how capacitors function in dc circuits. In subsequent studies of capacitance in ac (sinusoidal) circuits and waveshaping circuits the basics learned in this chapter will be drawn upon. The following chapter summary contains the important points and concepts that should be learned before proceeding.

Chapter Summary

1. The electrostatic force between two charges has a magnitude which is directly proportional to the product of the two charges and inversely proportional to the square of the distance between the charges (Coulomb's law).
2. The electric field associated with a charge is represented by lines of force. These lines of force always extend outward from a positive charge and in toward a negative charge.
3. The symbol for electrical field intensity is ξ and has units of volts/meter.
4. The electrical field intensity in the area between two charged plates equals the potential difference between the plates divided by the distance of plate separation.
5. The force that an electric field with intensity ξ exerts on a charge Q is equal to $Q \times \xi$.
6. A capacitor is an electrical device which consists of two metal conductor plates separated by an insulator (dielectric).
7. A capacitor has the ability to hold or store charges that have been placed on its plates.
8. A charged capacitor acts as a voltage source.
9. In a capacitor, the voltage between its plates is proportional to the amount of charge on the plates.
10. Capacitance C is a measure of how much charge is required to produce a given voltage between the plates of a capacitor.
11. Capacitance $C = Q/V$ so that its units are coulombs/volt which is commonly called a farad, F.
12. When a capacitor is being charged from a constant voltage source, a transient (temporary) current will flow until the capacitor is charged up to the source voltage, after which no further current will flow and the capacitor will act like an open circuit.

13. A capacitor's voltage cannot change instantaneously, but must build up (charge) or decrease (discharge) gradually.

14. When a capacitor is being charged from a constant voltage source, the charging current is maximum at the instant the switch is closed and gradually decreases as the capacitor charges.

15. The rate of change of capacitor voltage ($\Delta V/\Delta t$) is directly proportional to the charging current and inversely proportional to capacitance.

16. For an RC circuit, one time constant is equal to $\tau = RC$. An RC circuit will complete its transient response and reach steady state in a time interval equal to $5\,\tau$.

17. In an RC circuit all currents and voltages will complete 63 percent of their total ultimate change in one time constant.

18. In a series RC circuit the capacitor voltage will always charge or discharge toward a voltage equal to the input voltage, which it will reach in $5\,\tau$.

19. In a simple RC circuit, the currents and voltages will obey the exponential expressions below when a sudden change in the input occurs.
$$i = I_F + (I_0 - I_F)\epsilon^{-t/\tau}$$
$$v = V_F + (V_0 - V_F)\epsilon^{-t/\tau}$$

20. The equivalent capacitance of series capacitors is given by $1/C_{eq} = 1/C1 + 1/C2 + \ldots$ etc. C_{eq} is always less than the smallest series capacitor.

21. For capacitor series the applied voltage divides in inverse proportion to the capacitance values.

22. For capacitors in series the applied voltage divides in inverse proportion to the capacitance values.

23. If a capacitor is connected to any network containing only constant dc sources, the voltage which the capacitor will charge to can be found by treating the capacitor as an open circuit.

24. If a capacitor C is connected to a linear, resistive network with only dc sources, the capacitor will charge toward a voltage equal to the network's E_{Th} with a time constant $\tau = R_{Th} \times C$.

25. The value of capacitance is proportional to the area of the plates, inversely proportional to the separation of the plates, and proportional to the dielectric constant.

26. Dielectric breakdown occurs in a capacitor when the capacitor's plate voltage is increased beyond the capacitor's maximum voltage rating.

27. An ideal capacitor has an infinite leakage resistance. A practical capacitor has a large, but finite, leakage resistance.

28. The energy stored by a capacitor is equal to $CV^2/2$.

29. A capacitor does not dissipate energy (except for a minute amount in its leakage resistance).

30. The numerous applications of capacitors include energy storage, timing circuits, filters, and waveshaping circuits.

31. Stray or parasitic capacitances are present between any two wire conductors and across PN junction devices such as diodes.

32. Capacitors are linear devices since voltage is proportional to charge and vice versa.

33. Capacitors are linear devices since voltage is proportional to charge and vice versa.

Questions/Problems

Section 13.2

13-1 State Coulomb's law of electrostatic force.

13-2 Which of the following cases represents the greatest value of electric field intensity:
 a. two metal plates separated by 0.1 cm with a potential difference of 6 V between the plates
 b. two plates separated by 2 cm with 20 V between the plates
 c. two plates separated by 0.03 cm with 3 V between the plates

Section 13.3

13-3 If the two charged plates in Fig. 13-4(a) were separated by an *imperfect* insulator, what would eventually happen?

13-4 If these two charged plates are gradually brought closer together, what do you expect will eventually happen?

13-5 Why does a charged capacitor act like a *poor* storage battery?

Section 13.4

13-6 a. A certain capacitor requires 50 μC of charge on its plates in order to develop a voltage of 20 V. What is the capacitor's capacitance?
 b. If the capacitor's capacitance were increased, would it take more than or less than 50 μC to produce 20 V?

13-7 A certain capacitor has $C = 100$ μF. How much charge must be placed on its plates to produce 50 V?

13-8 Convert the following into farads using scientific notation: (a) 47 pF, (b) 0.005 μF, (c) 0.47 μF, (d) 200 μF.

13-9 Which of the following represents the largest capacitance: (a) 22000 pF, (b) 0.3 μF, (c) 0.00003 F?

Section 13.5

13-10 Refer to Fig. 13-33. At the instant the switch is closed, explain why the capacitor voltage cannot become 10 V immediately. What is missing from this circuit diagram?

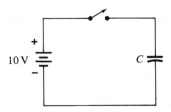

FIGURE 13-33

13-11 Explain why a momentary current will flow in Fig. 13-33 even though the capacitor plates are separated by an insulator.

13-12 Explain the process by which the capacitor in Fig. 13-33 eventually charges to 10 V.

13-13 The switch in Fig. 13-33 is closed long enough to fully charge C and then it is opened. What will happen to the capacitor voltage after the switch is opened? How can the capacitor voltage be made to decrease to zero?

13-14 Indicate which of the following statements are true:
    a. A capacitor consists of two insulators separated by a conductor.
    b. A capacitor's voltage cannot change instantaneously.
    c. When a capacitor is being charged, current flows through the insulator between the capacitor's plates.
    d. When a capacitor discharges, the discharge current is produced by the capacitor's stored charge.
    e. A 50-pF capacitor can store more charges than a 0.001-μF capacitor for the same voltage.

Section 13.6

13-15 In the circuit of Fig. 13-34, the switch is closed at $t = 0$. What is the value of charging current at $t = 0$?

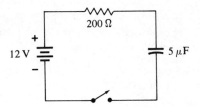

FIGURE 13-34

13-16 What voltage does the capacitor in Fig. 13–34 have to reach before the charging current has dropped to 20 mA?

Section 13.7

13-17 a. What is the initial rate of increase of the capacitor voltage in Fig. 13–34?
b. At what rate is the capacitor voltage increasing at the instant when the capacitor voltage is 8 V? Repeat for 10 V.
c. Sketch the waveforms of capacitor voltage, resistor voltage, and charging current.

13-18 Indicate how each of the following changes will affect the *initial* rate of increase of capacitor voltage in Fig. 13–34.
a. increase in supply voltage
b. increase in series resistance
c. decrease the capacitor

13-19 If a 100-μF capacitor is being charged (starting at 0 V) with a *constant* current of 2 mA, what will be its voltage after 2.5 s?

Sections 13.8–13.9

13-20 In the circuit of Fig. 13–34, the capacitor will become fully charged in approximately 5 ms. Indicate the effect each of the following changes will have on this charging time:
a. an increase in resistance
b. a decrease in capacitance
c. halving the source voltage
d. doubling E_s and R; but halving C.

13-21 Determine the time constant for the circuit in Fig. 13–34. How long will it take the circuit to reach steady state?

13-22 In Fig. 13–34, what will be the value of the capacitor voltage 1 ms after the switch is closed? What will be the value of charging current at that same instant of time?

13-23 Consider the circuit in Fig. 13–35. The switch is closed at time $t = 0$. Sketch the capacitor voltage v_{xy} and resistor voltage v_{wx} waveforms accurately showing the values at $t = 60$ ns, 120 ns, 180 ns, 240 ns, and 300 ns.

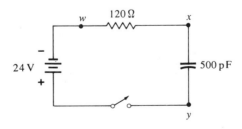

FIGURE 13-35

13-24 Repeat problem 13–23 except this time assume the capacitor was already initially charged to $v_{xy} = -6$ V before the switch was closed.

Section 13.10

13–25 In which of the following cases will the capacitor complete its discharge most rapidly:
  a. 10-μF capacitor with a 10-V charge discharging through a 1-kΩ resistor.
  b. the same capacitor with a 50-V charge discharging through a 1-kΩ resistor.
  c. a 50-pF capacitor with a 20-mV charge discharging through a 200-milliohm resistance.

13–26 A 2-μF capacitor is charged to 12 V. A 15-kΩ resistor is placed across the capacitor to discharge it.
  a. Sketch the waveforms of capacitor voltage and discharge current.
  b. Determine the initial discharge rate.
  c. Determine the discharge rate after 60 ms.
  d. Determine the capacitor voltage after 90 ms.

Section 13.11

13–27 Consider the circuit in Fig. 13–36. The capacitor is initially discharged ($v_{xy} = 0$). If the switch is closed at $t = 0$, determine:
  a. the capacitor voltage at $t = 0.6$ ms after the switch is closed
  b. at what time the resistor voltage will be 12 V
  c. the charging current at $t = 1.7$ ms

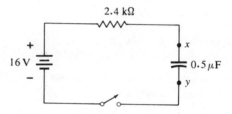

FIGURE 13–36

13–28 If the capacitor in Fig. 13–36 is already charged to 4 V before the switch is closed, what will be the capacitor voltage at $t = 0.9$ ms after the switch is closed? Sketch the capacitor and resistor voltage waveforms showing the time to reach steady state.

13–29 In the circuit of Fig. 13–37, the capacitor is initially discharged. The switch is thrown to position 2 at $t = 0$. It remains there until $t = 1$ ms and is then thrown to position 3 where it remains indefinitely.
  a. Sketch the capacitor voltage waveform.
  b. Determine the capacitor voltage at $t = 2$ ms. (1 ms after switch is thrown to position 3.)
  c. How much time does it take the capacitor voltage to reach 0 V after the switch is thrown to position 3?

13–30 Indicate how the answer to problem 13–29(b) will be affected (increase, decrease, or remain the same) for each of the following modifications of the circuit in Fig. 13–37:

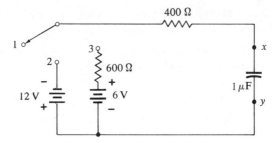

FIGURE 13-37

a. an increase in C
b. a decrease in the 12-V source
c. a decrease in the 6-V source
d. an increase in the 600-ohm resistor
e. capacitor initially charged (before $t = 0$) to a negative value ($v_{xy} = -V_0$).

Section 13.12

13-31 a. Write the exponential expression for the capacitor voltage in Fig. 13-15.
b. Use this expression to calculate the answers to Ex. 13.9 and compare to the results found using the time constant chart.

13-32 Repeat Ex. 13.10 by first writing the exponential equation for capacitor voltage.

13-33 Repeat problem 13-27 using the exponential equations for capacitor voltage, resistor voltage, and current.

13-34 Repeat problem 13-28 using the exponential equation for capacitor voltage.

Section 13.13

13-35 Find the equivalent capacitance in each of the cases in Fig. 13-38.

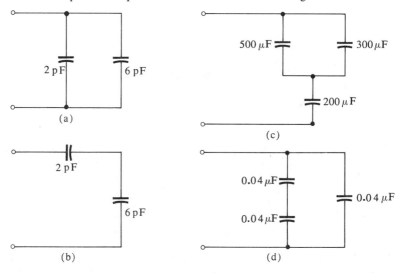

FIGURE 13-38

13–36 A 30-μF and 10-μF capacitor are connected in parallel across a 12-V source. Determine the total charge supplied by the source.

13–37 A 1-μF, 2-μF, and 3-μF capacitor are connected in series across a 300-V source. Determine the voltage across each capacitor.

Section 13.14

13–38 Determine the steady-state capacitor voltage in the circuits of Fig. 13–25(b) and (d).

13–39 The switch in Fig. 13–39 is closed at $t = 0$.
   a. Sketch the capacitor voltage waveform showing the time to complete its charge.
   b. If a capacitor is placed across the 4-kΩ resistor, what will be its steady-state voltage?

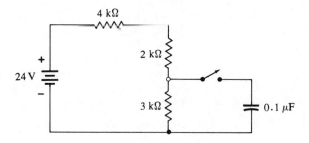

FIGURE 13–39

Section 13.15

13–40 Calculate the capacitance for two metal plates that are 1 cm on each side and are separated by 0.2 cm with a mica dielectric.

13–41 How will the capacitance in problem 13–40 be affected by the following changes:
   a. halve the area of the plates
   b. triple the separation distance
   c. change dielectric to air
   d. double the area, double the separation

13–42 A certain dielectric material has a dielectric strength of 10 kV/mil. What is the minimum plate separation that can be used with this dielectric, if the capacitor working voltage is to be 40 volts?

13–43 Suppose a certain circuit needs a 10-μF capacitor rated at 200 V, and all you have are 10-μF capacitors rated at 100 V. How could you combine these 100-V capacitors to meet the circuit requirement?

Section 13.16

13–44 Explain a procedure for determining the leakage resistance of a capacitor.

13–45 Consider the circuit of Fig. 13–40. The capacitor is supposed to be charged to the 20-V source voltage. However, a 3% voltmeter connected across the capacitor shows

an indication of only 15 V. Which of the following are possible causes for the discrepancy?
a. resistor value is out of tolerance
b. voltmeter is loading the circuit
c. source voltage is not at 20 V
d. capacitor is too large
e. voltmeter inaccuracy
f. capacitor is leaky

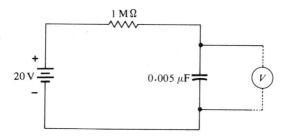

FIGURE 13-40

13-46 What procedure would you follow to determine the actual cause of the discrepancy in problem 13-45?

13-47 If the capacitor in Fig. 13-40 only charges to 12 V, what is its leakage resistance?

13-48 A 100-μF capacitor has a 2.3-MΩ leakage resistance. How long will it take a charge to completely leak off this capacitor?

Section 13.17

13-49 A 2-μF capacitor is charged to 120 V. How much energy will be dissipated in a 100-ohm resistor connected across the capacitor to discharge it? Repeat for a 1-MΩ resistor.

13-50 For the circuit of Fig. 13-36 how much energy is stored by the capacitor after it has been fully charged when the switch was closed? Why did the source have to supply *more* than this amount of energy during charging? Where did the extra energy go?

Sections 13.18-13.21

13-51 The circuit in Fig. 13-41 is used to produce a large pulse of current through the 10-ohm load. Assume the switch has been open for a long time.
a. Sketch the waveform of load current upon closure of the switch, showing its complete time duration.
b. How much energy is supplied to the load?
c. Where did this energy originally come from?
d. Why is the capacitor used in this application?

13-52 In Fig. 13-42 the input voltage source is a square wave with a period of 2 ms. Sketch the capacitor and resistor voltage waveforms for *three* cycles of the input. This is one example of a waveshaping circuit.

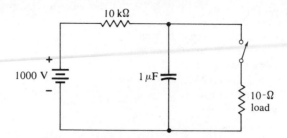

FIGURE 13–41

13–53 Change C to 1 µF and repeat problem 13–52.

13–54 Indicate which of the following statements pertain only to resistors, which pertain only to capacitors, and which pertain to both resistors and capacitors:
   a. are linear devices
   b. can dissipate considerable amounts of power
   c. can act as energy storage devices
   d. oppose sudden changes in voltage
   e. current is constant when connected to a dc source
   f. values depend on physical dimensions and type of material
   g. value doubles if area doubles
   h. an increase in value draws less current from the input source
   i. connecting them in parallel *increases* the total value
   j. is an open circuit in steady state

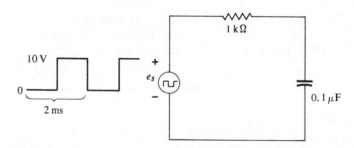

FIGURE 13–42

CHAPTER 14

Magnetism

14.1 Introduction

The three basic circuit properties are resistance, capacitance, and *inductance*. In this chapter we begin laying the groundwork for the study of inductance by briefly examining the concepts of *magnetism*. This study of magnetism is not only necessary for a thorough understanding of inductance, but it will also be of use to us in our later study of *transformers*.

A background in magnetism and magnetic effects will also aid in understanding many of the modern devices which utilize magnetic and electromagnetic principles. Such devices range from the common doorbell (buzzer) to the complex magnetic-core memories of a digital computer and include motors, generators, tape recorders, relays, speakers, telephones, and current meters to mention a few.

14.2 Some Basic Ideas

All of us have some basic concept of magnetism and magnetic effects. We know that a magnet has the ability to attract *magnetic materials*. Common magnetic materials are iron, steel, nickel, cobalt, and their alloys. Nonmagnetic materials such as wood, glass, paper, copper, and silver are not attracted by magnets. A magnet which is also made of a magnetic material can attract another magnetic material by *contact* or *at a distance*. Figure 14-1(a) illustrates attraction by contact while Fig. 14-1(b) illustrates attraction at a distance.

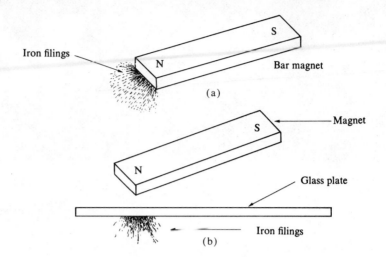

FIGURE 14-1 (a) *Magnetic attraction by contact.* (b) *Magnetic attraction at a distance.*

The situation in Fig. 14-1(b) also illustrates another important characteristic of magnetism whereby the force of a magnet can penetrate through a nonmagnetic material (glass) to attract a magnetic material (iron filings).

Most of us are also familiar with the difference between *permanent* magnets and *temporary* magnets. A permanent magnet retains its magnetic force for an indefinite period of time. A temporary magnet is magnetized only temporarily by a magnetic force. In Fig. 14-1(a) the iron filings are attracted to the permanent bar magnet. The magnetism of the bar magnet *induces* magnetism into the iron filings so that each iron filing becomes a little magnet that can attract other iron filings. These little magnets are temporary magnets, however, because they lose their magnetism when they are separated from the bar magnet.

Natural and Artificial Magnets The first magnetic material ever discovered is a type of iron ore called *magnetite* (also called *lodestone*). Pieces of magnetite are called *natural* magnets because they possess magnetic properties when found in their natural form. Natural magnets are no longer used because better magnets can now be produced artificially.

Artificial magnets are produced in a wide range of shapes and sizes. Three common types of artificial magnets are shown in Fig. 14-2. Artificial magnets are made from magnetic materials like steel by the process of *induction* similar to that described in Fig. 14-1(a) with the iron filings. Unlike the iron filings, however, steel will remain magnetized even after it is removed from the vicinity of the inducing magnet. In other words, the steel becomes a permanent magnet itself.

Some of the more common materials which are used to produce artificial permanent magnets include hard steels (like nickel-steel), cobalt, and alnico (an alloy of aluminum, nickel, iron, and cobalt). Soft iron is not used because it does not retain its magnetism.

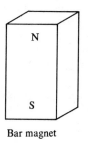 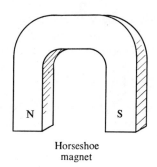 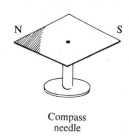

Bar magnet   Horseshoe magnet   Compass needle

FIGURE 14-2 *Examples of artificial magnets.*

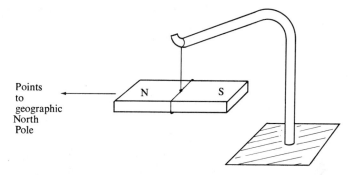

FIGURE 14-3 *Suspended magnet always points in a north-south direction.*

Another means of producing an artificial magnet uses electricity passed through a coil of wire wound around the magnetic material. This is an example of *electromagnetism*, which will be examined in detail later.

Magnetic Poles A magnet, natural or artificial, has two distinct regions called *poles*. These regions are usually at the extremities of the magnet. The effects of magnetism are not uniformly distributed over the surface of a magnet. The magnetic force is more concentrated at the poles.

If a magnet shaped like a bar or rod is suspended on a string so that it is free to rotate (Fig. 14-3), it will line itself up in a north-south geographic direction. The end of the magnet that points northward is called the *north-seeking pole* or more simply the *north pole* of the magnet. The other end points southward and is called the *south-seeking pole* or *south pole* for short. The reason that the suspended magnet points in the north-south direction is that the earth itself is a giant magnet whose poles are close to the geographic north and south poles. A fundamental law of magnetism states that:

Unlike magnetic poles attract each other, and like poles repel each other.

As such, it necessarily follows that the geographic North Pole is actually a south magnetic pole since it attracts the north magnetic pole of the suspended magnet. Similarly, the geographic South Pole is a north magnetic pole.

A compass is simply a small magnet which is free to rotate and thus indicates geographic directions. The north pole of the compass is usually the painted end and will always point to geographic north. Actually, the earth's magnetic poles do not exactly coincide with its geographic poles. In fact, the earth's magnetic poles randomly wander about. At present, the earth's north magnetic pole is about 500 miles from the true geographic North Pole.

14.3 Magnetic Field and Lines of Force

Like gravitational force and electrostatic force, magnetic force is a *field* force. An invisible magnetic force exists in the space surrounding a magnet. The space in the region around a magnet is called a *magnetic field*. This is similar to the electrostatic field associated with electric charges.

As was the case with electrostatic fields, it is useful to think of a magnetic field as consisting of *magnetic lines of force*. These magnetic lines of force are also referred to as *magnetic flux lines* or simply *magnetic flux*. Like electrostatic lines of force, magnetic lines of force are imaginary but are useful in representing the effects of magnetic fields.

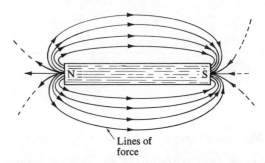

FIGURE 14-4 *Magnetic lines of force around a permanent bar magnet.*

Figure 14-4 illustrates the magnetic lines of force associated with a bar magnet. From this illustration, several important points can be ascertained.

Magnetic lines of force possess direction.

The arrows in the diagram indicate that the flux lines leave the magnet at the north pole and enter the magnet at the south pole.

Each line of force forms a continuous loop.

They do not begin at the north pole and end at the south pole, but instead continue from the south pole to the north pole inside the magnet to form continuous loops.

The direction of the lines of force was arbitrarily defined a long time ago as the direction in which the north pole of a compass needle will point if placed at any point along a line of force. In other words, we can take a compass and place it at various positions around the bar magnet and the compass needle will in each instance point in the direction of the flux lines. This is illustrated in Fig. 14-5.

Referring again to Fig. 14-4, it should be pointed out that the flux lines actually extend in all directions out into space. It is also true that the concentration of the flux lines is greatest near the poles and rapidly diminishes as distance from the poles increases. This can be verified by gradually moving a

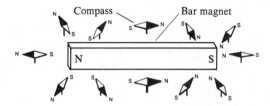

FIGURE 14-5 *Compass needle shows the direction of the flux lines around a bar magnet.*

compass further away from the magnet and noting that eventually the magnet will exert no visible force on the compass. Thus, we can state that

The concentration of flux lines is a direct measure of the strength of a magnetic field.

As another illustration of magnetic lines of force, Fig. 14-6 shows the flux distribution around a horseshoe-shaped magnet. Because of the different shape, this magnet's lines of force form a different pattern than those of a bar magnet. However, the characteristics are the same. That is, the flux direction is from north pole to south pole (external to the magnet) and the concentration of flux lines decreases as distance from the poles increases.

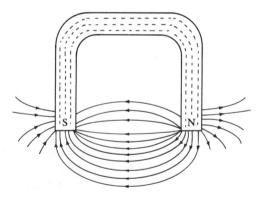

FIGURE 14-6 *Flux lines associated with a horseshoe magnet.*

As a final illustration of magnetic flux lines, consider the cases shown in Fig. 14-7. In part (a) of the figure, two magnetic north poles are shown with their resultant flux pattern. The two north poles produce flux lines which have a direction aimed at the space between the poles. The resultant flux distribution illustrates two more characteristics of flux lines:

Flux lines never intersect.

and

Flux lines pointing in the same direction tend to repel each other.

We can easily imagine that the flux lines from the two like poles appear to repel each other.

In Fig. 14-7(b) the flux distribution around two unlike magnetic poles is shown. The flux lines that leave the north pole combine with the flux lines that

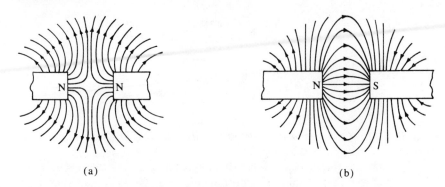

FIGURE 14-7 *Magnetic field around* (a) *two like poles and* (b) *two unlike poles.*

enter the south pole in the space between the poles. The result is an almost parallel pattern of flux lines in this region. This coming together of flux lines gives the impression of attraction between the two poles.

There are other characteristics of flux lines that will help us in our study of magnetism. These characteristics will be noted as they occur in the following sections.

14.4 Theory of Magnetism

Why are certain materials magnetic and others nonmagnetic? This is a natural question which should be considered in any study of magnetism. Strictly speaking, all natural substances are magnetic but to varying degrees. The type of strong magnetism observed in iron, nickel, and cobalt is called *ferromagnetism* (from the word "ferrum," which is Latin for "iron"). These substances are said to be *ferromagnetic*. Ferromagnetic materials are noticeably affected by any magnetic field. In other words, these ferromagnetic substances will be attracted by a magnet.

Materials which we think of as nonmagnetic can be classified as either *diamagnetic* or *paramagnetic* materials. Substances that are *slightly repelled* by a strong magnetic field are called diamagnetic. Examples of diamagnetic substances are glass, copper, wood, and water.

Substances that are *slightly attracted* by a strong magnetic field are classified as paramagnetic. Included in this category are materials such as air, liquid, oxygen, and aluminum.

In most of the discussions that follow we will not distinguish between diamagnetic and paramagnetic materials. Since such materials respond only weakly to a strong magnetic field, we will simply refer to them as nonmagnetic materials. Most of our study of magnetism will be concerned with ferromagnetic, or simply magnetic, substances.

Domain Theory of Magnetism One of the most prevalent theories behind magnetism is called the *domain theory*. This theory, which is very useful in helping us explain the characteristics of magnetic materials, was first proposed in 1852 by a German physicist named Weber. It was later modified by a Scottish physicist named Ewing in 1891. Since those early years modern physics has further modified it, but the basic idea is still the same.

Within every atom of a substance the orbiting electrons are also *spinning* as they revolve around the nucleus, much like the earth spins or rotates on its own axis as it revolves around the sun. This spinning charge represents a tiny current. As we shall see later, *a magnetic field is associated with any flow of current* (theory of electromagnetism). Thus, each spinning electron has a magnetic field and consequently each atom has a magnetic field associated with it.

In nonmagnetic materials, the magnetic fields from the various electrons of an atom oppose each other so that the net magnetic field is zero for each atom. In ferromagnetic materials, however, a number of electrons have their spin axes oriented in the same direction so that each atom has a net magnetic field. Groups of these atoms (containing about 10^{15} atoms) align themselves to form very small bar magnets called *domains*. A ferromagnetic substance can therefore be viewed as an arrangement of numerous tiny magnetic domains.

If we could observe the domain structure of an *unmagnetized* sample of iron, it might appear as shown in Fig. 14–8(a). The domain magnets are represented by small arrows with the heads of the arrows representing the north poles. For this unmagnetized condition, the domains are aligned in a random arrangement (not in any particular direction) so that the sample has no net magnetism.

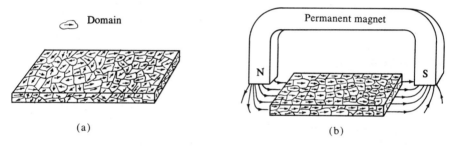

FIGURE 14–8 *Magnetic domains in an iron bar when* (a) *unmagnetized;* (b) *magnetized by the magnetic field of a permanent magnet.*

If this same iron sample is placed in a magnetic field (such as that of a permanent magnet) it will become magnetized by the process of induction discussed earlier. This process involves the actual movement of the small domains so that they align themselves with the magnetic field of the permanent magnet as illustrated in Fig. 14–8(b). The iron sample now possesses a net magnetic field of its own due to the uniform alignment of the domains. If the permanent magnet is removed, the domains in the iron sample will return to their random orientation [part (a) of Fig. 14–8] and the iron becomes unmagnetized again.

If the process stated here were repeated with a sample of hard steel, all of the domains would not reorient themselves when the permanent magnet was removed. Instead, many of them would maintain their uniform alignment so that the steel bar retains much of the magnetism and becomes a permanent magnet itself. We say that magnetic materials that behave like steel have a high *retentivity* of magnetism.

The domain theory also explains why when a permanent magnet is divided in half, the two halves both become permanent magnets. The domains in each half remain aligned in the original direction so that each half is a magnet.

14.5 Preferred Paths for Flux Lines

Figure 14–9(a) again shows the flux pattern around two unlike poles where we assume that the space around the poles contains air, a nonmagnetic substance. If a magnetic substance such as soft iron is placed between the two unlike poles, the flux pattern is affected as shown in Fig. 14–9(b). Many of the flux lines alter their normal paths to include the small iron bar. In other words, the flux prefers to travel the path through the iron bar rather than the air. This is because the iron is a better "conductor" of flux lines than is air or any nonmagnetic substance.

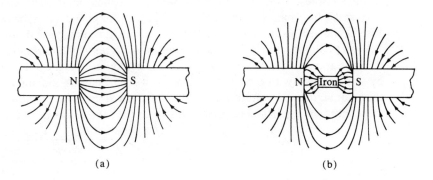

FIGURE 14–9 *Magnetic flux lines prefer to travel a path through a magnetic substance.*

This is analogous to the situation in electrical circuits where a low-valued resistor is placed in parallel with a large-valued resistor. Most of the electrical current will flow through the path of least resistance. In this magnetic situation, we can visualize the flux lines as being analogous to current (although they do not actually flow). The flux lines would rather travel through the path that has the least resistance to magnetic lines, namely the iron bar. The term "resistance" is actually not used in reference to magnetism. Instead we say that iron has a lower *reluctance* to the setting up of flux lines than does air. Reluctance will be defined more specifically later. Let us just say at this point that

Magnetic flux lines always prefer to take the path with the least magnetic reluctance.

If the iron bar is changed to a wooden bar, the flux pattern will be the same as if the wood were not there because wood is as nonmagnetic as air. In other words, any nonmagnetic material will not affect the flux lines. Any magnetic material will alter the flux pattern as in Fig. 14–9(b). However, some magnetic materials have a lower reluctance than other magnetic materials so that their effect will be even more pronounced.

It is important to note also that there is no magnetic "insulator" that will not pass flux lines. Even nonmagnetic materials such as air, wood, glass, etc., will pass magnetic lines of force although not as readily as iron or steel. The nonmagnetic materials are said to have a greater reluctance to flux lines than magnetic materials.

14.6 Magnetic Field Produced by Current Flow

The first connection between electricity and magnetism was discovered by Hans Oersted in 1820. He found that a wire carrying an electric current exerted an influence on a compass needle placed near the wire. In other words,

An electric current produces magnetism.

When we talk about the magnetism produced by the flow of current, we refer to it as *electromagnetism*. An illustration of the magnetism produced by a current flowing in a wire is shown in Fig. 14–10. The wire is shown threaded through a piece of cardboard so that the wire is perpendicular to the cardboard.

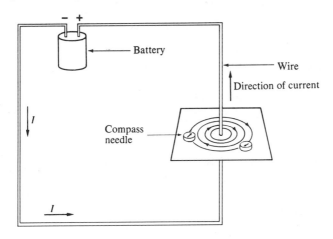

FIGURE 14–10 *Magnetic field around current-carrying wire.*

The magnetic lines of force produced by the current through the wire can be determined by placing a compass needle at various points around the wire. The resultant flux lines are as shown in the figure. Several important statements can be made concerning these lines of force:

(1) They form complete concentric circles around the wire.

(2) The separation between adjacent flux lines (circles) becomes greater as we move further from the wire. This indicates that the strength of the magnetic field decreases as distance from the wire increases so that eventually the compass will not be affected.

(3) The strength of the magnetic field (concentration of flux lines) will increase as the current in the wire increases, and vice versa.

(4) The flux pattern is the same at any point along the length of the wire, and is always in a plane perpendicular to the wire.

(5) The flux lines around a current-carrying wire have a specific direction. In Fig. 14–10 the upward flow of current produces concentric flux lines pointing in a clockwise direction. If the current is reversed, the direction of these flux lines will reverse and become counterclockwise. This can be verified with the compass needle.

Left-Hand Rule A quick method for determining the direction of the magnetic lines of force in a current-carrying wire is illustrated in Fig. 14–11. This method is called the left-hand rule and proceeds as follows: *Grasp the wire with the left hand so that the thumb points in the direction of electron flow in the wire; the fingers wrapped around the wire will then indicate the directions of the paths of the flux lines.* This method can be used to verify the flux direction in Fig. 14–10.

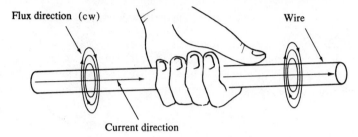

FIGURE 14–11 *Illustration of left-hand method for determining flux direction in a current-carrying wire.*

This left-hand rule applies only to *electron flow*. If the conventional current flow is used, then the same procedure can be followed except the *right* hand is used to determine the direction of the flux lines.

Cross and Dot Conventions To avoid perspective drawings when showing the flux lines around a conducting wire, we can use a method which looks at the cross section of the wire. This is illustrated in Fig. 14–12. In part (a) of the figure the small circle with a cross inside it represents the cross-sectional view of a wire. The cross symbolizes the tail feathers of the current arrow representing current flow in the wire. In other words, current is flowing into the page away from the observer. Using the left-hand rule, the direction of the concentric flux lines is ccw as shown.

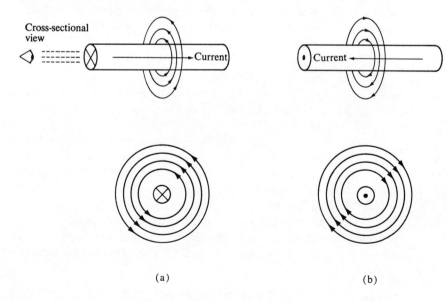

FIGURE 14–12 *Cross-sectional view of flux around a wire.*

Magnetism 347

In part (b) of the figure, the dot inside the small circle symbolizes the point of the current arrow indicating that current is flowing out of the page toward the observer. Once again the left-hand rule determines the direction of flux lines as cw.

The view of the flux patterns presented in Fig. 14–12 gives us another way of visualizing the magnetic field around a conducting wire. As the current in the wire increases from zero, a flux loop of very small diameter forms at the center of the conductor. As the current increases further, this flux loop expands in size and another small loop forms at the center of the conductor. Further increases in current produce more flux loops which expand outward from the wire. This action is very similar to the effect observed when a stone is thrown into a pool of water causing circular ripples that expand outward from the point where the stone hits the water. Passing a larger current through the wire increases the number of flux loops and also increases the distance at which they expand (the same effect as throwing the stone into the pool of water with a greater force).

When the current in the wire is held constant, the flux pattern remains stationary. A decrease in current causes the flux loops to collapse back toward the center of the wire. Of course, with zero current there will be no lines of force around the wire.

14.7 Magnetic Field Around a Solenoid (Coil)

If a current-carrying wire such as those in Figs. 14–10, 14–11, and 14–12 is bent into the shape of a loop, the same circular lines of magnetic force surround the wire as when it was straight [see Fig. 14–13(a)]. Examination of the field around the conducting loop shows that all the lines of force pass through the center of the loop in the *same* direction. By bending the straight wire into a loop, we have strengthened the magnetic field by concentrating the flux into a smaller area (the inside of the loop).

Another view is illustrated in part (b) of Fig. 14–13. Here we can see that all the flux lines enter the loop from one side and leave from the other, thus creating a north pole on one side of the wire loop and a south pole on the other.

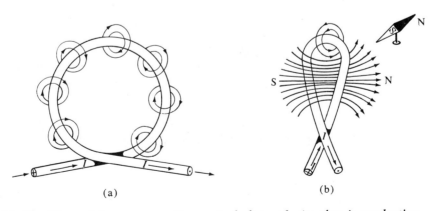

FIGURE 14–13 *Lines of force surrounding a single loop of wire that is conducting current.*

Solenoids The magnetic field can be concentrated even further by winding the conducting wire around a cardboard cylinder core so that several loops are formed as shown in Fig. 14–14. Such an arrangement is called a *coil* or *solenoid*. Since the current in each loop is traveling in the same direction around the circumference of the solenoid, then each loop contributes a magnetic flux in the same direction in the center of the solenoid (right to left). The individual flux contributions of each loop reinforce and add to each other so that the net magnetic field inside the solenoid is directed from right to left. The result is a flux pattern such as that shown in Fig. 14–14.

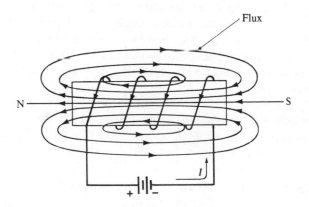

FIGURE 14–14 *Magnetic field around a current-carrying solenoid.*

As the figure shows, the flux pattern produced by a current-carrying solenoid is very similar to that of a permanent bar magnet (compare with Fig. 14–4). The solenoid also acts as a magnet with the north pole being the end out of which the flux lines emerge (the left end in our example). This type of magnet is called an *electromagnet* because it is the flow of electrons (current) which produces the magnetic field. As long as current flows through the coil, the solenoid has all the properties of a magnet. The strength of this electromagnet is easily increased by either increasing the current through the coil or by increasing the number of turns or loops wound around the core.

Left-Hand Rule for Solenoids If we wish to determine the magnetic polarity of a solenoid quickly, we can invoke another left-hand rule which proceeds as follows:

Grasp the solenoid with the left hand so that the fingers encircle the coil in the direction of current flow, and extend the thumb; the thumb will then point toward the north pole of the electromagnet.

This rule is illustrated in Fig. 14–15(a) and (b). Note that a reversal of current causes a reversal of the magnetic poles for a given solenoid.

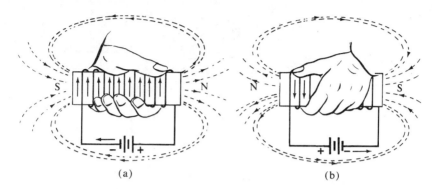

FIGURE 14–15 *Illustration of left-hand rule for determining magnetic poles of a solenoid.*

14.8 Solenoids With Ferromagnetic Cores

The solenoid in Fig. 14–14 is called an *air-core solenoid* or *air-core coil* because the coil is wound on a hollow nonmagnetic cylinder. The magnetic field of the air-core solenoid is relatively weak unless a tremendously large current is pushed through the coil. If a piece of magnetic material, usually soft iron, is used as the core, the magnetic properties of the solenoid are greatly enhanced, resulting in a greater number of flux lines. The reason for this increase in magnetic strength is because the soft iron has a lower *reluctance* than air and therefore provides a better path for flux lines.

Since the iron core is a ferromagnetic substance, its internal magnetic domains will align themselves with the magnetic field produced by the solenoid. This temporary magnetization of the iron core increases the number of flux lines in the solenoid and therefore the magnetic strength of the electromagnet. An iron-core electromagnet will have several thousand times the strength of a comparable air-core electromagnet.

Another important difference between the air-core and iron-core solenoids is in the distribution of their magnetic fields, as Fig. 14–16 illustrates. In the air-core solenoid a large number of flux lines do *not* leave the coil at the north pole, but somewhere in between. In the iron-core solenoid the majority of the flux lines extend the whole length of the core and leave at the north pole and enter at the south pole. The iron core provides a low reluctance path which the flux prefers to follow. In the air-core solenoid the flux is at its maximum in the center of the coil and decreases at the poles.

Any ferromagnetic material can be used as the core for an electromagnet. However, soft iron is generally used because of its low *retentivity*, which means that it will lose most of its magnetism when the current in the coil stops flowing. This is a very important property when the electromagnet is to be used in

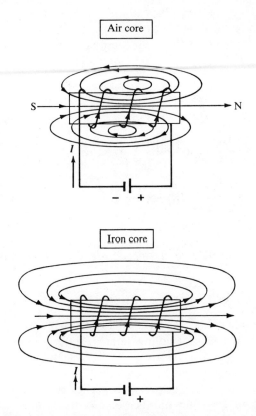

FIGURE 14–16 *Comparison of iron-core and air-core solenoid flux patterns.*

applications such as relays, motors, generators, and transformers. In such applications it is necessary that the amount of flux varies in proportion to the coil current.

If a material such as hard steel is used as the core of the solenoid, the result will be different. Because of the steel's high retentivity, it will retain a great portion of *residual magnetism* when the coil current stops flowing. This is one method of artificially producing a permanent magnet.

Saturation of an Electromagnet When a ferromagnetic material is used as the core of a solenoid, the amount of flux lines through the solenoid increases greatly. As the current through the coil is increased, the flux around each loop or turn of the coil also increases. This increase in the coil's magnetic field causes more of the domains of the core material to align themselves and contribute to the overall magnetic field. This process continues until all of the domains in the core have become aligned with the magnetic field. At this point we say that the core is *saturated* because it can contribute no additional flux lines even if the coil current is increased further. The concept of saturation will be discussed in more detail when we encounter *B-H* curves.

Toroidal and Rectangular Cores In many electromagnetic applications, a coil is wound around an iron core that is shaped so that the magnetic flux lines can follow a *complete* path through the low-reluctance iron. Two such examples are shown in Fig. 14–17. The core in part (a) of the figure is shaped in the form of a doughnut and is called a *toroid*, while the core in (b) is rectangular in shape. In both cases, when current is allowed to flow in the coil, the resulting magnetic lines of force will follow the path of the iron core and will not travel through the air outside the core. In other words, all of the flux lines will be concentrated inside the core material. Practically speaking, there will always be a few flux lines that do emerge from the core and travel a path that includes the surrounding air. This is called *leakage flux*.

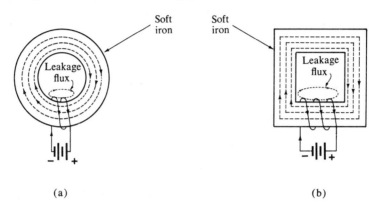

FIGURE 14–17 (a) *Toroid core.* (b) *Rectangular core.*

These types of cores find wide usage in such applications as transformers, magnetic amplifiers, and computer core memories. We will discuss them in more detail later in this chapter.

14.9 Magnetic Flux, Φ

Our discussion of magnetism and electromagnetism up to this point has been essentially a qualitative one. Starting in this section we will define the various important magnetic *quantities*, their basic units, and their interrelationships so that a quantitative feel for magnetism can be developed. In addition we will draw analogies between the magnetic quantities and their electrical counterparts.

The first magnetic quantity to be considered is one which we have been using all along in our discussions of magnetic fields, namely *magnetic lines of force* or *magnetic flux*. The letter symbol for the quantity of magnetic flux is the Greek letter Φ (phi). In the SI system of units, the basic unit for magnetic flux is the *weber*, abbreviated Wb. Thus, if we write

$$\Phi = 2 \text{ Wb}$$

this indicates that a flux of 2 webers is present.

A *weber* is actually a relatively large unit for flux. One weber equals 10^8 lines of flux. That is,

$$1 \text{ Wb} = 10^8 \text{ lines}$$

The microweber (μWb) is a smaller unit that is often used.

$$1 \ \mu\text{Wb} = 1 \times 10^{-6} \ (\text{Wb}) = 1 \times 10^{-6} \times (10^8 \ \text{lines})$$
$$= 10^2 \ \text{lines}$$
$$1 \ \mu\text{Wb} = 100 \ \text{lines}$$

A typical small magnet (1 lb) produces a flux of around 50 μWb, which corresponds to 5000 lines of flux.

Flux is Analogous to Electric Current We know that electric current must have a complete closed path in order to flow. Similarly, flux lines always form closed loops or paths. In electricity the paths for current constitute an *electric circuit*. Similarly, we can define the paths traced out by magnetic flux lines as *magnetic circuits*. Thus, we can state that

Magnetic flux follows a magnetic circuit in the same manner that current follows an electric circuit.

The analogy between current and magnetic flux is not an exact one because magnetic flux does not actually *flow*. However, the similarities are many and it will help us to a better understanding of electromagnetism.

14.10 Flux Density, B

In previous discussions it was stated that the strength of a magnetic field depends on the *concentration* of flux lines. That is, a magnetic field is stronger if its flux lines are concentrated into a smaller area. To illustrate, consider Fig. 14–18. In part (a) of the figure we have 10 flux lines passing through a cross-sectional area of 1 cm $\times$ 1 cm = 1 cm^2 = 10^{-4} m^2. In (b) we have 10 flux lines passing through a larger cross-sectional area of 2 cm $\times$ 2 cm = 4×10^{-4} m^2. Thus, the flux is more concentrated in (a) and therefore represents a stronger magnetic field than in (b).

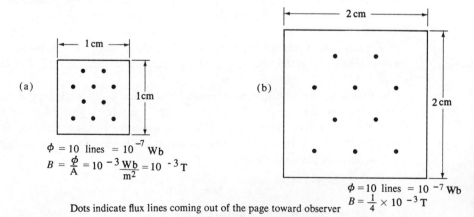

Dots indicate flux lines coming out of the page toward observer

FIGURE 14–18 *Difference between flux, Φ, and flux density, B.*

The flux concentration is more often called the *flux density* and represents the *amount of flux per unit of cross-sectional area*. The letter symbol for flux density is B. Since B equals flux per unit area, then it follows that

$$B = \frac{\Phi}{A} \qquad (14\text{--}1)$$

where Φ is the total flux in Wb passing through a cross-sectional area A in square meters (m²). From this relationship it is apparent that flux density is given in units of webers/square meter or Wb/m². However, the SI chooses to use *teslas* (T) as the basic unit for flux density. That is,

$$1 \text{ tesla} = 1 \text{ T} = 1 \text{ Wb/m}^2$$

We will use teslas and Wb/m² interchangeably.

EXAMPLE 14.1 Determine the flux density for each of the cases in Fig. 14–18.

Solution: (a) The flux is equal to 10 lines. We saw earlier that 100 lines equal 1 μWb. Thus, 10 lines represent 0.1 μWb. The cross-sectional area is 1 cm × 1 cm = 1 cm² = 10^{-4} m². Thus, using Eq. (14–1) we have

$$B = \frac{0.1 \text{ μWb}}{10^{-4} \text{ m}^2}$$

$$= \frac{10^{-7} \text{ Wb}}{10^{-4} \text{ m}^2} = 10^{-3} \frac{\text{Wb}}{\text{m}^2}$$

$$B = 10^{-3} \text{ T}$$

(b) Φ is the same, but A is now increased to 4 cm² = 4×10^{-4} m² so that

$$B = \frac{0.1 \text{ μWb}}{4 \times 10^{-4} \text{ m}^2} = \frac{1}{4} \times 10^{-3} \text{ T}$$

The same flux spread out over a larger area produces a smaller flux density.

Since the flux density (concentration) is a direct measure of magnetic strength, the flux density B is sometimes called the *magnetic field strength*. However, this term has fallen out of usage in order to avoid confusion with a similarly named magnetic quantity to be introduced later. We will use the term *flux density* throughout our coverage of magnetism.

Permanent magnets and electromagnets are often described in terms of their flux density. A bar magnet might have a flux density B equal to 10^{-3} Wb/m² = 10^{-3} T at its poles. The earth's magnetic field has a flux density roughly equal to 10^{-5} Wb/m² = 10^{-5} T. A large laboratory magnet could produce $B = 5$ Wb/m² = 5 T.

EXAMPLE 14.2 A certain bar magnet has a cross-sectional area of 9 cm². The flux density at the poles of this magnet is 0.0025 T. How much flux is emanating from this magnet's north pole?

Solution: Since $B = \Phi/A$, then it is also true that

$$\Phi = B \times A$$

In this example, $B = 0.0025$ T = 2.5×10^{-3} Wb/m² and

$A = 9$ cm² $= 9 \times 10^{-4}$ m². Thus,

$\Phi = 2.5 \times 10^{-3}$ Wb/m² $\times 9 \times 10^{-4}$ m² $= 2.25 \times 10^{-6}$ Wb
$= $ **2.25 μWb**

EXAMPLE 14.3 A certain air-core solenoid consists of 100 turns of wire wound around a cardboard cylinder. With 1 A of current through the coil, the flux produced in the center of the solenoid is only 0.4 μWb. Determine the flux density if the core has a diameter of 2 cm.

Solution: $A = \dfrac{\pi D^2}{4} = \dfrac{\pi \times (0.02 \text{ m})^2}{4}$

$= 0.000314$ m²

$= 3.14 \times 10^{-4}$ m²

$\Phi = 0.4$ μWb (given)

$B = \dfrac{\Phi}{A} = \dfrac{0.4 \times 10^{-6} \text{ Wb}}{3.14 \times 10^{-4} \text{ m}^2}$

$= 1.27 \times 10^{-3} \dfrac{\text{Wb}}{\text{m}^2} = $ **1.27 × 10⁻³ T**

EXAMPLE 14.4 If the cardboard core in Ex. 14.3 is replaced by a soft-iron toroidal core, the flux density increases by a factor of 1000. What will be the amount of flux in the toroidal core if it has the same cross-sectional area as the cardboard core?

Solution: The flux density has increased by 1000 times. This means that there is 1000 times as much flux passing through the same area. Thus, the new value for Φ will be 1000 times that which was present in the air-core solenoid.

$\Phi = 1000 \times 0.4$ μWb
$= $ **400 μWb** (40,000 lines)

Since flux Φ is analogous to current I, then flux density B is analogous to *current density*. Current density is a quantity that is not used very often in electrical circuit work. It is used quite often in the study of electromagnetic theory and is given the symbol J. By definition, $J = I/A$ representing current per unit area.

14.11 Magnetomotive Force, $\mathscr{F}$

We know that an electric current will flow in a circuit when a potential difference or voltage is applied to the circuit. This applied voltage is often called an *electromotive force*, emf. In a like manner, we have seen that magnetic flux lines will be produced in a magnetic circuit when a current is passed through a wire or coil. Furthermore, we have seen that an increase in the coil current or in the number of turns in the coil causes the flux to increase. We can say that the flux has been established by a *magnetomotive force*, mmf, that depends on the current and the number of turns of wire in the coil. In other words, magnetomotive force equals the product of *amperes* × *turns*. Stated formally

$$\mathcal{F} = N \times I \tag{14-2}$$

where $\mathcal{F}$ is the symbol for mmf, N is the number of turns, and I is the current in amperes. The basic units for mmf are ampere-turns, A-t, as defined by Eq. (14–2).

An mmf applied to a magnetic circuit produces flux in the same manner that an emf produces current in an electric circuit.

Thus, we can say that mmf is analogous to emf. Neither of them is actually a force in the strictest sense; instead they respectively represent the necessary applied energy needed to establish flux in a magnetic circuit and current in an electric circuit.

EXAMPLE 14.5 Calculate the magnitude of the mmf for each case in Fig. 14–19.

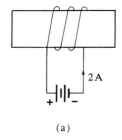
(a)

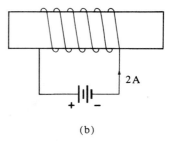
(b)

FIGURE 14–19

Solution: (a) In Fig. 14–19(a) there are *three* turns of wire wound around the core so that $N = 3$. Thus,

$$\begin{aligned}\mathcal{F} &= N \times I \\ &= 3 \text{ turns} \times 2 \text{ A} \\ &= 6 \text{ A-t}\end{aligned}$$

(b) Here there are *six* turns in the coil, which produces

$$\begin{aligned}\mathcal{F} &= 6 \times 2 \\ &= 12 \text{ A-t}\end{aligned}$$

14.12 Magnetic Field Intensity, H

Refer to Fig. 14–20 where two rectangular soft-iron cores of different lengths are shown. In each case, a coil of *five* turns is wound around the core and a current of *ten* amperes is flowing through the coil. Thus, for both cores the applied mmf is 50 A-t. Using the left-hand rule, the direction of flux will be cw for both cores.

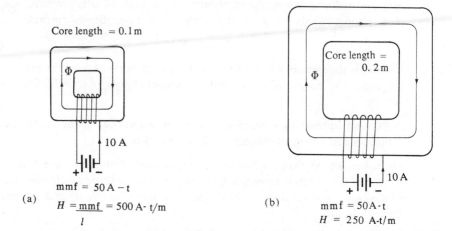

FIGURE 14-20 *The same mmf applied to cores of different magnetic path lengths produces different magnetic field intensities (H) and therefore different amounts of flux.*

The core in part (a) of the figure provides a magnetic circuit path length of 0.1 m while the core in part (b) provides a path length of 0.2 m. Both cores have the same cross-sectional area. This means that the flux produced in (a) travels a *shorter* path than the flux produced in (b). *For the same applied mmf, the flux will be greater for the core with the shorter length.*

This is analogous to the electric circuit situation where the same voltage (emf) is applied to two wires of different lengths. The shorter wire has *less* resistance and therefore will pass *more* current than the longer wire, for the same voltage.

Thus, the same mmf applied to cores of the same material and cross-sectional area, but of different lengths, will produce different amounts of flux Φ and flux density $B = \Phi/A$.

In order to take into account the magnetic path length, we can define a new magnetic quantity called *magnetic field intensity*, which will equal *mmf per unit length*. The symbol for magnetic field intensity is H. The defining relationship for H is

$$H = \frac{\mathcal{F}}{l} = \frac{NI}{l} \tag{14-3}$$

where $\mathcal{F}$ is the applied mmf and l is the path length in meters. The basic units for H can be determined from this expression as ampere-turns per meter (A-t/m).

EXAMPLE 14.6 Determine the values of H for the two cases in Fig. 14-20.

Solution: (a) For Fig. 14-20(a) we have

$$H = \frac{NI}{l} = \frac{(5 \times 10)\text{A-t}}{0.1 \text{ m}} = 500 \frac{\text{A-t}}{\text{m}}$$

(b) For Fig. 14–20(b) we have

$$H = \frac{NI}{l} = \frac{(5 \times 10)\text{A-t}}{0.2} = 250 \frac{\text{A-t}}{\text{m}}$$

The magnetic field intensity H is useful because it can be shown that *for the same H applied to cores of the same material, the flux density will be the same in each core*. This point will be discussed further in our investigation of magnetic permeability.

14.13 Magnetic Permeability, μ

When a certain value of H is applied to a core, this specifies how much magnetic field intensity is available to produce magnetic flux. The actual amount of flux which the applied H can produce depends on the core material. Every material has a characteristic called its *magnetic permeability*, which is a measure of how much *flux density*, B, can be established in it for a specified value of field intensity, H. Stated mathematically,

$$B = \mu \times H \qquad (14\text{–}4)$$

where μ is the letter symbol for magnetic permeability.

Equation (14–4) can also be written as

$$\mu = \frac{B}{H} \qquad (14\text{–}5)$$

which shows that μ is the ratio of the flux density to the magnetic field intensity that produces it. A larger value of permeability indicates a greater flux density for a given H. The units for μ can be determined from (14–5) as

$$\text{units for } \mu = \frac{B}{H} = \frac{\text{Wb/m}^2}{\text{A-t/m}} = \frac{\text{T}}{\text{A-t/m}}$$

The permeability of air and most nonmagnetic materials has a value of $4\pi \times 10^{-7} = 1.26 \times 10^{-6}$ and is given the symbol μ_o. Thus,

$$\mu_o = 1.26 \times 10^{-6} \frac{\text{T}}{\text{A-t/m}}$$

EXAMPLE 14.7 A 10-cm long air-core coil consists of 200 turns of wire. What will be the flux density in the coil if a current of 0.25 A flows in the coil?

Solution: $H = \dfrac{NI}{l} = \dfrac{200 \text{ t} \times 0.25 \text{ A}}{0.1 \text{ m}}$

$= 500$ A-t/m

$$B = \mu_o H$$
$$= 1.26 \times 10^{-6} \frac{T}{A\text{-}t/m} \times 500 \text{ A-t/m}$$
$$B = 6.3 \times 10^{-4} \text{ T}$$

Note that the A-t/m units cancel, leaving the units of teslas (T) for the flux density.

Relative Permeability, μ_r The permeability of magnetic materials can range from 50 to 80,000 times that of air. The *relative* permeability μ_r of a magnetic material is simply the ratio of its absolute magnetic permeability, μ, to the permeability of air, μ_o. That is,

$$\mu_r = \frac{\mu}{\mu_o} \tag{14-6}$$

For example, the relative permeability of cobalt is 60, which means it has a permeability of 60 times that of air. μ_r has no units but is a pure number.

EXAMPLE 14.8 If the core in Fig. 14–20(a) is made of silicon iron ($\mu_r = 7000$), determine the flux density in the core.

Solution: The value of H in Fig. 14–20(a) was previously determined to be 500 A-t/m. To determine B we have to know the value of μ. Since

$$\mu_r = \frac{\mu}{\mu_o}$$

then we have,

$$\mu = \mu_r \times \mu_o$$
$$= 7000 \times 1.26 \times 10^{-6}$$
$$= 8.8 \times 10^{-3} \frac{T}{A\text{-}t/m}$$

Thus,

$$B = \mu H$$
$$= 8.8 \times 10^{-3} \times 500$$
$$= 4.4 \text{ T}$$

The greater the permeability of a material, the better it is as a "conductor" of magnetic flux lines, and vice versa. Thus, μ is analogous to electrical conductivity, σ (Greek letter "sigma"). σ is the *reciprocal* of resistivity ρ (see chap. 12) and is a measure of a material's ability to conduct electrical current. A low resistivity ρ means a high conductivity σ and a better conductor of electricity.

14.14 B-H Curves

The value of permeability for air and other nonmagnetic materials is essentially constant at a value equal to μ_o. The permeability of ferromagnetic substances, however, is not constant but can vary as the applied H varies. In other words, the flux density in such materials is not *directly* proportional to the magnetic field intensity. This is much like the situation with a *nonlinear* resistor where the current is not directly proportional to the applied voltage.

For such materials it is often helpful to plot a curve showing how flux density B varies with magnetic field intensity H. Such a curve is called a *B-H curve* (or *magnetization curve*). An example of a B-H curve for cast iron is shown in Fig. 14–21. The value of H is easily varied by varying the current in the coil (by varying R). The resultant values of flux density B produce the B-H curve as shown.

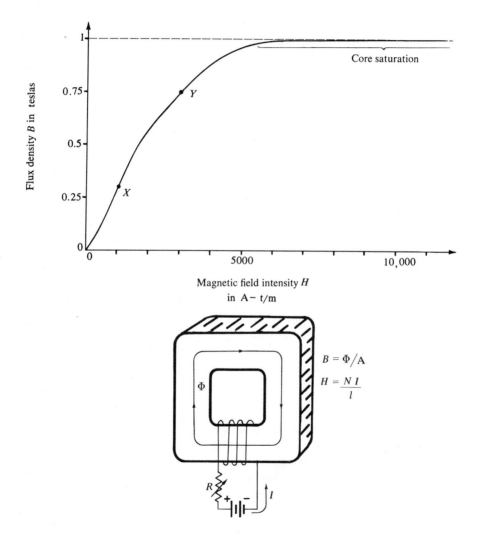

FIGURE 14–21 B-H *magnetization curve for cast iron.*

It is apparent that the curve is nonlinear, indicating that B does not vary in proportion to H. For example, the curve shows that $B = 0.3$ T, when $H = 1000$ A-t/m (see point x on the curve). If H is *tripled* in value to 3000 A-t/m, the value of B does not triple but increases to only 0.75 T (point y).

360 Chapter Fourteen

Because the B-H curve for cast iron is nonlinear, the permeability μ will be different at different points on the curve. For example,

$$\mu(\text{at point } x) = \frac{B}{H}$$

$$= \frac{0.3 \text{ T}}{1000 \text{ A-t/m}}$$

$$= 3 \times 10^{-4} \frac{\text{T}}{\text{A-t/m}}$$

$$\mu(\text{at point } y) = \frac{0.75 \text{ T}}{3000 \text{ A-t/m}} = 2.5 \times 10^{-4} \frac{\text{T}}{\text{A-t/m}}$$

The value of permeability decreases for higher values of H because B does not increase proportionally with H. In fact, at values of H above 5000 A-t/m, approximately, the value of B increases at a very slow rate, as indicated by the relatively flat slope of the curve. This effect whereby a small change in flux density occurs for a large change in field intensity is called *saturation*. As we discussed earlier, saturation occurs when most of the magnetic domains of the material align themselves with the applied field so that very little additional flux can be produced by increasing the applied mmf.

Many electronic applications of magnetic materials, such as transformers, inductors, speakers, and magnetic amplifiers, require that the core material does not saturate or even operate near saturation. This is because such applications require that the core's flux variations should follow the current variations (H variations). In these applications the core is operated on the linear portion of its B-H curve.

14.15 Hysteresis

The B-H curve in Fig. 14–21 shows us only part of the behavior of ferromagnetic materials. We are now going to take a closer look at B-H curves and the phenomenon of *hysteresis*.

Consider the B-H curve in Fig. 14–22(a). Note that it is a closed curve which is symmetrical about the origin. Also, note that negative as well as positive values of H and B are represented on the graph, the negative values representing a *reversal* of current in the coil and the accompanying *reversal* of the magnetic field.

The section of the curve 0 *abc* is similar to the B-H curve in Fig. 14–21, and represents the curve which would be followed by an initially unmagnetized core. If the core is initially unmagnetized and no current is supplied to the coil, then $H = 0$ and no net flux is produced in the core because the magnetic domains in the core are randomly oriented. This is illustrated in part (b) of the figure.

If current is supplied to the coil and is gradually increased, the flux density B in the core will follow the curve 0 *abc* until H reaches 3000 A-t/m or more and the core is saturated [see part (c) of the figure].

At point c on the curve, all the domains in the core are aligned with the applied magnetic field so that B is at its maximum value.

If the current is now gradually reduced, H will return to zero. However, the flux density B does not return along the original path *cba* 0 but follows the

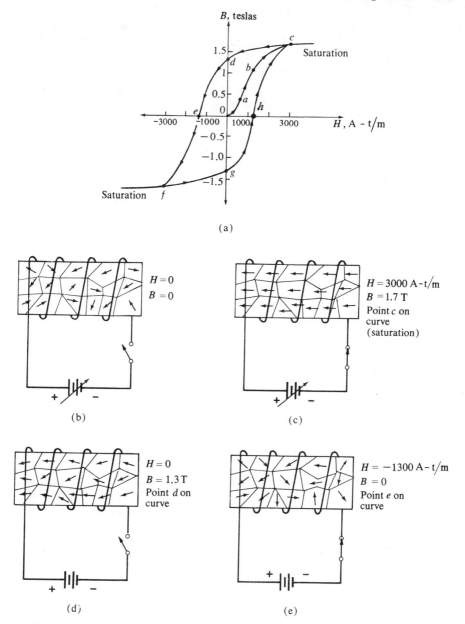

FIGURE 14-22 *Following a hysteresis curve.*

path to point *d* instead. In other words, when *H* is returned to zero, there is still a significant flux density in the core (*B* = 1.3 T at point *d*) representing a *residual* or *remanent magnetism*. The value of *B* at point *d* is called the *remanent flux density*, B_R. For this curve, $B_R = 1.3$ T.

The remanent magnetism is due to the fact that, once aligned, the magnetic domains do not return exactly to their original positions [Fig. 14-22(d)] when the magnetizing force (*H*) is removed. It's as if the domains were forced to move against an internal friction among the atoms of the magnetic material.

This effect is called *hysteresis*. The word *hysteresis* means "to lag behind," which describes how the change in B lags behind the change in H.

We can now reverse the current through the coil [see part (e) of the figure]. Since the current is flowing through the coil in the opposite direction, it will attempt to magnetize the core in the opposite direction. For these negative values of H, as we increase the current the flux density follows the curve from d to e. At point e, the magnetic field intensity has become large enough in the negative direction to nullify the remanent magnetism that was present at point d. This value of H is called the coercive force, H_c. For this curve $H_c = -1300$ A-t/m.

If the reverse current is increased beyond this point, the domains become aligned in the reverse direction (right to left) until the core is saturated in the reverse direction (point f on the curve). If the reverse current is then reduced to zero, the core will again retain some remanent magnetism (point g on the curve), but in the opposite polarity.

Finally, we can reverse the current back to its original direction. Increasing the current will then bring the core back to a demagnetized condition (point h), and eventually to saturation at point c. Thus, the complete curve for this ferromagnetic material is a closed curve and is called a *hysteresis curve* or *loop*.

Hysteresis Loss The hysteresis of a ferromagnetic substance is due to some of the domains not wanting to return to their original orientation and having to be forced to by a reversed magnetic field intensity ($-H_c$). Thus, energy is required to rotate these domains and this energy shows up as heat (sort of an internal friction). If an *alternating current* (ac) is applied to the coil, then the energy is taken from the ac source which is responsible for the domains having to rotate. The higher the frequency of the ac in the coil, the more rapidly the domains have to rotate and, therefore, the greater the *hysteresis energy loss*.

It turns out that the amount of energy loss in one complete magnetization cycle (once around the hysteresis loop) is measured by the area inside the loop. In applications where current in the coil will be continuously or often changing, such as in transformers or electromagnets, it is desirable that the hysteresis loss be kept to a minimum. There is no ideal ferromagnetic material that has no hysteresis loss. However, very soft and pure iron has a narrow hysteresis loop with a low remanent flux density (B_R) and a small coercive force (H_c) for demagnetization. Thus, it is commonly used in such applications. Its B-H curve is like the one in Fig. 14–21.

On the other hand, if we wish to construct a permanent magnet, we look for a ferromagnetic material with a high residual magnetism requiring a high coercive force to demagnetize it. Permanent magnets are made from a wide variety of materials, but all contain some form of ferromagnetic substance. Some of the popular permanent magnetic materials include Alnico 5 (an alloy of aluminum, nickel, cobalt, and copper) and hard cobalt steel.

In some applications a permanent magnet is desired which is not a metallic material, as in cases where a high electrical resistance is required. In such applications a *ceramic* magnet material can be used. This relatively new type of material is made from the oxides of certain metals (iron, manganese, magnesium). These oxides are called *ferrites*. Ferrite materials find useful application in memory banks of many modern digital computers.

14.16 Ohm's Law for Magnetic Circuits

As discussed earlier, magnetic flux Φ is analogous to electric current I, and magnetomotive force $\mathcal{F}$ is analogous to electromotive force (voltage) E. In an electric circuit, for a given emf, the current depends on the circuit's *resistance* R. Likewise in a magnetic circuit, for a given mmf, the flux depends on the circuit's opposition. For a magnetic circuit, this opposition is called *reluctance* and is given the symbol $\mathcal{R}$ (script R).

For a resistive electric circuit, we have shown that $I = E/R$, corresponding to Ohm's law. A similar law may be expressed for magnetic circuits as follows:

$$\Phi = \frac{\mathcal{F}}{\mathcal{R}} \tag{14-7}$$

This says that flux is *directly* proportional to mmf (the stimulus producing the flux) and *inversely* proportional to reluctance, which is the opposition to flux lines.

The units for reluctance can be established by rearranging (14–7):

$$\mathcal{R} = \frac{\mathcal{F}}{\Phi} \tag{14-8}$$

which indicates A-t/Wb as the units for reluctance.

EXAMPLE 14.9 A coil with 250 turns is wound around an iron core. With a coil current of 2 A, the flux in the core is 100 μWb. Calculate the reluctance of the core.

Solution:
$$\mathcal{R} = \frac{\text{mmf}}{\Phi} = \frac{N \times I}{\Phi}$$
$$= \frac{(250 \times 2) \text{ A-t}}{100 \times 10^{-6} \text{ Wb}} = 5 \times 10^6 \text{ A-t/Wb}$$

EXAMPLE 14.10 The same coil with an air core produces a flux of only 0.2 μWb. What is its reluctance?

Solution: Since Φ is only 0.2 μWb, a decrease by a factor of 500, the reluctance must be 500 times greater. Thus,

$$\mathcal{R} = 500 \times 5 \times 10^6 \text{ A-t/Wb} = 2.5 \times 10^9 \text{ A-t/Wb}$$

These examples emphasize the fact that ferromagnetic materials have a much *lower* reluctance than nonmagnetic materials.

Factors Determining $\mathcal{R}$ For a given magnetic circuit, the reluctance or opposition to flux lines depends on the geometrical dimensions of the circuit and on the material in the circuit. These are the same factors which determined the resistance of a resistor. The relationship between reluctance and these factors is summarized in the formula below:

$$\mathcal{R} = \frac{1}{\mu} \times \frac{l}{A} \tag{14-9}$$

where μ is the permeability of the material, l is the length of magnetic path in meters, and A is the cross-sectional area of the path in square meters.

As with resistance, the reluctance is proportional to length and inversely proportional to cross-sectional area. It is also interesting to note that $\mathcal{R}$ is inversely proportional to permeability, μ. Recall that μ is a *characteristic of the material* and is greater for ferromagnetic materials. Thus, a larger μ would result in a lower reluctance for a given circuit. Reluctance is a characteristic of the magnetic circuit and depends on μ as well as the circuit dimensions.

EXAMPLE 14.11 Consider the magnetic circuit in Fig. 14–23. The core has an average path length of 20 cm and a cross-sectional area of 5 cm². The core material is soft iron with $\mu_r = 500$. Calculate the core's reluctance and determine the coil current needed to produce $\Phi = 10\ \mu\text{Wb}$.

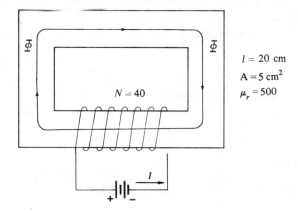

FIGURE 14–23

Solution: (a) First of all, the absolute permeability of the iron core must be calculated.

$$\mu = \mu_r \times \mu_o$$
$$= 500 \times 1.26 \times 10^{-6}$$
$$= 6.3 \times 10^{-4}\ \frac{\text{T}}{\text{A-t/m}} \quad (14\text{-}6)$$

Now we can calculate reluctance as

$$\mathcal{R} = \frac{l}{\mu A}$$

$$= \frac{0.2\ \text{m}}{6.3 \times 10^{-4}\ \frac{\text{T}}{\text{A-t/m}} \times 0.0005\ \text{m}^2}$$

$$= \mathbf{6.35 \times 10^5\ A\text{-}t/Wb}$$

(b) Since $\Phi = \text{mmf}/\mathcal{R}$, then we can write

$$\text{mmf} = \Phi \times \mathcal{R}$$
$$= (10 \times 10^{-6} \times 6.35 \times 10^5)\text{A-t}$$
$$= 6.35\ \text{A-t}$$

Since mmf = $N \times I$ and $N = 40$ (given in Fig. 14–23), then

$$40 \times I = 6.35 \text{ A-t}$$

or

$$I = 0.159 \text{ A}$$

Effect of an Air Gap in Magnetic Circuits In many applications of magnetic circuits, the lines of force are not always traveling through a low-reluctance ferromagnetic material. Magnetic circuits which are part of motors, generators, relays, loudspeakers, etc., usually consist of low reluctance ferromagnetic material. However, the flux lines in these circuits often have to cross an *air gap* to complete their loops. This gap is necessary in the case of motors, meters, and loudspeakers in order to permit space in which moving parts can move. Even in the case of stationary magnetic equipment, an air gap may be purposely inserted into the magnetic circuit. The purpose here is to *increase* the total reluctance of the circuit and thereby prevent the ferromagnetic portion of the circuit from saturating when a large current flows through the coil.

Figure 14–24(a) is an example of a magnetic circuit which includes an air gap. The core is the same one in Fig. 14–23 except that a small gap of 0.1 cm has been cut into it. Note that any flux which is produced in the core by an applied mmf must also flow through the air gap. We can say that the *air gap is in series with the ferromagnetic core*. As might be expected, we can treat this *series magnetic circuit* in much the same way that we treat a *series electrical circuit*.

Since Φ must travel through the core and the air gap, then the total reluctance in the circuit path is equivalent to the reluctance of the core *plus* the reluctance of the gap. We can represent this schematically as shown in Fig. 14–24(b) just like a series electric circuit. The mmf (equal to $N \times I$) is like the applied voltage. $\mathcal{R}_{core}$ is the reluctance of the iron core, $\mathcal{R}_{gap}$ is the reluctance of the air gap; they are like series resistors. Carrying the analogy further, we can say that

$$\Phi = \frac{\text{mmf}}{\text{total reluctance}}$$
$$= \frac{N \times I}{\mathcal{R}_{core} + \mathcal{R}_{gap}} \qquad (14\text{–}10)$$

Thus, in order to determine Φ for a given mmf we have to know the reluctance of both the core and the air gap.

EXAMPLE 14.12 For the magnetic circuit in Fig. 14–24(a) determine: (a) the reluctance of the core, (b) the reluctance of the air gap, (c) the amount of current needed to produce a flux of 10 μWb.

Solution: (a) The reluctance of the iron core was determined in Ex. 14.11 to be 6.35×10^5 A-t/Wb when the core was 20-cm long. The 0.1-cm air gap removed from the core will not change this value significantly. Thus,

$$\mathcal{R}_{core} \simeq 6.35 \times 10^5 \text{ A-t/Wb}$$

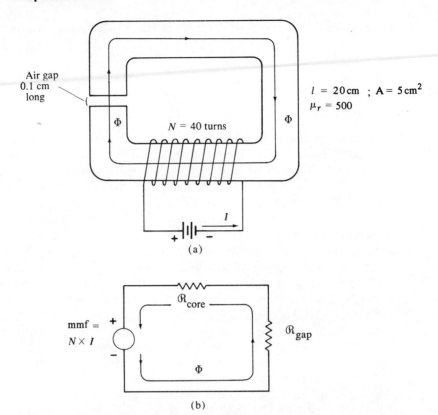

FIGURE 14-24 *Magnetic circuit with air gap can be treated like a series circuit.*

(b) To find $\mathcal{R}_{gap}$ we use (14-9) with

$$\mu_o = 1.26 \times 10^{-6}, l = 0.1 \text{ cm} = 10^{-3} \text{ m, and}$$
$$A = 5 \text{ cm}^2 = 5 \times 10^{-4} \text{ m}^2$$

Thus,

$$\mathcal{R}_{gap} = \frac{10^{-3}}{1.26 \times 10^{-6} \times 5 \times 10^{-4}}$$
$$= \mathbf{1.59 \times 10^6 \text{ A-t/Wb}}$$

(c) Rearranging (14-10) we have

$$\begin{aligned}
\text{mmf} &= \Phi \times \text{total reluctance} \\
&= 10 \times 10^{-6} \text{ Wb} \times (6.35 \times 10^5 + 1.59 \times 10^6) \text{ A-t/Wb} \\
&= 10^{-5} \times (0.635 + 1.59) \times 10^6 \text{ A-t} \\
&= 22.3 \text{ A-t}
\end{aligned}$$

Thus,

$$N \times I = 22.3 \text{ A-t}$$

or

$$I = \frac{22.3}{40} = \mathbf{0.577 \text{ A}}$$

This compares to a current of only 0.159 A needed to produce the same flux in the core *without* the air gap (Ex. 14.11). Thus, even a very small air gap significantly increases the reluctance of a magnetic circuit.

Chapter Summary

1. A magnet can attract magnetic materials by contact or at a distance.
2. A permanent magnet retains its magnetic force indefinitely; a temporary magnet is magnetized only as long as it is under the influence of another magnetic force.
3. Artificial permanent magnets are made from magnetic materials by the process of induction (exposing the magnetic material to a magnetic force).
4. The north pole of a magnet is that pole which points north when the magnet is made free to rotate. The south pole is the opposite pole.
5. Unlike magnetic poles attract each other; like magnetic poles repel each other.
6. Magnetic force is a field force that can be represented by lines of force called magnetic flux.
7. Magnetic lines of force always form continuous loops that leave a magnet at its north pole and enter a magnet at its south pole.
8. The concentration of magnetic flux lines is a direct measure of the strength of a magnetic field.
9. Flux lines never intersect.
10. Flux lines pointing in the same direction tend to repel each other.
11. Any substance which is noticeably affected by a magnetic field is said to be ferromagnetic.
12. Substances that are slightly repelled by a strong magnetic field are called *diamagnetic;* those that are slightly attracted by a strong magnetic field are called *paramagnetic*. Diamagnetic and paramagnetic materials are considered to be nonmagnetic.
13. Ferromagnetic materials can be viewed as consisting of many small atomic magnets called domains.
14. According to the domain theory of magnetism the domains in a ferromagnetic material always try to align themselves with any externally applied magnetic field.
15. Magnetic lines of force prefer to take the path of least reluctance. Reluctance is low for magnetic materials, high for nonmagnetic materials.
16. An electric current produces a magnetic field that consists of concentric flux loops that lie in a plane which is perpendicular to the current path and have a direction that can be determined using the left-hand rule (Fig. 14-11).
17. Magnetism which is produced by the flow of current is called electromagnetism.

18. The lines of force produced around a current-carrying wire can be concentrated into a smaller area by bending the wire into a loop or, better still, into several loops wound around a core (solenoid).

19. A current-carrying solenoid produces a flux pattern similar to a bar magnet and is therefore called an electromagnet. Its magnetic polarity can be determined using the left-hand rule for solenoids (Fig. 14-15).

20. The strength of an electromagnet can be increased by using a ferromagnetic material as the core.

21. As the coil current of an electromagnet increases, more flux lines are produced and the strength of the magnet increases until the core becomes saturated. Core saturation occurs when all of the magnetic domains in the core material have aligned themselves with the coil's magnetic field.

22. Table 14-1 summarizes the various magnetic quantities.

23. Magnetic flux, Φ, is analogous to electric current.

24. Flux density B is equal to flux per unit area, and is analogous to electric current density.

25. An mmf applied to a magnetic circuit produces flux in the same manner that an emf produces current in an electric circuit.

26. Magnetic field intensity, H, is equal to mmf per unit length of the magnetic circuit.

TABLE 14-1

Quantity	Symbol	Units	Electrical Analog
Flux	Φ	Weber (Wb)	I
Flux density	$B = \Phi/A$	Wb/m² = teslas (T)	$J = I/A$
Magnetomotive force	$\mathcal{F}$	Ampere-turns (A-t)	emf, voltage
Field intensity	$H = \mathcal{F}/l$	A-t/m	$\mathcal{E}$; field strength
Permeability	$\mu = B/H$	T/(A-t/m)	σ; conductivity
Relative μ	$\mu_r = \mu/\mu_o$	none, pure number	
Reluctance	$\mathcal{R} = \mathcal{F}/\Phi$	A-t/Wb	R
Permeance	$\rho = 1/\mathcal{R}$	Wb/A-t	G

27. Every substance has a magnetic permeability, μ, which is a measure of the substance's ability to establish a magnetic flux, and is analogous to electrical conductivity σ.

28. All magnetic materials have nonlinear B-H curves because of the saturation effect that occurs as the applied H is increased.

29. Hysteresis in a ferromagnetic material is caused by some of the domains not wanting to return to an unmagnetized orientation and having to be forced to by a reversed magnetic field intensity.

30. Hysteresis energy loss occurs because of the energy required to rotate the domains in a ferromagnetic material.

31. Ohm's law for a magnetic circuit can be stated as Φ = mmf/total reluctance.

32. An air gap in a magnetic circuit acts as a large reluctance in series with the low reluctance of the ferromagnetic core.

Questions/Problems

Sections 14.2–14.3

14–1 Name some magnetic substances.

14–2 Explain the difference between natural and artificial magnets.

14–3 Name some nonmagnetic substances.

14–4 How would you determine the poles of a magnet by using a magnetic compass?

14–5 Why is the earth's North Pole considered to be a south magnetic pole?

14–6 Sketch the arrangement of magnetic lines of force around a bar magnet.

14–7 Repeat problem 14–6 for a horseshoe magnet.

Sections 14.4–14.5

14–8 Indicate which of the following statements are true:
   a. Magnetic flux lines have direction.
   b. Magnetic flux lines cannot pass through all materials.
   c. Magnetic poles are regions of a magnet where the flux lines are most concentrated.
   d. Unlike poles repel each other.
   e. Flux lines begin at a certain point and terminate at a certain point.
   f. Flux lines do not intersect each other.
   g. A magnet which produces a greater amount of flux *always* has a stronger magnetic force.
   h. The strength of a magnet is independent of distance from the magnet.
   i. Diamagnetic materials are strongly magnetic.
   j. Paramagnetic materials are weakly attracted by a magnetic field.
   k. Ferromagnetic domains exist only in magnetic materials.
   l. Flux lines can be seen through a microscope.

14–9 Explain the domain theory of magnetism.

14–10 Use the domain concept to explain how a soft iron bar behaves when it is exposed to a magnetic field. What happens when the magnetic field is removed?

14–11 Repeat problem 14–10 for a steel bar which becomes a permanent magnet.

14–12 Show how the flux pattern of a horseshoe magnet will be affected if a small iron bar is placed between the poles of the magnet.

14–13 Repeat problem 14–12 for a small glass bar.

Section 14.6

14–14 Define electromagnetism.

14–15 Comment on the truth or falsity of each of the following statements concerning the magnetic field around a current-carrying wire:
   a. The flux lines are concentric circles.

b. The strength of the magnetic field depends *only* on the distance from the wire.
c. The direction of the flux lines will always be in a clockwise direction.
d. There will be no magnetic field if the current is zero.
e. The *amount* of flux will increase if the wire is bent into a loop.

14–16 Consider the circuit in Fig. 14–25. Determine the direction of the magnetic field around wire *A-B* and wire *C-D*. Sketch the flux lines around each wire.

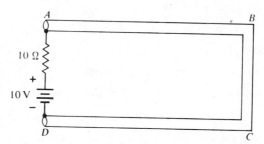

FIGURE 14–25

14–17 Figure 14–26(a) shows the cross-sectional view of a current-carrying wire. Sketch the lines of force around the wire.

FIGURE 14–26

14–18 Consider again the circuit in Fig. 14–25. Imagine that you were an observer sitting at point *D* looking toward point *C*. Use the dot-cross convention and sketch the cross-sectional view of the wire *C-D* and its flux pattern.

14–19 Repeat problem 14–18 for wire *A-B* as viewed from point *A*.

14–20 Use the rippling water analogy to help explain the change in flux as the current in a wire increases.

Sections 14.7–14.8

14–21 Each of the coils in Fig. 14–27 is wound around a cardboard core. Sketch the lines of force for each solenoid and determine the magnetic poles.

14–22 In Fig. 14–28 the magnetic poles of the electromagnets (solenoids) are given. Determine the polarity of the battery in each case.

14–23 Explain what happens to a solenoid's magnetic field when a ferromagnetic core is used to wind the coil around.

14–24 Why is soft iron a desirable core material for electromagnets?

14–25 Use the domain theory to explain how a steel bar can be turned into a permanent magnet by electromagnetism.

14–26 Two identical wire coils are wound around two cores, one made of soft iron and

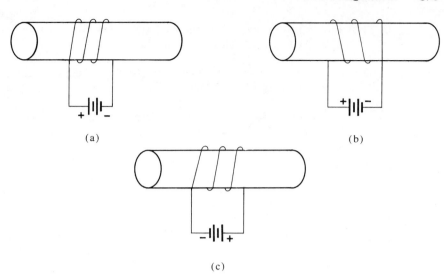

FIGURE 14-27

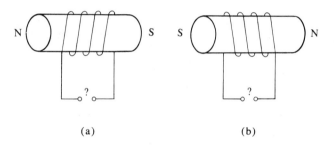

FIGURE 14-28

the other of nickel. For the same coil current, the iron-core solenoid produces *seven* times as much flux as the nickel-core solenoid
a. Which solenoid is a stronger electromagnet?
b. Which material has a *greater* reluctance, iron or nickel?
c. How can the nickel-core solenoid be made to produce more flux without changing the core material?

14-27 Explain magnetic *saturation* in an iron-core electromagnet.

14-28 Sketch a toroidal-core solenoid and its associated magnetic field.

Sections 14.9–14.10

14-29 A certain magnet produces 8,000 flux lines. How much is the flux Φ in webers and microwebers?

14-30 How is magnetic flux Φ similar to electric current?

14-31 A horseshoe magnet has poles with a cross-sectional area of 4 cm². If the magnet produces 6000 μWb, calculate the flux density B.

14-32 A certain electromagnet has a cross-sectional area of 10 cm² and produces a flux density $B = 0.2$ T. How much flux is produced?

14-33 Comment on the truth or falsity of each of the following statements:
  a. A magnet with a flux of 2000 μWb has more strength than one with a flux of 100 μWb.
  b. Flux density decreases as distance from the magnetic pole increases.
  c. The only way to increase B is to increase the amount of flux.

Sections 14.11–14.12

14-34 What is the magnetic quantity which is analogous to emf?

14-35 A 50-turn coil is wound around a soft-iron core. If 100 mA is passed through the coil, what is the mmf?

14-36 If the current in problem 14-35 is decreased to 40 mA, how can the mmf be maintained at the same value?

14-37 Which of the following cases represents the largest mmf:
  a. a 20-turn coil wound around a 10-cm iron core and passing a current of 3 A
  b. a 60-turn coil wound around a 20-cm iron core and passing a current of 1 A
  c. a 100-turn coil wound around a 30-cm cardboard core and passing a current of 0.75 A

14-38 Which of the cases in problem 14-37 represents the largest magnetic field intensity, H?

14-39 What is wrong, if anything, with the following statement: "Magnetic field intensity, H, is always greatest when the applied mmf is greatest."

Sections 14.13–14.15

14-40 (a) What will be the flux density for the core in question 14-37(c)? (b) Repeat for the core in 14-37(a) using $\mu_r = 500$ for iron.

14-41 Refer to the B-H curve in Fig. 14-21. Calculate μ and μ_r when $H = 4000$ A-t/m.

14-42 A certain cast-iron core has a length of 10 cm and a cross-sectional area of 6 cm². A coil of 150 turns is wound around the core and is passing a current of 4 A. Calculate: (a) mmf; (b) H; (c) B, using Fig. 14-21; and (d) Φ in the core.

14-43 Explain why the B-H curve in Fig. 14-21 flattens out for large values of H.

14-44 The *amount* of flux density B in an electromagnet depends on which of the following:
  a. the magnetic properties of the material
  b. the cross-sectional area of the magnet
  c. the applied mmf
  d. the length of the magnet
  e. the direction of the windings

14-45 Repeat problem 14-44 for flux Φ.

14-46 What causes hysteresis in a ferromagnetic substance?

14-47 A certain ferromagnetic material has a saturation flux density of 2 T and a remanent flux density (B_R) of 1 T. Another material has a saturation flux density of 1.5 T, and a B_R of 1.2 T. Which one would make a stronger permanent magnet?

14-48 Consider the hysteresis loops sketched in Fig. 14-29. Which one would be *most* desirable for use in a transformer? Which one would be *most* desirable in a permanent magnet?

14-49 Which curve in Fig. 14-29 represents the greatest hysteresis energy loss?

Section 14.16

14-50 Calculate the reluctance of the core in problem 14-42 first using Eq. (14-8) and then using (14-9).

14-51 A certain silicon-iron core has a length of 15 cm and a cross-sectional area of 3 cm². A coil of 200 turns and a current of 0.3 A are wound around the core. Calculate: (a) mmf; (b) the core's reluctance; (c) Φ in the core; (d) B in the core.

14-52 Repeat problem 14-51 but with a 0.25-cm air gap in the core.

14-53 Refer to the magnetic circuit in Fig. 14-30. A flux density of about 8 T is established in the core. For each of the following changes, indicate the effect each will have on the *flux density:*
  a. an increase in N
  b. a decrease in cross-sectional area A.
  c. an increase in core length
  d. a decrease in E_S
  e. an increase in μ_r of the core
  f. doubling N and doubling the core length

14-54 Repeat problem 14-53 for the effects on the core's *flux* Φ.

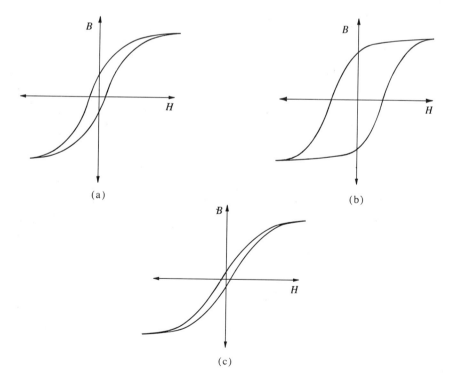

FIGURE 14-29

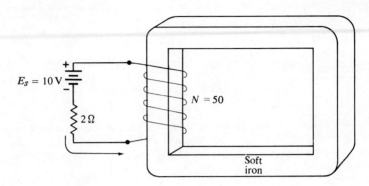

FIGURE 14-30

CHAPTER 15

Inductance

15.1 Introduction

After Oersted and others had shown that magnetism could be produced by electricity, many scientists began to look for the reverse effect whereby electric current could be produced by magnetism. Michael Faraday, in particular, undertook a series of experiments which led to the basic principles of *electromagnetic induction*. The importance of these principles is emphasized by the fact that generators, motors, inductors, transformers, and a host of other devices depend on electromagnetic induction effects.

In this chapter, we will examine the basic principles of electromagnetic induction. Then we will see how the electrical circuit property of *inductance* is produced and how it affects the flow of current in a circuit. After this chapter is completed, we will have become familiar with the three basic circuit properties: *resistance, capacitance, and inductance*.

15.2 Electromagnetic Induction

Figure 15-1 shows an arrangement for demonstrating electromagnetic induction. It is very similar to one used by Faraday in one of his basic experiments. The air-core coil is shown connected directly to a sensitive microammeter with no voltage source in the circuit. Under normal circumstances we would not expect any current to be flowing through the coil. Faraday's experiment consisted of pushing a bar magnet (shown in the figure) in and out of the coil. He found that a momentary pulse of current was registered on the meter whenever he moved one pole of the magnet quickly toward the coil. When he jerked the magnet away from the coil, a pulse of current was produced in the *opposite* direction, as indicated by a meter deflection in the reverse direction. The same effects occurred if the magnet was held stationary, and the coil was moved toward or away from the magnet. No current was observed as long as the coil and magnet were both held stationary.

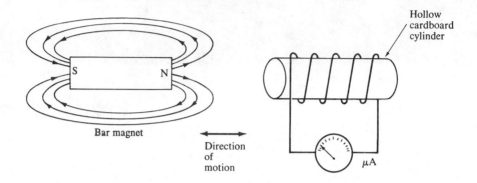

FIGURE 15-1 *Faraday's demonstration of electromagnetic induction.*

Faraday made some further observations. He noticed that the *magnitude* of the current pulse registered by the meter depended on the *speed* at which the magnet moved toward or away from the coil. He also found that the current pulse increased if the *number of turns* of wire on the coil was increased or if the *strength* of the bar magnet was increased. Faraday also noted that the *direction* of the coil current depended on which way the magnet was moving and also on which pole of the magnet was pushed into the coil. When the south pole of the magnet was moved toward the coil, the meter registered a current in the opposite direction.

It should be re-emphasized that in all cases the current in the coil persisted only momentarily when the magnet was in motion. When the magnet was held stationary either outside the coil or inside the coil, *there was no current flow*.

The astonishing result of this experiment and other similar ones performed by Faraday and his contemporaries is that a current was made to flow in a coil of wire with no *apparent* source of voltage present. Obviously, if a current is flowing in a wire, some source of electromotive force (emf) must have caused it. Faraday concluded from all of his experiments that an emf is *induced* in a loop or coil of wire whenever the number of magnetic lines of force (flux) passing through the loop was *changing*. If the circuit is closed (e.g., by connecting a current meter) the induced emf gives rise to a current flow.

Flux Linkages In his explanation of electromagnetic induction, Faraday utilized the concept of *flux linkages*. Whenever *one* flux line passes through or links *one* loop of wire, it is called *one* flux linkage. Figure 15-2(a) shows an example of a single flux line linking a single loop of wire. This represents one flux linkage. If two flux lines linked this single loop of wire it would represent two flux linkages. In part (b) of the figure, a single flux line is shown linking *two* loops or turns of wire. This also represents two flux linkages.

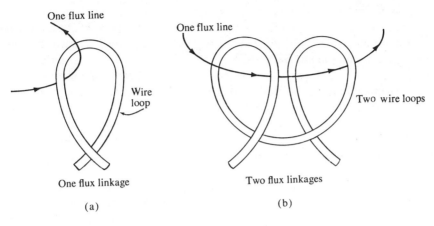

FIGURE 15-2 *Flux lines linking loops of wire (flux linkages).*

15.3 Faraday's Law

On the basis of his observations, Faraday proposed the following principle, which has become known as

Faraday's law of electromagnetic induction: An emf is induced in a loop or coil of wire whenever the number of flux linkages is changing; the magnitude of the emf is proportional to the rate at which the number of flux linkages is changing.

Let us refer to Fig. 15-3(a) and apply Faraday's law to the situation presented there.

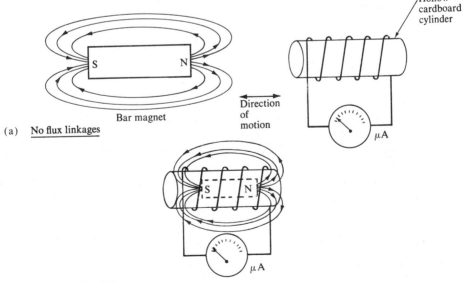

FIGURE 15-3 (a) *Bar magnet outside coil; no flux linkages.* (b) *Bar magnet all the way inside the coil produces large number of flux linkages.*

When the magnet is held far enough away from the coil, none of the flux lines from the magnet will *link* any of the loops of the coil. Thus, there are no flux linkages. As the north pole of the magnet is moved toward the coil, it is apparent that the flux lines will begin linking the turns of wire on the coil. Moving the magnet closer to the coil and eventually inside the coil will cause more of the flux lines to link more of the coil turns. Thus, *the number of flux linkages increases as the magnet is moved toward the coil.* According to Faraday's law, as the flux linkages increase, a voltage or emf is induced in each loop of wire which is being linked. Since the loops are in series, the individual loop emfs add up to produce a total coil emf which produces a pulse of current through the meter.

If the magnet is pushed inside the coil [Fig. 15–3(b)] and *held* there, the number of flux linkages will be at a maximum. However, no emf will be induced in the coil while the magnet is stationary because the number of flux linkages is *not changing*. In other words, the induced emf and current persist only while the number of flux linkages is increasing as the magnet moves toward the coil.

Faraday's law also states that the size of the induced emf is proportional to the rate at which the flux linkages are changing. This explains why the induced emf and current increase if the magnet is moved toward the coil at a faster speed. (Keep in mind that it is the relative speed between the magnetic field and the coil which is important; we could just as easily keep the magnet stationary and move the coil and produce the same results.)

When the magnet is inside the coil, and we pull it out, the number of flux linkages will *decrease* as the flux lines are moved further away from the coil. This *decrease in flux linkages* will induce an emf in the coil of the *opposite* polarity. Again, this induced emf will persist only while the number of flux linkages is decreasing as the magnet moves away from the coil. Once the magnet stops moving, or is far enough away so that no flux links the coil, then there will be no emf induced in the coil.

If the strength of the bar magnet is increased (more flux density), we can expect an increase in the number of flux linkages and therefore a larger induced emf according to Faraday's law.

15.4 Mutual Induction

In the example of electromagnetic induction presented in Fig. 15–1, motion was required to induce an emf in the coil. The motion was needed to produce a change in the number of flux linkages by moving the magnetic flux of the bar magnet closer to or further from the coil. Another common application of electromagnetic induction does not require any mechanical motion to change the flux linkages.

Consider Fig. 15–4 where a rectangular core of soft iron has two coils wound on it. The coil on the left is connected to a dc battery in series with a current-limiting resistor. The coil on the right is connected to a load resistor R_L. The coil connected to the battery is called the *primary* coil or winding, while the other coil is called the *secondary* winding.

When the switch in the primary winding is open, there will be no current through the coil and, therefore, no flux in the iron core. When the switch is closed, current flows in the primary winding in the direction shown. The cur-

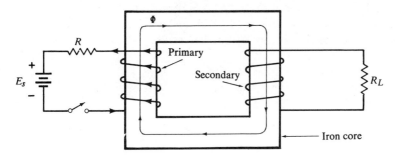

FIGURE 15-4 *Demonstration of mutual electromagnetic induction.*

rent in the primary coil produces a flux in the iron core. Because the iron is a low-reluctance material, all of the flux in the primary coil travels the iron path and *links* the secondary winding. This *increase* in flux linkages induces a momentary emf in the secondary winding which results in a pulse of current through the load resistor.

If the switch remains closed so that a *constant* primary current flows, there will be no emf in the secondary winding because the flux linkages will not be changing. However, if the battery voltage is suddenly *increased,* the primary current will *increase,* which in turn will *increase* the flux in the core and the flux linkages in the secondary coil, resulting in an induced emf in the secondary coil. *Decreasing* the battery voltage has the opposite effect; it *decreases* the primary current, which in turn *decreases* the flux in the core and the flux linkages in the secondary coil, resulting in an induced emf of the opposite polarity. Opening the switch in the primary coil reduces the current to zero and causes the flux in the core to collapse, thereby inducing an emf in the secondary coil.

To summarize, whenever the current in the primary winding *changes*, the flux in the core linking the secondary winding also *changes* and induces an emf in the secondary winding. As long as the primary current is steady, no emf will be induced in the secondary. Inducing a voltage in a secondary winding by a *changing* current in a primary winding is called *mutual induction.*

Mutual induction is the principle upon which the operation of all types of transformers is based. Mutual induction can sometimes have undesirable effects in situations where it is not purposely used but is present because of the proximity of current-carrying wires.

EXAMPLE 15.1 Indicate the effect each of the following changes will have on the emf induced in the *secondary* at the instant the switch is closed in Fig. 15-4. (a) increase R; (b) increase number of primary turns; (c) increase number of secondary turns; (d) use lower permeability (μ) core; (e) reverse the battery.

Solution: (a) A larger R in the primary circuit will reduce the primary current and produce less Φ. Thus, the amount of the change in Φ will be smaller. Since the induced emf depends on the *change* in flux linkages, the size of the emf will be *reduced*.

(b) An increase in primary turns means an increase in the magnetomotive force ($N \times I$) producing the flux in the core. Thus, more flux will be produced (provided the core is not saturated), resulting in more secondary flux linkages and a *larger* emf.

(c) A greater number of secondary turns will increase the number of flux linkages since the same amount of flux will be linking more turns. This will produce a *larger* secondary emf.

(d) A lower permeability core will result in *less* flux density for the same primary current, thereby resulting in fewer flux linkages and a *smaller* secondary emf.

(e) Reversing the battery will not affect the size of the induced emf, but it will cause it to be of opposite polarity since the flux will be in the opposite direction.

15.5 Mathematical Statement of Faraday's Law

The *magnitude* of the induced emf is directly proportional to the *rate of change* of flux linkages. Stated mathematically,

$$\text{emf} = \frac{\Delta \lambda}{\Delta t} \quad (15\text{--}1)$$

where λ represents the flux linkages, and $\Delta \lambda$ is the change in flux linkages that occurs in a time interval of Δt. Thus, $\Delta \lambda / \Delta t$ represents the rate of change of flux linkages.

In situations like that in Fig. 15-4, the flux Φ in the core links *all* of the secondary turns so that at any time, $\lambda = N \times \Phi$, where N is the number of turns in the secondary coil. Therefore, Eq. (15-1) for the induced emf becomes

$$\text{emf} = N \frac{\Delta \Phi}{\Delta t} \quad (15\text{--}2a)$$

where $\Delta \Phi / \Delta t$ represents the *rate of change* of flux in the core. This expression is more often written as

$$\text{emf} = N \frac{d\Phi}{dt} \quad (15\text{--}2b)$$

The d in $d\Phi$ and dt is an abbreviation for delta (Δ), which means a *change*.* We will continue to use $d\Phi$ and dt to represent a change in flux and a change in time respectively.

Equations [15-2(a)] and (b) are valid whenever *all* the flux Φ links *all* N turns of the coil. Using either [15-2(a)] or (b), the induced emf can be calculated in *volts* if $d\Phi$ is in *webers* and dt is in *seconds*.

EXAMPLE 15.2 The secondary coil in Fig. 15-4 has 10 turns. When the primary switch is closed, the flux in the core increases to 50 μWb in 1 ms. Calculate the magnitude of the induced emf in the secondary.

Solution: $\text{emf} = N \dfrac{d\Phi}{dt}$

where $N = 10$, $d\Phi = 50$ μWb, and $dt = 1$ ms.

---

* Mathematically speaking, d represents an infinitesimally small change.

$$\text{emf} = \frac{10 \times 50 \times 10^{-6}}{10^{-3}}$$
$$= 500 \times 10^{-3}$$
$$= \mathbf{0.5\ V}$$

EXAMPLE 15.3 With current flowing in the primary of Fig. 15-4, the core flux is 1000 μWb. The primary current is increased slightly so that Φ increases to 1050 μWb in 1 ms. What is the induced emf in the secondary?

Solution: We can again use $N\,d\Phi/dt$. The flux is changing from 1000 μWb to 1050 μWb, a change of 50 μWb. Thus, $d\Phi = 50$ μWb and $N \cdot d\Phi/dt$ is again equal to **0.5 volts**.

This emphasizes the fact that it is the amount by which the flux *changes* and not the actual amount of flux which determines the induced emf.

Induced emf Related to dI/dt In situations like Fig. 15-4, the flux in the core is produced by the primary current. From chap. 14 we know that

$$\Phi = \frac{N \times I}{\mathcal{R}} \qquad (14\text{-}7)$$

where N is the number of primary turns, I is the primary current, and $\mathcal{R}$ is the core's reluctance. If N and $\mathcal{R}$ are reasonably constant, then it is clear that Φ is proportional to I. Thus, any changes in I will produce proportional changes in Φ. Therefore, the rate of change of flux, $d\Phi/dt$, is directly proportional to the rate of change of current, dI/dt. For the situations where electromagnetic induction is due to a changing current rather than mechanical motion we can restate Faraday's law as:

> **The magnitude of the induced emf is directly proportional to the rate of change of current.**

This form of Faraday's law will be of more practical use to us in analyzing circuits that contain coils or transformers because it will be more convenient to deal with current rather than flux.

15.6 Lenz's Law

Up to now we have only talked about the *magnitude* of the induced emf, but have not shown how to determine its polarity. Faraday's law helps us to calculate the *magnitude* of the induced emf. Another important principle of electromagnetic induction is *Lenz's law*, which can be used to determine the *polarity* of the induced emf. Lenz's law is based on the principle of conservation of energy and can be stated as follows:

> **When a change in flux linkages induces an emf in a wire or a coil, the polarity of the induced emf will be such that any current flow produced by it will develop a magnetic field which opposes the original change in flux linkages.**

This statement of Lenz's law is quite a mouthful. It might make more sense if it were stated in several steps:

(1) A change in flux linkages will induce an emf in a coil; this emf will have a definite polarity.

(2) If the circuit is closed by connecting a meter or a load resistor across the coil, the induced emf will cause current to flow in the coil and circuit.

(3) The current flowing in the coil produces its own magnetic flux lines in the coil.

(4) The direction of these flux lines will always be such that they *oppose* the original *change* in flux (step 1) which induced the emf.

The key word in Lenz's law is *oppose*. It indicates that the force or impetus which is causing the original flux change will always meet with opposition.

As our first example of applying Lenz's law, let us consider the situation in Fig. 15–5. This is the same case considered earlier in our discussion of Faraday's law (Fig. 15–1) except that a resistor is shown in place of the current meter. If the bar magnet is made to move *toward* the coil, then according to Lenz's law the induced current in the coil must produce a magnetic field that *opposes* this motion. Thus, the coil current must produce a *north pole* on the left side of the coil as shown. This *induced* north pole will offer an opposition to the motion of the magnet's north pole. Using the left-hand rule we can establish the necessary coil current direction as indicated. Because of the coil's induced opposing north pole, it takes a certain amount of work (energy) to push the magnet into the coil. The mechanical energy expended in moving the magnet is converted to electrical energy, which produces the current in the coil and resistor.

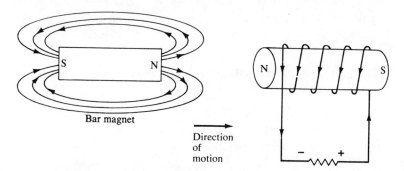

FIGURE 15–5 *Induced current in coil produces a north pole on left end opposing motion of bar magnet, illustrating Lenz's law.*

Consider the opposite case where the bar magnet is moved *away* from the coil. Then according to Lenz's law the induced pole at the left end of the coil must be a *south pole*. This induced south pole will attract the bar magnet's north pole in an attempt to keep it from moving away. For a south pole to be induced at the left end of the coil, the electron flow must be opposite to the direction shown in Fig. 15–5.

Lenz's Law Applied to Mutual Induction We can apply Lenz's law to situations like the one we discussed in Fig. 15–4 where the current in one coil can induce an emf in another coil by *mutual induction*. For convenience this situation is redrawn in Fig. 15–6(a). Before the switch is closed there will be no flux in the core and no emf induced in the secondary winding. When the switch is closed, the current flowing in the primary coil will produce cw flux in the core (labeled Φ_P). This flux links the secondary coil and induces an emf which produces a current in the load resistor.

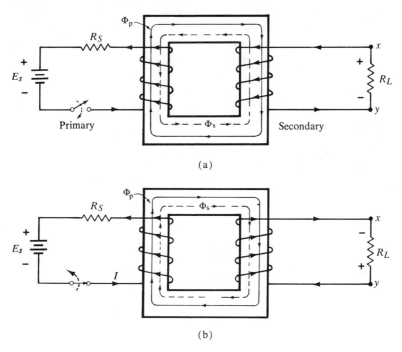

FIGURE 15–6 *Lenz's law applied to mutual induction:* (a) *primary current increasing;* (b) *decreasing.*

According to Lenz's law, the current in the secondary coil must be in such a direction that it produces a flux field that *opposes* the *change* in flux which produced the emf. Thus, the current flow in the secondary will be in the direction as shown. This direction of current will produce a flux in the secondary coil (using left-hand rule) that is in the ccw direction and is labeled Φ_S in the figure. Φ_S is in the *opposite* direction of Φ_P and therefore *opposes the increase in* Φ_P. The current in the secondary flows through R_L from y to x indicating that the induced emf makes x positive relative to y. Keep in mind that this induced emf and current will only be momentary as long as the primary current is increasing.

As the switch in the primary remains closed, the primary current will reach a constant value so that Φ_P is constant and no emf or current will be present in the secondary. If E_S is suddenly increased, the primary current will increase and Φ_P will *increase*. This will again induce an emf and current in the secondary with the polarity shown in Fig. 15–6(a). However, if the primary current is decreased by decreasing E_S or by opening the switch, the opposite effect will occur. The flux Φ_P will decrease so that the number of flux linkages in the secondary coil will be *decreasing*. As a result, the induced current in the secondary will have the direction indicated in Fig. 15–6(b). This direction of current will produce a flux Φ_S in the secondary coil that is in the cw direction, the *same* direction as Φ_P. Since Φ_S is in the same direction as Φ_P, it adds to Φ_P and attempts to keep the total flux from decreasing. In other words, the induced Φ_S *opposes the decrease* in Φ_P.

Note that the induced secondary flux Φ_S can be in either direction relative to Φ_P. Φ_S's direction will be opposite to Φ_P if Φ_P is trying to increase, and it will be the same as Φ_P if Φ_P is trying to decrease. In either case, Φ_S opposes the change that Φ_P is trying to make as dictated by Lenz's law. The secondary emf (and current) polarity is determined by choosing the polarity which will produce the correct direction for Φ_S.

Further applications of Lenz's law will be given in the problems at the end of the chapter. In addition, we will use it in the following discussion of self-inductance and inductors.

15.7 Self-Inductance

We have seen that when the magnetic flux linking a wire or a coil changes, an emf is induced in the circuit. This phenomenon of electromagnetic induction was explained in the previous sections using as examples the motion of a bar magnet with respect to a coil, and the mutual induction effect of the current in one coil on another coil. In these examples, we needed an external source of magnetic flux to induce an emf in a coil. Actually, as we shall see, this is not necessary because every wire and coil possesses the ability to induce an emf in *itself* by virtue of its own current flow. The process by which this occurs is called *self-induction*.

Consider Fig. 15-7 where a coil is shown with a current i flowing through it. At this point, we will not concern ourselves with how or what is producing this current; we will assume that an appropriate voltage or current source is present. If the current i is not changing, then the flux field it produces in the coil will be stationary. If the current i *increases*, the flux field will *expand;* if i *decreases*, the flux field will *collapse* (become less). Thus, as i changes, the flux linking the turns of the coil will change. This change in flux linkages will induce an emf in the coil. This process is called self-induction because the coil's own changing current is what induces the coil emf.

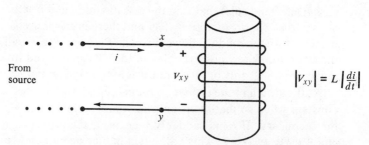

FIGURE 15-7

From our discussion of Faraday's law, we know that the *magnitude* of this self-induced emf is directly proportional to the rate of change of flux linking the coil ($d\Phi/dt$). However, since the flux in the coil depends directly on the current i, we can state that

Self-induced emf $\propto di/dt$ (rate of change of current).

The emf is proportional to di/dt, the rate of current change. The constant of proportionality is given the symbol L and is called *self-inductance* or simply *inductance*. Thus, we can write

$$\text{self-induced emf} = L \times \frac{di}{dt} \qquad (15\text{--}3)$$

This expression is concerned only with the *magnitude* of the emf. If we refer to Fig. 15–7, then, we can state that

$$|V_{xy}| = L \times \left|\frac{di}{dt}\right| \qquad (15\text{--}4)$$

where V_{xy} is the voltage across the coil. The vertical lines enclosing V_{xy} and di/dt in expression (15–4) indicate *magnitude only*. In other words, (15–4) can be used to determine the size of V_{xy} but not the polarity.

We can summarize the discussion at this point as follows:

The emf induced in a coil is proportional to the rate of change of current through the coil; the proportionality constant being L, the coil's inductance.

Note that although we have been referring to the coil voltage as an emf, we used the symbol V_{xy} rather than E_{xy} to represent the coil voltage. This is because the symbols E and e are reserved only for voltage sources (sources of emf). The coil is not, in the strictest sense, a source of voltage. Thus, we will use V and v to represent coil voltage the same as resistor and capacitor voltages.

A given coil will have a certain value of inductance L. The formula (15–4) serves to define the units of inductance. We can rearrange it as

$$L = \frac{|V_{xy}|}{|di/dt|} \qquad (15\text{--}5)$$

If V_{xy} is in *volts* (V) and di/dt is given in amperes/second (A/s), then L has the units

$$\frac{V}{A/s}$$

This derived unit is more simply called a *henry* and is given the symbol H. Thus, inductance, L, is given in units of henrys, H. From (15–5) we can see that

A coil has an inductance of 1 H if a current change (di/dt) of 1 A/s induces an emf of 1 volt in the coil.

15.8 The Inductor

Any current-carrying wire has inductance. However, a coil of wire has a greater amount of inductance because the flux will link more turns of wire and therefore produce a greater induced emf. A coil of wire which is purposely designed to display the property of inductance is often called an *inductor*. It is also referred to as a *choke* or simply a *coil*. The circuit symbols for an air-core inductor and an iron-core inductor are shown in Fig. 15–8. These symbols will be used throughout the text. Note that the symbol L is used for the inductor. This corresponds to R for resistance and C for capacitance.

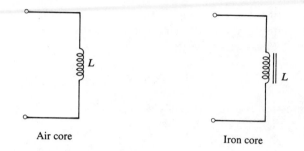

FIGURE 15-8 *Circuit symbols for inductors.*

The inductance of an inductor is a measure of its ability to induce an emf when its current is changing. An inductor with a larger value of L will induce a greater voltage across its terminals than a smaller L for the same change in current. There are many physical factors which determine the inductance value of a coil. We will discuss them in a later section. For now, we will attempt to gain a quantitative understanding of inductance by looking at several examples.

EXAMPLE 15.4 Refer to Fig. 15-7. The current i increases from 2 A to 6 A in a time interval of 0.5 s and induces a voltage of 12 V across the coil. Determine the inductance of the coil.

Solution: The rate of change of current di/dt is simply

$$\frac{di}{dt} = \frac{6\text{ A} - 2\text{ A}}{0.5\text{ s}} = \frac{4\text{ A}}{0.5\text{ s}}$$
$$= 8\text{ A/s}$$

Thus, using Eq. (15-5) to solve for L

$$L = \frac{12\text{ V}}{8\text{ A/s}} = 1.5\text{ H}$$

EXAMPLE 15.5 If the inductor of the previous example has its current change from 2 mA to 1 mA in a time interval of 10 μs, how much voltage will appear across the coil?

Solution: In this case

$$\frac{di}{dt} = \frac{1\text{ mA} - 2\text{ mA}}{10\text{ }\mu\text{s}} = \frac{-10^{-3}\text{ A}}{10^{-5}\text{ s}} = -100\text{ A/s}$$

$$\therefore \left|\frac{di}{dt}\right| = 100\text{ A/s}$$

Using Eq. (15-4) we can determine

$$|V_{xy}| = 1.5\text{ H} \times 100\text{ A/s} = 150\text{ V}$$

Note that in using this expression we use henries for L and A/s for di/dt in order to get volts for the result. Also note that the result is the magnitude of V_{xy} and does not tell us the polarity. As we shall see, the polarity can be determined using Lenz's law.

EXAMPLE 15.6 Two coils are subjected to the same rate of current change $di/dt = 300$ A/s. Coil x induces 0.6 V, while coil y induces 6 mV in response to this current change. Calculate the inductance for each coil.

Solution: Coil x

$$L_x = \frac{|V|}{|di/dt|} = \frac{0.6 \text{ V}}{300 \text{ A/s}} = \textbf{0.002 H}$$

This result can be expressed in the prefixed unit millihenries (mH). Since 1 mH = 0.001 H, then

$$L_x = 0.002 \text{ H}$$
$$= \textbf{2 mH}$$

Coil y

$$L_y = \frac{6 \text{ mV}}{300 \text{ A/s}} = \frac{6 \times 10^{-3} \text{ V}}{300 \text{ A/s}} = 2 \times 10^{-5} \text{ H}$$

This result can be expressed in microhenries (μH). Since 1 μH = 10^{-6} H, then

$$L_y = 2 \times 10^{-5} \text{ H} = 20.0 \times 10^{-6} \text{ H}$$
$$= \textbf{20.0 } \boldsymbol{\mu}\textbf{H}$$

EXAMPLE 15.7 A 50-mH coil has a voltage of 25 V across its terminals. What is the rate at which its current is changing?

Solution: Rearranging Eq. (15-4) we can solve for di/dt.

$$\left|\frac{di}{dt}\right| = \frac{|V_{xy}|}{L} = \frac{25 \text{ V}}{50 \text{ mH}}$$
$$= \frac{25 \text{ V}}{0.05 \text{ H}} = \textbf{500 A/s}$$

Note that the result is in A/s if we use volts and henries for V_{xy} and L respectively.

To induce a voltage in an inductor coil, its current must be *changing*. The actual value of the current is not important; but the rate at which it is changing determines the coil voltage. For example, consider these two cases: (1) the current in a 1-H coil changing from 1 A to 2 A in 1 s; (2) the current in a 1-H coil changing from 1000 A to 1001 A in 1 s. In both cases, the current *changes* by 1 A in 1 s so that $di/dt = 1$ A/s and the induced coil voltage will be 1 H $\times$ 1 A/s = 1 V. The fact that one coil is carrying 1000 times as much current makes no difference! (This, of course, assumes that the coil of wire has zero *resistance*.)

Practical Inductors It was stated that a *change in current* is necessary to produce a voltage across an inductor. A constant current would produce no inductor voltage. This is true for an *ideal* inductor. An ideal inductor is a coil of wire that has zero *resistance* to the flow of current. In practice, every inductor has some resistance. The amount depends mainly on the dimensions of the wire. Thus, we can expect that if a constant current flows through a practical inductor, there *will* be a voltage across the coil due to its resistance.

This effect can be taken into account by representing the practical inductor by an ideal inductor *in series* with a resistor. Figure 15-9 shows the comparison between an ideal inductor and a practical inductor when a constant I is flowing. For the ideal inductor, $V_{xy} = 0$ since I is not changing ($dI/dt = 0$). For the practical inductor, there will be no voltage across the inductor portion, but there will be a voltage $V = I \times R_c$ across the resistor portion. Since the inductive and resistive portions are distributed throughout the coil, this voltage would be measured across the coil terminals so that $V_{xy} = I \times R_c$.

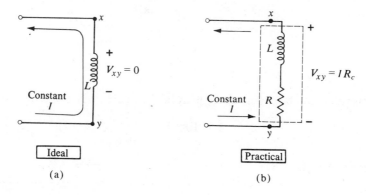

FIGURE 15-9 *Comparison of ideal and practical inductors.*

For the practical inductor, R_c usually represents the coil's wire resistance and can be easily measured with a dc ohmmeter across the inductor terminals. In many cases R_c is small enough to be neglected. In the work to follow we will often assume that the coil resistance is negligible. Whenever it must be considered, it will be included in the analysis or discussion.

EXAMPLE 15.8 A certain relay coil has 10 V across it when a *constant* 20-mA current passes through it. What is the coil's resistance?

Solution: Since the current is not changing, the 10 V across the coil must be due to the coil's resistance. Thus, the 10 V are across the R_c so that

$$R_c = \frac{10 \text{ V}}{20 \text{ mA}} = 500 \text{ }\Omega$$

15.9 Voltage Polarity Across an Inductance

We have seen how to determine the magnitude of the voltage across an inductance [Eqs. (15-3) and (15-4)]. Now we must turn our attention to the *polarity* of coil voltage. It is not as easy to determine as the polarity of voltage across a resistor. In a resistor, we know that the *electron* current always flows into the resistor's negative terminal and out of its positive terminal (see Fig. 15-10). If the value of the current changes, the resistor voltage will change since $v = iR$, but the voltage polarity will remain the same unless the current changes direction.

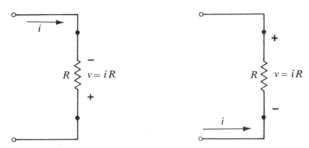

FIGURE 15-10 *Relationship between current direction and voltage polarity in a resistor.*

The situation is not as straightforward for an inductor. The inductor voltage is produced by a *change* in current; and according to Lenz's law the emf induced in the coil must be of such a polarity that it *opposes the change* that the current is trying to make. If the current increases, the coil voltage will have a polarity which tends to oppose this increase; if the current decreases, the coil voltage will have a polarity which tends to oppose this decrease. In other words, the change in current is *opposed* by the coil's induced emf. For this reason the coil's voltage is often referred to as a *counter emf* or *back emf*.

Figure 15-11 illustrates the details of applying Lenz's law to determine the polarity of the inductor voltage. Consider part (a) of the figure first. A *variable* dc voltage source is shown producing current through an inductor L. The source is shown in dotted lines because it might just as well be an ac source or other time-varying source or even a more complex network supplying current to the inductor. Note that the direction of current is into the inductor's terminal x. If the source voltage is suddenly *increased*, the current will increase. By Lenz's law, this will induce a coil voltage with the polarity as shown, negative at x and positive at y. The reason is that this polarity of the coil's counter emf acts in opposition to the source and therefore *opposes* the attempt to *increase* the current.

In part (b) of the figure, the same situation is shown with the source producing a current flow into terminal x of the coil. If the source voltage is suddenly *decreased*, the current will decrease. By Lenz's law, the coil's counter emf will then have the polarity shown, positive at x and negative at y. This polarity of counter emf acts to aid the source and therefore *opposes* the attempt to *decrease* the current.

In parts (c) and (d) of the figure, the source has been reversed so that the current is now flowing in the opposite direction. In (c) the current is increased and the resulting counter emf has the polarity shown [using the same reasoning as in (a)]. In (d) the current is decreased and the counter emf polarity is determined using the same reasoning as in (b).

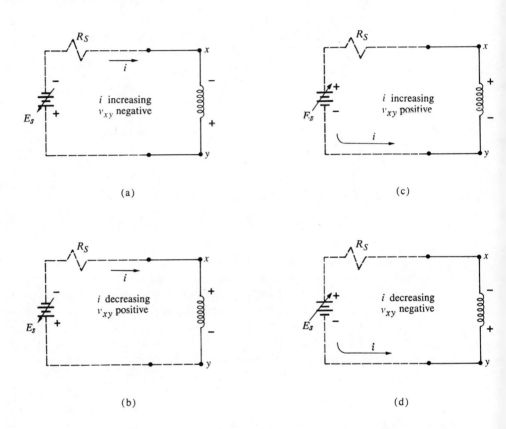

FIGURE 15-11 *The polarity of coil voltage depends on the current direction and how it is changing.*

Examination of these four cases indicates that current direction alone does not determine the coil voltage polarity. It depends on how the current is changing. If the *electron* current is *increasing*, then the end of the coil at which the current enters will be *negative* [cases (a) and (c)] and attempts to repel the electrons to keep the current from increasing. If the current is *decreasing*, then the end of the coil at which the current enters will be *positive* [cases (b) and (d)] in order to attract the electrons and keep the current from decreasing.

Because the counter emf of a coil always attempts to oppose any change in the coil current, we can state that

The principal circuit characteristic of an inductor is that it opposes any change in current.

A larger value of inductance will produce a larger counter emf; thus, it will offer a greater opposition to current changes.

Keep in mind that if a coil has no wire resistance, then it will offer no opposition to a *constant flow* of current; but it will always oppose any *change* in current. A practical coil always has some resistance so that it also offers some opposition to a dc current, but its opposition to changes in current is normally much greater. A comparison of inductance and resistance will be made later so that the difference between the two types of opposition will be clearer.

15.10 Factors Which Determine Inductance

As mentioned earlier, every conducting wire has a certain amount of inductance, but the inductance of a single wire is normally so small that it can be neglected. A coil of wire, however, can be constructed so that its inductance is significant. In terms of physical construction, the inductance of a coil is determined by two main factors.

(1) *The number of turns of wire in the coil.* A greater number N of turns increases the value of inductance L because a greater emf can be induced for a given change in current (recall that Faraday's law is basically emf $= N\, d\phi/dt$). Actually L increases in proportion to N^2, so that doubling N will increase L by four times.

(2) *The reluctance of the magnetic circuit.* The reluctance of the core determines how much flux will be produced for a given current. A larger value of reluctance means a smaller flux change for a given current change. The smaller flux change induces a smaller emf. Thus, a larger core reluctance will decrease the value of inductance and vice versa.

These two factors are summarized in the formula

$$L = \frac{N^2}{\mathcal{R}} \qquad (15\text{-}6a)$$

where N is given in turns (t), reluctance is given in amp-turns per weber (A-t/Wb) and L is in henries (H).

The formula for reluctance which was presented in chap. 14 can be used to replace $\mathcal{R}$ in the preceding equation. That is,

$$\mathcal{R} = \frac{l}{\mu A} \qquad (14\text{-}9)$$

so that

$$L = \frac{\mu N^2 \times A}{l} \qquad (15\text{-}6b)$$

where A is in square meters (m^2), l is in meters (m), and μ is the *absolute* permeability of the core material. This formula can be further modified by substituting

$$\begin{aligned} \mu &= \mu_r \times \mu_o \\ &= \mu_r \times 1.26 \times 10^{-6} \end{aligned} \qquad (14\text{-}6)$$

resulting in

$$L = \left(\frac{\mu_r N^2 A}{l} \times 1.26 \times 10^{-6}\right) \text{ H} \qquad (15\text{-}6c)$$

All of the formulas (15-6a, b, and c) are equivalent and can be used to calculate L for a given material and core dimensions.

EXAMPLE 15.9 An air-core coil with 50 turns is 20-cm long and has a cross-sectional area of 3 cm². Determine its inductance.

Solution: For this air-core coil, $\mu_r = 1$, $N = 50$, $A = 3 \times 10^{-4}$ m², and $l = 0.2$ m. Using 15-6(c)

$$L = \frac{1 \times 50^2 \times 3 \times 10^{-4} \times 1.26 \times 10^{-6}}{0.2} \text{ H}$$

$$= 4.725 \times 10^{-6} \text{ H} = \mathbf{4.725 \ \mu H}$$

EXAMPLE 15.10 Change the core in Ex. 15.9 to one made of soft iron ($\mu_r = 500$) and recalculate L.

Solution: Since $\mu_r = 500$, then according to (15-6c) the inductance L will be 500 times as great so that

$$L = 500 \times 4.725 \ \mu H$$
$$= 2362.5 \ \mu H$$
$$= \mathbf{2.3625 \ mH}$$

For the same core dimensions and coil turns, the iron-core inductor produces a much greater value of inductance than the air-core inductor.

15.11 Rise of Current in an Inductor

Refer to Fig. 15-12(a). We know that if we close the switch at time $t = 0$, a current will immediately begin to flow through the resistor with a value equal to 10 V/10 Ω = 1 A. The resistor offers an opposition to the flow of current and it is this resistor value that determines how much current will flow for a given applied voltage. However, the resistor does not prevent the current from instantaneously rising from 0 to 1 A as the switch is closed.

If we now insert a 1-H inductor in series with the resistor, the situation becomes that shown in part (b) of the figure. We will assume that any coil resistance is included in the 10-Ω resistance. Since the total resistance in the circuit is 10 Ω, we would expect that once again if the switch is closed a current of 1 A will flow. However, the presence of the inductor will not allow the current to rise to 1 A instantaneously. Instead, the current will rise *gradually* to 1 A as shown in Fig. 15-12(b). When the switch closes, the current is attempting to make a rapid change; the inductor opposes this rapid change and the current is forced to rise more slowly.

The inductor's opposition is in the form of a counter emf. As the current i is increasing, the voltage induced in the coil will be negative at y, positive at x.

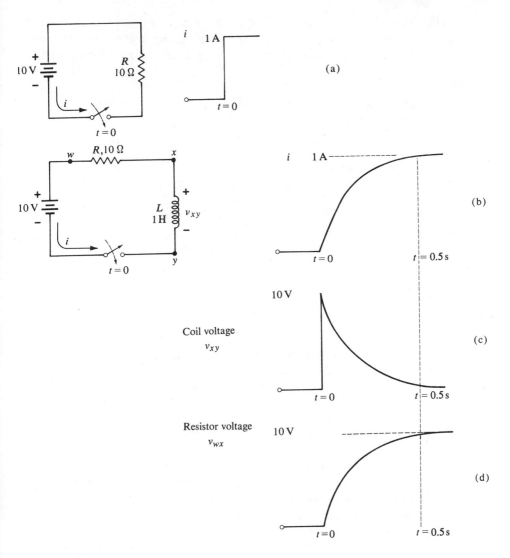

FIGURE 15–12 *Presence of inductance in a circuit prevents a rapid rise in current.*

This counter emf opposes the source voltage and slows down the rise of current.

Circuit Waveforms We will now analyze the operation of the circuit in Fig. 15–12(b) and investigate its waveforms. The analysis will proceed in a step-by-step fashion.

(1) At $t = 0$ the switch is closed and the 10-V source is applied to the circuit.

(2) According to Kirchhoff's voltage law, the *sum* of the resistor voltage, v_{wx}, and the coil voltage, v_{xy}, must equal the source voltage at any instant of time. That is,

$$v_{wx} + v_{xy} = 10 \text{ V}$$

(3) At the instant the switch is closed, the current will try to start flowing. However, the coil prevents the current from changing instantaneously by developing a counter emf which is equal to the source voltage. That is,

$$v_{xy}(@\ t = 0) = E_s = 10 \text{ V}$$

(4) This counter emf opposes the 10-V source so that there is no net voltage across the resistor to produce current. Thus,

$$v_{wx}(@\ t = 0) = 0 \quad \text{and} \quad i(@\ t = 0) = 0$$

Even though the current is zero at the instant the switch closes, it will immediately begin to increase and it is this increase which induces the counter emf in the coil.

(5) The initial rate at which the current will begin to increase can be determined by using

$$v_{xy} = L \frac{di}{dt}$$

Since $v_{xy} = 10$ V and $L = 1$ H, then

$$\frac{di}{dt}(@t = 0) = \frac{v_{xy}}{L} = \frac{10 \text{ V}}{1 \text{ H}} = 10 \text{ A/s}$$

Note that this initial rate is the *slope* of the i waveform at $t = 0$ and does not depend on the value of R.

(6) The current begins to increase at this initial rate. However, as the current increases, the voltage across the resistor must also increase since $v_{wx} = i \times R$. As the resistor voltage increases, the inductor voltage must *decrease* in order to satisfy KVL (step 2). Refer to the waveforms in Fig. 15–12(c) and (d).

(7) A lower coil voltage must mean that the rate at which the current is rising is becoming smaller. Thus, the current continues rising but at a slower rate. Refer to the waveform for i in Fig. 15–12(b).

(8) As the current continues rising, the resistor voltage keeps increasing and the coil voltage keeps decreasing. Since the coil voltage is decreasing, the rate at which the current is rising keeps getting slower and slower ($di/dt = v_{xy}/L$).

(9) This process continues until the coil voltage reaches 0 V. At this point all of the source voltage is across the resistor and the current is at $10 \text{ V}/10 \text{ }\Omega = 1$ A, and

$$\frac{di}{dt} = \frac{v_{xy}}{L} = 0 \quad \text{(no further change in current)}$$

The circuit will then be in its *steady state*, where it will remain indefinitely.

(10) As the waveforms in Fig. 15–12(b)–(d) show, the *transient interval* is 0.5 s for this circuit before steady state is reached. In steady state:

$v_{xy} = 0$ V	(coil voltage)
$v_{wx} = 10$ V	(resistor voltage)
$i = 1$ A	(equals E_s/R)

Upon closing the switch in Fig. 15–12(b) the current starts at zero and begins to gradually rise. It initially rises at a fast rate which gradually becomes slower and slower until the current reaches its steady-state value equal to E_s/R.

When the switch closes, a voltage equal to the source voltage appears across the coil. This voltage gradually decreases to zero as the current approaches its steady-state value.

The resistor voltage starts out at zero and its waveform follows the current waveform exactly until it equals E_s in steady state.

In comparison with the circuit containing no inductor, the circuit with the inductor causes the current to build up gradually to its final value. This is because

An inductor will not allow its current to change instantaneously from one value to another.

15.12 Time Constant of an RL Circuit

The circuit we have been considering is a series combination of a resistor R and an inductor L. We will refer to this circuit as a series RL circuit. We saw in the preceding analysis that the current in an RL circuit takes time to reach its final value. As with the RC circuits discussed in chap. 13, we can use the concept of a circuit *time constant* to help us analyze RL circuits.

Recall that the symbol for time constant is τ. For an RL circuit the time constant is given by

$$\tau = \frac{L}{R} \qquad (15\text{–}7)$$

when L is in henries, R is in ohms, and τ is in seconds.

EXAMPLE 15.11 Calculate the time constant of the RL circuit in Fig. 15–12(b).

Solution: $\tau = \dfrac{L}{R} = \dfrac{1 \text{ H}}{10 \text{ }\Omega} = \mathbf{0.1 \text{ s}}$

Everything which was said concerning the time constant of an RC circuit can be applied to the RL circuit. Thus, the RL circuit will essentially reach its steady state after 5 τ, five time constants.

EXAMPLE 15.12 For the circuit of Fig. 15–12(b) determine the time to reach steady state.

Solution: $5 \times \tau = 5 \times 0.1 \text{ s} = \mathbf{0.5 \text{ s}}$

This agrees with the waveforms in Fig. 15–12.

For the RL circuit the time constant and therefore the time to reach steady state are *directly* proportional to the value of inductance. A larger inductance offers more opposition to the increase in current and therefore causes the current to build up more slowly. The time constant is *inversely* proportional to the resistance value. A larger resistance value will decrease the steady-state current. For the same source voltage and inductance value, the current will have the same initial rate of increase (step 5) and will therefore reach this lower value of steady-state current faster. Thus, a larger R indicates a shorter τ and a shorter time to reach steady state. Note that this is opposite to an RC circuit where a larger R causes a longer τ and a slower approach to steady state.

EXAMPLE 15.13 For the circuit of Fig. 15–12(b) what will be the value of the current 0.1 s after the switch is closed? What will be the coil voltage?

Solution: For this circuit $\tau = 0.1$ s. Thus, 0.1 s represents a time interval equal to 1 τ. We learned in chap. 13 that all circuit voltages and currents will complete 63% of their total ultimate change in one time constant. This holds true for RL circuits as well as RC circuits.

The current is initially zero and is heading for 1 A. In 0.1 s it will be at $63\% \times 1\text{ A} = \textbf{0.63 A}$.

The coil voltage is initially 10 V and is heading for 0 V. In 0.1 s it will be at $10\text{ V} - 63\% \times 10\text{ V} = \textbf{3.7 V}$.

15.13 Time-Constant Graph

For the RL circuit we can utilize a *time-constant graph* exactly like the one we used for RC circuits. The graph is repeated in Fig. 15–13 for convenience. It plots the *percentage of total change* versus *number of time constants*. It is used in the same way that it was in chap. 13 with the RC circuits. The following examples illustrate.

EXAMPLE 15.14 Consider the circuit in Fig. 15–14(a). The switch is closed at $t = 0$.
  (a) Sketch the waveforms of inductor voltage and current.
  (b) What is the initial rate of current rise?
  (c) What is the value of inductor current at $t = 30\ \mu\text{s}$?
  (d) What is the inductor voltage at $t = 45\ \mu\text{s}$?
  (e) What is the rate of current rise at $t = 45\ \mu\text{s}$?

Solution: (a) At $t = 0$, the current is zero and gradually rises to a steady-state value of

$$i_{\text{s-s}} = \frac{32\text{ V}}{8\text{ k}\Omega} = 4\text{ mA}$$

The time constant τ is

$$\tau = \frac{L}{R} = \frac{120\text{ mH}}{8\text{ k}\Omega} = \frac{120 \times 10^{-3}}{8 \times 10^{3}} = 15 \times 10^{-6}\text{ s}$$
$$= 15\ \mu\text{s}$$

Thus, the current reaches steady state in $5\ \tau = 75\ \mu\text{s}$. Its waveform is drawn in Fig. 15–14(b).

At $t = 0$, the inductor voltage jumps up to equal the source voltage, 32 V, as it opposes the instantaneous rise in current. The coil voltage gradually decreases to 0 V, which it reaches in $5\ \tau = 75\ \mu\text{s}$. Its waveform is sketched in

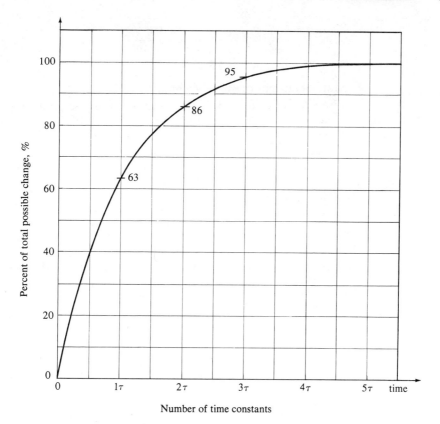

FIGURE 15–13 *Time-constant graph.*

part (c) of the figure.

(b) At $t = 0$, the coil counter emf = 32 V. Thus,

$$\frac{di}{dt}(@t = 0) = \frac{32 \text{ V}}{120 \text{ mH}} = 267 \text{ A/s}$$

Since this is the instant of maximum coil emf, then di/dt is maximum at this instant.

(c) 30 μs represents *two* time constants. In 2 τ the inductor current will have made 86% of its total change of 4 mA. Thus,

$$i(@ \ t = 30 \ \mu s) = 0.86 \times 4 \text{ mA}$$
$$= 3.44 \text{ mA}$$

See point *x* on the current waveform.

(d) 45 μs represents 3 τ. In 3τ the coil voltage will have moved 95% of the way from 32 V toward zero. Thus,

$$v_{coil} (@ \ t = 45 \ \mu s) = 32 \text{ V} - 0.95 \times 32 \text{ V} = \textbf{1.6 V}$$

(e) At $t = 45$ μs, the coil voltage is only 1.6 V. This counter emf is produced by a di/dt of

$$\frac{di}{dt}(@t = 45\ \mu s) = \frac{1.6\ V}{120\ mH}$$
$$= 13.33\ A/s$$

Note that this rate of current rise is substantially smaller than it was at $t = 0$. This accounts for the flattening out of the i waveform as steady state is approached.

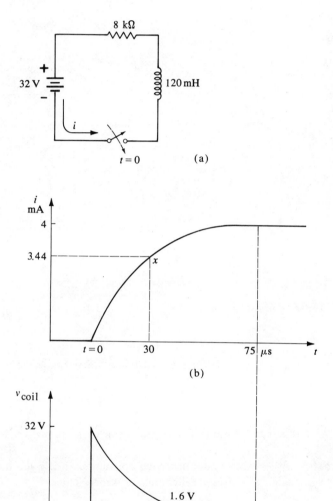

FIGURE 15–14

EXAMPLE 15.15 For the circuit of Fig. 15–14, find the value of current at $t = 20\ \mu s$.

Solution: τ for this circuit was previously calculated as 15 μs. Thus $t = 20\ \mu s$

represents $20/15 = 1.33$ time constants. Referring to the graph in Fig. 15–13, we see that after $1.33\,\tau$ the current will have completed about 73% of its total change. Since i is heading toward 4 mA from zero, then

$$i(@\ t = 20\ \mu s) = 0.73 \times 4\ \text{mA}$$
$$= \mathbf{2.92\ mA}$$

EXAMPLE 15.16 For the circuit of Fig. 15–14, how long will it take before the coil voltage drops to 12 V?

Solution: The coil voltage is decreasing from 32 V to 0 V, a total change of 32 V. When the coil is at 12 V, it will have completed 20 V of this change or $20/32 = 62.5\%$. Referring to the graph in Fig. 15–13, it can be seen that a 62.5% change takes place in a time interval of approximately $1\,\tau$. Thus, the coil voltage will be at 12 V at $t = \tau = \mathbf{15\ \mu s}$.

EXAMPLE 15.17 Consider the circuit in Fig. 15–15(a). Initially both switches are open so that no current flows. Switch $S1$ is closed at $t = 0$ and remains closed as the coil current builds up. At $t = 1$ ms, switch $S2$ is closed. Determine the inductor current at $t = 3$ ms (2 ms after $S2$ is closed).

Solution: *Step 1:* With $S1$ closed and $S2$ open, the total resistance in the circuit is 3 kΩ. Thus, the current i will start at zero and build up toward 24 V/3 kΩ = 8 mA with a time constant $\tau = 6$ H/3 kΩ = 2 ms.

At $t = 1$ ms, the current will have been building up for $0.5\,\tau$. According to the graph in Fig. 15–13, a change of 40% occurs in $0.5\,\tau$. Thus, at $t = 1$ ms the current will be

$$i(@\ t = 1\ \text{ms}) = 0.4 \times 8\ \text{mA}$$
$$= 3.2\ \text{mA}$$

Step 2: When $S2$ is closed at $t = 1$ ms, the coil current will be 3.2 mA since it cannot change instantaneously [see part (b) of the figure]. $S2$ shorts out the 500-ohm resistor so that the circuit resistance is now only 2.5 kΩ. With $S2$ closed, the current will now head for a steady-state value of 24 V/2.5 kΩ = 9.6 mA with a new value of $\tau = 6$ H/2.5 kΩ = 2.4 ms.

Step 3: Thus, when $S2$ is closed, the current is at 3.2 mA and heads for 9.6 mA, a total *change* of 6.4 mA. To find the current value 2 ms after $S2$ is closed, we first determine that 2 ms represents $2.0/2.4 = 0.83\,\tau$. According to the time-constant graph in Fig. 15–13, the current will complete 57% of its change in $0.83\,\tau$.

Thus, the additional current will be

$$\text{additional } i = 57\% \times 6.4\ \text{mA} = 3.65\ \text{mA}$$

The current value 2 ms after $S2$ is closed ($t = 3$ ms) will be 3.65 mA greater than its value at $t = 1$ ms. Therefore,

$$i(@\ t = 3\ \text{ms}) = 3.2\ \text{mA} + 3.65\ \text{mA} = \mathbf{6.85\ mA}$$

The complete waveform for i is shown in part (c) of the figure.

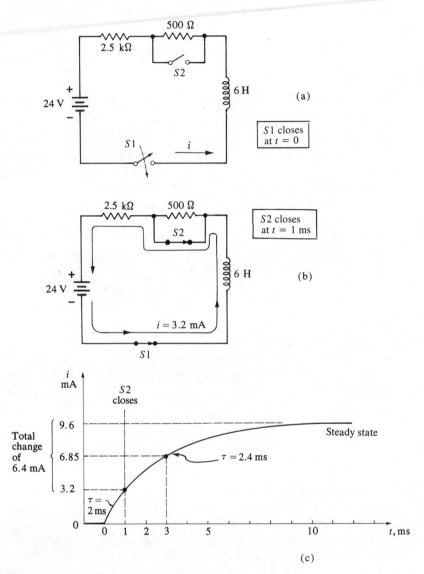

FIGURE 15–15

15.14 Exponential Form Applied To RL Circuits

The exponential form which was introduced in chap. 13 for RC circuits can be readily applied to simple RL circuits. The exponential expressions are the same except that $\tau = L/R$ instead of $R \times C$.

$$v = V_F + (V_0 - V_F)\epsilon^{-t/\tau} \quad (13\text{–}6)$$

$$i = I_F + (I_0 - I_F)\epsilon^{-t/\tau} \quad (13\text{–}7)$$

Of course, when using these expressions the intial and final values of voltage and current are found using the characteristics of RL circuits. The following examples will illustrate.

EXAMPLE 15.18 Write the exponential expression for i in Fig. 15–14. Then find the value of i at $t = 20$ μs.

Solution: $I_0 = 0$; $I_F = \dfrac{32 \text{ V}}{8 \text{ k}\Omega} = 4$ mA; $\tau = \dfrac{120 \text{ mH}}{8 \text{ k}\Omega} = 15$ μs

Substituting these values into the general expression (13–7) we have

$$i = (4 - 4\epsilon^{-t/15\,\mu s}) \text{ mA}$$

To find i at $t = 20$ μs, simply set $t = 20$ μs in this expression and evaluate i.

$$i = 4 - 4\epsilon^{-20/15} = 4 - 4\epsilon^{-1.33}$$
$$= 4 - 4 \times (0.2645) = \mathbf{2.94 \text{ mA}}$$

Compare to the result obtained in Ex. 15.15 using the time constant graph. This mathematical solution is more accurate.

EXAMPLE 15.19 Write the exponential expression for the coil voltage in Fig. 15–14. Then use it to determine how long it takes the coil voltage to drop to 12 V.

Solution: $V_0 = 32$ V; $V_F = 0$ V; $\tau = \dfrac{L}{R} = 15$ μs.

Substitute these values into the general expression (13–6) to obtain

$$v_{\text{coil}} = 0 + (32 - 0)\epsilon^{-t/15\,\mu s} = 32\epsilon^{-t/15\,\mu s} \text{ (V)}$$

Set $v_{\text{coil}} = 12$ V and solve for t.

$$12 = 32\epsilon^{-t/15\,\mu s}$$

or

$$\epsilon^{-t/15\,\mu s} = \frac{12}{32} = 0.375$$

$$\frac{-t}{15\,\mu s} = \ln(0.375) = -0.98$$

$$\therefore t = \mathbf{14.7 \text{ }\mu s}$$

Compare this result to the more approximate answer obtained in Ex. 15.16.

15.15 Comparison of RL and RC Circuits

It might be beneficial at this point to stop and consider the differences that exist between the RC circuit considered in chap. 13 and the RL circuit. These differences can best be pointed out by examining the various waveforms that occur when a dc source is applied to each circuit. These are shown in Fig. 15–16. The following comparisons can be made.

(1) The switch is closed at $t = 0$ in both circuits. In the *RL* circuit, the *coil current* must build up slowly from zero because of the coil's opposition to any increase in current. In the *RC* circuit, the *capacitor voltage* must build up slowly because the capacitor opposes a change in voltage.

(2) In the *RL* circuit the current is maximum in steady state ($t \geq 5\tau$) and zero at $t = 0$. In the *RC* circuit the current is maximum at $t = 0$ and zero in steady state.

(3) In both cases the resistor voltage waveform is the same shape as the current waveform since $v_R = i \times R$.

(4) The value of τ for the *RL* circuit is L/R while for the *RC* circuit it is

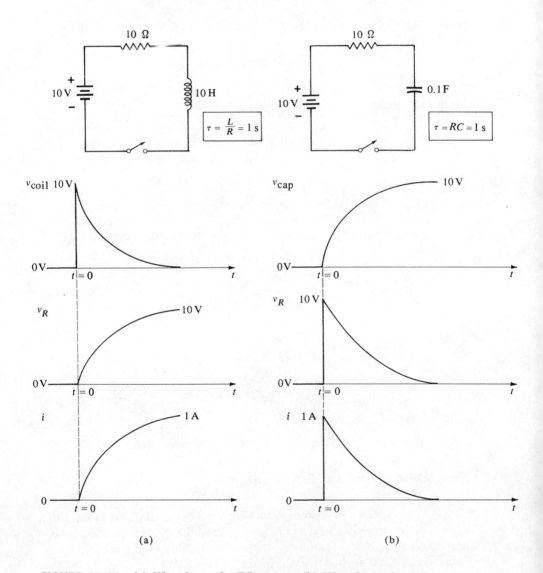

FIGURE 15–16 (a) *Waveforms for* RL *circuit.* (b) *Waveforms for* RC *circuit.*

$R \times C$. In both circuits steady state is reached after 5 τ.

(5) If both circuits have the same value of τ and the same value of source voltage, then v_{coil} for the RL circuit will be the same as v_R for the RC circuit; and v_R for the RL circuit will be the same as the v_{cap} waveform for the RC circuit.

"Charging" an Inductor In the RC circuit of Fig. 15-16(b), we say that the capacitor is being *charged* from 0 V to 10 V. After 5 τ it is fully charged with the number of electric charges on its plates needed to produce a 10-V potential difference across its plates. Although it is not strictly correct, we can use the term *charging* when we refer to the current building up in the inductor in Fig. 15-16(a). Instead of being charged with electric charges, an inductor is "charged" with magnetic flux. As the current increases, the flux in the coil increases until steady state is reached and the current and flux are constant. The term *charging an inductor* is used to describe the action of current (and flux) buildup in a coil.

Situation When R = 0 Although it is impossible to have zero resistance in either of the circuits of Fig. 15-16, we can investigate what would happen in each circuit if R is made smaller and smaller. First, in the RC circuit, as R is made smaller the circuit τ becomes smaller and the capacitor charging time is reduced. If R were zero, the capacitor would instantaneously charge to 10 V when the switch closed.

In the RL circuit, as R is made smaller the circuit τ becomes larger and the inductor charging time is increased. In addition, the smaller R will increase the value of steady-state current. If $R = 0$, τ would become infinite and the coil current would never reach steady state but would increase indefinitely because there would be no resistance to limit the current flow. With $R = 0$, the coil voltage would always equal the 10-V source voltage. This constant coil voltage would indicate a constant rate of current rise since

$$\frac{di}{dt} = \frac{v_{coil}}{L} = \frac{10 \text{ V}}{10 \text{ H}} = 1 \text{ A/s}$$

Thus, with no resistance in the RL circuit the current would rise steadily and indefinitely at this rate.

15.16 Discharging an Inductor

Consider the RL circuit in Fig. 15-17(a). With the switch closed, the inductor current will rise to 10 V/100 Ω = 100 mA in 5 τ = 25 ms. This current will flow indefinitely as long as the switch remains closed. What happens if the switch is suddenly opened?

When the switch is opened, the circuit current *quickly* drops from 100 mA to zero since there is no apparent closed path for current. This sudden decrease in current will induce a large counter emf in the coil. If we assume that it takes

1 μs for the switch contacts to open, then the current will be decreasing at a rate of

$$\frac{di}{dt} = \frac{100 \text{ mA}}{1 \text{ μs}} = 100{,}000 \text{ A/s!}$$

The coil emf will equal ($L\ di/dt$) or

$$\begin{aligned}\text{induced emf} &= 500 \text{ mH} \times 100{,}000 \text{ A/s} \\ &= 0.5 \times 10^5 \text{ V} \\ &= \mathbf{50{,}000 \text{ V!}}\end{aligned}$$

This large induced voltage is a result of the rapid collapse of the flux in the coil as the current drops rapidly to zero. The polarity of this induced emf is as shown in Fig. 15–17(b); note that it opposes the *decrease* in current by acting in the same direction as the source.

In practice, this 50,000-V coil emf would not be developed because, as can be seen from the figure, it would mean that there would have to be 50,000 V across the open switch contacts. This voltage is much more than enough to cause electrical *breakdown* of the insulation (usually air) between the switch contacts. When breakdown occurs, current will be able to flow through the switch insulation as the switch contacts are being separated. The electrons jumping the gap from one switch contact to the other produce a perceptible glow called *arcing*.

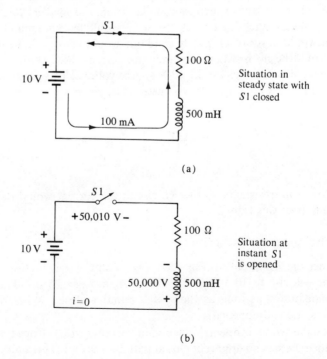

FIGURE 15–17 *Interrupting current flow in an inductor produces a large coil emf.*

As the switch contacts continue to separate, the electrons are not able to jump the gap so the current eventually becomes zero. However, because of the insulator arcing, the current does not decrease so rapidly and the smaller di/dt results in a smaller coil emf.

Continuous closing and opening of the switch can eventually damage the switch contacts because of the arcing. If the switch is a solid-state switch (like a transistor), damage is even more probable since such devices are usually not meant to operate in breakdown. For this reason it is usually undesirable to suddenly interrupt the flow of current in an inductor. The coil emf produced when current is suddenly cut off is often called an *"inductive kick."* We will presently look at a method for reducing inductive kick when a coil has its current reduced to zero.

Using a Discharge Resistor In Fig. 15–18(a) we have the same circuit of Fig. 15–17 except that a 1-kΩ resistor is placed in parallel with the coil. We will assume that the 100 ohms represent the coil's wire resistance so that the 1 kΩ is shown connected across the series combination of 100 ohms and 500 mH. The 10-V source is now across both parallel branches. In steady state, then, 100 mA will again flow through the branch containing the coil. In addition, 10 V/1 kΩ = 10 mA will flow through the 1-kΩ branch. The sum of these two currents, 110 mA, is of course the total current produced by the 10-V source.

If the switch is now opened [part (b) of the figure] the source current will suddenly drop to zero. However, the 100 mA which was flowing through the inductor will *not* have to suddenly drop to zero because it can flow around the loop through the 1-kΩ resistor as shown. What force keeps this current flowing after the source has been removed from the circuit? The counter emf induced in the coil as its current attempts to decrease will act as a voltage source in an attempt to keep the current flowing. At the instant the switch is opened, the coil emf will force its 100-mA current to keep flowing around the loop containing the 1-kΩ resistor. The coil emf will be 110 V as shown because it causes this 100 mA of current to flow through the series combination of 100 Ω and 1 kΩ. That is, 100 mA × 1.1 kΩ = 110 V.

The coil can maintain the flow of current by virtue of the energy which was stored in its magnetic field when the source was producing 100 mA through the coil. This energy is limited, and the current will gradually decrease to zero as the coil's magnetic field collapses. In other words, the coil will temporarily act as a voltage source and allows the current to decrease gradually. This *discharging* of the coil is very similar to the discharging of a charged capacitor.

The coil current will follow the discharge curve shown in part (c) of the figure. The discharge time constant τ is again given by $\tau = L/R$, which in this case is 500 mH/1,100 Ω = 0.46 ms. Thus, the complete coil current decay from 100 mA to zero will take 5 τ = 2.3 ms.

The 1-kΩ resistor serves as a *discharge resistor*, which provides a path for the coil current to flow when the switch is suddenly opened. The fact that the coil current is not forced to drop to zero rapidly, reduces the counter emf (inductive kick). In our example, it is only 110 V. Using KVL around the *w-x-y-w* loop in Fig 15–18(b), we can see that there will be only 110 V across the switch contacts upon opening. This is a great reduction from the 50,000 V encountered earlier when we did not use the discharge resistor. The coil voltage decays to 0 V in approximately 5 τ as shown in Fig. 15–18(c).

EXAMPLE 15.20 For the circuit in Fig. 15–18 what will be the effect on the circuit operation if the 1-kΩ resistor were increased in value?

Solution: Increasing the size of the discharge resistor has two major effects. First, the induced coil emf will have to be larger in order to push 100 mA

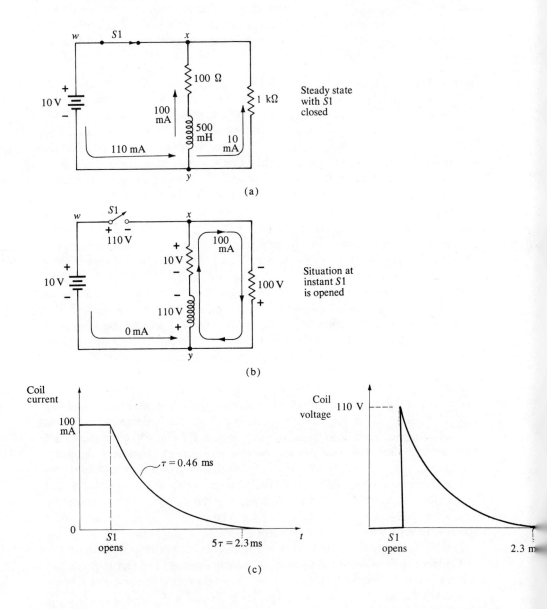

FIGURE 15–18 *Using a discharge resistor to reduce inductive kick by providing a path for coil current.*

through the larger resistor. For example, if $R = 2$ kΩ then the coil emf would have to be 100 mA $\times$ 2,100 Ω = 210 V. This increase in the inductive kick will of course be felt across the switch contacts.

Secondly, the larger R will decrease τ so that the coil current will decay more rapidly. This makes sense because a larger coil emf indicates that the current is changing more rapidly (larger di/dt).

As the example shows, a larger discharge resistor produces a greater inductive kick and causes a more rapid decay of the coil current. The opposite is true for a smaller discharge resistor. A smaller discharge resistor will reduce the inductive kick and will cause a slower decay because of the larger τ.

We can summarize this section by concluding that an inductor will not allow its current to be cut off instantaneously. It will always generate a counter emf which attempts to keep its current flowing. Unless a discharge path is provided, the counter emf can become very large and can cause damage to the switch contacts. Fortunately, the inductive kick is only momentary until the current reaches zero.

15.17 Using the Inductive Kick

As we saw in the last section, a very large coil voltage can be induced when its current is suddenly interrupted and there is no discharge path. This large inductive kick can serve a useful purpose in certain applications. A familiar example is the induction coil present in most automobile ignition systems. In this application the current in a coil is periodically interrupted by cam-driven switch contacts (referred to as a breaker point). The large induced emf is fed to spark plugs to produce the arc or spark needed to cause fuel combustion.

15.18 Radiated Transients

The large emf induced in a coil whose current is being interrupted can produce a tremendously strong electric field surrounding the coil. This electric field essentially radiates from the coil in all directions. As such, it can often affect the operation of other circuits and circuit components in close proximity to the coil. For this reason, it is usually undesirable to place sensitive solid-state circuitry in the same vicinity as relays, motors, or other inductive elements which are liable to radiate transient electric fields. If it cannot be avoided, usually some type of shielding is used to protect the sensitive elements.

15.19 Energy Stored in an Inductor

We saw in sec. 15.16 that an inductor can act as a temporary source of emf to produce current after the voltage source is removed from a circuit. The current which it produces causes energy to be dissipated in the form of heat in the circuit resistances. Where does this energy come from? This energy is present in the magnetic field of the coil and it was originally supplied by the voltage source which produced the current through the coil.

In other words, a current in the coil produces flux in the coil. This magnetic flux represents stored magnetic energy which can be used when the voltage source is removed from the circuit. As a coil discharges through a resistor, this magnetic energy is converted to electrical energy, which is dissipated as heat. An inductor itself does not dissipate energy. *An inductor stores energy in its magnetic field.*

The amount of energy stored in an inductor depends on the amount of current flowing and on the size of the inductance. A larger current means more flux and a stronger coil magnetic field. Thus, the stored energy will be greater for larger currents.

A larger value of inductance means that the inductor will discharge more slowly and therefore maintain the flow of current for a longer time. This requires more energy. Thus, stored energy must be greater for larger values of inductance.

The actual formula for energy stored in an inductor is:

$$\text{Energy} = \tfrac{1}{2} LI^2 \text{ (joules)} \tag{15-8}$$

In this formula, L is in henries, I is in amperes, and the resultant energy is in joules.

Consider the circuit in Fig. 15-19. When the switch is closed, current gradually increases towards its steady-state value of 24 V/200 Ω = 120 mA. As the current i increases, more and more energy is being dissipated in the resistor; also more and more energy is being stored in the inductor ($\tfrac{1}{2}LI^2$). When the current reaches its steady-state value of 120 mA, the energy stored in the coil will increase no further since the current is constant. The value of this energy will be

$$\begin{aligned} \tfrac{1}{2}LI^2 &= \tfrac{1}{2} \times 100 \text{ mH} \times (120 \text{ mA})^2 \\ &= 0.5 \times 0.1 \times (0.12)^2 \\ &= 0.00072 \text{ joule} \\ &= 7.2 \times 10^{-4} \text{ J} \end{aligned}$$

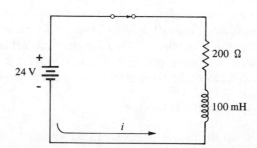

FIGURE 15-19

Since current is still being supplied by the source in steady state, then the source is still supplying energy but none of it is being stored in the inductor. Instead, it is being lost in the resistor as heat.

Inductance 409

15.20 Comparison of C and L Energy Storage

In a capacitor, energy is stored in the form of charges resting on the capacitor's plates. As long as current flows, the capacitor will continue charging and its energy will increase. When current stops flowing the capacitor's stored energy will remain fixed. A capacitor will store the energy without the need for current to continue flowing. It is a *static* type of energy storage and the energy is stored in the electrostatic field between the capacitor's plates.

In an inductor, energy is stored in the form of flux lines linking the coil. To maintain the magnetic field a current must be present. As such, the inductor's stored energy will be zero if no current flows. An inductor, then, possesses a *dynamic* type of energy storage which is present in its magnetic field.

15.21 Steady-State Solution of Circuits Containing R, L, and C

If a constant dc source (or sources) is applied to a network which contains more than one L or C, it is possible to determine the steady-state behavior of the circuit by replacing all Ls and Cs by their *steady-state equivalents* and then solving the circuit.

In steady state we know that an ideal capacitor will act as an *open circuit* to dc. In steady state an ideal inductor will act as a *short circuit* to dc. Consider the circuit in Fig. 15-20(a). It contains one inductor and one capacitor. We will assume that the inductor's wire resistance is included in the 2-kΩ resistor shown. To determine the steady-state values for this circuit, the inductor is replaced by a short circuit and the capacitor is replaced by an open circuit as redrawn in Fig. 15-20(b). This simple circuit is now easy to analyze. As can be seen, 4 mA of current flow through the coil and 8 V are present between terminals w and y (across the capacitor).

This same procedure can be followed no matter how many Ls and Cs are present in the circuit. Unfortunately there is no easy way to determine the time to reach steady state for the waveforms of current and voltage. Such analyses require techniques that are beyond the scope of this text and usually beyond the abilities of all but the most experienced professional circuit analysts.

In some cases, it is more accurate to replace a *practical* inductor by its dc wire resistance if the wire resistance is not already included in the circuit. It is sometimes necessary to replace a *practical* capacitor by its leakage resistance R_l in dc steady state. In many instances, the leakage resistance of a capacitor is so large that it makes little difference if it is assumed to be an open circuit. However, if capacitors are connected in series, it is their respective values of R_l that will determine their voltages in steady state. This will be illustrated in one of the problems at the end of the chapter.

Keep in mind that the preceding discussion is only valid for steady-state calculations when the sources are constant dc sources.

15.22 Combinations of Inductors

Inductors connected in *series* can be combined into one equivalent inductor by *adding* the individual inductances. That is,

$$L_{eq} = L1 + L2 + L3 \ldots \text{(for series } L\text{s)} \tag{15-9}$$

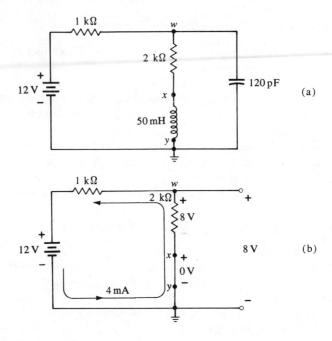

FIGURE 15-20 *Replacing coil by a short circuit and capacitor by an open circuit in steady state.*

This is the same manner in which series resistors are combined. As such, inductors in series produce a *larger* total inductance.

Inductors connected in *parallel* can be combined into one equivalent inductor by using the reciprocal formula

$$\frac{1}{L_{eq}} = \frac{1}{L1} + \frac{1}{L2} + \frac{1}{L3} \ldots \text{(for parallel } L\text{s)} \qquad (15\text{--}10)$$

This is the same relationship used for parallel resistors. For two Ls in parallel this reduces to

$$L_{eq} = \frac{L1 \times L2}{L1 + L2} \qquad (15\text{--}11)$$

which is the familiar product-over-sum formula. Thus, for inductors in parallel, the total equivalent inductance is *less than* any of the individual inductors.

Chapter Summary

1. When one flux line links one loop of wire this is called a flux linkage.
2. An emf is induced in a loop or coil of wire whenever the number of flux linkages is changing; the magnitude of the emf is proportional to the rate at which the number of flux linkages is changing (Faraday's law).

3. Mutual induction occurs when a changing current in one coil or winding (primary) induces an emf in another winding (secondary).
4. If the same flux Φ links all N turns of a coil, then the emf induced by a change in Φ will equal $N\, d\Phi/dt$.
5. If a change in flux is being caused by a change in current, then the emf induced in a coil is directly proportional to the rate of change of this current (Faraday's law).
6. When a change in flux linkages induces an emf in a wire or a coil, the polarity of the induced emf will be such that any current flow produced by it will develop a magnetic field which *opposes* the original change in flux linkages (Lenz's law).
7. Faraday's law determines the magnitude of an induced emf; Lenz's law determines the polarity of the induced emf.
8. The process by which an emf is induced in a coil by virtue of its own *changing* current is called self-induction.
9. The magnitude of the self-induced emf in a coil is equal to $L\, di/dt$ where L is the coil's inductance.
10. The unit for inductance is the henry (H).
11. A practical inductor possesses resistance as well as inductance.
12. In an inductor, the induced voltage across its terminals is of such a polarity that it opposes the change in current (Fig. 15–11).
13. If the electron current through an inductor is increasing (decreasing), then the end of the inductor at which the current enters will be of negative (positive) polarity.
14. The principal circuit characteristic of an inductor is that it opposes any change in current.
15. Inductance is proportional to the square of the number of coil turns N, and inversely proportional to the reluctance of the core.
16. An inductor will not allow its current to change instantaneously from one value to another.
17. When a voltage is applied to a series RL circuit, the current starts at zero and builds up gradually to a steady-state value of E_s/R.
18. The time constant τ for an RL circuit is equal to L/R. The currents and voltages in an RL circuit will reach steady state in 5 τ.
19. When current through an inductor is interrupted suddenly the inductor generates a large counter emf (inductive kick) which attempts to keep the current flowing.
20. If the inductor is provided with a discharge path for its current, the inductive kick can be substantially reduced.
21. An inductor will discharge through a resistive circuit with a discharge τ equal to L/R.
22. An inductor can store energy in its magnetic field. At any instant of time its stored energy will be equal to $\tfrac{1}{2}Li^2$.
23. In steady state an ideal inductor can be replaced by a short circuit and an ideal capacitor by an open circuit if the circuit contains only dc sources.
24. Table 15–1 is a comparison of the *three* basic circuit properties: resistance R;

capacitance C; and inductance L. At this point the student should have a reasonably clear understanding of the basic differences listed in the table.

25. In a simple RL circuit, the currents and voltages in response to a sudden input change will obey the exponential equations:

$$v = V_F + (V_0 - V_F)\epsilon^{-t/\tau}$$
$$i = I_F + (I_0 - I_F)\epsilon^{-t/\tau}$$

TABLE 15-1

Resistance, R	Capacitance, C	Inductance, L
$i = v/R$	$Q = Cv$ $i = C\left(\dfrac{dv}{dt}\right)$	$v = L\left(\dfrac{di}{dt}\right)$
Opposes the flow of current	Opposes any sudden *change* in voltage with $\tau = RC$	Opposes any sudden *change* in current with $\tau = L/R$
Dissipates energy as heat $\left(p = i^2 R = \dfrac{v^2}{R}\right)$	Stores energy $= \tfrac{1}{2}Cv^2$	Stores energy $= \tfrac{1}{2}Li^2$
$R_{eq} = R1 + R2 + \cdots$	$\dfrac{1}{C_{eq}} = \dfrac{1}{C1} + \dfrac{1}{C2} + \cdots$	$L_{eq} = L1 + L2 + \cdots$
	Series Combinations	
$\dfrac{1}{R_{eq}} = \dfrac{1}{R1} + \dfrac{1}{R2} + \cdots$	$C_{eq} = C1 + C2 + \cdots$	$\dfrac{1}{L_{eq}} = \dfrac{1}{L1} + \dfrac{1}{L2} \cdots$
	Parallel Combinations	
Ideal R has fixed resistance for any input voltage. *Practical R* will have its resistance change at higher voltages.	*Ideal C* dissipates no energy. *Practical C* has leakage resistance R_l which does dissipate some power.	*Ideal L* has no resistance and dissipates no energy. *Practical L* dissipates energy due to its winding resistance and also due to hysteresis losses in the core.
R acts the same during transient interval and in steady state	C opposes *changes* in v during transient interval and acts as open circuit in dc steady state	L opposes *change* in i during transient interval and acts as short circuit in dc steady state.

Questions/Problems

Sections 15.2–15.3

15-1 Consider the experiment illustrated in Fig. 15-1. When the bar magnet is quickly pushed inside the coil, the meter momentarily registers a current of 20 μA. For each of the following modifications indicate whether the meter's reading will be *greater* than 20 μA, *lower* than 20 μA, or *unchanged* if the experiment is repeated.

a. Use a stronger bar magnet.
b. Keep the magnet stationary and quickly push the coil over the magnet.
c. Increase the number of turns of wire in the coil.
d. Push the magnet inside the coil at a slower rate.
e. Wind the coil tighter so that the turns are closer together.

15-2 Indicate the truth or falsity of each of the following statements.
a. Reversing the poles of the magnet in Fig. 15-1 will produce a current pulse of the opposite polarity in the coil.
b. An induced voltage is present whenever flux links a coil.
c. Four flux lines linking two turns of wire represent *eight* flux linkages.
d. In Fig. 15-3 as the magnet is pulled away from the coil, a current will be induced in the coil.
e. The current will persist as long as the magnet is moving away from the coil, no matter how far away it gets.
f. A decrease in flux linkages will always induce an emf of opposite polarity from an increase in flux linkages.

Sections 15.4–15.5

15-3 Refer to Fig. 15-21 where a rectangular soft-iron core has three coils wound on it. A dc source is connected to the primary winding, which has 100 turns. The two secondary windings have 50 turns and 200 turns respectively. (a) When the primary switch is closed, which secondary winding will have the largest emf induced in it? (b) If 1 V appears across secondary #2, how much voltage appears across secondary #1?

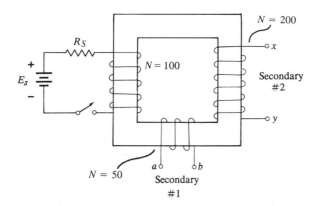

FIGURE 15-21

15-4 A bar magnet is moved toward a single loop of wire such that the flux linking the loop increases from 0 to 1000 μWb in 10 ms.
a. What is the magnitude of the emf induced in the wire?
b. If the magnet is moved at *twice* the speed, what will the emf be?
c. Repeat (b) if *two* closely wound loops are used.

15-5 Which of the following will produce the largest induced emf in a coil of 100 turns?
a. A flux of 30 μWb increasing to 40 μWb in 5 ms
b. A flux of 1000 μWb decreasing to 990 μWb in 5 ms
c. A flux of 1 μWb increasing to 2 μWb in 10 ms

414 Chapter Fifteen

15-6 Refer again to Fig. 15-21. Assume $E_s = 6$ V, and $R_s = 3$ ohms. Also assume that the reluctance of the core is 10^4 A-t/Wb.
  a. Calculate the amount of flux which will be present in the core after the primary switch is closed.
  b. If it takes 2 ms for this flux to build up in the core after the switch is closed, calculate the approximate emf induced in each secondary coil as the flux builds up.

Section 15.6

15-7 Refer to Fig. 15-22(a). The arrow shows that the permanent magnet is approaching the coil. Indicate the direction of the induced current in the coil.

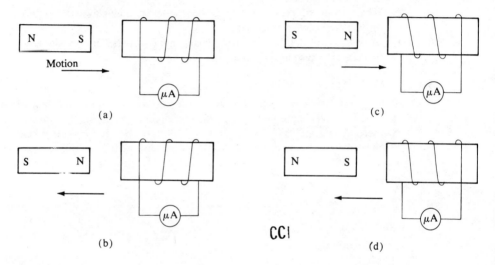

FIGURE 15-22

15-8 Repeat 15-7 for Figs. 15-22(b)-(d).

15-9 Refer to Fig. 15-23. The unlabeled bar magnet is moved toward the coil and induces a coil current in the direction shown. Label the poles of the magnet.

15-10 Refer again to Fig. 15-21. Determine the polarity of the emf induced in each secondary winding when the primary switch is closed. (Hint: it may be helpful to connect a resistor load across each secondary winding.)

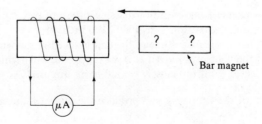

FIGURE 15-23

Inductance   415

Sections 15.7–15.8

15-11 Explain how a changing current through a coil will induce an emf in the coil.

15-12 Determine which of the following cases produce the largest coil voltage:
   a. a 3-H coil whose current changes from 1 A to 2 A in 0.2 s
   b. a 15-mH coil whose current increases from 10 mA to 15 mA in 5 ms
   c. a 3-H coil whose current changes from 101 A to 102 A in 0.2 s
   d. a 20-mH coil whose current changes from 5 mA to 7 mA in 2 μs
   e. a 50-mH coil with a constant current of 100 A

15-13 Why would a loop of wire have more inductance than a straight wire conductor?

15-14 Which of the following values of inductance will produce the largest emf for a given change in current: 10 mH; 0.005 H; 3200 μH; 3×10^2 mH; $15 \times 10^4 \mu$H?

15-15 A 50-mH inductor is connected in a dc circuit where a *constant* 200 mA is flowing through it. However, a voltage of 1 V is measured across the coil. Explain.

15-16 Calculate the coil resistance of the inductor in problem 15–15. Draw the circuit representation for this practical inductor.

15-17 A certain coil produces a voltage of 12 V when its current changes at the rate of 6 A/s. What is the coil's inductance?

Section 15.9

15-18 Indicate the truth or falsity of each of the following statements:
   a. Current always flows from negative to positive through a resistor.
   b. Current always flows from negative to positive through an inductor.
   c. A constant current produces no voltage in a practical inductor.
   d. Inductance is an opposition to the flow of current.
   e. The induced emf across a coil *always* opposes the polarity of the source.

15-19 Refer to Fig. 15–24(a). The current i is increasing at the rate of 12 mA/μs. Determine the polarity and magnitude of v_{xy}.

15-20 In Fig. 15–24(b) the current i decreases from 300 μA to 200 μA in 2 μs. Determine the polarity and magnitude of v_{xy}.

15-21 Determine the rate of change of current in Fig. 15–24(c). Is i increasing or decreasing?

15-22 Repeat for 15–24(d) and (e).

Section 15.10

15-23 A certain iron-core inductor has an inductance of 100 mH. Indicate how the inductance value will be affected by each of the following changes:
   a. Increase the number of coil turns.
   b. Use a lower permeability core.
   c. Increase the length of the core's magnetic path.
   d. Decrease the core cross-sectional area.
   e. Remove the iron core.

15-24 Calculate the inductance of an air-core coil which is 12-cm long, has a diameter of 1 cm, and has 250 turns.

Sections 15.11–15.13

15–25 Refer to the circuit in Fig. 15–25. The switch is closed at $t = 0$. Determine:
  a. i at $t = 0$
  b. di/dt at $t = 0$
  c. time to reach steady state
  d. i in steady state
  e. di/dt in steady state

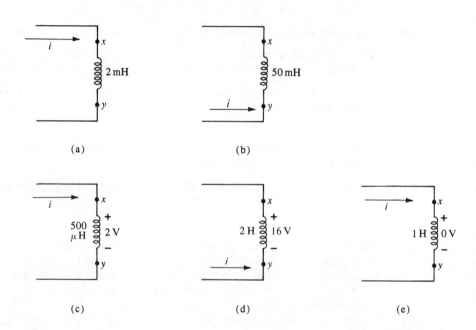

FIGURE 15–24

15–26 Sketch waveforms of i, v_R, and v_{coil} for the circuit of Fig. 15–25.

15–27 For the circuit in Fig. 15–25 determine:
  a. the current 0.2 ms after the switch is closed
  b. the coil voltage 0.3 ms after the switch is closed
  c. the amount of time it takes for the current to reach 6 mA

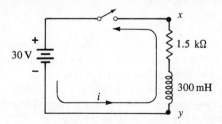

FIGURE 15–25

15–28 Consider the circuit in Fig. 15–26. Initially $S1$ is open and $S2$ closed. At $t = 0$, $S1$ is closed. One second later, $S2$ is open. Find the value of current one second after $S2$ is open.

Section 15.14

15-29 (a) Redo Ex. 15.14 using the exponential expressions for coil current and voltage.
  (b) Redo problem 15-27 using the exponential expressions.

15-30 A *relay* is an electromagnet that is used to open or close switch contacts by virtue of its magnetic force. Figure 15-27 shows the diagram of a relay that controls switch contacts *A-B*. When the 12-V source is connected to the relay coil, the coil current gradually builds up and the strength of the electromagnet increases. Eventually the magnet's strength is sufficient to pull switch *A-B* closed. At this point we say that the relay is *pulled-in*.

The relay coil has a resistance of 250 ohms and an inductance of 100 mH. The current needed to pull-in the relay is 20 mA. How long after the 12 V is applied to the relay coil will the relay pull-in? Use the exponential form for i.

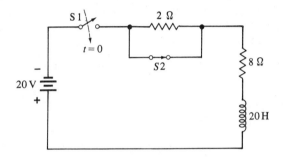

FIGURE 15-26

Section 15.15

15-31 Indicate which of the following statements refer to an *RC* circuit and which ones refer to an *RL* circuit, in response to an applied 10-V dc source:
  a. The current is maximum at $t = 0$.
  b. The current increases most rapidly at $t = 0$.
  c. The current reaches steady state in 5 τ.
  d. The time to reach steady state increases as *R* is decreased.
  e. No current flows after 5 τ.
  f. 10 V appear across *R* in steady state.
  g. Steady state will never be reached if $R = 0$.

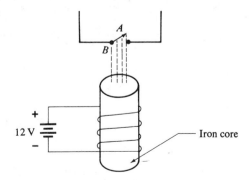

FIGURE 15-27

Section 15.16

15-32 Refer to Fig. 15–25. The 1.5-kΩ resistor represents the coil's resistance. The switch is closed long enough for the current to reach steady state. Then, the switch is suddenly opened. If it takes 1 μs for the switch to open, calculate the coil's inductive kick.

15-33 Place a 10-kΩ resistor across the coil (terminals x-y) in Fig. 15–25 and repeat problem 15–32. Also, sketch the coil current waveform and determine how long it takes to decay to zero.

15-34 Change the 2-ohm resistor in Fig. 15–26 to a 32-ohm resistor and repeat problem 15–28. Sketch the current waveform.

Section 15.19

15-35 A 30-mH coil is being discharged through a 2-kΩ resistor. The initial coil current is 50 mA.
  a. How long will it take the current to decay to 20 mA?
  b. How much energy will be dissipated in the resistor during the complete discharge?

15-36 Indicate how each of the following changes will affect the answers to 15–35(a) and (b):
  a. Increase the coil inductance.
  b. Decrease the resistor value.
  c. Decrease the initial current value.

15-37 What is the total energy stored in the coil of Fig. 15–18 when $S1$ is closed? What happens to this energy when $S1$ is opened?

Section 15.21

15-38 For each of the circuits in Fig. 15–28 determine all the currents and voltages in steady state.

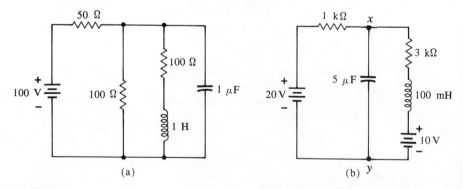

FIGURE 15-28

15-39 In Fig. 15–29 the upper capacitor has a leakage resistance of 10 MΩ and the lower capacitor has a leakage resistance of 15 MΩ. Determine the voltages which each capacitor will charge to in steady state.

15-40 Suggest a method which can be used to modify the circuit in Fig. 15–29 so that each capacitor will charge to 100 V (half of E_s).

Section 15.22

15–41 Three isolated inductors $L1 = 2$ mH, $L2 = 5$ mH, and $L3 = 10$ mH are connected in series. Determine their total equivalent inductance.

15–42 If the inductors in problem 15–41 are connected in parallel, determine L_{eq}.

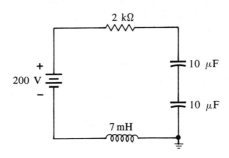

FIGURE 15-29

CHAPTER 16

The Sinusoidal Waveform (AC)

16.1 Introduction

In our earlier discussion of waveforms in chap. 12 the *sinusoidal waveform* (sine wave) was introduced as one of the many types of ac waveforms that can exist in electric circuits. In actuality, the sine wave is the most important and useful waveform in electronics and in most other scientific or technological fields. Briefly, the reasons for the importance of sine waves in electronics include those listed here.

(1) Most power utility companies generate and distribute ac power in the form of sinusoidal voltages.

(2) All types of time-varying waveforms can be shown to consist of a series of sine waves. As such, any complex waveform can be analyzed on the basis of its sinusoidal content. Audio and radio waveforms are examples of this.

(3) The sine wave is a useful test waveform for experimentally or analytically determining the operation of many types of circuits. The sine wave is unique in that it is the only waveform that can be applied to the input of *any* linear circuit to produce an output that is an undistorted version of the input.

(4) It is relatively easy to generate sine waves either electronically or electromechanically. Electronic sine wave generators or oscillators are used to produce sine waves with frequencies up to the GHz (10^9 Hz) range. Electromechanical generators use the principle of electromagnetic induction to produce a sinusoidal voltage in a coil which is rotating in a magnetic field.

Because of its numerous applications, the sine wave will be studied in detail in this chapter. In subsequent chapters we will investigate the operation of circuits containing resistance, inductance, and capacitance in response to sine wave inputs.

Although there are numerous types of ac waveforms, when we talk about ac we are usually referring to sine waves. In this chapter and most of those to follow, we will be concentrating on the sine wave. For this reason we will often use the term *ac voltage* when we mean sinusoidal voltage, as is customary in most texts. The term sine wave will still be used frequently throughout the text so that we do not forget what we are talking about.

A study of sine waves and the circuits which are driven by sine wave inputs is more meaningfully accomplished if certain mathematical techniques are employed. In particular, the use of trigonometry and trigonometric functions (sine, cosine, tangent) is a necessity in dealing with sine wave (ac) circuits. Although it is assumed that the reader has some background in trigonometry and can use trigonometric tables and/or the calculator to find the values of trig functions, we will begin the study of the sine wave with a brief review of some basic trigonometry concepts.

16.2 Rotating Vectors, Angles, and Quadrants

In Fig. 16–1(a), the axes of the familiar x-y coordinate systems are drawn and labeled. The intersection of the x and y axes is labeled O for *origin*. In part (b) of the figure a *vector*, OA, is drawn so that it lies on the $+x$-axis. A vector is a straight line that has a certain length and points in a certain direction. Vectors are often used to represent physical quantities like force and velocity which have both a magnitude and a direction. Later we will use vectors to represent currents and voltages in ac circuits.

The vector OA in Fig. 16–1(b) lies right on the $+x$-axis. In other words, there is no *angle* between the vector and the $+x$-axis. *The $+x$-axis will be our reference axis or $0°$ axis from which all angles will be measured.* If we keep one end of the vector on the origin and rotate the vector in a counterclockwise (ccw) direction, there will now be an angle between the vector and the $+x$-axis. Let us call this angle θ (Greek letter *theta*). Figures 16–1(c)-(f) depict some possible positions for the vector.

In part (c) the angle θ which the vector makes with the $+x$-axis is less than $90°$. If the vector is rotated so that it lies on the $+y$-axis, then θ would be $90°$. The area between the $+x$-axis and the $+y$-axis is called the *first quadrant* (quadrant I). It should be clear that whenever the vector lies in this quadrant, the angle θ will be between zero and $90°$. That is

$$0° \leq \theta \leq 90° \text{ (in quadrant I)}$$

In part (d) of the figure, the vector has been rotated into the second quadrant (quadrant II), which lies between the $+y$-axis and the $-x$-axis. In this quadrant the angle θ will be between $90°$ and $180°$.

$$90° \leq \theta \leq 180° \text{ (in quadrant II)}$$

If the vector is made to lie on the $-x$-axis, then θ would be $180°$. In other words, the $-x$-axis is $180°$ from the $+x$-axis. The vector has been rotated through one half of a full circle.

The Sinusoidal Waveform (AC) 423

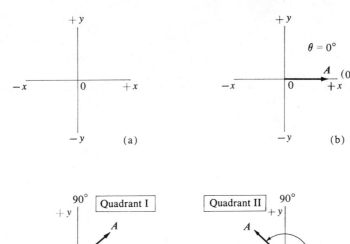

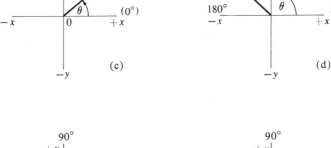

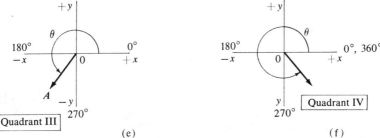

FIGURE 16-1 *Vector rotating about the origin of an x-y coordinate system.*

In part (e) of the figure, the vector has been rotated into the third quadrant (quadrant III), which lies between the −x-axis and the −y-axis. In this quadrant θ will be between 180° and 270°.

$$180° \leq \theta \leq 270° \text{ (in quadrant III)}$$

If the vector is rotated so that it lies on the −y-axis, then θ would equal 270°. The vector will have been rotated through three quarters of a full circle.

Finally, in part (f) the vector has been rotated into the fourth quadrant (quadrant IV), which lies between the +x-axis and the −y-axis. In this quadrant θ will be between 270° and 360°.

$$270° \leq \theta \leq 360° \text{ (in quadrant IV)}$$

If the vector is rotated so that it again lies on the +x-axis, then θ would be 360°. The vector will have been rotated through a complete circle starting at the $+x$-axis [part (b) of the figure] and rotating ccw until it again lies on the $+x$-axis. Whenever the vector makes one complete revolution, it will have gone through 360° of rotation.

Rotation Greater Than 360° In Fig. 16-2(a) we have shown the vector OA rotated ccw 45° from its 0° position on the $+x$-axis. In Fig. 16-2(b) we have shown the same vector rotated ccw through a total angle of 405°. It is obvious from the figure that in both cases the vector OA ends up in the same position in quadrant I. In part (b) the vector was rotated through 405°, but 405° = 360° + 45°. Thus, the first 360° of rotation simply brought it back to its starting position. The remaining 45° rotation brought it to its final position in quadrant I where it makes an angle of 45° with the $+x$-axis.

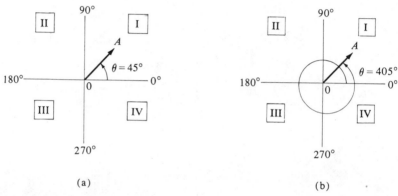

FIGURE 16-2 *If a vector is rotated through one or more complete revolutions (360°), it will end up in the same position.*

Since the vector OA ended up in the same position in both cases, then we can say that $\theta = 405°$ is the same as $\theta = 45°$ as far as the vector's position is concerned. Using the same reasoning, we can conclude that $\theta = 765°$ is the same as $\theta = 45°$ since $765° = 360° + 360° + 45°$ (two complete rotations plus 45° more).

EXAMPLE 16.1 Determine which quadrant the vector will end up in if it is rotated ccw through the following angles: (a) 480°; (b) 920°.

Solution: (a) Since $480° = 360° + 120°$, the vector will end up in quadrant II making an angle of 120° with the $+x$-axis. Thus, $\theta = 480°$ is the same as $\theta = 120°$.

(b) Since $920° = 360° + 360° + 200°$, the vector will end up in quadrant III just as if it were rotated through only 200° relative to the $+x$-axis. Thus, $\theta = 920°$ is the same as $\theta = 200°$.

Negative Angles In our discussion so far we have concerned ourselves with rotation in a ccw direction from the $+x$-axis. This direction is arbitrarily chosen as the *positive* direction of rotation and the resulting angles of rotation are *positive angles*. Thus, in Fig. 16-2, for example, $\theta = +45°$ in (a) and $\theta = +405°$ in (b).

Rotation in the cw direction is considered *negative* and the resulting angles are *negative angles*. Figure 16–3 shows two examples of this. In (a) the vector OA has been rotated 45° in the cw direction. Thus, $\theta = -45°$. In (b) the vector has been rotated 120° in the cw direction so that $\theta = -120°$.

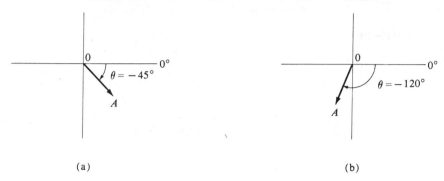

FIGURE 16–3 *Examples of negative angles.*

Looking at Fig. 16–3(a) it should be clear that the vector OA could also have been rotated $+315°$ in the ccw direction from the reference axis to reach the same position shown. Similarly in (b) the vector could have been rotated $+240°$ ccw to reach the same position shown. This means that we can stipulate the position of the vector in at least two ways. We can give its *positive* angle relative to the $+x$-axis (ccw rotation) or its *negative* angle relative to the $+x$-axis (cw rotation). Either representation would be correct.

EXAMPLE 16.2 Indicate the position of the vector in Fig. 16–2(a) using both positive and negative angle representations.

Solution: In (a) the vector's position could be described as 45° rotation in the ccw direction so that $\theta = +45°$. Or it could be described as 315° rotation in the cw direction so that $\theta = -315°$.

EXAMPLE 16.3 A vector is rotated through an angle of $\theta = -430°$ relative to the $+x$-axis. Sketch its position on the x-y coordinates and then represent it by an equivalent positive angle.

Solution: The vector is shown in Fig. 16–4(a). Since $-430° = -360° - 70°$,

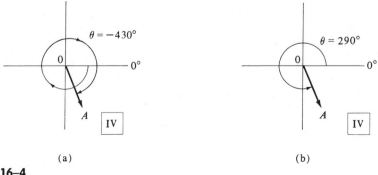

FIGURE 16–4

the vector has been rotated in the cw direction through one complete (360°) revolution and then 70° more to end up in the position shown.

The vector's position could also have been reached by rotating it 290° in the ccw direction. Thus, its angle could be $\theta = +290°$ [part (b) of the figure].

Labeling of Axes (Polar Coordinates) Since we will be dealing so much with angles in our subsequent work, it will be more useful to label our x-y coordinate axes as shown in Fig. 16–5. The $+x$-axis is labeled as 0° since it represents the reference line from which we will measure our angles. The $+y$-axis is labeled as 90° because a vector would have to rotate 90° from the $+x$-axis in order to lie on the $+y$-axis. For similar reasons, the $-x$-axis is labeled as 180° and the $-y$-axis is labeled as 270°. When the axes are labeled as such, we call them *polar coordinate axes*.

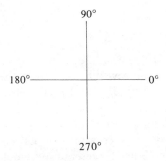

FIGURE 16–5 *Polar coordinate axes.*

Relative Position of Vectors Figure 16–6 shows two vectors, OA and OB. Vector OA is at an angle of 60° while vector OB is at an angle of 45° relative to the 0° reference axis. We can therefore say that OA is at an angle of $+15°$ relative to OB since OA is rotated 15° further in the ccw direction. Equivalently, we can say that OB is $-15°$ relative to OA. In the work to follow we will have occasion to use the terms *lead* and *lag* when referring to vectors (and sine waves). Thus, we can say that vector OA *leads* vector OB by 15°; or we can say that vector OB *lags* (behind) vector OA by 15°. When two vectors are at different angles, we say that they are *out of phase*.

16.3 Finding the Sine, Cosine, and Tangent of Any Angle

In the course of our investigation of sine waves and ac circuits, we will often need to find the *sine*, *cosine*, or *tangent* of an angle. It is a relatively easy matter to find the sine, cosine, or tangent of an *acute* angle (less than 90°) by using trig tables or electronic calculator. If we are dealing with an angle greater than 90°, the procedure is a little more complicated.* The steps listed here and the subsequent examples will illustrate how to determine the sine of *any* angle θ.

(1) On polar coordinate axes draw a vector from the origin so that it is at an angle θ relative to the 0° axis.

---

* Unless we have an electronic calculator.

The Sinusoidal Waveform (AC)   427

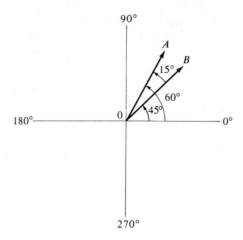

FIGURE 16–6 *Two vectors shown at different angular positions (out of phase).*

(2) Find the angle which this vector makes with the *closest* horizontal axis. Call this angle α. Angle α will always be an *acute* angle.
(3) Find sin α from trig tables or calculator.
(4) If the vector lies in quadrants I or II, then sin θ = sin α; if the vector lies in quadrants III or IV, then sin θ = −sin α (in other words, sin θ is negative in quadrants III and IV).

EXAMPLE 16.4 Determine: (a) sin 60°; (b) sin 120°; (c) sin 240°; and (d) sin 300°.

Solution: (a) In Fig. 16–7(a) a vector is drawn at an angle θ = 60° on the

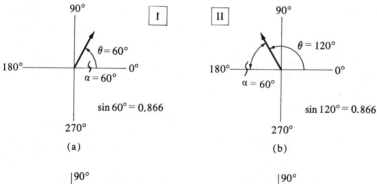

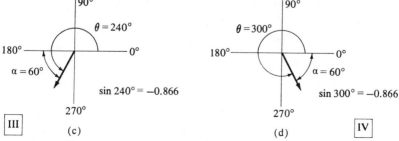

FIGURE 16–7 *Finding the sine of any angle.*

polar coordinate axes (step 1). The horizontal axis which is closest to the vector is the 0° axis (+x-axis). The angle which the vector makes with this axis is also 60°. In this case $\alpha = \theta = 60°$ (step 2). From the tables in Appendix A or from calculator

$$\sin \theta = \sin \alpha = \sin 60° = 0.866$$

Obviously, the sine of 60° could have been determined without going through this lengthy procedure. The same is true for any acute angle θ. This procedure, however, is necessary to determine the sine of angles greater than 90°.

(b) In Fig. 16–7(b) a vector is drawn so that it is at an angle $\theta = 120°$ on the polar axes (step 1). The horizontal axis which is closest to the vector is the 180° axis. Thus, as the diagram shows, $\alpha = 60°$ (step 2). Sin $\alpha = \sin 60° = 0.866$ (step 3)

Since the angle $\theta = 120°$ is in quadrant II, then $\sin \theta = \sin \alpha = 0.866$ (step 4)

(c) In part (c) of the figure the vector is drawn at an angle $\theta = 240°$. The closest horizontal axis is the 180° axis so that $\alpha = 60°$.

$$\sin \alpha = \sin 60° = 0.866$$

Since $\theta = 240°$ is in quadrant III, then $\sin \theta = -\sin \alpha = -0.866$.

(d) The reader should verify that $\sin 300° = -0.866$ using Fig. 16–7(d).

When finding the trig function of any angle greater than 90°, it is wise to get in the habit of sketching it on polar coordinates as was done in the preceding example. A graphical representation of the angle usually gives you a firmer understanding of what you are doing.

Cos θ and Tan θ The cosine or tangent of any angle can be found in a manner exactly like that outlined for the sine. The only difference is in the polarities of the trig functions in the different quadrants (step 4). Table 16–1 lists the polarities of each of the three trig functions for the four quadrants. As the table shows, the sine is positive in quadrants I and II while the cosine is positive in I and IV. The tangent alternates polarity starting positive in the first quadrant. It is only necessary to remember the polarities of the sine and cosine because the following identity is always true:

$$\tan \theta = \frac{\sin \theta}{\cos \theta}$$

Since $\tan \theta$ is equal to the quotient of $\sin \theta$ over $\cos \theta$, then $\tan \theta$ will be positive whenever $\sin \theta$ and $\cos \theta$ have the same polarity (quadrants I and III), and negative whenever $\sin \theta$ and $\cos \theta$ have opposite polarities (quadrants II and IV).

TABLE 16–1

	Quadrant I	Quadrant II	Quadrant III	Quadrant IV
Sin θ	+	+	−	−
Cos θ	+	−	−	+
Tan θ	+	−	+	−
	$0 \leq \theta \leq 90°$	$90° \leq \theta \leq 180°$	$180° \leq \theta \leq 270°$	$270° \leq \theta \leq 360°$

EXAMPLE 16.5 Determine cos 195° and tan 195°.

Solution: The same steps will be followed that were followed in finding the sine of an angle except that the polarity will be determined using Table 16–1. In Fig. 16–8 a vector is drawn on polar coordinate axes at an angle of 195° (step 1). The horizontal axis closest to the vector is the 180° axis. The angle between the vector and this axis is 15°. Thus, $\alpha = 15°$ (step 2).

$$\cos \alpha = \cos 15° = 0.966 \quad \text{(step 3)}$$
$$\tan \alpha = \tan 15° = 0.268$$

Referring to Table 16–1, the cosine function is negative in quadrant III and the tangent function is positive. Therefore,

$$\cos 195° = -\cos \alpha = -0.966$$
$$\tan 195° = \tan \alpha = 0.268$$

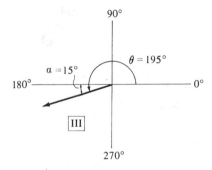

FIGURE 16–8

16.4 Radians

The unit of angular measure that we have been using thus far is the *degree*. Another unit of angular measure that is used in conjunction with sine waves is the *radian* (abbreviated rad). The radian is a larger unit than the degree. The relationship between the two is

$$2\pi \text{ radians} = 360 \text{ degrees} \quad (16\text{--}1)$$

In other words, 360°, which is one complete circle, is equivalent to 2π radians. Thus we can also say that

$$1 \text{ radian} = \frac{180°}{\pi} = 57.3° \quad (16\text{--}2)$$

Since 360° equals 2π radians, then 180° (half a circle) includes π radians, 90° (quarter circle) includes $\pi/2$ radians, and so forth. Table 16–2 lists some of the angles with their sizes expressed in degrees and radians.

TABLE 16-2

Degrees	0	30	45	60	90	180	270	360
Radians	0	$\pi/6$	$\pi/4$	$\pi/3$	$\pi/2$	π	$3\pi/2$	2π

EXAMPLE 16.6 Convert $\theta = 155°$ to radians.

Solution: 1 radian = 57.3°. Thus,

$$\theta = \frac{155°}{57.3°/\text{rad}}$$

$$\theta = 2.7 \text{ rad}$$

EXAMPLE 16.7 Convert $\theta = 0.37\pi$ rad to degrees.

Solution: Since 1 radian = $\frac{180°}{\pi}$,

then

$$\theta = 0.37\pi \text{ rad} = 0.37\pi \times \frac{180°}{\pi}$$

$$\theta = 66.6°$$

16.5 Plot of the Sine Function

A sine wave is the result of plotting the values of sine θ against the values of θ ranging from 0° to 360°. A convenient way to derive the sine values is through the use of a *rotating vector*. In Fig. 16-9(a) vector OA is resting at the 0° position on the polar coordinate axes. We will assign a length of 1 unit to this vector and we are going to observe what happens as the vector is allowed to rotate ccw at a uniform rate, just like the second hand on a clock but in the opposite direction. This rotating vector is given a special name. It is called a *phasor* and we will use this term henceforth.

As this phasor rotates through various angles θ, the distance above or below the horizontal axis of its endpoint A represents the sine θ. To illustrate, in part (b) of the figure we see the unit phasor at an angle of $\theta = 30°$. The dotted line AB represents the distance from point A to the horizontal axis; this distance is called the *vertical projection* of the phasor. To find the length of AB we can take advantage of the fact that the triangle OAB is a right triangle. For a right triangle we know that

$$\sin \theta = \frac{\text{length of opposite side}}{\text{length of hypotenuse}}$$

The Sinusoidal Waveform (AC)   431

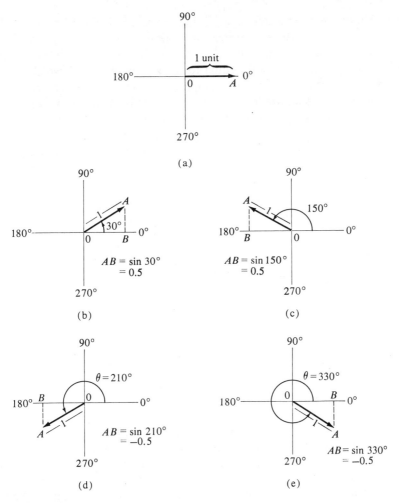

FIGURE 16-9 *Unit rotating vector (phasor) has a vertical projection which is equal to sin θ.*

For triangle OAB, $\theta = 30°$, the hypotenuse is $OA = 1$ unit, and the opposite side is AB. Thus,

$$\sin 30° = \frac{AB}{OA} = \frac{AB}{1} = AB$$

Since sine $30° = 0.5$, then $AB = 0.5$ units.

For any other angle θ in the first quadrant, the vertical projection of the phasor would equal sine θ. The same is true for angles in any of the other quadrants. For example, part (c) of Fig. 16-9 shows the phasor at an angle $\theta = 150°$. The reader can verify that $AB = \sin 150° = 0.5$ units.

Part (d) of the figure shows the phasor in quadrant III at an angle $\theta = 210°$. Once again, it is easy to see that the length of AB is equal to $\sin 210° = -0.5$. The negative sign reflects the fact that OA is now *below* the horizontal axis so that its vertical projection AB is negative.

Similar reasoning can be applied to the case in part (e) for $\theta = 330°$ in quadrant IV. Once again the vertical projection is negative.

To summarize, as the unit phasor is rotated ccw from 0°, its vertical projection length (AB) is always equal to the sine of the angle which the phasor is making at that instant. The complete 360° cycle of phasor rotation can be represented as shown in Fig. 16–10. The left-hand portion of the diagram shows the phasor at various angular positions in its cycle. For each angle θ, the vertical projection of the phasor (which represents the sine of the angle) is projected directly over to the right in the graph above the angle corresponding to θ.

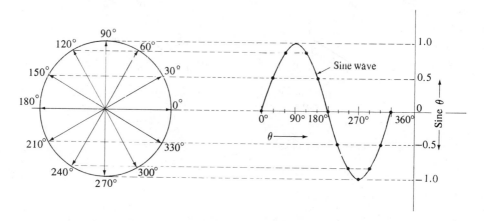

FIGURE 16–10 *Sine wave produced by rotating phasor.*

After the series of points has been projected, they can be connected with a smooth curve. This curve is a sinusoidal curve (sine wave). Actually, to obtain a more accurate sine-wave curve, more points are needed. To obtain more points, the phasor can be placed at different angles and its vertical projection can be projected over to the right-hand graph above the appropriate angle to obtain another plotting point. The sine-wave curve in Fig. 16–10 is a plot of sine θ versus θ.

The following points concerning the sine-wave curve in Fig. 16–10 are of special significance.

(1) The sine curve begins at 0° where $\sin \theta = 0$ and increases in the positive direction (has a positive slope).

(2) The maximum *positive* value of the sine curve occurs at $\theta = 90°$ where $\sin \theta = 1$. This is called the *positive peak* of the sine wave.

(3) The 180° point is half way through the cycle. At this point $\sin \theta$ is again equal to zero but the curve is decreasing (negative slope).

(4) The maximum *negative* value of the sine wave occurs at $\theta = 270°$ where $\sin \theta = -1$. This is the *negative peak* of the sine wave.

(5) At the end of the cycle (360°) the sine wave is back to zero ($\sin 360° = 0$). This corresponds to one complete revolution of the phasor so that the phasor is back to its starting position at 0°. Another complete revolution of the phasor will trace out a second sine curve.

(6) The sine-wave curve is symmetrical about the horizontal axis. That is, it goes through the same values between 180° and 360° as it does between 0° and 180° except that the values are negative between 180° and 360°.

From our discussion concerning Fig. 16–10 we can now say that a sine wave is *generated* by a phasor rotating ccw through 360°. The phasor doesn't actually physically *generate* the sine wave, but it would be to our advantage to think of it in this way. As we shall see, it will be very convenient to represent sine waves of voltage and current by the phasors which "generate" them.

16.6 Sine Wave of Voltage (and Current)

The phasor in Fig. 16–10 was assigned an arbitrary length of 1. As such, the sine curve which it generated had a maximum or peak value of 1. Since our interest is in sine waves of voltage and current, we can assign a length to the phasor in units of voltage or current. Figure 16–11 shows a *voltage* phasor with a length represented by E_M volts. As this phasor rotates it will generate a sine wave which has a maximum positive value of E_M volts at 90° and a maximum negative value of $-E_M$ volts at 270° as shown.

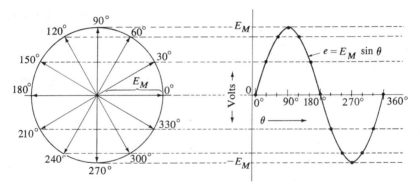

FIGURE 16–11 *Voltage sine wave produced by a rotating voltage phasor with length* E_M.

The resulting sine wave is now a plot of voltage versus angle θ. The vertical axis is marked off in **volts**. For the first half cycle (0° to 180°) the voltage is positive; for the second half cycle (180° to 360°) the voltage is negative. The *peak values* of the voltage sine wave occur at 90° and 270°.

Expression for Voltage Sine Wave The sine wave of voltage in Fig. 16–11 varies in direct proportion to $\sin \theta$ according to the following expression

$$e = E_M \sin \theta \tag{16-3}$$

In this expression e represents the voltage at any instant of time and depends on the angle θ at that instant of time. The symbol e, which is referred to as the *instantaneous* voltage, is written as a *lowercase* symbol because it represents a voltage which is changing with time. E_m represents the maximum or peak value of the voltage sine wave.

EXAMPLE 16.8 A certain voltage sine wave has a peak amplitude of 40 V. What will be the instantaneous voltage at: (a) the 45° point in the cycle; (b) the 225° point in the cycle?

Solution: (a) $E_m = 40$ V so that

$$e = 40 \sin \theta$$
$$= 40 \sin 45°$$
$$= 40 \times 0.707$$
$$= 28.28 \text{ V}$$

(b) $e = 40 \sin 225°$

The angle 225° is in quadrant III so that the sine is negative. Thus,

$$e = 40 \sin 225°$$
$$= 40(-\sin 45°)$$
$$= 40 \times (-0.707)$$
$$= -28.28 \text{ V}$$

A sine-wave curve like that in Fig. 16–11 can be plotted by using the expression (16–3) to calculate the value of e for several values of θ. The points thus obtained can be connected to produce the sine wave. The more points used, the more accurate the plot. For example, let us assume we have a voltage sine wave that has $E_M = 100$ V. Figure 16–12 shows the results of taking the expression $e = 100 \sin \theta$ and substituting values of θ at 30° intervals.

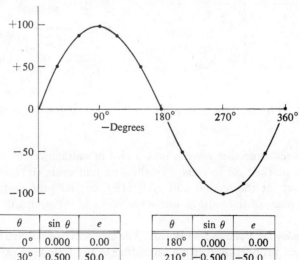

θ	$\sin \theta$	e
0°	0.000	0.00
30°	0.500	50.0
60°	0.866	86.6
90°	1.000	100.0
120°	0.866	86.6
150°	0.500	50.0
180°	0.000	0.00

θ	$\sin \theta$	e
180°	0.000	0.00
210°	−0.500	−50.0
240°	−0.866	−86.6
270°	−1.000	−100.0
300°	−0.866	−86.6
330°	−0.500	−50.0
360°	0.000	0.00

FIGURE 16–12 *Plotting voltage sine wave from* $e = 100 \sin \theta$.

Everything we have said about voltage sine waves can be repeated for current sine waves. A current sine wave can be said to be generated by a rotating phasor that has a length of I_M (amps). The resulting current sine wave will obey the expression

$$i = I_M \sin \theta \qquad (16\text{–}4)$$

where i is the instantaneous value of current and I_M is the maximum or peak value. Both voltage and current sine waves will be present in most of the ac circuits we are to study.

16.7 Period and Frequency

We have seen that a voltage phasor making one complete 360° (or 2π radians) revolution will generate one cycle of a voltage sine wave. Up to now we have not mentioned the rate at which the phasor is rotating. Let us say that it takes the phasor an amount of time, T, to complete one 360° cycle. This time, T, is called the *period* of the cycle.

If our phasor starts at 0° at time $t = 0$, then after a time interval equal to one period (T) it will have generated one cycle of a sine wave. At time $t = T$ it will be back at 0°. If it continues rotating, it will have completed its second cycle at time $t = 2T$ and so on. Figure 16–13 shows two cycles of a voltage sine wave that are generated by a phasor that rotates one cycle in a time interval equal to T.

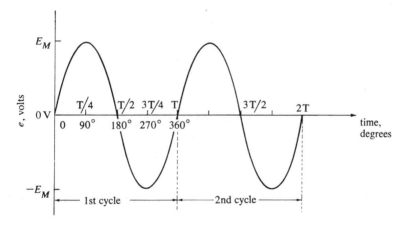

FIGURE 16–13 *Two cycles of a sine wave.*

Notice that the horizontal axis is marked off in units of time as well as degrees. Each full cycle of the sine wave contains 360° and takes an amount of time equal to T. Each half cycle contains 180° and takes an amount of time equal to $T/2$. Each quarter cycle contains 90° and takes an amount of time equal to $T/4$; and so on.

EXAMPLE 16.9 A certain sine wave has a peak voltage of 36 V and a period of 12 ms. What is its instantaneous voltage 1 ms after the start of its cycle?

Solution: Since $T = 12$ ms, then 1 ms represents 1/12 of a complete cycle or 30° of phasor rotation. Thus,

$$e = E_m \sin \theta$$
$$= (36 \text{ V}) \sin 30°$$
$$= 36 \times 0.5 = \mathbf{18 \text{ V}}$$

It is also of interest to know how many cycles of a sine wave occur in 1 second. The number of cycles per second is called *frequency*. If one cycle occurs in T seconds, then the number of cycles that occur in 1 second is simply 1 s/T s or $1/T$. Thus,

$$f = \frac{1}{T} \tag{16-5}$$

In Ex. 16.9, the sine wave had a period of 12 ms. The frequency of that sine wave is

$$f = \frac{1}{12 \text{ ms}} = 83.3 \text{ cycles/second}$$

Although the basic unit of frequency is the *cycle per second*, the SI system of units prefers to use the hertz (abbreviated Hz). Thus, $f = 83.3$ Hz.

EXAMPLE 16.10 What is the frequency of a sine wave that has a period of 2 μs?

Solution: $f = \dfrac{1}{T} = \dfrac{1}{2 \text{ μs}} = 5 \times 10^5 \text{ Hz} = \mathbf{500 \text{ kHz}}$

EXAMPLE 16.11 What is the period of one cycle of 60-Hz line voltage?

Solution: $f = \dfrac{1}{T}$

which can be transposed so that

$$T = \frac{1}{f}$$
$$= \frac{1}{60} = 0.0166 \text{ s} = \mathbf{16.6 \text{ ms}}$$

EXAMPLE 16.12 Determine the frequency of each of the sine waves in Fig. 16–14.

The Sinusoidal Waveform (AC)

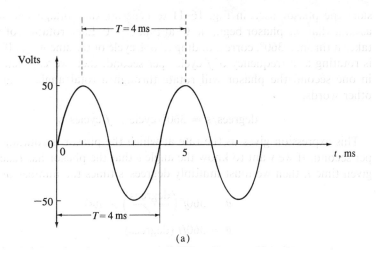

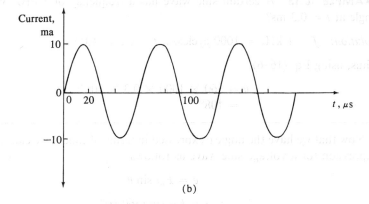

FIGURE 16–14

Solution: In (a) the period is seen to be 4 ms. Thus,

$$f = \frac{1}{4 \text{ ms}} = 250 \text{ Hz}$$

In (b) the length of one cycle is seen to be 60 μs. Thus,

$$f = \frac{1}{60 \text{ μs}} = \frac{10^6}{60} = 1.6 \times 10^4 \text{ Hz} = \mathbf{16 \text{ kHz}}$$

Note that the period of one cycle can be measured as the distance between the same points in two consecutive cycles.

16.8 Sine Wave Expressed in Terms of *t*

Previously, we expressed the instantaneous voltage of a sine wave as $e = E_M \sin \theta$. This expression tells us the voltage for any angle θ. It will also be useful to express the instantaneous voltage in terms of *time*. To do this, consider the phasor used in Fig. 16–11 to generate the voltage sine wave. Let's assume that the phasor begins at 0° at time $t = 0$. Each rotation of the phasor takes it through 360°, corresponding to one cycle of the sine wave. If the phasor is rotating at a frequency of f cycles per second, then we can determine that in one second the phasor will rotate through a total angle of $360° \times f$. In other words,

$$\text{degrees/s} = 360°/\text{cycle} \times f \text{ cycles/s}$$

This expression gives us the rate at which the phasor is rotating in degrees per second. If we want to know the angle θ that the phasor has reached at any given time t, then we must multiply degrees/s times the number of seconds t:

$$\theta = 360f \left(\frac{\text{degrees}}{\cancel{s}}\right) \times t\cancel{(s)}$$

$$\theta = 360ft \text{ (degrees)} \qquad (16\text{–}6)$$

EXAMPLE 16.13 A certain sine wave has a frequency of 1 kHz. What is its angle at $t = 0.3$ ms?

Solution: $f = 1$ kHz $= 1000$ cycles/s; $t = 0.3 \times 10^{-3}$ s

Thus, using Eq. (16–6)

$$\theta = 360 \times 1000 \times 0.3 \times 10^{-3}$$
$$= \mathbf{108°}$$

Now that we have the angle θ expressed in terms of time t, we can rewrite the expression for a voltage sine wave as follows:

$$e = E_M \sin \theta \qquad (16\text{–}3)$$

$$\therefore e = E_M \sin (360 ft)° \qquad (16\text{–}7)$$

We can now use this last expression to determine the instantaneous voltage of any sine wave at any instant of time if we know its peak voltage E_M and its frequency f.

EXAMPLE 16.14 Write the expression for the sine wave in Fig. 16–14(a). Then determine the instantaneous voltage at $t = 1.4$ ms.

Solution: For this sine wave $f = 250$ Hz [Ex. (16.12)] and $E_M = 50$ V. Using Eq. (16–7),

$$e = 50 \sin (360 \times 250\, t)°$$
$$= \mathbf{50 \sin (9 \times 10^4\, t)°}$$

Substituting $t = 1.4$ ms in this expression:

$$e = 50 \sin (9 \times 10^4 \times 1.4 \times 10^{-3})°$$
$$= 50 \sin 126° = 50 \sin 54°$$
$$= 50 \times 0.809$$
$$e = \mathbf{40.45 \text{ V}}$$

Sine-Wave Expression Using Radians In the equation $e = E_M \sin(360\, ft)°$, the quantity in parentheses represents the value of the angle θ in *degrees* at any instant of time. In dealing with ac circuits in later chapters it will be necessary to express θ in *radians* rather than degrees. Since $360° = 2\pi$ radians, then

$$\theta = (360\, ft)° = (2\pi\, ft) \text{ radians}$$

Thus, the expression for e can be written as

$$e = E_M \sin(2\pi\, ft) \qquad (16\text{-}8)$$

where the angle is now in radians.

EXAMPLE 16.15 A 60-Hz sine wave has $E_M = 150$ V. Express this sine wave in terms of radians.

Solution: $e = 150 \sin(2\pi \times 60 \times t)$
$ = 150 \sin(120\pi\, t)$
$ = \mathbf{150\ sin\ (377\ t)}$

Equation (16-8) can also be expressed as

$$e = E_M \sin \omega t \qquad (16\text{-}9)$$

where ω (greek letter omega) is defined as

$$\omega = 2\pi f \qquad (16\text{-}10)$$

ω is called the *angular velocity* or *angular frequency* of the sine wave. If we think back to the rotating phasor, ω represents its angular velocity in radians per second. That is,

$$\omega = 2\pi\, \frac{\text{radians}}{\text{cycle}} \times f\, \frac{\text{cycles}}{\text{s}}$$
$$ = 2\pi f \text{ radians/s}$$

Equations (16-8) and (16-9) are the most common ways of expressing a voltage (or current) sine wave. These equivalent expressions use radians as the angular measure. Therefore, when we are using these expressions we often have to convert from radians to degrees.

EXAMPLE 16.16 A current sine wave has the expression $i = 20 \sin(377\, t)$mA. What is the instantaneous current at $t = 1$ ms?

Solution: Substituting $t = 10^{-3}$ s,

$$i = 20 \sin(377 \times 10^{-3}) \text{ mA}$$
$$ = 20 \sin(0.377 \text{ rad}) \text{ mA}$$

But,

$$0.377 \text{ rad} = 0.377 \times 57.3°$$
$$\phantom{0.377 \text{ rad}\ } = 21.6°$$

Thus,
$$i = 20 \sin (21.6°) \text{ mA}$$
$$= (20 \times 0.37) \text{ mA}$$
$$= 7.4 \text{ mA}$$

16.9 Amplitude of a Sine Wave

In chap. 12 we saw that the amplitude of a sine wave could be expressed in either of two ways. The *peak amplitude* is the value of the sine wave at its positive (or negative) peak relative to zero. In our discussions thus far, E_M and I_M have represented the peak amplitudes of voltage and current sine waves respectively. The *peak-to-peak amplitude* is the vertical measurement from the positive peak value to the negative peak value. As indicated in Fig. 16-15, the *peak-to-peak value is always twice the peak value.*

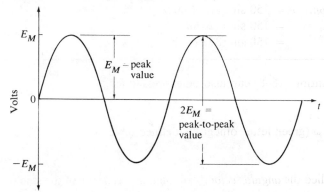

FIGURE 16-15 *Peak and peak-to-peak amplitudes of a sine wave.*

In the next section we will discover a third way of expressing the sine-wave amplitude. This third way of doing it is the most widely used for reasons which will become apparent later. After we have seen this third method we will show the comparisons between all three.

16.10 Effective (rms) Value of a Sine Wave

If a voltage sine wave is applied to a linear resistor R, the current through the resistor will also be a sine wave. This is illustrated in Fig. 16-16(a). The expression for the applied voltage is

$$e_s = E_M \sin (2\pi ft)$$

Since $i = e_s/R$, then the expression for i is

$$i = \frac{e_s}{R} = \left(\frac{E_M}{R}\right) \sin (2\pi ft)$$

Clearly, the peak value of the current waveform is $I_M = E_M/R$.

Because the current sine wave follows *exactly* the variations of the applied voltage sine wave, we say that the two waveforms are *in phase* with one another. As seen in Fig. 16-16(a), the waveforms reach their positive peaks simul-

The Sinusoidal Waveform (AC)   441

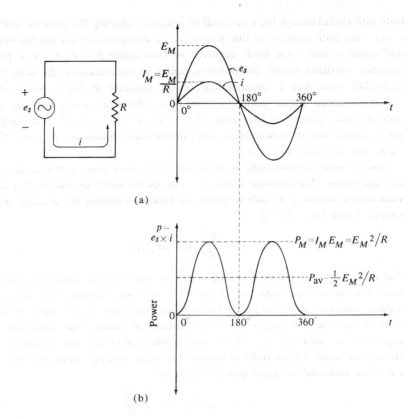

FIGURE 16-16 *Voltage current and power waveforms in a resistor.*

taneously and their negative peaks simultaneously and, in fact, all corresponding points on both sine waves are reached at the same instant of time. The waveforms are exactly *in phase*.

As the voltage and current vary, the *power* which is being applied to the resistor from the source also varies since $p = v \times i$ at any instant of time. The expression for the instantaneous power p can be obtained by multiplying the expressions for e_s and i. Thus,

$$p = e_s \times i$$
$$= E_M \sin(2\pi ft) \times \frac{E_M}{R} \sin(2\pi ft) \qquad (16\text{-}11)$$
$$= \frac{[E_M \sin(2\pi ft)]^2}{R}$$

Expression (16-11) can be plotted for various value of t resulting in the graph shown in Fig. 16-16(b). This graph represents the waveform of instantaneous power p plotted versus time. This graph could also be obtained by taking the *product* of the e_s and i values at each instant of time [from their waveforms in (a)] and plotting these values versus t.

Notice that the power waveform is exactly the same during the positive half cycle of *the applied voltage* as it is during the negative half cycle. This is as it should be because the current and voltage waveforms are exactly the same in both half cycles except for a reversal in polarity. During the positive half cycle e_s and i are both positive so that $p = e_s \times i$ is also *positive;* during the negative half cycle e_s and i are both negative so that again $p = e_s \times i$ is a *positive* quantity. In other words, the power supplied to the resistor is the same during each half cycle and is independent of the polarity of the applied voltage.

The maximum power point P_M takes place when the current and voltage are both at their peaks. At this point, $P_M = E_M \times I_M = E_M^2/R$. The minimum power points are of course when the current and voltage are at zero. Thus, p varies between zero and P_M.

Since p varies continuously in ac circuits, it is often more useful to talk about *average power*. The average power P_{av} can be obtained by calculating the instantaneous power p at various points in time and finding the average of these values. From Eq. (16–11)

$$p = \frac{E_m^2}{R} [\sin (2\pi ft)]^2$$

The term $2\pi ft$ represents the angle θ of the sine wave at different instants of time. We can let this angle vary from 0° to 360° and calculate the value of p at regular intervals, say every 15°. These values can then be averaged by adding them all up and dividing by the total number of values. The result will be an *approximate* value for P_{av}. If we use smaller intervals and therefore more values, the result will be more accurate. The exact average value of power turns out to be one half the peak power.* That is,

$$P_{av} = \frac{1}{2} E_M I_M = \frac{1}{2} \frac{E_M^2}{R} \qquad (16\text{–}12)$$

EXAMPLE 16.17 What is the average power dissipated in a 10-ohm resistor when an ac voltage of 20-V peak amplitude is applied to it?

Solution: In this case, $E_M = 20$ V and $R = 10$ ohms. Using Eq. (16–12) we have

$$P_{av} = \frac{1}{2} \times \frac{(20 \text{ V})^2}{10 \text{ }\Omega}$$
$$= 20 \text{ W}$$

Expression (16–12) indicates that the average power produced by a sine wave with peak value E_M is only *half* as much as that produced by a dc source of the same value. If we had a dc source with a voltage, $E_{dc} = E_M$, it would supply power equal to E_M^2/R. Let us now concern ourselves with finding out what value of dc voltage will produce the *same power* as a sine wave with a peak value of E_M.

The power supplied by a dc source E_{dc} to a resistor R is given by E_{dc}^2/R. If this power is to equal the power supplied by the ac source with peak value E_M, then

---

*Calculation of P_{av} is left as an exercise for the student.

$$\frac{E_{dc}^2}{R} = \frac{E_M^2}{2R}$$

Cancelling the Rs and solving for E_{dc}:

$$E_{dc} = \frac{1}{\sqrt{2}} E_M = \frac{E_M}{1.414} = 0.707 \, E_M$$

This last expression indicates that a dc source of only 7.07 V produces the same power as an ac source with a peak value of 10 V. This result should not be surprising since an ac voltage is at its peak point only momentarily and is usually below its peak point.

To illustrate, if we applied an ac voltage of 100-V peak to a light bulb, the power dissipation in the bulb's filament would produce a certain amount of heat causing the filament to glow with a certain amount of brightness. To obtain the same average lamp brightness, a dc voltage of only 70.7 V need be applied to the lamp since it will produce the same amount of power. What we are saying, then, is that:

A sinusoidal voltage with peak value E_M has the same effect as a dc voltage equal to 0.707 E_M as far as average power is concerned.

EXAMPLE 16.18 A 300-V-peak sine wave is applied to a 100-ohm resistor. What dc voltage will supply the same average power when applied to this resistor?

Solution: $E_{dc} = 0.707 \, E_M = 0.707 \times 300 \text{ V} = 212.1 \text{ V}$

To verify this result calculate the power in both cases. For the sine wave

$$P_{av} = \frac{1}{2} \frac{E_M^2}{R} = \frac{(300)^2}{2 \times 100} = \frac{9 \times 10^4}{2 \times 10^2} = 450 \text{ W}$$

For the dc source

$$P_{av} = \frac{E_{dc}^2}{R} = \frac{(212.1)^2}{100} = \frac{4.5 \times 10^4}{10^2} = 450 \text{ W}$$

Thus, 212.1 V dc produces the same power as 300-V peak ac.

Effective rms Value of Sine Waves We have seen that a voltage sine wave with peak voltage E_M is *effective* as a dc voltage of 0.707 E_M in producing power. Thus, we can define the *effective value* of a sine wave to be

$$E_{eff} = \frac{1}{\sqrt{2}} E_M = 0.707 \, E_M \qquad (16\text{--}13a)$$

The *effective value* is also often called the *rms value;* rms stands for *root-mean-square*, which is the mathematical procedure by which we determine the effective value. The rms value is the same as the effective value. Thus,

$$E_{rms} = E_{eff} = \frac{1}{\sqrt{2}} E_M = 0.707 \, E_M \qquad (16\text{--}13b)$$

Effective value and rms value are used interchangeably in the field of electronics. However, rms is more commonly used so we will continue to use it here, keeping in mind that it represents the *effective* value of the waveform.

The rms value of a voltage sine wave is the third way of expressing its amplitude. *rms* is used more often in the field of electronics than is *peak* or *peak-to-peak* values. In fact, if an ac voltage value is given, it is always assumed that the voltage given is the rms value. For example, the 60-Hz power line voltage is usually given as 110 V. This 110 V is the rms value of the line voltage.

EXAMPLE 16.19 What is the peak voltage of the 60-Hz power line? What is the peak-to-peak voltage?

Solution: Rearranging Eq. (16–13b) to solve for E_M, we have

$$E_M = \sqrt{2}\, E_{\text{rms}} = 1.414\, E_{\text{rms}}$$

For the 60-Hz line voltage, $E_{\text{rms}} = 110$ V. Thus,

$$E_M = 1.414 \times 110 \text{ V} = \mathbf{155.5 \text{ V}}$$

The peak-to-peak value is twice this amount.

$$\text{peak-to-peak} = 2\, E_M = \mathbf{311 \text{ V}}$$

EXAMPLE 16.20 An ac voltage observed on an oscilloscope has its peak-to-peak voltage measured as 32 V. What is the rms value of the waveform?

Solution: Since $2\, E_M = 32$ V, then $E_M = 16$ V. Thus,

$$E_{\text{rms}} = 0.707 \times 16 \text{ V} = \mathbf{11.3 \text{ V}}$$

The rms value is always $0.707/2 = 0.3535$ *times the peak-to-peak value.*

EXAMPLE 16.21 Write the expression for a voltage sine wave that has a frequency of 200 Hz and an rms value of 80 V.

Solution: The general sine-wave expression is

$$e = E_M \sin (2\pi f t)$$

f is given as 200 Hz. E_M can be determined from the rms value given

$$E_M = \sqrt{2}\, E_{\text{rms}}$$
$$= 1.414 \times 80 \text{ V} = 113.1 \text{ V}$$

Thus,

$$e = 113.1 \sin (2\pi \times 200\, t)$$
$$\mathbf{\mathit{e} = 113.1 \sin (1256\, \mathit{t})}$$

We have now seen three ways of representing the amplitude of a sine wave.

These three representations are compared in Fig. 16–17. The formulas relating the three are also given.

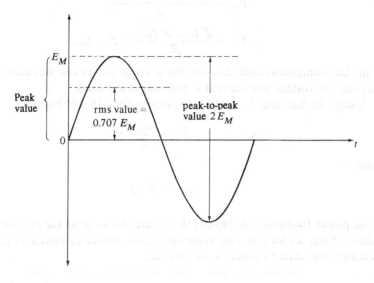

FIGURE 16–17 *Various ways of expressing sine-wave amplitude.*

rms Values for Current A sine wave of current also has an rms value that is related to its peak value, I_M. The relationship is given as

$$I_M = \sqrt{2}\, I_{rms} \tag{16-14}$$

or conversely,

$$I_{rms} = \frac{I_M}{\sqrt{2}} = 0.707\, I_M \tag{16-15}$$

These are the same relationships that were derived for voltage.

EXAMPLE 16.22 What is the rms current in a 6-ohm resistor that has a peak voltage of 30 V applied to it?

Solution: With $E_M = 30$ V, the peak current I_M will be 30 V/6 Ω = 5 A. Thus,

$$I_{rms} = 0.707\, I_M$$
$$= 3.54 \text{ A}$$

16.11 Power Calculations Using rms Values

The usefulness of rms values lies in their application to power calculations. In the last section the average power from a sine-wave source was determined to be

$$P_{av} = \frac{E_M \times I_M}{2} \tag{16-12}$$

Since $E_M = \sqrt{2}\, E_{\text{rms}}$ and $I_M = \sqrt{2}\, I_{\text{rms}}$, then

$$P_{\text{av}} = \frac{\sqrt{2}\, E_{\text{rms}} \times \sqrt{2}\, I_{\text{rms}}}{2}$$

$$P_{\text{av}} = \frac{2\, E_{\text{rms}} \times I_{\text{rms}}}{2} = E_{\text{rms}}\, I_{\text{rms}} \qquad (16\text{--}16)$$

This last expression indicates that the average power can be calculated as the product of voltage and current if rms values are used.

Using the fact that $I_{\text{rms}} = E_{\text{rms}}/R$, expression (16–16) becomes

$$P_{\text{av}} = \frac{E_{\text{rms}}^2}{R} \qquad (16\text{--}17)$$

and

$$P_{\text{av}} = I_{\text{rms}}^2 R \qquad (16\text{--}18)$$

The power formulas (16–16) to (16–18) are the same as the dc power relationships. Thus, as long as rms values are used, power calculations in ac circuits employ the same formulas as dc circuits.

EXAMPLE 16.23 The 60-Hz power line voltage is applied to a 22-ohm load. How much power is delivered to the load?

Solution: $E_{\text{rms}} = 110$ V; therefore, $I_{\text{rms}} = 110\text{ V}/22\text{ ohms} = 5$ A.

$$P_{\text{av}} = E_{\text{rms}} \times I_{\text{rms}} = 110 \times 5 = \mathbf{550\ W}$$

We will often be using rms values to represent the amplitude of ac voltage and current. For this reason, we will abandon the use of the rms subscripts for convenience, and simply use capital letters E and V for rms voltage and I for rms current. Whenever the subscript M is used, this indicates *maximum* or peak value. When the subscript M is not present, the value is assumed to be rms. As an example, for the 60-Hz power line voltage, $E = 110$ V (rms value) and $E_M = 156$ V (maximum peak value). Table 16–3 lists the various voltage and current symbols used for ac.

TABLE 16–3

Symbol	Meaning	Relationship
e	Instantaneous voltage	$e = E_M \sin 2\pi f t$
i	Instantaneous current	$i = I_M \sin 2\pi f t$
E_M	Peak voltage	$E_M = E\sqrt{2}$
I_M	Peak current	$I_M = I\sqrt{2}$
E	rms voltage	$E = E_M/\sqrt{2} = 0.707\, E_M$
I	rms current	$I = I_M/\sqrt{2} = 0.707\, I_M$

EXAMPLE 16.24 A 6.3-V ac source produces 2.2 A in a vacuum tube filament. How much power is being dissipated by the filament?

Solution: The values of voltage and current given are assumed to be rms values. Using rms values we can calculate the average power.

$$P_{av} = E \times I = 6.3 \text{ V} \times 2.2 \text{ A} = \mathbf{13.86 \text{ W}}$$

Most ac voltmeters and ammeters indicate rms values unless otherwise indicated. However, these meters are usually designed to read the true rms values *only* for sine waves. Thus, if any other type of ac waveform is measured by the meter, the meter indication will be wrong. To measure rms values of *all* types of waveforms, a *true rms* meter is used.

Before leaving the topic of rms values, it should be stated that the relationships between rms, peak, and peak-to-peak values we have used thus far are valid *only* for sine waves. We cannot apply the same formulas to other ac waveforms like square waves.

16.12 Phase Angles

When an ac voltage is applied to a resistor or a pure resistive circuit, the current and voltage sine waves are always *in phase*. Many of the circuits which we will encounter contain inductance and capacitance. In such circuits, as we shall see, the current and voltage sine waves will *not* usually be *in phase*; they will *not* reach corresponding points in their cycles at the same time. To compare two sine waves of the *same frequency* but which are not in phase, we use a quantity called *phase angle*. The symbol for phase angle is ϕ (Greek letter phi, not to be confused with theta, θ).

Consider the situation shown in Fig. 16–18(a). Two phasors labeled E_A and E_B are shown resting at the 0° and 90° positions respectively. Let us assume that these are the starting positions of the two phasors at time $t = 0$ as they both begin to rotate ccw at the *same* frequency. Phasor E_A will generate the familiar sine-wave shown at the right in the figure and labeled sine wave A. Phasor E_B will also generate a sine wave with the same period and frequency of sine wave A. However, since phasor E_B started out at the 90° point to begin with, sine wave B (shown dotted) starts out at its maximum positive voltage point. The complete 360° cycle for sine wave B begins at its maximum positive voltage point and ends at its maximum positive voltage point.

Thus, sine wave B is at its positive peak (*x*) at $t = 0$ while sine wave A doesn't reach its positive peak (*y*) until after 90° of rotation. In other words, sine wave A is *lagging* behind sine wave B by an angular distance of 90°. Any point in the sine-wave A waveform occurs 90° later than the corresponding point in the B waveform. The angular difference between phasors A and B is called the *phase angle* ϕ. In this case $\phi = 90°$. Because of the phase angle $\phi = 90°$, the two generated sine waves are *out of phase* by 90°.

Phasor E_A is 90° behind phasor E_B *at all times*, as they both rotate at the same frequency. We can say that phasor E_A lags phasor E_B by 90°, so that sine wave A *lags* sine wave B by 90°. Equivalently, we can say that phasor E_B *leads* phasor

448 Chapter Sixteen

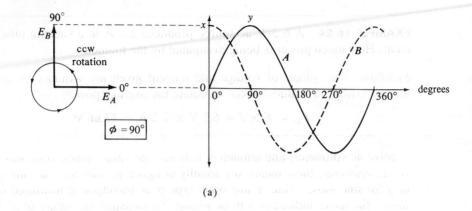

(a)

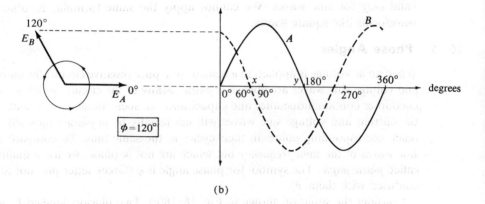

(b)

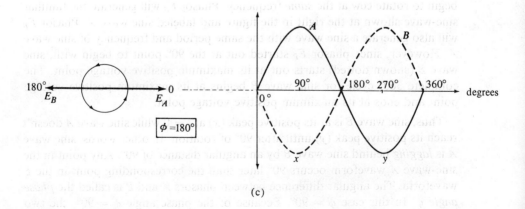

(c)

FIGURE 16–18 *Illustration of various phase angles between two phasors and their respective sine waves.*

E_A by 90° so that sine wave B leads sine wave A by 90°. Either way of saying it means the same thing.

Since phasors E_A and E_B are rotating at the *same* frequency, the phase angle difference between sine waves A and B will be maintained throughout the complete cycle and in all successive cycles. *Phase angles are not normally used for sine waves of different frequencies.*

Two phasors can be out of phase by anywhere from 0° to 360°. Figure 16–18(b) shows the two phasors drawn so that E_B leads E_A by 120°, resulting in the generated sine waves shown at the right. Sine wave B reaches zero (point x) after only 60° rotation; sine wave A reaches the corresponding point in its cycle (point y) after 180° rotation, thus showing that sine wave A *lags* behind sine wave B by 120° ($\phi = 120°$).

Part (c) of Fig. 16–18 shows the case where the phasors and sine waves are 180° out of phase ($\phi = 180°$). For this case it is arbitrary as to whether we say sine wave A lags 180° behind B or whether sine wave B leads A by 180°. Notice for the case where $\phi = 180°$ that sine wave B goes through its complete *negative* half cycle while sine wave A goes through its complete *positive* half cycle and vice versa. If the two sine waves had exactly the same amplitude and we added the two sine waves together, they would cancel each other out and the result would be zero. Waveforms that are 180° out of phase are said to be of completely *opposite phase.*

Phase Angles Greater Than 180° Figure 16–19 shows phasor E_A at 0° and phasor E_B at 210°. For this situation we could say that E_B leads E_A by 210°. It is also correct to say that E_B lags E_A by 150°. Either way correctly describes the phasor positions.

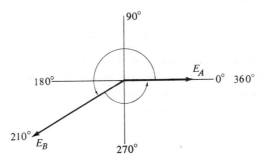

FIGURE 16–19

16.13 Determining Phase Angle From Waveforms

Phase angles play an important part in measurements made on ac circuits. The phase angle between two waveforms is usually determined by viewing the waveforms on an oscilloscope. As such, it is necessary that we be able to determine ϕ directly from waveform diagrams. To illustrate, refer to Fig. 16–20. Here two sine waves of different amplitudes but the same frequency are shown. The waveforms are not in phase. The period of the two waveforms is clearly seen to be $T = 1$ ms.

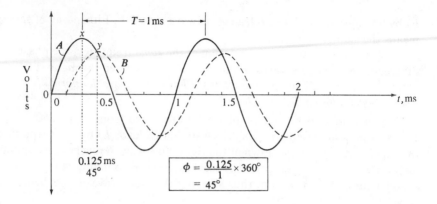

FIGURE 16-20 *Determining phase angle from waveforms.*

As the waveforms indicate, sine wave B reaches its positive peak *later* in time than sine wave A. Thus, sine wave A *leads* sine wave B. To determine the phase angle in degrees the following steps are performed.

(1) Pick a convenient point on sine wave A. Find the corresponding point during the same cycle on sine wave B. In the diagram the peak points x and y are chosen.

(2) Determine the time difference between these points along the horizontal axis. For our example this distance is 0.125 ms as shown in the figure.

(3) Determine what fraction of the total period this time difference constitutes. In our example $T = 1$ ms, thus, the time difference is 1/8 of a period.

(4) Multiply this fraction by 360° to determine the phase angle. In our example,

$$\phi = \frac{1}{8} \times 360° = 45°$$

Thus, sine wave A *leads* sine wave B by 45°.

The method just described simply measures the time difference between the waveforms and determines what portion of the total 360° cycle this time difference constitutes. The phase angle difference between sine waves is a more useful description of the relationship between two sine waves than is a simple time difference.

0° Phase Angle Sine waves that are in phase do not have a phase angle difference. They reach corresponding points in their cycles at the same time. Thus, $\phi = 0°$ for wave forms that are in phase. Waveforms that have any measureable phase difference are *out of phase*.

16.14 Phasor Representation of Phase Angle

The phase angle of one sine wave can be specified only with respect to another as reference. It makes no sense to say "sine wave A has a phase angle of 25°." We must say "sine wave A has a phase angle of 25° relative to sine wave B." We can use phasors to depict the phase angle difference between two or more sine waves.

For example, suppose we have sine wave A *lagging* sine wave B by 30°. There are two ways we can represent this situation using phasors. Figure 16-21 shows both possibilities. In (a) the phasor E_A (corresponding to sine wave A) is chosen as the reference phasor and is placed at 0°. Then phasor E_B (representing sine wave B) is placed at 30° *counterclockwise* indicating that sine wave B leads A by 30°. We can equivalently say that sine wave B is +30° with respect to sine wave A.

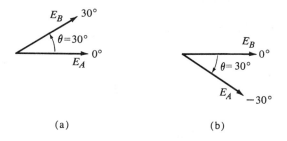

FIGURE 16-21 *Phase angle of* 30°. (a) E_A *is reference and* E_B *leads* E_A *by* 30°. (b) E_B *is reference and* E_A *lags* E_B *by* 30°.

In part (b) of the figure, phasor E_B is chosen as the reference and is placed at 0°. Then phasor E_A is placed 30° *clockwise* indicating that sine wave A lags B by 30°. Another way to express this is to say that sine wave A is −30° with respect to sine wave B.

In representing phase angles using phasors, positive angles are ccw from 0° and correspond to *leading* phase angles; negative angles are cw from 0° and represent *lagging* phase angles. The reference determines whether the phase angle is considered leading or lagging. However, the phase difference is not actually changed by the method used to represent it.

EXAMPLE 16.25 Three sine waves, A, B, and C, have the same frequency and are related as follows: sine wave A has an rms amplitude $E_A = 10$ V; sine wave B has an amplitude $E_B = 15$ V and leads sine wave A by 90°; sine wave C has $E_C = 20$ V and lags sine wave B by 30°. Sketch the phasor representation of the three sine waves.

Solution: If we choose sine wave A as the reference, then the correct phasor diagram is shown in Fig. 16-22. Phasor E_A is drawn at 0° as the reference. The *length* of this phasor is given as 10 V, representing the rms value of sine wave A.

Phasor E_B is drawn at +90° with respect to E_A. The length of E_B is 15 V, representing the amplitude of sine wave B. Drawn to scale, the length of E_B is 1.5 times that of E_A.

Phasor E_C has to lag E_B by 30°, which means that it leads E_A by 60°. It is shown in the figure at +60° with respect to E_A and with a length of 20 V.

452 Chapter Sixteen

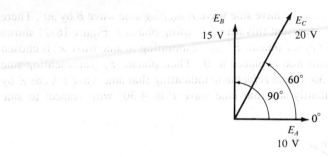

FIGURE 16-22

As this example illustrates, phasors can be used to represent both the amplitude of sine waves and the relative phase angles between sine waves. This phasor representation will be shown to be useful in later work when we have to consider the addition and subtraction of sine waves. As we shall see, these operations will be carried out using *vector* addition and subtraction performed on the phasors, which are essentially vectors representing sine waves.

16.15 Including ϕ in the Sine-Wave Expression

The basic equation for a voltage sine wave A is $e_A = E_{MA} \sin(2\pi ft)$ (or $e_A = E_{MA} \sin \omega t$). This expression represents a sine wave that is at its 0° point at $t = 0$. This can be verified by substituting $t = 0$ into the expression. A second sine wave B that *leads* sine wave A will not be at its 0° point at $t = 0$. It will be at some angle ϕ representing its phase-lead angle relative to sine wave A. The expression for sine wave B can be written as

$$e_B = E_{MB} \sin(2\pi ft + \phi) \quad \begin{Bmatrix} \text{leading} \\ \text{angle} \end{Bmatrix}$$

where ϕ represents the *constant* angle by which sine wave B leads sine wave A. Comparing the expressions for e_A and e_B shows that at any time t the angle of e_B is larger than that of e_A by an amount equal to ϕ.

If sine wave B *lags* sine wave A by an angle ϕ, then its expression will be

$$e_B = E_{MB} \sin(2\pi ft - \phi) \quad \begin{Bmatrix} \text{lagging} \\ \text{angle} \end{Bmatrix}$$

This shows that at any time t, the angle of e_B is smaller than the angle of e_A by an amount equal to ϕ.

EXAMPLE 16.26 Sine wave A has a peak amplitude of 50 V and a frequency of 60 Hz. Sine wave B has a peak amplitude of 30 V at the same frequency and leads sine wave A by 45°.
(a) Write the expression for both waveforms.
(b) Determine e_A and e_B at $t = 1$ ms.

Solution: (a) $e_A = 50 \sin (2\pi \times 60 \times t)$
$ = \mathbf{50 \sin (377\ t)}$

$e_B = 30 \sin (377\ t + \phi)$

Since $\phi = 45°$ is in degrees, we must convert it to radians since $2\pi\ ft$ is in radians.

$$45° = 45° \times \frac{\pi}{180°} \text{ radians} = \frac{\pi}{4} \text{ radians}$$

Thus,

$$e_B = \mathbf{30 \sin (377\ t + \pi/4)}$$

(b) At $t = 1$ ms,

$e_A = 50 \sin (377 \times 10^{-3}) = 50 \sin (0.377 \text{ radians})$
$ = 50 \times \sin (21.6°) = \mathbf{18.5\ V}$

$e_B = 30 \sin (377 \times 10^{-3} + \pi/4)$
$ = 30 \sin (0.377 + \pi/4)$
$ = 30 \sin (21.6° + 45°)$
$ = 30 \sin (66.6°) = \mathbf{27.5\ V}$

Case Where $\phi = 180°$ If $\phi = 180° = \pi$ radians

$$e_A = E_{MA} \sin (2\pi\ ft)$$

and

$$e_B = E_{MB} \sin (2\pi\ ft + \pi)$$

But, $\sin (\theta + \pi) = -\sin \theta$ for all values of θ. Thus,

$$e_B = -E_{MB} \sin 2\pi\ ft$$

This shows that e_B is of the exact *opposite* polarity as e_A. In other words, for $\phi = 180°$ the waveform e_B is of exact opposite polarity as e_A. Thus, a phase angle difference of 180° is the same as inverting the polarity.

Chapter Summary

1. On the *x-y* coordinate plane, the +*x*-axis will be our reference or 0° axis from which all angles will be measured.

2. The *x-y* coordinate plane is divided into four quadrants (see Fig. 16–1).

3. When a vector is rotated through one or more complete revolutions (360°) it will end up in the same position.

4. A vector rotated ccw about the origin will rotate through a positive angle; if it is rotated cw the angle is negative.

5. Two vectors are out of phase when they are at different angles.

6. To find the sine (cosine or tangent) of any angle θ, draw a vector at the angle θ on polar coordinate axes and find the sine (cosine or tangent) of the angle this vector makes with the closest horizontal axis. Assign polarity according to Table 16–1.

7. A radian is another unit of angular measure. 1 radian = 57.3° and 2π radians = 360°.

8. A sine wave can be generated by a rotating vector (phasor) by plotting the vertical projection of the phasor versus angle θ.

9. The positive and negative peaks of a sine wave occur at the 90° and 270° points respectively.

10. Each 360° rotation of the phasor generates a complete sine wave cycle.

11. The period T of a sine wave is the amount of time it takes for one cycle (one complete rotation of the phasor).

12. The frequency f of a sine wave is the number of cycles that occur per second ($f = 1/T$).

13. The basic unit for frequency is cycles/s but the SI unit is the hertz (Hz).

14. A voltage sine wave can be expressed using any of the following:
$e = E_M \sin(360 ft)° = E_M \sin(2\pi ft) = E_M \sin(\omega t)$

15. The angular frequency ω is equal to $2\pi f$ radians/s.

16. When a voltage sine wave is applied to a resistor, the instantaneous power is maximum when the applied voltage is at its peaks. $P_M = E_M \times I_M$.

17. The average power P_{av} equals one half the peak power.

18. A sinusoidal voltage with peak value E_M produces the same average power in a resistive load as a dc voltage equal to $E_M/\sqrt{2}$.

19. The rms (effective) value of a voltage sine wave is equal to $E_M/\sqrt{2}$. The rms value of a current sine wave is equal to $I_M/\sqrt{2}$.

20. The following formulas can be used to calculate average power dissipation in a resistor when a sinusoidal voltage is applied:
$$P_{av} = E_{rms} \times I_{rms} = \frac{E_{rms}^2}{R} = I_{rms}^2 \times R$$

21. The symbols for the rms values of ac voltage and current which will be used throughout the text are simply E and I.

22. The phase angle difference between two sine waves is the angular difference between corresponding points in the cycles of the two sine waves.

23. The length of a phasor represents sine-wave amplitude, while the angle of the phasor represents the phase angle with respect to the 0° reference. Leading phase is shown with positive (ccw) angles; lagging phase is shown with negative (cw) angles.

24. Waveforms that are 180° out of phase are said to have completely opposite phase.

Questions/Problems

Section 16.2

16-1 On a set of *x-y* coordinates sketch vectors rotated about the origin through each of the following angles relative to the $+x$-axis:
(a) 75°; (b) 795°; (c) −285°; (d) 135°; (e) −135°; (f) 240°; (g) 300°; (h) −60°; (i) 90°; (j) 180°; (k) −180°; (l) 1440°.

16-2 Two vectors *OA* and *OB* are out of phase by 90°. Vector *OA* is lagging behind *OB*, and is at an angle of 120°. Sketch both vectors on polar coordinate axes.

16-3 Three vectors *OA*, *OB*, and *OC* are situated such that: *OA* is at an angle of −30°; *OB* lags *OA* by 45°; and *OC* leads *OB* by 90°. Sketch the vectors on the polar coordinate axes and determine the angle between *OA* and *OC*.

16-4 Two vectors are out of phase by 30°. If they are both rotated 180°ccw, what will be the *phase angle* between them?

Section 16.3

16-5 Find the sine of each of the following angles: 50°, 110°, 197°, 313°, −110°.

16-6 Find the cosine and tangent of each of the following angles: 36°, 152°, 226°, 317°, −65°.

Section 16.4

16-7 Convert the following:
 a. 39° = _____ rad
 b. 117° = _____ rad
 c. 1.5 rad = _____ degrees
 d. $\pi/5$ rad = _____ degrees
 e. -3π rad = _____ degrees

Sections 16.5–16.6

16-8 For each of the following statements fill in the blanks with the correct word or number:
 a. A sine wave begins at $\theta = 0°$ and increases in the _____ direction.
 b. The positive peak of a sine wave will occur at $\theta =$ _____.
 c. The negative peak will occur at $\theta =$ _____.
 d. The sine curve has a negative slope between $\theta =$ _____ and $\theta =$ _____.

16-9 A certain current sine wave has a peak amplitude of 32 mA. What will be the instantaneous current at:
 a. the 30° point in the cycle
 b. the 250° point in the cycle

16-10 A phasor with a length $E_M = 120$ V is rotating ccw at the rate of 2 ms per cycle. Sketch two cycles of the resulting voltage sine wave.

16-11 A certain sine wave has $E_M = 80$ V, $T = 100$ μs. At what angle θ will it be after 20 μs? What will be its instantaneous voltage at this point?

Section 16.7

16-12 Determine the frequency of the sine waves of the two previous problems.

16-13 A certain sine wave reaches its positive peak at $t = 2.5$ ms. What is the frequency of this sine wave?

16-14 What is the period of a 200-kHz sine wave?

Section 16.8

16-15 Refer to Fig. 16-14(b). Write the expression for instantaneous current i. Then determine the value of i at $t = 100$ μs.

16-16 A certain 1-kHz sine wave reaches -70 V at $t = 0.6$ ms. Determine the peak value of this sine wave.

16-17 A certain sine wave is expressed as $e = 12 \sin (4000 \, t)$. Determine the frequency of this sine wave. Determine its angular velocity ω.

16-18 Refer to the waveform in Fig. 16-23. For this waveform determine values for the following:
(a) E_M; (b) T; (c) f; (d) ω.

Sections 16.9–16.11

16-19 A voltage sine wave with $E_M = 50$ V is applied to a 10-ohm resistor. Plot the instantaneous power waveform by taking points at 30° intervals. Using the values of power at these various points calculate the approximate average power P_{av}. Compare your result to the exact value $P_{av} = \frac{1}{2}E_m^2/R$.

16-20 A sine wave of voltage produces a peak current of 12 mA in a 3-kΩ resistor. What is the average power dissipation in the resistor?

16-21 What value of dc voltage will produce the same average power as a 150-V peak sine wave?

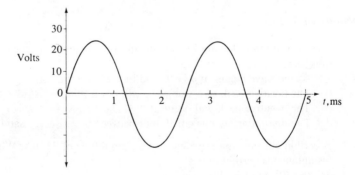

FIGURE 16-23

16-22 Determine the rms values of the waveforms in Fig. 16-14.

16-23 A 600-V peak-to-peak sine wave is applied to a 12-kΩ load. Determine the rms values of voltage and current.

16-24 A 60-Hz voltage sine wave is applied to a 10-ohm light bulb. (a) Determine the rms value of the voltage if the bulb dissipates 40 W. (b) Write the expression for the instantaneous current in the light bulb.

16–25 What is the peak-to-peak amplitude of a current sine wave that produces 2 W of power in a 500-ohm resistor?

16–26 Determine the average power dissipation in problem 16–23.

16–27 What current will an ac ammeter show when connected in series with a 32-ohm load that is dissipating a *peak* power of 288 W?

16–28 A 160-W soldering iron is operated from the 60-Hz power line. What is the resistance of the soldering iron?

Sections 16.12–16.15

16–29 Determine the approximate phase angle difference between the sine waves in Fig. 16–24. Express your result in the form "sine wave B leads (or lags) sine wave A by $\phi°$."

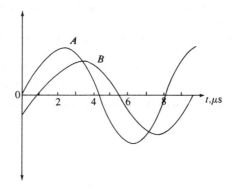

FIGURE 16–24

16–30 Sketch two sine waves that each have a frequency of 10 Hz and their phase angle difference is 120°.

16–31 Three sine waves having the same frequency (20 kHz) are related as follows: sine wave A has an rms amplitude of 100 V; sine wave B has an amplitude of 50 V and lags sine wave A by 50°; sine wave C has an amplitude of 150 V and leads sine wave B by 120°.
  a. Sketch the phasor representation of the three sine waves.
  b. What is the phase of C relative to A?

16–32 Sine wave A has a frequency of 100 Hz and an rms amplitude of 100 mA. Sine wave B has the same frequency, has an rms amplitude of 200 mA, and lags sine wave A by 60°.
  a. Write the expression for each sine wave.
  b. Calculate i_A and i_B at $t = 0.35$ ms.

16–33 Refer to the phasor diagram in Fig. 16–25. The voltage values shown represent rms values. Write the expression for the sine waves represented by each phasor.

16–34 Answer true or false to each of the following statements:
  a. Two sine waves are always in phase if they reach zero at the same time.
  b. A 180° phase angle indicates that two sine waves have completely opposite phase.
  c. If A leads B by 120°, then B lags A by 240°.

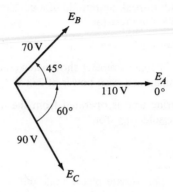

FIGURE 16–25

    d. If $e_2 = 10 \sin(1000\,t + \pi/2)$ and $e_1 = 5 \sin(1000\,t + \pi/4)$, then e_2 leads e_1 by 45°.
    e. As two phasors rotate at the same frequency, their relative phase angle gradually increases.
    f. The rms value of a sine wave increases with an increase in phase angle.
    g. If sine wave A leads B by 240°, we can also say that A lags B by 120°.

16–35 On the same time axis, sketch the sine waves described in problem 16–31.

CHAPTER 17

Opposition in AC Circuits

17.1 Introduction

In this chapter we will discuss the three forms of opposition to the flow of alternating current: resistance, inductive reactance, and capacitive reactance. We will find out how inductors and capacitors respond to the application of sinusoidal (ac) voltages.

17.2 AC Voltage Applied to Resistor

We have already seen that a voltage sine wave applied to a resistor will produce a sine wave of current that is in phase with the applied voltage. This is shown in Fig. 17–1. The peak value of the current sine wave is easily determined using Ohm's law

$$I_M = \frac{E_M}{R} \qquad (17\text{--}1)$$

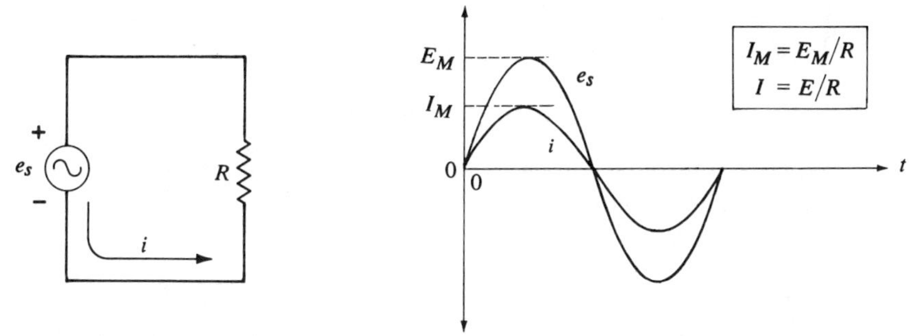

FIGURE 17–1 *Sine wave applied to a resistor.*

460   Chapter Seventeen

The rms value of current can also be determined from Ohm's law if the rms voltage is used

$$I = \frac{E}{R} \tag{17-2}$$

The resistor opposes ac current with the same amount of opposition that it opposes dc current.

For any kind of applied voltage, the resistor behaves according to Ohm's law. The resistor's resistance R is the same for any applied voltage. If a sine wave is applied to the resistor, the resistance R will be the same for any sine wave frequency. As we shall see, this characteristic is *not* true of inductors and capacitors.

17.3 Inductance Opposes Change in Current

Before we discuss how an inductor behaves when a sinusoidal voltage is applied to it, let us review the effect that an inductor has in the circuit shown in Fig. 17-2. We know from the discussion in chap. 15 that after the switch is closed, the current in this circuit will gradually build up to its final value E_s/R. Because of the inductor, the current cannot instantaneously jump to its steady-state value, but reaches this value after five time constants (5τ).

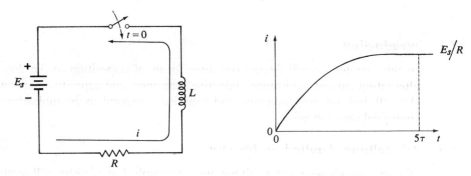

FIGURE 17-2 *Buildup of current is slow due to presence of inductor.*

In this circuit, the inductor opposes the rapid *change* in current that is trying to occur at the instant the switch closes. In steady state when the current is no longer *changing*, the inductor offers no opposition. In steady state the current is limited solely by the resistance R. Thus, *an inductor opposes a changing current but does not oppose a constant dc current.*

17.4 Inductive Opposition to AC: Reactance

When a sinusoidal voltage is applied to a circuit, it produces an alternating current that is continually changing. An inductor will offer opposition to this changing current. Thus, it can be stated that

Inductors oppose the flow of alternating current.

While the opposition of a resistor is called *resistance*, the opposition of an inductor to ac is called *inductive reactance* or simply *reactance*. The symbol for reactance is X. As we shall see, capacitors also possess reactance called *capacitive reactance*. To distinguish between the two types of reactance we use the subscripts L and C. Thus, X_L represents inductive reactance and X_C represents capacitive reactance.

Inductive reactance X_L is a measure of an inductor's opposition to alternating current. It is the counterpart of resistance in a resistor. In fact, X_L is also measured in units of ohms, and Ohm's law can be applied to an inductor in an ac circuit. If a sinusoidal voltage with a peak amplitude E_M is applied to an inductor, the current through the inductor will be a sine wave whose amplitude is given by

$$I_M = \frac{E_M}{X_L} \qquad (17\text{-}3)$$

The rms value of the current can be determined using the similar relationship

$$I = \frac{E}{X_L} = \frac{0.707\, E_M}{X_L} \qquad (17\text{-}4)$$

These equations compare to the Ohm's law equations for the resistor (17–1 and 17–2).

EXAMPLE 17.1 The 110-V, 60-Hz line voltage is applied to an inductor that has a reactance of 500 ohms. What is the *peak value* of inductor current?

Solution: The rms current is found using (17–4):

$$I = \frac{E}{X_L} = \frac{110\text{ V}}{500\ \Omega} = 0.22 \text{ A (rms)}$$

The peak current I_M is thus

$$I_M = \sqrt{2} \times I = 1.414 \times 0.22 \text{ A}$$
$$= 0.31 \text{ A (peak)}$$

Formula For X_L The value of inductive reactance can be determined from the formula

$$X_L = 2\pi \times f \times L = \omega \times L \qquad (17\text{-}5)$$

where f is the frequency of the applied voltage in Hz, L is the inductance in henries, and X_L is in ohms.

Examination of this formula indicates that X_L is *proportional* to inductance. This is reasonable since a larger value of inductance will produce more opposition to a changing current (ac). The formula also indicates that X_L is *proportional* to frequency. This is reasonable since a higher frequency of alternating current means that the current is going through more changes per second, and this faster changing current will be opposed more by the inductive reactance.

The fact that X_L is proportional to frequency means that the inductor offers *more* opposition to ac as the applied frequency is increased. This is a great deal different than a resistor whose opposition is always equal to R for all ac frequencies as well as dc.

EXAMPLE 17.2 Determine in which circuit of Fig. 17–3 the lamp will be brightest, and in which circuit it will be least bright. Assume $R_{lamp} = 100\ \Omega$.

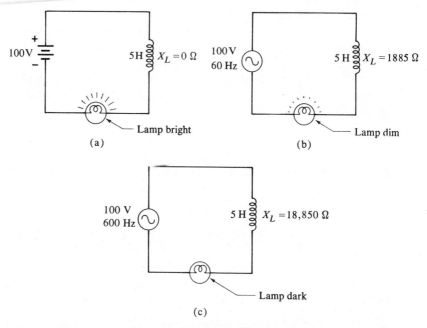

FIGURE 17–3 *Inductor offers more opposition at higher frequencies*

Solution: In (a), the input source is 100 V dc. In a dc circuit an inductor offers no opposition to current (in steady state). If we neglect the resistance of the coil windings, then we can consider the inductor to be a short circuit in this situation. Equivalently, we can say that the inductor's reactance X_L is zero for dc. This is verified by formula (17–5). Since dc represents *zero* frequency or $f = 0$, then

$$X_L = 2\pi f L = 2\pi \times 0 \times L = 0 \text{ ohms}$$

Thus, in (a) the current would be 100 V/100 ohms = 1 A just as if the inductor were not present. In (b) the input is a 60-Hz, 100-V source. The 5-H coil has a reactance of

$$X_L = 2\pi \times 60 \times 5$$
$$= 1885 \text{ ohms at 60 Hz}$$

Thus, in this case the coil offers a great deal of opposition to current. The current will have an approximate rms value of 100 V/1885 Ω = 53 mA (neglecting the lamp's resistance).

In (c) the input frqeuency has been increased to 600 Hz. This will cause the 5-H coil to have a greater opposition to current.

$$X_L = 2\pi \times 600 \times 5$$
$$= 18,850 \text{ ohms at 600 Hz}$$

Thus, the current in this case will be *ten* times less than in (b).

Clearly, then, the lamp will be brightest in the circuit in (a) because the inductor offers no opposition to dc current. In (c) the lamp current is smallest because of the larger X_L at 600 Hz. Thus, the lamp will be least bright in (c).

EXAMPLE 17.3 Calculate the X_L of a 15-mH coil at 40 kHz and at 10 MHz.

Solution: At $f = 40$ kHz,

$$X_L = 2\pi fL$$
$$= 2\pi \times 40 \times 10^3 \times 15 \times 10^{-3}$$
$$= 3768 \ \Omega$$

At $f = 10$ MHz,

$$X_L = 2\pi \times 10^7 \times 15 \times 10^{-3}$$
$$= 942 \ k\Omega$$

EXAMPLE 17.4 Calculate the X_L of a 200-μH coil at $f = 40$ kHz and 10 MHz.

Solution: At $f = 40$ kHz,

$$X_L = 2\pi fL$$
$$X_L = 2\pi \times 4 \times 10^4 \times 200 \times 10^{-6}$$
$$= 50.3 \text{ ohms}$$

At $f = 10$ MHz,

$$X_L = 2\pi \times 10^7 \times 200 \times 10^{-6}$$
$$= 12.56 \text{ kilohms}$$

These two examples illustrate the fact that X_L is proportional to inductance and is proportional to frequency. Example 17.4 shows that even a small value of inductance like 200 μH can have a considerable opposition to current at high frequencies (10 MHz).

EXAMPLE 17.5 What size inductance has a 10-kΩ reactance at a frequency of 20 kHz?

Solution: In this problem we know X_L and f and must find the value of L. We can solve formula (17–5) for L to obtain

$$L = \frac{X_L}{2\pi f} \qquad (17\text{–}6)$$

For the values given,

$$L = \frac{10 \times 10^3}{2\pi \times 20 \times 10^3}$$

$$\therefore L = 0.08 \text{ H} = \mathbf{80 \text{ mH}}$$

The inductance of a coil can very often be determined by applying an ac voltage and measuring the amount of ac current that results. The ratio of E/I will produce the value of X_L, and L can be found from (17-6).

EXAMPLE 17.6 A coil with negligible wire resistance has a sine wave with $E = 110$ V and $f = 60$ Hz applied to it. The coil current is measured to be $I = 15$ mA. Determine the value of L.

Solution: $X_L = \dfrac{E}{I} = \dfrac{110 \text{ V}}{15 \text{ mA}}$

$= 7.33 \text{ k}\Omega$

$$L = \dfrac{X_L}{2\pi f} = \dfrac{7.33 \times 10^3}{2\pi \times 60} = \mathbf{19.4 \text{ H}}$$

EXAMPLE 17.7 A certain 1-H coil has a coil resistance of 150 ohms. At what frequency will the inductive reactance have the same value as the coil resistance?

Solution: In this problem we know $L = 1$ H and we must find the value of f needed to produce a reactance $X_L = 150$ ohms. We can rearrange formula (17-5) to solve for f:

$$f = \dfrac{X_L}{2\pi L} \qquad (17\text{-}7)$$

For the values given,

$$f = \dfrac{150}{2\pi \times 1} = \mathbf{24 \text{ Hz}}$$

EXAMPLE 17.8 A 50-mH coil has a 10-kHz ac current flowing through it whose rms value is measured as 2.4 mA. Determine the *peak* value of coil voltage.

Solution: At $f = 10$ kHz,

$$X_L = 2\pi fL$$
$$= 2\pi \times 10^4 \times 50 \times 10^{-3} = 3140 \text{ }\Omega$$

Since according to Ohm's law, $I = E/X_L$, then we can also write $E = I \times X_L$. Thus

$$E = 2.4 \text{ mA} \times 3.14 \text{ k}\Omega$$
$$= 7.54 \text{ V}$$

This is the rms value of coil voltage since we used the rms value of current. To find the peak voltage we use

$$E_M = \sqrt{2}\, E = \mathbf{10.6 \text{ V}}$$

Opposition in AC Circuits   465

To summarize our discussion of inductive opposition to current we can make the following statements.

(1) An inductance can have an appreciable opposition (X_L) to current in ac circuits. Furthermore, the higher the frequency of ac, and the larger the inductance, the greater is the opposition.

(2) The Ohm's law relationship $I = E/X_L$ can be applied to an inductor in an ac circuit.

(3) There is no X_L for steady state dc ($f = 0$). In this case the inductor offers no opposition to current except for its wire resistance.

These characteristics have numerous applications in practical circuits. Inductance is often used where it is desired to have a high opposition to alternating current but very little opposition to constant direct current. It is also used where it is desired to have more opposition to higher frequency ac than to lower frequency ac. The fact that X_L varies with frequency gives inductors an advantage over resistors in applications which require a circuit to respond differently to different frequencies. This characteristic will become evident in our subsequent studies of *filters* and *resonant* circuits.

17.5 Series and Parallel Inductive Reactances

Reactance represents opposition to ac in *ohms*. Thus, inductive reactances in series or parallel are combined the same way as ohms of resistance. This is illustrated in Fig. 17–4. With two inductors in series, the total inductive reactance is the sum of the individual reactances. That is,

$$X_{LT} = X_{L1} + X_{L2} \qquad (17\text{–}8)$$

For the general case of more than two series inductors

$$X_{LT} = X_{L1} + X_{L2} + X_{L3} \ldots \text{etc.} \qquad (17\text{–}9)$$

With two inductors in parallel, the total reactance is given by the *product-over-sum* formula

$$X_{LT} = \frac{X_{L1} \times X_{L2}}{X_{L1} + X_{L2}} \qquad (17\text{–}10)$$

For more than two parallel inductors the reciprocal formula is used

$$\frac{1}{X_{LT}} = \frac{1}{X_{L1}} + \frac{1}{X_{L2}} + \frac{1}{X_{L3}} \ldots \qquad (17\text{–}11)$$

These relationships are the same ones used to calculate series and parallel resistances.

EXAMPLE 17.9 Figure 17–5 shows a 160-Hz source connected to two series inductors. The source voltage is 100 V rms. Determine: (a) total inductive reactance; (b) rms current; (c) rms voltage across each inductor.

Solution: (a) $X_{L1} = 2\pi f L_1 = 2\pi \times 160 \times 1 = 1 \text{ k}\Omega$

$X_{L2} = 2\pi \times 160 \times 3 = 3 \text{ k}\Omega$

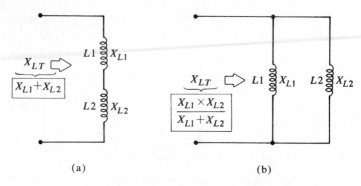

FIGURE 17–4 *Inductive reactances in series or parallel combine the same way as resistances.*

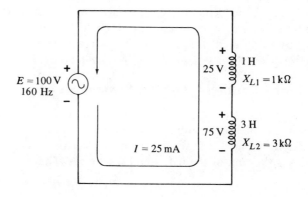

FIGURE 17–5

Thus,

$$X_{LT} = X_{L1} + X_{L2} = 4 \text{ k}\Omega$$

(b) $I = \dfrac{E}{X_{LT}} = \dfrac{100 \text{ V}}{4 \text{ k}\Omega} = 25 \text{ mA}$

Since this is a series circuit, the same current flows through each inductor.

(c) $V_1 = I \times X_{L1}$ $V_2 = I \times X_{L2}$
    $= 25 \text{ mA} \times 1 \text{ k}\Omega = 25 \text{ V}$ $= 25 \text{ mA} \times 3 \text{ k}\Omega = 75 \text{ V}$

These are the rms voltages across each inductor. Note that $V_1 + V_2 = E$. Also note that the larger inductor has the larger ac voltage.

EXAMPLE 17.10 Figure 17–6 shows the same source connected to the same two inductors in parallel. Determine: (a) total inductive reactance; (b) rms value of total source current; (c) current through each inductor.

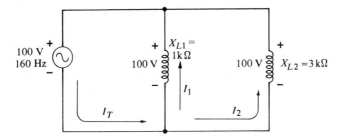

FIGURE 17-6

Solution: (a) $X_{LT} = \dfrac{X_{L1} \cdot X_{L2}}{X_{L1} + X_{L2}} = \dfrac{1 \text{ k}\Omega \times 3 \text{ k}\Omega}{4 \text{ k}\Omega} = 0.75 \text{ k}\Omega$

(b) $I_T = \dfrac{E}{X_{LT}} = \dfrac{100 \text{ V}}{0.75 \text{ k}\Omega} = 133.3 \text{ mA}$

(c) $I_1 = \dfrac{E}{X_{L1}} = \dfrac{100 \text{ V}}{1 \text{ k}\Omega} = 100 \text{ mA}$

$I_2 = \dfrac{E}{X_{L2}} = \dfrac{100 \text{ V}}{3 \text{ k}\Omega} = 33.3 \text{ mA}$

Note that since the inductive reactances are in parallel with the source, each one has a voltage equal to the source voltage. Also note that $I_1 + I_2 = I_T$ and the smaller inductor has the larger current.

17.6 Inductive Phase Angle

Up to this point we have talked about inductive reactance X_L as the opposition to ac. We have treated it much like resistance in using it to calculate current in a circuit. Unlike resistance, however, in an inductor *the current and voltage will not be in phase*. The phase angle difference between the current and voltage in an inductor occurs because the inductor voltage is always proportional to the rate of change of current (di/dt). For a pure inductor (zero wire resistance) it can be stated that:

The current through the inductor lags the voltage across the inductor by 90° ($\pi/2$ radians) at all frequencies.

Figure 17-7 shows the waveforms for current and voltage in an inductor. The current waveform is a normal sine curve that begins at 0° and reaches its positive peak at 90°. The voltage sine wave is at its positive peak at $t = 0$. The current waveform is *lagging* the voltage by 90° since it reaches its positive peak 90° later than the voltage waveform.

In order to see why this 90° phase difference is present, we must recall that in an inductor the coil voltage is proportional to the rate at which the current is changing. It is the changing current which induces the coil voltage. Thus, at all times we have

$$|v_L| = L \left|\frac{di}{dt}\right| \tag{15-4}$$

The instantaneous inductor voltage is equal to L times the instantaneous rate of current change.

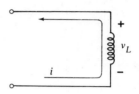

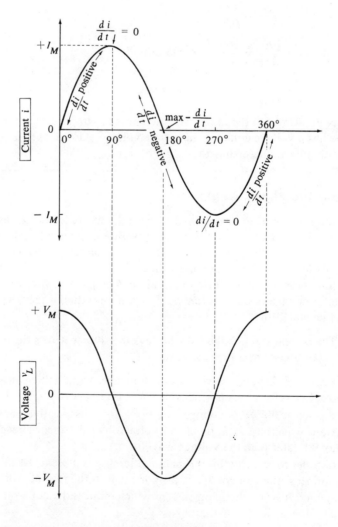

FIGURE 17-7 Current lags voltage by 90° in an inductor.

Situation at 0° If we look at the current sine wave at 0°, we can see that the *slope* of the curve is steepest at this point. Also, the slope is positive (*i* increasing) at this point. Thus, the rate of change of current di/dt has its *maximum positive* value at 0°. Since *i* is increasing, the coil voltage will have the polarity shown in the figure (recall Lenz's law from chap. 15). Furthermore, since di/dt is maximum at this point, the coil voltage v_L will be maximum. Thus, $v_L = V_M$ when the current is at 0° as shown.

Between 0° and 90° Between 0° and 90° the slope of the current curve becomes gradually less and less. This means that di/dt is decreasing from its maximum value at 0°. With the current changing at a slower and slower rate, the induced coil voltage will also gradually become less and less. Eventually at 90° the current waveform reaches its positive peak at which point di/dt is equal to zero. The inductor voltage will then be at zero volts [$v_L = L(di/dt) = 0$].

Between 90° and 180° After the 90° point, the current *i* begins to decrease. The slope of the curve is negative and this negative slope becomes gradually steeper in going from the 90° to the 180° points. This indicates that di/dt is negative on this region of the curve and it reaches its maximum negative value at 180°.

Since *i* is now decreasing (di/dt negative), the induced coil voltage will reverse polarity so that v_L becomes negative. Furthermore, v_L will increase in the negative direction because the rate of change of current is increasing in the negative direction. At the 180° point, the current is decreasing at its maximum rate so that the coil voltage has its maximum negative value.

Between 180° and 270° After the 180° point, the current continues decreasing but the rate at which it decreases becomes less and less until the 270° peak point where $di/dt = 0$ again. Thus, v_L remains negative in this interval but its value becomes less and less negative (increases) until 270° where $v_L = 0$ again.

Between 270° and 360° After 270°, the current begins to increase again. The rate at which it is increasing becomes gradually greater until at 360° the value of di/dt is again at its maximum positive value. Since the current is now increasing, v_L reverts back to its positive polarity. And since di/dt increases to its maximum value, v_L also increases to its maximum value.

At the 360° point the current is back to zero and the voltage is at its positive peak. This is the same situation that began at 0°, thus completing one full cycle. The next cycle of the current sine wave will produce a voltage cycle identical to the first voltage cycle. Thus, both waveforms will be at the *same frequency* but the *i* waveform *lags* the v_L waveform by 90° or, equivalently, the v_L waveform *leads* the *i* waveform by 90°. *This same 90° phase difference occurs no matter what the frequency.*

17.7 Complete Inductor Response to AC

When a sine wave of voltage is applied to an inductor, the inductor opposes current due to its reactance X_L. The current that does flow will always lag the applied voltage by 90°. The amplitude of the current depends on the amplitude of the applied voltage and inductive reactance.

EXAMPLE 17.11 Consider Fig. 17–8(a) where the voltage sine wave shown is applied to a 200-mH inductor. Determine the current waveform.

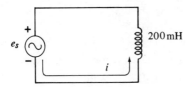

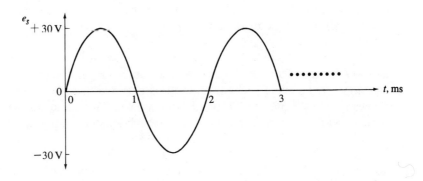

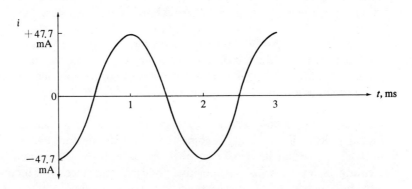

FIGURE 17-8

Solution: The frequency of the applied voltage can be determined from $f = 1/T$. Clearly, $T = 2$ ms. Thus, $f = 1/0.002 = 500$ Hz. The inductive reactance is therefore

$$X_L = 2\pi \times 500 \times 200 \times 10^{-3}$$
$$= 628 \text{ ohms}$$

The peak voltage is $E_M = 30$ V. Thus, the peak current is

$$I_M = \frac{E_M}{X_L} = \frac{30 \text{ V}}{628 \text{ }\Omega} = 47.7 \text{ mA}$$

The complete waveform for i is drawn in part (b) of the figure. Its peak amplitude is 47.7 mA and it is lagging the voltage waveform by 90° (one-quarter of a cycle).

Expression for Inductor Current The expression for a sine wave voltage source is

$$e_s = E_M \sin(2\pi ft) \tag{17-12a}$$

If this voltage is applied to an inductor, the current i through the inductor will be a sine wave with its expression given by

$$i = \underbrace{\frac{E_M}{X_L}}_{I_M} \sin(2\pi ft - \pi/2) \tag{17-12b}$$

This expression reflects the fact that the maximum current will be E_M/X_L. The expression also includes the effect of the 90° phase difference. Since 90° is $\pi/2$ radians, the current sine wave angle will always be $\pi/2$ radians less than the voltage sine wave angle. Thus, the $\pi/2$ term is subtracted from $2\pi ft$ in the expression.

EXAMPLE 17.12 Write the expressions for e_s and i of Fig. 17-8. Determine the value of i at $t = 1$ ms.

Solution: $e_s = E_M \sin(2\pi ft)$
$= 30 \sin(2\pi \times 500\ t)$
$e_s = \mathbf{30 \sin(3140\ t)\ volts}$

$i = \mathbf{47.7 \sin(3140\ t - \pi/2)\ mA}$

At $t = 1$ ms,

$i = 47.7 \sin(3140 \times 10^{-3} - \pi/2)$
$= 47.7 \sin(3.14 - \pi/2) = 47.7 \sin(\pi - \pi/2)$
$= 47.7 \sin(\pi/2) = 47.7 \sin(90°)$
$= \mathbf{47.7\ mA}$

The expression for inductor voltage and current can also be written with the current sine wave chosen as the reference. Thus, we would have

$$i = I_M \sin(2\pi ft) \tag{17-13a}$$

and

$$e_s = E_M \sin(2\pi ft + \pi/2) \tag{17-13b}$$

The $+\pi/2$ term in the expression for e_s indicates that the voltage sine wave is 90° *ahead* of the current sine wave.

Phasor Representation of Inductor e and i If the voltage across an inductor is a sine wave with an rms value V_L, we can represent this on a phasor diagram as a phasor with length equal to V_L. We can choose the voltage sine wave as our reference so that the voltage phasor is at 0°. This is shown in Fig. 17-9(a). The inductor current will lag the applied voltage by 90°. Thus, we can represent the current sine wave by a phasor with a length of $I_L = V_L/X_L$ at an angle of $-90°$ as shown in the same figure.

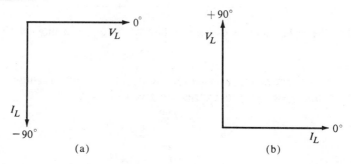

FIGURE 17-9 *Equivalent phasor representations of inductor current and voltage.*

If we wish to choose the coil current as reference, the phasor representation is shown in part (b) of the figure with I_L at 0° and V_L at +90°, leading I_L by 90°. Both representations in Fig. 17-9 are correct. It is arbitrary as to which sine wave is chosen as the reference.

17.8 A Capacitor Blocks Direct Current

Consider the circuit in Fig. 17-10 where a constant dc source is applied to a capacitor. When the switch is closed a current will flow from the source to charge the capacitor. This flow of current is only momentary since it lasts only long enough for the capacitor to charge up to the source voltage. This charging takes an amount of time equal to 5 τ, after which no appreciable current will flow (assuming a nonleaky capacitor dielectric). Thus, when a constant dc voltage is applied to a capacitor, current stops flowing when the capacitor voltage stops changing. *In steady state a capacitor blocks the flow of direct current (dc).*

$$\frac{dv_{xy}}{dt} = \frac{i}{C} \qquad (13\text{-}4)$$

In order for the capacitor voltage to change, a charge must be flowing to or from the capacitor plates.

17.9 Capacitive Opposition to AC

Figure 17-11 shows a sinusoidal voltage source e_s connected directly to a capacitor. As a result, the voltage across the capacitor plates must at all times equal e_s. Since e_s is a sine wave which is continually changing, the capacitor voltage is also continually changing. When e_s is increasing, the capacitor voltage is increasing (charging); and when e_s is decreasing, the capacitor voltage is decreasing (discharging). In order for the capacitor to charge as e_s increases, then a charging current must flow to the capacitor. Conversely, in order for the capacitor to discharge as e_s decreases, a discharge current must flow away from the capacitor. Thus, the current must continually reverse its direction; it is alternating current.

Opposition in AC Circuits   473

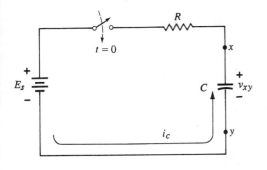 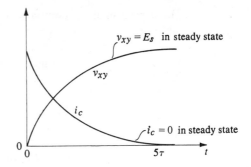

FIGURE 17–10 *In steady state capacitor blocks dc.*

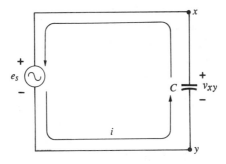

FIGURE 17–11 *Capacitor connected to ac source.*

The amplitude of the alternating current that flows when e_s is connected to the capacitor will depend on the value of the capacitance. Regardless of the size of the capacitor, its voltage must change at the same rate as e_s. Since $dv/dt = i/C$ for a capacitor, then a larger capacitance C will require a larger current i for the same rate of voltage change dv/dt. A smaller capacitance value will not require as much current for the same dv/dt. In other words, the amount of current needed to charge and discharge a capacitor at a certain rate is directly proportional to C.

For a given source voltage, then, the size of the capacitance will determine the amount of current. As C increases, the current increases and vice versa. Thus, we can say that

A larger capacitor opposes ac less than a smaller capacitor.

The opposition which a capacitor presents to the flow of alternating current is called *capacitive reactance* and is given the symbol X_C. Like its inductive counterpart X_L, capacitive reactance is measured in ohms. If a sinusoidal voltage with a peak amplitude E_M is applied to a capacitor the capacitor current will be a sine wave whose amplitude is given by

$$I_M = \frac{E_M}{X_C} \tag{17-14}$$

The rms value of current can be determined from

$$I = \frac{E}{X_C} = \frac{0.707 \, E_M}{X_C} \tag{17-15}$$

These relationships compare to the similar formulas (17-3) and (17-4) for the inductor.

Formula For X_C The value of capacitive reactance X_C can be determined from the formula

$$X_C = \frac{1}{2\pi f C} = \frac{1}{\omega C} \tag{17-16}$$

where f is in Hz, C is in farads, and X_C is in ohms.

Examination of this formula indicates that X_C is *inversely proportional* to capacitance. This follows from the previous discussion where it was shown that a larger capacitor opposes ac less than a smaller capacitor. The formula also indicates that X_C is *inversely proportional* to frequency. This is reasonable since a higher frequency means that the capacitor voltage is changing at a faster rate; thus, the capacitor charging and discharging current would have to become correspondingly higher ($i = C\, dv/dt$) in order to follow the applied voltage.

The fact that X_C is inversely proportional to frequency means that the capacitor offers *less* opposition to ac as the applied frequency is increased. This is just opposite to the behavior of inductors.

EXAMPLE 17.13 Determine in which circuit of Fig. 17-12 the lamp will be brightest, and in which case it will be least bright.

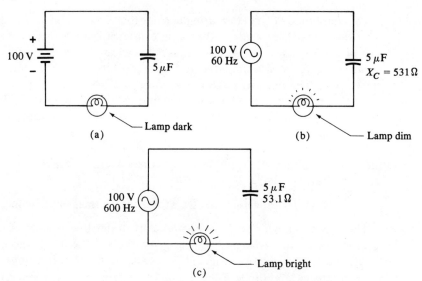

FIGURE 17-12 *Capacitor offers less opposition at higher frequencies.*

Solution: In (a) the source is 100 V dc. The capacitor will block dc current so that we can consider the capacitor to be an open circuit to dc. Equivalently, we can say that the capacitive reactance X_C is *infinite* for dc. That is,

$$X_C = \frac{1}{2\pi f C} = \frac{1}{2\pi \times 0 \times C} = \frac{1}{0} = \infty$$

for $f = 0$

In (b) the same capacitor will offer less opposition since f is now 60 Hz. The capacitive reactance is

$$X_C = \frac{1}{2\pi f C} = \frac{1}{2\pi \times 60 \times 5 \times 10^{-6}} = 531 \ \Omega$$

With this value of X_C, a substantial current will flow through the lamp.

In (c) the frequency has been increased to 600 Hz, so that X_C will decrease to 53.1 ohms. Thus, the lamp current will increase. Clearly, then, the lamp will be brightest in circuit (c) and least bright in (a).

EXAMPLE 17.14 Calculate the X_C of a 0.05-μF capacitor at 400 Hz and at 10 kHz.

Solution: At $f = 400$ Hz,

$$X_C = \frac{1}{2\pi f C}$$

$$= \frac{1}{2\pi \times 400 \times 0.05 \times 10^{-6}}$$

$$X_C = 8 \ k\Omega$$

At $f = 10$ kHz

$$X_C = \frac{1}{2\pi \times 10^4 \times 0.05 \times 10^{-6}}$$

$$X_C = 320 \ \Omega$$

EXAMPLE 17.15 Calculate the X_C of a 25-μF capacitor at 400 Hz and then at 10 kHz.

Solution: At $f = 400$ Hz,

$$X_C = \frac{1}{2\pi \times 400 \times 25 \times 10^{-6}} = 16 \ \Omega$$

At $f = 10$ kHz,

$$X_C = \frac{1}{2\pi \times 10^4 \times 25 \times 10^{-6}} = 0.64 \ \Omega$$

These last two examples illustrate that X_C decreases as f increases, and as C increases. Example 17.15 also shows that a relatively large capacitance (25 μF) will have a very low reactance at higher frequencies.

EXAMPLE 17.16 What size capacitance has a reactance of 160 Ω at 10 kHz?

Solution: In this problem we know X_C and f and are asked to find C. We can solve formula (17–16) for C with the result

$$C = \frac{1}{2\pi f X_C} \qquad (17\text{–}17)$$

For the values given,

$$C = \frac{1}{2\pi \times 10^4 \times 160}$$
$$= 10^{-7} \text{ F} = 0.1 \text{ μF}$$

The capacitance of a capacitor can often be determined by applying an ac voltage and measuring the amount of ac current that results. The ratio of E/I will produce the value of X_C, and C can be found from (17–17).

EXAMPLE 17.17 A capacitor has a voltage sine wave applied to it with $E = 12$ V and $f = 1$ kHz. The capacitor current is measured to be 3 mA. What is the value of C?

Solution: $X_C = E/I = 12 \text{ V}/3 \text{ mA} = 4 \text{ k}\Omega$

$$C = \frac{1}{2\pi f X_C} = \frac{1}{2\pi \times 10^3 \times 4 \times 10^3} = 0.04 \text{ μF}$$

EXAMPLE 17.18 At what frequency will a 2-μF capacitor have a reactance of 800 ohms?

Solution: Formula (17–16) can be solved for f with the result

$$f = \frac{1}{2\pi X_C C} \qquad (17\text{–}18)$$

For the values given,

$$f = \frac{1}{2\pi \times 800 \times 2 \times 10^{-6}}$$
$$= 100 \text{ Hz}$$

EXAMPLE 17.19 A 500-pF capacitor has a 200-kHz current whose rms value is measured as 10 mA. Determine the *peak* value of capacitor voltage.

Solution: $X_C = \dfrac{1}{2\pi f C} = \dfrac{1}{2\pi \times 200 \times 10^3 \times 500 \times 10^{-12}} = 1.6 \text{ k}\Omega$

Since $I = V/X_C$, then $V = I \times X_C$ so that

$$V = 10 \text{ mA} \times 1.6 \text{ K}\Omega$$
$$= 16 \text{ V}$$

This is the rms value of voltage. The peak value V_M is $\sqrt{2}V = 22.56$ V.

To summarize our discussion of capacitive opposition to current we can make the following statements.

(1) A capacitor offers infinite opposition to current in a dc circuit ($f = 0$).
(2) In an ac circuit, the capacitor's opposition X_C decreases as the frequency is increased and as the capacitance is increased.
(3) The Ohm's law relationship $I = V/X_C$ can be applied to a capacitor in an ac circuit.

These characteristics have numerous applications in practical circuits. Capacitance is often used where it is desired to block direct current but not alternating current. It is also used where it is desired to have more current opposition at low ac frequencies than at higher ac frequencies. Like inductors, capacitors are used in filters and resonant circuits.

17.10 Series and Parallel Capacitive Reactances

Capacitive reactances are combined in series and parallel in the same manner as inductive reactances. This is illustrated in Fig. 17–13. For the general case of capacitors in *series*, the total capacitive reactance is equal to the sum of the individual capacitive reactances.

$$X_{CT} = X_{C1} + X_{C2} + X_{C3} \ldots \text{etc.} \qquad (17\text{–}19)$$

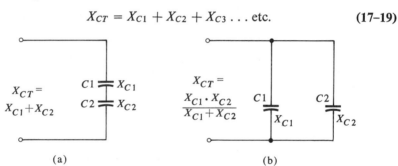

FIGURE 17–13 *Capacitive reactances combine in the same manner as inductive reactances and resistors.*

For two *parallel* capacitive reactances the total reactance is given by

$$X_{CT} = \frac{X_{C1} X_{C2}}{X_{C1} + X_{C2}} \quad \{\text{product-over-sum}\} \qquad (17\text{–}20)$$

For more than two parallel capacitive reactances, the reciprocal formula is used

$$\frac{1}{X_{CT}} = \frac{1}{X_{C1}} + \frac{1}{X_{C2}} + \frac{1}{X_{C3}} \ldots \text{etc.} \qquad (17\text{–}21)$$

It is important to realize that capacitive reactances combine just like resistors and inductive reactances and not like capacitors (capacitors in *parallel* add together). This is because X_C is inversely proportional to C.

EXAMPLE 17.20 Figure 17–14 shows a 160-Hz source connected to two series capacitors. The source voltage is 100 V rms. Determine: (a) total capacitive reactance; (b) rms current; (c) rms voltage across each capacitor.

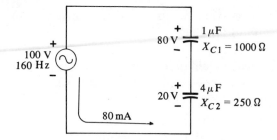

FIGURE 17-14

Solution: (a) $X_{C1} = \dfrac{1}{2\pi f C_1} = 1000 \;\Omega$

$X_{C2} = \dfrac{1}{2\pi f C_2} = 250 \;\Omega$

$X_{CT} = 1000 + 250 = \mathbf{1250 \;\Omega}$

(b) $I = \dfrac{E_s}{X_{CT}} = \dfrac{100 \text{ V}}{1250 \;\Omega} = \mathbf{80 \text{ mA}}$

(c) $V_1 = I X_{C1} = 80 \text{ mA} \times 1 \text{ k}\Omega$
$= \mathbf{80 \text{ V}}$
$V_2 = I X_{C2} = \mathbf{20 \text{ V}}$

These are the rms voltages across each capacitor. Note that $V_1 + V_2 = E$. Also note that the *larger* capacitor has the *smaller* voltage, since it has a smaller reactance.

EXAMPLE 17.21 Figure 17–15 shows the same source connected to the same capacitors in parallel. Determine: (a) total capacitive reactance; (b) rms value of total current; (c) current through each capacitor.

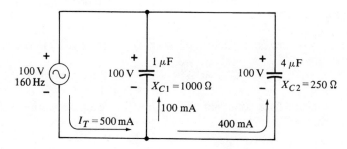

FIGURE 17-15

Solution: (a) $X_{CT} = \dfrac{X_{C1} X_{C2}}{X_{C1} + X_{C2}} = \dfrac{1000 \times 250}{1250} = \mathbf{200 \;\Omega}$

(b) $I_T = \dfrac{E_s}{X_{CT}} = \dfrac{100 \text{ V}}{200 \;\Omega} = \mathbf{500 \text{ mA}}$

(c) $I_1 = \dfrac{E}{X_{C1}} = \dfrac{100 \text{ V}}{1 \text{ k}\Omega} = 100 \text{ mA}$

$I_2 = \dfrac{E}{X_{C2}} = \dfrac{100 \text{ V}}{250 \text{ }\Omega} = 400 \text{ mA}$

Note that each capacitor has the full source voltage across its terminals. Also note that $I_1 + I_2 = I_T$ and the *larger* capacitor has the *larger* value of current since it has a lower reactance.

17.11 Capacitive Phase Angle

When a capacitor is used in an ac circuit it can be stated that:

> The capacitor current leads the capacitor voltage by 90° ($\pi/2$ radians) at all frequencies.

This characteristic is opposite to that of an inductor where *i lags v* by 90°.

Figure 17–16 shows the waveforms for voltage and current in a capacitor. The voltage waveform is a normal sine curve that begins at 0° at $t = 0$, and reaches its positive peak at 90°. The current sine wave is at its positive peak at $t = 0$. The current waveform is *leading* by 90° since it reaches its positive peak 90° earlier than the voltage.

In order to see why this 90° phase difference is present, we must recall that the capacitor current is proportional to the rate at which its voltage is changing.

$$i_c = C \dfrac{dv_c}{dt} \qquad (13\text{--}5)$$

The capacitor voltage v_c is exactly the same as the applied voltage sine wave.

Situation at 0° If we look at the capacitor voltage sine wave at 0° it can be seen that the *slope* of the curve is steepest at this point. Also, the slope is positive (v_c increasing) at this point. Thus, the *rate of change* of capacitor voltage dv_c/dt has its *maximum positive* value at 0°. Since the capacitor is charging at its maximum rate, the charging current i_c must be maximum at this point. Thus, $i_c = I_M$ when the voltage is at 0° as shown.

Between 0° and 90° Between zero and 90° the slope of the voltage curve becomes gradually less and less, which means that dv_c/dt is decreasing from its maximum value at 0°. With the capacitor charging at a slower and slower rate, the charging current i_c must be getting smaller and smaller. Eventually at 90° the voltage reaches its positive peak at which point dv_c/dt is equal to zero. Thus, at 90° the current will be zero ($i = C\, dv/dt = C \times 0$).

Between 90° and 180° After the 90° point, the voltage begins to decrease. The slope of the curve is negative and this negative slope becomes gradually steeper in going from 90° to 180°. This means that the capacitor is discharging at a gradually increasing rate. Thus, the current i_c reverses. This discharge current reaches its maximum value when the voltage is at 180° (maximum negative dv_c/dt).

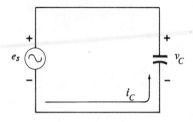

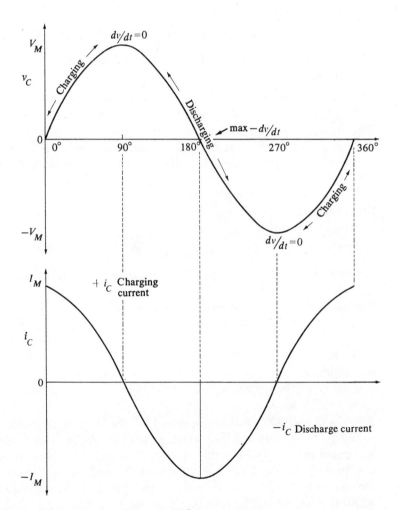

FIGURE 17–16 *Current leads voltage by 90° in a capacitor.*

Between 180° and 270° After the 180° point, the voltage continues decreasing but the rate at which it decreases becomes less and less until the 270° peak point where $dv_c/dt = 0$ again. In this interval the current remains negative (discharge) but its value becomes less and less negative (increases) until 270° where $i_c = 0$ again.

Between 270° and 360° After 270° the voltage begins to increase again. The rate at which it is increasing becomes gradually greater until at 360° the value of dv_c/dt is again at its maximum value. Since the capacitor is now charging again, the current is positive and it increases to its maximum value at 360°.

At the 360° point the voltage is back to zero and the current is at its positive peak. This is the same situation that began at 0°, thus completing one full cycle. The next full cycle of voltage sine wave will produce a current cycle identical to the first current cycle. Thus, both waveforms will be at the *same frequency* but the i_c waveform *leads* the v_c waveform by 90° or, equivalently, the v_c waveform *lags* the i_c waveform by 90°. *This same 90° phase difference occurs at all frequencies.*

17.12 Complete Capacitor Response to AC

When a sine wave of voltage is applied to a capacitor, the capacitor opposes current due to its reactance X_C. The amplitude of the current depends on the applied voltage and the capacitive reactance. The phase angle of the current will always be 90° leading with respect to the voltage.

EXAMPLE 17.22 Consider Fig. 17–17(a) where the voltage sine wave shown is applied to a capacitor. Determine the current waveform.

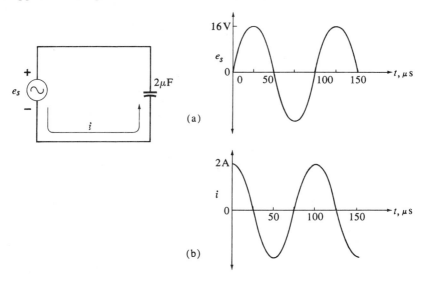

FIGURE 17–17

Solution: The frequency of the applied voltage can be found from $f = 1/T$. Clearly, $T = 100$ μs. Thus, $f = 1/(100 \text{ μs}) = 10$ kHz. The capacitive reactance is therefore

$$X_C = \frac{1}{(2\pi)(10^4)(2 \times 10^{-6})} \approx 8 \text{ Ω}$$

482   Chapter Seventeen

The peak voltage is $E_M = 16$ V. Thus, the peak current is $I_M = 16\,\text{V}/8\,\Omega = 2$ A. The complete i_c waveform is shown in (b). Its peak amplitude is 2 A and it leads the voltage by 90°.

Expression for Capacitor Current   The expression for a sine-wave voltage source is

$$e_s = E_M \sin(2\pi ft) \tag{17-22a}$$

If this voltage is applied to a capacitor, the capacitor current will be a sine wave with its expression given by

$$i = \underbrace{\frac{E_M}{X_C}}_{I_M} \sin(2\pi ft + \pi/2) \tag{17-22b}$$

This expression reflects the fact that the peak current will be E_M/X_C, and also includes the effect of the 90° phase difference. Since 90° is $\pi/2$ radians, the current sine wave angle will always be $\pi/2$ radians more than the voltage sine wave angle. Thus, the $+\pi/2$ term is present in the current expression.

EXAMPLE 17.23   Write the expression for e and i of Fig. 17–17.

Solution: $e_s = E_M \sin 2\pi ft$
$\qquad\quad = 16 \sin (2\pi \times 10^4\, t)$
$\qquad\quad = 16 \sin (62{,}800\, t)$ volts
$\qquad i = 2 \sin (62{,}800\, t + \pi/2)$ amps

The expression for capacitor voltage and current can also be written with the current sine wave as reference.

$$i = I_M \sin(2\pi ft) \tag{17-23a}$$

and

$$v_c = V_M \sin(2\pi ft - \pi/2) \tag{17-23b}$$

The $-\pi/2$ term indicates that the voltage is lagging the current by 90°.

Phasor Representation of v and i   If the voltage across a capacitor is a sine wave with an rms value V_c, we can represent this on a phasor diagram as a phasor with length equal to V_c. We can choose the voltage sine wave as our reference so that the voltage phasor is at 0°. This is shown in Fig. 17–18(a). The capacitor current will lead the applied voltage by 90°. Thus, we can represent the current sine wave by a phasor with a length of $I_C = V_c/X_C$ at an angle of $+90°$ as shown in the figure.

If we wish to choose the capacitor current as the reference, the phasor representation is shown in part (b) of the figure with I_C at 0° and V_c at $-90°$, lagging I_C by 90°. Both representations in Fig. 17–18 are correct. It is arbitrary as to which sine wave is chosen as the reference.

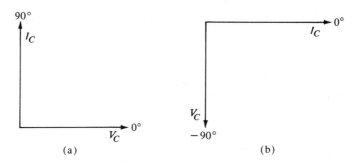

FIGURE 17-18 *Equivalent phasor representations of capacitor current and voltage.*

Chapter Summary

1. A resistor opposes ac current with the same amount of opposition that it opposes dc current.
2. An inductor opposes a changing current (ac) but does not oppose a constant dc current.
3. The opposition of an inductor to *ac* is called its reactance X_L.
4. The formula for inductive reactance is $X_L = 2\pi fL$ ohms.
5. The peak amplitude of current in an inductor is determined from $I_M = E_M/X_L$. The rms amplitude is determined from $I = E/X_L$.
6. Since X_L is proportional to frequency, the inductor offers more opposition to current as the frequency of the applied voltage is increased.
7. Inductive reactances in series (parallel) are combined in the same manner as resistors in series (parallel).
8. The current sine wave through an inductor *lags* the voltage sine wave across the inductor by 90° ($\pi/2$ radians) at all frequencies.
9. A capacitor blocks constant direct current.
10. The opposition of a capacitor to ac is called its reactance X_C.
11. The formula for capacitive reactance is $X_C = 1/(2\pi fC)$.
12. The peak current in a capacitor is given by $I_M = E_M/X_C$. The rms current is given by $I = E/X_C$.
13. The capacitor offers less opposition to current as the frequency of the applied voltage is increased.
14. Capacitive reactances in series (parallel) are combined in the same manner as resistors in series (parallel).
15. The current sine wave in a capacitor *leads* the voltage across the capacitor by 90° ($\pi/2$ radians) at all frequencies.

16. Table 17-1 summarizes the ac characteristics of the elements R, L, and C.

TABLE 17-1

Element	R	L	C
Basic units	ohms—Ω	henries—H	farads—F
Opposition to dc	resistance R $\left(I = \dfrac{E}{R}\right)$	zero (short circuit)	infinite (open circuit)
Opposition to ac	"	reactance X_L $\left(I = \dfrac{E}{X_L}\right)$	reactance X_C $\left(I = \dfrac{E}{X_C}\right)$
Effects of frequency	none: resistance = R at all f.	X_L increases with f: $X_L = 2\pi f L$	X_C decreases with f: $X_C = \dfrac{1}{2\pi f C}$
Series combination	$R_T = R_1 + R_2$	$X_{LT} = X_{L1} + X_{L2} \ldots$	$X_{CT} = X_{C1} + X_{C2} \ldots$
Parallel combination	$R_T = \dfrac{R_1 R_2}{R_1 + R_2}$	$X_{LT} = \dfrac{X_{L1} X_{L2}}{X_{L1} + X_{L2}}$	$X_{CT} = \dfrac{X_{C1} X_{C2}}{X_{C1} + X_{C2}}$
Phase angle	i in phase with e	i lags e by 90°	i leads e by 90°
Voltage and current expressions	$e = E_M \sin 2\pi ft$ $i = \dfrac{E_M}{R} \sin 2\pi ft$	$i = \dfrac{E_M}{X_L} \sin (2\pi ft - \pi/2)$	$i = \dfrac{E_M}{X_C} \sin (2\pi ft + \pi/2)$
Phasor diagram	I, E at 0°	E at 0°, I at −90°	I at 90°, E at 0°

Questions/Problems

Sections 17.2–17.4

17-1 In which circuit in Fig. 17-19 will the lamp be brightest? In which circuit will the lamp be least bright?

17-2 Change the 100-ohm resistor to a 100-mH inductor for each of the circuits in Fig. 17-19 and repeat problem 17-1.

17-3 Calculate the X_L of a 150-mH coil at each of the following frequencies: (a) 40 Hz; (b) 1 kHz; (c) 40 kHz; (d) 1 MHz.

Opposition in AC Circuits   485

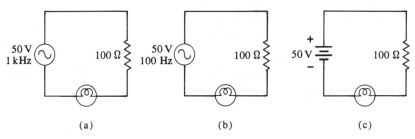

FIGURE 17–19

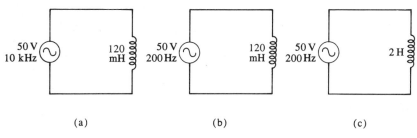

FIGURE 17–20

17–4   Calculate the X_L of 50-μH coil at each of the frequencies given in problem 17–3.

17–5   Determine the rms current in each of the circuits of Fig. 17–20.

17–6   What value of inductance would produce an rms current of 1 mA in the circuit of Fig. 17–20(a).

17–7   What is the inductance of a coil that passes a current of 1 mA when connected to a 12-V, 10-MHz source?

17–8   A certain coil has a reactance of 1000 ohms at a certain frequency. If its inductance is cut in half and the frequency is doubled, what will be the new value of X_L?

17–9   At what frequency will a 400-mH inductance have a reactance equal to the resistance of a 500-kΩ resistor?

17–10 What is the voltage across a 100-μH coil that is passing a 15-μA, 500-kHz current?

Section 17.5

17–11 For the circuit of Fig. 17–21(a) determine: (a) rms current and (b) voltage across each coil.

17–12 For the circuit of Fig. 17–21(b) determine: (a) total reactance; (b) source current I_T; (c) current through each coil.

17–13 Give two differences and one similarity in comparing L and R with respect to their opposition to alternating current.

Sections 17.6–17.8

17–14 A 10-mH coil has a 60-Hz current flowing through it with an instantaneous value given by $i = 50 \sin (377t)$ mA. Draw the waveforms of coil current and coil voltage on the same time axis.

17-15 A certain voltage sine wave behaves according to the expression $e_s = 150 \sin (25120t)$. This voltage is applied to a 10-mH coil. Draw the waveforms of coil voltage and current on the same time axis showing amplitudes of each.

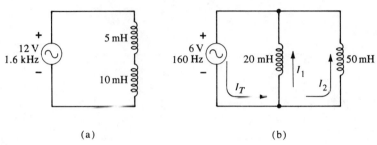

(a)   (b)

FIGURE 17–21

17-16 Write the expression for the coil voltage of problem 17–14.

17-17 Write the expression for the coil current of problem 17–15. Calculate the instantaneous current at $t = 100$ μs.

17-18 Draw the phasor diagram for problem 17–14.

17-19 Draw the phasor diagram for problem 17–15.

17-20 Comment on the truth or falsity of each of the following statements:
   a. An inductor will pass more current at 10 Hz than at 1 Hz.
   b. The phase difference between coil current and voltage is greater at high frequencies than at low frequencies.
   c. Connecting inductors in series increases the opposition to alternating current.
   d. Connecting coils in series increases the phase angle difference between current and voltage.
   e. Connecting coils in parallel decreases the opposition to alternating current.

Sections 17.8–17.9

17-21 Change the 100-ohm resistor to a 10-μF capacitor for each circuit of Fig. 17–19 and repeat problem 17–1.

17-22 Calculate the X_C of a 0.15-μF capacitor at each of the following frequencies: (a) 40 Hz; (b) 1 kHz; (c) 40 kHz; (d) 1 MHz.

17-23 Calculate X_C for a 50-pF capacitor at each of the frequencies of problem 17–22.

17-24 Determine the rms current in each of the circuits of Fig. 17–22.

17-25 What value of capacitance would produce an rms current of 1 mA in the circuit of Fig. 17–22(b).

17-26 What is the capacitance of a capacitor that passes a current of 50 μA when connected to a 6-V, 60-Hz source?

17-27 A certain capacitor has a reactance of 1000 ohms at a certain frequency. If its capacitance is doubled and the frequency is doubled, what will be the new X_C?

17-28 At what frequency will a 40-μF capacitor have the same reactance as a 100-mH inductor?

17-29 What is the voltage across a 0.001-μF capacitor that has a 12-μA, 500-Hz current?

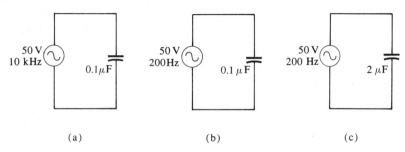

FIGURE 17-22

Section 17.10

17-30 Change the coils in Fig. 17-21(a) to 5-μF and 10-μF capacitors and repeat problem 17-11.

17-31 Change the coils in Fig. 17-21(b) to 20-μF and 50-μF capacitors and repeat problem 17-12.

Sections 17.11–17.12

17-32 Give two differences and one similarity in comparing C and R with respect to their opposition to ac.

17-33 Repeat problem 17-32 for C and L.

17-34 A 0.033-μF capacitor has a 60-Hz current whose instantaneous value is given by 10 sin (377t) μA. Draw the waveforms of capacitor current and voltage on the same time axis.

17-35 A voltage sine wave $e = 150 \sin (6280\ t)$ is applied to a 0.0047-μF capacitor. Draw the waveforms of capacitor voltage and current on the same time axis showing amplitudes.

17-36 Write the expression for the voltage in problem 17-34.

17-37 Write the expression for the current in problem 17-35.

17-38 Draw the phasor diagram for problem 17-34.

17-39 Draw the phasor diagram for problem 17-35.

17-40 For each of the following statements, indicate which one (or more) of the devices R, L, or C the statement pertains to.
  a. Doubling the amplitude of the applied ac voltage doubles the current.
  b. The phase difference between i and v does not change with changes in frequency.

c. Current increases as frequency increases.
d. Two devices in parallel produce more opposition to current.
e. It will conduct direct current.
f. Opposition to ac increases at high frequencies.
g. Current and voltage phasors are at right angles.
h. It becomes a short circuit at high frequencies.

17-41 A 20-V, 1-kHz source is applied to a circuit that consists of the parallel combination of a 500-ohm resistor, a 500-ohm X_L, and a 500-ohm X_C. On the same time axis draw the waveforms of the applied voltage and the currents through each element.

17-42 Draw a phasor diagram representing the currents and voltage of the previous problem. Use the source voltage as reference.

17-43 A 10-kHz voltage is applied to a certain device. The waveforms of voltage and current are displayed on the oscilloscope and appear as shown in Fig. 17-23(a). From these waveforms determine whether the device is a resistor, capacitor, or inductor and its value.

17-44 Repeat 17-43 for the waveforms in Figure 17-23(b)

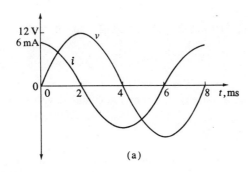
(a)

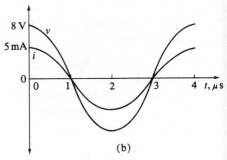
(b)

FIGURE 17-23

CHAPTER 18

RC Circuit Response to AC

18.1 Introduction

In chap. 13 we investigated how circuits containing both *resistance* and *capacitance* responded to the sudden application of a dc voltage. We saw that after *five* circuit time constants had passed, the circuit reached a *steady state*. In steady state, all the circuit currents and voltages had reached constant (steady) values. In this chapter we examine the response of *RC* circuits to the application of a sinusoidal voltage. As in the dc case, the *RC* circuit reaches a steady state after 5τ. In ac steady state, however, all the circuit currents and voltages will be sine waves. In our work with ac circuits in this and subsequent chapters we will only be interested in the steady-state response and will not concern ourselves with the transient interval that occurs before steady state is reached.

18.2 Review of X_C and R

RC circuits contain both resistance and capacitance. In the last chapter we saw how resistance behaves when an ac voltage is applied to it. Resistance opposes ac current with an opposition equal to R ohms. The ac current in a resistance is always in phase ($\Phi = 0°$) with the ac voltage across the resistor. This is shown in Fig. 18-1(a). The resistor voltage is labeled v_R. The resistor current is i_R. The waveform diagram and the phasor diagram show the 0° phase difference between v_R and i_R.

We also saw that capacitance opposes ac current with an opposition equal to X_C ohms. The capacitor current always *leads* the capacitor voltage by 90° ($\pi/2$ rad). This is shown in Fig. 18-1(b) where the capacitor voltage is labeled v_c and the current is i_c.

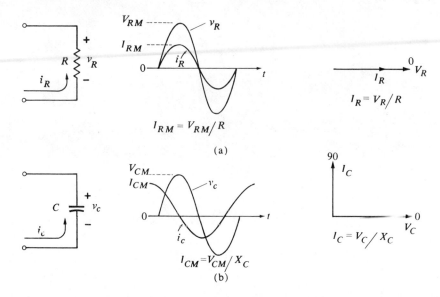

FIGURE 18-1 *Behavior of* (a) *resistor and* (b) *capacitor in alternating current circuit.*

In Fig. 18-1 the *peak* values of the resistor and capacitor voltages are shown on the waveform diagrams as V_{RM} and V_{CM} respectively. The subscript M indicates maximum or peak value. The same is true for the peak currents I_{RM} and I_{CM}. The phasor diagrams shown in the figure use the rms values of voltage and current. For example, the resistor phasor diagram uses V_R and I_R as the phasor magnitudes. The absence of the M subscript means that these are rms values. The same is true for the capacitor phasors V_C and I_C. In most of the subsequent work with phasor diagrams we will use rms values unless otherwise indicated.

18.3 X_c and R in Series

Figure 18-2(a) shows an ac source connected to a circuit that contains a capacitive reactance X_C in series with a resistor R. The current from the source is the *same* through each of the circuit components. Since i flows through both X_C and R, then both elements will contribute to the total opposition to current. Since $X_C = 100$ ohms and $R = 100$ ohms for the circuit shown, we might expect the total opposition to be 200 ohms. As we shall see, this is *not* the case.

Part (b) of Fig. 18-2 shows the various circuit waveforms for the circuit values chosen. The following points can be ascertained from these waveforms.

(1) The current waveform is in phase with the resistor voltage waveform v_R. The peak voltage across the resistor is equal to the peak current 1.41 A × 100 ohms = 141 V. The rms value is $V_R = 0.707 \times 141$ V = 100 V.

(2) The capacitor voltage waveform v_c *lags* the current waveform by 90°. Therefore, it also *lags* v_R by 90°. The peak capacitor voltage is equal to the peak current 1.41 A × 100 ohms = 141 V. The rms value is $V_C = 0.707 \times 141$ V = 100 V.

(3) The source voltage waveform at any instant of time is equal to the algebraic sum of the resistor voltage and capacitor voltage. This, of course, is a

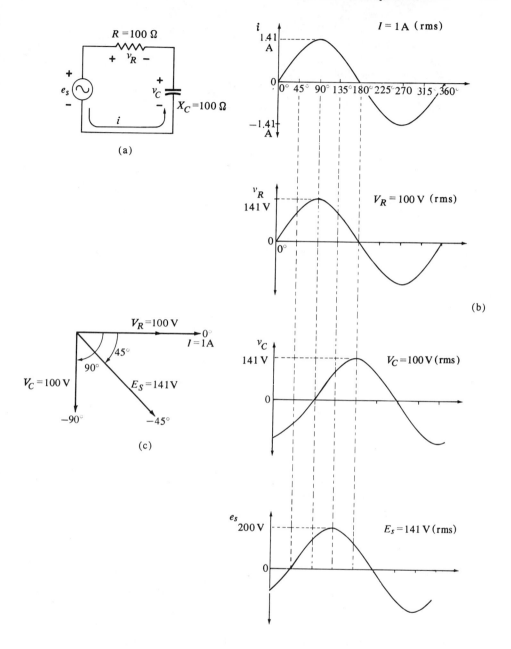

FIGURE 18-2 (a) *Circuit with X_C and R in series.* (b) *Waveform.* (c) *Phasor diagram.*

result of Kirchhoff's voltage law. That is, $e_s = v_R + v_C$. The e_s waveform can be obtained by adding the values of v_R and v_C at each instant of time.

(4) The peak source voltage is 200 V. This does *not* equal the sum of the peak voltages across R and X_c. This is because the peaks of the v_C and v_R waveforms do not occur at the same time since these waveforms are 90° out of phase. When v_R is at its peak of 141 V, the v_C waveform is at zero volts and vice versa.

The peak source voltage occurs at 135° when v_R and v_C are both 100 V so that $e_s = 200$ V(peak).

(5) The source voltage reaches its peak at 135°. Thus, e_s *lags* the current i by 45°. Or, equivalently, i *leads* e_s by 45°. In other words, the phase difference between the circuit current and the applied voltage is neither 0° nor 90°. Keep in mind, however, that i is still in phase with v_R and leads v_C by 90°.

Phasor Diagram The phasor diagram in Fig. 18-2(c) shows the phase relationships described. Since the current is common to all the circuit elements, the current is used as the reference sine wave and its phasor I is drawn at 0°. The magnitude of the current phasor is shown as 1 A. This is the rms value of current (0.707 × 1.41 A). The resistor voltage phasor V_R is also at 0° since it is in phase with the current. $V_R = 100$ V, the rms voltage of the v_R waveform. The capacitor voltage phasor V_C is drawn at $-90°$ relative to the current phasor and also has an rms value of 100 V. The source voltage phasor E_s is drawn so that it *lags* the current phasor by 45° and has an rms magnitude $E_s = 141$ V (0.707 × 200). Note that $E_s \neq V_R + V_C$.

Combining v_R and v_C Using Vector Addition As shown in Fig. 18-2(b), we can obtain the applied source waveform e_s by arithmetically adding the v_R and v_C waveforms point by point. We saw that because v_R and v_C are out of phase, the peak or rms values of e_s *cannot* be obtained by adding the peak or rms values of v_C and v_R. In other words, *when adding two series voltages that are out of phase, we have to take into account the phase difference.*

There is a much easier method of adding two out-of-phase sine waves than the point-by-point addition described earlier. This easier method uses the phasor representation of the sine waves. The phasors for each sine wave are drawn on a phasor diagram. Instead of adding the sine waves we now add their respective phasors. The resultant phasor represents the sine wave that is the sum of the two added sine waves. Since phasors are actually *vectors* with magnitude and direction, we have to use *vector addition* to add the phasors.

Figure 18-3 shows the procedure for determining the sum of the v_R and v_C waveforms of Fig. 18-2(b). In part (a) of the figure the phasors for the capacitor and resistor voltage are drawn with their lengths proportional to their respective rms values. The procedure for adding the two phasors begins by moving one of the phasors so that its tail coincides with the arrowhead of the other phasor. In part (b) of the figure, the V_C phasor has been moved so that its tail is at the arrowhead of the V_R phasor. Note that the V_C phasor still points in the *same* direction and still makes a 90° angle with V_R even after it has been moved.

Once the phasors are connected head to tail as in Fig. 18-3(b), the sum or *resultant* of the two phasors is obtained by drawing a phasor from the tail of V_R to the arrowhead of V_C as shown in part (c) of the figure. This resultant phasor is the vector sum of the V_C and V_R phasors and represents the source voltage. This resultant phasor is therefore labeled E_S. If V_C and V_R are drawn to scale, then the magnitude and angle of E_S can be measured using a ruler and protractor. A more mathematical method is described here.

RC Circuit Response to AC

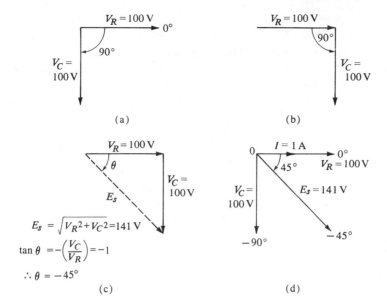

FIGURE 18-3 *Finding the total applied voltage by vector addition of resistor and capacitor voltage phasors.*

The magnitude of the E_S phasor can be found using the Pythagorean theorem since the phasors V_C, V_R, and E_S form a right triangle. E_S is the hypotenuse. Therefore

$$E_S = \sqrt{V_R^2 + V_C^2} \qquad (18\text{-}1)$$

For $V_C = V_R = 100$ V, the rms value of source voltage is therefore

$$E_S = \sqrt{100^2 + 100^2} = \sqrt{2 \times 10^4} = 141 \text{ V}$$

In Fig. 18-3(c), the angle θ represents the phase angle by which V_R *leads* E_s. Remember V_R is at 0° because it is in phase with the circuit current which was chosen as the reference phasor. Thus, θ also represents the phase angle by which the circuit current I leads the source voltage E_s. The value of θ can be determined using a bit of right-triangle trigonometry.

$$\tan \theta = \frac{\text{opposite}}{\text{adjacent}}$$

$$= -\left(\frac{V_C}{V_R}\right) \qquad (18\text{-}2)$$

The negative sign is used because θ will be negative since it is *clockwise* relative to the 0° reference. For our example $V_C = V_R$ so that $\tan \theta = -1$. Thus, $\theta = -45°$. The negative sign indicates θ is clockwise by 45° relative to the 0° reference.

Thus, the source voltage has an rms magnitude of 141 V and *lags* the circuit current (and V_R) by 45°. This result agrees with that obtained in Fig. 18-2. The formulas (18-1) and (18-2) for the magnitude and phase of the source voltage can be used for *any* series RC circuit. However, the student should not abandon the phasor diagram simply because the formulas are available. The phasor diagram is useful in helping to visualize the circuit phase relationships.

Part (d) of Fig. 18-3 shows the I, V_R, V_C, and E_S phasors all drawn from the origin (O). This representation gives a complete picture of the circuit phase relationships.

What we have shown so far is that

In a series RC circuit the phasor representing the source voltage is equal to the vector sum of the V_R and V_C phasors.

EXAMPLE 18.1 A series RC circuit contains $R = 2$ kΩ and $X_C = 1$ kΩ. (a) Find the source voltage needed to produce an rms current of 5 mA in the circuit; (b) determine the phase angle between the source voltage and current; (c) determine the phase angle between the source voltage and the capacitor voltage; and (d) find the phase angle between the source voltage and resistor voltage.

Solution: (a) The circuit diagram is shown in Fig. 18-4(a). Since $I = 5$ mA we can calculate the rms values of the resistor and capacitor voltages.

$$V_R = 5 \text{ mA} \times 2 \text{ k}\Omega = 10 \text{ V}$$
$$V_C = 5 \text{ mA} \times 1 \text{ k}\Omega = 5 \text{ V}$$

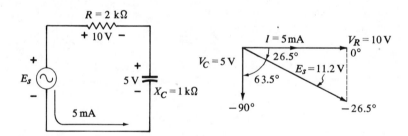

FIGURE 18-4

To find E_S we must add the V_R and V_C phasors using vector addition. These phasors are drawn in part (b) of the figure along with the resultant E_S. The rms magnitude of E_S is

$$E_S = \sqrt{V_R^2 + V_C^2}$$
$$= \sqrt{10^2 + 5^2} \approx \mathbf{11.2 \text{ V}}$$

(b) $\tan \theta = -\left(\dfrac{V_C}{V_R}\right) = -\left(\dfrac{5 \text{ V}}{10 \text{ V}}\right) = -0.5$

$\therefore \theta = -26.5°$ as shown on the phasor diagram. The current *leads* E_S by 26.5°. In a series RC circuit, *the current always leads the applied voltage*.

(c) From the phasor diagram it can be seen that V_C lags E_S by $90° - 26.5° = 63.5°$. *In a series RC circuit, the capacitor voltage always lags the source voltage.*

(d) The resistor voltage is in phase with the current. Thus, V_R leads E_S by $26.5°$. *In a series RC circuit the resistor voltage always leads the source voltage.*

18.4 Impedance of a Series RC Circuit

In the last section we saw how to combine voltages in a series RC circuit using vector addition of the voltage phasors. We have not as yet seen how to determine the magnitude of the current that flows in the circuit. To find this current we must know the *total opposition* of the circuit. The RC circuit has two types of opposition, a resistance R and a reactance X_C both measured in ohms. It might appear that we could determine the total opposition by simply adding the two oppositions numerically as we would if two resistors were in series. Unfortunately, this is not the case because the resistance and reactance effects are not in phase. Again, vector addition must be used, as we shall see.

Whenever resistance and reactance are combined in the same circuit, the total opposition to ac is called the circuit's impedance. The symbol for *impedance* is Z and its unit is ohms. When an rms voltage E_S is applied to a circuit that has a total impedance Z, we can use Ohm's law to calculate the rms current. That is,

$$I = \frac{E_s}{Z} \tag{18-3}$$

This relationship can be rearranged as

$$E_S = IZ \tag{18-4}$$

which indicates that the source voltage equals the current times the circuit impedance.

Finding Z For a Series RC In Fig. 18–3 we constructed a phasor diagram of the voltages in a series RC circuit. The voltage phasor triangle of Fig. 18–3(c) is redrawn in Fig. 18–5(a). Note that each phasor in the voltage triangle has a magnitude proportional to I. Thus, if we divide each phasor magnitude by I, we still have the same right triangle reduced in scale by the factor I. This new triangle is shown in Fig. 18–5(b). As can be seen, the sides of this triangle are R, X_C, and Z. Thus, it is called an *impedance triangle*.

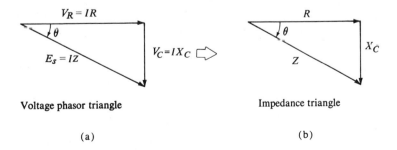

FIGURE 18–5 *Impedance Z is vector sum of* R *and* X_C.

In the impedance triangle R, X_C, and Z are drawn as vectors. These vectors have the same directions as their respective voltage phasors. From the impedance triangle it can be seen that the total circuit impedance Z is the *vector* sum of R and X_C. Using the Pythagorean theorem again, we have

$$Z = \sqrt{R^2 + X_C^2} \tag{18-5}$$

This relationship allows us to calculate the total opposition of any *series RC* circuit. It is important to realize that we cannot simply *add* R and X_C to obtain Z. That is, $Z \neq R + X_C$.

The angle θ in the impedance triangle is the same as the angle θ in the voltage triangle. Using the impedance triangle the value of θ can be calculated from

$$\tan \theta = -\left(\frac{X_C}{R}\right) \tag{18-6}$$

The negative sign is necessary because θ is a negative angle relative to the 0° reference.

EXAMPLE 18.2 A series RC circuit contains $R = 120$ ohms and $C = 2\ \mu\text{F}$. A 500-Hz, 40-V source is applied to the circuit. Determine: (a) the circuit impedance; (b) the circuit current; and (c) the various circuit voltages.

Solution: (a) $X_C = 1/2\pi f C$

$$= \frac{1}{2\pi \times 500 \times 2 \times 10^{-6}} \approx 160\ \Omega$$

$$Z = \sqrt{R^2 + X_C^2}$$
$$= \sqrt{120^2 + 160^2} = 200\ \Omega$$

(b) $I = E_S/Z$

$$= \frac{40\ \text{V}}{200\ \Omega} = 0.2\ \text{A}$$

(c) $V_R = IR = 0.2\ \text{A} \times 120\ \text{ohms} = \mathbf{24\ V}$
$V_C = IX_C = 0.2\ \text{A} \times 160\ \text{ohms} = \mathbf{32\ V}$

EXAMPLE 18.3 Draw the phasor diagram showing the current and voltages for the previous example.

Solution: The phasor diagram is shown in Fig. 18-6(a). The current phasor is drawn as the 0° reference. The V_R phasor is also at 0°. The V_C phasor is at $-90°$ lagging relative to the current. The source voltage is the vector sum of V_R and V_C. That is,

$$E_S = \sqrt{24^2 + 32^2} = 40\ \text{V}$$

The phase angle θ can be obtained from

$$\tan \theta = -\left(\frac{V_C}{V_R}\right) = -\frac{32}{24} = -1.33$$

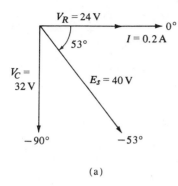
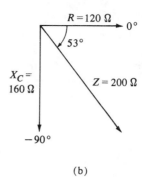

FIGURE 18-6

resulting in $\theta = -53°$. The negative sign is included to indicate that θ is clockwise relative to $0°$.

EXAMPLE 18.4 Draw the vector diagram showing R, X_C, and Z for the circuit of Ex. 18-2. Calculate θ.

Solution: The vector diagram is shown in Fig. 18-6(b). The resistance vector is at $0°$. The capacitive reactance vector is *always* drawn at $-90°$ relative to the resistance vector. The Z vector is the resultant of vectorially adding R and X_C. The angle θ is obtained from

$$\tan \theta = -\left(\frac{X_C}{R}\right) = -\frac{160}{120} = -1.33$$

so that $\theta = -53°$. This is the exact same angle determined using the voltage phasor diagram in Fig. 18-6.

We have seen in this section that

The impedance Z of a series RC circuit equals the vector sum of R and X_C.

18.5 Phasor Notation — Polar Form

The voltage and current phasors we have been using are characterized as having a magnitude and a phase angle. To completely describe a phasor we must give its magnitude, and its angle relative to the $0°$ reference axis. For example, in Fig. 18-6 the phasor which represents the source voltage has a magnitude of 40 V and a phase angle of $-53°$.

A special shorthand notation is often used to represent current and voltage phasors. A *current* phasor can be represented as

$$\bar{I} = I\underline{/\theta} \qquad (18\text{-}7)$$

In this expression, $\bar{I}$ is the symbol for the current phasor (vector). (The "bar" over the top of any quantity indicates that it is a vector quantity.) The I represents the rms value of the current and θ represents the phase angle of current relative to $0°$. Thus, if we have $\bar{I} = 10 \text{ mA}\underline{/25°}$ this would indicate that the I phasor has a magnitude of 10 mA and is at $25°$.

A similar representation is used for *voltage* phasors.

$$\overline{E} = E\underline{/\theta} \text{ (or } \overline{V} = V\underline{/\theta}) \tag{18-8}$$

In this expression $\overline{E}$ is the symbol for the voltage phasor. E is the rms magnitude of the voltage and θ is its phase angle. In Fig. 18-6(a), for example, we express the source voltage phasor as

$$\overline{E}_S = 40 \text{ V}\underline{/-53°}$$

When we represent phasors using the notations of (18-7) and (18-8), we are expressing these phasors in *polar form*. It is called polar form because from it we can accurately draw the phasor on a set of polar coordinate axes. We will employ the polar form to express phasors conveniently. Later we will find more uses for the polar form.

EXAMPLE 18.5 Refer to the phasor diagram in Fig. 18-4. Express each of the phasors in polar form.

Solution: $\overline{I} = 5 \text{ mA}\underline{/0°}$

$$\overline{V}_R = 10 \text{ V}\underline{/0°}$$
$$\overline{V}_C = 5 \text{ V}\underline{/-90°}$$
$$\overline{E}_S = 11.2 \text{ V}\underline{/-26.5°}$$

Note that the angle for $\overline{E}_S$ is negative since $\overline{E}_S$ *lags* $\overline{I}$.

EXAMPLE 18.6 Sketch the following phasors on a polar coordinate graph: (a) 100 V$\underline{/90°}$; (b) 50 V$\underline{/30°}$; (c) 75 V$\underline{/-60°}$.

Solution: The results are shown in Fig. 18-7. Note the magnitudes and angle of each phasor.

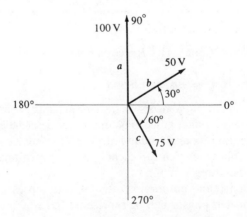

FIGURE 18-7

Earlier we found that in an *RC* series circuit the *vector* sum of the capacitor voltage and resistor voltage phasors was equal to the applied voltage phasor. We can express this as

$$\overline{E}_S = \overline{V}_R + \overline{V}_C \tag{18-9}$$

This expression is a vector expression and the addition sign stands for vector addition.

We also saw earlier that the magnitude of $\overline{E_S}$ is given by $E_S = \sqrt{V_R^2 + V_C^2}$ and the phase angle of E_S is given by $\tan\theta = -(V_C/V_R)$. Thus, $\overline{E_S}$ can be written as

$$E_S = \underbrace{\sqrt{V_R^2 + V_C^2}}_{\substack{\text{magnitude}\\ E_S}} \underbrace{/\theta = -\arctan(V_C/V_R)}_{\substack{\text{phase angle}\\ \theta}}$$

This last expression is simply a summary of the ideas presented in our earlier discussion (sec. 18.2). Note that θ is shown as a *negative* angle because E_S lags the reference current in a series RC circuit.

Vector Notation for Z R, X_C, and Z are not phasors. Recall that a phasor is a *rotating* vector that generates a sine wave. However, R, X_C, and Z can be represented as vectors as we saw in Fig. 18–5(b). We also saw earlier that for a series RC circuit the impedance equals the vector sum of the resistance and capacitive reactance. Thus, we have

$$\overline{Z} = \overline{R} + \overline{X_C} \qquad (18\text{--}10)$$

where $\overline{R} = R/0°$ and $\overline{X_C} = X_C/{-90°}$ are the polar forms of these vectors.

The magnitude of the impedance vector is given by Eq. (18–5) as $Z = \sqrt{R^2 + X_C^2}$ and the phase angle is given by (18–6) as $\tan\theta = -(X_C/R)$. Thus, $\overline{Z}$ can be written as

$$\overline{Z} = Z/\theta$$
$$= \sqrt{R^2 + X_C^2}/\theta = -\arctan(X_C/R)$$

Note that θ is once again negative because the $\overline{Z}$ vector is clockwise with respect to the 0° reference (Fig. 18–5). This last expression simply expresses the magnitude and angle of the impedance vector in polar form. *The angle of the impedance vector is always negative for a capacitive circuit.*

EXAMPLE 18.7 Write the polar form expression for the impedance in Fig. 18–6(b).

Solution: $\overline{Z} = 200\ \Omega/{-53°}$

Note that the impedance angle is negative. This is always the case for series RC circuits.

When expressing the impedance vector in polar form Z/θ, the angle θ represents the phase angle between the current and the applied voltage. In other words, the impedance Z/θ causes the current to be θ degrees out of phase with E_S. If θ is positive, $\overline{I}$ lags $\overline{E_S}$. If θ is negative, $\overline{I}$ leads $\overline{E_S}$.

Multiplying Vectors in Polar Forms Let two vectors $\overline{A}$ and $\overline{B}$ be represented in polar form as $\overline{A} = A/\theta$ and $\overline{B} = B/\phi$. The *product* of the two vectors is found as follows

$$\overline{A} \times \overline{B} = A/\theta \times B/\phi = (A \times B)/\theta + \phi \qquad (18\text{--}11a)$$

The procedure is to *multiply* the magnitudes of the two vectors ($A \times B$) to obtain the resultant magnitude, and to *add* the phase angles of the two vectors ($\theta + \phi$) to obtain the resultant phase angle.

EXAMPLE 18.8 Since the source voltage equals the current times the circuit impedance, then we can write in vector notation that $\overline{E_S} = \overline{I} \times \overline{Z}$. If $\overline{I}$ and $\overline{Z}$ are expressed in polar form, $\overline{E_S}$ can be obtained using the vector multiplication shown.

(a) Find $\overline{E_S}$ for $\overline{I} = 2$ A$/0°$ and $\overline{Z} = 50$ Ω$/-40°$
(b) Find $\overline{E_S}$ for $\overline{I} = 5$ mA$/45°$ and $\overline{Z} = 2$ kΩ$/-45°$
(c) Find $\overline{E_S}$ for $\overline{I} = 0.1$ A$/30°$ and $\overline{Z} = 10$ kΩ$/10°$

Solution: (a) $\overline{E_S} = 2$ A$/0° \times 50$ Ω$/-40°$
$= (2 \text{ A} \times 50 \text{ Ω})/0° - 40°$
$= 100$ V$/-40°$

(b) $\overline{E_S} = (5 \text{ mA} \times 2 \text{ kΩ})/45° - 45°$
$= 10$ V$/0°$

(c) $\overline{E_S} = (0.1 \text{ A} \times 10 \text{ kΩ})/30° + 10°$
$= 1$ k V$/40°$

Dividing Vectors in Polar Form If we want to divide vector $\overline{A}$ by vector $\overline{B}$ the quotient is found as follows

$$\frac{\overline{A}}{\overline{B}} = \frac{A/\theta}{B/\phi} = \left(\frac{A}{B}\right)/\theta - \phi \qquad (18\text{--}11b)$$

The procedure is to divide the magnitudes of the two vectors (A/B) to obtain the resultant magnitude, and to subtract the angle of the denominator from the angle of the numerator ($\theta - \phi$) to obtain the resultant phase angle.

EXAMPLE 18.9 (a) Find $\overline{I}$ if $\overline{E_S} = 10$ V$/0°$ and $\overline{Z} = 2$ Ω$/-30°$

(b) Find $\overline{I}$ if $\overline{E_S} = 60$ V$/20°$ and $\overline{Z} = 2$ Ω$/-30°$

(c) Find $\overline{Z}$ if $\overline{E_S} = 14$ V$/0°$ and $\overline{I} = 3.5$ mA$/+45°$

Solution: (a) $\overline{I} = \dfrac{\overline{E_S}}{\overline{Z}} = \dfrac{10 \text{ V}/0°}{2 \text{ Ω}/-30°} = \dfrac{10 \text{ V}}{2 \text{ Ω}} /0° - (-30°) = 5$ A $/30°$

(b) $\overline{I} = \dfrac{60 \text{ V}/20°}{2 \text{ Ω}/-30°} = 30$ A$/50°$

(c) $\overline{Z} = \dfrac{\overline{E_S}}{\overline{I}} = \dfrac{14 \text{ V}/0°}{3.5 \text{ mA}/+45°} = \dfrac{14 \text{ V}}{3.5 \text{ mA}} /0° - 45°$
$= 4$ kΩ$/-45°$

EXAMPLE 18.10 The impedance of a certain series RC circuit is 500 $\Omega/\!-\!60°$. A source voltage of 15 V$/0°$ is applied to the circuit. Find $\bar{I}$.

Solution: $\bar{I} = \dfrac{\bar{E}_S}{\bar{Z}} = \dfrac{15 \text{ V}/0°}{500 \text{ }\Omega/\!-\!60°} = 30 \text{ mA}/\!+\!60°$

Note that $\bar{I}$ has a positive angle indicating that it *leads* E_S by 60° while $\bar{Z}$ has a negative angle. Because of the division, the negative angle for $\bar{Z}$ becomes positive for $\bar{I}$.

18.6 Finding R and C From Z

If the impedance of a series RC circuit is known, then it is an easy matter to determine the values of R and C. Refer to Fig. 18-8 where the impedance vector diagram for an RC circuit is drawn. If $\bar{Z} = Z/\theta$ is known, then R and X_C can be determined using the trigonometry of right triangles. From the diagram we have

$$\text{sine } \theta = \dfrac{\text{opposite}}{\text{hypotenuse}} = \dfrac{X_C}{Z}$$

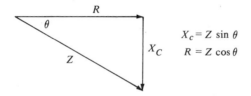

$X_C = Z \sin \theta$
$R = Z \cos \theta$

FIGURE 18-8 *Impedance relationships for* RC *circuit.*

so that,

$$X_C = Z \sin \theta \qquad (18\text{-}12)$$

Similarly, we have

$$\cos \theta = \dfrac{\text{adjacent}}{\text{hypotenuse}} = \dfrac{R}{Z}$$

so that

$$R = Z \cos \theta \qquad (18\text{-}13)$$

EXAMPLE 18.11 A series RC circuit has $\bar{Z} = 320 \text{ }\Omega/\!-\!30°$ at a frequency of 500 Hz. Find R and C.

Solution: $R = Z \cos \theta$
$= 320 \times \cos 30°$
$= 277 \text{ }\Omega$

$X_C = Z \sin \theta$
$= 320 \times \sin 30°$
$= 160 \text{ }\Omega$

Since $X_C = \dfrac{1}{2\pi fC}$, then $C = \dfrac{1}{2\pi fX_c}$ so that

$$C = \dfrac{1}{2\pi \times 500 \times 160} \approx 2 \ \mu F$$

Note that for these calculations it is not necessary to include the negative sign for the angle since we are only interested in the magnitude of R and X_C.

EXAMPLE 18.12 A 50-V sine wave is to be applied to a series RC circuit so that a current of 10 mA flows. It is also desired that the current lead the voltage by 20°. What values of R and X_C should be used?

Solution: With $E_S = 50$ V and $I = 10$ mA, $Z = 50$ V/10 mA $= 5$ kΩ. Also, since I leads E_S by 20°, then $\theta = -20°$. Thus,

$$R = Z \cos \theta = 5 \text{ k}\Omega \times \cos 20°$$
$$= 4.7 \text{ k}\Omega$$

$$X_C = Z \sin \theta = 5 \text{ k}\Omega \times \sin 20°$$
$$= 1.71 \text{ k}\Omega$$

18.7 Variation of Z and θ With Frequency

In an RC circuit, the capacitive reactance portion of the total impedance will vary with frequency since $X_C = 1/2\pi fC$. As such, the impedance Z and the phase angle θ will also be different at different frequencies. This can be graphically illustrated using the impedance vector diagrams of Fig. 18–9.

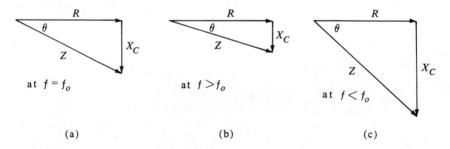

FIGURE 18–9 *Vector diagrams showing how Z and θ change as X_C changes with frequency.*

In part (a) of the figure a certain resistance R and a certain X_C produce an impedance Z and phase angle θ at some frequency f_o. If the applied frequency is *increased*, the resistance R will not change, but X_C will *decrease*. The effect is shown in part (b) of the figure. The smaller X_C results in a smaller phase angle θ and a smaller total impedance Z. As the frequency is increased further, X_C will become smaller and smaller until eventually θ would be almost 0° and Z would be approximately equal to R.

RC Circuit Response to AC

If the applied frequency is *decreased*, then X_C will *increase* producing the result shown in part (c) of the figure. The larger X_C causes both the phase angle θ and the impedance Z to increase. If the frequency is decreased further, X_C will become larger and larger until eventually θ would be almost $-90°$ and Z would be approximately equal to X_C.

These same effects can be shown using the relationships developed earlier. Since we have

$$Z = \sqrt{R^2 + X_C^2} \qquad (18\text{–}5)$$

it is easily seen that Z will increase as X_C increases and it will decrease as X_C decreases. At very *high* frequencies X_C will become very small (compared to R) and its effect can be neglected. If X_C is set equal to zero in the expression, it can be seen that $Z = \sqrt{R^2} = R$. At very *low* frequencies X_C will become very large compared to R so that the effect of R can be neglected. If R is set equal to zero, then $Z = \sqrt{X_C^2} = X_C$.

EXAMPLE 18.13 A 120-Ω resistor is in series with a 1-μF capacitor. Determine Z at $f = 1$ kHz. Repeat for $f = 10$ Hz and $f = 100$ kHz.

Solution: At $f = 1$ kHz,

$$X_C = \frac{1}{2\pi \times 10^3 \times 10^{-6}} = 160 \text{ Ω}$$

Thus,

$$Z = \sqrt{120^2 + 160^2} = 200 \text{ Ω}$$

At $f = 10$ Hz, X_C will be 100 times larger at 16 kΩ. Thus,

$$Z = \sqrt{120^2 + 16{,}000^2}$$
$$\approx 16 \text{ kΩ} \; (= X_C)$$

At $f = 100$ kHz, X_C will be 100 times smaller than its value at 1 kHz.

$$Z = \sqrt{120^2 + 1.6^2}$$
$$\approx 120 \text{ Ω} \; (= R)$$

We also know that θ can be calculated from

$$\tan \theta = -\left(\frac{X_C}{R}\right) \qquad (18\text{–}6)$$

At higher frequencies, the ratio X_C/R becomes very small and approaches zero. Thus, $\tan \theta \simeq 0$ so that $\theta \simeq 0°$ at *high* frequencies. At lower frequencies, X_C/R becomes very large and approaches infinity. Thus, $\tan \theta \rightarrow -\infty$ so that $\theta \simeq -90°$ at *low* frequencies. Remember, θ is the angle of E_S relative to I.

EXAMPLE 18.14 Determine the phase angle for the three cases of the previous example.

Solution: At $f = 1$ kHz, $X_C = 160$ Ω. Thus,

$$\tan\theta = -\frac{X_C}{R} = \frac{-160}{120} = -1.33$$

so that $\theta = -53°$

At $f = 10$ Hz, $X_C = 16$ kΩ. Thus,

$$\tan\theta = -\left(\frac{16{,}000}{120}\right) = -133$$

so that $\theta = -89.6°$ ($\simeq -90°$)

At $f = 100$ kHz, $X_C = 1.6$ Ω. Thus,

$$\tan\theta = -\frac{1.6}{120} = -0.0133$$

and $\theta = -0.76°$

Graphs of Z and θ Versus Frequency

For a given *RC* circuit we can calculate Z and θ over a wide range of frequencies using Eqs. (18–5) and (18–6). If we take many frequency values and calculate Z and θ the results can be expressed in graphical form as in Fig. 18–10. The graph of Z vs f shows that the impedance decreases with frequency. At higher frequencies, Z levels off at a value equal to R. At lower frequencies, Z increases rapidly as f decreases. The graph of θ versus f shows that θ is approximately $-90°$ at the lower frequencies and decreases to $0°$ at the higher frequencies.

From these graphs and the discussion that preceded them, we can conclude that at lower frequencies a series *RC* circuit behaves almost purely capacitive since $Z \simeq X_C$ and $\theta \simeq -90°$. At the lower frequencies the current will lead the applied voltage by 90°. Similarly, at the higher frequencies a series *RC* circuit behaves almost purely resistive since $Z \simeq R$ and $\theta \simeq 0°$. At high frequencies the current will be *in phase* with the applied voltage. At in-between frequencies the phase angle is between 0° and $-90°$ and the circuit is a combination of resistive and capacitive.

Transition Frequency f_T For a series *RC* circuit, a special significance is attached to the frequency at which the capacitive reactance is equal to the resistance. For the time being, we will call this frequency the transition frequency f_T. At frequencies below f_T the value of X_C exceeds the value of R; at frequencies above f_T the value of X_C is less than R. The value of f_T can be determined for any *RC* circuit by setting $X_C = 1/2\pi fC$ equal to R and solving for f_T. The result is

$$f_T = \frac{1}{2\pi RC} \qquad (18\text{--}13)$$

At $f = f_T$, the capacitive reactance equals the resistance. As such, the circuit phase angle at this frequency is $-45°$ since $\tan\theta = -(X_C/R) = -1$. At frequencies above f_T the phase angle is less than 45°. At frequencies below f_T θ is greater than 45° (see θ graph in Fig. 18–10).

Since $X_C = R$ at $f = f_T$, then the impedance Z at this frequency is

$$Z = \sqrt{R^2 + X_C^2} = \sqrt{R^2 + R^2} = \sqrt{2R^2}$$
$$= \sqrt{2}\,R \text{ (at } f = f_T)$$

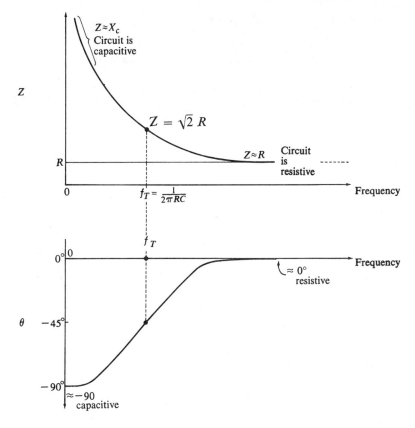

FIGURE 18–10 Plots of Z and θ variations with frequency.

Also, with $X_C = R$ there will be equal voltages across the resistor and capacitor since $V_C = IX_C$ and $V_R = IR$. At frequencies below f_T, X_C is larger than R so that more voltage will be across the capacitor. Conversely, at frequencies above f_T, more voltage will be across R since X_C is less than R.

EXAMPLE 18.15 In the circuit of Fig. 18–11(a) the applied voltage is $E_S = 10$ V rms. Determine the resistor and capacitor voltages at the following applied voltage frequencies: $f = f_T$, $f = 10\,f_T$, and $f = 0.1\,f_T$.

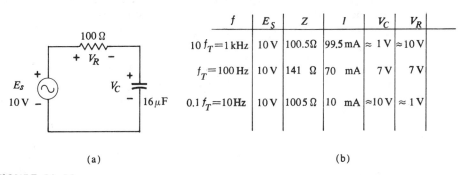

	f	E_S	Z	I	V_C	V_R
	$10\,f_T=1$ kHz	10 V	100.5 Ω	99.5 mA	≈ 1 V	≈ 10 V
	$f_T=100$ Hz	10 V	141 Ω	70 mA	7 V	7 V
	$0.1\,f_T=10$ Hz	10 V	1005 Ω	10 mA	≈ 10 V	≈ 1 V

(a) (b)

FIGURE 18–11

Solution: (a) For the circuit values shown,

$$f_T = \frac{1}{2\pi RC} = \frac{1}{2\pi \times 10^2 \times 16 \times 10^{-6}}$$
$$= 100 \text{ Hz}$$

At $f = f_T$, $X_C = R = 100 \, \Omega$. Thus,

$$Z = \sqrt{100^2 + 100^2} = 141 \, \Omega$$

The circuit current is therefore $I = 10 \text{ V}/141 \, \Omega = 70$ mA. Thus, $V_R = 70$ mA $\times$ 100 Ω = 7 V and V_C = 70 mA $\times$ 100 Ω = 7 V. This shows that $V_R = V_C$ at $f = f_T$.

(b) $10 f_T = 10 \times 100$ Hz $= 1000$ Hz. At $f = 1000$ Hz, X_C will be ten times *smaller* than its value at 100 Hz. Thus,

$$X_C = 10 \, \Omega$$
$$Z = \sqrt{100^2 + 10^2} = 100.5 \, \Omega$$

and

$$I = 10 \text{ V}/100.5 \, \Omega = 99.5 \text{ mA}$$

so that

$$V_R = 99.5 \text{ mA} \times 100 \, \Omega = \mathbf{9.95 \text{ V}} \approx \mathbf{10 \text{ V}}$$
$$V_C = 99.5 \text{ mA} \times 10 \, \Omega = \mathbf{0.995 \text{ V}} \approx \mathbf{1 \text{ V}}$$

(c) $0.1 f_T = 0.1 \times 100 = 10$ Hz. At $f = 10$ Hz, $X_C = 1000 \, \Omega$. Thus,

$$Z = \sqrt{100^2 + 1000^2} = 1005 \, \Omega$$

and

$$I = 10 \text{ V}/1005 \, \Omega = 9.95 \text{ mA}$$

so that

$$V_R = 9.95 \text{ mA} \times 100 \, \Omega = \mathbf{0.995 \text{ V}} \approx \mathbf{1 \text{ V}}$$

and

$$V_C = 9.95 \text{ mA} \times 1000 \, \Omega = \mathbf{9.95} \approx \mathbf{10 \text{ V}}$$

The results of the three cases are tabulated in part (b) of Fig. 18–11. Note that at a frequency of $10 f_T$ most of the applied voltage is across R and at a frequency of $0.1 f_T$ most of the applied voltage is across C.

EXAMPLE 18.16 Design a series RC circuit that has a transition frequency of 12 kHz and draws a maximum current of 50 mA from a 6-V source.

Solution: First of all, we can use the formula for f_T to solve for $R \times C$. That is,

$$f_T = \frac{1}{2\pi RC} = 12 \times 10^3 \text{ Hz} \qquad (18\text{--}13)$$

so that
$$RC = \frac{1}{2\pi \times 12 \times 10^3} = 1.32 \times 10^{-5}$$

To determine values for R and C we must utilize the second specification, namely that the current should not exceed 50 mA at any frequency. The current will be greatest at the higher frequencies when X_C is almost zero and $Z \simeq R$.

Thus, at high frequencies ($> 10 f_T$) the current will be $I = E_S/R$. Setting this equal to 50 mA and solving for R, we have

$$\frac{6 \text{ V}}{R} = 50 \text{ mA}$$

or $R = 120 \text{ }\Omega$. This is the value of R which will draw a maximum current of 50 mA.

Since $RC = 1.32 \times 10^{-5}$, we can solve for C.

$$C = \frac{1.32 \times 10^{-5}}{R} = \frac{1.32 \times 10^{-5}}{120} = \mathbf{0.11 \text{ }\mu F}$$

EXAMPLE 18.17 A certain series RC circuit is to be designed to handle input frequencies in the range of 50 Hz to 50 kHz. It is required that the circuit phase shift not exceed 1° for this range of frequencies. If $R = 5000 \text{ }\Omega$, choose an appropriate value for C.

Solution: The phase shift will be greatest at the lower frequencies. Therefore we will concern ourselves with the minimum frequency 50 Hz. If we pick C so that $\theta = 1°$ at $f = 50$ Hz, then at any higher frequency θ will be even smaller. If $\theta = 1°$ then

$$\tan \theta = \tan 1° = \frac{X_C}{R} \text{ (at } f = 50 \text{ Hz)}$$

From calculator or tables, $\tan 1° = 0.0175$. Thus,

$$\frac{X_C}{R} = 0.0175$$

or $X_C = 0.0175 \, R = 87.5$ ohms at $f = 50$ Hz.

Since $X_C = 1/2\pi fC$, then $C = 1/2\pi f X_C$ so that

$$C = \frac{1}{2\pi \times 50 \times 87.5} = \mathbf{36.3 \text{ }\mu F}$$

This value of C represents the minimum value that will work. Any larger value of C will make X_C lower at 50 Hz and therefore reduce θ even further.

18.8 Decibels, dB

We pause now in our discussion of *RC* circuits to introduce the *decibel* unit of measure. An understanding of the decibel will help us in our study of *RC filters* in this chapter and other filters in subsequent chapters.

Gain and Attenuation If a voltage sine wave with a peak value of 1 V is applied to the input of an amplifier circuit, the amplifier output might be a 20-V sine wave at the same frequency. Because the output voltage is 20 times greater than the input voltage, we say that the amplifier has a *gain* of 20. Modern electronic amplifiers are available with gains ranging as high as 1,000,000.

If a 1-V sine wave is applied to a circuit and the circuit output is only, say, 0.5 V, then the circuit is said to have *attenuated* (reduced) the input voltage by a factor of 0.5. The *attenuation factor* is 0.5.

Decibel The voltage *gain* or *attenuation* of a circuit is equal to the ratio of output voltage to input voltage. That is, V_{out}/V_{in}. Since gain and attenuation are ratios of volts to volts, they have no conventional units. However, a commonly used measure of gain and attenuation used in the study of amplifiers and filters is the *decibel*, abbreviated dB.

For a circuit with an input voltage V_{in} and output voltage V_{out}, its gain (or attenuation) can be expressed in dB as

$$\text{gain} = 20 \log_{10} (V_{out}/V_{in}) \qquad (18\text{--}14)$$

For example, if $V_{in} = 1$ V and $V_{out} = 20$ V the gain will be

$$20 \log_{10} (20) = 20 \times 1.3$$
$$= 26 \text{ dB}$$

If $V_{in} = 1$ V and $V_{out} = 0.5$ V the attenuation in decibels is

$$20 \log_{10} (0.5) = 20 \times (-0.3)$$
$$= -6 \text{ dB}$$

Whenever V_{out}/V_{in} is greater than one, the number of dB is *positive*. Whenever the voltage ratio is less than one (attenuation), the number of dB is *negative* since the log of a fraction is always negative.

For the case where $V_{out} = V_{in}$ the gain in decibels will be

$$20 \log_{10} (1) = 20 \times 0 = 0$$

Thus, 0 db represents a voltage ratio of 1, meaning $V_{out} = V_{in}$.

The following table lists the dB values for voltage ratios which are powers of 10. Note that in each case the dB value is equal to 20 times the exponent of the power of ten.

V_{out}/V_{in}	dB
10^{-6}	$-120 \text{ dB} = (20 \times -6)$
10^{-3}	$-60 \text{ dB} = (20 \times -3)$
10^{-2}	$-40 \text{ dB} = (20 \times -2)$
10^{-1}	$-20 \text{ dB} = (20 \times -1)$
10^{0}	$0 \text{ dB} = (20 \times 0)$
10^{1}	$20 \text{ dB} = (20 \times 1)$
10^{2}	$40 \text{ dB} = (20 \times 2)$
10^{3}	$60 \text{ dB} = (20 \times 3)$
10^{6}	$120 \text{ dB} = (20 \times 6)$

The characteristic of the decibel unit that makes it so useful is the way in which successive gains combine to give an overall gain. To find the overall gain of a complicated electronic system, we merely *add* up the decibel gain figures for each of the component stages. To illustrate, Fig. 18–12(a) shows two amplifiers $A1$ and $A2$ connected so that the output of $A1$ is fed to the input of $A2$. The output of $A2$ is the final output. $A1$ has a gain of 20 dB and $A2$ has a gain of 12 dB. The overall gain of the combination is equal to 20 dB + 12 dB = 32 dB. Thus, the decibel gain between V_{in} and V_{out} is 32 dB.

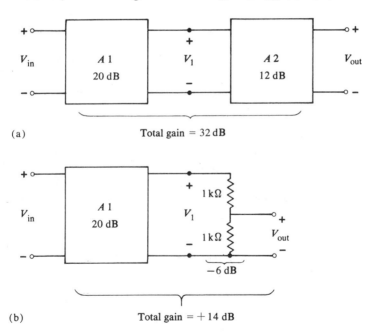

FIGURE 18–12 *Successive decibel gains are added to determine overall gain.*

Another example is shown in part (b) of the figure. There, the output of $A1$ is fed to a voltage divider that has an attenuation factor of 0.5 (-6 dB). Thus, the overall decibel gain between V_{in} and V_{out} is 20 dB $-$ 6 dB $=$ 14 dB.

EXAMPLE 18.18 (a) Convert the following voltage ratios to dB: 0.25; 5; 100; 10,000; 0.707.

(b) Convert the following dB gains to voltage ratios: -20 dB; $+30$ dB.

Solution: (a) $20 \log_{10} (0.25) = 20 \times (-0.6) = $ **-12 dB**
$20 \log_{10} (5) = 20 \times 0.69 = $ **13.8 dB**
$20 \log_{10} (100) = 20 \times 2 = $ **40 dB**
$20 \log_{10} (10{,}000) = 20 \times 4 = $ **80 dB**
$20 \log_{10} (0.707) = 20 \times (-0.15) = $ **-3 dB**

Note that the number of dB does not change much as the voltage ratio varies over a wide range.

(b) $20 \log_{10} (V_{\text{out}}/V_{\text{in}}) = -20$ dB

$$\log_{10} (V_{\text{out}}/V_{\text{in}}) = \frac{-20}{20} = -1$$

Thus, $V_{\text{out}}/V_{\text{in}} = 10^{-1} = $ **0.1**
Similarly,

$$20 \log_{10} (V_{\text{out}}/V_{\text{in}}) = +30 \text{ dB}$$
$$\log_{10} (V_{\text{out}}/V_{\text{in}}) = 1.5$$

thus,

$$V_{\text{out}}/V_{\text{in}} = \textbf{31.6}$$

18.9 Filters — General

Electronic circuits often contain voltages over a wide range of frequencies. As examples, the input to an audio circuit can have frequencies ranging from very low (10 Hz) to very high (20 kHz). The audio detector portion of a radio contains both audio frequencies (10 Hz − 20 kHz) and radio frequencies (1000 kHz). The rectifier in a dc power supply produces a dc voltage with an ac voltage (ripple) superimposed on the dc level.

In applications where many frequencies of voltage exist, it is often necessary to design circuits which will pass or reject one frequency or group of frequencies. Such circuits are called *electrical filters*. The operation of three types of electrical filters is illustrated in Fig. 18–13. In part (a) of the figure the filter is

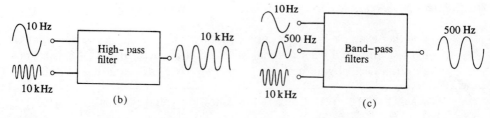

FIGURE 18–13 *Various types of electrical filters.*

called a *low-pass filter* because it passes the lower frequencies (10 Hz) from its input to its output without any attenuation, while it rejects or attenuates the higher frequencies (10 kHz).

Part (b) shows a *high-pass filter* that passes higher frequencies but attenuates the lower frequencies. Part (c) represents a *band-pass filter* that attenuates both the lower frequencies (10 Hz) and higher frequencies (10 kHz) but passes the medium (mid-range) frequencies (500 Hz). There are many other types of filters that we will encounter. However, for now we will be interested in the three shown in Fig. 18–13 since they can be easily constructed using resistance and capacitance.

18.10 RC Circuit as a High-Pass Filter

A series *RC* circuit can function as a high-pass filter if the output voltage is taken across the *resistor*. This is shown in Fig. 18–14. The input voltage E_S is applied across the series combination, and the output is taken as the voltage across *R*.

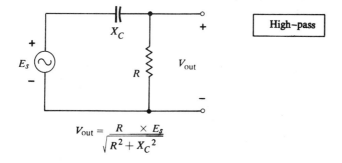

FIGURE 18–14 *RC circuit arranged to act as a high-pass filter.*

To determine the output voltage V_{out} we can utilize Ohm's law.

$$I = \frac{E_S}{Z}$$

Thus,

$$V_{out} = V_R = R \times I = \frac{R}{Z} \times E_S \qquad (18\text{--}15)$$

This same result can be obtained using a voltage divider relationship that is similar to the one for resistive circuits except impedances are used instead of resistors.

$$V_{out} = \frac{Z_{out}}{Z_{total}} \times V_{in} = \frac{R}{Z} \times E_S \qquad (18\text{--}16)$$

Here $Z_{out} = R$ (output is taken across *R*), Z_{total} is the total equivalent impedance *Z* of the series circuit, and $V_{in} = E_S$.

In any case, the output V_{out} is given by Eq. (18–15). Since $Z = \sqrt{R^2 + X_C^2}$, this can be rewritten as

$$V_{out} = \left(\frac{R}{\sqrt{R^2 + X_C^2}}\right) \times E_S \qquad (18\text{--}17)$$

At high-input frequencies the capacitive reactance X_C will become very small (almost a short circuit at high enough frequencies). As a result, most of the input voltage E_S will appear across the resistor. This can be seen in Eq. (18–17) if we set $X_C = 0$:

$$V_{out} = \left(\frac{R}{\sqrt{R^2 + 0}}\right) \times E_S \approx E_S \text{ (at high frequencies)}$$

At very low frequencies X_C becomes very large ($X_C = 1/2\pi fC$). As a result, most of the input voltage will appear across the capacitor and very little will appear across R. Of course, at zero frequency (dc) the resistor voltage will be zero since the capacitor blocks dc current. As the frequency increases, the capacitor allows more and more current to flow, resulting in a greater V_R output.

At the transition frequency $f_T = 1/2\pi RC$ the voltages across R and X_C will be equal since $X_C = R$. From Eq. (18–17) we can calculate V_{out} for this situation as

$$V_{out} = \left(\frac{R}{\sqrt{R^2 + R^2}}\right) E_S = \left(\frac{R}{\sqrt{2R^2}}\right) E_S = \frac{1}{\sqrt{2}} \times E_S$$
$$= 0.707 \, E_S \text{ (at } f = f_T\text{)}$$

At frequencies above f_T, V_{out} increases; at frequencies below f_T, V_{out} decreases toward zero.

Frequency-Response Curve The characteristics described here can be summarized in a graph called a *frequency-response curve*. This graph shows how the attenuation of the high-pass filter varies with frequency. Most filter frequency-response curves plot the attenuation in *decibels* and use a log scale for the frequency. Such a frequency-response curve is shown in Fig. 18–15 for the accompanying RC high-pass filter.

This curve could be obtained experimentally by applying input voltages at various frequencies and measuring the ratio of the output voltage to input voltage for each frequency. This ratio is converted to dB and is plotted versus frequency. These attenuation ratios can also be obtained from Eq. (18–17) using $E_S = V_{in}$.

$$V_{out} = \frac{R}{\sqrt{R^2 + X_C^2}} \times V_{in} \qquad (18\text{--}17)$$

so that

$$\frac{V_{out}}{V_{in}} = \frac{R}{\sqrt{R^2 + X_C^2}} \qquad (18\text{--}18a)$$

In decibels,

$$20 \log_{10}\left\{\frac{V_{out}}{V_{in}}\right\} = 20 \log_{10}\left\{\frac{R}{\sqrt{R^2 + X_C^2}}\right\} \qquad (18\text{--}18b)$$

This ratio of course varies with frequency because of X_C.

In examining the curve in Fig. 18–15 several features should be pointed out.

(1) The transition frequency f_T is 100 Hz for the circuit values shown. At frequencies above $10 f_T$ (1 kHz) the attenuation is essentially 0 dB, which indicates $V_{out} = V_{in}$.

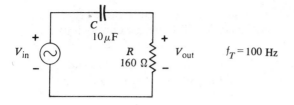

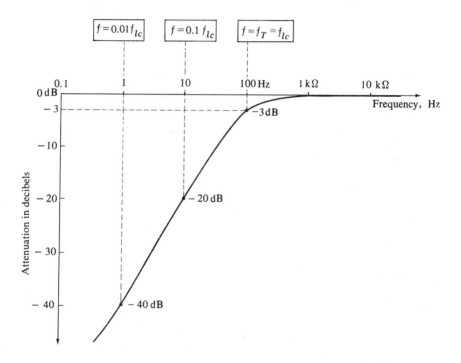

FIGURE 18–15 RC *high-pass filter and its frequency-response curve.*

(2) At frequencies below 100 Hz (f_T) the attenuation increases rapidly in the negative direction. At 10 Hz (0.1 f_T) the attenuation is -20 dB, representing an attenuation ratio of 0.1, which indicates that $V_{out} = 0.1\ V_{in}$ at this frequency. At 1 Hz (0.01 f_T) the attenuation is -40 dB and so on.

(3) At a frequency equal to f_T (100 Hz) the attenuation is -3 dB, representing an attenuation ratio of 0.707. That is, 20 $\log_{10}$ (0.707) $= -3$ dB. In other words, the attenuation is down 3 dB at $f = f_T$. The transition frequency f_T is often called the *low cutoff frequency* and is given the symbol f_{lc}. The f_{lc} represents the frequency *below* which the filter greatly attenuates the input voltage. At frequencies above f_{lc} ($= f_T$), the filter passes most of the input voltage. Because the attenuation is -3 dB at $f = f_{lc}$, it is also often called the *lower 3 dB frequency*.

All series *RC* high-pass filters have a frequency-response curve with the exact same shape as that shown in Fig. 18–15. Varying the *R* and *C* values results in a different value of low cutoff frequency ($f_{lc} = f_T = 1/2\pi\ RC$) and simply shifts the curve left or right. The value of f_{lc} is chosen to satisfy the requirements for which the filter is being designed. The following example illustrates.

EXAMPLE 18.19 Design an *RC* filter that will attenuate any frequencies *below* 120 Hz by at least −20 dB (attenuation ratio of 0.1). The filter is to draw a maximum of 50 mA from a 10-V source.

Solution: This design requires a high-pass filter since the low frequencies have to be blocked. The attenuation of an *RC* high-pass filter will be −20 dB at $f = 0.1 f_{lc}$. Thus,

$$0.1 f_{lc} = 120 \text{ Hz}$$
$$f_{lc} = 1200 \text{ Hz}$$

and

$$f_{lc} = f_T = \frac{1}{2\pi RC} = 1200 \text{ Hz}$$

so that

$$RC = \frac{1}{2\pi \times 1200} = 1.32 \times 10^{-4}$$

Since the circuit is to draw a maximum of 50 mA at 10 V,

$$R = \frac{10 \text{ V}}{50 \text{ mA}} = 200 \text{ }\Omega$$

Thus

$$200 \times C = 1.32 \times 10^{-4}$$
$$C = \mathbf{0.66 \text{ }\mu F}$$

The desired high-pass filter then has $R = 200 \text{ }\Omega$, $C = 0.66 \text{ }\mu F$, and $f_{lc} = 1200$ Hz.

Coupling Network An *RC* high-pass filter is often used to couple the output of an amplifier or pulsing circuit to the input of another circuit as illustrated in Fig. 18–16. The function of the filter in these applications is to block low frequencies (usually dc) from reaching the input of the second circuit while allowing the relatively high frequencies to pass unattenuated from the output of the first circuit to the input of the second. Used in this arrangement, the high-pass filter is often called an *RC coupling network*.

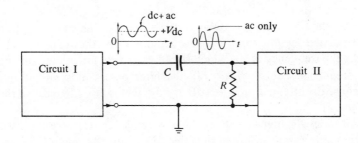

FIGURE 18–16 *RC coupling network (high-pass filter).*

As shown in the figure, the output of circuit I contains an ac voltage superimposed on a dc level, $+V_{dc}$. In other words, the sine wave is not symmetrical above and below zero volts, but instead is symmetrical above and below $+V_{dc}$. The $+V_{dc}$ represents a *constant bias level* that is not allowed to reach the input of circuit II because the high-pass filter blocks dc. The sine wave portion of the signal, however, is passed by the high-pass filter to circuit II provided that the low cutoff frequency f_{lc} is chosen so that it is less than the frequency of the sine wave.

18.11 RC Circuit as a Low-Pass Filter

Since the capacitor voltage is equal to the input voltage at low frequencies, the series RC circuit can function as a low-pass filter if the output voltage is taken across the *capacitor* as shown in Fig. 18–17. Because the output is V_C, it follows that at low frequencies most of the input E_S will appear at the output. At high frequencies X_C decreases rapidly so that it essentially acts as a short circuit (compared to R) and V_C will be very low.

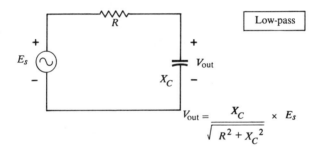

FIGURE 18–17 *RC circuit arranged to act as a low-pass filter.*

The output voltage V_{out} can be determined using a voltage divider formula derived in the same manner as for the high-pass filter except that V_{out} is taken across X_C. Thus,

$$V_{out} = \left(\frac{X_C}{\sqrt{R^2 + X_C^2}}\right) \times E_S \qquad (18\text{–}19)$$

At the lower input frequencies the capacitive reactance X_C will become very large compared to R. Thus, R^2 will be small compared to X_C^2 and can be neglected in (18–19) so that

$$V_{out} \approx \left(\frac{X_C}{\sqrt{0 + X_C^2}}\right) \times E_S \approx E_S \text{ (at low frequencies)}$$

which indicates very little attenuation at low frequencies.

At high frequencies X_C decreases to a small value compared to R so that most of E_S appears across R and very little reaches the output V_{out}. At high enough frequencies X_C is almost a short circuit and $V_C \simeq 0$.

It should be obvious that the RC low-pass filter is the same circuit as the RC high-pass filter except that the outputs are taken across different points in each case. Thus, it can also be stated for the low-pass filter that $V_C = V_R$ at $f = f_T$.

From (18-19) we can calculate V_{out} at the transition frequency since $X_C = R$ at this frequency.

$$V_{out} = \left(\frac{X_C}{\sqrt{R^2 + X_C^2}}\right) E_S = \left(\frac{R}{\sqrt{R^2 + R^2}}\right) E_S = 0.707\ E_S \ (\text{at}\ f = f_T)$$

At frequencies $>f_T$, V_{out} decreases and at frequencies $<f_T$, V_{out} increases.

Confusion Between High-Pass and Low-Pass There is an easy way to remember which RC circuit arrangement is a low-pass filter and which one is a high-pass filter. We know that a capacitor opposes rapid changes in its voltage. This means that the capacitor voltage will not be able to follow the rapid changes in input voltage at high frequencies, but it can follow the more slowly changing low frequencies. Thus, V_C will be greater at low frequencies and therefore V_R will be greater at high frequencies.

Frequency-Response Curve The frequency-response curve for an RC low-pass filter is shown in Fig. 18-18. Notice that the R and C values in the circuit are

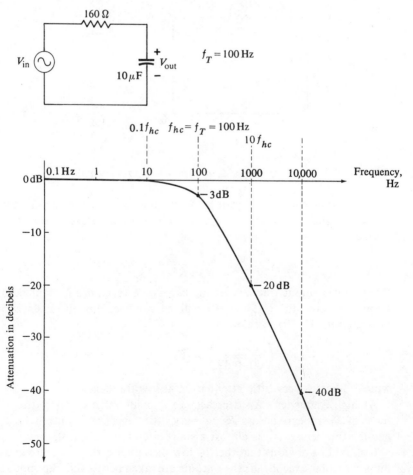

FIGURE 18-18 RC *low-pass filter and its frequency-response curve.*

the same as those used in the high-pass filter of Fig. 18–15. The graph in 18–18 shows how the attenuation of the low-pass filter varies with the applied frequency. Once again the attenuation is plotted in dB and a log scale is used for frequency.

The attenuation can be measured experimentally by applying various input frequencies, or it can be calculated from Eq. (18–19) with $E_S = V_{in}$. That is,

$$V_{out} = \left(\frac{X_C}{\sqrt{R^2 + X_C^2}}\right) V_{in}$$

so that

$$\frac{V_{out}}{V_{in}} = \left(\frac{X_C}{\sqrt{R^2 + X_C^2}}\right) \quad (18\text{–}20)$$

This ratio of course varies with frequency because of X_C.

The frequency-response curve in Fig. 18–18 has several important features.

(1) At frequencies below approximately $0.1 f_T$ (10 Hz) the attenuation is essentially 0 dB ($V_{out} \simeq V_{in}$).

(2) At frequencies above f_T, the attenuation increases rapidly in the negative direction. At 1000 Hz (10 f_T) the attenuation is -20 dB. ($V_{out}/V_{in} = 0.1$). At 10,000 Hz the attenuation is -40 dB and so on.

(3) At a frequency equal to f_T the attenuation is -3 dB as it also was for the high-pass filter. This represents an attenuation ratio of 0.707. For a low-pass filter the transition frequency f_T is often called the *high-cutoff frequency* and is given the symbol f_{hc}. This f_{hc} represents the frequency *above* which the filter greatly attenuates the input voltages. At frequencies below f_{hc} the filter passes most of the input voltage. Because the attenuation is -3 dB at $f = f_{hc}$, it is also often called the *upper-3-dB frequency*.

All series RC low-pass filters have a frequency-response curve with the exact same shape as that shown in Fig. 18–18. Varying the R and C values results in a different value of high-cutoff frequency ($f_{hc} = f_T = 1/2\pi RC$) and simply shifts the curve left or right. The value of f_{hc} is chosen to satisfy the requirements for which the filter is being designed.

Low-Pass Filter at dc (f = 0) No matter what values of R and C are chosen, *a series RC low-pass filter will always pass dc with no attenuation.* This is because the capacitor will charge up to the value of the dc input leaving no voltage across R, and no dc current. Of course, as we saw in chap. 13 a *leaky* capacitor will not necessarily charge up to the total source voltage.

EXAMPLE 18.20 Design an RC filter that will attenuate any frequencies *above* 15 kHz by at least -20 dB. The filter is to draw a maximum current of 12 mA from a 120-V source.

Solution: This design requires a low-pass filter since the high frequencies have to be blocked. The attenuation of an RC low-pass filter will be -20 dB at $f = 10 f_{hc}$. Thus,

$$10 f_{hc} = 15 \text{ kHz}$$

Therefore,
$$f_{hc} = f_T = \frac{1}{2\pi RC} = 1500 \text{ Hz}$$
so that
$$RC = 1.06 \times 10^{-4}$$
Since the current has to be a maximum of 12 mA at 120 V,
$$R = \frac{120 \text{ V}}{12 \text{ mA}} = 10 \text{ k}\Omega$$
Thus,
$$10^4 \times C = 1.06 \times 10^{-4}$$
$$\therefore C = \mathbf{0.01 \ \mu F}$$

The required low-pass filter then has $R = 10 \text{ k}\Omega$, $C = 0.01 \ \mu\text{F}$, and $f_{hc} = 1500 \text{ Hz}$

An RC low-pass filter is often used in situations where ac and dc are present at the same time but only the dc is desirable. An example of this is in dc power supplies where a pulsating type of signal containing both a constant dc level and an ac portion (ripple) must be filtered so that only the dc appears at the output. Figure 18–19 illustrates this situation. The input and output waveforms show the effect of the low-pass filter in passing the dc but blocking the ac provided that the filter's high-cutoff frequency f_{hc} is less than the frequency of the ac.

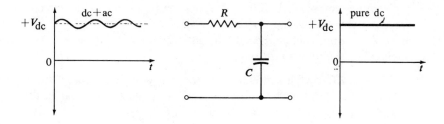

FIGURE 18–19 *Low-pass filter used to pass dc and block ac.*

18.12 Low-Pass and High-Pass Comparison

Now that we have seen how a series RC circuit can be arranged as either a low-pass or high-pass filter, it might be useful to compare the two arrangements with the help of Fig. 18–20. It is interesting to note that for given R and C values, the value of f_{hc} for the low-pass filter is the same as f_{lc} for the high-pass filter. This should seem reasonable because both f_{hc} and f_{lc} are equal to the transition frequency f_T, which is the frequency at which $V_C = V_R$. At frequencies on either side of f_T more of the voltage will be across one or the other circuit elements.

18.13 R and X_C in Parallel

Figure 18–21(a) shows an ac source connected to a circuit that contains a capacitive reactance X_C in parallel with a resistor R. Because the elements are

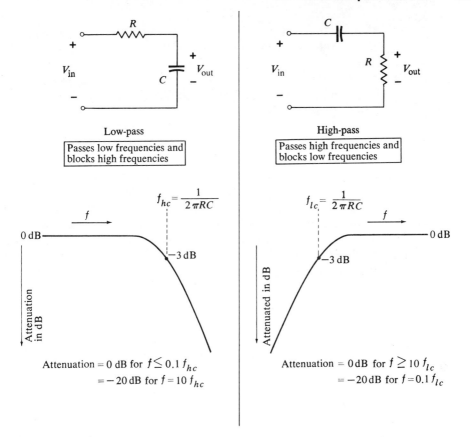

FIGURE 18-20 *Comparison of* RC *low-pass and high-pass circuits.*

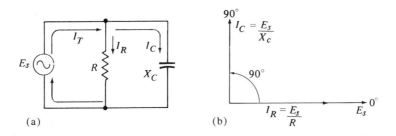

FIGURE 18-21 *In parallel RC circuit* I_R *and* I_C *are 90° out of phase.*

connected in parallel, each one has a voltage across it equal to the applied voltage E_S. In other words, $V_C = V_R = E_S$.

The current through each branch will depend on the value of E_S and on the amount of opposition in each branch. Thus,

$$I_R = \frac{E_S}{R} \quad \text{and} \quad I_C = \frac{E_S}{X_C} \tag{18-21}$$

The total current I_T is the rms ac current supplied by the source. We might be tempted to say that I_T can be found by adding the rms currents through each branch. After all, this is the procedure we followed in our study of parallel resistive circuits. Unfortunately, this *cannot* be done here because of the phase difference between I_C and I_R.

The current I_R will be *in phase* with the applied voltage E_S. The current I_C will lead by 90° the applied voltage (see phasor diagram in figure). Thus I_C leads I_R by 90°. Since I_R and I_C are out of phase, we must resort to a *vector* addition to find the total resultant current I_T. This is exactly the same situation we encountered when combining V_R and V_C in the series RC circuit (sec. 18.3).

Finding I_T To find the total current phasor $\overline{I_T}$ we must combine the $\overline{I_R}$ and $\overline{I_C}$ phasors using vector addition. To help do this we can draw the phasor diagram showing the phasors representing $\overline{E_S}$, $\overline{I_C}$, and $\overline{I_R}$. This is done in Fig. 18-22(a). In drawing this phasor diagram, the phasor E_S is chosen as the *reference* phasor at 0°. This is because E_S is common to all the branches in a parallel circuit. Recall that in the series RC circuit, the circuit current I was chosen as the reference phasor because it was common to all the series elements.

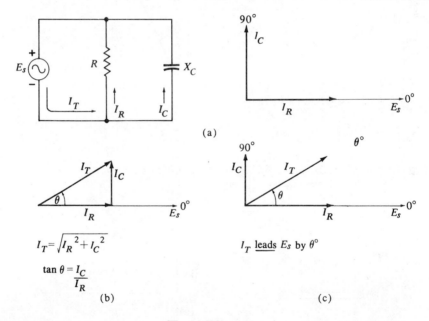

FIGURE 18-22 *Vector addition of $\overline{I_C}$ and $\overline{I_R}$ produces total current phasor $\overline{I_T}$.*

The phasor representing $\overline{I_R}$ is drawn in phase with $\overline{E_S}$ at 0°. The magnitude of the resistor current phasor is, of course, E_S/R. The phasor representing $\overline{I_C}$ is drawn at a phase angle of $+90°$ relative to E_S (capacitor current leads capacitor voltage by 90°). The magnitude of the capacitor current phasor is, of course, E_S/X_C.

The phasor representing the total source current $\overline{I_T}$ is obtained by adding the $\overline{I_C}$ and $\overline{I_R}$ phasors vectorially. This is shown in part (b) of Fig. 18-22. Since $\overline{I_C}$ and $\overline{I_R}$ are at right angles to each other, we can use Pythagorean's theorem to determine the magnitude of $\overline{I_T}$. Thus,

$$I_T = \sqrt{I_R^2 + I_C^2} \qquad (18\text{-}22)$$

The phasor right triangle in Fig. 18–22(b) can also be used to determine the angle θ.

$$\tan \theta = \frac{I_C}{I_R} \quad (18\text{–}23)$$

As the diagram shows, θ is the angle by which the total current leads the applied voltage E_S. The complete phasor diagram is shown in part (c) of the figure.

EXAMPLE 18.21 A parallel RC circuit has $R = 3$ kΩ and $X_C = 4$ kΩ. (a) Find the total current for $E_S = 12$ V; (b) determine the phase angle between the total current and the applied voltage; (c) find the phase angle between the capacitor current and the total current.

Solution: (a) The circuit diagram is shown in Fig. 18–23(a). Since $E_S = 12$ V we can calculate the rms values of I_C and I_R.

$$I_C = \frac{12 \text{ V}}{4 \text{ k}\Omega} = 3 \text{ mA} \quad \text{and} \quad I_R = \frac{12 \text{ V}}{3 \text{ k}\Omega} = 4 \text{ mA}$$

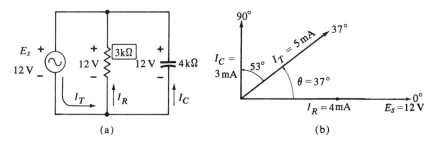

FIGURE 18–23

To find I_T we must add these currents vectorially. The phasors are drawn in part (b) of the figure. The *magnitude* of $\overline{I_T}$ is

$$I_T = \sqrt{I_R^2 + I_C^2}$$
$$= \sqrt{4^2 + 3^2} = 5 \text{ mA} \quad (18\text{–}22)$$

(b) From the phasor diagram

$$\tan \theta = \frac{I_C}{I_R} = \frac{3 \text{ mA}}{4 \text{ mA}} = 0.75$$

so that $\theta = 37°$. Thus, $\overline{I_T}$ leads $\overline{E_S}$ by 37°. In the parallel RC circuit the total current *always* leads the applied voltage.

(c) It is obvious from the phasor diagram that I_C leads I_T by

$$90° - 37° = 53°$$

In the parallel RC circuit as was the case for the series RC circuit, the total current *leads* the applied voltage. However, in the parallel circuit the phase angle θ is *positive* while in the series circuit the phase angle θ is *negative*. The reason for this is the fact that the reference is different in each case. In the series circuit the current was chosen as the 0° reference; in the parallel circuit the applied voltage was chosen as the 0° reference. One must always be aware of the reference used in each case.

18.14 Impedance of Parallel RC Circuit

The total current in a parallel *RC* circuit can be determined as shown in the last section. Once $\overline{I_T}$ is known it is relatively simple to find the total impedance $\overline{Z}$ of the *RC* parallel combination. Since Z is the total opposition to current:

$$\overline{Z} = \frac{\overline{E_S}}{\overline{I_T}} \qquad (18\text{--}24)$$

For example, for the circuit of Fig. 18–23 the applied voltage is 12 V and I_T was determined to be 5 mA. Thus,

$$Z = \frac{12 \text{ V}}{5 \text{ mA}} = 2.4 \text{ k}\Omega$$

This impedance is the combined opposition of the 3-kΩ resistor in parallel with 3-kΩ capacitive reactance. *For the parallel* RC *circuit $\overline{Z}$ is not equal to the vector sum of $\overline{R}$ and $\overline{X_C}$, because $\overline{R}$ and $\overline{X_C}$ are not in series.* $\overline{Z}$ is found by finding $\overline{I_T}$ and then using (18–24).

EXAMPLE 18.22 What is the total Z of a 600-Ω resistor in parallel with a 300-Ω X_C?

Solution: To find Z we must apply a voltage to the parallel *RC* circuit and determine I_T. Since no voltage was specified, we can assume any value we wish. A good value to use in this example is 600 V so that the branch currents come out as whole numbers.

For $E_S = 600$ V, we have

$$I_R = \frac{600 \text{ V}}{600 \text{ }\Omega} = 1 \text{ A}$$

$$I_C = \frac{600 \text{ V}}{300 \text{ }\Omega} = 2 \text{ A}$$

Thus,

$$I_T = \sqrt{I_R^2 + I_C^2} = \sqrt{1 + 4} = 2.23 \text{ A}$$

The impedance is therefore

$$Z = \frac{E_S}{I_T} = \frac{600 \text{ V}}{2.23 \text{ A}} = 268 \text{ }\Omega$$

Keep in mind that this value of Z is only the impedance at the particular frequency at which $X_C = 300$ Ω. As f changes, X_C will change causing Z to take on different values.

Impedance Vector in Polar Form We can express the impedance as a vector in polar form like we did in the series *RC* circuit. In Fig. 18–24 the phasor diagram for the parallel RC circuit is shown. The magnitude of the total current phasor is I_T; its phase angle is $\theta°$ relative to the 0° reference. Thus, we can express the total current phasor as

$$\overline{I_T} = I_T\underline{/\theta}$$

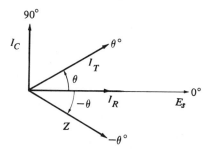

FIGURE 18-24 *Impedance vector has negative angle for parallel RC circuit.*

The source voltage phasor is $\overline{E_S} = E_S\underline{/0°}$. Since $\overline{Z} = \overline{E_S}/\overline{I_T}$, we have

$$\overline{Z} = \frac{E_S\underline{/0°}}{I_T\underline{/\theta}} = \frac{E_S}{I_T}\underline{/0° - \theta}$$

$$= Z\underline{/-\theta}$$

This shows that the impedance vector for the parallel RC circuit has a *negative* angle. This was also true for the series RC circuit. In general, it can be stated that

For a series or parallel RC circuit or any capacitive circuit, the impedance has a negative angle ($Z\underline{/-\theta}$).

EXAMPLE 18.23 A certain parallel RC circuit has an impedance of 1.2 k$\Omega\underline{/-30°}$. Determine the R and X_C values.

Solution: For the parallel RC circuit, $\overline{Z}$ is *not* the vector sum of $\overline{X_C}$ and $\overline{R}$ as it was for the series circuit. To find R and X_C we can apply a voltage to the circuit and calculate I_T. From I_T we can determine I_R and I_C from which we can determine R and X_C.

For convenience, let $\overline{E_S} = 12$ V$\underline{/0°}$. Thus,

$$\overline{I_T} = \frac{\overline{E_S}}{\overline{Z}} = \frac{12 \text{ V}\underline{/0°}}{1.2 \text{ k}\Omega\underline{/-30°}} = 10 \text{ mA}\underline{/30°}$$

This $\overline{I_T}$ phasor is shown in Fig. 18-25. It is clear that I_R is related to I_T by

$$I_R = I_T \cos\theta$$
$$= 10 \text{ mA} \cos(30°) = 8.66 \text{ mA}$$

Similarly,

$$I_C = I_T \sin\theta$$
$$= 10 \text{ mA} \sin(30°) = 5 \text{ mA}$$

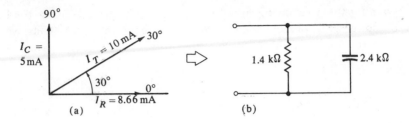

FIGURE 18–25

Thus,

$$R = \frac{E_S}{I_R} = \frac{12 \text{ V}}{8.66 \text{ mA}} = 1.4 \text{ k}\Omega$$

Similarly,

$$X_C = \frac{E_S}{I_C} = \frac{12 \text{ V}}{5 \text{ mA}} = 2.4 \text{ k}\Omega$$

18.15 Effect of Frequency on Parallel RC

Since X_C depends on frequency, we can expect that the behavior of the parallel RC circuit will be different at different frequencies. At very low frequencies, X_C will become very large compared to R. As such, the capacitor current I_C will be very small compared to the resistor current I_R. Thus, the total current I_T is essentially equal to I_R. That is,

$$I_T \approx I_R \quad \text{(at very low frequencies)}$$

This situation is depicted in the phasor diagram in Fig. 18–26(a). Since the total current is composed mainly of the resistive current I_R, the phase angle between I_T and E_S is approximately zero at low frequencies.

Also at very low frequencies, we can state that impedance is approximately equal to R since

$$Z = \frac{E_S}{I_T} \simeq \frac{E_S}{I_R} = R \quad \text{(at very low } f\text{)}$$

Of course, at $f = 0$ (dc) the capacitor draws no current (in steady state) and it acts as an open circuit.

At very high frequencies X_C will become smaller and smaller compared to R. In fact, at high enough frequencies X_C will essentially be a short circuit. As such, the capacitor current I_C will be much greater than I_R and $I_T \simeq I_C$ (at very high frequencies).

This situation is depicted in part (b) of Fig. 18–26. As can be seen there, because I_T is composed mainly of capacitive current I_C, the phase angle is approximately 90°.

Also, at very high frequencies we can state that the impedance is approximately equal to X_C since

$$Z = \frac{E_S}{I_T} \simeq \frac{E_S}{I_C} = X_C \quad \text{(at very high } f\text{)}$$

RC Circuit Response to AC   525

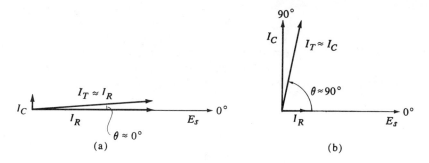

FIGURE 18-26 *Situation at* (a) *very low frequencies and* (b) *very high frequencies.*

Of course, as f increases, X_C approaches zero as does Z.

Figure 18-27 shows a graphical representation of how Z and θ vary with frequency. At low frequencies Z is almost pure resistance; at high frequencies Z is almost pure capacitance.

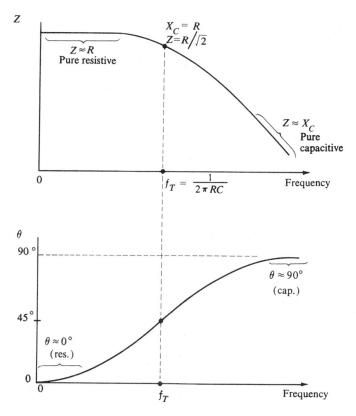

FIGURE 18-27 *Plots of Z and θ variations with frequency for a parallel RC circuit.*

Transition Frequency f_T A parallel RC circuit also has a transition frequency $f_T = 1/2\pi\ RC$. At this frequency, as in the series RC circuit, the value of X_C is equal to R. Since $X_C = R$, then $I_C = I_R$ at the transition frequency. Thus,

$$\tan \theta = \frac{I_C}{I_R} = 1$$

so that $\theta = 45°$. At frequencies $> f_T$, the value of X_C is less than R so that $I_C > I_R$ and the circuit is more capacitive ($\theta > 45°$). At frequencies $< f_T$ the value of X_C is greater than R so that $I_C < I_R$ and the circuit is more resistive ($\theta < 45°$).

18.16 Bypass Capacitor

We saw earlier how a capacitor in series with a resistor will prevent direct current from flowing through the resistor but will pass alternating current to the resistor. There are many applications in electronics in which it is required that dc current be allowed to flow through a resistor while ac is not allowed. One method for accomplishing this is shown in Fig. 18-28.

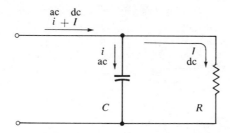

FIGURE 18-28 *Shunt capacitor acts as a bypass for ac current.*

In this circuit the capacitor C in parallel with R acts as a *bypass* capacitor. A *bypass* is a parallel or *shunt* path used to shunt current away from R. The bypass capacitor is chosen so that its reactance X_C is much less than R at the ac frequency of interest. In this way, the greater part of the ac current will flow through X_C and not through R. For dc, of course, no current will flow through C and all of it will flow through R.

The result is practically zero ac voltage across R because of the bypass action of C. The dc voltage across R is unaffected by C. For proper bypass action, C must be chosen such that X_C is much smaller than R by at least a factor of ten.

EXAMPLE 18.24 Choose a capacitor that will act as an effective bypass across a 2-kΩ resistor over a frequency range of 100 Hz to 10 kHz.

Solution: If the capacitor is to be an effective bypass it must have a low enough X_C at the lowest frequency of operation, which is 100 Hz in this case. Since $R = 2$ kΩ, we can choose $X_C = R/10 = 200$ Ω at 100 Hz. Thus, we have

$$X_C = 1/2\pi fC = 200 \text{ Ω}$$

or

$$C = \frac{1}{2\pi f X_C} = \frac{1}{2\pi \times 100 \times 200} \simeq 8 \text{ μF}$$

This value of C will have a reactance of only 200 Ω at 100 Hz so that it effectively bypasses R. At the higher frequencies of operation X_C is even smaller, thereby providing even better bypass action. At 10 kHz, X_C is only 2 Ω.

18.17 Comparison of Series RC and Parallel RC

In both series *RC* and parallel *RC* circuits, the total current *leads* the source voltage by a phase angle between 0° and 90°. The impedance of either circuit is of the form $Z/\!-\!\theta$. As such, as far as a source is concerned, it is impossible to distinguish between the two circuits at a given frequency. However, as we have seen, the two circuits behave differently with changes in frequency. Although the impedances of each circuit *decrease* as *f increases*, the phase angle of the parallel *RC* circuit approaches 90° at high frequencies, while the phase angle of the series *RC* circuit approaches 0° at high frequencies. In addition, at high frequencies the parallel circuit impedance approaches zero, while the impedance of the series circuit approaches a value equal to *R*.

Although the impedances of each circuit *increase* as frequency *decreases*, the phase angle of the parallel *RC* circuit approaches 0° at low frequencies while the phase angle of the series *RC* circuit approaches 90° at low frequencies. In addition, at low frequencies the parallel circuit impedance approaches a value equal to *R* while the impedance of the series circuit increases indefinitely ($Z \simeq X_C$).

These comparisons are summarized in Table 18-1. Note also from the table that for each circuit the phase angle is 45° at the transition frequency.

TABLE 18-1 *Comparison of Series and Parallel RC Circuits*

	Parallel *RC*	Series *RC*
Z at low frequencies ($f < 0.1\, f_T$) θ at low frequencies	$\simeq R$ $\simeq 0°$ } pure res.	$\simeq X_C$ $\simeq 90°$ } pure cap.
Z at high frequencies ($f > 10\, f_T$) θ at high frequencies	$\simeq X_C$ $\simeq 90°$ } pure cap.	$\simeq R$ $\simeq 0°$ } pure res.
Z at $f = f_T$ (transition frequency) θ at $f = f_T$	$R/\sqrt{2}$ 45°	$\sqrt{2}\, R$ 45°

Chapter Summary

1. In a series *RC* circuit the current is the same through each circuit element.

2. In a series *RC* circuit the amplitude (peak or rms) of the applied voltage does *not* equal the sum of the amplitudes of the resistor and capacitor voltages.

3. In a series *RC* circuit the phasor representing the source voltage is equal to the *vector* sum of the phasors representing the resistor and capacitor voltages: $\overline{E_S} = \overline{V_R} + \overline{V_C}$ (KVL).

4. In a series *RC* circuit, the current always *leads* the applied voltage, the resistor voltage always *leads* the applied voltage, and the capacitor voltage always *lags* the applied voltage.

5. Whenever resistance and reactance are combined in a circuit, the total opposition is called the impedance Z.

6. For a given circuit the impedance Z is equal to the ratio of applied voltage to current (E_S/I) and is measured in ohms.

7. For a series RC circuit the total impedance is equal to the *vector* sum of resistance and reactance. That is, $\overline{Z} = \overline{R} + \overline{X_C}$.

8. Table 18–2 summarizes the basic relationships for a *series* RC circuit.

TABLE 18–2

$$Z = \sqrt{R^2 + X_C^2}$$

$$I = \frac{E_S}{Z}$$

$$V_R = I \times R$$

$$V_C = I \times X_C$$

$$\sqrt{V_R^2 + V_C^2} = E_S$$

$$\theta = -\arctan\left(\frac{X_C}{R}\right) = -\arctan\left(\frac{V_C}{V_R}\right)$$

Note: θ is angle by which I leads E_S

9. Any phasor or vector can be expressed in the polar form $\overline{A} = A\underline{/\theta}$ where $\overline{A}$ is the vector or phasor, A is its magnitude, and θ is its angle relative to the 0° reference.

10. If $\overline{A} = A\underline{/\theta}$ and $\overline{B} = B\underline{/\phi}$, then $\overline{A} \times \overline{B} = (A \times B)\underline{/\theta+\phi}$ and $\overline{A}/\overline{B} = (A/B)\underline{/\theta-\phi}$.

11. In a series RC circuit, the impedance Z and phase angle θ both decrease as the applied frequency increases.

12. The transition frequency f_T of an RC circuit is the frequency at which $X_C = R$. The value of f_T is $1/2\pi RC$.

13. At very low frequencies ($< 0.1\, f_T$) a series RC circuit has $\theta \simeq 90°$, $Z \simeq X_C$, and behaves as almost purely capacitive.

14. At very high frequencies ($> 10\, f_T$) a series RC circuit has $\theta \simeq 0°$ and $Z \simeq R$, and behaves as almost purely resistive.

15. At the transition frequency the phase shift of a series RC is 45° and $V_C = V_R = 0.707\, E_S$.

16. The gain or attenuation ratio of a circuit is equal to V_out/V_in. This ratio can be expressed in decibel units (dB) as $20\log_{10}(V_\text{out}/V_\text{in})$.

17. Positive dBs indicate gain while negative dBs indicate attentuation. Zero dB represents $V_\text{out}/V_\text{in} = 1$.

18. Circuits which are designed to pass or reject one frequency or group of frequencies are called filters.

19. A low-pass filter passes low frequencies (including dc) and blocks or rejects high frequencies. A high-pass filter passes high frequencies and blocks low frequencies. A band-pass filter blocks lower and higher frequencies and passes medium (mid-range) frequencies.

20. A series RC circuit can be used as a high-pass filter if the output is taken as the resistor voltage V_R. It can be used as a low-pass filter if the output is taken as the capacitor voltage V_C.

21. The frequency-response curve of a filter is a plot of attenuation in dB versus frequency.

22. The low-cutoff frequency f_{lc} of a high-pass filter is the frequency below which the filter greatly attenuates the input voltage. $f_{lc} = f_T = 1/2\pi RC$. At $f = f_{lc}$ the attenuation is -3 dB.

23. The high-cutoff frequency f_{hc} of a low-pass filter is the frequency above which the filter greatly attenuates the input voltage. f_{hc} also equals f_T. At f_{hc} the attenuation is -3 dB.

24. A complete comparison between the series RC low-pass and high-pass filters is given in Fig. 18-20.

25. In a parallel RC circuit, the voltage is the same across each element.

26. In a parallel RC circuit, the *amplitude* (peak or rms) of the total current does *not* equal the sum of the amplitudes of the individual branch currents.

27. In a parallel RC circuit, the total current phasor equals the *vector* sum of the resistor current and capacitor current phasors: $\overline{I_T} = \overline{I_R} + \overline{I_C}$ (KCL).

28. In a parallel RC circuit, I_T always *leads* the applied voltage.

29. The total opposition of a parallel RC circuit is called its impedance Z. It is determined by the ratio of the source voltage to the total current: $Z = E_S/I_T$.

30. For a parallel RC circuit (or any capacitive circuit) the impedance vector has a negative angle: $\overline{Z} = Z\underline{/-\theta}$.

31. Table 18-3 summarizes the relationships for a parallel RC circuit.

TABLE 18-3

$$I_R = \frac{E_S}{R}$$

$$I_C = \frac{E_S}{X_C}$$

$$I_T = \sqrt{I_R^2 + I_C^2}$$

$$\theta = \arctan\left(\frac{I_C}{I_R}\right)$$

Note: θ is angle by which I_T leads E_S;

$$Z = \frac{E_S}{I_T}$$

32. In the parallel RC circuit, the impedance Z decreases and phase angle θ increases as frequency *increases*.

33. At very low frequencies ($< 0.1 f_T$) a parallel RC circuit has $Z \simeq R$, $\theta \approx 0°$, and behaves as almost purely resistive.

34. At the transition frequency f_T the phase shift θ is 45° and the branch currents are equal in a parallel RC circuit.

35. At very high frequencies ($> 10\, f_T$) a parallel RC circuit has $Z \approx X_C$, $\theta \approx 90°$, and behaves as almost purely capacitive.

36. Series and parallel RC circuits are compared in Table 18–1 (sec. 18.17).

Questions/Problems

Sections 18.2–18.3

18–1 In a certain series RC circuit, the resistor voltage is measured at 86.6 V and the capacitor voltage is measured at 50 V. The frequency is 100 Hz. Plot the waveforms of v_R and v_C versus time. Plot the waveforms of source voltage e_s by taking points on the v_R and v_C waveforms at 30° intervals and *algebraically* adding them.

18–2 Determine the approximate amplitude of e_s and its phase relative to v_R from the e_s waveform plotted in problem 18–1.

18–3 Determine the amplitude and phase of e_s for the situation in problem 18–1 using phasors and performing vector addition.

18–4 In a series RC circuit why don't the rms values of V_C and V_R add up to the rms value of the input E_S?

18–5 A series RC circuit contains $R = 500\ \Omega$ and $C = 1\ \mu F$: (a) find the source voltage needed to produce an rms current of 200 mA at 160 Hz; (b) draw a complete phasor diagram showing all currents and voltages.

Sections 18.4–18.5

18–6 Draw the impedance vector diagram for problem 18–5 and determine $\overline{Z}$ (magnitude and angle).

18–7 Express each of the phasors of problem 18–5 in polar form.

18–8 An RC circuit has $\overline{Z} = 1000\ \Omega/\!-20°$ at a certain frequency. If the applied voltage is 60 V, draw the phasors for $\overline{I}$ and $\overline{E_S}$.

18–9 Some of the following statements are *false*. Modify the false statements to make them correct.
   a. In a series RC circuit the current always leads the applied voltage by 90°.
   b. For a series RC circuit, V_R is measured to be 4 V and V_C is measured as 3 V; the ac source voltage must be 7 V.
   c. An X_C of 50 Ω in series with an R of 50 Ω has the same opposition to current as a single R or X_C of 100 Ω.
   d. The resistor voltage in a series RC circuit is always in phase with the applied voltage.
   e. If $\overline{Z} = 10\ \Omega/\!-30°$, then $R = 10\ \Omega$ and $X_C = 10\ \Omega$.

Sections 18.6–18.7

18–10 Refer to the circuit in Fig. 18–29(a). The box contains a series RC circuit. It is necessary to determine the values of R and C without getting inside the box. In other words, measurements can only be made external to the box at the terminals x-y.

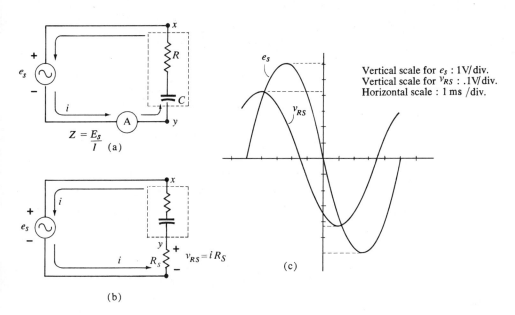

FIGURE 18-29

The first step is to apply a known voltage to the circuit and measure the current flow with an ac ammeter as shown. The ratio of the applied voltage to current will give us the magnitude of the impedance Z.

Once Z is known, we can use the arrangement shown in part (b) of the figure to determine the phase angle θ. A sampling resistor R_S is placed in series with the RC circuit so that the current i will produce a voltage across R_S that can be viewed on an oscilloscope. The value of R_S is chosen at least 10 times smaller than the calculated Z so that its effect on the total impedance will be negligible.

   a. In the circuit of Fig. 18–29(a), the source voltage is 12 V p-p at 100 Hz; the ammeter reads 6 mA. Determine the impedance Z of the series RC circuit (answer: $Z = 707\ \Omega$).

   b. Using $R_S = 50\Omega$, the source voltage is applied to the circuit in Fig. 18–29(b). The waveforms of e_s and v_{RS} appear on the oscilloscope as shown in part (c) of the figure. From these waveforms determine the phase angle θ. (answer: $\theta = -54°$)

   c. From the values of Z and θ determine the values of R and C.

18-11 A sine wave voltage at a certain frequency is applied to a series RC circuit resulting in a certain Z and θ. Indicate how Z and θ will change (increase, decrease, no change) as each of the following changes is made in the circuit:
   a. a decrease in frequency
   b. an increase in capacitance
   c. a decrease in capacitance
   d. an increase in input amplitude
   e. a two-fold increase in R and a halving of C

18-12 In each of the following determine the missing vector quantity using vector multiplication or division:
 a. $\bar{I} = 6\underline{/0°}, \bar{Z} = 50\ \Omega\underline{/-15°}, \bar{E_S} = ?$
 b. $\bar{I} = 20\underline{/10°}, \bar{Z} = 3\ \Omega\underline{/-70°}, \bar{E_S} = ?$
 c. $\bar{Z} = 4\ k\Omega\underline{/-50°}, \bar{E_S} = 100\ V\underline{/0°}, \bar{I} = ?$
 d. $\bar{E_S} = 50\ V\underline{/0°}, \bar{I} = 2\ mA\underline{/30°}, \bar{Z} = ?$

18-13 A 20-kΩ resistor is in series with a 0.5-μF capacitor. (a) Calculate Z and θ at the following frequencies: 1 Hz, 10 Hz, 100 Hz, and 1000 Hz. (b) Plot these values using a *log scale* for frequency. (c) Determine f_T and calculate Z and θ at this frequency.

18-14 Describe a simple method for determining f_T for a series RC circuit using only a variable frequency generator and an ac voltmeter.

18-15 The table below shows measurements taken on the current in a series RC circuit at various frequencies. From these measurements determine the values of R and C.

f	E_S	I
20 Hz	12 V	0.48 mA
100 Hz		2.35 mA
500 Hz		8.5 mA
10000 Hz		12 mA
15000 Hz		12 mA

18-16 Design a series RC circuit that has $f_T = 300$ Hz and draws a current of 2 mA when a 100-V, 10-kHz voltage is applied to it.

18-17 A series RC circuit is to operate at frequencies in the range of 100 Hz-600 kHz. It is required that the current lead the applied voltage by at *least* 60° throughout this range. Choose an appropriate value of C if R is 1.5 kΩ.

18-18 Refer to Fig. 18-30. The two blocks represent transistor amplifiers. The input to amplifier $A1$ is an audio voice signal from a microphone pickup. This audio signal contains frequencies in the range from 20 Hz to 10 kHz. The output of $A1$ is connected to the input of amplifier $A2$ through a *coupling capacitor* C_c. The output of $A1$ contains dc as well as amplified audio ac. The coupling capacitor is used to prevent the flow of dc current from the output of $A1$ to the input of $A2$ while allowing the audio signals to reach the input of $A2$.

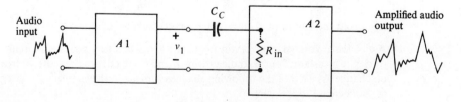

FIGURE 18-30

A series RC circuit is formed by the coupling capacitor and R_{in}, the input

RC Circuit Response to AC 533

resistance of $A2$. For a given value of R_{in}, the value of C_c must be chosen so that this RC circuit produces very little phase shift at frequencies in the audio range. Any appreciable phase shift at any of the audio frequencies will cause distortion in the output of $A2$.

With $R_{in} = 20$ kΩ, choose a value for C_c so that θ is less than $2°$ for the range of frequencies which are being amplified.

Section 18.8

18-19 Convert the following gain and attenuation ratios to decibel units: 12, 0.01, 7000, 0.3, 10^5, 10^{-3}.

18-20 Convert the following db to V_{out}/V_{in} ratios: 14 dB, -60 dB, -3 dB, 120 dB.

18-21 Three circuits, A, B, and C, are connected so that the output of circuit A is the input to B, and the output of B is the input to C. Circuit A has a gain of 36 dB; circuits B and C each have an attenuation of -8 dB. If an input voltage of 10 V is applied to the input of A, determine the voltage at the output of C.

Sections 18.9–18.10

18-22 Design an RC filter that attenuates all frequencies below 10 kHz by at least -20 dB. The value of C is not to exceed 1000 pF.

18-23 Sketch the frequency-response curve for the RC filter of problem 18–22.

18-24 Consider the circuit in Fig. 18–16. Assume $R = 10$ kΩ and $C = 0.05$ μF. Indicate which of the following frequencies will be passed from the output of circuit I to the input of circuit II with essentially no attenuation: $f = 0$, 20 Hz, 330 Hz, 4 kHz, 200 kHz.

Sections 18.11–18.12

18-25 Design an RC filter that attenuates all frequencies *above* 16 kHz by at least -20 dB. The circuit must draw a maximum current of 20 mA when the input is 10 V.

18-26 Sketch the frequency-response curve for the filter of problem 18–25.

18-27 Consider the circuit of Fig. 18–17. Assume $R = 10$ kΩ and $C = 0.05$ μF. Indicate which of the following frequencies will be passed from input to output with essentially no attenuation: $f = 0$, 20 Hz, 220 Hz, 4 kHz, 200 kHz.

18-28 A certain RC filter has input voltages of various frequencies applied to it. For each frequency the input amplitude is 10 V and the output amplitude is recorded in the data table below. From the data table:
   a. Determine what type of filter it is.
   b. Determine its cut-off frequency.
   c. Determine the values of R and C.

f	E_S	V_{out}	I
10 Hz	10 V	9.95 V	0.1 mA
100 Hz	10 V	9.9 V	$\simeq$1.0 mA
1000 Hz	10 V	7.1 V	7.1 mA
10000 Hz	10 V	1.0 V	$\simeq$10.0 mA
100000 Hz	10 V	0.1 V	$\simeq$10.0 mA

18–29 Each of the voltage signals (waveforms) in Fig. 18–31 is applied to an RC low-pass filter with $R = 400\,\Omega$ and $C = 0.08\,\mu F$. Sketch the filter output for each case.

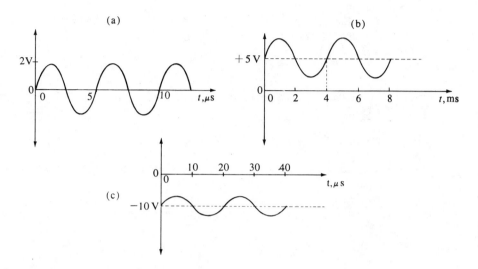

FIGURE 18–31

18–30 Repeat problem 18–29 for a high-pass filter with $f_{lc} = 5$ kHz.

Sections 18.13–18.15

18–31 A parallel RC circuit has $R = 4$ kΩ and $X_C = 2$ kΩ. Find: (a) I_T if $E_S = 32$ V; (b) phase angle between I_T and E_S; (c) impedance $\overline{Z}$ (polar form).

18–32 For the circuit of the previous problem indicate what would happen to I_T and θ for each of the following circuit changes.
    a. an increase in frequency
    b. a decrease in R
    c. an increase in C
    d. a decrease in E_S
    e. adding another capacitor in parallel

18–33 When a 50-V, 200-Hz voltage is applied to a parallel RC circuit, the total current is 8 mA and leads the source voltage by 26.5°. Determine the values of R and C.

18–34 Determine the R and X_c values of a parallel circuit that has $\overline{Z} = 500\underline{/-45°}$.

18–35 Determine the impedance and phase angle of a parallel circuit with $R = 160\,\Omega$ and $C = 1\,\mu F$ at the following frequencies: 100 Hz, 500 Hz, 1 kHz, 5 kHz, 10 kHz. Plot these Z and θ values versus f.

Sections 18.16–18.17

18–36 Choose a capacitor that will effectively bypass ac current around a 500-Ω resistor over a frequency range of 150 Hz to 1 MHz.

18–37 The data in the table below are taken for a certain RC circuit. From the data determine: (a) if it is a series or parallel circuit, and (b) the R and C values.

Frequency	e_s	i_t
10 Hz	20 V	10 mA
100 Hz		10.1 mA
1 kHz		14 mA
10 kHz		101 mA
100 kHz		1.0 A
1 MHz		10 A

18-38 Consider each of the following statements. For each one indicate whether it pertains to a series RC circuit, a parallel RC circuit, or both.
 a. Z decreases as frequency increases
 b. becomes more capacitive as frequency increases
 c. becomes more resistive as frequency decreases
 d. total current is always limited by R
 e. total current leads applied voltage
 f. capacitor current leads total current
 g. can be used as voltage filter circuits
 h. transition frequency $f_T = 1/2\pi RC$
 i. I_T leads E_S by 45° at transition frequency
 j. voltages across R and X_C are equal at all times
 k. blocks current at very low frequencies
 l. acts as a short to very high frequencies
 m. applied voltage is vector sum of component voltages
 n. total current is vector sum of component currents
 o. impedance vector always has a negative angle
 p. capacitor voltage lags applied voltage
 q. decreasing C causes the circuit to become more capacitive
 r. increasing R causes the circuit to become more capacitive

CHAPTER 19

RL Circuit Response to AC

19.1 Introduction

In chap. 15 we investigated how circuits containing both *resistance* and *inductance* responded to the sudden application of a dc voltage. We saw that after *five* circuit time constants had passed, the circuit reached a steady-state condition where all the circuit currents and voltages had reached constant (steady) values. In this chapter we will investigate the response of *RL* (resistance-inductance) circuits to the application of a sinusoidal voltage. In doing so we will be interested only in the steady-state response (after 5τ) where all the circuit currents and voltages are sine waves.

Much of what will be done concerning *RL* circuits will be similar to the concepts covered in the last chapter on *RC* circuits. We will find that the *RL* circuits in many cases behave exactly opposite to their *RC* counterparts. This fact should make analysis of *RL* circuits somewhat easier to perform.

19.2 Review of X_L

We know ac current in an inductor is always 90° out of phase with the inductor voltage. This is shown in Fig. 19–1. Note that i_L *lags* v_L by 90°. We also know that an inductor opposes ac with an opposition equal to X_L ohms. Thus, the peak value of inductor current I_{LM} is equal to V_{LM}/X_L where V_{LM} is the peak voltage.

The phasor diagram relating the inductor current and voltage is also drawn in the figure. Here rms values are used for current and voltage. Thus, $I_L = V_L/X_L$ where I_L and V_L are rms values. Of course, the value of X_L is given by $X_L = 2\pi f L$ and is dependent on the frequency.

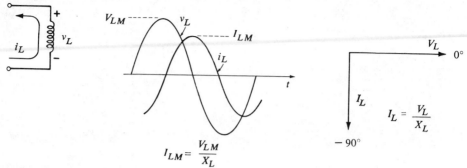

FIGURE 19-1 *Behavior of inductor in alternating current circuit.*

19.3 X_L and R in Series

Figure 19-2 shows an ac source connected to a circuit that contains an inductive reactance X_L in series with a resistor R. The current from the source is the *same* through each of the circuit components. Since i flows through both R and X_L, then both elements contribute to the total opposition to current. Since $X_L = 100$ ohms and $R = 100$ ohms for the circuit shown, we might expect the total opposition (impedance) to be 200 ohms. As we shall see, this is *not* the case.

Part (b) of Fig. 19-2 shows the various circuit waveforms for the circuit values chosen. The following points can be ascertained from these waveforms.

(1) The current waveform i is in phase with the resistor voltage waveform v_R. The peak resistor voltage is equal to the peak current $1.41 \text{ A} \times 100 \text{ }\Omega = 141$ V. The rms value is $V_R = 0.707 \times 141 \text{ V} = 100$ V.

(2) The inductor voltage waveform v_L *leads* the current waveform by 90°. It also *leads* v_R by 90°. The peak inductor voltage is $1.41 \text{ A} \times 100 \text{ }\Omega = 141$ V. The rms value is $V_L = 0.707 \times 141 \text{ V} = 100$ V.

(3) The source voltage waveform at any instant of time is equal to the algebraic sum of the resistor and inductor voltages. This, of course, is a result of Kirchhoff's voltage law. That is, $e_S = v_R + v_L$. The e_S waveform can be obtained by adding the values of v_L and v_R at each instant of time.

(4) The peak source voltage is 200 V. This does *not* equal the sum of the peak voltages across R and X_L. This is because the peaks of the v_R and v_L waveforms do not occur at the same time since these waveforms are 90° out of phase. When v_R is at its peak of 141 V, the v_L waveform is at 0 V and vice versa. The peak source voltage occurs at 45° when v_R and v_L are both 100 V so that $e_s = 200$ V.

(5) The source voltage reaches its peak at 45°. Thus, e_s *leads* the current i by 45°. Equivalently, i *lags* e_s by 45°. Thus, the phase angle between the circuit current and the applied voltage is neither 0° nor 90°. Keep in mind, however, that i is still in phase with v_R and lags v_L by 90°.

Phasor Diagram The phasor diagram in Fig. 19-2(c) shows the phase relationships. Since it is a series circuit, the current i is common to all the circuit elements and is therefore used as the reference phasor. The magnitude of the

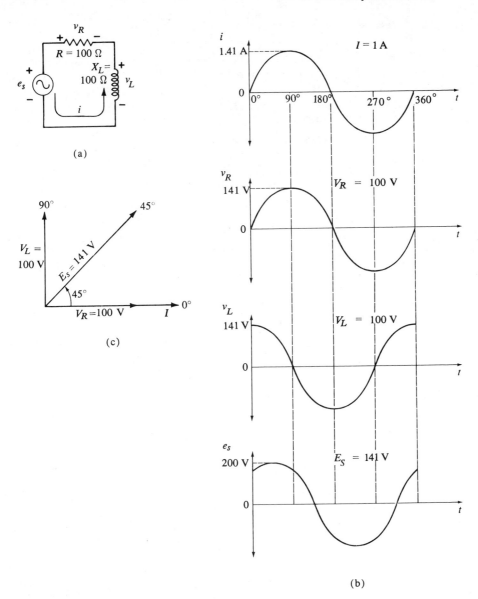

FIGURE 19-2 (a) *Circuit with* X_L *and R in series.* (b) *Waveforms.* (c) *Phasor diagram.*

current phasor is $I = 1$ A (rms) and is drawn at 0°. The resistor voltage phasor V_R is also at 0° and $V_R = 100$ V. The inductor voltage phasor V_L is drawn at +90° relative to the current phasor and $V_L = I \times X_L = 100$ V. The source voltage phasor E_S is drawn so that it *leads* the current phasor by 45° and has an rms magnitude $E_S = 141$ V. Note that $E_S \neq V_R + V_L$.

Combining v_R and v_L Using Vector Addition As shown in Fig. 19-2(b), we can obtain the applied voltage waveform e_s by arithmetically adding the v_R and v_L waveforms point-by-point. Because v_R and v_L are 90° out of phase, the peak or rms values of e_s cannot be obtained by adding the peak or rms values of v_L and v_R. In other words, *when we are adding two series voltages that are out of phase, we have to take into account the phase difference.*

An easy way to add v_R and v_L while taking their phase difference into account is to use vector addition exactly the way it was used in the series RC circuit. Figure 19-3 shows the procedure for determining the sum of the v_R and v_L waveforms of Fig. 19-2(b). In part (a) of the figure the phasors for the inductor and resistor voltages are drawn with $V_L = V_R = 100$ V and V_L leading V_R by 90°. To add $\mathbf{V_L}$ and $\mathbf{V_R}$ we use the same method employed in the RC circuit; $\mathbf{V_L}$ is moved so that its tail coincides with $\mathbf{V_R}$'s arrowhead [part (b) of figure].

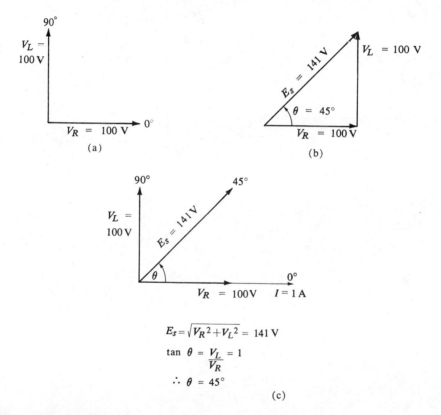

FIGURE 19-3 *Finding the total applied voltage by vector addition of resistor and inductor voltage phasors.*

The vector sum of $\overline{\mathbf{V_L}}$ and $\overline{\mathbf{V_R}}$ is obtained by drawing the resultant vector from the tail of $\overline{\mathbf{V_R}}$ to the head of $\overline{\mathbf{V_L}}$. This resultant phasor is the $\overline{\mathbf{E_S}}$ phasor. That is, $\overline{\mathbf{E_S}} = \overline{\mathbf{V_L}} + \overline{\mathbf{V_R}}$ (vector sum). The magnitude of the E_S phasor can be found from

$$E_S = \sqrt{V_R^2 + V_L^2} \qquad (19\text{-}1)$$

For $V_R = V_L = 100$ V, the rms value of the source voltage is

$$E_S = \sqrt{100^2 + 100^2} = 141 \text{ V}$$

In the same phasor diagram in Fig. 19-3(b) the angle θ represents the phase angle by which E_S leads V_R. Since V_R is in phase with the current, then θ also represents the phase angle by which E_S leads I. The value of θ can be determined using

$$\tan \theta = \left(\frac{V_L}{V_R}\right) \tag{19-2}$$

For our example,

$$\tan \theta = \frac{100 \text{ V}}{100 \text{ V}} = 1$$

Thus, $\theta = 45°$. Note that θ is positive relative to the 0° reference.

The total source voltage phasor E_S has an rms amplitude of 141 V and *leads* the current (and V_R) by 45°. The complete phasor diagram with all the circuit phasors (I, V_R, V_L, and E_S) drawn from the origin is shown in part (c) of Fig. 19-3. This diagram is a complete picture of the circuit phase relationships.

Formulas (19-1) and (19-2) can be used to find the magnitude and phase angle of the source voltage in *any* series RL circuit. However, the student should not abandon the phasor diagram simply because these formulas are available. The phasor diagram is useful in helping to visualize the circuit phase relationships.

To summarize this section, we have shown that *in a series* RL *circuit the phasor representing the source voltage is equal to the vector sum of the $\overline{V_R}$ and $\overline{V_L}$ phasors.*

EXAMPLE 19.1 A series RL circuit contains $R = 2$ kΩ and $X_L = 1$ kΩ.
(a) Find the source voltage needed to produce an rms current of 5 mA in the circuit.
(b) Determine the phase angle between E_S and I.
(c) Determine the phase angle between E_S and V_L.
(d) Determine the phase angle between E_S and V_R.

Solution: (a) The circuit diagram is shown in Fig. 19-4(a). Since $I = 5$ mA we can calculate the rms values of the resistor and inductor voltages.

$$V_R = 5 \text{ mA} \times 2 \text{ k}\Omega = 10 \text{ V}$$

$$V_L = 5 \text{ mA} \times 1 \text{ k}\Omega = 5 \text{ V}$$

To find E_S we must add the V_R and V_L phasors using *vector* addition. These phasors are drawn in part (b) of the figure along with the resultant E_S. The rms value of E_S is

$$E_S = \sqrt{V_R^2 + V_L^2} = \sqrt{10^2 + 5^2}$$
$$= 11.2 \text{ V}$$

(b) $\tan \theta = \left(\dfrac{V_L}{V_R}\right) = \dfrac{5 \text{ V}}{10 \text{ V}} = 0.5$

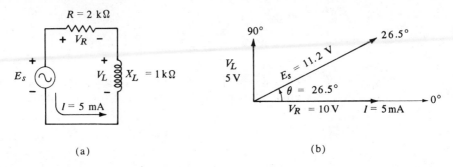

(a) (b)

FIGURE 19-4

therefore, $\theta = 26.5°$ as shown on the phasor diagram. The current I lags E_S by 26.5°. *In a series RL circuit, the current always lags the applied voltage.*

(c) From the phasor diagram it can be seen that V_L leads E_S by $90° - 26.5° = 63.5°$. *In a series RL circuit, the inductor voltage always leads the source voltage.*

(d) The resistor voltage is in phase with I. Thus, V_R lags E_S by 26.5°. *In a series RL circuit $\overline{V_R}$ always lags the applied voltage.*

19.4 Impedance of a Series RL Circuit

The impedance Z of a series RL circuit is the total opposition that it presents to alternating current. If Z is known, then the circuit current can be determined from

$$I = \frac{E_S}{Z} \qquad (19\text{-}3)$$

This relationship can also be arranged as

$$E_S = IZ \qquad (19\text{-}4)$$

The procedure for determining Z for a series RL circuit is similar to what we did for the series RC circuit.

In Fig. 19-5(a) the phasor diagram of the voltages in the series RL circuit is redrawn from Fig. 19-3(b). Note that in this voltage phasor triangle each phasor has a magnitude proportional to I. Thus, if we divide each phasor's magnitude by I, we still will have the same right triangle reduced in scale by the factor I. This new triangle is drawn in part (b) of the figure. As can be seen, the sides of this *impedance* triangle are R, X_L, and Z.

The $\overline{R}$, $\overline{X_L}$, and $\overline{Z}$ vectors have the same directions as their respective voltage phasors. From the impedance triangle, it is seen that Z is the vector sum of R and X_L ($\overline{Z} = \overline{R} + \overline{X_L}$). Thus, the magnitude of the impedance is

$$Z = \sqrt{R^2 + X_L^2} \qquad (19\text{-}5)$$

This formula allows us to calculate the impedance of any series RL circuit. It is important to realize that we cannot simply *add* R and X_L to obtain Z.

The angle θ in the impedance triangle is the same as the angle θ in the voltage triangle. Using the impedance triangle the values of θ can be found from

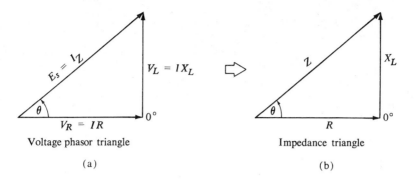

FIGURE 19-5 *The impedance Z is the vector sum of R and X_L.*

$$\tan \theta = \left(\frac{X_L}{R}\right) \qquad (19\text{-}6)$$

Recall that θ is the phase angle by which the current *lags* E_S in the series *RL* circuit.

EXAMPLE 19.2 A 200-ohm resistor is in series with a 24-mH inductor. A 1000-Hz, 50-V source is applied to the circuit.
(a) Determine the circuit impedance.
(b) Find the circuit current and voltages.
(c) Draw the complete circuit phasor diagram.
(d) Determine θ.
(e) Express $\bar{I}$, $\overline{E_S}$, and $\overline{Z}$ in polar form.

Solution: (a) $X_L = 2\pi f L = 2\pi \times 10^3 \times 24 \times 10^{-3} = 150\ \Omega$.

$$Z = \sqrt{R^2 + X_L^2} = \sqrt{200^2 + 150^2} = \mathbf{250\ \Omega}$$

(b) $I = E_S/Z$

$$= \frac{50\ \text{V}}{250\ \Omega} = \mathbf{0.2\ A}$$

$$V_R = IR = 0.2\ \text{A} \times 200\ \Omega = \mathbf{40\ V}$$
$$V_L = IX_L = \mathbf{30\ V}$$

(c) The phasor diagram is shown in Fig. 19–6. The current phasor is at 0° as is the V_R phasor. The V_L phasor is at 90° since the inductor voltage *leads* its current by 90°. The source voltage phasor is the vector sum of V_R and V_L.

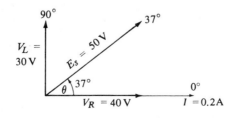

FIGURE 19-6

(d) The angle θ is the phase angle between E_S and I.

$$\tan \theta = \left(\frac{V_L}{V_R}\right) = \frac{30 \text{ V}}{40 \text{ V}} = 0.75$$

$$\therefore \theta = 37°$$

This value could also be found from

$$\tan \theta = \left(\frac{X_L}{R}\right) = \frac{150 \text{ }\Omega}{200 \text{ }\Omega} = 0.75$$

(e) $\bar{I} = 0.2 \text{ A}\underline{/0°}$
$\overline{E_S} = 50 \text{ V}\underline{/37°}$

Thus,

$$\bar{Z} = \frac{\overline{E_S}}{\bar{I}} = \frac{50\underline{/37°}}{0.2\underline{/0°}} = \left(\frac{50}{0.2}\right)\underline{/37° - 0°}$$

$$= 250 \text{ }\Omega\underline{/37°}$$

EXAMPLE 19.3 A series RL circuit draws 15 mA of current from a 120-V, 60-Hz source. The current lags the voltage by 60°. Determine the values of R and L.

Solution:

$$Z = \frac{120 \text{ V}}{15 \text{ mA}} = 8 \text{ k}\Omega$$

Since I lags E_S by 60°, then $\theta = 60°$. Thus,

$$\bar{Z} = 8 \text{ k}\Omega\underline{/60°}$$

To find R and X_L we can utilize the impedance triangle for the RL circuit [Fig. 19–5(b)]. Since Z and θ are known, then R and X_L can be determined using trigonometry.

$$R = Z \cos \theta$$
$$= 8 \text{ k}\Omega \times 0.5 = \mathbf{4 \text{ k}\Omega}$$

$$X_L = Z \sin \theta$$
$$= 8 \text{ k}\Omega \times 0.866 = \mathbf{6.9 \text{ k}\Omega}$$

To find L, we use the fact that $X_L = 2\pi f L$, so that

$$L = \frac{X_L}{2\pi f} = \frac{6900}{2\pi \times 60} = \mathbf{18.3 \text{ H}}$$

19.5 The Practical Inductor

We know from previous discussions that a *practical inductor* possesses both inductance and resistance. We represent a practical inductor by an inductance

L in series with a resistance R_e. The resistance R_e represents the *effective resistance* of the coil. At *low* frequencies R_e is simply the ohmic resistance of the wire which makes up the coil. Its value can be easily measured with a dc ohmmeter placed across the coil. In other words, at low frequencies

$$R_e = R_{dc}$$

where R_{dc} is the coil's dc wire resistance.

At frequencies above around 10 kHz, several effects occur which cause the coil's *effective* resistance R_e to increase significantly as frequency is increased. The factors that make the R_e of the coil more than its wire resistance include the *skin effect, eddy currents,* and *hysteresis losses*.

Skin Effect The skin effect occurs when a conductor is carrying current at high frequencies (> 10 kHz). This effect is caused by the tendency of high frequency current to flow at the surface (skin) of a conductor and not throughout the entire cross section of the conductor. Explanation of the skin effect requires the use of some of the concepts previously described in chap. 15.

It will be remembered from chap. 15 that inductance is the property by which a conductor opposes any change in the amount of magnetic flux which links it (Fig. 19–7). Since magnetic flux is produced by the flow of current in the wire, then in an ac circuit the flux Φ will change at the same frequency as the current. Thus, at higher frequencies of ac the flux Φ changes more rapidly. As such, the inductance offers a greater opposition to these more rapid flux changes. This is why the inductive opposition to ac (inductive reactance X_L) increases as frequency increases.

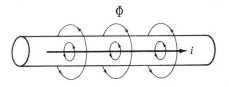

FIGURE 19–7 *Magnetic flux around a current-carrying wire.*

The flux lines linking the conductor actually extend outward from the *center* of the wire. Thus, more flux lines link the center of the wire than the outside of the wire. In other words, the flux is more concentrated at the center of the conductor. As a result, the inductive opposition will be greater at the center than it is on the outer portion of the wire. This will cause the electrons to flow near or on the surface of the conductor since there is less inductive opposition there. This effect reduces the effective cross-sectional area through which the current flows, and therefore causes an increase in the resistance of the conductor (recall R is inversely proportional to A; $R = \rho l/A$). This skin effect is usually insignificant at low frequencies. As the frequency is increased (above a few kHz) the effect on R_e becomes significant. For example, a 100-μH coil with a resistance of 10 ohms at dc or low frequencies might have its effective resistance increased to 120 ohms at 2 MHz.

Eddy Currents In any inductor with an iron core, the changing flux in the core will induce an emf in the core itself (recall Faraday's law). Since iron is a conductor, current will flow in the core as a result of this induced emf. This current is called an *eddy current* since it flows in circular paths throughout the cross section of the core (see Fig. 19-8).

These eddy currents can result in a waste of power in the form of heat, equal to I^2R where I is the amount of eddy current and R is the resistance of the iron core. The magnitude of this power loss depends on the type of core used. If the core is a material like wood or air (poor conductors of electricity), then the eddy currents will be very small as will the power losses.

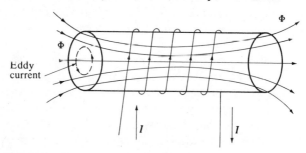

FIGURE 19-8 *View of iron-core inductor showing eddy currents.*

Unfortunately, however, air-core coils have much less inductance than iron-core coils. The use of air-core coils is limited to radio-frequency applications (> 10 kHz) where required values of inductance are in the µH or low-mH range.

To minimize power losses due to eddy currents in iron-core coils while maintaining large values of inductance, the core can be constructed of: (1) laminated sheets insulated from one another; or (2) powdered-iron granules; or (3) certain ceramic materials called ferrites that have high values of magnetic permeability µ but are poor conductors of electricity.

For a given core material, the eddy currents will increase as the frequency of the current flowing through the coil increases. The higher frequency coil current produces more rapid flux changes (greater $d\Phi/dt$) causing a larger induced voltage in the core and therefore larger eddy currents.

The effect of the increased eddy current power loss at higher frequencies is to make an inductor appear to have a greater effective resistance R_e, since it is resistance (not inductance) which dissipates power. Thus, the eddy current effect and the skin effect both act to increase R_e as frequency is increased.

Hysteresis The third effect is due to the *hysteresis* present in most ferromagnetic materials. As we saw in our discussion in chap. 14, hysteresis is caused by a sort of atomic friction which prevents the magnetic domains from completely reversing when the flux in the core is reversed. When ac is passed through a coil, the flux in the core will be reversing many times per second. As it does so, heat is produced in the core as a result of the atomic friction (hysteresis). This heat represents lost power and its effect is to make the inductor appear to have a greater effective resistance. At higher frequencies of ac the hysteresis power loss becomes greater because the magnetic domains in the iron core have to reverse their directions more times per second. Inductors with cores of non-ferromagnetic materials (air, wood, etc.) do not suffer from hysteresis effects.

Measuring L and R_e. Since an inductor's effective resistance R_e is dependent on frequency, its value must be measured when the inductor is carrying an ac current at its normal operating frequency. For example, if a certain *rf* coil is to be used at a frequency of 500 kHz, then R_e should be measured at that frequency. This means that R_e cannot be measured by simply using a dc ohmmeter unless the inductor is to operate at very low frequencies ($\ll$10 kHz).

Figure 19-9 shows one method which can be used to determine the values of L and R_e for a practical inductor. L and R_e are shown enclosed in dotted lines since they are inseparable and are therefore not individually accessible. The coil terminals x and y are accessible. The series resistor R_S is inserted in the circuit deliberately and is used to measure the amount of current flow. The value of R_S is chosen so that the current will produce a measurable voltage across R_S. The procedure is as follows.

(1) A known source voltage is applied at the frequency of interest.

(2) The voltage across the coil v_{coil} is displayed on an oscilloscope. The voltage across R_S is also displayed on the oscilloscope, so that the phase difference θ between v_{coil} and v_{RS} can be measured (see waveform in figure).

(3) The value of θ thus obtained also represents the phase angle between v_{coil} and i, since i is in phase with v_{RS}.

(4) Using Ohm's law, $I = V_{RS}/R_S$.

(5) The magnitude of the coil impedance (includes X_L and R_e) is therefore $Z_{\text{coil}} = V_{\text{coil}}/I$.

(6) Knowing Z_{coil} and θ, the impedance triangle can be drawn as shown in the figure. From this triangle, X_L and R_e can be calculated.

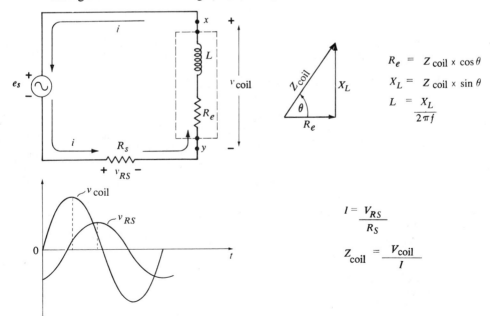

FIGURE 19-9 *Method for determining* L *and* R_e *for a practical coil.*

EXAMPLE 19.4 The setup of Fig. 19-9 is used to measure L and R_e for a certain coil at 100 kHz. The phase angle between v_{coil} and v_{RS} is measured as 82°. The rms voltages are measured as $V_{coil} = 12$ V and $V_{RS} = 3$ V. The value of R_S used is 1.5 kΩ. Determine L and R_e.

Solution: $I = \dfrac{V_{RS}}{R_S} = \dfrac{3 \text{ V}}{1.5 \text{ k}\Omega} = 2 \text{ mA}$

$$Z_{coil} = \dfrac{V_{coil}}{I} = \dfrac{12 \text{ V}}{2 \text{ mA}} = 6 \text{ k}\Omega$$

Thus, $\overline{Z}_{coil} = 6 \text{ k}\Omega \underline{/82°}$. From the impedance triangle,

$$R_e = Z_{coil} \times \cos\theta$$
$$= 6 \text{ k}\Omega \times 0.14 = \mathbf{840\ \Omega}$$

$$X_L = Z_{coil} \times \sin\theta$$
$$= 6 \text{ k}\Omega \times 0.99 = \mathbf{5.94 \text{ k}\Omega}$$

The value of L is then determined as

$$L = \dfrac{X_L}{2\pi f} = \dfrac{5.94 \times 10^3}{2\pi \times 10^5} \approx \mathbf{9.5 \text{ mH}}$$

Because of its effective resistance, a practical inductor will not have its current lagging the applied voltage by exactly 90°. The total impedance of the coil, Z_{coil}, equals the vector sum of X_L and R_e and this impedance will have an angle *less* than 90°. That is $\overline{Z}_{coil} = Z_{coil}\underline{/\alpha}$ where $\alpha < 90°$. Thus, I_L will lag the applied voltage by an angle equal to α. Of course, if R_e is much less than X_L, then α will be approximately 90° and the inductor will essentially act as a pure inductance.

19.6 Frequency Effects on Series RL

Since X_L increases with frequency, the impedance Z and phase angle θ will change with changes in the applied frequency. This can be graphically illustrated using the impedance diagrams of Fig. 19-10.

The impedance triangle in (a) shows a resistance R and a reactance X_L producing an impedance Z and phase angle θ at some frequency f_o. If the applied frequency is *increased* [part (b) of the figure], the R will remain the same but X_L will *increase*. This causes the phase angle θ and impedance Z to increase. If the frequency is increased even further, X_L would become larger and larger until eventually θ would be almost 90° and Z would approximately be equal to X_L.

The effect of a *decrease* in frequency is shown in part (c) of the figure. X_L has decreased causing a smaller phase angle and a smaller Z. If f is decreased further, X_L becomes smaller and smaller so that θ approaches 0° and Z would be approximately equal to R.

These same effects can be shown using the relationships developed earlier. Since we have

$$Z = \sqrt{R^2 + X_L^2} \qquad (19\text{–}5)$$

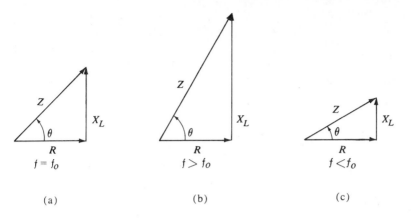

FIGURE 19-10 *Impedance diagrams show how Z and θ vary with frequency.*

it is easily seen that Z will increase as X_L increases with frequency. At very high frequencies X_L becomes very large compared to R so that the effect of R can be neglected and $Z \simeq \sqrt{X_L^2} = X_L$. At very low frequencies X_L becomes very small (compared to R) so that its effect can be neglected and $Z \simeq \sqrt{R^2} = R$.

We also know that

$$\tan \theta = \frac{X_L}{R} \tag{19-6}$$

At higher frequencies the ratio X_L/R becomes very large and approaches ∞. Thus, $\tan \theta \to \infty$ and $\theta \simeq 90°$, at high frequencies. At lower frequencies X_L/R approaches zero so that $\tan \theta \to 0$ and $\theta \simeq 0°$.

Graphs of Z and θ Versus f Figure 19-11 contains the graphs showing how Z and θ vary for a series RL circuit as the frequency is varied. These graphs can be obtained using equations (19-5) and (19-6) to calculate Z and θ over a wide range of frequencies. The graph of Z versus f shows that the impedance *increases* with frequency. At low frequencies Z levels off at a value equal to R. At higher frequencies, Z increases rapidly as f increases. The graph of θ versus f shows θ at $0°$ at low frequencies and increasing to $90°$ at high frequencies.

From these graphs and the preceding discussion we can conclude that at low frequencies the series RL circuit behaves almost purely resistive since $Z \simeq R$ and $\theta \simeq 0°$. At low frequencies the circuit current will be in phase with the applied voltage. At high frequencies, the series RL circuit behaves almost purely inductive since $Z \simeq X_L$ and $\theta \simeq 90°$. At high frequencies the circuit current *lags* the applied voltage by $90°$. At in-between frequencies, θ is between $0°$ and $90°$ and the circuit is a combination resistive and inductive.

Transition Frequency f_T At a frequency called the transition frequency f_T, the X_L portion of a series RL circuit is equal to R. At frequencies below f_T the value of X_L is smaller than R; at frequencies above f_T the value of X_L is greater than R. The value of f_T can be determined for the RL circuit by setting $X_L = R$ and solving for f_T.

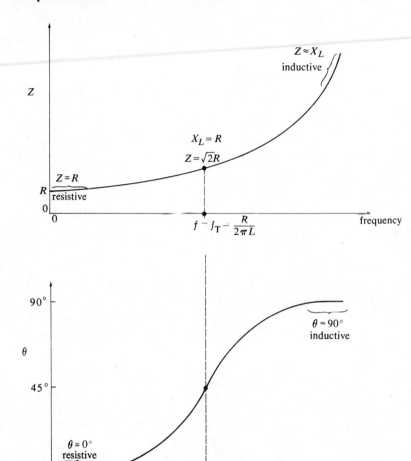

FIGURE 19–11 *Plots of Z and θ versus frequency.*

$$X_L = 2\pi f_T L = R$$

so that

$$f_T = \frac{R}{2\pi L} \tag{19-7}$$

At $f = f_T$ the circuit phase angle is 45° since $\tan\theta = X_L/R = 1$. At frequencies above f_T the phase angle is greater than 45° as the circuit becomes more inductive. Below f_T, θ is less than 45° as the circuit becomes less inductive. The impedance Z at the transition frequency is determined by

$$Z = \sqrt{R^2 + X_L^2} = \sqrt{R^2 + R^2}$$
$$= \sqrt{2}\,R$$

RL Circuit Response to AC 551

EXAMPLE 19.5 For the circuit of Fig. 19-12(a) determine the resistor and inductor voltages at the following frequencies: $f = f_T$; $f = 10 f_T$; and $f = 0.1 f_T$.

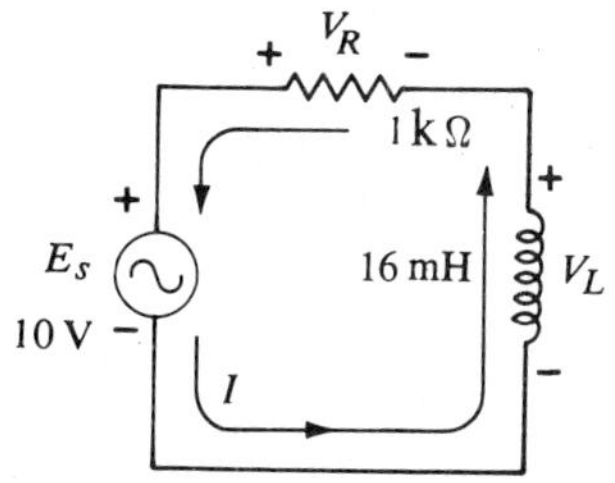

f		Z	I	V_L	V_R
$0.1\,f_T =$	1000 Hz	1 kΩ	10 mA	1 V	10 V
$f_T =$	10000 Hz	1.4 kΩ	7 mA	7 V	7 V
$10\,f_T =$	100000 Hz	10 kΩ	1 mA	10 V	1 V

(a)   (b)

FIGURE 19-12

Solution: (a) $f_T = \dfrac{R}{2\pi L} = \dfrac{10^3}{2\pi \times 16 \times 10^{-3}} = 10$ kHz

At $f = f_T$, $X_L = R = 1000$ Ω. Thus

$$Z = \sqrt{2}\,R = 1410\ \Omega$$

The current I is therefore 10 V/1410 Ω $\simeq$ 7 mA. Thus,

$$V_R = 7\text{ mA} \times 1000\ \Omega = 7\text{ V}$$
$$V_L = 7\text{ mA} \times 1000\ \Omega = 7\text{ V}$$

Note that $\sqrt{V_R^2 + V_L^2} \simeq 10$ V $= E_S$.

(b) At $f = 10\,f_T$ the value of X_L will be 10 times larger than its value at $f = f_T$. Thus,

$$X_L = 10\text{ k}\Omega$$
$$Z = \sqrt{1\text{ k}\Omega^2 + 10\text{ k}\Omega^2} = 10.05\text{ k}\Omega \approx 10\text{ k}\Omega$$

Therefore,

$$I \simeq \dfrac{10\text{ V}}{10\text{ k}\Omega} = 1\text{ mA}$$
$$V_R = 1\text{ mA} \times 1\text{ k}\Omega = \mathbf{1\text{ V}}$$
$$V_L = 1\text{ mA} \times 10\text{ k}\Omega = \mathbf{10\text{ V}}$$

(c) At $f = 0.1\,f_T$ the value of X_L will be only 0.1×1 kΩ $= 100$ Ω. Thus,

$$Z = \sqrt{1\text{ k}\Omega^2 + 100^2} = 1005\ \Omega \simeq 1\text{ k}\Omega$$
$$I = 10\text{ V}/1\text{ k}\Omega = 10\text{ mA}$$
$$V_R = 10\text{ mA} \times 1\text{ k}\Omega = \mathbf{10\text{ V}}$$
$$V_L = 10\text{ mA} \times 100\ \Omega = \mathbf{1\text{ V}}$$

The results of the three cases are tabulated in part (b) of Fig. 19–12. Note that at a frequency of $10 f_T$ most of the voltage is across the inductor, while at a frequency of $0.1 f_T$ most of the voltage is across the resistor.

EXAMPLE 19.6 A certain series RL circuit has $L = 300\ \mu\text{H}$. The circuit is required to operate at frequencies in the range of 100 kHz to 2 MHz. Over this range of frequencies, it is desired to keep the circuit phase angle below 3°. Determine an appropriate value for R.

Solution: The greatest phase angle will occur at the highest frequency of operation. Therefore R should be chosen so that $\theta \leq 3°$ at $f = 2$ MHz. At 2 MHz, the inductive reactance is

$$X_L = 2\pi \times 2 \times 10^6 \times 300 \times 10^{-6}$$
$$= 3.8\ \text{k}\Omega$$

Since $\tan \theta = X_L/R$, then

$$\tan 3° = \frac{3.8\ \text{k}\Omega}{R}$$

Thus,

$$R = \frac{3.8\ \text{k}\Omega}{0.053} = 71.7\ \text{k}\Omega$$

Any value of $R \geq 71.7\ \text{k}\Omega$ will be okay since a greater value of R will cause θ to be even smaller.

19.7 RL Circuit as a High-Pass Filter

A series RL circuit can function as a high-pass filter if the output voltage is taken across the *inductor*. This is shown in Fig. 19–13 where the input voltage E_S is applied across the series combination and the output voltage is taken across X_L. (Assume the inductor's own effective resistance is small enough to be neglected.)

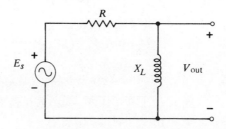

FIGURE 19–13 *Series RL circuit used as a high-pass filter.*

The output V_{out} can be determined using either Ohm's law or the voltage divider formula. In either case,

$$V_{out} = \frac{X_L}{Z} \times E_S$$

$$= \left(\frac{X_L}{\sqrt{R^2 + X_L^2}}\right) \times E_S \qquad (19\text{–}8)$$

At high frequencies the inductive reactance X_L will become very large compared to R. As a result, most of the input voltage will appear across the inductor. This can be seen in (19–8) if we neglect the resistance of R.

$$V_{out} \simeq \frac{X_L}{\sqrt{X_L^2}} \times E_S \simeq E_S \quad \text{(at high frequencies)}$$

At very low frequencies X_L becomes very small (approaches a short circuit). Thus, for low enough frequencies most of the input voltage will appear across R. Of course, at zero frequency (dc) the voltage across X_L will be zero since the inductor will be a short circuit.

At the transition frequency f_T, the voltages across R and X_L will be equal since $X_L = R$. From (19–8) we can determine V_{out} at $f = f_T$.

$$V_{out} = \left(\frac{X_L}{\sqrt{R^2 + X_L^2}}\right) \times E_S = \left(\frac{R}{\sqrt{R^2 + R^2}}\right) \times E_S = \frac{\cancel{R}}{\sqrt{2}\cancel{R}} \times E_S$$

$$= 0.707 \, E_S \text{ (at } f = f_T)$$

At frequencies above f_T, V_{out} increases; at frequencies below f_T, V_{out} decreases toward zero.

Frequency-Response Curve Figure 19–14 shows the frequency-response curve for an RL high-pass filter. This curve shows how the attenuation of this filter varies with frequency. Once again the attenuation ratio V_{out}/V_{in} is given in dB and a log scale is used for frequency. That is,

$$\text{Attenuation in dB} = 20 \log_{10}\left(\frac{V_{out}}{V_{in}}\right)$$

$$= 20 \log_{10}\left(\frac{X_L}{\sqrt{R^2 + X_L^2}}\right) \qquad (19\text{–}9)$$

Of course, since X_L varies with frequency, the attenuation will also vary. The curve in Fig. 19–14 can be obtained either experimentally or by using Eq. (19–9) for various frequencies.

In examining the curve in Fig. 19–14 several features should be pointed out.

(1) At frequencies above $10 f_T$ (1 kHz) the attenuation is essentially 0 dB, which means $V_{out} = V_{in}$.

(2) At frequencies below f_T (100 Hz) the attenuation increases rapidly in the negative direction. At $0.1 f_T$ (10 Hz), the attenuation is -20 dB representing a ratio of 0.1 ($V_{out} = 0.1 \, V_{in}$) at this frequency. At 1 Hz ($0.01 f_T$) the attenuation is -40 dB ($V_{out} = 0.01 \, V_{in}$) and so on.

(3) At $f = f_T$ the attenuation is -3 dB, representing an attenuation ratio of 0.707. In other words, the attenuation is down 3 dB at $f = f_T$. Recall from our

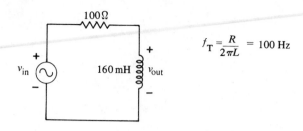
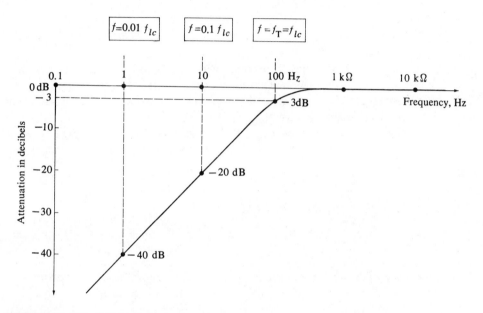

FIGURE 19-14 RL *high-pass filter and its frequency-response curve.*

discussion of *RC* filters that the transition frequency for a high-pass filter is also called the *low cutoff frequency* f_{lc} (also the lower 3-dB frequency).

All series *RL* high-pass filters have a frequency-response curve with the same shape as the one in Fig. 19-14. Varying the *R* and *L* values results in different values of the low cutoff frequency ($f_{lc} = f_T = R/2\pi L$), and simply shifts the curve left or right. The value of f_{lc} is chosen to satisfy the requirements for which the filter is being designed.

EXAMPLE 19.7 Design an *RL* filter that will attenuate any frequencies *below* 120 Hz by at least −20 dB. The filter is to draw a maximum current of 50 mA from a 10-V source.

Solution: The attenuation of an *RL* high-pass filter will be −20 dB at $f = 0.1 f_{lc} = 0.1 f_T$. Thus,

$$0.1 f_{lc} = 120 \text{ Hz}$$

$$f_{lc} = 1200 \text{ Hz} = \frac{R}{2\pi L}$$

Since $I_{max} = 50$ mA at 10 V, then

$$R = \frac{10 \text{ V}}{50 \text{ mA}} = 200 \text{ }\Omega$$

Thus,

$$\frac{200 \text{ }\Omega}{2\pi L} = 1200 \text{ Hz}$$

Solving for L,

$$L = \frac{200}{2\pi \times 1200} = 27 \text{ mH}$$

19.8 RL Circuit as a Low-Pass Filter

As we have seen, in a series RL circuit most of the applied voltage will appear across R at low frequencies. As such, we can use the series RL circuit as a *low-pass* filter if the output voltage is taken across R. This arrangement is shown in Fig. 19–15.

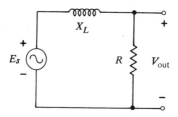

FIGURE 19–15 *Series* RL *low-pass filter.*

The output voltage V_{out} is determined using

$$V_{out} = \left(\frac{R}{\sqrt{R^2 + X_L^2}}\right) \times E_S \qquad (19\text{-}10)$$

At low frequencies X_L is almost zero and can be neglected compared to R. Thus,

$$V_{out} \simeq \frac{R}{\sqrt{R^2}} \times E_S \simeq E_S \qquad \text{(at } low \text{ frequencies)}$$

which indicates very little attenuation at low frequencies. At high frequencies X_L becomes very large compared to R so that very little voltage reaches the output. At high enough frequencies $V_{out} \simeq 0$.

The RL low-pass filter is the same circuit as the RL high-pass filter except that the outputs are taken across different points in each case. Thus, it can also be stated for the low-pass filter that $V_L = V_R$ at $f = f_T$. From Eq. (19-10) we can calculate V_{out} at the transition frequency since $X_L = R$.

$$V_{out} = \left(\frac{R}{\sqrt{R^2 + X_L^2}}\right) \times E_S = \left(\frac{R}{\sqrt{R^2 + R^2}}\right) \times E_S = \frac{1}{\sqrt{2}} \times E_S$$
$$= 0.707 \ E_S \text{ (at } f = f_T)$$

At frequencies $>f_T$, V_{out} decreases and at frequencies $<f_T$, V_{out} increases.

Frequency-Response Curve The frequency-response curve for an *RL* low-pass filter is shown in Fig. 19–16. Note that the R and L values are the same as those used in the high-pass filter of Fig. 19–14. The curve in Fig. 19–16 shows how the low-pass filter attenuation in dB varies with frequency. The attenuation in dB is found from

$$20 \log_{10} \left(\frac{V_{out}}{V_{in}} \right) = 20 \log_{10} \left(\frac{R}{\sqrt{R^2 + X_L^2}} \right) \tag{19-11}$$

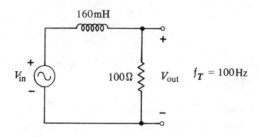

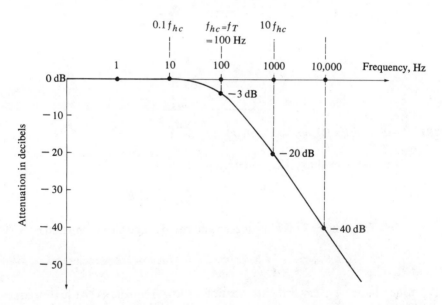

FIGURE 19–16 *RL low-pass filter and its frequency-response curve.*

The value of attenuation varies because X_L varies with frequency.

Several important features can be pointed out concerning the frequency-response curve of Fig. 19–16.

(1) At frequencies below $0.1 f_T$ (10 Hz) the attenuation is essentially 0 dB ($V_{out} = V_{in}$).

(2) Above $f = f_T$ (100 Hz) the attenuation increases rapidly in the negative direction. At $f = 10 f_T$ (1000 Hz) the attenuation is -20 dB ($V_{out}/V_{in} = 0.1$). At 10 kHz the attenuation is -40 dB and so on.

(3) At $f = f_T$ the attenuation is -3 dB ($V_{out}/V_{in} = 0.707$) as it also was for the high-pass filter. For the low-pass filter f_T is often called the *high cutoff frequency* f_{hc} (also the upper 3-dB frequency).

All series *RL* low-pass filters have a frequency-response curve with the same shape as that in Fig. 19–16. Varying the *R* and *L* values results in a different value of f_{hc} and shifts the curve left or right. The value of f_{hc} is chosen to satisfy the requirements for which the filter is being designed.

Low-Pass Filter at dc Like the *RC* low-pass filter, the *RL* low-pass filter *always* passes dc with no attenuation. This is because the inductor acts as a short circuit (after 5 τ) so that all the source voltage is across *R*. Of course, if the coil's own effective resistance R_e is significant compared to *R*, then for dc V_{out} will be less than E_S because of the voltage divider action between R_e and *R*.

EXAMPLE 19.8 Design an *RL* filter that will attenuate any frequencies above 15 kHz by at least -20 dB. Use a 30-mH inductor.

Solution: Since the higher frequencies are to be attenuated, a low-pass filter is required. The attenuation will be -20 dB at $f = 10 f_{hc}$. Thus,

$$10 f_{hc} = 15 \text{ kHz}$$

This gives

$$f_{hc} = \frac{R}{2\pi L} = 1500 \text{ Hz}$$

$$\therefore R = 1500 \times 2\pi \times 30 \times 10^{-3} \approx \mathbf{284 \ \Omega}$$

19.9 Comparison of RC and RL Filters

We have seen how both *RC* and *RL* circuits can be used as filters. Figure 19–17 is a comparison of the two types of circuits. It is important to note the differences between the two. For the *RC* circuit, the resistor voltage is used as the *high-pass* output while for the *RL* circuit, the resistor voltage is the *low-pass* output. In the *RC* circuit the reactance voltage is used as the *low-pass* output while in the *RL* circuit the reactance voltage is the *high-pass* output.

The *RC* and *RL* filters have the same general frequency-response curves. This can be verified by comparing Figs. 18–15 and 19–14 for high-pass filters and comparing Figs. 18–18 and 19–16 for low-pass filters. Either type of filter circuit (*RC* or *RL*) can be used for a given application. However, because of the wider range of capacitor values that are available and because of the inductor losses at high frequencies (due to R_e), *RC* filters are more common than *RL* filters.

An application where an *RL* filter is more useful than an *RC* filter occurs when a resistive load is to be protected from high-frequency voltages (Fig. 19–18). In such applications, the inductor is referred to as a *choke*. A choke is

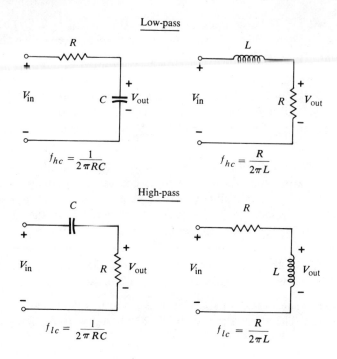

FIGURE 19–17 *Comparison of RC and RL filter circuits.*

an inductance placed in series with a resistor to prevent any high frequencies in the input voltage from developing any appreciable voltage across the resistor. When used as a choke, the inductor value is chosen so that X_L is at least *ten* times greater than R_L over the range of frequencies which are to be choked (kept from reaching R_L).

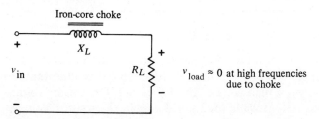

FIGURE 19–18 *When used as a high-frequency choke, the inductor has $X_L \gg R_L$ so that $V_R \simeq 0$ at the high frequencies.*

19.10 R and X_L in Parallel

The analysis of a parallel combination of R and X_L proceeds along the same lines as our previous analysis of parallel RC circuits. Figure 19–19(a) shows an ac source connected to a parallel RL circuit so that $V_L = V_R = E_S$. Thus,

$$I_R = \frac{E_S}{R} \quad \text{and} \quad I_L = \frac{E_S}{X_L} \tag{19-12}$$

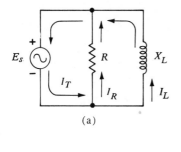
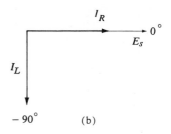
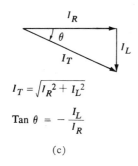
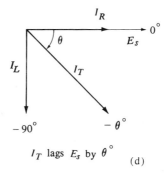

FIGURE 19–19 (a) *Parallel RL circuit.* (b) *Phasor diagram.* (c) *Find* I_T *phasor.* (d) *Complete phasor diagram showing* I_T.

(We will assume that the effective resistance R_e of the inductor is small enough to be neglected.)

The total current I_T is the total current supplied by the source. The value of I_T can be found by taking the *vector* sum of I_R and I_L. That is, $\overline{I_T} = \overline{I_R} + \overline{I_L}$. Part (b) of Fig. 19–19 shows the phasor representations of $\overline{E_S}$, $\overline{I_R}$, and $\overline{I_L}$. $\overline{E_S}$ is used as the reference phasor at 0° since $\overline{E_S}$ is common to all the circuit elements. $\overline{I_R}$ is shown in phase with $\overline{E_S}$ at 0°. $\overline{I_L}$ will *lag* $\overline{E_S}$ by 90°; therefore its phasor is drawn at −90°. Thus, $\overline{I_L}$ also lags $\overline{I_R}$ by 90°.

If I_R and I_L are known, the total current can be determined using vector addition as illustrated in part (c) of Fig. 19–19. The magnitude of $\overline{I_T}$ is easily obtained using Pythagorean's theorem

$$I_T = \sqrt{I_R^2 + I_L^2} \qquad (19\text{–}13)$$

As the phasor diagram in 19–19(c) shows, the $\overline{I_T}$ phasor *lags* $\overline{I_R}$ by the angle θ. In other words I_T is at a negative phase angle relative to the 0° reference. The value of θ can be determined using

$$\tan \theta = \left(-\frac{I_L}{I_R}\right) \qquad (19\text{–}14)$$

The negative sign indicates that θ is a lagging (negative) phase angle.

The complete phasor diagram for the parallel *RL* circuit is shown in Fig. 19–19(d). It shows that the total current I_T lags the applied source voltage by the angle θ. This is the same relationship that existed in the series *RL* circuit. *In any inductive circuit, the current delivered by the source lags the source voltage.*

Chapter Nineteen

EXAMPLE 19.9 A parallel RL circuit has $R = 2$ kΩ and $X_L = 1.5$ kΩ (a) Find I_T for $E_S = 120$ V; (b) determine θ.

Solution: (a) The circuit is drawn in Fig. 19–20(a).

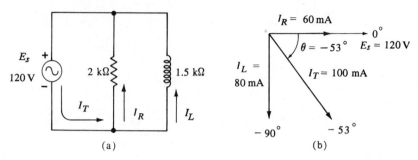

FIGURE 19–20

Since $E_S = 120$ V we can calculate I_R and I_L.

$$I_R = \frac{120 \text{ V}}{2 \text{ k}\Omega} = 60 \text{ mA}$$

$$I_L = \frac{120 \text{ V}}{1.5 \text{ k}\Omega} = 80 \text{ mA}$$

To find $\overline{I_T}$ we must add $\overline{I_R}$ and $\overline{I_L}$ vectorially. The phasors are drawn in part (b) of the figure.

$$I_T = \sqrt{I_R^2 + I_L^2}$$
$$= \sqrt{60^2 + 80^2} = \textbf{100 mA}$$

(b) From the phasor diagram

$$\tan \theta = -\left(\frac{I_L}{I_R}\right) = -1.33$$

so that $\theta = -53°$. Thus, I_T lags E_S by 53°.

In the parallel RL circuit as was the case for the series RL circuit, the total current lags the applied voltage. However, in the parallel circuit the phase angle θ is *negative* while for the series circuit θ is *positive*. The reason for this is that our reference is different in each case. In the series circuit the current was chosen as the 0° reference; in the parallel circuit the applied voltage was chosen as the 0° reference. One must be aware of the reference used in each case.

19.11 Impedance of Parallel RL Circuit

The total current in a parallel RL circuit can be determined as shown in the last section. Once I_T is known it is relatively easy to find the total impedance Z of the parallel combination. Since Z is the total opposition of the circuit:

$$Z = \frac{E_S}{I_T} \tag{19-15}$$

For example, in the circuit of Fig. 19–20 the applied voltage is 120 V and I_T was determined to be 100 mA. Thus,

$$Z = \frac{120 \text{ V}}{100 \text{ mA}} = 1.2 \text{ k}\Omega$$

This impedance is the combined opposition of the 2-kΩ resistor and the 1.5-kΩ inductive reactance. For the parallel RL circuit Z is *not* equal to the vector sum of R and X_L, because R and X_L are *not* in series. Instead, Z is found by first calculating I_T and then using Eq. (19–15).

EXAMPLE 19.10 What is the total Z of a 600-ohm resistor in parallel with 400-ohm inductive reactance?

Solution: To find Z we must apply a voltage to the parallel RL circuit and determine I_T. For convenience, a source voltage of 1200 V can be used. Thus,

$$I_R = \frac{1200 \text{ V}}{600 \text{ }\Omega} = 2 \text{ A}$$

$$I_L = \frac{1200 \text{ V}}{400 \text{ }\Omega} = 3 \text{ A}$$

$$I_T = \sqrt{I_R^2 + I_L^2} = \sqrt{4 + 9} = 3.6 \text{ A}$$

The impedance is therefore

$$Z = \frac{E_S}{I_T} = \frac{1200 \text{ V}}{3.6 \text{ A}} = 333 \text{ }\Omega$$

Keep in mind that this is the impedance value only at the frequency at which $X_L = 400$ ohms. As f changes, X_L will change causing Z to take different values.

Polar Form of Z We can express the impedance of a parallel RL circuit as a vector in polar form. In Fig. 19–21 the phasor diagram for the parallel RL circuit is shown. The source voltage phasor is $E_S/\underline{0°}$ and the total current phasor is $I_T/\underline{-\theta}$. Thus,

$$\overline{Z} = \frac{\overline{E_S}}{\overline{I_T}} = \frac{E_S/\underline{0°}}{I_T/\underline{-\theta}} = \frac{E_S/\underline{0° - (-\theta)}}{I_T}$$

$$= Z/\underline{+\theta}$$

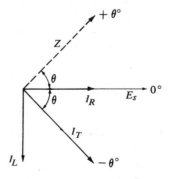

FIGURE 19–21 *Impedance vector has a positive angle for a parallel RL circuit.*

This shows that the impedance vector for the parallel *RL* circuit has a *positive angle*. This was also true for the series *RL* circuit. In general, then, it can be stated that *for a series or parallel* RL *circuit or any inductive circuit, the impedance vector has a positive angle* $(Z/+\theta)$.

EXAMPLE 19.11 A parallel *RL* circuit has $\overline{Z} = 15\ \Omega/26.5°$. Determine the *R* and X_L values.

Solution: $\overline{Z}$ is *not* the vector sum of *R* and X_L in a parallel circuit. To find *R* and X_L we can apply a voltage to the circuit and calculate I_T. From I_T we can determine I_R and I_L from which we can then determine *R* and X_L.

For convenience, let $\overline{E_S} = 15\ V/0°$. Thus,

$$\overline{I_T} = \frac{\overline{E_S}}{\overline{Z}} = \frac{15\ V/0°}{15\ \Omega/+26.5°}$$

$$= 1\ A/-26.5°$$

This $\overline{I_T}$ phasor is shown in Fig. 19-22. I_R and I_L can be found as follows:

$$I_R = I_T \cos(26.5°)$$
$$= 1\ A \times 0.9 = 0.9\ A$$

$$I_L = I_T \sin(26.5°)$$
$$= 1\ A \times 0.45 = 0.45\ A$$

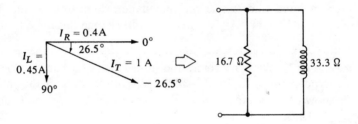

FIGURE 19-22

Thus,

$$R = \frac{E_S}{I_R} = \frac{15\ V}{0.9\ A} = 16.7\ \text{ohms}$$

$$X_L = \frac{E_S}{I_L} = \frac{15\ V}{0.45\ A} = 33.3\ \text{ohms}$$

19.12 Frequency Effects on Parallel RL

Figure 19-23 contains graphs which show how Z and θ vary with frequency in a parallel *RL* circuit. As the graphs show, at low frequencies the impedance Z is essentially equal to X_L, since X_L will be much lower than R at low frequencies. The phase angle at low frequencies will be $-90°$ because the inductive current I_L will be much larger than I_R, thus making I_T almost purely inductive current.

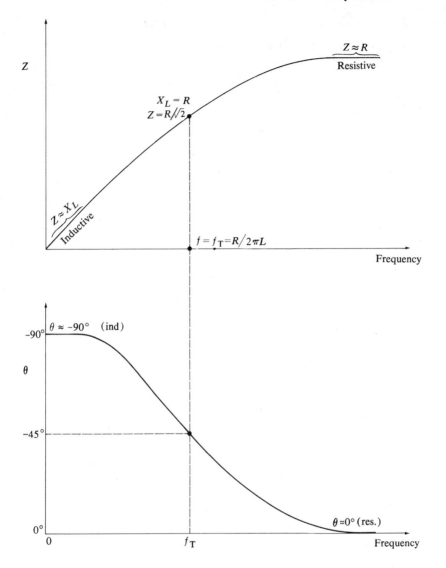

FIGURE 19–23 Z and θ variations with frequency for a parallel RL circuit.

At high frequencies, Z is essentially equal to R since X_L will become very large compared to R and will draw negligible current. θ will approach 0° at high frequencies because the resistor current I_R will be much larger than I_L, thus making I_T almost purely resistive current.

At the transition frequency $f = f_T$, the X_L and R branches draw equal currents. Thus, $\theta = -45°$ since $\tan \theta = -(I_L/I_R) = -1$. At $f > f_T$, the value of X_L is larger than R so that $I_R > I_L$ and the circuit is more resistive ($\theta < -45°$). At $f < f_T$, the value of X_L is less than R so that $I_L > I_R$ and the circuit is more inductive ($\theta > -45°$).

In summary, Z increases and θ decreases with increases in frequency for a parallel *RL* circuit. Recall that in a *series* RL circuit Z increases with frequency but θ also increases with frequency (see Fig. 19–11).

19.13 Comparison of Series RL and Parallel RL

In both series and parallel RL circuits, the total current *lags* the applied voltage by a phase angle between 0° and 90°. The impedance of either circuit is of the form $Z\underline{/+\theta}$. As such, as far as the source is concerned, it is impossible to distinguish between the two circuits at a given frequency. However, as we have seen, the two circuits behave differently with changes in frequency. The phase angle of the series RL approaches 90° at *high* frequencies while the phase angle of the parallel RL approaches 0° at high frequencies. In addition, at high frequencies the parallel circuit impedance approaches R while the series circuit impedance increases indefinitely.

At *low* frequencies the phase angle of the series RL approaches 0° while the parallel circuit phase angle approaches 90°. In addition, at low frequencies the parallel circuit impedance approaches 0 while the series circuit impedance approaches R. All of these comparisons are summarized in Table 19–1. Note also from the table that $\theta = 45°$ at the transition frequency.

TABLE 19–1 *Comparison of Series and Parallel* RL

	Parallel RL	Series RL
Z at low frequencies ($f < 0.1\,f_T$) θ at low frequencies	$\simeq X_L$ $\simeq 90°$ } inductive	$\simeq R$ $\simeq 0°$ } resistive
Z at high frequencies ($f > 10\,f_T$) θ at high frequencies	$\simeq R$ $\simeq 0°$ } resistive	$\simeq X_L$ $\simeq 90°$ } inductive
Z at $f = f_T$ θ at $f = f_T$	$R/\sqrt{2}$ 45°	$\sqrt{2}\,R$ 45°

19.14 Practical Inductor in a Parallel RL Circuit

In our study of the parallel RL circuit, we have assumed that the inductor's effective series resistance R_e was small enough to neglect. Although this is often a valid assumption, there are many cases where R_e will have a significant effect. Figure 19–24 shows a parallel RL circuit with a practical inductor.

The inductive branch of the parallel circuit is no longer a *pure* inductance. Thus, I_L will *not* lag E_s by 90°. The X_L and R_e series combination will have an impedance $Z_L\underline{/\alpha}$ where the phase angle α will be something *less* than 90°. Thus, the phasor for I_L will be at an angle of $-\alpha$ relative to the source voltage. This is shown in the phasor diagram of Fig. 19–24(b). Note that $\mathbf{I_L} = \mathbf{E}_s/\mathbf{Z_L}$.

The total current $\overline{I_T}$ delivered by the source is equal to the vector sum of $\overline{I_R}$ and $\overline{I_L}$. This vector addition is performed in part (c) of the figure by placing the $\overline{I_R}$ and $\overline{I_L}$ phasors head-to-tail. The vector triangle formed by $\overline{I_T}$, $\overline{I_R}$, and $\overline{I_L}$ is *not* a right triangle because $\overline{I_L}$ is not at 90° relative to the 0° reference. As such, $\overline{I_T}$ cannot be calculated by using Pythagorean's theorem. The method for calculating $\overline{I_T}$ is somewhat more complicated. Later we will study the mathematical procedure for solving problems such as this.

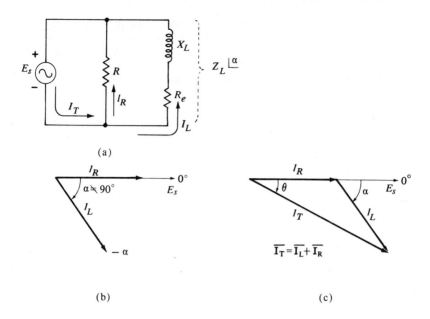

FIGURE 19-24 *The resistance of a practical inductor causes the current in the inductor branch to lag E_S by less than 90°.*

Chapter Summary

1. In a series *RL* circuit the amplitude (rms or peak) of the applied voltage does *not* equal the sum of the amplitudes of the resistor and inductor voltages.

2. In a series *RL* circuit the E_S phasor is equal to the vector sum of the V_R and V_L phasors: $\overline{E_S} = \overline{V_R} + \overline{V_L}$.

3. In a series *RL* circuit, $\bar{I}$ and $\overline{V_R}$ always *lag* $\overline{E_S}$, and $\overline{V_L}$ always leads $\overline{E_S}$.

4. For a series *RL* circuit the total impedance is equal to the *vector* sum of R and X_L. That is, $\overline{Z} = \overline{R} + \overline{X_L}$.

5. Table 19-2 summarizes the basic relationships for a series *RL* circuit.

TABLE 19-2

$$Z = \sqrt{R^2 + X_L^2}$$
$$\bar{I} = \overline{E_S}/\overline{Z}$$
$$V_R = IR$$
$$V_L = IX_L$$
$$E_S = \sqrt{V_R^2 + V_C^2}$$
$$\theta = \arctan\left(\frac{X_L}{R}\right) = \arctan\left(\frac{V_L}{V_R}\right)$$

Note: θ is angle by which I lags E_S.

6. A practical inductor possesses both inductance and resistance. The *effective resistance* R_e of a coil depends on several factors: wire resistance, skin effect, eddy currents, and hysteresis. The value of R_e usually is higher at higher frequencies.

7. In a series RL circuit, the impedance Z and phase angle θ increase as the applied frequency increases.

8. The transition frequency of an RL circuit is the frequency at which $X_L = R$. The value of f_T is $R/2\pi L$.

9. At very low frequencies ($f < 0.1\ f_T$) a series RL circuit has $\theta \simeq 0°$, $Z \simeq R$, and behaves as purely resistive.

10. At very high frequencies ($f > 10\ f_T$) a series RL circuit has $\theta \simeq 90°$, $Z \simeq X_L$, and behaves as purely inductive.

11. At $f = f_T$, the phase shift of a series RL circuit is 45° and $V_L = V_R = 0.707\ E_S$.

12. A series RL circuit can be used as a high-pass filter if the output is taken as the inductor voltage. It can be used as a low-pass filter if the output is taken as the resistor voltage.

13. The low cutoff frequency f_{lc} of an RL high-pass filter is the frequency below which the filter greatly attenuates the input voltage. $f_{lc} = f_T = R/2\pi L$. At f_{lc} the attenuation is -3 dB.

14. The high cutoff frequency f_{hc} of an RL low-pass filter is the frequency above which the filter greatly attenuates the input. f_{hc} also equals f_T. At f_{hc} the attenuation is -3 dB.

15. In a parallel RL circuit, the total current equals the vector sum of the resistor and inductor branch currents: $\overline{I_T} = \overline{I_R} + \overline{I_L}$.

16. In a parallel RL circuit (or any inductive circuit), I_T always *lags* the applied voltage.

17. For a parallel RL circuit (or any inductive circuit) the impedance vector is of the form $Z/{+\theta}$.

18. Table 19–3 summarizes the basic relationships for a parallel RL circuit.

TABLE 19–3

$$I_R = \frac{E_S}{R};\ I_L = \frac{E_S}{X_L}$$

$$I_T = \sqrt{I_R^2 + I_L^2}$$

$$\theta = \arctan\left(\frac{I_L}{I_R}\right)$$

Note: θ is angle by which I_T lags E_S.

$$\overline{Z} = \frac{\overline{E_S}}{\overline{I_T}}$$

19. For the parallel RL circuit, Z *increases* and θ *decreases* as frequency *increases*.

20. At very low frequencies ($< 0.1\, f_T$) the parallel RL circuit has $Z \simeq X_L$ and $\theta \simeq -90°$, and behaves as almost purely inductive. At very high frequencies ($> 10\, f_T$) it has $Z \simeq R$ and $\theta \simeq 0°$, and behaves purely resistive.

21. Table 19-1 (sec. 19.13) gives a comparison of series and parallel RL circuits.

22. If the inductor in a parallel RL circuit has a significant effective resistance R_e, the current through the inductive branch will lag E_s by less than 90°.

Questions/Problems

Sections 19.2–19.3

19-1 In a series RL circuit a VOM is used to measure the rms values of $V_R = 12$ V and $V_L = 5$ V. Why isn't the rms source voltage E_S equal to 17 V?

19-2 Determine E_S for the previous problem. Also determine the phase angle between the circuit current and E_S.

19-3 A series RL circuit has $R = 500$ ohms and $L = 1$ H. (a) Find the source voltage required to produce $I = 200$ mA at 160 Hz; (b) draw a complete phasor diagram showing all voltage and current phasors.

Section 19.4

19-4 Draw the impedance vector diagram for the circuit of the previous problem and determine $\overline{Z}$ (magnitude and angle).

19-5 An RL series circuit has $\overline{Z} = 1000\, \Omega/\underline{30°}$ at a certain frequency. If the applied voltage is 60 V, draw the phasors representing $\overline{I}$ and $\overline{E_S}$.

19-6 A 500-ohm resistor is in series with a 10-mH inductor. A 5-kHz, 24-V source is applied to the circuit.
  a. Determine the circuit impedance.
  b. Determine I, V_R, and V_L.
  c. Determine the phase angle between $\overline{I}$ and $\overline{E_S}$.
  d. Draw the complete phasor diagram.

Section 19.5

19-7 The setup of Fig. 19-9 is used to measure L and R_e for a certain coil. The waveforms of v_{coil} and v_{RS} are displayed on an oscilloscope and appear as shown in Fig. 19-25. If $R_S = 200$ ohms, determine L and R_e.

19-8 Some of the following statements are *false*. Modify them to make them correct.
  a. In a series RL circuit the current always lags the applied voltage by 90°.
  b. If $V_R = 5$ V and $V_L = 12$ V then the ac source voltage must be 17 V.
  c. An X_L of 20 ohms in series with an R of 20 ohms has the same opposition to current as a single 40-ohm resistor.
  d. The resistor voltage in a series RL circuit is always lagging the source voltage.

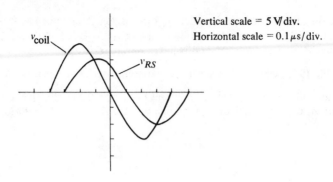

FIGURE 19-25

e. If a practical coil has $\overline{Z} = 1000$ ohms$/60°$, its effective resistance is 1000 ohms.
f. The total opposition of a practical coil increases with frequency only because X_L increases.
g. Eddy currents occur only in air-core coils.
h. Hysteresis effects produce a greater power loss at low frequencies.
i. The skin effect produces a greater R_e at high frequencies.

Section 19.6

19-9 A sine wave of voltage at a certain frequency is applied to a series *RL* circuit. Indicate how Z and θ will change (increase, decrease, no change) as each of the following changes is made in the circuit:
 a. a decrease in frequency
 b. an increase in resistance
 c. a decrease in inductance
 d. an increase in input amplitude
 e. a two-fold increase in R and a two-fold increase in inductance

19-10 A 5-kΩ resistor is in series with a 100-mH inductor. (a) Calculate Z and θ at $f = 80$ Hz, 800 Hz, 8 kHz, and 80 kHz; (b) plot these values versus f (use log scale for f); (c) determine f_T.

19-11 Describe a method for determining f_T for a series *RL* circuit using only a variable frequency generator and an ac voltmeter.

19-12 A series *RL* circuit is to operate at frequencies in the range of 100 Hz-600 kHz. It is required that the current lags the source voltage by at least 30° over this range. If $R = 1000$ ohms, choose an appropriate value for L.

19-13 Design a series *RL* circuit that has $f_T = 15$ kHz and draws a current of 10 mA when a 6-V, 100-Hz voltage is applied to it.

Sections 19.7-19.8

19-14 Choose appropriate values for an *RL* high-pass filter which is to attenuate all frequencies below 200 Hz by at least -20 dB. Sketch the filter's response curve.

19-15 A low-pass *RL* filter has $R = 1200$ ohms and $L = 0.2$ H. Indicate which of the following frequencies will be passed with virtually no attenuation: $f = 0$, 100 Hz, 2 kHz, 40 kHz. Sketch the response curve for this filter.

19-16 A certain RL filter has measurements made on it that are recorded in the following data table. From the data determine:
a. type of filter (low- or high-pass)
b. filter cutoff frequency
c. values of R and L

f	E_S	V_{out}	I
20 Hz	20 V	0.2 V	$\sim$5 mA
200 Hz		2.0 V	$\sim$5 mA
2 kHz		14.1 V	3.5 mA
20 kHz		$\sim$20.0 V	$\sim$0.5 mA
200 kHz		$\sim$20.0 V	0.05 mA

19-17 Each of the signals in Fig. 19-26 is applied to a high-pass filter with $R = 190$ ohms and $L = 10$ mH. Sketch the filter output for each case.

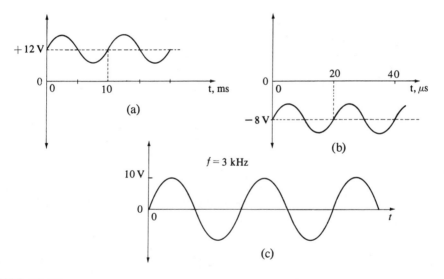

FIGURE 19-26

19-18 Repeat 19-17 for a low-pass filter with the same R and L values.

Section 19.9

19-19 What value of inductor can be used as a *choke* for a 10-kΩ resistive load at frequencies > 50 kHz?

19-20 We have seen that both series RC and series RL circuits can be arranged to operate as high-pass or low-pass filters. Very often the output of a filter is connected to a load resistance R_L (see Fig. 19-27). For each of the four filter circuits discussed in chaps. 18 and 19, comment on the effects of R_L on the filter operation (attenuation, cutoff frequency).

Sections 19.10–19.14

19-21 A parallel RL circuit has $R = 4$ kΩ and $X_L = 2$ kΩ. Find (a) I_T if $E_S = 32$ V; (b) θ; (c) $\overline{Z}$ (polar form).

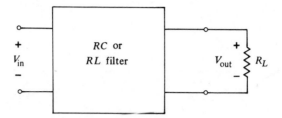

FIGURE 19-27

19-22 For the circuit of the previous problem indicate what would happen to I_T and θ for each of the following circuit changes:
   a. a decrease in frequency
   b. an increase in R
   c. an increase in L
   d. a decrease in E_S
   e. adding another resistor in parallel
   f. adding another inductor in parallel

19-23 For the circuit of problem 19-21 sketch the graphs of Z and θ versus frequency being sure to show the values at the transition frequency.

19-24 When a 50-V, 200-Hz voltage is applied to an unknown circuit, the total current is 8 mA and *lags* the source voltage by 26.5°. Determine R and L values for a *series RL* circuit equivalent to the unknown circuit.

19-25 Repeat 19-24 using a *parallel RL* equivalent circuit.

19-26 The data in the table below are taken for a certain *RL* circuit. From the data determine: (a) if it is a series or a parallel *RL* circuit; (b) the R and L values.

Frequency	E_S	I_T
10 Hz	20 V	10 A
100 Hz		1 A
1 kHz		101 mA
10 kHz		14.1 mA
100 kHz		10.1 mA
1 MHz		10 mA

19-27 Consider each of the following statements. For each one indicate which types of circuit it pertains to (series *RC*, parallel *RC*, series *RL*, parallel *RL*, or any combination of these):
   a. Z decreases as f increases
   b. draws its maximum current at low frequencies
   c. total current always lags applied voltage
   d. total current is always limited by R
   e. I_T is out of phase with E_S by 45° at the transition frequency
   f. f_T increases as R increases

g. impedance vector always has a positive angle
h. increasing R causes the circuit to become more inductive
i. blocks current at high frequency
j. $Z \simeq R$ at high frequencies
k. $Z \simeq R$ at low frequencies
l. I_T always leads I_R
m. V_R always lags E_S

CHAPTER 20

Vector Algebra for AC Circuits

20.1 Introduction

In our analysis of simple *RC* and *RL* circuits it was necessary to deal with vector quantities in order to solve the circuit. The vector operations in those circuits were simple enough to perform using right-triangle trigonometry. Unfortunately, as we begin to analyze more complex ac circuits, the vector operations also become somewhat more complex. As such, this chapter will introduce a mathematical technique which can be used to analyze any ac circuit. This technique, which utilizes the *j operator*, is especially useful in analyzing series-parallel circuits that contain both resistance and reactance.

Although the mathematical techniques are helpful in analyzing complex circuits, care must be taken to insure that the mathematics does not obscure the important concepts of circuit behavior. This is one reason why the simple *RC* and *RL* circuits were analyzed using only a minimum of math. The mathematical tools developed in this chapter will be helpful in our subsequent study of ac circuits, but we will not rely solely on these techniques. The concepts of circuit impedance and phase angle developed in the last three chapters will be just as useful in helping us gain a good conceptual understanding of more complex circuits.

20.2 Vectors in Polar Form

We have been representing ac voltages and currents as *phasors*. The phasors are rotating vectors that trace out the sinusoidal waveforms of these voltages and currents. The phasor is represented as a vector with magnitude and angle. The magnitude corresponds to the amplitude of i or v; the angle represents the phase angle of i or v relative to the 0° reference. We have also been representing circuit opposition to ac as vectors (resistance, reactance, impedance). For these impedance vectors, the magnitude corresponds to the amount of opposition, and the angle corresponds to the phase angle of the impedance's voltage relative to the impedance's current.

Up to this point we have represented all of these current, voltage, and impedance vectors in *polar form*. Figure 20-1(a) shows the polar form of a vector $\bar{A}$ representing a certain quantity. The length of the vector is the *magnitude* of the quantity, and the angle θ which the vector makes with the 0° reference line is the phase angle of the quantity. In (a) the vector angle is positive θ since it is ccw relative to the 0° reference. In (b) the vector angle is $-\theta$ since it is cw relative to the 0° reference.

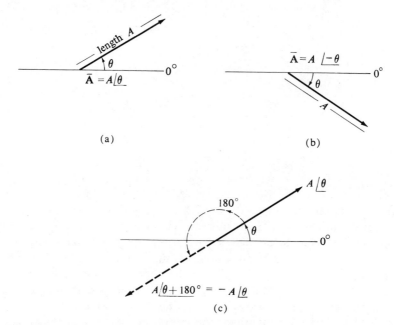

FIGURE 20-1 (a) *and* (b) *Polar representations of a vector.* (c) *Rotating a vector 180° is the same as making its magnitude negative.*

Negativeness of a Vector In part (c) of Fig. 20-1 the vector A/θ is again shown. Also shown is the same vector rotated 180° ccw. The new rotated vector has the same magnitude as A but its angle is $\theta + 180°$. The new vector is pointed in the *exact* opposite direction as the original vector. Thus, the new vector is the *negative* of the original vector. That is,

$$A/\theta + 180° = -(A/\theta) = -A/\theta \qquad (20\text{-}1)$$

In other words, *rotating a vector by 180° is the same as multiplying its magnitude by -1.*

There are two ways of indicating the negative of a vector: by (1) placing a minus sign in front of its magnitude, or (2) changing the angle by 180°. For example, suppose we have a voltage vector (phasor) $\overline{V} = 100 \text{ V}\underline{/20°}$. We can represent the negative of $\overline{V}$ as

$$-\overline{V} = -100 \text{ V}\underline{/20°}$$

or as

$$-\overline{V} = 100 \text{ V}\underline{/20° + 180°} = 100 \text{ V}\underline{/200°}$$

Both representations are equivalent.

20.3 Vectors in Rectangular Form

In addition to polar form, a vector can also be represented in *rectangular (Cartesian) form* by specifying its coordinates on a set of rectangular coordinate axes. In the study of mathematics these axes are usually called the horizontal axis (x axis) and the vertical axis (y axis). Figure 20–2(a) shows a vector drawn on a set of x-y rectangular coordinate axes. The tail of the vector is at the origin. The head of the vector is at a point that has the coordinates (3,4). Thus, the position of the head of the vector is 3 units in the $+x$ direction and 4 units in the $+y$ direction. Part (b) of the figure shows a vector with coordinates (3,−4) since its arrow is 3 units in the $+x$ direction and 4 units in the $-y$ direction.

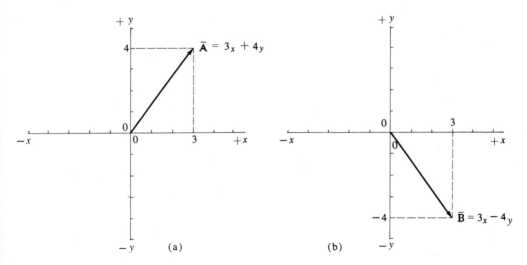

FIGURE 20-2 *Vectors expressed in rectangular form using x-y coordinate system.*

In mathematical studies the vector in (a) would be expressed as $\overline{A} = 3_x + 4_y$ and the vector in (b) would be expressed as $\overline{B} = 3_x - 4_y$. These representations give the position of the vector in rectangular form. If the x and y coordinates of a vector are known, then its exact position and length are specified. For example, the length of the vector in (a) is easily found using the Pythagorean theorem

$$\text{length} = \sqrt{(x \text{ coordinate})^2 + (y \text{ coordinate})^2}$$
$$= \sqrt{3^2 + 4^2} = 5 \qquad (20\text{-}2)$$

Similarly, the length of the vector in (b) is also 5 units.

In electrical applications, when vectors are expressed in rectangular form we do not use the familiar *x* and *y* axes. The horizontal axis is instead called the *real axis* and the vertical axis is called the *j axis* for reasons which will be explained later. The *j* axis is also called the *imaginary axis* since, as we shall see, *j* represents $\sqrt{-1}$, the imaginary number. Figure 20–3 shows several vectors expressed in rectangular form using the real- and *j*-axis coordinates. Note that the positive vertical axis is labeled as +*j* and the negative vertical axis as −*j*. The values along the vertical axis are marked off in units of *j*.

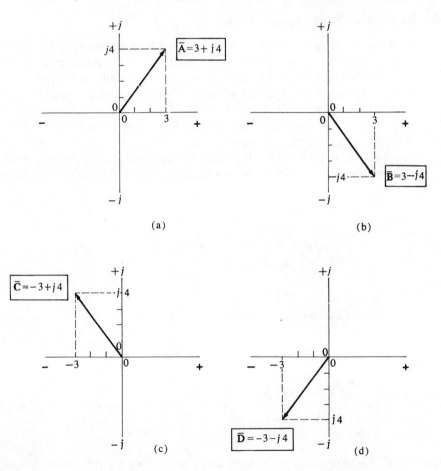

FIGURE 20-3 *Vectors in rectangular form using real axis (horizontal) and j axis (vertical).*

When expressing a vector using this rectangular coordinate system, it is written in the form

$$a + jb \tag{20-3}$$

where *a* is the coordinate along the *real axis* and *b* is the coordinate along the *j axis*. The symbol *j* is used to represent the vertical coordinates. For example, in Fig. 20–3(a) the vector $\overline{A}$ is 3 units along the +real axis and 4 units along the +*j* axis. Thus,

$$\overline{A} = 3 + j4$$

In (b) the vector $\overline{B}$ is 3 units along the +real axis and 4 units along the −j axis. Thus,

$$\overline{B} = 3 - j4$$

Note that the j term is negative since the vector extends in the −j direction.

In (c) the vector $\overline{C}$ is 3 units along the −real axis and 4 units along the +j axis. Thus,

$$\overline{C} = -3 + j4$$

Here the real term is negative since the vector points in the direction of the −real axis. In (d) the vector $\overline{D}$ is expressed as

$$\overline{D} = -3 - j4$$

since the vector extends 3 units along the −real axis and 4 units along the −j axis.

Thus, any vector can be represented in the form $(a + jb)$ where a is the coordinate along the real axis and can be either positive or negative and b is the coordinate along the j axis and can be either positive or negative.

20.4 More About j

Refer to Fig. 20–4(a) where a vector is shown with a magnitude of 3 units and is situated on the +real axis. Since this vector has no component in the vertical direction, it can be expressed as

$$\overline{A} = 3 \quad \text{(rectangular form)}$$

It can also be expressed in polar form using its magnitude (3) and its angle (0°). Thus,

$$\overline{A} = 3\underline{/0°} \quad \text{(polar form)}$$

These two representations for the vector $\overline{A}$ are equivalent.

If we take this vector $\overline{A}$ and rotate it 90° ccw, the vector in Fig. 20–4(b) results. This new vector $\overline{B}$ has a vertical component of $+j3$ and has no horizontal component. Thus, $\overline{B}$ can be expressed as

$$\overline{B} = +j3 \quad \text{(rectangular form)}$$

It can also be expressed in polar form as

$$\overline{B} = 3\underline{/90°} \quad \text{(polar form)}$$

since the +j axis is also the 90° axis relative to the 0° reference axis.

Thus, by rotating vector $\overline{A}$ by 90° ccw we essentially multiplied $\overline{A}$ by the factor j. It can be stated in general that

If a vector $\overline{A}$ represented in rectangular form is rotated 90° ccw, it is the same as multiplying $\overline{A}$ by the factor j.

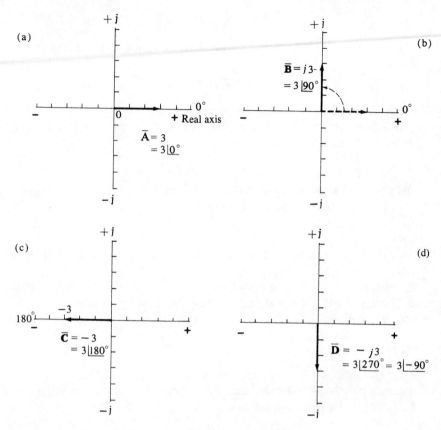

FIGURE 20-4 *Rotating a vector by 90° ccw is the same as multiplying it by j; each 90° rotation multiplies it by j.*

Let us now rotate the vector in (b) 90° ccw so that the vector in (c) results. This new vector $\overline{C}$ lies along the negative real axis so that

$$\overline{C} = -3 \quad \text{(rectangular form)}$$

or

$$\overline{C} = 3\underline{/180°} \quad \text{(polar form)}$$

Since $\overline{C}$ was obtained by rotating $\overline{B}$ by 90° ccw, then

$$\overline{C} = j \times \overline{B} = j \times j3 = j^2 \times 3$$

But as we see in Fig. 20-4(c), the vector $\overline{C} = -3$. Thus,

$$\overline{C} = j^2 \times 3 = -3$$

or

$$j^2 = -1$$

and

$$j = \sqrt{-1}$$

This verifies that j actually represents $\sqrt{-1}$, the imaginary number. In the study of mathematics the symbol i is used to represent $\sqrt{-1}$. In electronics we use j to avoid confusion with the symbol i for current.

If the vector in (c) is now rotated 90° ccw, the vector in (d) results. This new vector $\overline{D}$ lies along the $-j$ axis so that

$$\overline{D} = -j \times 3 \text{ (rectangular form)}$$

or

$$\overline{D} = 3\underline{/270°} = 3\underline{/-90°} \quad \text{(polar form)}$$

Since $\overline{D}$ is obtained by rotating $\overline{C}$ by 90° ccw, then

$$\overline{D} = j\,\overline{C}$$
$$= j(-3) = -j3$$

as the diagram in (d) shows.

If $\overline{D}$ is rotated 90° ccw we would be back to the situation in (a). That is

$$\overline{A} = j\,\overline{D} = j(-j3)$$
$$= -j^2 3$$
$$= -(-1)3 = 3$$

To summarize, each time a vector is multiplied by the factor j, it results in a vector that is $+90°$ relative to the original vector with the same magnitude. Conversely, if a vector is rotated 90° ccw, it is the same as multiplying the vector by j. Table 20-1 summarizes the steps shown in Fig. 20-4. Note that $j^2 = -1$.

TABLE 20-1

$$\overline{A} = 3 = 3\underline{/0°}$$
$$\overline{B} = j(3) = 3\underline{/90°}$$
$$\overline{C} = j(j3) = j^2 3 = -3 = 3\underline{/180°}$$
$$\overline{D} = j(j^2 3) = j^3 3 = -j3 = 3\underline{/270°} = 3\underline{/-90°}$$

If a vector is multiplied by $-j$, then it is the same thing as rotating it $-90°$ (cw). For example, if $\overline{A} = 3 = 3\underline{/0°}$ then

$$-j(\overline{A}) = -j3 = 3\underline{/-90°}$$

Repeated multiplications by the factor $-j$ cause corresponding $-90°$ rotations.

j operator The factor j is often called the *j operator*. When the j operator multiplies a real number, the combination represents a vector at $+90°$. Thus, the j operator causes a rotation of $+90°$. Similarly, the $-j$ operator causes a rotation of $-90°$. We will sometimes use the term j operator when referring to j since this term has become widely used in the field of electronics.

20.5 Representing R, X_L, X_C, and Z in Rectangular Form

In previous chapters we saw fit to represent reactances and impedances in polar form. Inductive reactance X_L was represented as $\overline{X_L} = X_L\underline{/90°}$. This vector is drawn in Fig. 20-5(a). Since $\overline{X_L}$ is at 90° it can be expressed in rectangular form as

$$\overline{X_L} = j X_L \quad (20\text{-}4)$$

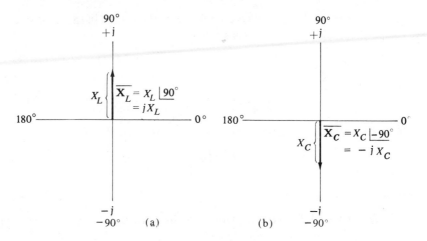

FIGURE 20-5 *Representing reactances in both polar and rectangular form.*

Capacitive reactance X_C was represented as $\overline{X_C} = X_C \underline{/-90°}$. This vector is shown in Fig. 20-5(b). Since $\overline{X_C}$ is at $-90°$ it can be expressed in rectangular form as

$$\overline{X_C} = -jX_C \qquad (20\text{-}5)$$

Resistance R was represented as $\overline{R} = R\underline{/0°}$. Since $\overline{R}$ is at 0°, then in rectangular form we simply have

$$\overline{R} = R \qquad (20\text{-}6)$$

Formulas (20-4) through (20-6) are the rectangular forms for representing reactance and resistance. These forms can be used in any type of circuit to help us calculate total impedance.

The total impedance of any circuit can be represented in polar form as $\overline{Z} = Z\underline{/\theta}$. The angle θ can be positive or negative. If θ is positive the circuit is inductive (current lags applied voltage); if θ is negative the circuit is capacitive (current leads applied voltage). The impedance can also be represented in rectangular form. To illustrate refer to Fig. 20-6(a) where a 3-ohm resistor is shown in series with a 4-ohm inductive reactance. The inductive reactance is written as $j4$ Ω. The total impedance of the series combination is obtained by *adding* $\overline{R}$ and $\overline{X_L}$.

$$\overline{Z} = \overline{R} + \overline{X_L} = (3 + j4) \text{ Ω}$$

This is the expression for $\overline{Z}$ in rectangular form. This vector is drawn in the figure. As can be seen, $\overline{Z}$ consists of a component of 3 ohms along the real axis and a component of 4 ohms along the j axis.

As another example consider the series RC circuit in Fig. 20-6(b). In this circuit the capacitive reactance is written as $-j4$. The total impedance of the series combination is obtained by *adding* $\overline{R}$ and $\overline{X_C}$. That is,

$$\overline{Z} = \overline{R} + \overline{X_C} = (3 - j4) \text{ Ω}$$

This vector is drawn in the figure. As can be seen, $\overline{Z}$ consists of a component of 3 ohms along the real axis and a component of 4 ohms along the $-j$ axis.

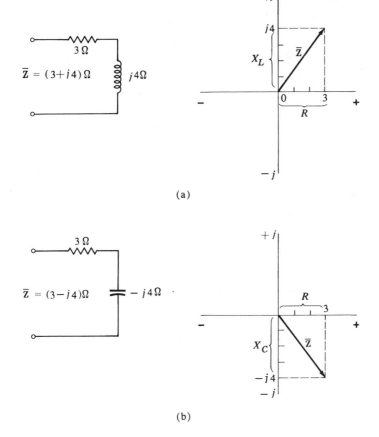

FIGURE 20–6 *Representing impedances in rectangular form.*

The above examples show that an impedance vector represented in rectangular form has a real component and a *j* component. *The real component always represents resistance. The j component always represents reactance.* Any impedance then has the general form

$$\bar{Z} = R \pm jX \qquad (20\text{–}5)$$

If Z is inductive, then

$$\bar{Z} = R + jX_L \qquad (20\text{–}6)$$

If it is capacitive, then

$$\bar{Z} = R - jX_C \qquad (20\text{–}7)$$

EXAMPLE 20.1 A certain network has an impedance $\bar{Z} = 100\,\Omega\underline{/30°}$. Express this impedance in rectangular form.

Solution: This vector is drawn in Fig. 20–7(a). To express $\bar{Z}$ in rectangular form we must determine its real component and its *j* component. Using trig we can determine the real component.

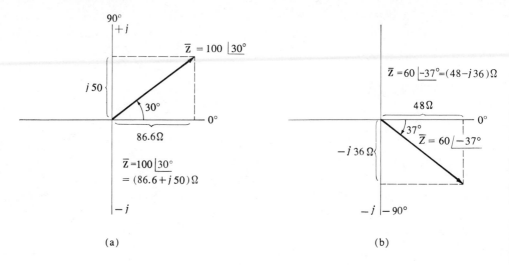

FIGURE 20-7

$$\text{real component} = Z \cos \theta = 100 \times \cos(30°)$$
$$= 86.6 \text{ ohms}$$

Similarly, the j component is obtained as

$$j \text{ component} = Z \sin \theta = 100 \times \sin(30°)$$
$$= 50 \ \Omega$$

Thus,

$$\overline{Z} = (86.6 + j50) \ \Omega$$

The real component (86.6 Ω) is pure resistance; the $j50$-Ω component is inductive reactance.

EXAMPLE 20.2 Express the impedance $\overline{Z} = 60 \ \Omega /\!\!-37°$ in rectangular form.

Solution: This vector is drawn in Fig. 20-7(b). Its real-axis component is

$$\text{real component} = Z \cos \theta$$
$$= 60 \times \cos(37°)$$
$$= 60 \times 0.8 = 48 \ \Omega$$

The j-axis component is

$$j \text{ component} = -Z \sin \theta$$
$$= -60 \times \sin(37°)$$
$$= -60 \times 0.6 = -36 \text{ ohms}$$

Thus,

$$\overline{Z} = (48 - j36) \text{ ohms}$$

Note that the angle of the impedance is negative so that the j component is negative. This indicates that the reactance portion is capacitive reactance $(-jX_C)$.

EXAMPLE 20.3 For the circuit in Fig. 20-8(a) express the total impedance vector in rectangular form.

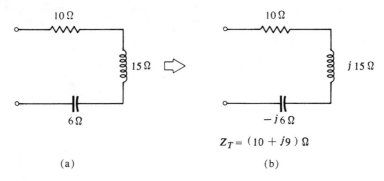

FIGURE 20-8

Solution: The circuit is redrawn in Fig. 20-8(b) using the j operator to help represent the inductive and capacitive reactances. Since the three elements are in series, the total impedance $\overline{Z_T}$ is given by

$$\overline{Z_T} = \overline{R} + \overline{X_L} + \overline{X_C}$$
$$= 10\,\Omega + j15\,\Omega - j6\,\Omega$$
$$= 10\,\Omega + j9\,\Omega$$

Note that the capacitive reactance *cancels* out some of the inductive reactance. This phenomenon is due to the fact that the capacitor voltage and inductor voltage are 180° out of phase in a series circuit. This will be discussed in more detail later when we study *RLC* circuits.

20.6 Currents and Voltages in Rectangular Form

Current and voltage phasors can also be expressed in rectangular form. The following examples illustrate.

EXAMPLE 20.4 In a certain ac circuit the voltage across the output is $\overline{V} = 10\text{ V}\underline{/-60°}$. Express this voltage phasor in rectangular form.

Solution: The voltage phasor is drawn in Fig. 20-9(a). To express $\overline{V}$ in rectangular form we must determine the real-axis and j-axis components of $\overline{V}$.

$$\text{real-axis component} = 10\text{ V} \times \cos(60°)$$
$$= 10\text{ V} \times 0.5 = 5.0\text{ V}$$
$$j\text{-axis component} = -10\text{ V} \times \sin(60°)$$
$$= -10\text{ V} \times 0.866 = -8.66\text{ V}$$

Thus,
$$\overline{V} = (5 - j8.66)\text{ V}$$

Note that since θ is negative, the j component is in the $-j$ direction.

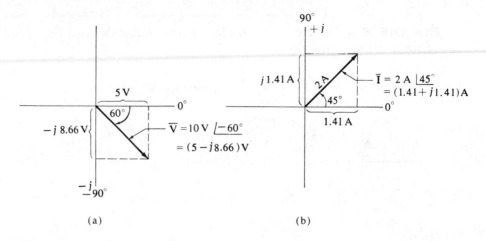

FIGURE 20-9

EXAMPLE 20.5 A voltage source of 120 V$\underline{/0°}$ is applied to a circuit with $\overline{Z}$ = 60 Ω$\underline{/-45°}$. Express the circuit current phasor in rectangular form.

Solution: $\overline{I} = \dfrac{\overline{E_s}}{\overline{Z}} = \dfrac{120\ V\underline{/0°}}{60\ \Omega\underline{/-45°}} = \left(\dfrac{120}{60}\right)\underline{/0° - (-45°)}$

$= 2\ A\underline{/+45°}$

This current phasor is drawn in Fig. 20-9(b). The real-axis and *j*-axis components of $\overline{I}$ are

$$\text{real component} = 2\ A\ \cos(45°)$$
$$= 1.41\ A$$
$$j\text{ component} = 2\ A\ \sin(45°)$$
$$= 1.41\ A$$

Thus, **$\overline{I}$ = (1.41 + j1.41) A**

20.7 Converting From Rectangular to Polar Form

If a vector is expressed in polar form we can convert it to rectangular form by finding its real component and *j* component using the technique shown in Exs. 20.1 to 20.5. Very often it becomes necessary to take a vector that is expressed in rectangular form and convert it to polar form. To illustrate, refer to Fig. 20-10(a) where a series *RC* circuit is shown.

The impedance in rectangular form is $\overline{Z} = (2 - j1)$ kΩ. This impedance vector is drawn in (b) by drawing a vector with a real-axis component of 2 kΩ and a *j*-axis component of $-j1$ kΩ. To express $\overline{Z}$ in polar form we must calculate the magnitude (Z) of the vector, and the angle θ. Using Pythagorean's theorem we can determine

$$Z = \sqrt{2\ k\Omega^2 + 1\ k\Omega^2} = 2.23\ k\Omega$$

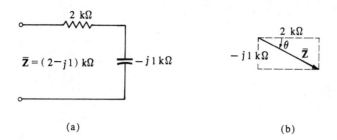

(a)   (b)

FIGURE 20-10 *Converting an impedance vector from rectangular form to polar form.*

The value of θ is expressed as

$$\tan\theta = -\left(\frac{1\ k\Omega}{2\ k\Omega}\right) = -0.5$$

$$\theta = -26.5°$$

This procedure should be very familiar to us since we used it extensively in the previous two chapters to calculate Z for series RC and series RL circuits. This procedure can be generalized for converting *any* vector from rectangular form to polar form as follows:

Any vector a + jb can be converted to polar form M/θ. The magnitude M and angle θ of the polar form are found using

$$M = \sqrt{(\text{real component})^2 + (j\ \text{component})^2}$$
$$= \sqrt{a^2 + b^2} \qquad (20\text{–}8)$$

$$\tan\theta = \frac{j\ \text{component}}{\text{real component}}$$
$$= \frac{b}{a} \qquad (20\text{–}9)$$

EXAMPLE 20.6 (a) Convert $\overline{E_S} = (-12 + j5)$ volts to polar form.
(b) Convert $-\overline{E_S}$ to polar form.

Solution: (a) The phasor $\overline{E_S}$ is drawn in Fig. 20–11(a). The magnitude of $\overline{E_S}$ is

$$E_S = \sqrt{12^2 + 5^2} = \sqrt{169} = 13\ \text{V}$$

The phase angle θ is in the second quadrant. To find θ we must first find the acute angle α

$$\tan\alpha = \left(\frac{5}{12}\right) = 0.416$$
$$\therefore \alpha = 22.6°$$

Thus, θ = 180° − α = 157.4°

and

$$\overline{E_S} = 13\ \text{V}/\underline{157.4°}$$

586   Chapter Twenty

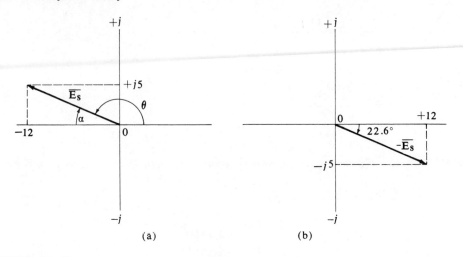

FIGURE 20-11

(b) $-\overline{E_s}$ is obtained by rotating the $\overline{E_s}$ phasor 180° (cw or ccw, it makes no difference) as shown in part (b) of Fig. 20-11. Thus, we can express $-\overline{E_s}$ as either

$$-\overline{E_s} = -13 \text{ V}\underline{/157.4°}$$

or

$$-\overline{E_s} = 13 \text{ V}\underline{/157.4° - 180°} = 13 \text{ V}\underline{/-22.6°}$$

If $-\overline{E_s} = 13 \text{ V}\underline{/-22.6°}$ is converted to rectangular form, we will find that $-\overline{E_s} = 12 - j5$, which shows that the real part of $-\overline{E_s}$ is the *negative* of the real part of $\overline{E_s}$, and the j part of $-\overline{E_s}$ is the negative of the j part of $\overline{E_s}$. That is,

$$\overset{\overline{E_s}}{\overbrace{-\overline{E_s} = -(-12 + j5)}}$$
$$= (12 - j5) \text{ volts}$$

EXAMPLE 20.7 Find the impedance of a series combination of a 120-ohm resistor and 1-μF capacitor at 1 kHz. Express Z in rectangular and polar form.

Solution: $\overline{Z} = \overline{R} + \overline{X_C} = R - jX_C$ (rectangular form)

$$X_C = \frac{1}{2\pi fC} = \frac{1}{2\pi \times 10^3 \times 10^{-6}} = 160 \text{ ohms}$$

therefore

$$\overline{Z} = (120 - j160) \text{ ohms}$$

To convert to polar form $Z\underline{/\theta}$

$$Z = \sqrt{120^2 + 160^2} = 200 \text{ ohms}$$

$$\tan \theta = -\frac{160}{120} = -1.33$$

$$\therefore \theta = -53°$$

Thus,
$$\overline{Z} = 200 \text{ }\Omega\underline{/-53°} = (120 - j160) \text{ }\Omega$$

20.8 Vector Operations

We have seen how to express current, voltage, and impedance vectors in both polar and rectangular form; a reasonable question at this point might be "why do we need both, since we solved the problems in the previous chapters using only the polar form?" The answer to this is simply that in dealing with more complicated circuits, certain calculations are easier when the vectors are in polar form and other calculations are easier in rectangular form. As such, it is necessary to be able to use both forms and to be able to convert between the two.

Multiplication and Division of Vectors In chap. 18 when the polar form was first introduced, we saw how to multiply vector quantities and divide vector quantities that are expressed in *polar form*. We could also spend time showing how to divide and multiply vectors that are expressed in *rectangular form*. However, the procedure for doing so is rather cumbersome and is very rarely used. In other words, *whenever we want to multiply or divide vector quantities, we will express these vectors in polar form* and then perform the desired operations. The following examples illustrate.

EXAMPLE 20.8 A 10-V source is applied to a series combination of $R = 100 \text{ }\Omega$ and $X_L = 173 \text{ }\Omega$. Determine the magnitude and phase angle of the current.

Solution: $\overline{I} = \dfrac{\overline{E_S}}{\overline{Z}}$

To divide $\overline{E_S}$ by $\overline{Z}$, both vectors have to be in polar form. Let $\overline{E_S} = 10 \text{ V}\underline{/0°}$. To find $\overline{Z} = Z\underline{/\theta}$ we have

$$Z = \sqrt{R^2 + X_L^2} = 200 \text{ }\Omega$$

$$\tan \theta = \left(\frac{X_L}{R}\right) = 1.73$$

$$\theta = 60°$$

Thus, $\overline{Z} = 200 \text{ }\Omega\underline{/60°}$ and

$$\overline{I} = \frac{10 \text{ V}\underline{/0°}}{200 \text{ }\Omega\underline{/60°}} = \left(\frac{10}{200}\right)\underline{/0° - 60°} = 50 \text{ mA}\underline{/-60°}$$

EXAMPLE 20.9 In a certain complex circuit, the current through one part of the circuit is $4 \text{ mA}\underline{/-45°}$. This current passes through an impedance $\overline{Z} = (3 + j6)\text{k}\Omega$. What is the voltage across $\overline{Z}$?

Solution: $\overline{V} = \overline{I} \times \overline{Z}$

To multiply these vectors we must have them both expressed in polar form. To find $\overline{Z} = Z\underline{/\theta}$ we have

$$\overline{Z} = 3 \text{ k}\Omega + j6 \text{ k}\Omega = Z\underline{/\theta}$$

Thus,
$$Z = \sqrt{3 \text{ k}\Omega^2 + 6 \text{ k}\Omega^2} = 6.7 \text{ k}\Omega$$
$$\tan \theta = \frac{j \text{ component}}{\text{real component}} = \frac{6 \text{ k}\Omega}{3 \text{ k}\Omega} = 2$$

therefore $\theta = 63.5°$ so that $\overline{Z} = 6.7 \text{ k}\Omega \underline{/63.5°}$.

Thus,
$$\overline{V} = 4 \text{ mA}\underline{/-45°} \times 6.7 \text{ k}\underline{/63.5°}$$
$$= (4 \text{ mA} \times 6.7 \text{ k}\Omega)\underline{/-45° + 63.5°}$$
$$\overline{V} = 26.8 \text{ V}\underline{/18.5°}$$

Addition and Subtraction of Vectors In chaps. 18 and 19 we often had to add phasors or vectors. In all those cases, however, the vectors which were being added were at right angles to one another so that right-triangle techniques could be used to find their resultant. In the more general case, we will have to be able to add and subtract *any* two vectors.

Whenever vectors are to be added or subtracted, they should be expressed in rectangular form.

Addition and subtraction are relatively simple processes when the vectors are in rectangular form. The basic process is as follows.

(1) The two vectors to be added (subtracted) are expressed in rectangular form $(a + jb)$.

(2) The real components of each vector are then added (subtracted) to give the real component of the resultant vector.

(3) The j components of each vector are added (subtracted) to give the j component of the resultant vector.

EXAMPLE 20.10 Perform the following vector operations:
(a) $(6 + j4) + (3 - j12)$; (b) $(36 - j10) - (36 - j7)$; (c) $10\underline{/30°} + (14 - j10)$; (d) $5\underline{/0°} + 5\underline{/+90°}$.

Solution: (a) $\quad (6 + j4)$
$\quad\quad\quad\quad +(3 - j12)$
$\quad\quad\quad\quad \overline{(6 + 3) + (j4 - j12)} = \mathbf{9 - j8}$

(b) $\quad (36 - j10)$
$\quad\quad -(36 - j7)$
$\quad\quad \overline{(36 - 36) + (-j10 + j7)} = \mathbf{0 - j3}$

(c) To add $10\underline{/30°}$ and $14 - j10$, we must first convert $10\underline{/30°}$ to rectangular form.
$$10\underline{/30°} = 10 \cos(30°) + j10 \sin(30°)$$
$$= 8.66 + j5$$

Thus,

$$\begin{array}{r}(8.66 + j5)\\+(14 - j10)\\\hline 22.66 - j5\end{array}$$

(d) $5\underline{/0°} = 5 + j0$
$5\underline{/90°} = 0 + j5$
$\overline{5 + j5}$

EXAMPLE 20.11 Consider the circuit in Fig. 20–12(a). Determine the total current I_T, and the total circuit impedance:

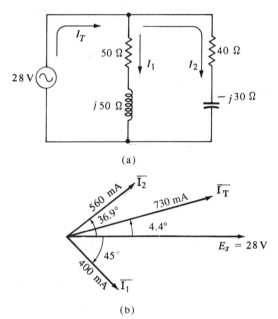

FIGURE 20–12

Solution: *Step (1)* Using KCL we know that

$$\overline{I_T} = \overline{I_1} + \overline{I_2}$$

To find $\overline{I_1}$ we must calculate the total impedance in branch 1.

$$\overline{Z_1} = (50 + j50) \ \Omega$$

Converting it to polar form

$$\overline{Z_1} = Z_1 \underline{/\theta}$$

where

$$Z_1 = \sqrt{50^2 + 50^2} = 70 \ \Omega$$

$$\theta = \text{arc tan}\left(\frac{50}{50}\right) = 45°$$

Thus,
$$\overline{Z}_1 = 70\ \Omega\underline{/45°}$$

and
$$\overline{I}_1 = \frac{28\ \text{V}\underline{/0°}}{70\ \Omega\underline{/45°}} = 400\ \text{mA}\underline{/-45°}$$

Step (2) To find $\overline{I}_2$ we must calculate the total impedance in branch 2.
$$\overline{Z}_2 = (30 - j40)\ \Omega$$

Converting to polar form
$$Z_2 = \sqrt{40^2 + 30^2} = 50\ \Omega$$
$$\theta = \text{arc tan} - \left(\frac{30}{40}\right) = -36.9°$$

Thus,
$$\overline{Z}_2 = 50\ \Omega\underline{/-36.9°}$$

and
$$\overline{I}_2 = \frac{28\ \text{V}\underline{/0°}}{50\ \Omega\underline{/-36.9°}} = 560\ \text{mA}\ \underline{/+36.9°}$$

Step (3) The phasors for $\overline{I}_1$ and $\overline{I}_2$ are drawn in part (b) of the figure. The two phasors are *not* at right angles. To add $\overline{I}_1$ and $\overline{I}_2$ we must break each vector down into its real and j components.

$$\overline{I}_1 = 400\ \text{mA}\underline{/-45°}$$
$$= (400\ \text{mA} \times \cos 45°) - j(400\ \text{mA} \times \sin 45°)$$
$$= 280\ \text{mA} - j280\ \text{mA}$$

Note that because the $\overline{I}_1$ phase angle is negative, its j component is negative.

$$\overline{I}_2 = 560\ \text{mA}\underline{/36.9°}$$
$$= (560\ \text{mA} \times \cos 36.9°) + j(560\ \text{mA} \times \sin 36.9°)$$
$$= 448 + j\,336\ \text{mA}$$

Thus,
$$\overline{I}_T = \begin{array}{r} 280\ \text{mA} - j280\ \text{mA} \\ +448\ \text{mA} + j336\ \text{mA} \\ \hline 728\ \text{mA} + j56\ \text{mA} \end{array}$$

This phasor can now be converted to polar form $I_T\underline{/\theta}$
$$I_T = \sqrt{728^2 + 56^2} = 730\ \text{mA}$$
$$\tan \theta = \frac{56}{728} = 0.077$$
$$\therefore \theta = 4.4°$$

Therefore,

$$I_T = 730 \text{ mA}/4.4°$$

The phasor for $\overline{I_T}$ is shown in the figure.

The total impedance is now easily found using Ohm's law

$$\overline{Z_T} = \frac{\overline{E_S}}{\overline{I_T}} = \frac{28 \text{ V}/0°}{730 \text{ mA}/4.4°} = 38.4 \text{ }\Omega/-4.4°$$

Note that the total impedance is *capacitive* (negative angle). This is because the total current is *leading* the source voltage by 4.4°.

EXAMPLE 20.12 Two impedances, $\overline{Z_1}$ and $\overline{Z_2}$, are in series. Their total impedance $\overline{Z_T} = 150 \text{ }\Omega/20°$. If $\overline{Z_1}$ is $85 \text{ }\Omega/-30°$ find $\overline{Z_2}$. Also determine if $\overline{Z_2}$ is capacitive or inductive.

Solution: $\overline{Z_T} = \overline{Z_1} + \overline{Z_2}$

Thus,

$$\overline{Z_2} = \overline{Z_T} - \overline{Z_1}$$

To subtract vectors they must be expressed in rectangular form:

$$\overline{Z_T} = 150 \text{ }\Omega/20°$$
$$= 150 \cos(20°) + j150 \sin(20°)$$
$$= 141 \text{ }\Omega + j51.3 \text{ }\Omega$$

$$\overline{Z_1} = 85 \text{ }\Omega/-30°$$
$$= 85 \cos 30° - j85 \sin 30°$$
$$= 73.6 \text{ }\Omega - j42.5 \text{ }\Omega$$

Therefore,

$$\overline{Z_2} = \quad (141 \text{ }\Omega + j51.3 \text{ }\Omega)$$
$$-(73.6 \text{ }\Omega - j42.5 \text{ }\Omega)$$
$$\overline{Z_2} = \quad 67.4 \text{ }\Omega + j93.8 \text{ }\Omega$$

In polar form

$$Z_2 = \sqrt{67.4^2 + 93.8^2} = 117 \text{ }\Omega$$

$$\tan \theta = \frac{93.8}{67.4} = 1.4$$

$$\therefore \theta = 55°$$

Thus, $\overline{Z_2} = 117 \text{ }\Omega/55°$. Since θ is positive, $\overline{Z_2}$ is **inductive**.

Chapter Summary

1. A vector $\overline{A}$ expressed in polar form is written as $\overline{A} = A/\theta$ where A represents the magnitude of the vector and the angle θ represents the angle which the vector makes with the 0° reference line.

2. Rotating a vector by 180° is the same as making its magnitude negative.

$$A\underline{/\theta} + 180° = -A\underline{/\theta}.$$

3. A vector can be represented in rectangular form by expressing its vertical and horizontal coordinates. In electrical applications the vertical axis is called the *j* axis or imaginary axis and the horizontal axis is the real axis.

4. In rectangular form, a vector is expressed as $(a + jb)$ where *a* represents the coordinate of the vector along the horizontal (real) axis and *b* represents the co-ordinate along the vertical (j) axis. The factor j in front of b indicates the vertical (j axis) component and is called the j operator.

5. If a vector $\overline{A}$ represented in rectangular form is rotated 90° ccw, it is the same as multiplying $\overline{A}$ by the factor j. Rotation by 90° cw is multiplication by $-j$.

6. The factor j is equal to $\sqrt{-1}$, the imaginary number.

7. Table 20–2 shows the representation of resistance, reactances, and impedance in both polar and rectangular form.

TABLE 20-2

	Polar Form	Rectangular Form
Resistance R	$R\underline{/0°}$	R (no j term)
Reactance X_L	$X_L\underline{/90°}$	jX_L (no real term)
Reactance X_C	$X_C\underline{/-90°}$	$-jX_c$ (no real term)
Impedance Z	$Z\underline{/\pm\theta}$	$R \pm jX$

8. A vector is expressed in polar form and can be converted to rectangular form by drawing the vector on polar coordinates and using trig to determine its real-axis and *j*-axis components.

9. A vector expressed in rectangular form can be converted to polar form by drawing the vector on rectangular coordinates and using trig to determine the vector length (magnitude) and angle. If the vector is $a + jb$, then its magnitude is

$$M = \sqrt{a^2 + b^2}$$

and its angle is

$$\theta = \arctan(b/a)$$

10. When vectors are to be multiplied or divided, the vectors are expressed in polar form and the operation is carried out using the polar form (as first learned in chap. 18).

11. When vectors are to be added or subtracted, the vectors are expressed in rectangular form and the operation is carried out using the rectangular form.

Questions/Problems

Sections 20.2–20.4

20-1 A certain current phasor $\overline{I} = 4$ mA$\underline{/32°}$. If this phasor is rotated 180° cw, express the resulting phasor in polar form in *two* ways.

20-2 Draw each of the following vectors on a set of rectangular coordinate axes: (a) (12-j6) volts; (b) (−60 + j45)mA; (c) (−50-j22) volts; (d) (114 + j96)kΩ; (e) (10 + j0)mA; (f) (0 − j62)Ω.

20-3 For each vector of problem 20-2 do the following: (a) rotate it 90° ccw and express the resulting vector in rectangular form; (b) rotate it 90° cw and repeat (a).

Sections 20.5–20.6

20-4 Consider the circuits shown in Fig. 20-13. For each circuit, write the expression for the total impedance vector $\overline{Z}$ in rectangular form.

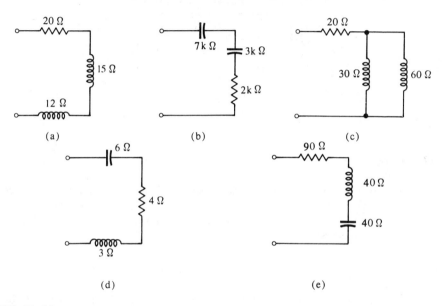

FIGURE 20-13

20-5 A certain circuit has $\overline{Z} = 3000/\underline{-18°}$. Express this impedance in rectangular form. Is the circuit capacitive or inductive?

20-6 A 1-kHz voltage of 72 V is applied to a circuit and results in a 120-mA current that lags the applied voltage by 66°. Express the circuit's impedance in rectangular form.

20-7 A certain circuit has $\overline{Z} = (6 + j10)$. If the frequency is doubled, what will be $\overline{Z}$?

20-8 Express the voltage and current phasors in rectangular form: (a) 24 V$/\underline{37°}$; (b) 100 A$/\underline{-70°}$; (c) 70 V$/\underline{160°}$

Section 20.7

20-9 Express the following vectors in polar form:
  a. $\overline{Z} = 20\ \Omega - j10\ \Omega$ b. $\overline{I} = 12\ A + j16\ A$
  c. $\overline{V} = -12\ V + j12\ V$ d. $\overline{I} = -j6$ mA

20-10 Express each of the impedances in Fig. 20-13 in polar form.

Section 20.8

20-11 Three vectors are given as $\overline{A} = 100\underline{/25°}$, $\overline{B} = 30\underline{/-25°}$, and $\overline{C} = 24 + j18$. Perform the following vector operations and give the result in polar form:
  a. $\overline{A} \cdot \overline{B}$
  b. $\overline{A} + \overline{B} + \overline{C}$
  c. $-\overline{B}/\overline{C}$
  d. $\overline{A} \cdot (\overline{C}/\overline{B}) + \overline{C}$

20-12 For the circuit of Fig. 20–14 determine the following:
  a. $\overline{Z_1}$ and $\overline{Z_2}$
  b. $\overline{I_1}$ and $\overline{I_2}$
  c. $\overline{I_T} = \overline{I_1} + \overline{I_2}$
  d. $\overline{Z_T}$

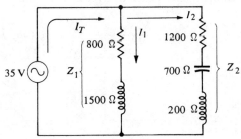

FIGURE 20-14

20-13 In the last problem the total impedance was determined by finding the total current. Another way to find the equivalent impedance of two *parallel* impedances makes use of the same *product-over-sum formula* that is used for two parallel resistors. That is,

$$\overline{Z_T} = \frac{\overline{Z_1} \cdot \overline{Z_2}}{\overline{Z_1} + \overline{Z_2}}$$

Use this formula to find $\overline{Z_T}$ for the circuit of Fig. 20–14. Express $\overline{Z_T}$ in polar form.

20-14 Two impedances, $\overline{Z_1} = 1000\underline{/-45°}$ and $\overline{Z_2} = 2000\underline{/65°}$, are connected in series across a 40-V source. Determine the magnitude and phase angle of the voltage across Z_2 with respect to the input voltage.

CHAPTER 21

RLC Circuits and Resonance

21.1 Introduction

In this chapter we will examine the behavior of circuits that contain all three types of ac opposition: resistance, inductive reactance, and capacitive reactance. Many interesting effects occur in these circuits because of the opposite effects that L and C have on alternating current. The most important of these effects, called *resonance*, will be studied in detail because of its many applications in communications equipment.

21.2 X_L and X_C in Series

It is virtually impossible to construct a circuit that contains only pure inductance and pure capacitance in series because of the presence of resistance in connecting wires, source resistance, and coil resistance. However, it will be more helpful to investigate the *pure* series LC circuit first, after which we will consider the effects of the added series resistance.

In an ac series combination of L and C [Fig. 21-1(a)] the opposition is due to inductive reactance and capacitive reactance only. If a source voltage E_s is applied to the circuit, a current I will flow. Using the current phasor as reference (0°), we can draw the circuit phasor diagram as shown in part (b) of the figure. The inductor voltage $V_L = IX_L$ *leads* the circuit current by 90°; the capacitor voltage $V_C = IX_C$ *lags* the current by 90°. As such, V_L and V_C are 180° out of phase.

Since the circuit is a series circuit, the source voltage $\overline{E_S}$ must equal the *vector* sum of $\overline{V_L}$ and $\overline{V_C}$. Assume for now that $X_L > X_C$ so that $V_L > V_C$. The vector

595

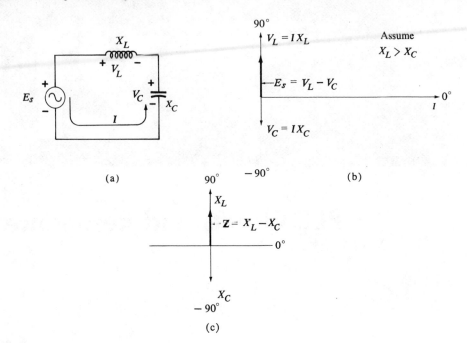

FIGURE 21-1 (a) *Series* LC *circuit.* (b) *Phasor diagram assuming* $X_L > X_C$. (c) *Impedance vector diagram.*

sum of the $\overline{V_L}$ and $\overline{V_C}$ will simply be the difference $\overline{V_L} - \overline{V_C}$ because $\overline{V_C}$ is negative relative to $\overline{V_L}$. Thus,

$$E_S = V_L - V_C \tag{21-1a}$$

as shown in the phasor diagram. Note that E_S leads the current I by 90°; this is because the circuit is *inductive* since $X_L > X_C$ and $V_L > V_C$.

The impedance Z of this series combination can be determined as follows:

$$E_s = V_L - V_C \tag{21-1}$$

$$IZ = IX_L - IX_C$$

$$\therefore Z = X_L - X_C \tag{21-2}$$

The total impedance is the *difference* $(X_L - X_C)$. This is also apparent from the impedance vector diagram in part (c) of the figure. This total impedance is *inductive* since $X_L > X_C$.

If the situation in Fig. 21-1 is reversed with X_C greater than X_L, then the source voltage E_S would be

$$E_S = V_C - V_L \tag{21-3}$$

and would *lag* the circuit current by 90° since the net voltage would be capacitive. Similarly, the total impedance would be

$$Z = X_C - X_L \tag{21-4}$$

and would be *capacitive*.

RLC Circuits and Resonance   597

In a series *LC* circuit the total impedance *Z* equals the difference between X_L and X_C. *Z* will be inductive (+90°) or capacitive (−90°) depending on which reactance is greater.

Thus, we see that X_L and X_C in *series* have a cancelling effect on each other as far as their *voltages* and their *total opposition* are concerned. The following examples will illustrate.

EXAMPLE 21.1 A 60-V ac source is applied to a series circuit containing $X_L = 100$ ohms and $X_C = 70$ ohms. For this circuit determine the current and the voltages across each element.

Solution: The circuit is drawn in Fig. 21–2(a). To find the circuit current, we must first find the total impedance. Since $X_L > X_C$, the net impedance is

$$Z = X_L - X_C = 30 \text{ ohms}$$

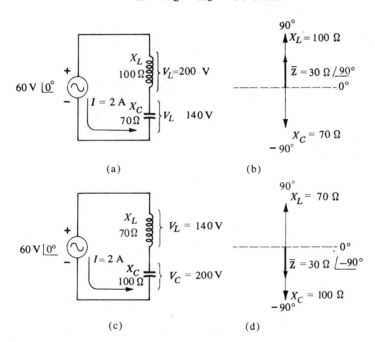

FIGURE 21-2

As the diagram in Fig. 21–2(b) shows, $\overline{Z} = 30 \text{ }\Omega/90°$, indicating that the total impedance is *inductive*. The 70 Ω X_C cancels out 70 Ω of the X_L, leaving 30 Ω net inductive reactance. Thus, the circuit acts just as if it were a simple 30-Ω *inductive* reactance.

Using Ohm's law the current will be

$$\overline{I} = \frac{\overline{E_s}}{\overline{Z}} = \frac{60 \text{ V}/0°}{30 \text{ }\Omega/90°} = 2 \text{ A}/\!-\!90°$$

Since *Z* is pure *inductive* reactance, this current will lag the source voltage by 90°.

The voltage V_L will be

$$V_L = IX_L = 2\text{ A} \times 100\text{ ohms}$$
$$= 200\text{ V}$$

The voltage V_C will be

$$V_C = IX_C = 2\text{ A} \times 70\text{ }\Omega$$
$$= 140\text{ V}$$

Note that the individual element voltages *exceed* the input source voltage. This unusual result does not violate Kirchhoff's voltage law however because, recall, these voltages are of completely opposite phase. Thus,

$$E_S = V_L - V_C = 200\text{ V} - 140\text{ V} = 60\text{ V}$$

EXAMPLE 21.2 Reverse the values of X_L and X_C and repeat the preceding example.

Solution: The revised circuit is drawn in Fig. 21-2(c). Since X_C is now greater than X_L, the total impedance is

$$Z = X_C - X_L = 30\text{ ohms}$$

As the diagram in part (d) of the figure shows, the impedance $\overline{Z} = 30\text{ }\Omega\underline{/-90°}$ indicating that the impedance is *capacitive*. Thus, the series combination acts as if it were a simple 30-Ω *capacitive* reactance.

The circuit current is therefore

$$\overline{I} = \frac{60\text{ V}\underline{/0°}}{30\text{ }\Omega\underline{/-90°}} = 2\text{ A}\underline{/+90°}$$

and this current *leads* the applied voltage by 90° since $\overline{Z}$ is pure *capacitive* reactance.

The voltages in the circuit are

$$V_L = IX_L = 2\text{ A} \times 70\text{ }\Omega = 140\text{ V}$$
$$V_C = IX_C = 2\text{ A} \times 100\text{ }\Omega = 200\text{ V}$$

Once again note that the element voltages are greater than E_S. Also note, however, that $E_S = V_C - V_L = 60\text{ V}$ as expected.

Using j operator The results of the preceding discussions can also be demonstrated using the *j* operator. In rectangular notation $\overline{X_L} = jX_L$ and $\overline{X_C} = -jX_C$. Thus, since $\overline{Z} = \overline{X_L} + \overline{X_C}$ we have

$$\overline{Z} = jX_L - jX_C = j(X_L - X_C)$$

If $X_L > X_C$, then $\overline{Z}$ will be of the form

$$\overline{Z} = j(X_L - X_C) = (X_L - X_C)\underline{/+90°}$$

which indicates a net *inductive* reactance. If $X_C > X_L$ then $\overline{Z}$ will be of the form

$$\overline{Z} = -j(X_C - X_L) = (X_C - X_L)/\!-90°$$

which indicates a net *capacitive* reactance.

21.3 Series RLC

Figure 21-3(a) shows the more practical situation of resistance, inductance, and capacitance in series. The resistance R represents the effects of the source resistance, the inductor's effective resistance, and any purposely inserted series resistor. The analysis of a series RLC circuit follows the same procedure as a series RL or series RC circuit once the reactances X_L and X_C are combined. In other words, we first combine X_L and X_C as we did in the last section to come up with a *total net reactance*, X. If $X_L > X_C$, then X will be inductive and we can analyze the circuit in the same manner as a series RL circuit [see (b) in figure]. If $X_C > X_L$, then X will be capacitive and we can analyze the circuit as a series RC circuit [see (c) in figure]. The following examples will illustrate.

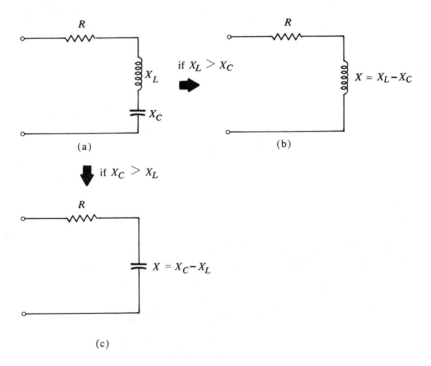

FIGURE 21-3 (a) *Series RLC circuit* (b) *becomes series RL circuit when* $X_L > X_C$; (c) *becomes series RC circuit when* $X_C > X_L$.

EXAMPLE 21.3 Consider the circuit in Fig. 21-4(a). For this circuit determine (a) total impedance; (b) current; (c) voltage across each element; (d) verify KVL.

Solution: (a) Part (b) of Fig. 21-4 shows the impedance vector diagram for the circuit. Since $X_L = 150\ \Omega$ and $X_C = 100\ \Omega$, then the *net reactance* of the

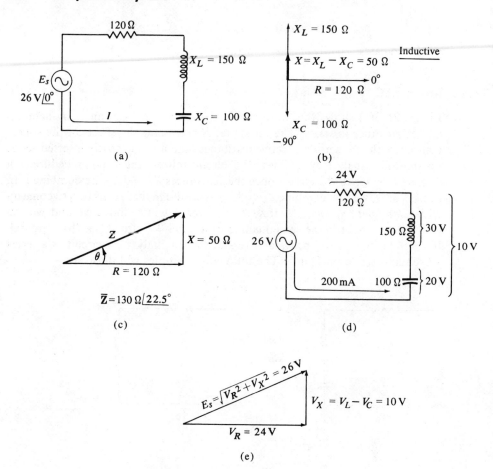

FIGURE 21-4

circuit is $X = 150 - 100 = 50$-ohm *inductive*. Thus, we can combine $R = 120$ ohms and $X = 50$ ohms as in a series RL circuit. The appropriate impedance triangle is shown in part (c) of the figure. From this triangle

$$Z = \sqrt{120^2 + 50^2} = 130 \text{ ohms}$$

$$\tan \theta = \frac{X}{R} = \frac{50}{120} = 0.416$$

$$\therefore \theta = 22.5°$$

Thus,

$$\overline{Z} = 130 \text{ }\Omega\underline{/22.5°} \text{ (inductive)}$$

We can also express $\overline{Z}$ in rectangular form as

$$\overline{Z} = 120 + j50 \text{ }\Omega$$

(b) With $\overline{E}_S = 26 \text{ V}\underline{/0°}$ and $\overline{Z} = 130 \text{ }\Omega\underline{/22.5°}$ the circuit current is

$$\overline{I} = \frac{26 \text{ V}\underline{/0°}}{130 \text{ }\Omega\underline{/22.5°}} = 0.2 \text{ A}\underline{/-22.5°} = 200 \text{ mA}\underline{/-22.5°}$$

Since the circuit impedance is inductive with a phase angle of 22.5°, the circuit current *lags* E_S by 22.5°.

(c) With $I = 200$ mA, we have (using Ohm's law)

$$V_R = 200 \text{ mA} \times 120 \text{ }\Omega = 24 \text{ V}$$
$$V_L = 200 \text{ mA} \times 150 \text{ }\Omega = 30 \text{ V}$$
$$V_C = 200 \text{ mA} \times 100 \text{ }\Omega = 20 \text{ V}$$

Note that $V_L > E_S$.

(d) To verify KVL we must *vectorially* add the element voltages to see if the sum equals E_S. First, we can combine V_L and V_C. Since they are 180° out of phase the net voltage across the *LC* combination is 30 V − 20 V = 10 V. This is the net reactance voltage $V_X = 10$ V.

Next, the resistor voltage $V_R = 24$ V is vectorially combined with the net reactance voltage $V_X = 10$ V as shown in part (e) of Fig. 21–4.

$$E_S = \sqrt{V_R^2 + V_X^2} = \sqrt{24^2 + 10^2}$$
$$= 26 \text{ V}$$

which verifies KVL.

EXAMPLE 21.4 Halve the frequency of the applied voltage for the circuit in Fig. 21–4(a) and determine its total impedance.

Solution: Halving of the applied frequency will cause X_L to decrease from 150 Ω to 75 Ω and X_C to increase from 100 Ω to 200 Ω. The revised circuit is drawn in Fig. 21–5(a). Since $X_C = 200$ Ω and $X_L = 75$ Ω, the net reactance is $X = 125$ ohms *capacitive* [see part (b) of figure].

Using the impedance triangle in part (c) of the figure we can calculate Z and θ.

$$Z = \sqrt{120^2 + 125^2} = 173 \text{ ohms}$$
$$\tan \theta = -\left(\frac{125}{120}\right) = -1.04$$
$$\therefore \theta \approx -45°$$

Thus,

$$\overline{Z} = 173 \text{ }\Omega\underline{/-45°} \text{ (capacitive)}$$

This can also be expressed in rectangular form as

$$\overline{Z} = 120 \text{ }\Omega - j125 \text{ }\Omega$$

21.4 Resonance in Series RLC Circuit

As the preceding examples illustrated, a series RLC circuit can behave either inductively or capacitively depending on the relative values of X_L and X_C, which in turn depend on the applied frequency. It seems reasonable to suspect that for given L and C values there must be a frequency where X_L equals X_C. When this situation exists in a *series RLC* circuit it is referred to as *series resonance*, and the circuit is called a *series resonant* circuit. The frequency at which resonance occurs is called the *resonant frequency* and is given the symbol f_r. The value of f_r can be calculated by setting $X_L = X_C$ as follows:

$$X_L = X_C$$

$$2\pi f_r L = \frac{1}{2\pi f_r C}$$

so that

$$f_r = \frac{1}{2\pi}\sqrt{\frac{1}{LC}} = \frac{1}{2\pi\sqrt{LC}} \qquad (21\text{-}5)$$

Thus, for any values of L and C the formula (21–5) gives the frequency at which $X_L = X_C$. Note that f_r does *not* depend on the value of the circuit resistance.

Since $X_L = X_C$ when the applied frequency equals the resonant frequency f_r, then it follows that at frequencies *below* f_r the value of X_L will be *less* than the value of X_C. This is because X_L decreases and X_C increases as frequency decreases. Conversely, at frequencies *above* f_r the value of X_L will be *greater* than X_C. These different conditions are illustrated in Fig. 21–5.

Part (a) of the figure shows the impedance vector diagram for $f < f_r$. Since X_C is greater than X_L the net reactance $X = X_C - X_L$. The total circuit Z is equal to the vector sum of R and X and is *capacitive* in this case.

Part (b) of the figure depicts the situation when f is increased to the resonant frequency f_r. Since $X_C = X_L$ the net reactance $X = 0$. As such, the total circuit Z is simply equal to the circuit resistance R. At resonance, then, the circuit impedance is purely resistive.

Part (c) of the figure shows the impedance vector diagram for frequencies greater than f_r. Here, X_L is greater than X_C so that the net reactance is $X = X_L - X_C$. The total circuit Z is the vector sum of R and X and is *inductive* in this case.

From this analysis it can be seen that the circuit impedance is at its *minimum* value at the resonant frequency. This occurs because the net reactance at resonance is zero. Above and below the resonant frequency the reactance is greater than zero, causing the impedance to increase. At resonance the total impedance is equal to the total resistance in the circuit. Since Z is a minimum at resonance, it follows that the circuit current will be a *maximum* at the resonant frequency. At resonance the current is limited solely by the series resistance, and is in phase with the applied voltage.

To summarize the characteristics of series resonance thus far, we can state that when $f = f_r$:

(1) The net circuit reactance is zero since $X_L = X_C$.
(2) The total impedance is a minimum and is equal to R.

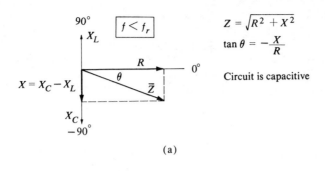

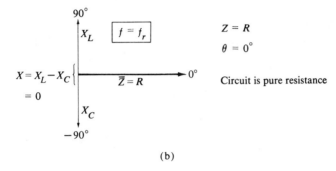

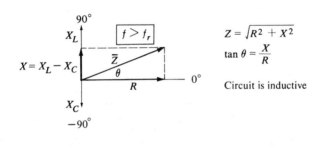

FIGURE 21–5 *Series* RLC *circuit conditions when* (a) $f < f_r$; (b) $f = f_r = 1/2\pi\sqrt{LC}$; *and* (c) $f > f_r$.

(3) The circuit current is a maximum and is in phase with the applied voltage.

EXAMPLE 21.5 For the circuit in Fig. 21–6(a) determine the circuit current and the element voltages at the following frequencies: (a) $f = f_r$; (b) $f = 0.9 f_r$; (c) $f = 1.1 f_r$.

Solution: (a) $f_r = \dfrac{1}{2\pi\sqrt{LC}} \approx 10$ kHz

At $f = f_r$, the values of X_C and X_L should be equal.

$$X_L = 2\pi f_r L = 5 \text{ k}\Omega$$

$$X_C = \frac{1}{2\pi f_r C} = 5 \text{ k}\Omega$$

Thus, net $X = 0$ and

$$Z = R = 100 \ \Omega$$

The circuit current is therefore

$$I = \frac{E_S}{Z} = \frac{10 \text{ V}}{100 \ \Omega} = \textbf{0.1 A}$$

and is in phase with E_S.

With $I = 0.1$ A we can find V_R, V_L, and V_C using Ohm's law:

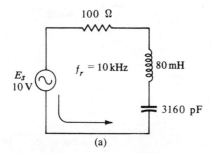

Freq.	X_C	X_L	Z	I	Phase angle of I relative to E_S	V_R	V_L	V_C
9 kHz	5.5 kΩ	4.5 kΩ	1050 Ω	10 mA	84° (Capacitive)	1 V	45 V	55 V
10 kHz f_r	5 kΩ	5 kΩ	100 Ω pure R	100 mA	0° (Resistive)	10 V	500 V	500 V
11 kHz	4.5 kΩ	5.5 kΩ	1050 Ω	10 mA	−84° (Inductive)	1 V	55 V	45 V

(b)

FIGURE 21-6

$$V_R = 0.1 \text{ A} \times 100 = \textbf{10 V}$$

$$V_L = 0.1 \text{ A} \times 5 \text{ k}\Omega = \textbf{500 V}$$

$$V_C = 0.1 \text{ A} \times 5 \text{ k}\Omega = \textbf{500 V}$$

All of these calculated values are entered into the middle row of the table in Fig. 21-6(b).

Note that the inductor and capacitor voltages are much larger than the input voltage. However, V_L and V_C are 180° out of phase with each other so that these voltages *cancel* leaving only 10-V net voltage. V_L and V_C are large because I is so large at resonance.

  (b) $f = 0.9 f_r = 9$ kHz

At $f = 9$ kHz:

$$X_L = 2\pi f L = 4.5 \text{ k}\Omega$$

$$X_C = \frac{1}{2\pi f C} = 5.5 \text{ k}\Omega$$

Thus, the net reactance is 1-kΩ *capacitive*, and

$$Z = \sqrt{R^2 + X^2}$$
$$= \sqrt{100^2 + 1000^2} = 1005 \text{ }\Omega$$

The circuit phase angle is

$$\tan \theta = \frac{X}{R} = \frac{1000}{100} = 10$$
$$\therefore \theta = 84°$$

The circuit current will lead E_S by 84°.

The value of the current at this frequency will be

$$I = \frac{10 \text{ V}}{1005 \text{ }\Omega} \simeq 10 \text{ mA}$$

Thus, we can determine the element voltages

$$V_R = 10 \text{ mA} \times 100 \text{ }\Omega = \mathbf{1 \text{ V}}$$
$$V_L = 10 \text{ mA} \times 4.5 \text{ k}\Omega = \mathbf{45 \text{ V}}$$
$$V_C = 10 \text{ mA} \times 5.5 \text{ k}\Omega = \mathbf{55 \text{ V}}$$

All of these values are tabulated in Fig. 21–6(b).

Note that at this lower frequency the circuit impedance is much greater (1050 Ω) than it was at resonance (100 Ω). Correspondingly, the current is much less (10 mA) than its value at resonance (100 mA). Note also that the V_C and V_L values are much lower than their values at resonance.

(c) $f = 1.1 f_r = 11$ kHz

The circuit values can be calculated at this frequency in the same manner as was done in (b). The calculations are left to the student. The results are entered in the bottom row of the table in Fig. 21–6(b).

The table shows that at this higher frequency X_L has increased from its value at resonance and X_C has decreased from its value at resonance. As such, the net reactance is 1-kΩ *inductive*, producing a total circuit impedance of 1050 Ω. This is again much higher than Z at resonance.

The current is 10 mA and *lags* E_S by 84° at this frequency. This current is again much smaller than the current at resonance. Accordingly, the values of V_C and V_L are lower than their resonant values.

21.5 Voltage Magnification at Resonance

The preceding example pointed out another important characteristic of series resonance:

The magnitude of the inductor and capacitor voltages at resonance are equal

and can be much greater than the input source voltage.

This characteristic is called *voltage magnification*. In the circuit of Fig. 21-6 the input voltage was 10 V and the capacitor (and inductor) voltage at resonance was 500 V. This represents a magnification factor of 500 V/10 V = 50.

For a series *RLC* circuit this voltage magnification factor is also called the circuit's *quality factor* and is given the symbol Q. For our circuit in Fig. 21-6 the value of Q was 50. Q is a unitless quantity since it is a ratio of volts to volts. Every *RLC* circuit has a certain value of Q which represents a figure of merit, a higher value of Q usually being more desirable than a lower value. Typical Q values range from 1 to 500 although in certain cases they may go much higher.

EXAMPLE 21.6 A certain *RLC* circuit has $Q = 250$. If the input voltage is 200 µV at $f = f_r$, what will be the capacitor voltage?

Solution: At $f = f_r$, the value of V_C (and V_L) is equal to the input voltage magnified by the factor Q. Thus, $V_C = 250 \times 200\ \mu V = 50{,}000\ \mu V = \mathbf{0.5\ V}$.

The quality factor Q can be determined for any series *RLC* circuit by measuring V_C (or V_L) at resonance and taking the ratio V_C/E_S (or V_L/E_S). That is,

$$V_L = V_C = Q \times E_S \quad \text{(at resonance)} \tag{21-6a}$$

so that

$$Q = \frac{V_L}{E_S} = \frac{V_C}{E_S} \tag{21-6b}$$

Since $V_L = IX_L$ and $V_C = IX_C$, then

$$Q = \frac{IX_L}{E_S} = \frac{IX_C}{E_S}$$

At resonance the value of I is determined only by the circuit resistance R since X_C and X_L cancel each other out. Thus, $I = E_S/R$ at resonance. Substituting this for I in the above expression yields

$$Q = \frac{X_C}{R} = \frac{X_L}{R} \tag{21-7}$$

These last expressions tell us that Q is also given by the ratio of the capacitive (or inductive) reactance at the resonant frequency to the circuit resistance. Since $X_C = X_L$ at resonance either of these expressions can be used to find Q.

EXAMPLE 21.7 A certain series *RLC* circuit has $E_S = 10$ mV, $R = 200\ \Omega$, $L = 40$ mH, and $C = 1580$ pF. Find the circuit Q and find V_L at resonance.

Solution: $Q = \dfrac{X_L}{R}$ \hfill (21-7)

To find X_L at resonance we need to find f_r first.

$$f_r = \frac{1}{2\pi\sqrt{LC}} = 20 \text{ kHz}$$

Thus,

$$X_L = 2\pi f_r L = 5 \text{ k}\Omega$$

and

$$Q = \frac{X_L}{R} = \frac{5 \text{ k}\Omega}{200 \text{ }\Omega} = 25$$

At resonance

$$\begin{aligned} V_L &= QE_S \\ &= 25 \times 10 \text{ mV} \\ &= 250 \text{ mV} \end{aligned} \quad (21\text{-}6a)$$

Frequency Selectivity The voltage magnification property of a series RLC circuit at resonance can be used to select one input frequency by providing much more voltage output at the resonant frequency compared with the frequencies above and below f_r. This is illustrated in Fig. 21–7 where a series RLC circuit is shown with the output voltage taken across the capacitor. The resonant frequency of the circuit is 1000 kHz and its Q is 75. The input to this circuit is a radio-frequency voltage signal whose frequency components range from 540 kHz to 1620 kHz. This type of signal is present at the antenna of any AM radio receiver.

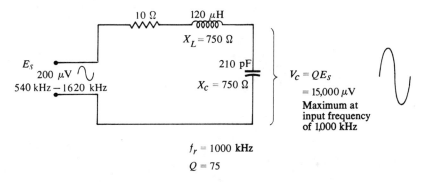

FIGURE 21-7 Series RLC circuit selects resonant frequency by producing maximum output voltage at resonant frequency.

Although the input signal contains a wide band of frequencies, the output voltage will be relatively low except at the resonant frequency of 1000 kHz. At the resonant frequency, $Q = 75$ and the output voltage will be 75 times the input voltage. In this way, the series RLC circuit selects one input frequency out of a band of input frequencies. In actuality, the output voltage will be high at frequencies which are *close to* f_r as well as at $f = f_r$. Exactly *how close* to f_r depends on the sharpness of the resonance. The *sharpness of resonance* concept will be discussed subsequently when we discuss *bandwidth*.

21.6 Series Resonance Curves

For a series *RLC* circuit, the variations of impedance and current with frequency can be shown graphically by plotting Z versus f, and I versus f for frequencies above, below, and at resonance. Figure 21–8 shows the general shape of these *response curves*. In part (a) of the figure we see how Z varies with frequency in the series *RLC* circuit. Note that Z is at its minimum value at the resonant frequency f_r. This minimum is of course equal to the circuit R. Above f_r the impedance increases rapidly and becomes *inductive*. Below f_r the impedance increases rapidly and becomes *capacitive*. Note that the curve is not entirely symmetrical around f_r, especially at higher frequencies.

In part (b) of the figure we see how I varies with frequency. At the resonant frequency the current is at its maximum value E_S/R. At frequencies below the resonant frequency, I decreases rapidly because of the increase in Z. At these lower frequencies the circuit is capacitive so that *I leads E_S*. At frequencies above the resonant frequency, I decreases rapidly because of the increase in Z.

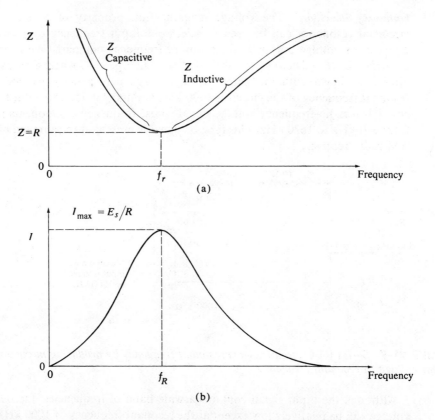

FIGURE 21–8 *Response curves showing variations with frequency in series* RLC. (a) *Impedance* Z. (b) *current* I.

At these higher frequencies the circuit is inductive so that *I lags E_S*. Note that the *I*-vs-*f* curve is not quite symmetrical around f_r. However, in the region near f_r we can assume that the curve is symmetrical.

Bandwidth We have been saying that a series *RLC* circuit is resonant at one frequency. This is true if we are considering the *maximum* resonance effect. In actuality, frequencies just above and just below f_r are also effective. At these near-resonance frequencies the circuit current will be close to maximum. Therefore, there is actually a small *band* of frequencies that produces resonance effects. This band of frequencies is centered around f_r. The width of this resonant band of frequencies is called the *bandwidth* of the resonant circuit.

Measuring Bandwidth Figure 21–9(b) shows the *I*-vs-*f* curve for the circuit in (a). The circuit's resonant frequency is 10 kHz. At resonance $I = E_S/R = 10 \text{ V}/100 \, \Omega = 100$ mA. At 9.9 kHz and at 10.1 kHz, the current is down to 70.7 mA. **The bandwidth (BW) of this circuit is measured as the difference between the two frequencies (f_1 and f_2) on either side of resonance at which the current is 70.7 percent of its maximum resonant value.** Thus,

$$\text{BW} = f_2 - f_1 \qquad (21\text{--}8)$$

For the circuit in Fig. 21–9(a) we have $f_1 = 9.9$ kHz and $f_2 = 10.1$ kHz.

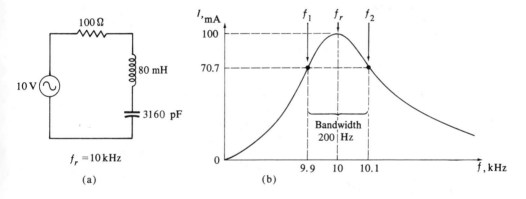

FIGURE 21–9 *Bandwidth of series resonant circuit is measured between f_1 and f_2, which are 70.7% response points.*

Therefore its bandwidth is

$$\text{BW} = 10.1 - 9.9 = 0.2 \text{ kHz} = 200 \text{ Hz}$$

The frequencies f_1 and f_2 are referred to by several different names. They are commonly called the *edge frequencies* because they define the edges of the bandwidth. They are also commonly called the *half-power* frequencies. The reason for this latter designation is because at these frequencies the power dissipation in the circuit resistance drops to one half of its value at resonance. To illustrate, for the circuit in 21–9(a) we have at resonance

$$P = I^2 R = (100 \text{ mA})^2 \times 100 \text{ ohms}$$
$$= 1 \text{ watt}$$

At f_1 or f_2 we have

$$P = (70.7 \text{ mA})^2 \times 100 \text{ ohms}$$
$$= 500 \text{ mW} = 0.5 \text{ W}$$

f_1 and f_2 are also sometimes referred to as the *cutoff* frequencies of the resonant circuit.

Relationship Between BW and Q The series *RLC* circuit is said to be resonant over a bandwidth of frequencies centered around f_r. Frequencies within the bandwidth will produce a current that is 70.7% or more of I_{max}. At frequencies outside the bandwidth the current drops rapidly. When the BW is narrow, the peak of the current response curve is much sharper than when the bandwidth is wider. This is illustrated in Fig. 21–10 by the *I*-vs-*f* response curves for three different *RLC* circuits. Each circuit has the same f_r = 10 kHz but a different value of quality factor *Q*.

As the curves show, the bandwidth is narrower for larger values of *Q*. In other words, as *Q* increases, the peak of the response curve becomes sharper and the bandwidth shrinks. As *Q* decreases, the peak of the response curve becomes broader (less sharp) and the bandwidth expands. In general, then,

The sharpness of resonance in a series *RLC* circuit increases for higher values of *Q*.

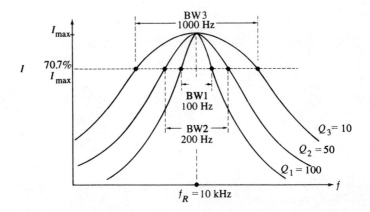

FIGURE 21–10 *BW of resonant circuit decreases for higher values of* Q *so that resonance becomes sharper.*

For higher values of *Q* the bandwidth is narrower, which means that the resonance effects are confined to a narrower band of frequencies. As such, the circuit will be more *selective* when used in applications like the one we previously discussed in Fig. 21–7. For this reason the quality factor *Q* is also referred to as the *selectivity* of a resonant circuit. Thus, a higher *Q* means not only a greater voltage magnification factor (recall $V_L = V_C = QE_S$ at resonance), but also a sharper frequency selectivity because of the narrower bandwidth. Usually $Q \geq 10$ is desirable for good selectivity.

There is a definite relationship between the resonant frequency f_r, BW, and *Q* of a series *RLC* circuit.

$$BW = \frac{f_r}{Q} \qquad (21\text{-}9)$$

This relationship shows that BW is proportional to f_r and inversely proportional to selectivity Q. If f_r and Q are known, the BW can be calculated. For example, we can calculate the BW for each of the cases in Fig. 21–10 as follows:

$$\left. \begin{array}{l} BW_1 = \dfrac{f_r}{Q_1} = \dfrac{10\text{ kHz}}{100} = 100\text{ Hz} \\[6pt] BW_2 = \dfrac{f_r}{Q_2} = \dfrac{10\text{ kHz}}{50} = 200\text{ Hz} \\[6pt] BW_3 = \dfrac{f_r}{Q_3} = \dfrac{10\text{ kHz}}{10} = 1000\text{ Hz} \end{array} \right\} \text{ As } Q \downarrow, BW \uparrow$$

Once the BW is known the edge frequencies f_1 and f_2 can be determined. The BW is essentially symmetrical around f_r so that f_1 will be lower than f_r by an amount equal to BW/2, and similarly f_2 will be higher than f_r by an amount equal to BW/2. For example, for the $Q = 10$ curve in Fig. 21–10, the edge frequencies are $f_1 = f_r - BW/2 = 10\text{ kHz} - 500\text{ Hz} = 9.5\text{ kHz}$ and $f_2 = 10\text{ kHz} + 500\text{ Hz} = 10.5\text{ kHz}$.

The following examples will further illustrate the interrelationships between f_r, BW, and Q.

EXAMPLE 21.8 For the circuit in Fig. 21–11(a) find the values of f_r, bandwidth, and Q.

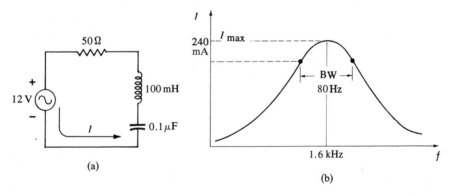

FIGURE 21–11

Solution:

$$f_r = \frac{1}{2\pi\sqrt{LC}} = 1.6\text{ kHz}$$

To find BW we must first determine Q.

$$Q = \frac{X_L}{R} = \frac{2\pi f_r L}{R} = \frac{2\pi \times 1.6 \times 10^3 \times 10^{-1}}{50} \approx 20$$

Therefore,

$$\text{BW} = \frac{f_r}{Q} = \frac{1600 \text{ Hz}}{20} = 80 \text{ Hz}$$

The I-vs-f curve for this circuit is sketched in part (b) of the figure.

EXAMPLE 21.9 How can the bandwidth of the circuit in Fig. 21–11 be doubled while keeping f_r at 1.6 kHz?

Solution: We want BW = 160 Hz and f_r = 1.6 kHz. Therefore

$$160 \text{ Hz} = \frac{1.6 \text{ kHz}}{Q}$$

so that $Q = 10$. We must reduce Q from 20 to 10 without changing f_r. Since $Q = X_L/R$, we can cut Q in half by doubling R to 100 Ω. This change in R will *not* affect f_r since f_r depends only on L and C.

EXAMPLE 21.10 Consider again the circuit in Fig 21–11. Indicate the effect each of the following circuit changes will have on the circuit BW: (a) increase ($\uparrow$) in L; (b) decrease ($\downarrow$) in C; (c) $\downarrow$ in R; (d) $\uparrow$ in input amplitude; (e) $\downarrow$ in input frequency.

Solution: (a) If L is increased, the resonant frequency will decrease since ($f_r = 1/2\pi\sqrt{LC}$). The increase in L also will cause Q to change since $Q = X_L/R = 2\pi f_r L/R$. But we cannot readily tell how Q will change since $f_r \downarrow$ and $L \uparrow$. To help us answer this question let's assume L is increased 4 times. Since f_r is inversely proportional to $\sqrt{L}$, then f_r will be cut in half as $L \uparrow$ by four times. Thus, with

$$Q = 2\pi f_r L/R$$

when $L \uparrow$ by four times and f_r decreases by a factor of one half, it is clear that Q will double. As such, the BW will decrease because BW $= f_r/Q$ and we have determined that $f_r \downarrow$ and $Q \uparrow$.

(b) Using a similar procedure, if C is $\downarrow$ by 4 times the value of f_r will double. Also, since $Q = 2\pi f_r L/R$ the value of Q will double. Thus, the BW will not change since BW $= f_r/Q$ and both f_r and Q change by the same amount keeping their ratio constant.

This interesting result indicates that the value of C does not affect the circuit bandwidth.

(c) A decrease in R will not affect f_r, but it will produce a larger Q. As such, the BW will decrease in inverse proportion to Q's increase. This means that the value of R should be kept small if a narrow BW, high selectivity circuit is required.

(d)–(e) Any changes in the input amplitude or frequency do not affect the circuit BW. BW, as well as f_r and Q, are dependent *only* on the R, L, and C values and not on the input source.

21.7 Q of a Coil

We know that any practical coil possesses an effective resistance, R_e, which we can think of as being in series with the coil's inductance X_L (see Fig. 21–12). This coil resistance is part of the total resistance in a series RLC circuit. As we have seen, the Q of a series resonant circuit is inversely proportional to R. For this reason, if it is desired that the resonant circuit have a high Q and narrow bandwidth, then the value of R must be kept as low as possible. This in turn means that the coil's effective resistance should be kept small.

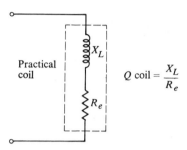

FIGURE 21–12 *The Q of a coil depends on its effective resistance.*

In this respect we can talk about the *quality* of a coil as being the ratio of its reactance (at the resonant frequency) to its effective resistance. That is

$$Q_{coil} = \frac{X_L}{R_e} = \frac{2\pi f_r L}{R_e} \qquad (21\text{--}10)$$

For example, if a coil with an inductance of 10 mH and an effective resistance of 5 ohms is to be used in a resonant circuit with $f_r = 1000$ Hz, the Q of the coil would be

$$Q_{coil} = \frac{2\pi f_r L}{R_e} = \frac{2\pi \times 1000 \times 10 \text{ mH}}{5 \text{ }\Omega} = 12.6$$

A high value of Q for a coil indicates a low value of R_e. A coil is considered to be a *high-Q* coil if it has $Q > 10$. The Q of practical coils generally never exceeds 1000. It might appear from formula (21–10) that the coil Q would increase considerably at the higher resonant frequencies which might be used in radio-frequency circuits. This is not true, however, because it may be recalled that the coil's effective resistance will increase considerably at higher frequencies due to hysteresis, eddy currents, and skin effects (chap. 19). This increase in R_e offsets the increase in X_L at higher values of f_r, thereby preventing Q from increasing without limit.

When a coil with a certain Q is used in a series RLC circuit whose other resistance is negligible compared to the coil's resistance, the *total circuit Q* will be the same as the coil's Q because $R = R_e$. For this reason, a high-Q coil must be used if the resonant circuit's Q is to be kept high. *The Q of the series resonant circuit can never be greater than the Q of the coil in the circuit.* Of course, if the circuit's other resistances (source resistance, wiring resistance, etc.) are significant, then the Q of the circuit will be *less* than the coil's Q because the circuit R will be greater than R_e.

EXAMPLE 21.11 A series RLC circuit has $R = 18$ ohms, $L = 214$ μH, and $C = 0.003$ μF. The total series resistance of 18 ohms consists of 16-ohm source resistance and 2-ohm coil resistance.

(a) Calculate f_r and Q; (b) discuss how Q may be increased *without* changing f_r.

Solution: (a) $f_r = \dfrac{1}{2\pi\sqrt{LC}} = 200$ kHz

$$Q = \frac{X_L}{R} = \frac{2\pi \times 200 \text{ kHz} \times 214 \text{ μH}}{18 \text{ Ω}} = \frac{270 \text{ Ω}}{18 \text{ Ω}}$$

$$= 15$$

(b) The circuit Q may be increased by reducing R. This change in R will not affect f_r. To reduce R we can try to find a coil with a higher Q (lower R_e). For the coil given we have

$$Q_{\text{coil}} = \frac{X_L}{R_e} = \frac{270 \text{ Ω}}{2 \text{ Ω}} = 135$$

In practice, we might be able to find a coil with a higher Q. Even so, the best we can ever hope to do is to reduce R_e to zero, which will still leave us with a total R of 16 Ω in the circuit and a Q of only 270 Ω/16 Ω = 17.

Another approach we can take to increase Q is to think about decreasing the source resistance. If this is possible, the circuit Q will be increased without having to find a higher-Q coil.

21.8 Measuring f_r of Series Resonant Circuits

The resonant frequency of a series RLC circuit can be experimentally measured by varying the frequency of the input source while monitoring the circuit current. If the frequency is started out low and gradually increased, the current will increase rapidly. The current will reach its maximum value at f_r and will begin to decrease as the frequency goes above f_r. The value of f_r can also be determined by monitoring the capacitor voltage V_c as the input frequency is varied. The capacitor voltage will reach its maximum value at $f = f_r$.

Either of these methods will give satisfactory results for high-Q circuits ($Q > 10$). For low-Q circuits the resonance effects are spread out over a greater bandwidth so that the exact value of f_r would be difficult to determine by monitoring I or V_C. For such cases the value of f_r can be more accurately determined by utilizing the fact that at resonance the circuit phase angle is 0°. Since I is in phase with E_S at resonance, it follows that V_C will *lag* E_S by exactly 90° at $f = f_r$. Thus, the E_S and V_C waveforms can be monitored (using an oscilloscope) so that their relative phase difference can be monitored as frequency is varied. The waveforms will be exactly 90° out of phase when the resonant frequency is reached. This procedure will give a more accurate measurement of f_r for low-Q circuits because the circuit phase angle increases rapidly on either side of resonance.

Measuring L or C Using Resonance Values of L or C can be accurately determined using the effects of series resonance. An unknown L can be inserted in series with a known value of C and the resonant frequency measured using one of the methods already outlined. Once f_r is known, L can be calculated from the formula for f_r. For example, assume that an unknown L is placed in series with $C = 1\ \mu F$, and f_r is measured to be 13 kHz. The formula for f_r is

$$f_r = \frac{1}{2\pi\sqrt{LC}} \quad (21\text{-}5)$$

which can be solved for L as follows:

$$f_r^2 = \frac{1}{(2\pi)^2 LC}$$

$$L = \frac{1}{(2\pi f_r)^2 C} \quad (21\text{-}11)$$

With $f_r = 13$ kHz and $C = 1\ \mu F$ the value of L can be determined as

$$L = \frac{1}{(2\pi \times 13 \times 10^3)^2 \times 10^{-6}} = 150\ \mu H$$

A similar technique can be used to measure an unknown capacitor by placing it in series with a known inductance. Formula (21–11) can be transposed to solve for C:

$$C = \frac{1}{(2\pi f_r)^2 L} \quad (21\text{-}12)$$

21.9 Measuring Q of a Series Resonant Circuit

There are two basic methods for experimentally measuring the Q of a series RLC circuit. The first method utilizes the voltage magnification effect that occurs at resonance. Since

$$V_C = Q \times E_S \quad (21\text{-}6a)$$

if E_S and V_C are measured at $f = f_r$, the value of Q can be calculated using $Q = V_C/E_S$.

The second method utilizes the fact that the circuit bandwidth is related to Q by

$$BW = \frac{f_r}{Q} \quad (21\text{-}9)$$

Thus, Q can be determined as

$$Q = \frac{f_r}{BW}$$

if the values of f_r and BW are experimentally measured. The BW is measured by determining the two edge frequencies f_1 and f_2 [see Fig. 21–9(b)]. Recall that at these frequencies the circuit current is 70.7 percent of its maximum value at resonance.

21.10 Tuning a Series Resonant Circuit

Tuning is a commonly used term that represents the process of *changing* the resonant frequency by varying either L or C. To illustrate, Fig. 21–13 shows a series RLC circuit with a variable capacitance. As C is varied, the resonant frequency of the circuit will vary so that the resonance effects can be made to occur at different frequencies. If the input signal consists of voltages at various different frequencies, the variable capacitor can be used to *tune* the circuit so that it will be resonant at *one* of these input frequencies. Thus, the circuit output voltage (across C) will be maximum at the selected frequency.

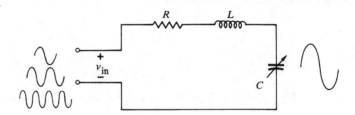

FIGURE 21–13 *Tuning a series resonant circuit with a variable C so that output voltage is maximum at one of the selected input frequencies.*

A common application of this tuning technique is in tuning a radio receiver to the carrier frequency of the desired station. The tuning is done by rotating the plates of a variable capacitor to change the value of C. As C is varied, the f_r of the circuit is varied until f_r equals the desired carrier frequency. The output of the resonant circuit then consists mainly of a voltage at the desired carrier frequency. The carrier frequencies of the other stations will not appear in the output if the circuit BW is made narrow enough. As far as these other carrier frequencies are concerned, the resonant circuit is *mistuned* since it is tuned to another resonant frequency. This tuning technique is used in AM and FM radio receivers; it is also used in television receivers to tune in the desired channel.

EXAMPLE 21.12 The frequency range for AM radio broadcasting is 540 kHz to 1620 kHz. A tuning circuit like that in Fig. 21–13 is to be used to tune in any frequency over this range. If $L = 239$ μH, over what limits must the variable capacitor be varied?

Solution: The value C has to be varied such that f_r can be varied between 540 kHz and 1620 kHz. We can use formula (21–12) to calculate the required value of C for $f_r = 540$ kHz:

$$C = \frac{1}{(2\pi f_r)^2 L} = 360 \text{ pF} \quad (\text{at } f_r = 540 \text{ kHz})$$

For $f_r = 1620$ kHz, C will have to be *decreased*. We could also use formula (21-12) to calculate the required C for this frequency. However, we can utilize the fact that f_r is inversely proportional to the square root of C. Thus, since 1620 kHz is three times 540 kHz, then C will have to be nine times smaller than its value at 540 kHz. Thus,

$$C = \frac{360 \text{ pF}}{9} = 40 \text{ pF} \; (f_r = 1620 \text{ kHz})$$

The total range of the variable C has to be **40 pF to 360 pF**.

21.11 R, X_C and X_L in Parallel

In chaps. 18 and 19 we analyzed parallel RC and parallel RL circuits by calculating the branch currents from which I_T and Z were then determined. The same procedure can be used to analyze a parallel RLC circuit. Figure 21–14(a) shows a parallel RLC circuit driven by a sine-wave voltage source. Since E_S, R, X_C, and X_L are known, we can easily determine the currents through each branch. We will assume the input source to be the reference sine wave; that is, $\overline{E_S} = 100 \text{ V}/\underline{0°}$. Thus,

$$\overline{I_R} = \frac{\overline{E_S}}{\overline{R}} = \frac{100 \text{ V}/\underline{0°}}{100 \text{ }\Omega/\underline{0°}} = 1 \text{ A}/\underline{0°}$$

$$\overline{I_L} = \frac{\overline{E_S}}{\overline{X_L}} = \frac{100 \text{ V}/\underline{0°}}{50 \text{ }\Omega/\underline{90°}} = 2 \text{ A}/\underline{-90°}$$

$$\overline{I_C} = \frac{\overline{E_S}}{\overline{X_C}} = \frac{100 \text{ V}/\underline{0°}}{80 \text{ }\Omega/\underline{-90°}} = 1.25 \text{ A}/\underline{90°}$$

The phasor diagram for the circuit is drawn in part (b) of the figure showing I_R in phase with E_S, I_C leading E_S by 90°, and I_L lagging E_S by 90°.

The total source current I_T can be found by using KCL: $\overline{I_T}$ is the *vector* sum of the individual branch currents. That is,

$$\overline{I_T} = \overline{I_R} + \overline{I_L} + \overline{I_C}$$

Part (c) of the figure shows how this vector addition can be performed. Note that I_C and I_L are 180° out of phase, so that they will have a canceling effect on each other. Since $I_L > I_C$ in this example, these two currents can be combined to give the *net reactive* current I_X as

$$I_X = I_L - I_C$$
$$= 2 \text{ A} - 1.25 \text{ A} = 0.75 \text{ A}$$

This I_X will be inductive current since $I_L > I_C$. Thus, $\overline{I_X} = 0.75 \text{ A}/\underline{-90°}$. (This procedure for combining I_L and I_C to find net I_X is the counterpart of the method we used in the *series RLC* circuit to combine V_L and V_C to obtain the net V_X).

The total source current can now be found by finding the vector sum of I_X and I_R as shown in Fig. 21–14(c). The magnitude of I_T is

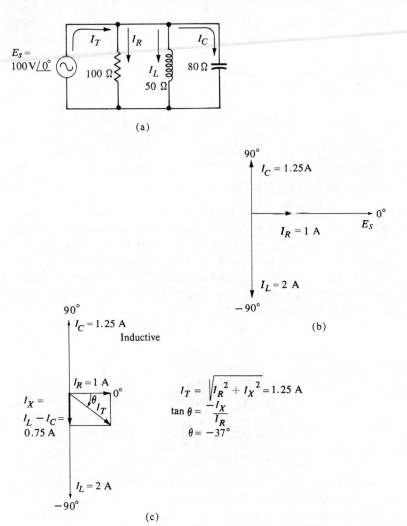

FIGURE 21-14 (a) *A parallel* RLC *circuit.* (b) *Phasor diagram.* (c) *Finding* I$_T$.

$$I_T = \sqrt{I_R^2 + I_X^2}$$
$$= \sqrt{1^2 + 0.75^2} = 1.25 \text{ A}$$

The phase angle θ by which I_T lags E_S is given by

$$\tan \theta = -\left(\frac{I_X}{I_R}\right) = -\left(\frac{0.75 \text{ A}}{1 \text{ A}}\right) = -0.75$$

so that $\theta = -37°$. Thus, $\overline{I_T} = 1.25$ A$/\underline{-37°}$. It is interesting to note that the total current of 1.25 A is *less than* the inductive branch current. This is because part of I_L was cancelled by I_C, which is 180° out of phase with I_L.

Determining Z$_T$ The total equivalent impedance of the three parallel branches can be determined using Ohm's law:

$$\overline{Z_T} = \frac{\overline{E_S}}{\overline{I_T}} = \frac{100 \text{ V}/0°}{1.25 \text{ A}/-37°} = 80 \text{ }\Omega/37°$$

This total impedance is *inductive* since I_T lags E_S. It should be noted that Z_T will be inductive whenever $X_L < X_C$ because the smaller X_L will cause I_L to be greater than I_C, therefore causing I_T to be inductive. (This is opposite to a series RLC circuit where Z_T will be inductive if $X_L > X_C$).

Analysis Using j operator The preceding analysis could also have been performed using the *j* operator techniques introduced in chap 20.

$$\overline{I_R} = 1 \text{ A}/0° = 1 + j0 \quad \text{amps}$$
$$\overline{I_L} = 2 \text{ A}/-90° = 0 - j2 \quad \text{amps}$$
$$\overline{I_C} = 1.25 \text{ A}/90° = 0 + j1.25 \text{ amps}$$
$$\therefore \overline{I_T} = 1 - j0.75 \text{ amps}$$

Converting this from rectangular to polar form

$$I_T = \sqrt{1^2 + 0.75^2} = 1.25 \text{ A}$$
$$\tan \theta = -\left(\frac{0.75}{1}\right) \rightarrow \theta = -37°$$

Therefore, $\overline{I_T} = 1.25 \text{ A}/-37°$ as determined earlier.

Thus, $\overline{Z_T} = 100 \text{ V}/0°/1.25 \text{ A}/-37° = 80 \text{ }\Omega/37°$. $\overline{Z_T}$ can be converted to its rectangular form

$$\overline{Z_T} = Z_T \cos \theta + j Z_T \sin \theta$$
$$= (80 \times 0.8) + j(80 \times 0.6)$$
$$= 64 \text{ }\Omega + j48 \text{ }\Omega$$

The total impedance, then, is equivalent to a 64-ohm resistance in series with a 48-ohm X_L.

Situation With $X_C < X_L$ Consider now the case where the capacitive reactance is less than the inductive reactance. The analysis follows the same procedure as used for the circuit of Fig. 21–14. Of course, since $X_C < X_L$ the capacitive branch current I_C will be greater than the inductive current I_L. As such, I_T will be *capacitive* and will *lead* E_S. For example, if the values of X_L and X_C are reversed in the circuit of Fig. 21–14 the results would be

$$\overline{I_T} = 1.25 \text{ A}/37° \text{ (capacitive)}$$

and

$$\overline{Z_T} = 80 \text{ }\Omega/-37° \text{ (capacitive)}$$

The student should verify these results.

Situation With $X_C = X_L$ For the case where $X_C = X_L$, the magnitudes of I_C

620 Chapter Twenty-one

and I_L will be equal. Therefore, I_X will be zero since the two reactive currents will exactly cancel. For this condition, the total current will simply equal I_R, the resistor current, and the circuit will behave as *purely resistive*. For example, if $X_L = X_C = 50\ \Omega$ in the circuit of Fig. 21–14, the I_C and I_L currents would both be 2A. Thus,

$$\overline{I_T} = 1\text{ A} - j2\text{ A} + j2\text{ A} = 1\text{ A} \qquad \text{(pure resistive)}$$

and

$$\overline{Z_T} = \frac{100\text{ V}\underline{/0°}}{1\text{ A}\underline{/0°}} = 100\ \Omega \text{ (same as } R\text{)}$$

The three conditions $X_L > X_C$, $X_C > X_L$, and $X_L = X_C$ are summarized in Fig. 21–15.

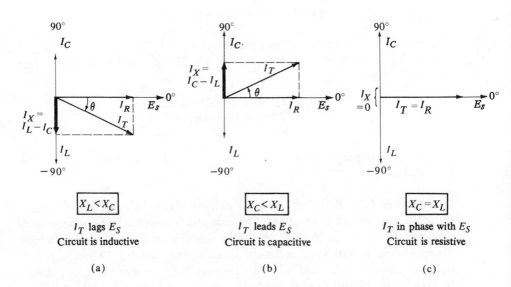

FIGURE 21-15 *Phasor relationships in parallel* RLC *circuit for* (a) $X_L < X_C$; (b) $X_C < X_L$; (c) $X_C = X_L$.

21.12 Ideal Parallel Resonant Circuit

We previously saw that a series resonant circuit is formed by connecting an inductor and capacitor in series. Resistance is also present in the series resonant circuit usually as a result of the coil's effective resistance R_e. A *parallel* resonant circuit can be formed by connecting a coil and capacitor in parallel as shown in Fig. 21–16. For now we will assume that the coil is *ideal* and has zero R_e. In accordance with this, the circuit will be called an *ideal parallel resonant circuit*. This circuit is also commonly referred to as a *tank circuit*. We will use both designations henceforth.

The resonant frequency f_r of the ideal parallel resonant circuit is the frequency at which $X_L = X_C$. This is the same condition that determined f_r for the series resonant circuit. As such, for this ideal parallel resonant circuit f_r is also

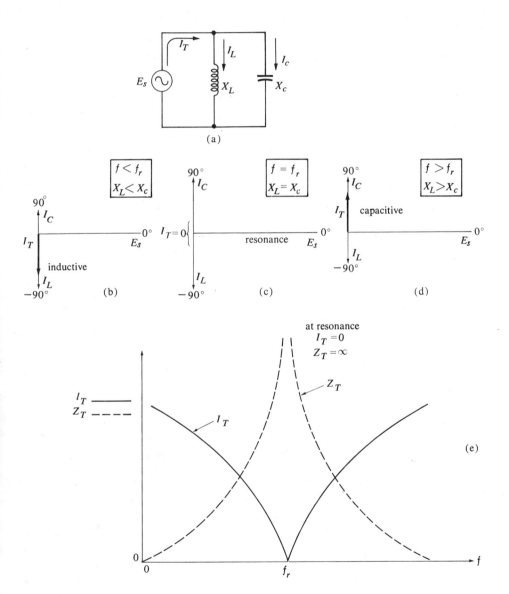

FIGURE 21–16 (a) *Ideal parallel resonant (tank) circuit.* (b) *Phasor diagrams for $f < f_r$,* (c) $f = f_r$, *and* (d) $f > f_r$. (e) I_T *and* Z_T *response curves.*

given by

$$f_r = \frac{1}{2\pi\sqrt{LC}} \tag{21-5}$$

At this frequency, $X_L = X_C$. Below this frequency $X_L < X_C$. Above this frequency $X_L > X_C$. Thus, as the input frequency is varied the current through each branch and therefore the total current will vary. The phasor diagrams in Fig. 21–16(b)-(d) illustrate these current variations. In (b) the current phasors are shown for a frequency below resonance. Since $X_L < X_C$, the inductive branch current I_L is larger than the capacitive branch current I_C. The total current I_T is therefore equal to $I_L - I_C$ and is *inductive*.

If the frequency is increased to the resonant frequency f_r, the situation in (c) results. The I_L and I_C branch currents will be equal since $X_L = X_C$. Thus, I_T must equal *zero*. In other words, I_L and I_C exactly cancel each other to produce *zero* net current.

Increasing the frequency further produces the results shown in (d). Since $X_L > X_C$, the I_C branch current will exceed the I_L branch current. The total current is therefore equal to $I_C - I_L$ and is *capacitive*.

A complete picture of how I_T varies with frequency can be shown graphically by sketching the I_T-vs-f response curve. This curve is shown in part (e) of Fig. 21–16 (solid curve). At frequencies above and below f_r, the current increases rapidly; but at $f = f_r$ the current value is exactly *zero*.

Total Impedance Z_T The total equivalent impedance of the parallel LC combination can be found at any frequency by simply taking the ratio E_S/I_T. The curve of Z_T vs f is sketched in Fig. 21–16(e) (dotted curve), showing how Z_T varies with frequency. At resonance, the value of Z_T is *infinite* (open circuit) since no net current is drawn from the source. Above and below resonance, Z_T drops in value since I_T is increasing.

At this point, we can summarize the resonance effects for the *ideal* parallel LC circuit.

(1) At resonance $X_L = X_C$ so that $I_L = I_C$ and the total current is zero.

(2) Z_T is infinite indicating that the tank circuit acts like an open circuit at resonance.

How? How is it possible to have current flowing in each element of the parallel LC circuit at resonance without any current being drawn from the source? This is a reasonable question in light of what we have previously learned about current: namely, that a source of energy is necessary to produce the flow of charges (current). In trying to resolve this paradox we must realize that the circuit, to begin with, is ideal and can never be realized in practice. We are studying it only because it will help us better understand the more practical circuit which contains the coil's effective resistance. In the practical circuit we will find that the source current at resonance is small but is *not* zero.

EXAMPLE 21.13 For the ideal tank circuit in Fig. 21–17(a) determine (a) f_r and (b) I_C, I_L, and I_T at resonance.

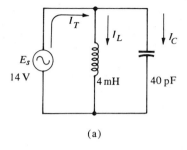 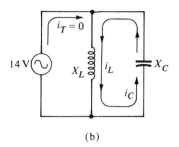

FIGURE 21-17 (a) *Ideal tank circuit.* (b) *At resonance* $I_T = 0$ *but a 1.4-mA current circulates around* LC *tank circuit.*

Solution: (a) $f_r = \dfrac{1}{2\pi\sqrt{LC}} = 400$ kHz

(b) At $f = f_r$,

$$X_C = X_L = 2\pi f_r L = 10 \text{ k}\Omega$$

Thus,

$$\overline{I_L} = \dfrac{14 \text{ V}\underline{/0°}}{10 \text{ k}\Omega\underline{/90°}} = 1.4 \text{ mA}\underline{/-90°}$$

$$\overline{I_C} = \dfrac{14 \text{ V}\underline{/0°}}{10 \text{ k}\Omega\underline{/-90°}} = 1.4 \text{ mA}\underline{/90°}$$

$$\begin{aligned}\overline{I_T} &= \overline{I_L} + \overline{I_C} \\ &= 1.4\underline{/-90°} + 1.4\underline{/90°} \\ &= -j1.4 + j1.4 \\ &= 0\end{aligned}$$

As expected, I_C and I_L cancel each other because they are 180° out of phase. In other words, at any instant of time the currents in the two branches will have the same value but will be flowing in *opposite* directions. This is illustrated in Fig. 21-17(b) where the instantaneous currents are shown. Since $i_T = 0$, the i_L and i_C currents actually circulate around the parallel LC tank circuit. This *circulating* current at resonance is 1.4 mA (rms) for the circuit values given.

21.13 Practical Parallel Resonance

A *practical* parallel resonant tank circuit is shown in Fig. 21-18(a). The branch containing the inductor also includes the coil's effective resistance. The inclusion of this effective resistance will have a significant effect on the characteristics at resonance. We saw that for the *ideal* case I_L and I_C canceled at resonance since $X_L = X_C$. In this practical circuit I_L and I_C will *not* quite cancel because: (1) the total impedance of the inductive branch will be slightly greater than X_L due to the presence of R_e, and (2) I_L will not lag E_S by exactly 90° due to R_e so that I_L and I_C will not be exactly 180° out of phase.

Although I_C and I_L will not exactly cancel at the resonant frequency f_r, the total current I_T will be very small if the resistance R_e is small compared to X_L. In other words, if the Q of the coil is large (≥ 10), the value of I_T will reach a *minimum* (but not zero) value at resonance. As such, the parallel circuit's impedance will be *maximum* (but not infinite) at resonance. We can illustrate these points by making calculations on the circuit in Fig. 21–18(a).

Calculating I_C and I_L With $L = 4$ mH and $C = 40$ pF the resonant frequency was previously calculated as 400 kHz. At this frequency $X_L = X_C = 10$ kΩ. With $\overline{E_S} = 14$ V$/0°$ this gives for $\overline{I_C}$

$$\overline{I_C} = \frac{14 \text{ V}/0°}{10 \text{ k}\Omega/-90°} = 1.4 \text{ mA}/90°$$

To determine $\overline{I_L}$ we must first calculate the total impedance of the X_L and R_e series combination. Let's call this impedance Z_L. Thus,

$$\overline{Z_L} = \overline{X_L} + \overline{R_e}$$
$$= \sqrt{10 \text{ k}\Omega^2 + 100 \text{ }\Omega^2} \Big/ \left(\arc\tan \frac{10 \text{ k}\Omega}{100}\right)°$$
$$\approx 10 \text{ k}\Omega/89.4°$$

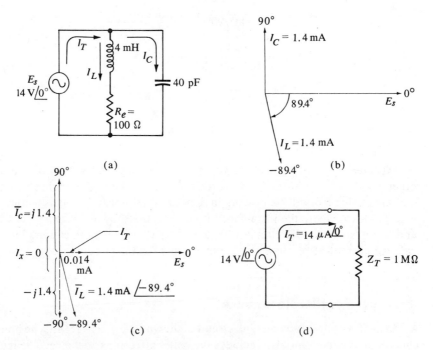

FIGURE 21–18 (a) *Practical parallel resonant circuit* ($Q > 10$). (b) *Phasor diagram at resonance.* (c) $\overline{I_L}$ *broken down into its rectangular components.* (d) *At resonance Z_T is purely resistive.*

RLC Circuits and Resonance

Note that R_e is so small compared to X_L that $Z_L \simeq X_L$.
The current I_L is therefore found to be

$$\overline{I_L} \approx \frac{14 \text{ V}\angle 0°}{10 \text{ k}\Omega\angle +89.4°} \approx 1.4 \text{ mA}\angle -89.4°$$

Note that I_L has approximately the same magnitude as I_C, but its phase angle is not quite $-90°$. The phasor diagram in Fig. 21–18(b) summarizes the situation at resonance.

Calculating $\overline{I_T}$ and $\overline{Z_T}$ The total circuit current is obtained by finding the vector sum of $\overline{I_L}$ and $\overline{I_C}$. The process is not so simple as for the ideal case where I_L and I_C were of completely opposite phase. Recall from chap. 20 that we can add two vectors, if we first convert them to *rectangular* form.

$$\begin{aligned}\overline{I_L} &= 1.4 \text{ mA}\angle -89.4° \\ &= 1.4 \text{ mA} \times \cos(-89.4°) + j1.4 \text{ mA} \sin(-89.4°) \\ &\approx .014 \text{ mA} - j1.4 \text{ mA} \\ \overline{I_C} &= \phantom{.014 \text{ mA}} 0 + j1.4 \text{ mA} \end{aligned}$$

$$\overline{I_T} = \overline{I_L} + \overline{I_C} \approx 0.014 \text{ mA} + j0 = 0.014 \text{ mA}\angle 0°$$

Thus, $\overline{I_T}$ is only 0.014 mA and is *in phase* with the source voltage.

We can also view these operations graphically in part (c) of Fig. 21–18. The rectangular form for $\overline{I_L}$ indicates that its components consist of 0.014 mA along the 0° axis and 1.4 mA along the $-90°$ ($-j$) axis. The component along the 0° axis is called the *in-phase* component since it is in phase with source voltage. The component along the $-90°$ axis is called the *out-of-phase* component (or *quadrature* component) since it is out of phase with E_S.

The out-of-phase component of $\overline{I_L}$ essentially cancels $\overline{I_C}$ to leave zero net reactive current. This leaves only 0.014 mA$\angle 0°$, the *in-phase* component, as the total source current. The value of Z_T can now be calculated as

$$\overline{Z_T} = \frac{\overline{E_S}}{\overline{I_T}} = \frac{14 \text{ V}\angle 0°}{0.014 \text{ mA}\angle 0°} = 1 \text{ M}\Omega\angle 0°$$

As stated earlier this total impedance is very large at resonance. It is also *purely resistive* [see Fig. 21–18(d)]. Recall that Z_T for the *series* resonant circuit was also purely resistive at resonance.

For frequencies above and below resonance I_T will increase and Z_T will decrease. Table 21–1 shows results for 360 kHz and 440 kHz. The calculations at these frequencies are performed in the same manner as those already done and are left as an exercise for the reader. Note that Z_T is *resistive* only at f_r. Above f_r, Z_T is capacitive, and below f_r, Z_T is inductive.

TABLE 21–1

f	I_L	I_C	I_T	Z_T	
360 kHz	1.54 mA	1.26 mA	0.28 mA	50 kΩ	(inductive)
f_r = 400 kHz	$\simeq$1.4 mA	1.4 mA	0.014 mA	1 MΩ	(resistive)
440 kHz	1.26 mA	1.54 mA	0.28 mA	50 kΩ	(capacitive)

Current Magnification — Q For the circuit in Fig. 21–18 at resonance, I_T is only 0.014 mA while the I_L and I_C branch currents are each 1.4 mA. This means that the inductor and capacitor *circulating* current is 100 times larger than the source current. This *current magnification* is the counterpart of the voltage magnification in the series resonant circuit. Once again the *magnification factor* is equal to the circuit Q. That is,

$$I_L = I_C = Q\, I_T \text{ (at resonance)} \tag{21-13}$$

For the circuit in Fig. 21–18, the circuit Q is the same as the coil's Q, which is given by

$$Q_{\text{circuit}} = Q_{\text{coil}} = \frac{X_L}{R_e} = \frac{10 \text{ k}\Omega}{100 \text{ }\Omega} = 100$$

Thus, the current magnification is 100, which agrees with our previous calculations.

Impedance Magnification The total impedance at resonance is much greater than the impedance of either branch. In fact, Z_T is related to X_L and X_C at resonance by the following expression

$$Z_T = Q\, X_L = Q X_C \text{ (at resonance)} \tag{21-14}$$

In other words, the *total* impedance at resonance is Q times the impedance (reactance) of either branch. This makes sense since the *total* current is Q times smaller than the current of either branch. This relationship can be checked for the circuit in Fig. 21–18. Since $X_L = X_C = 10$ kΩ at resonance and $Q = 100$, then

$$Z_T = 100 \times 10 \text{ k}\Omega = 1 \text{ M}\Omega$$

as previously determined.

I_T and Z_T Response Curves Figure 21–19 contains a graphical presentation of how I_T and Z_T vary with frequency in a *practical* parallel resonant circuit. These curves have essentially the same shape as those obtained for the *ideal* parallel resonant circuit in Fig. 21–16(e). The major differences occur at resonance: Z_T is maximum, *but not infinite,* and I_T is minimum, *but not zero,* at f_r for the practical circuit. Note that these response curves are just the opposite of those obtained for the series resonant circuit (Fig. 21–6).

EXAMPLE 21.14 A parallel resonant circuit consists of a coil with $L = 200$ μH and $R_e = 60$ ohms and a 130-pF capacitor. Determine: (a) f_r; (b) Q; (c) Z_T; (d) I_T, I_C, and I_L. Assume $E_S = 50$ mV.

Solution: (a) $f_r = 1/2\pi\sqrt{LC} \simeq 1$ MHz

(b) $Q = \dfrac{X_L}{R_e} = \dfrac{2\pi f_r L}{R_e} = \dfrac{1260 \text{ }\Omega}{60 \text{ }\Omega} = 21$

(c) $Z_T = Q\, X_L = 21 \times 1260$ $\Omega = 26.5$ kΩ

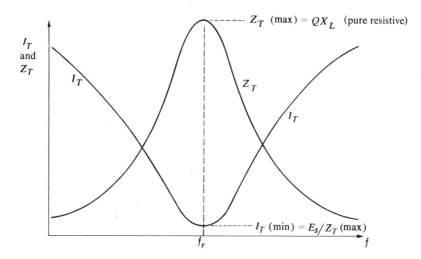

FIGURE 21-19 I_T- and Z_T-vs-f response curves for the practical parallel resonant circuit.

(d) $I_T = \dfrac{E_S}{Z_T} = \dfrac{50 \text{ mV}}{26.5 \text{ k}\Omega} = 1.9 \text{ μA}$

$I_C = I_L = QI_T = 21 \times 1.9 \text{ μA} = 40 \text{ μA}$

Summary Thus Far The principal effects of a practical parallel resonant circuit can be summarized as follows: (1) at resonance $X_L = X_C$; (2) at resonance, Z_T is at its maximum value so that I_T is at its minimum value; (3) Z_T is pure resistance so that I_T is in phase with E_S.

We have said that these effects occur at the resonant frequency $f_r = 1/2\pi\sqrt{LC}$. Actually this is only true if the circuit Q is greater than 5, preferably 10. In other words, for a high-Q circuit these resonant effects all occur at the same frequency. For a low-Q circuit ($Q < 5$), effects (2) and (3) occur at slightly *lower* frequencies than f_r. Since most parallel resonant circuits are designed with a high Q, we will not concern ourselves with the low-Q situation. It may be recalled that for the series resonant circuit all the resonant effects occurred at exactly the same f_r regardless of the Q value.

21.14 Selectivity and Bandwidth

The fact that a parallel resonant circuit has maximum impedance at resonance makes it useful in applications where a high impedance is required over a narrow band of frequencies. One such application is shown in Fig. 21–20. In this circuit, the tank circuit is connected between terminals y and z. A 100-kΩ resistor is connected in series with the source and the parallel resonant circuit, which is the one previously analyzed in Fig. 21–18. This circuit will behave as follows:

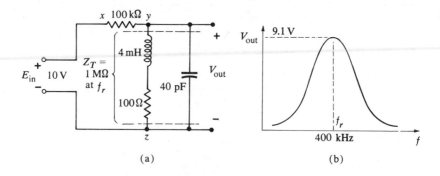

FIGURE 21-20 *Parallel resonant tank circuit (y-z) used in a band-pass filter.*

(1) At the resonant frequency of the tank circuit (400 kHz), the impedance between points y-z will be at its maximum value (1 MΩ) and will be purely resistive. As such, *most* of the input voltage E_{in} will appear across the tank circuit. In fact, using the voltage divider formula,

$$V_{out} = \left(\frac{1 \text{ M}\Omega}{1 \text{ M}\Omega + 100 \text{ k}\Omega}\right) \times E_{in}$$
$$= 0.91 \, E_{in} = 9.1 \text{ V} \quad \text{(at resonance)}$$

(2) At frequencies above and below resonance, the tank circuit impedance decreases rapidly. As such, most of the input voltage will appear across the series 100-kΩ resistor and very little voltage will appear across the tank circuit. In other words, V_{out} will decrease as the tank circuit impedance decreases on either side of f_r.

(3) The output V_{out} will therefore vary with frequency according to the graph in Fig. 21-20(b). For frequencies at or near resonance, V_{out} will contain most of E_{in}, with only a slight attenuation. Thus, we can call this circuit a *band-pass filter* since it passes input frequencies over a certain band.

In the application explained here, the resonant tank circuit was used to produce a maximum output voltage over a narrow band of frequencies. This *frequency selectivity* of the resonant tank circuit is due to the fact that its impedance is maximum at $f = f_r$ and decreases on either side of f_r. As was the case for the series resonant circuit, the sharpness of the frequency selectivity depends on the shape of the response curves, which in turn depends on the circuit Q (selectivity). Figure 21-21 shows how the shape of the Z_T-vs-f response curve varies for different Q values. For higher values of Q the peak is much sharper at resonance and the impedance drops off quickly above and below resonance. For lower Q values the peak is much broader indicating that the resonance effects are spread out over a wider frequency band.

We can define the *bandwidth* of the parallel resonant circuit similarly to the series resonant circuit. The BW is measured between the two edge frequencies at which Z_T has dropped to 0.707 of its maximum resonance value. As the curves in Fig. 21-21 show, the BW is smaller for larger values of Q, and vice versa. This characteristic was also true for the series resonant circuit and the same relationship is valid for the parallel circuit; that is,

$$BW = \frac{f_r}{Q} \qquad (21\text{-}9)$$

EXAMPLE 21.15 (a) Determine the BW of the tank circuit portion of the circuit in Fig. 21–20; (b) how can the BW be decreased for better selectivity without changing f_r?; (c) how can the BW be increased?

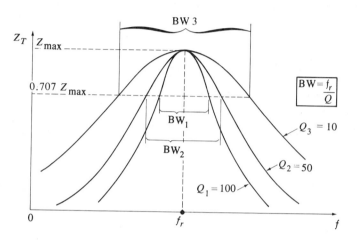

FIGURE 21–21 *Sharpness of resonance peak increases for higher values of Q.*

Solution: (a) Q for this circuit $= X_L/R_e$ was previously calculated as 100. Thus,

$$BW = \frac{f_r}{Q} = \frac{400 \text{ kHz}}{100} = 4 \text{ kHz}$$

The BW, then, extends from 398 kHz to 402 kHz.

(b) To decrease BW, Q must be increased. We know that $Q = X_L/R_e$ where R_e is the coil's effective resistance. We could try to find a coil with a smaller R_e. If this is not possible, then we could use a *larger* value of inductance L so that X_L would be greater. However, to maintain the same value of $f_r = 1/2\pi\sqrt{LC}$, the value of C must be proportionally *decreased*.

For example, if we *double* L to 8 mH then $Q = X_L/R_e = 200$ so that BW = 400 kHz/200 = 2 kHz. To maintain f_r = 400 kHz, C must therefore be cut by a *half* from 40 pF to 20 pF.

(c) To increase BW, Q must be decreased. This is easily accomplished by *adding* resistance in series with the coil. This will increase the effective value of R_e and therefore decrease Q without affecting f_r. In this case the circuit Q will be *less* than the actual coil Q, which is still X_L/R_e.

Tuning The parallel resonant circuit can be *tuned* to a particular resonant frequency by varying L or C. For many tuning applications a variable capacitor is used to vary f_r over a particular range. For example, in the circuit of Fig. 21–20 we saw that maximum V_{out} occurred at frequencies close to $f_r = 400$ kHz. By varying C, the circuit can be tuned to another frequency for maximum output.

21.15 Measurements on a Parallel Resonant Circuit

A technician must often make measurements on a resonant circuit to determine actual values for f_r, Q, and BW. The measured values will, in general, not be exactly the same as the values which can be calculated from the values of L, C, and R_e. This is mostly because of the fact that accurate values for L, C, and R_e are rarely known.

Measuring f_r The resonant frequency of a parallel tank circuit can be determined experimentally by monitoring the total current I_T as the input frequency is varied. Since Z_T is maximum at resonance, I_T will reach a minimum at $f = f_r$. When making this measurement a sensitive current meter should be used. This is especially true for relatively low Q circuits where the resonance effects are not as pronounced and are spread over a wider bandwidth.

Measuring Q Q can usually be measured by measuring the current magnification at resonance. Since

$$I_L = Q \, I_T \tag{21-13}$$

then

$$Q = \frac{I_L}{I_T}$$

The currents I_L and I_T can be measured and Q calculated from this relationship. When measuring I_L, care must be taken that the current meter resistance is much smaller than R_e, otherwise the additional resistance would cause the circuit Q to decrease, giving an erroneous result.

Q can also be experimentally determined by first determining the circuit BW. Then Q can be calculated from

$$Q = \frac{f_r}{BW}$$

Measuring BW To measure the circuit BW, the total current I_T must be monitored. At resonance Z_T is maximum so that $I_T = E_S/Z_T$ is minimum. At the edge frequencies Z_T will have *decreased* to 0.707 of this *maximum* value. Thus, I_T will have *increased* to $1/0.707 = 1.4$ of its *minimum* resonant value. For example, suppose $E_S = 10$ V and $Z_T = 100$ kΩ at resonance. Then, $I_T = 10$ V/100 kΩ = 0.1 mA at resonance. At the edge frequencies $Z_T = 0.707 \times 100$ kΩ = 70.7 kΩ so that $I_T = 10$ V/70.7 kΩ = 0.14 mA. After determining the edge frequencies f_1 and f_2 at which I_T increases to 1.4 times its minimum value, the BW is simply equal to $f_2 - f_1$.

21.16 Damping a Parallel Resonant Circuit

Up to now we have seen that for the practical parallel resonant circuit, the circuit Q is the same as the Q of the coil in the inductive branch. If a resistor R_P is put in parallel (shunt) with the tank circuit, the effect is to *reduce* the circuit Q. This effect is called *damping* the resonant circuit. R_P may represent the resistance of a load which the tank circuit is driving, or it could represent an actual resistor purposely inserted to reduce Q.

This reduction in Q can be demonstrated by referring to Fig. 21–22 where the circuit which we analyzed in Fig. 21–18 is redrawn with $R_P = 1$ MΩ connected in parallel. Recall that the original circuit Q was calculated as 100. The insertion of R_P across the 14-V source means that there will be a 14-μA current (14 V/1 MΩ) through R_P at resonance. This current, I_R, will of course be in phase with E_S. We already saw that the tank circuit draws a current of 0.014 mA = 14 μA in phase with E_S at resonance. Thus, I_T at resonance is now $I_{\text{tank}} + I_R = 14\ \mu\text{A} + 14\ \mu\text{A} = 28\ \mu\text{A}$.

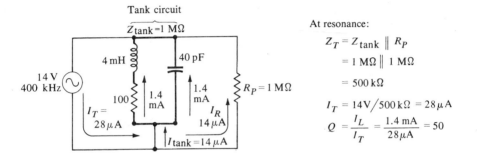

FIGURE 21–22 *Shunt resistor* R_P *produces a smaller* Z_T *and larger* I_T *at resonance, thereby reducing* Q *(damping).*

This new value for I_T can also be determined by combining the tank circuit impedance ($Z_{\text{tank}} = 1$ MΩ) with $R_P = 1$ MΩ in parallel to find the new value of Z_T.

$$Z_T = Z_{\text{tank}} \parallel R_P = 1\ \text{M}\Omega \parallel 1\ \text{M}\Omega = 500\ \text{k}\Omega$$

Thus, $I_T = 14$ V/500 kΩ = 28 μA.

Since $I_L = I_C = 1.4$ mA, as before, the current magnification I_L/I_T has been reduced (from 100) to 1.4 mA/28 μA = 50. Thus, Q is now only 50, a reduction of 50 percent caused by the presence of R_P. Of course, the lower Q will increase the circuit BW. The resonant frequency will not be affected since that is determined solely by L and C. If R_P is made smaller, it will draw more in-phase current to add to I_T at resonance and thereby reduce Q further.

Earlier we said that Q could be reduced by increasing the effective resistance R_e in series with the inductor. Now we have seen that adding a shunt resistor R_P has the some effect. In general, both R_e and R_P are present and must be taken into account in calculating Q. The following relationship can be used to calculate Q in the general case:

$$Q = \frac{X_L}{R_e + X_L^2/R_P} \tag{21-15}$$

For the example in Fig. 21–22, we have

$$Q = \frac{10 \text{ k}\Omega}{100 \text{ }\Omega + (10 \text{ k}\Omega)^2/1 \text{ M}\Omega} = \frac{10000}{100 + 100} = 50$$

If R_P is very large, then the X_L^2/R_P term in the denominator of Eq. (21-15) can be neglected and the formula reduces to $Q = X_L/R_e$, which is the coil Q. The effect of R_P can be neglected when the current it draws is much less than the resonant tank current. In other words, if $R_P \geq 10 Z_{\text{tank}}$, then $I_R \leq 0.1 \, I_{\text{tank}}$ and R_P can be neglected so that the total circuit Q will be the same as the coil Q.

EXAMPLE 21.16 A certain resonant tank circuit has $f_r = 1$ MHz, $L = 160 \, \mu$H, and $Q_{\text{coil}} = 52$. (a) What minimum value of R_P will have negligible effect on Q? (b) Find Q for $R_P = 100$ kΩ.

Solution: (a) With no R_P, the total tank circuit impedance at resonance is

$$Z_{\text{tank}} = QX_L = 52 \times (2\pi \times 1 \text{ MHz} \times 160 \, \mu\text{H})$$
$$= 52 \times 1 \text{ k}\Omega = 52 \text{ k}\Omega$$

Thus, R_P can be made as small as 10×52 kΩ = **520 kΩ** with little or no effect on Q.

(b) The new value of Q can be found using Eq. (21-15). R_e must be determined first using the original $Q = 52$, which is the coil Q. That is,

$$Q_{\text{coil}} = 52 = \frac{X_L}{R_e} = \frac{1 \text{ k}\Omega}{R_e}$$

so that $R_e = 1000/52 \approx 19 \, \Omega$. Thus, the circuit Q with $R_P = 100$ kΩ included is

$$Q = \frac{X_L}{R_e + X_L^2/R_P} = \frac{1000}{19 + 10} = \textbf{34.4}$$

The new value of Q could also be found using $Z_T = QX_L$. We have $Z_T = Z_{\text{tank}} \parallel R_P = 52 \text{ k}\Omega \parallel 100 \text{ k}\Omega \approx 34.4$ kΩ. We also have $X_L = 1$ kΩ. Thus,

$$34.4 \text{ k}\Omega = Q \times 1 \text{ k}\Omega$$

or

$$Q = \textbf{34.4}$$

Chapter Summary

1. When a capacitor and inductor are in series in an ac circuit, their respective voltages are 180° out of phase.

2. Because of this 180° phase difference, the effects of X_L and X_C try to cancel each other in a series circuit. The net reactance is equal to the *difference* between X_L and X_C.

3. In a series circuit if $X_L > X_C$, the circuit behaves as an inductive circuit; if $X_C > X_L$ the circuit behaves as a capacitive circuit.

4. When a capacitor and inductor (ideal) are in parallel in an ac circuit, their respective currents are 180° out of phase. As a result, these currents try to cancel each other and the net current is the difference between I_C and I_L.

5. In a parallel RLC circuit with $X_L > X_C$, the circuit is capacitive; if $X_L < X_C$ the circuit is inductive.

6. Table 21–2 provides a comparison between series resonance and parallel resonance. The principal difference is that series resonance produces maximum current and minimum Z at f_r, while parallel resonance produces minimum current and maximum Z at f_r.

TABLE 21–2

Series Resonance (any Q)	Parallel Resonance ($Q > 5$)
a. $X_L = X_C$ at $f_r = 1/2\pi\sqrt{LC}$	a. $X_L = X_C$ at $f_r = 1/2\pi\sqrt{LC}$
b. I is *maximum* at f_r and is in-phase with E_S	b. I_T is *minimum* at f_r and is in phase with E_S
c. Impedance Z is *minimum* at f_r; $Z = R$ (pure resistive)	c. Impedance Z_T is *maximum* at f_r; $Z_T = QX_L$ (pure resistive)
d. Voltage magnification: $V_C = V_L = QE_S$	d. Current magnification: $I_L = I_C = QI_T$
e. Quality factor (selectivity): $Q = X_L/R$, also $Q = V_C/E_S = V_L/E_S$	e. $Q = X_L/R_e$, also $Q = I_L/I_T = I_C/I_T = Z_T/X_L = Z_T/X_C$
f. Bandwidth: BW $= f_2 - f_1 = f_r/Q$ At edge frequencies f_1 and f_2, $I = 0.707\, I_{max}$	f. BW $= f_2 - f_1 = f_r/Q$ At f_1 and f_2, $Z_T = 0.707\, Z_{max}$ and $I_T = 1.4\, I_{min}$
g. Circuit is capacitive at frequencies below f_r and inductive above f_r	g. Circuit is inductive at frequencies below f_r and capacitive above f_r
h. To decrease Q: add resistance in series	h. To decrease Q: add resistance in series with coil or place resistor R_P in parallel with tank circuit

Questions/Problems

Sections 21.2–21.3

21-1 A 50-V, 1-kHz source is applied to a series circuit containing $L = 40$ mH and $C = 1$ µF. The inductor's effective resistance is 30 ohms and the source resistance is 60 ohms. Calculate: (a) circuit current and phase angle; and (b) voltages across each element.

21-2 Reduce the frequency to 600 Hz and repeat problem 1.

21-3 How is it possible for the V_L and V_C voltages in a series RLC circuit to be larger than E_S? Doesn't this defy KVL? Explain.

Sections 21.4–21.5

21-4 Determine the resonant frequency for the circuit of problem 1. Then, calculate Z, I, V_L, and V_C at $f = f_r$.

21-5 Use *two* methods to calculate Q for the circuit of problem 1.

21-6 A series RLC circuit has $Q = 120$. If the input voltage is 10 mV at $f = f_r$, determine V_R, V_C, and V_L at resonance.

21-7 In a certain series resonant circuit, V_C is measured as 320 V at resonance when $E_S = 4$ V. If $L = 2$ μH and $C = 3$ pF determine: (a) Q, (b) f_r, (c) total circuit resistance.

21-8 What value of L is necessary with a C of 0.001 μF to produce series resonance at 100 kHz?

21-9 The following statements refer to a *series RLC* circuit. Indicate which statements are true. Correct each false statement by correcting the italicized portions:
 a. Resonance occurs when $X_L = R$.
 b. At resonance the circuit phase angle is *maximum*.
 c. At frequencies below resonance the current *leads* the applied voltage.
 d. If R is doubled, f_r will *double* and Q will be *halved*.
 e. A Q of 50 will produce a *greater* voltage magnification at resonance than a Q of 25.
 f. Doubling L and doubling C will cause f_r to be doubled.
 g. Increasing E_S will increase V_C at resonance, thereby *increasing Q*.
 h. An increase in input frequency will *always* cause Z to decrease and I to increase.
 i. At frequencies above f_r the inductance voltage is *greater than* the capacitor voltage.
 j. V_C and V_L can never be less than V_R.

Section 21.6

21-10 A series *RLC* circuit has $R = 20$ ohms, $L = 4$ mH, and $C = 0.1$ μF. For $E_S = 12$ V sketch the *I*-vs-*f* response curve showing the values of: (a) f_r; (b) edge frequencies f_1 and f_2; (c) I_{max}; (d) bandwidth.

21-11 For the circuit of problem 21-10 consider each of the following changes: (a) $R \downarrow$; (b) $L \uparrow$; (c) $C \downarrow$; (d) $E_S \uparrow$. For *each* of these indicated changes sketch the new *I*-vs-*f* curve showing the effects caused by the change.

21-12 A 10-V variable-frequency source is applied to each of three different series *RLC* circuits in order to obtain data to plot their *I*-vs-*f* curves. The resulting curves are shown in Fig. 21-23. Each curve has the *same* resonant frequency. From these curves determine:
 a. which circuit has highest Q
 b. which circuit has greatest BW
 c. which circuit has largest R
 d. which circuit has largest L
 e. which circuit has largest C

21-13 Consider the circuit in Fig. 21-7. It is desired that the output exhibit resonance effects over a range of frequencies from 950 kHz to 1050 kHz. Show how this can be done by changing the value of R.

Sections 21.7–21.10

21-14 A 200-μH coil has a Q of 250 at $f_r = 800$ kHz. What is the coil's effective resistance?

21-15 Describe a procedure for experimentally determining the effective resistance of a coil by using it in a series resonant circuit.

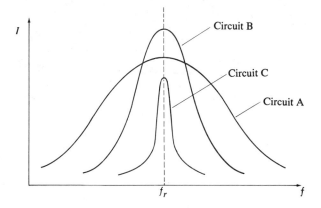

FIGURE 21-23

21-16 A series *RLC* circuit has edge frequencies of $f_1 = 220$ kHz and $f_2 = 230$ kHz. Determine the circuit Q.

21-17 Consider the tuned resonant circuit in Fig. 21-13. The input signal contains frequencies in the range from 540 kHz to 1620 kHz spaced at intervals of 10 kHz apart (i.e., 540 kHz, 550 kHz, 560 kHz, etc.). Assume C is adjusted so that $f_r = 800$ kHz. Also assume $L = 239$ μH. What minimum circuit Q is required if the output is to contain *only* the desired 800-kHz frequency. (Hint: assume that any frequencies outside the BW will not appear significantly in the output.) What value of R is required?

21-18 Calculate the range of capacitor values required for an FM tuner with $L = 0.2$ μH to tune through frequencies from 88 MHz to 108 MHz.

21-19 Indicate which of the following statements concerning *series* resonant circuits are *always* true:
   a. In a series resonant circuit the circuit Q is never greater than the coil Q.
   b. A large BW indicates a high-Q circuit.
   c. If two series resonant circuits have the same edge frequencies, then they have the same f_r and Q.
   d. A larger value of C will tune the circuit to a higher frequency.
   e. Increasing the input voltage will increase the output voltage *only* at resonance.
   f. The impedance curve and current curve have the same general shape.
   g. A circuit with a higher Q is more selective in its frequency response.
   h. A low-Q circuit will exhibit resonance effects over a more narrow range of frequencies than a high-Q circuit.
   i. The value of C does not affect the BW.
   f. *Any* high-Q circuit has a narrower BW than *any* low-Q circuit.

21-20 Using the basic formulas for Q and f_r derive a new formula for Q which contains only L, C, and R.

21-21 Consider the series resonant circuit of Fig. 21-24. (a) Calculate V_C at resonance with $R = 0$. (b) The capacitor is rated to handle a maximum voltage of 600 V. What is the minimum R needed in order to ensure that V_C does not exceed 600 V?

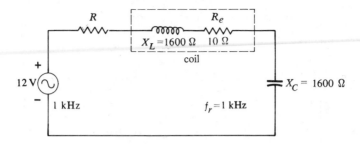

FIGURE 21-24

Sections 21.11–21.12

21-22 A 2-kΩ resistor, 0.03-μF capacitor, and a 30-mH inductor are connected in parallel across an ac source. Determine I_R, I_L, I_C, and $\overline{I_T}$ and $\overline{Z_T}$ at the following frequencies: (a) 4 kHz; (b) 5.3 kHz; (c) 6 kHz. (Assume $E_S = 20$ V.)

21-23 In a parallel *RLC* circuit, how is it possible that I_L and I_C can be larger than I_T?

Sections 21.13–21.14

21-24 A parallel resonant circuit consists of a coil with $L = 500$ μH and $R_e = 50$ ohms in parallel with $C = 125$ pF. If $E_S = 16$ V determine: (a) f_r; (b) I_T and Z_T at resonance; (c) Q.

21-25 For the circuit of the previous problem, indicate the effect each of the following changes will have on f_r and Q:
   a. increase R_e b. decrease L
   c. increase E_S (d) increase C

21-26 A certain parallel resonant circuit contains a coil with $Q = 80$. When a 32-V source is applied at the resonant frequency of 50 kHz, the total source current is 20 μA. Determine the values of L, R_e, and C.

21-27 If the L, R_e, and C of the previous problem are connected *in series* across the source, what will be the values of f_r, Q, and Z_T? Compare these values to those for the parallel circuit in problem 21-26.

21-28 For the circuit in Fig. 21-25 sketch the I_T and Z_T response curves showing values for the following: edge frequencies f_1 and f_2, bandwidth, Z_{max}, and I_{min}.

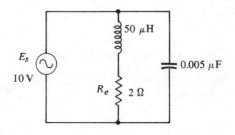

FIGURE 21-25

Sections 21.15–21.16

21-29 (a) Show how the response curves for the circuit in Fig. 21–25 will be affected if R_e is *increased*. (b) Show how the response curves will be affected if a resistance R_P is placed across the tank circuit.

21-30 A parallel resonant circuit has its f_r measured as 200 kHz. At $f = 192$ kHz the value of I_T is increased to 1.4 times its resonant value. What is the circuit Q?

21-31 A 10-V variable-frequency source is applied to each of three parallel resonant circuits in order to obtain data to plot their I_T-vs-f curves. The resulting curves are shown in Fig. 21–26. Each curve has the same f_r. From these curves determine:

   a. which circuit has highest Q
   b. which circuit has largest BW
   c. which circuit has smallest L
   d. which circuit has largest C

21-32 A resonant tank circuit has edge frequencies of 115 kHz and 125 kHz. Determine the circuit Q.

21-33 Why does the insertion of a parallel resistance R_P across a resonant tank circuit decrease the circuit Q? How does adding R_P affect f_r and bandwidth?

21-34 A certain tank circuit contains a coil with $Q = 150$ and has $f_r = 1$ MHz. The tank circuit impedance is 300 kΩ at resonance. What is the smallest value of parallel resistance which can be tolerated if the circuit BW is not to exceed 10 kHz?

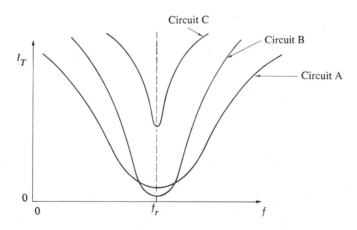

FIGURE 21-26

21-35 For each of the following indicate whether the statement refers to series resonance, parallel resonance, both, or neither:
   a. f_r changes as circuit resistance is made smaller
   b. $X_L = X_C$ at resonance
   c. behaves inductively for $f < f_r$
   d. maximum current flows at resonance

e. large Q means narrow BW and sharper tuning
f. circuit Q can be greater than coil Q
g. capacitor voltage is maximum at resonance
h. impedance is maximum at resonance
i. I is in phase with E_S at resonance
j. f_r doubles if L or C is cut in half
k. total impedance is zero at high frequencies
l. Bandwidth can be narrowed by decreasing circuit resistance
m. $Z_T = QX_L$

21-36 The box shown in Fig. 21-27 is known to contain a resonant circuit. The components in the box are not visible, so that it is not known whether the box contains a series resonant or parallel resonant circuit. Describe a method which can be used to experimentally determine the type of resonant circuit, the value of f_r, and the circuit Q. No measurements can be made inside the box.

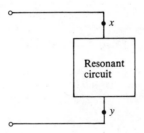

FIGURE 21-27

21-37 A certain parallel resonant circuit is being tested in the lab by a technician. He makes calculations based on the values of L, R_e, and C and determines that f_r should be 150 kHz and Q should be 60. He then proceeds to experimentally measure f_r and Q using the setup shown in Fig. 21-28. Ammeter A_1 is used to monitor the total current I_T and ammeter A_2 is used to monitor the inductor current I_L. The input source is a variable frequency generator. Its frequency is varied until I_T reaches a minimum value to determine f_r. Once f_r is reached, Q is calculated as I_L/I_T at that frequency. The oscilloscope monitors the tank circuit voltage. Using this procedure the technician measures $f_r \simeq 150$ kHz, and $Q = 40$. In other words, the measured Q is off by 33 percent from the calculated value. Which of the following could be possible causes for this discrepancy? Explain each choice.

a. resistance of A_1
b. resistance of A_2
c. input resistance of the oscilloscope
d. inaccuracy of capacitor value used in calculations
e. inaccuracy of R_e value used in calculations
f. value for E_s was too large

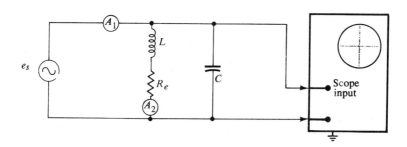

FIGURE 21-28

21-38 Using formulas (21-7) and (21-9), derive an expression for resonant circuit bandwidth in terms of R and L. This formula shows that BW depends on L and R but not on C. What is the significance of this fact in an application like that shown in Fig. 21-13?

21-39 Figure 21-29 shows two arrangements for using tuned circuits to produce an output voltage which will be maximum over a narrow band of frequencies. Which of the two arrangements would be better if the input source has a large source resistance? Which would be better if the input source has a low source resistance? Explain.

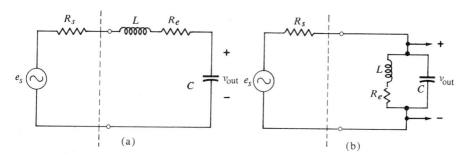

FIGURE 21-29

CHAPTER 22

Filters

22.1 Introduction

Filters are electrical circuits which are designed to prevent the passage of certain frequencies of input voltage while permitting the passage of other frequencies. We have already seen how series RC and series RL circuits can be used as certain types of filters. In this chapter we will look at more examples of filtering. An exhaustive study of electrical filters, however, would require a volume in itself. As such, we will limit our study to the basic types of filters in an effort to develop a thorough understanding of filtering principles.

Filter circuits are necessary because most electronic circuits and systems usually contain voltage signals which are not simple dc or simple sine waves. Instead these signals are often made up of many sine waves of different frequencies. For example, the audio signals in a radio can contain many frequencies that cover a range typically from 10 Hz to 20 kHz; the audio detector in a radio contains both radio frequencies (>500 kHz) and audio frequencies. Another example is the output of an amplifier stage which contains a dc voltage level with an ac signal superimposed or "riding" on it.

In such applications where the voltage signals have different frequency components, it is often necessary to either pass or reject one of these frequencies or group of frequencies by using a suitable filter. Because a filter circuit must behave differently for different frequencies, then it is necessary that any filter circuit contains inductance and/or capacitance since these elements are frequency-dependent.

22.2 Filter Types

Most filter circuits fall into one of the following categories: (1) *low pass;* (2) *high pass;* (3) *band pass;* or (4) *band reject.* The basic operation of these filter types is illustrated in Fig. 22–1.

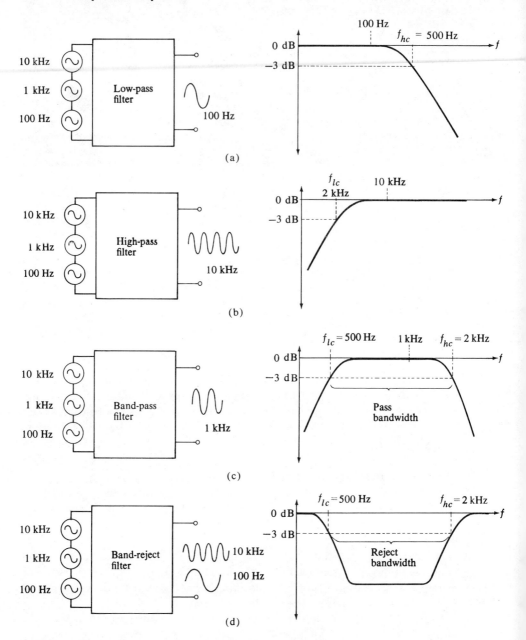

FIGURE 22–1 *Illustration of operation of basic filter types.*

Low-Pass Filter Low-pass filter operation is illustrated in part (a) of Fig. 22–1. A low-pass filter will pass lower frequencies and will attenuate (block) higher frequencies. In the figure, three ac voltage sources are shown connected in series to the input of the low-pass filter. Thus, the *total* input to the filter consists of three superimposed sine waves at frequencies of 100 Hz, 1 kHz, and 10 kHz. The filter output, however, shows only a sine wave of 100 Hz. The filter has attenuated the 1-kHz and 10-kHz input sine waves while passing the 100-Hz input.

The attenuation-versus-frequency plot for this low-pass filter is also shown in the figure. It shows 0-dB attenuation for low frequencies and negative-dB attenuation at higher frequencies. Recall, that 0 dB represents a V_{out}/V_{in} ratio of *one* while $-$dB indicates a ratio less than one. The high-frequency cutoff f_{hc} (upper 3-dB frequency) is shown as 500 Hz. Thus, frequencies *below* 500 Hz will be passed with very little attenuation. This includes 0 Hz which, of course, is dc.

For our example, the 1-kHz and 10-kHz inputs will be attenuated by the filter. In practice, the attenuation is not complete so that these frequencies will be present to some extent in the filter output. The attenuation becomes greater (more negative dB) at frequencies much greater than f_{hc}.

High-Pass Filter High-pass filter operation is illustrated in part (b) of Fig. 22-1. A high-pass filter will pass higher frequencies and will attenuate lower frequencies. The same three input frequencies are applied to the high-pass filter resulting in an output at only 10 kHz. The filter has attenuated the 100-Hz and 1-kHz input sine waves while passing the 10-kHz input.

The attenuation-versus-frequency graph for this high-pass filter shows 0 dB (no attenuation) at high frequencies and negative dB at lower frequencies. The low-frequency cutoff f_{lc} is shown as 2 kHz. Thus, frequencies *above* 2 kHz are passed with very little attenuation. For our example, the 100-Hz and 1-kHz inputs are below 2 kHz and will therefore be attenuated. A *high-pass filter also attenuates any dc present in the input.*

Band-Pass Filter Band-pass filter operation is illustrated in part (c) of Fig. 22-1. A band-pass filter will attenuate both lower frequencies and higher frequencies, but will pass frequencies in between. For example, the same three input frequencies are shown applied to the band-pass filter resulting in an output at only 1 kHz. The filter has attenuated the lower input frequency (100 Hz) and the higher input frequency (10 kHz).

The attenuation-versus-frequency plot for this band-pass filter shows 0 dB (no attenuation) at middle frequencies and negative dB at lower and higher frequencies. The band-pass filter has both a low-frequency cutoff ($f_{lc} = 500$ Hz) and a high-frequency cutoff ($f_{hc} = 2$ kHz). Frequencies between these two cutoff frequencies are considered to be passed by the filter. The *pass bandwidth* for the band-pass filter is equal to ($f_{hc} - f_{lc}$) as shown.

Band-Reject Filter Band-reject filter operation is illustrated in part (d) of Fig. 22-1. A band-reject filter will pass both lower frequencies and higher frequencies, but will attenuate (reject) frequencies in between. For example, the same three input frequencies are shown applied to the band-reject filter, resulting in an output that contains *both* 100 Hz and 10 kHz, but *not* 1 kHz. The filter has rejected the 1-kHz voltage but has passed the lower frequency (100 Hz) and higher frequency (10 kHz) voltages.

The attenuation-versus-frequency plot for this band-reject filter shows 0 dB at lower and higher frequencies and negative dB at frequencies in the middle. The band-reject filter has both a low-frequency cutoff ($f_{lc} = 500$ Hz) and a high-frequency cutoff ($f_{hc} = 2$ kHz). Frequencies between these two cutoff frequencies are considered to be attenuated (rejected) by the filter. The *reject bandwidth* equals ($f_{hc} - f_{lc}$).

EXAMPLE 22.1 Figure 22-2 shows a filter system which is used to filter an audio signal into three frequency ranges. The audio input contains frequencies in the range from 20 Hz to 20 kHz. Determine the range of frequencies contained in each filter output.

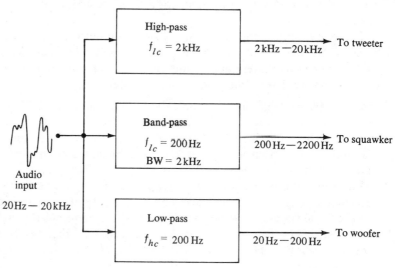

FIGURE 22-2

Solution: (a) The filter in part (a) of the figure is a high-pass filter with a cutoff frequency of 2 kHz, which means that it attenuates frequencies *below* 2 kHz. Thus, its output will contain only audio frequencies above 2 kHz. In other words, frequencies between 2 kHz and 20 kHz will be present at the filter output. These frequencies are fed to a *tweeter*, which is a speaker which is designed to reproduce the higher frequencies of sound.

(b) The filter in part (b) of the figure is a band-pass filter with a low-frequency cutoff of 200 Hz and a bandwidth of 2 kHz. This means that it passes frequencies between 200 Hz and 200 + 2000 = 2200 Hz. Thus, its output will contain only audio frequencies in the range 200-2.2 kHz. These frequencies are fed to a *squawker*, which is a speaker which reproduces the mid-range sound frequencies.

(c) The filter in part (c) is a low-pass filter with a cutoff frequency of 200 Hz, which means that it attenuates frequencies *above* 200 Hz. Thus, its output will contain only audio frequencies below 200 Hz. In other words, frequencies between 20 Hz and 200 Hz will be present at the filter's output. These frequencies are fed to a *woofer*, which is a speaker designed to reproduce the lower frequencies of sound.

22.3 Sharpness of Filter Response

A filter is designed to pass frequencies over a certain range (pass-band) and attenuate frequencies outside this range (stop-band). Ideally, a filter would either pass frequencies completely or attenuate them completely. In practice, no real filter behaves ideally. In other words, a practical filter will never completely attenuate all frequencies outside its pass-band (bandwidth). In addition, the transition from the pass-band to the stop-band is a gradual one. The *sharpness* of this transition may be different for different filter circuits.

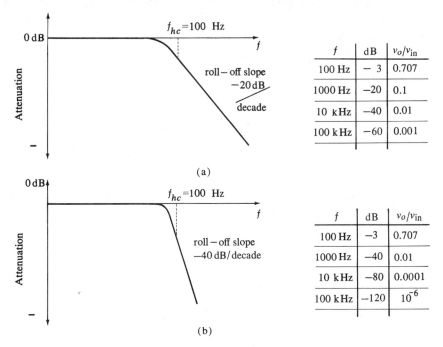

FIGURE 22-3 *Two low-pass frequency-response curves. The curve in (b) shows a sharper (steeper) cutoff than the curve in (a).*

For example, refer to Fig. 22-3, which shows the frequency-response curves for two different types of low-pass filter. For the response curve in (a), the attenuation is shown as increasing at the rate of -20 *dB/decade* for frequencies greater than the 100-Hz cutoff frequency. This means that for every tenfold (decade) increase in f, the attenuation increases by -20 dB. The table next to the curve gives several values of attenuation at various frequencies. Note that -20 dB represents an attenuation ratio of 0.1, -40 dB represents 0.01, etc.

Examination of the frequency-response curve in (b) shows that the attenuation increases at the rate of -40 *dB/decade* for frequencies greater than the cutoff frequency. This means that for each tenfold increase in frequency, the attenuation increases by -40 dB. This steeper slope represents a sharper filter response. This filter will discriminate between frequencies above f_{hc} and below f_{hc} more effectively than the filter in (a).

The filter represented in (a) is said to have a response curve that has a *roll-off* of -20 dB/decade; the curve in (b) has a roll-off of -40 dB/decade. All filters have response curves that roll off at some multiple of -20 dB/decade. A filter with a faster roll-off has a sharper frequency response and its operation more closely approaches the ideal filter operation. In general, filters that employ *both* inductance and capacitance have a sharper response than those that employ only one or the other.

A filter's roll-off slope can also be expressed as so many dB per *octave*. An octave is a change in frequency by a factor of two. For example, if a low-pass filter has a roll-off of -6 dB/octave this means that for every increase in frequency by a factor of two, the attenuation increases by -6 dB. A filter's roll-off can be expressed in either dB/decade or dB/octave. The equivalent representations are

$$-20 \text{ dB/decade} = -6 \text{ dB/octave}$$
$$-40 \text{ dB/decade} = -12 \text{ dB/octave}$$
$$-60 \text{ dB/decade} = -18 \text{ dB/octave}$$
$$\text{etc.} \qquad\qquad \text{etc.}$$

22.4 Low-Pass Filter Circuits

We have already seen how RC and RL circuits can be used as low-pass filters. Figure 22–4 shows these two types of low-pass filter circuits. In the RC low-pass filter, the higher frequency input voltages are kept from reaching the output because of the *shunting* action of the capacitor. At the higher frequencies X_C approaches *zero* ohms, essentially *shorting out* the output so that $V_{out} \simeq 0$. In the RL low-pass filter, the higher frequency input voltages are kept from reaching the output because of the *blocking* action of the inductor. At the higher frequencies X_L becomes very large; as a result the circuit current is blocked so that very little voltage appears across R, and $V_{out} \approx 0$.

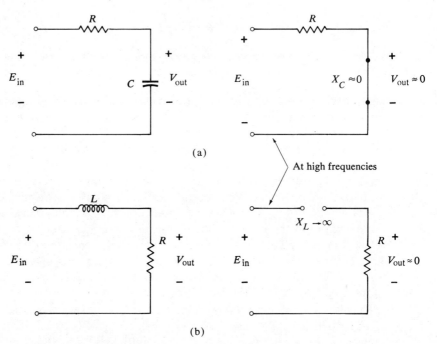

FIGURE 22–4 (a) RC *low-pass and* (b) RL *low-pass filters.*

These two low-pass filters illustrate the two basic filter actions: *shunting* and *blocking*. Shunting refers to shorting out of the output by a reactive element

(*L* or *C*) placed in *parallel* (shunt) with the output so that $V_{out} \simeq 0$ over a certain frequency range. Blocking refers to a reactive element placed in *series* with the output so that it acts as a high impedance over a certain frequency range preventing any of the input from reaching V_{out}. In the *RC* low-pass filter, *C* acts as the shunting element; in the *RL* low-pass, *L* acts as the blocking element.

RLC Low-Pass Filter The low-pass filter shown in Fig. 22–5 combines the capacitor shunting effect of the *RC* low-pass with the series inductor blocking effect of the *RL* low-pass to produce a sharper filter response. At higher frequencies the capacitor acts as a low impedance shunt and the inductor acts as a large blocking impedance. These effects combine to reduce V_{out} at high frequencies. The high-frequency cutoff f_{hc} for this filter is equal to f_r, the circuit's natural resonant frequency. The value of *R* is usually chosen to reduce the circuit *Q* to around one so that V_{out} will not rise sharply at resonance. Recall that at resonance $V_{out} = V_C = Q E_{in}$. If it is desired that $V_{out} = V_{in}$ then *Q* should be equal to one.

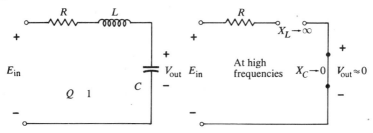

FIGURE 22–5 RLC *low-pass filter*.

Because of the combined effect of the *L* and *C*, this filter's response is much sharper than the *RC* or *RL* low-pass filter. In other words, the attenuation increases more rapidly at frequencies above the cutoff frequency for the *RLC* filter. In fact, the *RLC* low-pass has a -40-dB/decade roll-off while the *RL* and *RC* filters have only a -20-dB/decade roll-off.

Applications of Low-Pass Filters Low-pass filters are used to separate low frequencies from high frequencies. A common example is the filtering of the detector output of an AM receiver where the audio freqeuncies are passed while the *rf* carrier frequencies are blocked. Another common example is the filtering required to separate a dc voltage from a signal that contains both dc and ac.

EXAMPLE 22.2 Figure 22–6(a) shows an *RC* low-pass filter being driven from a source e_s that contains a dc component of 10 V and a 1-kHz ac component of 6 V_{p-p}. The waveform for e_s as shown in the figure shows the 6-V_{p-p} sine wave *riding* on the 10-V dc level. Rather than alternating above and below zero volts, this sine wave alternates above and below 10 V. This combination of dc and ac is often referred to as *pulsating* (changing) dc to distinguish it from pure dc. For this circuit determine the seatdy-state output voltage.

Solution: The pulsating dc input can be represented in the circuit diagram by a series combination of a 10-V dc source and a 6-V_{p-p}, 1-kHz ac source. This is

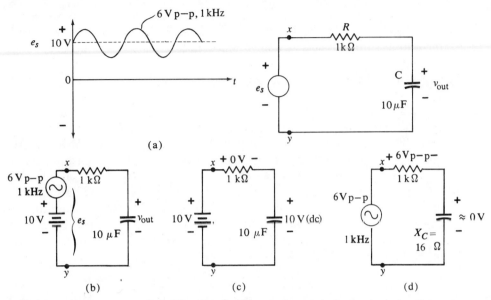

FIGURE 22-6 *Low-pass RC filter used to filter out pure dc from a pulsating dc signal.*

shown in part (b) of Fig. 22-6. The combination of these two sources produces the desired pulsating dc voltage between points x and y.

Since the circuit now contains *two* voltage sources, we can employ the method of *superposition* to determine the circuit voltages. Recall that the superposition method consists of solving the circuit using *one* source at a time. In part (c) of the figure only the dc source is shown; the ac source has been deenergized (its voltage set to zero). The *RC* circuits' response to the 10-V dc input is easily determined since we know that the capacitor will act as an open circuit in steady state and will charge up to the full 10 V. This of course means $V_R = 0$.

In part (d) of the figure only the ac source is considered. To determine the *RC* circuit's response to this 1-kHz voltage, we must first calculate X_C.

$$X_C = \frac{1}{2\pi \times 10^3 \times 10^{-5}} \approx 16 \, \Omega$$

Since X_c is so small compared to *R*, it should be obvious that most of the ac input voltage will appear across *R* and very little across *C*. Without making any further calculations we can safely approximate $V_{out} \approx 0$.

The results of this analysis can be summarized as follows: (1) the low-pass filter completely passes the dc portion of the input signal to the output (capacitor voltage); (2) it completely attenuates the ac portion of the input so that the output contains no ac.

Cascading Low-Pass Filters If a single low-pass filter does not provide a sufficient degree of filtering, then it is possible to connect two or more low-pass filters so that their filtering actions combine to give a better overall filter action. In Fig. 22-7(a), two low-pass filters are connected so that the output of one serves as the input of the other. This is called *cascading* the two filter circuits.

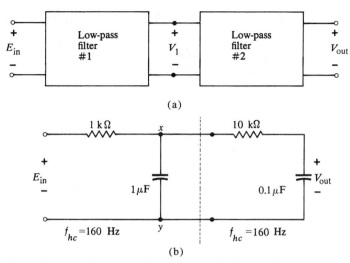

FIGURE 22–7 *Cascading two low-pass filters to provide better filtering.*

The voltage signal to be filtered is fed to the input of the first low-pass filter. This first filter attenuates the high frequencies of the input E_{in} so that its output V_1 contains only a small amount of high-frequency ac. This output V_1 is then fed to the second low-pass filter which takes V_1 and further attenuates the already diminished high-frequency content. This additional high-frequency attenuation provided by the second low-pass filter produces a final output V_{out} which contains much less high-frequency ac than would have been possible using only one filter section.

Figure 22–7(b) shows two RC low-pass filters connected in cascade to provide better filtering. The two filters are both designed to have the same high-frequency cutoff f_{hc}. Since $f_{hc} = 1/2\pi RC$, then the $R \times C$ products of the two filters have to be the same. The first filter has $R = 1$ kΩ and $C = 1$ μF so that $R \times C = 10^{-3}$ and $f_{hc} = 160$ Hz. The second filter has $R = 10$ kΩ and $C = 0.1$ μF so that $RC = 10^{-3}$ and $f_{hc} = 160$ Hz. The reason for choosing a 10 times larger R and a 10 times smaller C for the second filter is to prevent the second filter from affecting the operation of the first filter. The much larger R and smaller C (larger X_C) of the second filter presents a large total impedance connected across the output of the first filter. This large impedance will not *load* down the first filter. As such, the first filter will act as it normally would with nothing connected to its output.

With both filters designed for the same high-frequency cutoff of 160 Hz, the result of cascading the two filters is to provide a sharper cutoff response than can be obtained with one filter. Since each filter has a roll-off of -20 dB/decade past f_{hc}, the cascaded combination will have a roll-off of $(-20$ dB$) + (-20$ dB$) = -40$ dB/decade. In other words, cascading the filters produces an attenuation roll-off equal to the sum of each filter's roll-off. This of course produces sharper filter action.

22.5 High-Pass Filter Circuits

We have previously discussed RC and RL high-pass filters. Figure 22–8(a) shows these two types of high-pass filter circuits. In the RC high-pass, the lower frequency input voltages are *blocked* by the capacitor so that they cannot reach the output. In the *RL* high-pass, the lower frequencies are *shunted* by the inductor, essentially shorting out the output. Both of these filters have a roll-off of -20 dB/decade for frequencies below f_{lc}. In other words, the filter attenuation increases by -20 dB for each tenfold *decrease* in frequency below the low-frequency cutoff.

RLC High-Pass Filter The *RLC* high-pass filter shown in part (b) of Fig. 22–8 combines the capacitor blocking effect of the *RC* high-pass with the inductor shunting effect of the *RL* high-pass to produce a sharper filter response. Because of this combined effect of *L* and *C*, the attenuation of this filter increases rapidly at frequencies below the low-frequency cutoff f_{lc}. The *RLC* high-pass has a -40 dB/decade roll-off, which is twice as sharp as the *RL* or *RC* high-pass filters. f_{lc} for this filter is equal to f_r, its resonant frequency. Once again $Q \approx 1$ for proper operation.

Applications of High-Pass Filters High-pass filters are used to separate high-frequency ac from low-frequency ac, and for separating ac from dc. A common application of a high-pass filter is in the coupling between amplifier stages. The output of an amplifier typically consists of an ac signal superimposed on a dc bias level. However, when the output of an amplifier is being fed to the input of a second amplifier, it is usually desired that only the ac portion of the signal be allowed to pass to the second amplifier. In other words, the dc portion of the first amplifier's output is to be blocked from reaching the second amplifier's input without attenuating the ac signal. Figure 22–9 shows an *RC* high-pass network used as a coupling circuit in such an application.

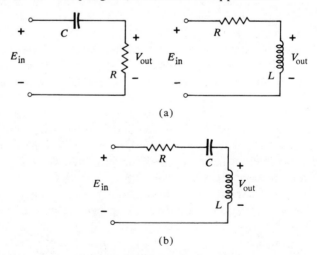

FIGURE 22–8 (a) RC *and* RL *high-pass filters.* (b) RLC *high-pass filters.*

The *R* and *C* values are chosen so that the filter's low-frequency cutoff is lower than any ac frequencies which are to be passed from *A*1 to *A*2. The capacitor will of course provide perfect attenuation of the dc on the output of *A*1. Usually, *R* represents the input resistance of amplifier *A*2. Problem 22–8 at the end of this chapter is devoted to this particular circuit.

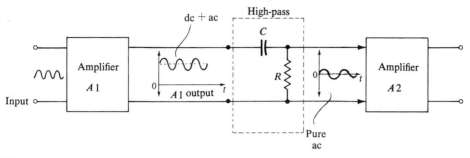

FIGURE 22-9 Use of a high-pass network as coupling between amplifiers.

22.6 Band-Pass Filter Circuits

A band-pass filter must block frequencies above and below a certain band of frequencies. There are two basic methods for achieving band-pass filter action. One method is simply to combine a basic high-pass filter with a basic low-pass filter; the other method uses some form of resonant circuit.

RC Band-Pass Filter Figure 22-10 shows an RC band-pass filter. It consists of a low-pass filter stage followed by a high-pass filter stage. Intuitively, then, it appears that we can analyze this band-pass filter by simply analyzing the individual stages. This is not quite true however unless we can guarantee that the high-pass components $C2$ and $R2$ do not produce a loading effect on the low-pass filter. In other words, the series combination of $C2$ and $R2$ is in parallel with $C1$; in order to treat the low-pass filter independently of the high-pass filter, the impedance of the $C2$-$R2$ combination must be at least ten times greater than the reactance of $C1$. This can be guaranteed if we specify that $C2$ should always be less than $0.1 \times C1$ so that $X_{C2} > 10 X_{C1}$ at all frequencies; then, we can assume a negligible loading effect of $C2$ and $R2$ upon $C1$.

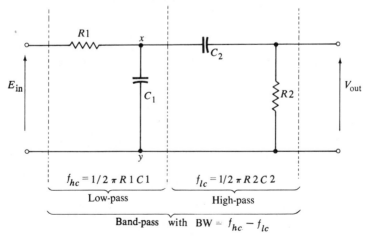

FIGURE 22-10 RC band-pass filter.

The low-pass filter consisting of $R1$ and $C1$ will attenuate the higher frequencies of the input while passing the lower frequencies. The cutoff frequency of this low-pass stage is of course simply

$$f_{hc} = 1/(2\pi R_1 C_1)$$

Input frequencies below f_{hc} will be passed from the input terminals to x-y, the output of the low-pass filter. These lower frequencies are then fed to the high-pass filter. The high-pass filter will attenuate any of these frequencies which are below its cut-off frequency

$$f_{lc} = 1/(2\pi R_2 C_2)$$

Thus, the only input frequencies which will be passed to the output are those which pass through both filters. In other words, frequencies which are below f_{hc} and above f_{lc}. The band-pass filter bandwidth is therefore equal to $f_{hc} - f_{lc}$. Obviously, in order for the circuit to function as a band-pass filter, the R and C values must be chosen so that f_{hc} is greater than f_{lc}. In fact for a well defined bandwidth, f_{hc} must be at least *twice* f_{lc}.

EXAMPLE 22.3 Design a band-pass filter that will pass frequencies below 20 kHz and has a bandwidth of 18 kHz.

Solution: $f_{hc} = 1/(2\pi R_1 C_1) = 20$ kHz
therefore,

$$R_1 C_1 = \frac{1}{2\pi f_{hc}} = \frac{1}{2\pi \times 20 \times 10^3} = 8 \times 10^{-6}$$

We can choose any values of R_1 and C_1 to satisfy this requirement. Thus, we can use $R_1 = $ **1 kΩ** so that

$$C_1 = \frac{8 \times 10^{-6}}{10^3} = 8 \times 10^{-9} = \mathbf{8000 \text{ pF}}$$

Since BW = 18 kHz, then f_{lc} must be 20 kHz $-$ 18 kHz = 2 kHz. Thus,

$$f_{lc} = 1/2\pi R_2 C_2 = 2 \text{ kHz}$$

so that

$$R_2 C_2 = 1/2\pi f_{lc} = \frac{1}{2\pi \times 2000} = 80 \times 10^{-6}$$

The value of C_2 must be chosen at least *ten* times smaller than C_1 to prevent loading. Let us choose $C_2 = $ **500 pF**. Thus,

$$R_2 = \frac{80 \times 10^{-6}}{C_2} = \frac{80 \times 10^{-6}}{500 \times 10^{-12}}$$
$$= 0.16 \times 10^6 \, \Omega = \mathbf{160 \text{ k}\Omega}$$

Resonant Band-Pass Filters Tuned resonant circuits provide a convenient means of filtering a band of frequencies, especially radio frequencies where relatively small values of L and C can be used for resonance. A resonant circuit produces filtering action by virtue of the fact that it exhibits certain properties over a band of frequencies centered around resonance. As we know, the width of this band of frequencies depends on the circuit Q, a higher Q produces a narrower bandwidth.

Figure 22–11(a) shows a band-pass filter circuit which employs series resonance. Inductor L and capacitor C constitute the series resonant components and are connected in series with a load resistor R_L. The coil's own resistance is not included; we will assume it is much smaller than R_L. The output voltage is taken across R_L. As we know, the series LC combination will have *minimum* impedance at and near its resonant frequencies. Thus, at these frequencies most of the applied voltage will appear across R_L. At frequencies outside the bandwidth of the LC circuit, the impedance of the LC combination increases rapidly; this will block the input voltage from reaching R_L. This action means that V_{out} will be maximum over a band of frequencies around the f_r of the series LC combination.

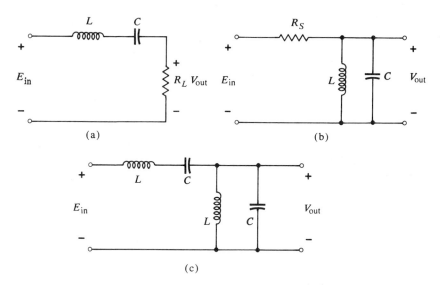

FIGURE 22–11 *Resonant band-pass filters.*

Figure 22–11(b) shows a band-pass filter which utilizes a parallel resonant tank circuit. The filter output is taken across the tank circuit. At frequencies near resonance the tank circuit impedance Z_T is maximum. The series resistor R_S and the tank circuit impedance form a voltage divider. Thus, frequencies near f_r will produce maximum voltage across the output and minimum voltage across R_S. This, of course, is band-pass operation.

Sharper band-pass filtering can be provided by combining the series and parallel resonant circuits. Figure 22–11(c) shows one such arrangement. The series resonant circuit is in series with the resonant tank circuit. The L and C values are chosen so that both resonant circuits have the same resonant frequency f_r. Thus, at frequencies at and near f_r the impedance of the series resonant circuit will be minimum and the tank circuit impedance will be maximum. This will produce maximum output voltage across the tank circuit. Away from resonance, the series LC impedance increases while the tank circuit impedance decreases. This causes the output voltage to decrease rapidly, providing sharper cutoff outside the filter bandwidth. In each of the resonant band-pass filters, a larger Q produces a narrower bandwidth.

22.7 Band-Reject Filter Circuits

Effective band-reject filters can also be constructed using tuned resonant circuits. Figure 22–12 shows three of these circuits. They may be recognized as the same as the band-pass circuits of Fig. 22–11 except that the outputs are taken across the *opposite* portions of the circuit. In the circuit in (a) the series LC has a very low impedance at resonance; thus, V_{out} will be *low* at frequencies near f_r. At frequencies away from resonance the LC impedance increases rapidly; this causes V_{out} to increase, thus producing band-reject operation. Similar analysis will show that the circuits in (b) and (c) also produce band-reject operation. In (c) a sharper band-reject characteristic is produced by using series and parallel resonant circuits tuned to the same f_r. Band-reject filters are also often referred to as *band-stop* filters or *wavetraps*.

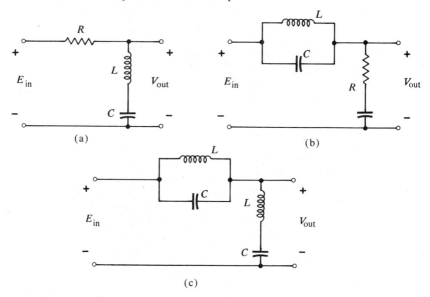

FIGURE 22–12 *Resonant band-reject filters.*

22.8 Phase-Shifting Networks

Occasionally in electronics a circuit is required which will take an input voltage at a certain frequency and will produce an output voltage at the same frequency but which is out of phase with the input by a certain phase angle. When a circuit is used for such a purpose, it is called a *phase-shifting network*. Two circuits which are often used are the simple RC low-pass and high-pass circuits.

Figure 22–13 shows an RC low-pass circuit used as a phase-shifting circuit. The output voltage in this circuit will always *lag* the input voltage. The amount of phase lag can be determined as follows. Assume the input voltage is $E_S/\underline{0°}$. The circuit impedance is

$$\overline{Z} = Z/\underline{-\theta}$$

where

$$\tan \theta = \left(\frac{X_C}{R}\right) \quad \text{and} \quad Z = \sqrt{R^2 + X_C^2}$$

The angle for $\overline{Z}$ is negative since the circuit is capacitive.

$$\overline{I} = \frac{E_S\underline{/0°}}{Z\underline{/-\theta}} = \left(\frac{E}{Z}\right)\underline{/+\theta}$$

Thus, $\overline{V}_{out}$ can be found as

$$\overline{V}_{out} = \overline{I} \times \overline{X_C} = \frac{E_S}{Z}\underline{/+\theta} \times X_C\underline{/-90°}$$

$$\overline{V}_{out} = \left(\frac{E_S X_C}{Z}\right)\underline{/\theta - 90°}$$

Thus, $\overline{V}_{out}$ *lags* $\overline{E_S}$ by a phase angle equal to

$$\text{phase lag} = \theta - 90°$$

$$= \left[\arctan\left(\frac{X_C}{R}\right) - 90°\right] \quad (22\text{-}1)$$

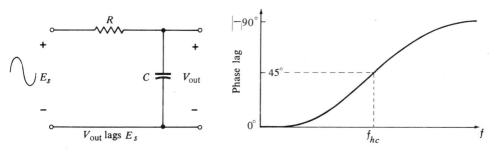

FIGURE 22-13 RC *low-pass circuit acts as a phase-lag network to shift the phase of the input sine wave.*

θ can have a value anywhere between 0° and 90° depending on the values of X_C and R. The phase lag therefore can be as much as $-90°$ when $\theta = 0°$. At low frequencies $X_C \gg R$ so that $\theta \simeq 90°$ and phase lag $\simeq 0°$. At high frequencies $X_C \simeq 0$ so that $\theta \simeq 0°$ and phase lag $\simeq -90°$. At $f = f_{hc}$, $X_C = R$ so that $\theta = 45°$ and phase lag $= -45°$. The variation of phase lag with frequency is shown in the figure.

When used as a phase-shifting network, the *RC* low-pass circuit is called a *phase-lag network* or simply a *lag network*. Any low-pass filter exhibits phase-lag properties and can be used as a lag network.

EXAMPLE 22.4 Choose values for R and C to produce a phase lag of $-30°$ at an input frequency of 25 kHz. Find the filter attenuation at this frequency.

Solution: phase lag $= -30° = \theta - 90°$

Thus, $\theta = 60°$. Since $\tan 60° = X_C/R$, then $X_C/R = 1.73$ so that $X_C = 1.73R$. Let's use $R = 1$ kΩ. This gives us $X_C = 1.73 \times 1$ k$\Omega = 1.73$ kΩ at 25 kHz. Thus,

$$X_C = \frac{1}{2\pi f C} = 1.73 \text{ k}\Omega$$

or

$$C = \frac{1}{2\pi f X_C} = 3680 \text{ pF}$$

The low-pass filter's attenuation at this frequency can be calculated as follows:

$$V_{out} = \left(\frac{X_C}{Z}\right) \times E_S = \frac{1.73 \text{ k}\Omega \times E_S}{\sqrt{1 \text{ k}\Omega^2 + 1.73 \text{ k}\Omega^2}} = 0.87 \, E_s$$

Thus, the attenuation is **0.87**.

Figure 22–14 shows an *RC* high-pass circuit used as a *phase-lead network*. The output $\overline{V}_{out}$ always leads the input voltage. The amount of phase lead is equal to the circuit phase angle θ as shown below:

$$\overline{V}_{out} = \overline{I} \times \overline{R} = \left(\frac{E_S}{Z}\right)\underline{/+\theta} \times R\underline{/0°} = \left(\frac{E_S R}{Z}\right)\underline{/+\theta}$$

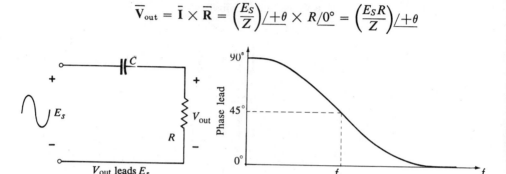

FIGURE 22–14 RC *high-pass circuit acts as a phase-lead network.*

Thus, $\overline{V}_{out}$ *leads* $\overline{E_S}$ by

$$\text{phase lead} = \theta = \arctan\left(\frac{X_C}{R}\right) \tag{22-2}$$

This phase lead can have a value anywhere from 0° to 90°. At low frequencies $X_C \gg R$ and phase lead $\simeq 90°$. At high frequencies $X_C \simeq 0$ and phase lead $\simeq 0°$. At $f = f_{lc}$, $X_C = R$ so that phase lead = 45°. A plot of phase lead versus f is shown in the figure.

The two phase-shifting networks that we have looked at can shift the phase of the input by as much as 90°. If a greater amount of phase shift is required, a number of *RC* circuits can be connected in cascade. One of the problems at the end of this chapter will illustrate this technique.

Chapter Summary

1. A low-pass filter attenuates high frequencies which exceed f_{hc}, the filter's high-frequency cutoff.

2. A high-pass filter attenuates frequencies below f_{lc}, the low-frequency cutoff.

3. A band-pass filter attenuates frequencies which fall *outside* its bandwidth; its bandwidth extends from f_{lc} to f_{hc}.

4. A band-reject filter attenuates frequencies that fall *within* its reject bandwidth between f_{lc} and f_{hc}.

5. A filter's response is said to be sharper if it's attenuation curve rolls off at a steeper slope outside the filter's pass band. A roll-off of -40 dB/decade is sharper than one of -20 dB/decade.

6. The two basic filter actions are *shunting* and *blocking*. Shunting refers to shorting out the output over a particular frequency range by an element (or elements) in *parallel* with the output. Blocking refers to preventing certain frequencies of voltage from reaching the output by placing an element (or elements) in *series* with the output.

7. Filters which use both L and C elements usually provide a sharper response than those which use only L or C.

8. For sharper filtering it is possible to connect filters in cascade so that the input voltage must pass through more than one filter before reaching the output.

9. In general, low-pass filters can also be used as phase-lag networks, and high-pass filters can be used as phase-lead networks.

Questions/Problems

Sections 22.2–22.3

22-1 Sketch the response curves for each of the following filters:
   a. high-pass with $f_{lc} = 500$ Hz
   b. low-pass with $f_{hc} = 10$ kHz
   c. band-pass filter with $f_{lc} = 10$ kHz and bandwidth $= 50$ kHz
   d. band-reject filter with $f_{hc} = 500$ Hz and bandwidth $= 480$ Hz

22-2 An audio signal containing frequencies in the range of 50 Hz-15 kHz is applied to the input of each of the filters in problem 22-1. Determine the frequency content of the filter outputs in each case.

22-3 A 10-kHz voltage is applied to the three low-pass filters described here. In which filter will this voltage have the greatest attenuation? (a) $f_{hc} = 5$ kHz, -40-dB/decade roll-off; (b) $f_{hc} = 1$ kHz, -20 dB/decade; (c) $f_{hc} = 500$ Hz, -60 dB/decade.

22-4 Which of the filters in problem 22-3 has the sharpest frequency response?

22-5 A certain voltage signal contains frequencies in the range from 100 Hz-100 kHz. This signal is to be passed through a filter network such that the output voltage signal contains only the following frequencies: 1 kHz to 5 kHz, and 20 kHz to 50 kHz. By combining two or more of the basic types of filters, show a *block* diagram of a filter system that will perform this filter action. Indicate the cutoff frequencies for each filter used.

Section 22.4

22-6 A pulsating dc voltage consists of a 100 mV p-p 120-Hz signal riding on a 12-V dc level. This signal is applied to each of the low-pass filters in Fig. 22-15. In each case determine the dc and ac components contained in output v_{out}. Which filter is better?

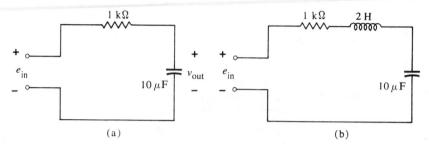

FIGURE 22-15

22-7 Determine the cutoff frequency and the roll-off of the filter in Fig. 22-16. Why aren't the R values and C values of each stage chosen to be the same?

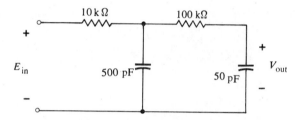

FIGURE 22-16

Section 22.5

22-8 In the circuit shown in Fig. 22-9, the amplifiers are audio amplifiers. (a) If the input audio frequencies range from 50 Hz-15 kHz, determine an appropriate value for C. Assume that the input resistance to A2 is $R = 1.5$ kΩ. (b) If C is chosen too small, how will it affect the circuit operation? (c) What do you think limits the maximum value you can make C?

Sections 22.6–22.7

22-9 Design an RC band-pass filter which will attenuate frequencies above 3 kHz and has a bandwidth of 2500 Hz.

22-10 A resonant band-pass circuit such as that shown in Fig. 22-11(b) is to be used to pass frequencies in a narrow band around 455 kHz. (a) If $L = 300$ μH choose an appropriate C value. (b) What should be the Q of the tank circuit if V_{out} is to be at least 0.9 E_{in} at resonance? Assume $R_s = 10$ kΩ.

22-11 Figure 22-17 shows several different filters. Some of them are modifications of, or combinations of, those covered in the chapter. Identify each filter as low-pass, high-pass, band-pass, or band-reject. Explain your answers.

Section 22.8

22-12 Design an RC phase-lead network that will provide a 60° phase shift at $f = 50$ kHz.

22-13 An electronic circuit called an *RC phase-shift oscillator* is shown in Fig. 22-18. The circuit consists of an amplifier whose output is fed to a phase-shifting network. The output of the phase-shifting network is in turn fed back to serve as the input to the amplifier. Thus, there is no external input voltage to the circuit (of course, there is the dc source used to bias the amplifier). This arrangement will function as a sine-wave oscillator producing a sine wave at the amplifier output. The frequency of the generated sine wave is determined by the phase-shift network. The sine-wave frequency will be the frequency at which the phase-shift network produces a 180° phase shift (the theory behind this cannot be explained here).

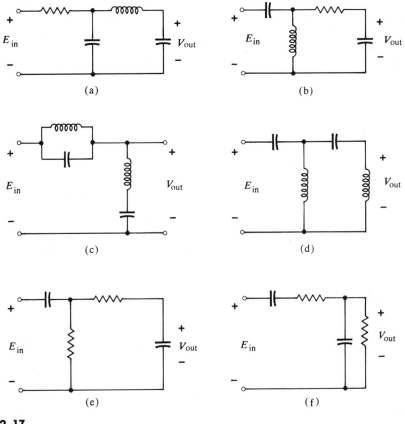

FIGURE 22-17

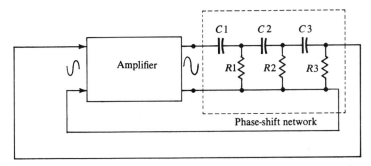

FIGURE 22-18

Thus, for a desired sine-wave frequency, the circuit values must be chosen to provide exactly 180° phase shift at that frequency. Since one RC section can only provide a maximum phase shift of 90°, it is necessary to cascade more than one RC section. To obtain 180° phase shift, three sections can be used with each section contributing 60° phase shift.

Choose values for each R and C to obtain the desired 180° phase shift at $f = 1$ kHz. Make sure that the second stage does not load down the first stage and that the third stage doesn't load down the second. Use $R1 = 1$ kΩ.

22-14 The oscillator in problem 22-13 can be used as a variable frequency oscillator by making $R1$ a variable resistor. If $R1$ is increased in value, what will then happen to the oscillator frequency (recall, oscillations occur only at frequency at which phase shift = 180°).

22-15 The band-pass filter in Fig. 22-19 is designed so that $f_{lc} = 2$ kHz and $f_{hc} = 20$ kHz. A technician constructs this circuit in the lab and makes measurements on its response by using a VOM to measure E_{in} and V_{out} at various frequencies. His measurements verify that $f_{hc} \approx 20$ kHz; however, the measurements show that f_{lc} is 4 kHz. Which of the following could be reasons for the discrepancy (explain each choice)?
a. effect of E_{in}'s source resistance (not shown)
b. C1 was out of tolerance
c. loading effect of VOM's input resistance

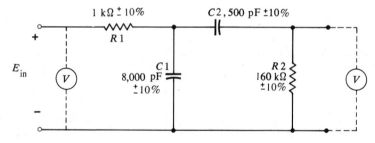

FIGURE 22-19

22-16 Repeat 22-15 except this time assume that f_{lc} measured okay but f_{hc} measured to be 12 kHz.

CHAPTER 23

AC Network Analysis

23.1 Introduction

In this chapter we will look at techniques which can be used to analyze the more complex impedance networks that may occur in electronic circuitry. The techniques which are presented will parallel those which we studied in the earlier chapters on resistive circuits. As we shall see, the major difference here will be that we are dealing with impedances rather than simple resistors so that we must consider phase angles as well as magnitudes.

23.2 Impedances in Series

Figure 23–1 shows a circuit where impedances Z_1 and Z_2 are connected in series across an ac source E_S. Z_1 and Z_2 can each represent any complex impedance (i.e., any combination of resistors, capacitors, and inductors). These

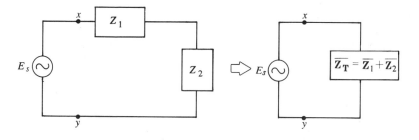

FIGURE 23–1 *Impedances in series are added vectorially to find total impedance.*

impedances can be represented in either polar form or rectangular form. In order to find the total series impedance Z_T seen by the source, we must add the Z_1 and Z_2 impedance vectors. That is,

$$\overline{Z}_T = \overline{Z}_1 + \overline{Z}_2 \qquad (23\text{--}1)$$

EXAMPLE 23.1 (a) Find $\overline{Z}_T$ for the circuit in Fig. 23-2. (b) Determine $\overline{I}$.

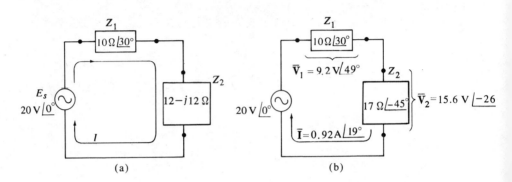

FIGURE 23-2

Solution: (a) To add Z_1 and Z_2 they must both be expressed in rectangular form. Thus,

$$\overline{Z}_1 = 10/\underline{30°} = 10\cos(30°) + j10\sin(30°)$$
$$= (8.66 + j5)\Omega$$
$$\overline{Z}_2 = (12 - j12)\Omega$$
$$\overline{Z}_T = \overline{Z}_1 + \overline{Z}_2 = (20.66 - j7)\Omega$$

Converting $\overline{Z}_T$ to polar form $Z_T/\underline{\theta}$

$$Z_T = \sqrt{20.66^2 + 7^2} = 21.8\ \Omega$$
$$\theta = \arctan\left(\frac{-7}{20.66}\right) = -19°$$

Thus, $\overline{Z}_T = 21.8\ \Omega/\underline{-19°}$ and is therefore capacitive.

(b) $\overline{I} = \overline{E}_S/\overline{Z}_T = 20\ \text{V}/\underline{0°}\,/21.8\ \Omega/\underline{-19°}$

$$\overline{I} = 0.92\ \text{A}/\underline{19°}$$

Voltage Divider Rule The voltage divider method which we used in analyzing series resistive circuits can also be applied to series impedance circuits. For $\overline{Z}_1$ and $\overline{Z}_2$ in series, the voltage across $\overline{Z}_1$ will be

$$\overline{V}_1 = \left(\frac{\overline{Z}_1}{\overline{Z}_1 + \overline{Z}_2}\right) \times \overline{E}_S \qquad (23\text{–}2a)$$

Similarly,

$$\overline{V}_2 = \frac{\overline{Z}_2}{\overline{Z}_1 + \overline{Z}_2} \times \overline{E}_S \qquad (23\text{–}2b)$$

These two voltage divider formulas are *vector* formulas where both the Zs are expressed as vectors.

EXAMPLE 23.2 Find $\overline{V}_1$ and $\overline{V}_2$ in Fig. 23-2 using the voltage divider formulas.

Solution: We have already determined $\overline{Z}_T = 21.8\ \Omega\underline{/-19°}$.

Thus,

$$\overline{V}_1 = \frac{\overline{Z}_1}{\overline{Z}_T} \times \overline{E}_S = \left(\frac{10\ \Omega\underline{/30°}}{21.8\ \Omega\underline{/-19°}}\right) \times 20\ V\underline{/0°}$$

$$= \left(\frac{10 \times 20}{21.8}\right)\underline{/49°}$$

$$\therefore\ \overline{V}_1 = 9.2\ V\underline{/49°}$$

$$\overline{V}_2 = \frac{17\ \Omega\underline{/-45°}}{21.8\ \Omega\underline{/-19°}} \times 20\ V\underline{/0°}$$

$$= \left(\frac{17 \times 20}{21.8}\right)\underline{/-26°}$$

$$\overline{V}_2 = 15.6\ V\underline{/-26°}$$

Note that the sum of the *magnitudes* of the individual impedance voltages does *not* equal the input 20 V. This of course is because the voltages are not in phase. The student can verify that

$$\overline{V}_1 + \overline{V}_2 = \overline{E}_S$$

to satisfy KVL.

To summarize this section we can conclude that series impedances are treated the same way as series resistances except that all math operations are *vector* operations. Thus, the series impedances have to be expressed in rectangular form in order to add them to find Z_T.

23.3 Impedances in Parallel

Figure 23-3 shows several impedances connected in parallel across an ac source E_S. To find the total impedance Z_T seen by the source, we can use the same basic formulas that we used for resistive circuits. That is,

$$\frac{1}{\overline{Z}_T} = \frac{1}{\overline{Z}_1} + \frac{1}{\overline{Z}_2} + \frac{1}{\overline{Z}_2} \ldots \text{etc.} \qquad (23\text{--}3)$$

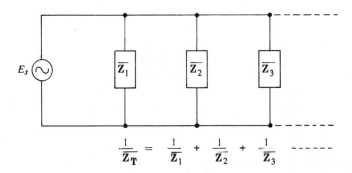

FIGURE 23-3 *Impedances in parallel are combined similar to resistors in parallel.*

where each parallel impedance is represented vectorially.

For the common case of *two* parallel impedances this relationship can be reduced to

$$\overline{Z}_T = \frac{\overline{Z}_1 \overline{Z}_2}{\overline{Z}_1 + \overline{Z}_2} \qquad (23\text{--}4)$$

which is the same "product-over-sum" form used for two parallel resistors.

EXAMPLE 23.3 (a) Determine $\overline{Z}_T$ for the circuit in Fig. 23-4. (b) Find $\overline{I}_1$, $\overline{I}_2$, and $\overline{I}_T$.

Solution: (a) $\overline{Z}_T = \dfrac{\overline{Z}_1 \overline{Z}_2}{\overline{Z}_1 + \overline{Z}_2}$

To add $\overline{Z}_1$ and $\overline{Z}_2$ we must convert them to rectangular form.

$$\begin{aligned}
\mathbf{Z}_1 &= 5 \text{ k}\Omega \underline{/60°} \\
&= 5 \text{ k}\Omega \cos(60°) + j5 \text{ k}\Omega \sin(60°) \\
&= 2.5 \text{ k}\Omega + j4.33 \text{ k}\Omega \\
\overline{Z}_2 &= 10 \text{ k}\Omega \underline{/-37°} \\
&= 10 \text{ k}\Omega \cos(-37°) + j10 \text{ k}\Omega \sin(-37°) \\
&= 8 \text{ k}\Omega - j6 \text{ k}\Omega
\end{aligned}$$

Thus, $\overline{Z}_1 + \overline{Z}_2 = 10.5 \text{ k}\Omega - j1.67 \text{ k}\Omega$
$= 10.6 \text{ k}\Omega \underline{/-9°}$ (in polar form)

$$\therefore \overline{Z}_T = \frac{(5 \text{ k}\Omega \underline{/60°})(10 \text{ k}\Omega \underline{/-37°})}{10.6 \text{ k}\Omega \underline{/-9°}}$$

$$= \mathbf{4.7 \text{ k}\Omega \underline{/32°}}$$

AC Network Analysis   665

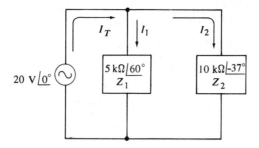

FIGURE 23-4

(b) $\overline{I}_1 = \dfrac{\overline{E}_s}{\overline{Z}_1} = \dfrac{20 \text{ V}\underline{/0°}}{5 \text{ k}\Omega\underline{/60°}} = 4 \text{ mA}\underline{/-60°}$

$\overline{I}_2 = \dfrac{\overline{E}_s}{\overline{Z}_2} = \dfrac{20 \text{ V}\underline{/0°}}{10 \text{ k}\Omega\underline{/-37°}} = 2 \text{ mA}\underline{/37°}$

$\overline{I}_T = \dfrac{\overline{E}_s}{\overline{Z}_T} = \dfrac{20 \text{ V}\underline{/0°}}{4.7 \text{ k}\Omega\underline{/32°}} = 4.3 \text{ mA}\underline{/-32°}$

The total source current $\overline{I}_T$ could also be found using Kirchhoff's current law (KCL) which states that

$$\overline{I}_T = \overline{I}_1 + \overline{I}_2$$

The student can verify that KCL is satisfied.

EXAMPLE 23.4 Figure 23-5 shows a parallel resonant tank circuit. Its resonant frequency is 400 kHz. (a) Find Z_T using the methods developed in the resonance chapter; (b) find Z_T using the parallel impedance formulas (23-4).

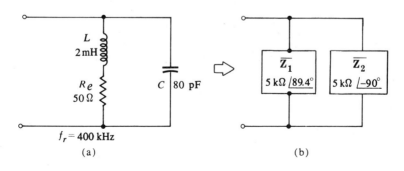

FIGURE 23-5

Solution: (a) $X_L = 2\pi f_R L = 5$ kΩ

$$Q = \frac{X_L}{R_e} = \frac{5 \text{ k}\Omega}{50 \text{ }\Omega} = 100$$

$$Z_T = QX_L = 100 \times 5 \text{ k}\Omega = 500 \text{ k}\Omega$$

We know from our work on resonance that this impedance is almost purely resistive at resonance. Thus,

$$Z_T \approx 500 \text{ k}\Omega/0°$$

(b) We can call the two parallel branches of the tank circuit Z_1 and Z_2 as shown in part (b) of the figure. Z_1 is the series combination of 5-kΩ X_L and 50-Ω R_e. Thus,

$$\overline{Z_1} = 50 \text{ }\Omega + j5 \text{ k}\Omega$$

$$\approx 5 \text{ k}\Omega\; \underline{/89.4°} \text{ (in polar form)}$$

$\overline{Z_2}$ is simply the capacitor's impedance

$$\overline{Z_2} = -j5 \text{ k}\Omega = 5 \text{ k}\Omega\underline{/-90°}$$

Thus,

$$\overline{Z_T} = \frac{\overline{Z_1}\,\overline{Z_2}}{\overline{Z_1} + \overline{Z_2}} = \frac{5 \text{ k}\Omega\underline{/89.4°} \times 5 \text{ k}\Omega\underline{/-90°}}{(50 + j5 \text{ k}\Omega) - j5 \text{ k}\Omega}$$

$$= \frac{25 \times 10^6 \underline{/-0.6°}}{50}$$

$$= 500 \text{ k}\Omega \underline{/-0.6°}$$

which agrees closely with the result in (a).

23.4 Conductance, Susceptance, and Admittance

In dealing with parallel impedances it is sometimes more useful to work with quantities in *siemens* rather than ohms. To illustrate, for three resistors in parallel we have

$$\frac{1}{R_T} = \frac{1}{R_1} + \frac{1}{R_2} + \frac{1}{R_3}$$

Recall from our early discussion of resistance that a quantity called conductance, G, is equal to the reciprocal of resistance. That is, $G = 1/R$ and has the units of $1/\Omega$ = siemens (S). Thus, for our three parallel resistors we have

$$G_T = G_1 + G_2 + G_3 \tag{7-3}$$

so that the total conductance equals the sum of the individual conductances in *parallel*.

Susceptance When dealing with capacitors or inductors in ac circuits we can likewise define susceptance, B, as being the reciprocal of reactance. That is,

$$\overline{B} = \frac{1}{\overline{X}} \qquad (23\text{-}5)$$

where the units of $\overline{B}$ are also 1/ohms = siemens. Since reactance $\mathbf{X}$ is actually a vector quantity, the susceptance $\mathbf{B}$ is also a vector quantity.

If we are dealing with capacitive reactance $\overline{X_C}$, then we can find the *capacitive susceptance* $\overline{B_C}$ as

$$\overline{B_C} = \frac{1}{\overline{X_C}} = \frac{1}{X_C\underline{/-90°}} = \left(\frac{1}{X_C}\right)\underline{/+90°} \qquad (23\text{-}6a)$$

$$= j\left(\frac{1}{X_C}\right) = jB_C \qquad (23\text{-}6b)$$

Similarly, for *inductive susceptance* $\overline{B_L}$, we have

$$\overline{B_L} = \frac{1}{\overline{X_L}} = \frac{1}{X_L\underline{/90°}} = \left(\frac{1}{X_L}\right)\underline{/-90°} \qquad (23\text{-}7a)$$

$$= -j\left(\frac{1}{X_L}\right) = -jB_L \qquad (23\text{-}7b)$$

Note that $\overline{B_C}$ has the opposite angle of $\overline{X_C}$ because of the reciprocal relationship. The same is true for $\overline{B_L}$ and $\overline{X_L}$.

Susceptances in *parallel* can be combined in the same manner as conductances. That is,

$$\overline{B_T} = \overline{B_1} + \overline{B_2} + \overline{B_3} \ldots \qquad (23\text{-}8)$$

Using Eq. (23-8) the total susceptance equals the sum of the individual parallel susceptances.

EXAMPLE 23.5 Four reactances are connected in parallel. Their values are $X_{C1} = 10$ ohms, $X_{C2} = 20$ ohms, $X_{L1} = 5$ ohms, $X_{L2} = 25$ ohms. Find total parallel reactance by using susceptances.

Solution: Convert each reactance to a susceptance.

$$\overline{B_{C1}} = \frac{1}{\overline{X_{C1}}} = \frac{1}{10\underline{/-90°}} = 0.1 \text{ S}\underline{/90°}$$

$$= 100 \text{ mS}\underline{/90°}$$

$$= j100 \text{ mS}$$

Similarly,

$$\overline{B_{C2}} = \frac{1}{20\underline{/-90°}} = j50 \text{ mS}$$

$$\overline{B_{L1}} = \frac{1}{5\underline{/90°}} = 200 \text{ mS}\underline{/-90°} = -j200 \text{ mS}$$

$$\overline{B_{L2}} = \frac{1}{25\underline{/90°}} = -j40 \text{ mS}$$

Total susceptance is thus

$$\overline{B_T} = \overline{B_{C1}} + \overline{B_{C2}} + \overline{B_{L1}} + \overline{B_{L2}}$$
$$= j100 + j50 - j200 - j40$$
$$= -j90 \text{ mS} = 90 \text{ mS}/\underline{-90°}$$

This total susceptance is inductive as shown by the $-j$. To convert back to total reactance $\overline{X_T}$ we have

$$\overline{X_T} = \frac{1}{\overline{B_T}} = \frac{1}{90 \text{ mS}/\underline{-90°}}$$
$$= 11.1 \text{ }\Omega/\underline{90°} \text{ (inductive)}$$

Admittances The reciprocal of resistance is conductance; the reciprocal of reactance is susceptance. The reciprocal of impedance is called admittance and is given the symbol Y. That is,

$$\overline{Y} = \frac{1}{\overline{Z}} \text{ siemens} \qquad (23\text{-}9)$$

Whereas impedance Z represents a measure of opposition to current, admittance Y represents a measure of ability to conduct current. That is,

$$\overline{I} = \frac{\overline{E}}{\overline{Z}} = \overline{E} \times \frac{1}{\overline{Z}} = \overline{E} \times \overline{Y}$$

which means that a larger $\overline{I}$ will result for a larger $\overline{Y}$.

EXAMPLE 23.6 A 40-ohm resistor is in series with a 30-ohm X_L. Find the impedance and admittance of this series combination.

Solution: $\overline{Z} = 40 + j30$
$$= 50 \text{ }\Omega/\underline{37°}$$

Thus,

$$\overline{Y} = \frac{1}{\overline{Z}} = \frac{1}{50/\underline{37°}} = 0.02 \text{ S}/\underline{-37°} = 20 \text{ mS}/\underline{-37°}$$

$\overline{Y}$ can be converted to rectangular form as

$$\overline{Y} = 20 \cos(37°) - j20 \sin(37°)$$
$$= (16 - j12) \text{ mS}$$

Note that the angle for $\overline{Y}$ has the opposite polarity of the angle for $\overline{Z}$. As such, the j term for $\overline{Y}$ has the opposite sign of the j term for Z. This is always true.

When impedances are connected in parallel, we may use admittances to help find total impedance. Since

$$\frac{1}{\overline{Z_T}} = \frac{1}{\overline{Z_1}} + \frac{1}{\overline{Z_2}} + \frac{1}{\overline{Z_3}} + \ldots \text{ etc.} \qquad (23\text{-}3)$$

then,

$$\overline{Y}_T = \overline{Y}_1 + \overline{Y}_2 + \overline{Y}_3 + \ldots \text{ etc.} \qquad (23\text{--}10)$$

Thus, the total admittance equals the sum of individual admittances connected in *parallel*.

EXAMPLE 23.7 $\overline{Z}_1 = 10 + j10$ and $\overline{Z}_2 = 20\,\Omega\underline{/30°}$ are connected in parallel. Find $\overline{Z}_T$ using admittances.

Solution: $\overline{Y}_1 = \dfrac{1}{\overline{Z}_1}$

We must convert $\overline{Z}_1$ to polar form to use in this formula.

$$\overline{Z}_1 = 14.1\underline{/45°}$$

Thus,

$$\overline{Y}_1 = \dfrac{1}{14.1\underline{/45°}} = 70.7 \text{ mS}\underline{/-45°}$$

$$\overline{Y}_2 = \dfrac{1}{\overline{Z}_2} = 50 \text{ mS}\underline{/-30°}$$

$$\overline{Y}_T = \overline{Y}_1 + \overline{Y}_2$$

To perform this sum, Y_1 and Y_2 must be converted to rectangular form.

$$\overline{Y}_1 = 70.7 \text{ mS}\underline{/-45°}$$
$$= (50 - j50) \text{ mS}$$
$$\overline{Y}_2 = 50 \text{ mS}\underline{/-30°}$$
$$= (43.3 - j25) \text{ mS}$$
$$\therefore \overline{Y}_T = (93.3 - j75) \text{ mS}$$
$$= 120 \text{ mS}\underline{/-39°}$$

$$\overline{Z}_T = \dfrac{1}{\overline{Y}_T} = \dfrac{1}{120 \times 10^{-3}\underline{/-39°}}$$
$$= \mathbf{8.33\ \Omega\underline{/39°}}$$

Series and Parallel Equivalences The general rectangular form for an impedance is

$$\overline{Z} = R \pm jX \qquad (23\text{--}11)$$

where the j term will be $+jX_L$ if the impedance is inductive or $-jX_C$ if the impedance is capacitive. The real term R of course represents resistance. The total impedance of any network, simple or complex, can be expressed in this form. Thus, any total impedance can be represented by a resistance R *in series* with a reactance X. For example, suppose a certain complex circuit has $\overline{Z}_T = 100\,\Omega\underline{/60°}$. Converting to rectangular form,

$$\overline{Z}_T = (86.6 + j50)\Omega$$

Thus, Z_T can be viewed as consisting of an 86.6-ohm resistor connected in series with a 50-ohm inductive reactance. *For any complex linear network, the total impedance can be viewed as a simple series combination of an R and an X.*
In a like manner, the general rectangular form for an admittance is

$$\overline{Y} = G \pm jB$$

where the j term will be $+jB_C$ for a capacitive circuit or $-jB_L$ for an inductive circuit. The real term G represents conductance. The total admittance of any network can be expressed in this form. Thus, any total Y_T can be represented by a conductance G *in parallel* with a susceptance B. For example, suppose a certain circuit has $\overline{Y_T}$ = 120 mS$/-39°$ (ex. 23.7). Converting to rectangular form,

$$\overline{Y_T} = (93.3 - j75) \text{ mS}$$

Thus, $\overline{Y_T}$ can be viewed as consisting of a 93.3-mS conductance in parallel with a 75-mS inductive susceptance. *In general, then, any circuit's total admittance can be viewed as a simple parallel combination of a G and a B.*

These two ways of viewing any complex circuit are illustrated in Fig. 23–6.

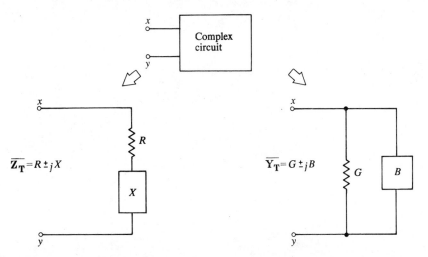

FIGURE 23–6 *Any complex circuit can be represented as a simple series circuit or as a simple parallel circuit.*

Both viewpoints are equivalent. Whichever one is used depends on which one is more convenient for a given problem or situation.

EXAMPLE 23.8 A certain network has $\overline{Z_T}$ = 8 k$\Omega$$/22°$. (a) Draw an equivalent series circuit representing this network; (b) draw an equivalent parallel circuit.

Solution: $\overline{Z_T}$ = 8 k$\Omega$$/22°$
= 8 kΩ cos (22°) + j8 kΩ sin (22°)
= 7.4 kΩ + j3 kΩ

Thus the equivalent series circuit is shown in Fig. 23-7(a) as a 7.4-kΩ resistor and a 3-kΩ inductive reactance.

$$\mathbf{Y_T} = \frac{1}{\mathbf{Z_T}} = \frac{1}{8 \text{ k}\Omega/22°} = 0.125 \text{ mS}/{-22°}$$

$$= 125 \text{ }\mu\text{S}/{-22°}$$

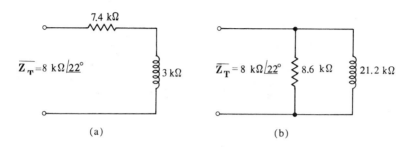

FIGURE 23-7

Converting to rectangular form

$$\mathbf{Y_T} = (116 - j47) \text{ }\mu\text{S}$$

Thus the equivalent parallel circuit consists of a 116 μS conductance and a 47-μS *inductive* susceptance. The conductance can be converted to a resistance.

$$R = \frac{1}{G} = \frac{1}{116 \times 10^{-6}} = 8.6 \text{ k}\Omega$$

and the inductive susceptance can be converted to a reactance

$$X_L = \frac{1}{B_L} = \frac{1}{47 \times 10^{-6}} = 21.2 \text{ k}\Omega$$

This equivalent parallel circuit is shown in Fig. 23-7(b).

Note that both of these equivalent circuits contain resistance and inductance, although different values. This is because the original network is inductive ($\mathbf{Z_T}$ = 8 kΩ/22°). If the original network were capacitive, the equivalent circuits would both contain an R and X_C.

23.5 Series-Parallel Impedances

The analysis of ac circuits that contain both series and parallel impedances is performed using the same techniques which we used in analyzing series-parallel resistor circuits in chap. 8. Of course, here we are dealing with complex impedances rather than pure resistances so that the arithmetic is much more cumbersome. Rather than developing formal rules or steps, the following example will serve to illustrate the techniques used in analyzing a series-parallel network.

EXAMPLE 23.9 Determine the source current for the circuit of Fig. 23–8(a).

Solution: The procedure is to find the total impedance as seen by the source so that the source current can be calculated. Z_2 and Z_3 are in parallel and can be combined as

$$\overline{Z}_2 \| \overline{Z}_3 = \frac{\overline{Z}_2 \times \overline{Z}_3}{\overline{Z}_2 + \overline{Z}_3}$$

$\overline{Z}_2$ and $\overline{Z}_3$ have to be converted to rectangular form so that they can be added.

$$\overline{Z}_2 = 50\ \Omega\underline{/-70°} = (17.1 - j47)\Omega$$
$$\overline{Z}_3 = 100\ \Omega\underline{/45°} = (70.7 + j70.7)\Omega$$

Thus,

$$\overline{Z}_2 + \overline{Z}_3 = (87.8 + j23.7)\Omega$$
$$= 91\ \Omega\underline{/15°}$$

Therefore,

$$\overline{Z}_2 \| \overline{Z}_3 = \frac{50\underline{/-70°} \times 100\underline{/45°}}{91\underline{/15°}}$$
$$= 55\ \Omega\underline{/-40°}$$

The equivalent circuit is shown in part (b) of Fig. 23–8. The total impedance seen by the source is therefore

$$\overline{Z}_T = \overline{Z}_1 + \overline{Z}_2 \| \overline{Z}_3$$
$$= 75\ \Omega\underline{/53°} + 55\ \Omega\underline{/-40°}$$
$$= (45 + j60) + (42.1 - j34.4)$$
$$= (87.1 + j25.6)\Omega$$
$$= 90.7\ \Omega\underline{/16.4°}$$

Thus, the total source current is

$$\overline{I}_S = \frac{\overline{E}_S}{\overline{Z}_T} = \frac{50\ V\underline{/0°}}{90.7\ \Omega\underline{/16.4°}}$$
$$= \mathbf{0.55\ A\underline{/-16.4°}}$$

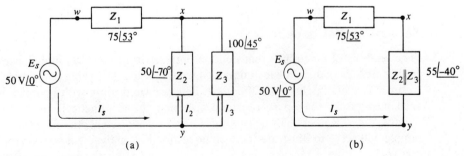

FIGURE 23–8

23.6 The Superposition Method

Any circuit containing only linear elements and more than one source can be analyzed using the *principle of superposition*. Recall that the method of superposition simply means that the circuit is analyzed by considering only one source at a time. The results of each analysis are combined (superimposed) to give the total solution.

To illustrate, consider the circuit in Fig. 23–9(a). It contains two ac sources $\overline{E}_1$ and $\overline{E}_2$. Suppose it is desired to find the current through the load impedance $\overline{Z}_L$. The steps involved in the superposition process are illustrated in (b)-(d) of the same figure. In (b) the $\overline{E}_2$ source has been de-energized and replaced by a short circuit (it is assumed that $\overline{E}_2$'s internal impedance is included in $\overline{Z}_2$). This circuit is now a single-source series-parallel circuit which can be analyzed to find the current through $\overline{Z}_L$ produced by the source $\overline{E}_1$. This current is labeled $\overline{I}_{L1}$ since it is that portion of the total $\overline{I}_L$ which is due to the $\overline{E}_1$ source alone.

In (c) the same process is carried out, this time with $\overline{E}_1$ replaced by a short. The current through $\overline{Z}_L$ due to the $\overline{E}_2$ source is labeled $\overline{I}_{L2}$. The total solution [part (d) of figure] for the total current through $\overline{Z}_L$ is obtained by combining the results of (b) and (c). That is,

$$\overline{I}_L = \overline{I}_{L1} + \overline{I}_{L2}$$

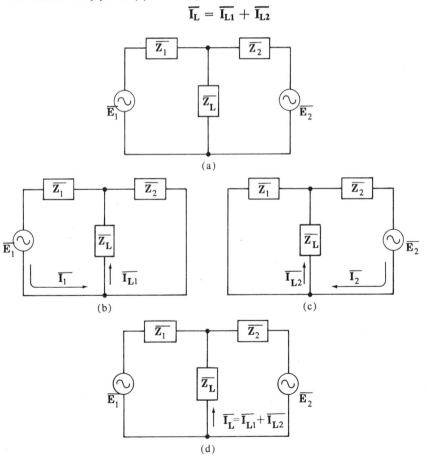

FIGURE 23–9 *Illustration of steps involved in using the method of superposition.*

Note that the combination is a vector addition since $\overline{I}_{L1}$ and $\overline{I}_{L2}$ will not generally be in phase.

Sources With Different Frequencies If a network contains two or more sources that are not all the same frequency, the superposition method can still be used provided that the circuit impedances are calculated at the frequency of the source which is being used. In other words, as each source is considered one at a time, the impedance values in the circuit must correspond to the frequency of the particular source being considered. The most typical example of this situation occurs in circuits where a dc source and an ac source are used at the same time.

Figure 23–10(a) shows a network which is present in certain types of voltage amplifiers. The dc source V_{GG} is used to provide a dc *bias* level for the amplifier input voltage. The ac source represents the signal to be amplified; this signal is coupled to the amplifier input via the coupling capacitor. The R_{in} represents the amplifier's input resistance. The amplifier input terminals are x-y. The total amplifier input voltage (V_{xy}) consists of a dc component due to V_{GG} plus an ac component due to e_s.

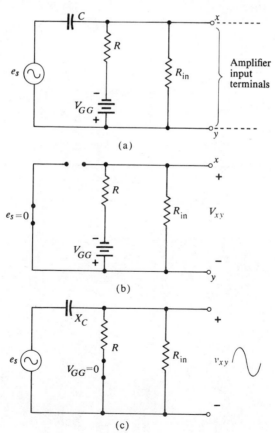

FIGURE 23–10 *Superposition method applied to a network containing a dc source and an ac source.*

Parts (b) and (c) of the figure show the steps involved in determining the total amplifier input voltage using the superposition method. In (b) the ac source has been de-energized so that the dc solution can be obtained. For dc, of course, the coupling capacitor acts as an open circuit (neglecting leakage). The dc voltage V_{xy} is easily calculated from the simplified circuit. In (c) the dc source has been de-energized so that the ac solution can be determined. The reactance X_C of the coupling capacitor is calculated at the frequency of the ac source. The ac voltage v_{xy} is calculated using any of the methods previously discussed. Usually in an amplifier, X_C is made small enough to insure that most of the e_s signal reaches the amplifier input.

23.7 Thevenin's Theorem

In chap. 10 we found that we could replace any linear resistive network with its *Thevenin equivalent circuit* consisting of a voltage source in series with a resistance. According to Thevenin's theorem this technique can be extended to impedance networks with ac sources. The Thevenin equivalent circuit for such networks will consist of an ac source E_{Th} in series with a Thevenin equivalent impedance Z_{Th}. This is illustrated in Fig. 23–11. There are situations where Z_{Th} will be purely resistive, but in general it will be a complex impedance.

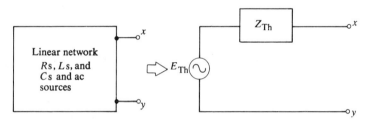

FIGURE 23–11 *Thevenin equivalent circuit for impedance networks.*

The equivalent source E_{Th} is calculated or measured as the voltage between terminals x-y with no load connected to the network. The Thevenin impedance Z_{Th} is determined as the total equivalent impedance seen looking into terminals x-y with all the network sources de-energized. The Thevenin equivalent circuit essentially represents a *practical* ac voltage source whose effect on any load connected to its terminals x-y will be the same as the effect produced by the original network on the load.

EXAMPLE 23.10 The circuit in Fig. 23–12(a) is a simple *RC* low-pass filter. It is desired to calculate the filter's output voltage for many different loads on the filter output. The process can be somewhat simplified by finding the Thevenin equivalent of the filter at its output terminals. Find this equivalent circuit for $f = 1$ kHz.

Solution: The Thevenin equivalent source E_{Th} is found by calculating V_{xy} with no load connected. At 1 kHz, X_C will be

$$X_C = \frac{1}{2\pi \times 10^3 \times 10^{-6}} \approx 160 \; \Omega$$

Using the voltage divider method:

$$E_{Th} = \overline{V}_{xy} = \left(\frac{-j160}{120 - j160}\right) \times 10 \text{ V}\underline{/0°}$$

$$= \frac{160\underline{/-90°}}{200\underline{/-53°}} \times 10 \text{ V}\underline{/0°}$$

$$= 8 \text{ V}\underline{/-37°}$$

The Thevenin equivalent impedance is calculated as shown in Fig. 23–12(b).

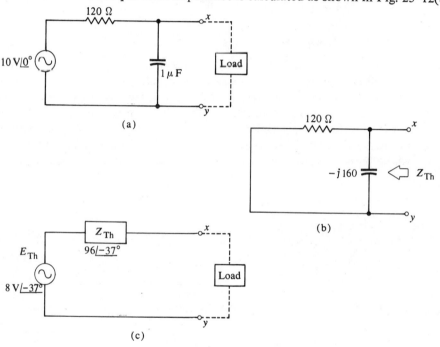

FIGURE 23–12

The source is de-energized ($E_S = 0$) and Z_{Th} is the parallel combination of 120 ohms and $-j160$ ohms.

$$\overline{Z}_{Th} = \frac{(120) \times (-j160)}{120 - j160}$$

$$= \frac{120\underline{/0°} \times 160\underline{/-90°}}{200\underline{/-53°}}$$

$$= 96 \text{ }\Omega\underline{/-37°}$$

The complete Thevenin equivalent is shown in part (c) of the figure. It is important to realize that this equivalent circuit is valid only for $f = 1$ kHz. If f is changed, both E_{Th} and Z_{Th} will be different because X_C will have changed. Now that the Thevenin equivalent has been determined, the various loads can be applied to the equivalent circuit to easily calculate the load voltage and current.

23.8 Norton Equivalent Circuits

The companion to the Thevenin equivalent circuit is the Norton equivalent circuit. The Norton equivalent for an ac network consists of a *current* source in parallel with an impedance (see Fig. 23–13). The Norton equivalent circuit essentially represents a *practical* ac current source with a current equal to I_N and a source impedance equal to Z_N. The value of I_N is determined by calculating or measuring the current through a short circuit placed across x-y. Z_N is found in the same way as Z_{Th}; in fact, they are identical:

$$Z_N \equiv Z_{Th} \tag{23–15}$$

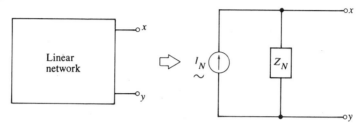

FIGURE 23–13 *Norton equivalent circuit for an ac network.*

The value of I_N is also related to E_{Th} for the same network

$$I_N = \frac{E_{Th}}{Z_{Th}} \tag{23–16}$$

Thus, if a network's Thevenin equivalent circuit is known, its Norton equivalent can be found using these two relationships.

The Norton current source I_N will represent an ac current (usually rms value) if the network being *Nortonized* is an ac network. This current will be supplied to any load connected to the Norton equivalent circuit. Of course, some of the current will flow through the parallel Z_N and will not be available for the load.

EXAMPLE 23.11 Determine the Norton equivalent network for the low-pass filter of Fig. 23–12. Then determine the current through a 100-ohm resistive load connected to x-y.

Solution: The value of I_N is determined by finding the current through a short circuit across the output terminals [see Fig. 23–14(a)]. It should be clear that all the current from the source will flow through the short. The value of this current is

$$\overline{I_N} = \frac{10 \text{ V}\underline{/0°}}{120 \text{ }\Omega\underline{/0°}} = 83.3 \text{ mA}\underline{/0°}$$

The value of Z_N is determined in the same manner as Z_{Th}. This value was previously calculated as

$$\overline{Z_N} = \overline{Z_{Th}} = 96 \text{ }\Omega\underline{/-37°}$$

The complete Norton equivalent is shown in Fig. 23–14(b).

In part (c) of the figure, the 100-ohm load is connected to the Norton equivalent circuit. The total parallel impedance at the x-y terminals is $R_L \parallel \overline{Z}_N$. Thus,

$$\overline{Z}_{xy} = 100\ \Omega\underline{/0°} \parallel 96\ \Omega\underline{/-37°}$$
$$= 51.6\ \Omega\underline{/-18.8°}$$

We have, therefore,

$$\overline{V}_{xy} = \overline{I}_N \times \overline{Z}_{xy} = 83.3\ \text{mA}\underline{/0°} \times 51.6\ \Omega\underline{/-18.8°}$$
$$= 4.3\ \text{V}\underline{/-18.8°}$$

This voltage is across R_L. Therefore,

$$\overline{I}_L = \frac{\overline{V}_{xy}}{R_L} = \frac{4.3\ \text{V}\underline{/-18.8°}}{100\ \Omega\underline{/0°}} = 43\ \text{mA}\underline{/-18.8°}$$

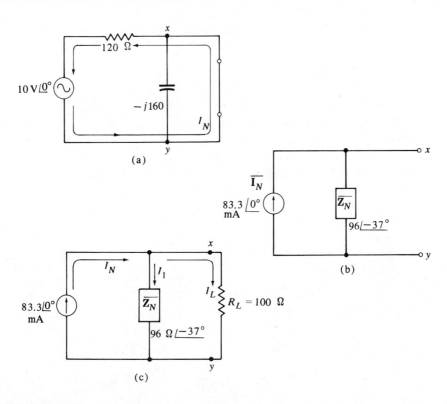

FIGURE 23–14

23.9 Delta (Δ)-Wye (Y) Conversion

In chap. 10 we studied a method for converting circuits from a Δ configuration to a Y configuration and vice versa. The conversion formulas presented there were based on pure resistive circuit elements. These same formulas can also be used for Δ-Y conversions involving *impedances* if we replace the Rs by Zs in all of the formulas. The conversion process using these formulas proceeds in the same manner as outlined in chap. 10. These conversions are illustrated in Fig.

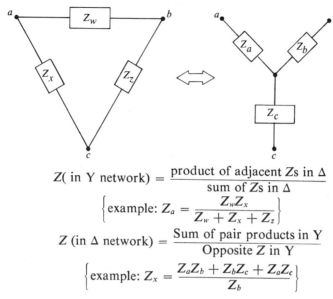

$$Z(\text{in Y network}) = \frac{\text{product of adjacent Zs in }\Delta}{\text{sum of Zs in }\Delta}$$

$$\left\{\text{example: } Z_a = \frac{Z_w Z_x}{Z_w + Z_x + Z_z}\right\}$$

$$Z(\text{in }\Delta\text{ network}) = \frac{\text{Sum of pair products in Y}}{\text{Opposite Z in Y}}$$

$$\left\{\text{example: } Z_x = \frac{Z_a Z_b + Z_b Z_c + Z_a Z_c}{Z_b}\right\}$$

FIGURE 23–15 Δ-Y *impedance conversion.*

23–15. *When using these conversion formulas, all of the impedances must be expressed in vector form.*

23.11 Bridge Circuits

Bridge circuits are extremely useful in impedance measuring instruments and in the sensing circuits of instrumentation and control systems. The general bridge circuit arrangement is shown in Fig. 23–16. The connection between points x and y is the so-called *bridge connection*. For now, this bridge connection will be left open. The impedances Z_1, Z_2, Z_3, and Z_4 can take on many forms, but initially we will treat them as general impedances.

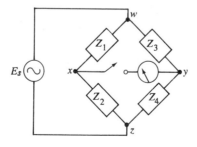

FIGURE 23–16 *General bridge circuit.*

In applications of bridge circuits, particular consideration is given to the condition under which the voltage between points x-y is exactly equal to zero.

When this condition exists the bridge circuit is said to be *balanced* or in the *balanced condition*. When the bridge is balanced, there will be no potential difference between points *x-y*. Thus, when the switch is closed in the bridge connection, there will be no voltage available to produce current through the current meter. The meter, then, will indicate zero current when the bridge is balanced.

With the bridge connection open, we can determine the basic condition necessary for the bridge to be balanced. At balance, the voltage difference between points *x* and *y* is zero. If so, then $\overline{V_{xz}}$ must equal $\overline{V_{yz}}$. We can determine $\overline{V_{xz}}$ and $\overline{V_{yz}}$ using the voltage divider method. $\overline{Z_1}$ and $\overline{Z_2}$ are in series across the source $\overline{E_S}$; therefore, $\overline{V_{xz}}$ (voltage across $\overline{Z_2}$) is given by

$$\overline{V_{xz}} = \left(\frac{\overline{Z_2}}{\overline{Z_1} + \overline{Z_2}}\right) \times \overline{E_S}$$

Similarly,

$$\overline{V_{yz}} = \left(\frac{\overline{Z_4}}{\overline{Z_3} + \overline{Z_4}}\right) \times \overline{E_S}$$

Setting $\overline{V_{xz}} = \overline{V_{yz}}$:

$$\left(\frac{\overline{Z_2}}{\overline{Z_1} + \overline{Z_2}}\right)\overline{E_S} = \left(\frac{\overline{Z_4}}{\overline{Z_3} + \overline{Z_4}}\right)\overline{E_S}$$

Transposing,

$$\overline{Z_2}(\overline{Z_3} + \overline{Z_4}) = \overline{Z_4}(\overline{Z_1} + \overline{Z_2})$$
$$\overline{Z_2}\overline{Z_3} + \overline{Z_2}\overline{Z_4} = \overline{Z_4}\overline{Z_1} + \overline{Z_4}\overline{Z_2}$$

Rearranging,

$$\frac{\overline{Z_2}}{\overline{Z_1}} = \frac{\overline{Z_4}}{\overline{Z_3}} \tag{23-16}$$

This expression is known as the *general bridge formula* and must hold true if the bridge is to be balanced.

The bridge formula simply indicates that the ratio of the impedances in branch *w-x-z* ($\overline{Z_2}/\overline{Z_1}$) must be the same as the ratio of the impedances in branch *w-y-z* ($\overline{Z_4}/\overline{Z_3}$) in order for bridge balance to occur. Note that these ratios are *vector* ratios so that both the magnitudes and angles of the impedances must be used. Also note that the source voltage E_S does not appear in formula (23-16). This means that the balance condition depends only on the impedance ratios and is not affected by the size of the input voltage. This makes sense, because if the bridge were balanced so that the voltage V_{xz} was the same as the voltage V_{yz}, any change in E_S would cause a proportionate change in both V_{xz} and V_{yz}, thereby maintaining the balanced condition.

EXAMPLE 23.12 For each of the following combinations of impedances determine if the bridge circuit will be balanced:

(a) $\overline{Z_1} = 100\ \Omega/\underline{40°}$; $\overline{Z_2} = 50\ \Omega/\underline{-11°}$; $\overline{Z_3} = 800\ \Omega/\underline{70°}$; $\overline{Z_4} = 400\ \Omega/\underline{10°}$

(b) $\overline{Z}_1 = 2\ \text{k}\Omega\underline{/30°}$; $\overline{Z}_2 = 8\ \text{k}\Omega\underline{/45°}$; $\overline{Z}_3 = 0.5\ \text{M}\Omega\underline{/-45°}$; $\overline{Z}_4 = 2\ \text{M}\Omega\underline{/-30°}$

Solution: (a) Using the basic bridge formula for the balanced condition

$$\frac{\overline{Z}_1}{\overline{Z}_2} = \frac{\overline{Z}_3}{\overline{Z}_4}$$

$$\frac{100\underline{/40°}}{50\underline{/-11°}} \stackrel{?}{=} \frac{800\underline{/70°}}{400\underline{/10°}}$$

$$2\underline{/51°} \neq 2\underline{/60°}$$

The *magnitudes* of the ratios are equal, but the *phase angles* are not. Thus, the condition for balance is *not* satisfied.

(b)

$$\frac{2\ \text{k}\Omega\underline{/30°}}{8\ \text{k}\Omega\underline{/45°}} \stackrel{?}{=} \frac{0.5\ \text{M}\Omega\underline{/-45°}}{2\ \text{M}\Omega\underline{/-30°}}$$

$$0.25\underline{/-15°} = 0.25\underline{/-15°}$$

Therefore, for these impedance values the bridge will be balanced. Note that the impedances in one branch are much larger than the impedances in the other branch; and the impedance angles in one branch are negative (capacitive), while in the other the angles are positive (inductive). This makes no difference; as long as the impedance *ratios* are the same for each branch, balance will occur.

Wheatstone Bridge One of the most common bridge circuits is the Wheatstone bridge, in which all the Zs are pure resistance. The Wheatstone bridge is often used to accurately determine the value of an unknown resistor. Figure 23–17 shows a typical Wheatstone bridge setup. In this arrangement $Z_1 = R_1$, $Z_2 = R_2$, $Z_3 = R_3$, and $Z_4 = R_X$ (the unknown resistor). The resistors $R1$ and $R3$ are fixed-value *precision* resistors. Resistor R_2 is a variable resistance whose value will be varied until the bridge is balanced. The source E_S can be either dc or ac for this pure resistance bridge.

The value of R_X is determined by adjusting R_2 until the bridge is balanced (no current through meter when bridge connection is closed). When balance occurs, the value of R_X can be calculated from the basic bridge formula

$$\frac{R_2}{R_1} = \frac{R_X}{R_3}$$

or

$$R_X = \frac{R_3 R_2}{R_1} \qquad (23\text{-}17)$$

The value of R_2 is usually read off a calibrated dial.

EXAMPLE 23.13 The bridge circuit in Fig. 23–17 uses a dc source of 20 V and has $R_3 = 50$ ohms and $R_1 = 18$ ohms. (a) If the bridge balances when R_2 is adjusted to 13.5 ohms, what is the value of R_X; (b) If E_S is doubled to 40 V, at what value of R_2 will the bridge balance?

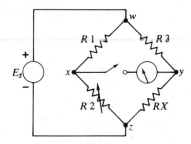

FIGURE 23-17 *Wheatstone bridge.*

Solution: (a)

$$R_x = \frac{R_3 R_2}{R_1} = \frac{50 \times 13.5}{18}$$
$$= 37.5 \ \Omega$$

(b) The value of E_S has no bearing on the bridge balance conditions. Thus, the bridge will still balance when $R_2 = 13.5$ ohms.

Capacitance Bridge   The bridge arrangement of Fig. 23–18 is a *capacitance bridge* which is used to measure the value of the unknown capacitance C_X. R_3 is a precision resistor and C_S is a precision capacitor. R_1 is a variable resistor used to obtain balance. In this arrangement

$$\overline{Z_1} = R_1;\ \overline{Z_2} = -jX_{CS}$$
$$\overline{Z_3} = R_3;\ \overline{Z_4} = -jX_{CX}$$

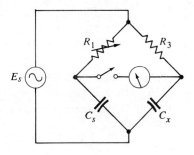

FIGURE 23-18 *Capacitance bridge*

At balance, $\overline{Z_1}/\overline{Z_2} = \overline{Z_3}/\overline{Z_4}$ so that

$$\frac{R_1}{-jX_{CS}} = \frac{R_3}{-jX_{CX}}$$

Substituting for

$$X_{CX} = 1/2\pi f C_X \quad \text{and} \quad X_{CS} = 1/2\pi f C_S$$

results in

$$\frac{R_1 2\pi f C_S}{-j} = \frac{R_3 2\pi f C_X}{-j}$$

Canceling like terms on each side

$$R_1 C_S = R_3 C_X$$

or

$$C_X = \frac{R_1 C_S}{R_3} \tag{23-18}$$

Thus, C_X can be determined from the values of R_1, R_3, and C_S at balance. Notice that the input source frequency does not appear in formula (23–18), thus indicating that the accuracy of the capacitance bridge is not dependent on the accuracy of the input source frequency.

Maxwell Inductance Bridge An inductance bridge similar to the capacitance bridge in Fig. 23–18 could be constructed to accurately measure an unknown inductance. To do so, however, requires having precision values of inductors which are expensive and bulky. Instead, the Maxwell bridge circuit shown in Fig. 23–19 can be used to balance the bridge by using a standard capacitor C_S. Notice that C_S is put in the upper part of the left-hand branch while L_X is in the lower part of the right-hand branch because of their opposite phase angles. At balance,

$$\frac{-jX_{CS}}{R_2} = \frac{R_3}{jX_{LX}}$$

$$\frac{-j}{2\pi f C_S R_2} = \frac{R_3}{j 2\pi f L_X}$$

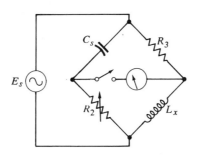

FIGURE 23–19 *Maxwell inductance bridge.*

Solving for L_X we have

$$L_X = \frac{R_3 R_2 C_S}{-j^2}$$

Recall that $j^2 = -1$. Thus,

$$L_X = R_3 R_2 C_S \tag{23-19}$$

and L_X can be determined from the values of R_3, R_2, and C_S (at balance). Once again note that f does not appear in this relationship.

A more practical form of the Maxwell bridge is shown in Fig. 23–20. Here the effective resistance of the unknown coil is shown as R_X in series with inductance

L_X. This series combination is $\overline{Z_4}$ of the bridge. Because of R_X the angle of $\overline{Z_4}$ will not be exactly 90°, but somewhat less. Since $\overline{Z_3} = R_3$, the ratio $\overline{Z_3}/\overline{Z_4}$ will have an angle between 0° and $-90°$. For balance, then, the ratio $\overline{Z_1}/\overline{Z_2}$ must have the same angle. With C_S alone (as in Fig. 23–19) the angle for $\overline{Z_1}/\overline{Z_2}$ would be exactly $-90°$. By putting adjustable resistor R_1 in parallel with C_S, the angle for $\overline{Z_1}$ can be made smaller than $-90°$. This will allow us to balance the phase angles of the ratios by adjusting R_1; the magnitudes will be balanced by adjusting R_2. The following example illustrates.

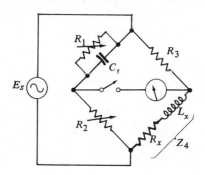

FIGURE 23–20 *More practical Maxwell bridge circuit.*

EXAMPLE 23.14 In the bridge circuit of Fig. 23–20, $R_3 = 100$ ohms and $C_S = 1$ μF. The input frequency is 1 kHz and balance occurs when $R_1 = 600$ ohms and $R_2 = 1$ kΩ. Determine the values of L_X and R_X.

Solution: $\overline{Z_1} = R_1 \parallel X_{CS} = 154.6\ \Omega\underline{/-75°}$

$\overline{Z_2} = R_2 = 1000\ \Omega\underline{/0°}$

$\overline{Z_3} = R_3 = 100\ \Omega\underline{/0°}$

$\overline{Z_4} = R_X + jL_X$

From the balance formula,

$$\overline{Z_4} = \frac{\overline{Z_3} \times \overline{Z_2}}{\overline{Z_1}} = \frac{100\underline{/0°} \times 1000\underline{/0°}}{154.6\underline{/-75°}}$$
$$= 647\ \Omega\underline{/75°}$$

Converting $\overline{Z_4}$ to rectangular form:

$$\overline{Z_4} = 168\ \Omega + j625\ \Omega$$

But, we know that

$$\overline{Z_4} = R_X + jX_{LX}$$

Thus,

$$R_X = \mathbf{168\ \Omega} \quad \text{and} \quad X_{LX} = 625\ \Omega$$

To find L_X,
$$L_X = \frac{X_{LX}}{2\pi f} = 100 \text{ mH}$$

There are general formulas available for the bridge circuit of Fig. 23–20 which express L_X and R_X in terms of the other circuit values. These are:

$$L_X = C_X R_2 R_3 \tag{23-19}$$

$$R_X = \frac{R_2 R_3}{R_1} \tag{23-20}$$

The formulas could have been used in Ex. 23.14. However, the general procedure followed in that example is applicable to any bridge balance problem. Note that formulas (23–19) and (23–20) are not dependent on f, which shows that the Maxwell bridge accuracy in Fig. 23–20 is also independent of the input source frequency accuracy.

Questions/Problems

Sections 23.2–23.3

23-1 Three impedances $\overline{Z_1} = 150 \text{ } \Omega\underline{/37°}$, $\overline{Z_2} = (50 - j50)\Omega$, and $\overline{Z_3} = 100 \text{ } \Omega\underline{/90°}$ are in series across a 50-V source. Calculate: $\overline{Z_T}$, $\overline{I_T}$, $\overline{V_1}$, $\overline{V_2}$, and $\overline{V_3}$.

23-2 Use the voltage divider rule to calculate $\overline{V_1}$ of problem 23–1.

23-3 $\overline{Z_1}$ and $\overline{Z_2}$ of problem 23–1 are connected in parallel across the 50-V source. Calculate: $\overline{Z_T}$, $\overline{I_1}$, $\overline{I_2}$, $\overline{I_T}$.

Section 23.4

23-4 Using admittances find $\overline{Y_T}$ and $\overline{Z_T}$ of the parallel combination of $\overline{Z_1}$, $\overline{Z_2}$, and $\overline{Z_3}$ of problem 23–1.

23-5 Calculate the susceptance of: (a) 100-pF capacitor at 1 MHz; (b) 6-mH inductor at 1 MHz.

23-6 What is the total susceptance of the *parallel* combination of the elements of problem 23–5?

23-7 $R = 100 \text{ } \Omega$, $X_C = 50 \text{ } \Omega$, and $X_L = 200 \text{ } \Omega$ are connected in parallel across a 10-V source. Using susceptances and admittances, find I_T delivered by the source.

23-8 $R = 300 \text{ } \Omega$ and $X_L = 400 \text{ } \Omega$ are in parallel. Convert this parallel combination to an equivalent series RL circuit.

Section 23.5

23-9 For the circuit in Fig. 23–21 determine the current and voltage for each impedance.

Section 23.6

23-10 In the circuit of Fig. 23-21 replace Z_4 by a second voltage source $E_{s2} = 12$ V$\underline{/-30°}$. Then use superposition to solve for the total current through Z_2.

23-11 Consider the amplifier input network shown in Fig. 23-10. Using the following values determine the total voltage across terminals x-y: $C = 0.1$ μF, $R = 1$ MΩ, $R_{in} = 5$ MΩ, $V_{GG} = 3$ V, $e_s = 12$ mV$_{p-p}$ at 1 kHz. Sketch the complete waveform for v_{xy} (dc and ac).

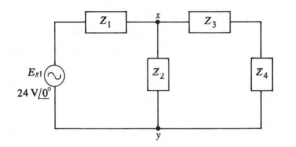

$\overline{Z}_1 = 25$ $\Omega\underline{/13°}$

$\overline{Z}_2 = 35$ $\Omega\underline{/45°}$

$\overline{Z}_3 = 50$ $\Omega\underline{/0°}$

$\overline{Z}_4 = 120$ $\Omega\underline{/-90°}$

FIGURE 23-21

23-12 Without making any calculations, show what will happen to the waveform of v_{xy} of the preceding problem for each of the following changes:
a. decrease in e_s frequency
b. increase in R_{in}
c. decrease in V_{GG}
d. decrease in C

Sections 23.7–23.8

23-13 Refer again to the circuit in Fig. 23-21. Remove Z_2 and Thevenize the circuit at terminals x-y. Use the Thevenin equivalent to then determine the current and voltage of Z_2. Compare to your results of problem 23-9.

23-14 Repeat problem 23-13 using the Norton's equivalent circuit.

23-15 The network in Fig. 23-21 is to have its Thevenin equivalent circuit determined experimentally in the laboratory. Describe a *laboratory* procedure for determining $\overline{E}_{Th}$ and $\overline{Z}_{Th}$ (note vector quantities) at terminals x-y. $\overline{E}_{Th}$ and $\overline{Z}_{Th}$ are *not* to be calculated from the circuit values, but rather from the results of measurements.

Section 23.9

23-16 Convert the Δ-network in Fig. 23-22(a) to an equivalent Y-network.

23-17 Convert the Y-network in Fig. 23-22(b) to an equivalent Δ-network.

Section 23.10

23-18 Consider the general bridge circuit of Fig. 23-16. The bridge is balanced with $\overline{Z}_1 = 100$ $\Omega\underline{/90°}$, $\overline{Z}_2 = 5$ $\Omega\underline{/0°}$, $\overline{Z}_3 = 2$ k$\Omega\underline{/0°}$. (a) Determine $\overline{Z}_4$; (b) if $f = 1$

kHz, determine the component value (R, L, or C) contained in Z_4; (c) if f is *increased* will the bridge remain balanced? Explain.

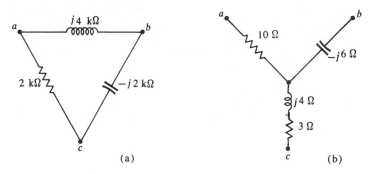

FIGURE 23-22

23-19 A Wheatstone bridge contains $R_1 = 500$ Ω and $R_3 = 200$ Ω. Balance is reached when $R_2 = 12.5$ Ω. What is the value of R_X?

23-20 Fig. 23-23 shows a Wheatstone bridge circuit used in a *light-sensing* application. The resistance arms R_2 and R_3 are photoconductive cells whose resistance *decreases* as the amount of light falling on the cells *increases*. Resistor R_1 is adjustable and is used to balance the bridge for a certain light condition on the photocells. At balance, of course, $V_{xy} = 0$. Once the bridge is balanced, it will remain so unless the amount of light falling on the photocells *changes*. This will change R_2 and R_3 and throw the bridge out of balance.

  a. Assume that under a certain light condition on both photocells the bridge is balanced with $R_4 = 400$ ohms. If R_2 and R_3 are identical photocells, calculate R_2 and R_3 under this light condition.
  b. If the light falling on the cells is now *increased*, R_2 and R_3 will drop to 150 ohms. Determine V_{xy} for this condition.
  c. If the light level *decreases*, R_2 and R_3 will increase to 250 ohms. Calculate V_{xy} for this condition.

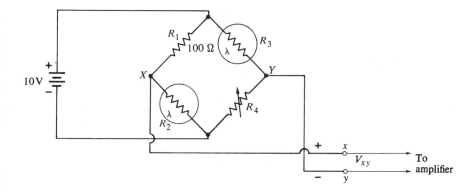

FIGURE 23-23

23-21 A capacitance bridge balances when $R_1 = 1$ kΩ, $R_3 = 600$ ohms, and $C_S = 500$ pF. What is C_X?

23-22 In the Maxwell bridge circuit in Fig. 23-20, the adjustable resistor R_1 was placed in *parallel* with C_1 so that the phase angle of $\mathbf{Z}_1$ could be balanced with the phase angle of $\overline{\mathbf{Z}_X}$. It would seem that we could also have placed R_1 in *series* with C_1 to accomplish the same purpose. However, with that arrangement the bridge balance would not be *independent* of frequency. That is, once the bridge was balanced, any change in f would unbalance it. Without making any calculations, explain why this would happen

23-23 Figure 23-24 shows a type of bridge circuit.
 a. By varying R_2 is it possible to balance this bridge? Why not?
 b. Add a component to this circuit that can be used to help achieve balance.

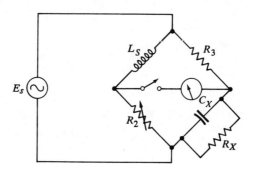

FIGURE 23-24

CHAPTER 24

Power in Sinusoidal Circuits

24.1 Introduction

In this chapter we will investigate the concept of power in circuits that are driven by sinusoidal voltage sources. Before doing so we will discuss the power characteristics of each of the three basic circuit elements: $R, L,$ and C.

24.2 Power in a Pure Resistance

In chap. 16 we investigated the nature of the power dissipation in a pure resistance when a sinusoidal voltage is applied to it. As we saw then, a sine wave of voltage across the resistor produces a sine wave of current which is exactly in phase with the voltage. As a result, the *instantaneous* power dissipation in the resistor reached its peak values when the applied voltage reached its peak values. The *average* power dissipation in the resistor is equal to

$$P_{av} = V_{rms} \times I_{rms} \qquad (24\text{–}1)$$

where rms values for the resistor voltage and current are used. Using Ohm's law, two other equivalent formulas can be derived:

$$P_{av} = V_{rms}^2/R \qquad (24\text{–}2)$$

and

$$P_{av} = I_{rms}^2 R \qquad (24\text{–}3)$$

EXAMPLE 24.1 A 100-Hz, 20-V waveform is applied to a 40-ohm resistor.
 (a) Determine the average power dissipation in the resistor.

(b) Determine the average rate at which energy is being lost in the resistor as heat.

(c) If f is increased to 10 kHz, how will that affect the answers to (a) and (b)?

Solution: (a) $P_{av} = \dfrac{V_{rms}^2}{R} = \dfrac{(20 \text{ V})^2}{40 \text{ }\Omega} = 10 \text{ W}$

(b) Recall that power represents *energy per unit time*. That is,

$$1 \text{ watt} = 1 \text{ joule/second}$$

Thus, if the resistor has an average power dissipation of 10 W, then it must be dissipating energy at the rate of **10 joules/second**. This energy is, of course, dissipated in the form of heat.

(c) Since R does not depend on frequency, then any change in the applied frequency will *not* affect the amount of power dissipated in the resistor.

True Power The power supplied by a source to a resistor represents energy being *consumed*; this energy can never be returned to the source but instead is consumed in the resistor as heat energy (or light energy as in the case of light bulbs). Thus, we will speak of the average power in a resistor as representing the *true power* that is drawn from the source and consumed. The symbol for true power is P_T. Thus, for a resistor we have

$$P_T = \text{(true power)} = V \times I = \dfrac{V^2}{R} = I^2 R \qquad (24\text{–}4)$$

where V and I are rms values.

24.3 Power in a Pure Inductance

When a sinusoidal voltage is applied to a pure inductance, the current through the inductor *lags* the voltage across the inductor by exactly 90° (see Fig. 24–1). We can determine the instantaneous power p in this pure inductance by multiplying $v \times i$ at any instant of time. If this is done at every instant of time, then a waveform of p vs t can be drawn. This is shown in Fig. 24–1 as the dotted curve.

Note that this instantaneous power waveform is *positive* during the first quarter of a cycle (0° to 90°) because both v and i are positive ($p = v \times i$). However, in the second quarter cycle (90° to 180°) the power waveform goes *negative* because v is negative and i is positive. Similarly, p is again positive in the interval from 180° to 270° and negative in the interval from 270° to 360°. Any successive cycles of the applied voltage will produce the same result.

What does this instantaneous power waveform show us? The power curve has two positive peaks and two negative peaks during one complete cycle of the applied voltage. *Positive* power means power consumed or absorbed by the circuit; *negative* power means power *returned* from the circuit back to the source. Thus, it is evident that a pure inductance in an ac circuit alternately takes from and returns to the source equal amounts of energy, so we can state that

In a pure inductance the average or true power is zero.

Power in Sinusoidal Circuits

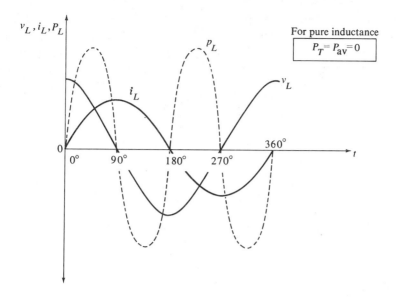

FIGURE 24-1 *Instantaneous voltage, current, and power waveform in a pure inductance.*

During one quarter cycle (0° — 90°) energy is supplied to the inductor from the source and stored in the coil's magnetic field; during the next quarter cycle (90°-180°) as the coil current decreases the magnetic field collapses and this energy is returned to the source. Consequently, the *true power* consumption is zero since no net power is actually consumed or burned up in a pure inductor. For a pure inductor, then, $P_T = 0$.

Reactive Power In the case of a pure resistance we are able to determine true power by simply multiplying the rms values of voltage and current. In the circuit of Fig. 24-2, the meters will read the rms values of inductor voltage and current. The product of these values $V_L \times I_L$ does *not* represent true power dissipation since the true power is *zero* for a pure inductor. Instead, this product is given the special name of *reactive power* and is given the symbol P_Q. Therefore, in a pure inductor

$$P_Q = V_L I_L \tag{24-5}$$

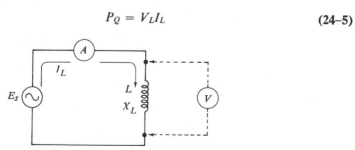

FIGURE 24-2 P_Q, *reactive power in a pure inductance, equals* $V_L \times I_L$.

where V_L and I_L are both rms values. Reactive power P_Q is not the same thing as true power P_T. Reactive power represents the amount of power being transferred back and forth between the source and the inductor, while true power is delivered from the source and dissipated in the circuit. In order to distinguish between true power and reactive power, the units for P_Q are not called *watts* but are called *vars*. Vars stands for *volt-amperes-reactive* and means the product of voltage and current in a reactance (inductive reactance in this case).

Formula (24–5) can be modified using the fact that $I_L = V_L/X_L$ to provide us with three power relationships for a pure inductance:

$$P_Q = V_L I_L = \frac{V_L^2}{X_L} = I_L^2 X_L \qquad (24\text{–}6)$$

These formulas are exactly the same as those for P_T in a resistor.

EXAMPLE 24.2 A 2-H pure inductance draws 0.3 A from a 60-Hz source. (a) What is the *true* power in the inductance? (b) What is the reactive power in the inductance? (c) How much power is dissipated as heat in the inductance?

Solution: (a) $P_T = 0$ for a pure inductance

(b) $X_L = 2\pi f L = 754 \; \Omega$
$P_Q = I_L^2 X_L = (0.3)^2 \times 754 =$ **67.9 vars**

(c) True power is dissipated power. In a pure inductor there is **zero** power dissipated as heat. (Of course, any practical inductor will have some resistance; the practical coil's resistance *will* dissipate true power).

24.4 Power in a Pure Capacitance

When a sinusoidal voltage is applied to a pure capacitance, the current *leads* the voltage by 90° (see Fig. 24–3). The waveform of instantaneous power p is obtained in the same manner as for the pure inductance. The power waveform (shown dotted) shows two positive peaks and two negative peaks during one complete cycle of the capacitor voltage. Thus, just as was the case for pure inductance, the pure capacitance in an ac circuit alternately takes from and returns to the source equal amounts of energy, so we can state that

In a pure capacitor the average or true power is zero.

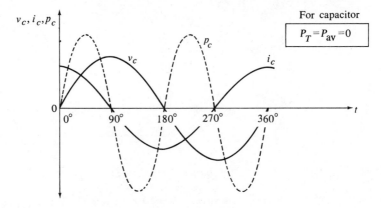

FIGURE 24–3 v, i, p *waveforms in a pure capacitance.*

During one quarter cycle (0°-90°) energy is supplied to the capacitor as it is charged to the peak of the source voltage; during the next quarter cycle (90°-180°) the capacitor voltage discharges to zero as this energy is returned to the source. Consequently, the *true power* consumption is zero since no net power is actually consumed or burned up in a pure capacitance; $P_T = 0$.

Reactive Power in a Capacitor The product of capacitor voltage and current (rms values) is also called reactive power P_Q and represents the amount of power being transferred back and forth between the source and the capacitor. To distinguish between reactive power for inductances (L) and capacitances (C) the two symbols P_{QL} and P_{QC} will be used. Thus, for a capacitor

$$P_{QC} = V_C I_C = \frac{V_C^2}{X_C} = I_C^2 X_C \qquad (24\text{-}7)$$

where I_C and V_C are rms values, and the units for P_{QC} are *vars* (volt-amperes-reactive).

EXAMPLE 24.3 A 20-μF pure capacitor is connected across a 100-V, 60-Hz line.
  (a) How much true power is dissipated in the capacitor?
  (b) How much is the capacitor's reactive power?

Solution: (a) In a pure capacitor no true power is dissipated as heat. Thus, $P_T = 0$. (Of course, any practical capacitor has leakage resistance associated with its dielectric. In most cases, the true power dissipated by this large resistance is negligible.)

  (b) $X_C = \dfrac{1}{2\pi f C} = 133 \ \Omega$

$$P_{QC} = \frac{E^2}{X_C} = \frac{(100)^2}{133} = 75.2 \text{ vars}$$

Summary of R, L, and C Power Table 24-1 summarizes the power relationships for each of the basic elements using rms values of I and V.

TABLE 24-1

Element	True Power	Reactive Power
Resistance	$P_T = VI = \dfrac{V^2}{R} = I^2 R$ (watts)	zero in a resistor
Inductance	P_T = zero	$P_{QL} = VI = \dfrac{V^2}{X_L} = I^2 X_L$ (vars)
Capacitance	P_T = zero	$P_{QC} = VI = \dfrac{V^2}{X_C} = I^2 X_C$ (vars)

24.5 Apparent Power

Most circuits contain combinations of R, L, and C. As such, the ac power in these circuits will not be totally true power or totally reactive power, but rather a combination of the two. For example, a source is applied to a complex impedance network in Fig. 24–4, causing a total current I_T to flow. In general, the product $E_S \times I_T$ will represent neither *true* power nor *reactive* power because the impedance network can contain both resistance and reactance. If we have only the E_S and I_T magnitudes to go by, then there is no way to determine the *true* power and the *reactive* power of the network.

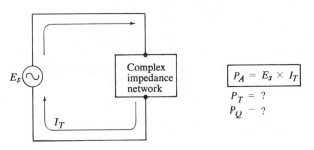

FIGURE 24–4 *Apparent power in a complex circuit equals the product of applied voltage and total circuit current.*

If the network was purely resistive, then the product $E_S I_T$ would represent the true power being delivered by the source and dissipated by the circuit. Thus, $E_S \times I_T$ represents what we call the *apparent power* of the circuit. It represents the amount of power which is *apparently* being consumed by the circuit. We need more information to be able to determine the *true* consumed power. In general, then, we can state that

In any ac circuit the product of the source voltage and the total source current is called the apparent power of the circuit.

The letter symbol which we shall use for apparent power is P_A. Because P_A is, in general, neither all true power (watts) or all reactive power (vars), we will express apparent power in units of *volt-amperes* (abbreviated VA). Thus,

$$P_A = E_S I_T \tag{24-8}$$

where E_S and I_T are rms values and the units for P_A are VA.

The complex network in Fig. 24–4 will have a total impedance $\overline{Z_T}$. Using the fact that $\overline{I_T} = \overline{E_S}/\overline{Z_T}$ provides us with two alternate expressions for P_A.

$$P_A = E_S I_T = \frac{E_S^2}{Z_T} = I_T^2 Z_T \tag{24-9}$$

These formulas use rms values for E_S and I_T, and only the magnitude of Z_T.

EXAMPLE 24.4 A 240-V ac source drives a network with $\overline{Z_T} = 800\ \Omega \underline{/60°}$.
(a) What is the apparent power in the circuit?
(b) If the impedance is $800\ \Omega \underline{/35°}$, how will that affect P_A?

Solution: (a) $P_A = \dfrac{E_S^2}{Z_T} = \dfrac{(240)^2}{800} = 72\ \text{VA}$

Power in Sinusoidal Circuits 695

(b) The *angle* of Z_T does not affect the magnitude of the current I_T supplied by the source. Therefore, it has no affect on P_A. Any impedance with a magnitude of 800 ohms will result in the *same* $P_A = 72$ VA.

It should be clear that in a pure resistive network the apparent power P_A will equal the total true power P_T, while in a pure reactive network the apparent power will equal the total reactive power P_Q. In the general case of a network which is both resistive and reactive, the apparent power will be a combination of the true power and reactive power of the network. This relationship between P_A, P_T, and P_Q is developed in the next section.

24.6 The Power Triangle

Any complex impedance network can be simplified by finding its total equivalent impedance $\overline{Z_T}$. This impedance will be of the form

$$\overline{Z_T} = R \pm jX$$

The reactive portion will be $+jX_L$ if the impedance is inductive and $-jX_C$ if the impedance is capacitive. In either case Z_T can be represented as a series combination of R and $\pm jX$.

Figure 24–5(a) shows the case where $\overline{Z_T}$ is inductive. For this inductive series circuit we can draw the impedance vector triangle shown in (b) with X_L and R as sides of the right triangle and Z_T as the hypotenuse. The angle θ is the angle of the impedance and also the angle by which the circuit current *lags* the applied voltage. If *each* side of the impedance triangle is multiplied by the square of the circuit current (I^2), the triangle in (c) results.

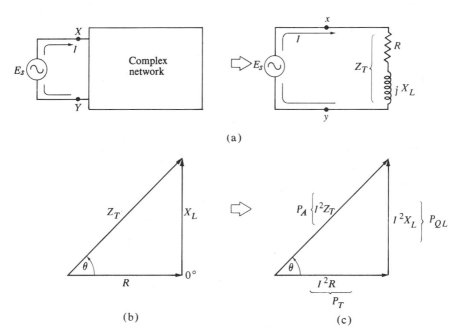

FIGURE 24–5 *Developing the power triangle for an inductive impedance network.*

Examination of this triangle reveals that: (1) it is similar to the impedance triangle in (b) since its sides are proportional to the sides of the impedance triangle; therefore, the angle θ is the same in both triangles; (2) the sides of the new right triangle represent true power (I^2R), reactive power (I^2X_L) and apparent power (I^2Z_T). Therefore,

$$P_A^2 = P_T^2 + P_{QL}^2 \qquad (24\text{--}10)$$

This expression gives us the relationship between apparent power, true power, and reactive power. The triangle in Fig. 24–5(c) is called a *power triangle*.

A similar power triangle can be developed for a capacitive network where $\overline{Z_T} = R - jX_C$. This is shown in Fig. 24–6. Note that since X_C is drawn at $-90°$, the capacitive reactive power $P_{QC} = I^2 \times X_C$ is at $-90°$ in the power triangle. This is opposite to P_{QL} in the inductive circuit power triangle. For the power triangle in Fig. 24–6, then, we have

$$P_A^2 = P_T^2 + P_{QC}^2 \qquad (24\text{--}11)$$

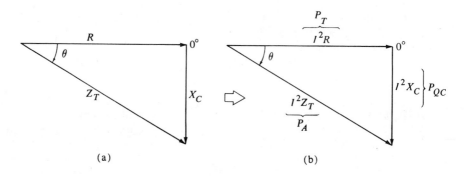

(a) (b)

FIGURE 24–6 *Power triangle for a capacitive impedance network.*

Formulas (24–10) and (24–11) can be replaced by the more general formula

$$P_A^2 = P_T^2 + P_Q^2 \qquad (24\text{--}12)$$

where P_Q can be either P_{QL} or P_{QC}.

EXAMPLE 24.5 A 24-V source is applied to a network that has $\overline{Z_T} = 100\ \Omega\underline{/30°}$. Calculate P_Q, P_T, and P_A for the circuit.

Solution: There are two methods for solving this problem.

Method I: $\overline{Z_T} = 100\ \Omega\underline{/30°} = \overbrace{86.6}^{R} + \overbrace{j50}^{X_L})\Omega$

With $E_S = 24$ V, the circuit current is

$$I_T = 24\text{ V}/100\ \Omega = 0.24\text{ A}$$

Thus,

$$P_A = I^2 Z_T = (0.24)^2 \times 100 = \mathbf{5.76\text{ VA}}$$

$$P_T = I^2 R = (0.24)^2 \times 86.6 = \mathbf{4.99\text{ W}}$$

Power in Sinusoidal Circuits

$$P_{QL} = I^2 X_L = (0.24)^2 \times 50 = \mathbf{2.88 \text{ vars}}$$

Note the different units for each power component. Also note that $P_A^2 = P_T^2 + P_{QL}^2$.

Method II: $I_T = E_S/Z_T = 0.24$ A

Thus,

$$P_A = E_S I_T = 24 \text{ V} \times 0.28 \text{ A}$$
$$= \mathbf{6.72 \text{ VA}}$$

The power triangle for this circuit is drawn in Fig. 24–7(a) with $P_A = 5.76$ VA. Note that $\theta = 30°$, the same as the impedance angle. To find P_T and P_{QL} we can use the trig relations for the right triangle.

$$P_T = P_A \cos(\theta) = 5.76 \times \cos(30°)$$
$$= \mathbf{4.99 \text{ W}}$$

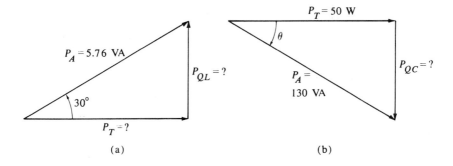

(a)   (b)

FIGURE 24–7

Similarly,

$$P_{QL} = P_A \sin(30°) = 5.76 \times 0.5$$
$$= \mathbf{2.88 \text{ vars}}$$

EXAMPLE 24.6 A capacitive impedance network draws 2 A from a 65-V source. The network dissipates 50 W of power. (a) Determine P_A, P_T, and P_{QC}; (b) determine $\overline{Z_T}$.

Solution: (a) Since $E_S = 65$ V and $I_T = 2$ A, then $P_A = 65 \times 2 = 130$ VA. Also, $P_T = 50$ W since true power is power which is dissipated (consumed). The power triangle for the circuit is drawn in Fig. 24–7(b). We know that

$$P_A^2 = P_T^2 + P_{QC}^2 \qquad (24\text{–}11)$$

Therefore,

$$P_{QC} = \sqrt{P_A^2 - P_T^2} = \sqrt{130^2 - 50^2}$$
$$= \mathbf{120 \text{ vars}}$$

(b) The magnitude of $\overline{Z_T}$ is 65 V/2 A = 32.5 ohms. The angle of $\overline{Z_T}$ is the same as the angle θ in the power triangle. From the triangle in Fig. 24-7(b),

$$\cos\theta = \frac{P_T}{P_A} = \frac{50}{130} = 0.385$$

therefore,

$$\theta = 67.35°$$

Thus, $\overline{Z_T} = 32.5\ \Omega\,/\!-\!67.35°$ (capacitive)

24.7 Power Factor

When an impedance load is connected to a source, the source supplies an apparent power P_A. The *true* power which the load consumes or converts to some other form of energy can never be greater than P_A, and is usually less than P_A. *The ratio of the true power to the apparent power is called the power factor of the load.* That is,

$$\text{PF} = \frac{P_T}{P_A} \qquad (24\text{-}13)$$

where PF is the abbreviation for power factor.

If we examine the power triangles in Figs. 24-5 and 24-6, we see that the ratio P_T/P_A is equal to the cosine of the angle θ. Thus, we can write

$$\text{PF} = \frac{P_T}{P_A} = \cos\theta \qquad (24\text{-}14)$$

where θ is the load's phase angle.

The PF is a unitless number which will always have a value between 0 and 1. The value of PF will be zero when $\theta = \pm 90°$ since $\cos(\pm 90°) = 0$. This is reasonable since $\theta = \pm 90°$ indicates a pure reactive circuit (90° inductive, $-90°$ capacitive), which contains no resistance and therefore dissipates no true power. Thus, $P_T = 0$ so that $\text{PF} = P_T/P_A = 0$.

On the other hand, the PF will equal unity when $\theta = 0°$, which indicates a pure resistive circuit. In such a circuit $P_Q = 0$ so that $P_A = P_T$ and $\text{PF} = P_T/P_A = 1$.

In order to distinguish between inductive loads and capacitive loads, we always say that an inductive load has a *lagging* power factor and a capacitor load has a *leading* power factor. An inductive load has a *lagging* PF since the load current will *lag* the applied voltage. The opposite is true for the capacitive load.

PF, then, represents a ratio of the true power consumed by a load to the apparent power delivered to the load by the source. PF is often expressed as a percentage as well as a decimal fraction.

EXAMPLE 24.7 Determine the power factor for each of the following load impedances: (a) 500 $\Omega/18°$; (b) 5 M$\Omega/18°$; (c) $(8 - j4)$ ohms; (d) 1 k$\Omega/0°$; (e) $-j3000\ \Omega$.

Solution: (a) Since $\overline{Z_T} = 500/18°$, the load is *inductive*. Therefore,

$$PF = \cos(18°) = \textbf{0.951 lagging}$$
$$\text{or} \quad \textbf{95.1\% lagging}$$

(b) Since $\overline{Z_T} = 5\ M\Omega\underline{/18°}$, the load is *inductive*. Therefore,

$$PF = \cos(18°) = \textbf{0.951 or 95.1\% lagging}$$

Notice that this PF is the same as in (a) even though the *magnitudes* of the impedances are quite different. It is the *angle* of the impedance which solely determines PF.

(c) $\overline{Z_T} = (8 - j4)\Omega$, which is capacitive. The angle of this impedance is found from

$$\tan\theta = \frac{-4}{8} = -0.5 \quad \text{therefore } \theta = -26.5°$$

$$PF = \cos(-26.5°) = \textbf{0.895 or 89.5\% leading}$$

(d) $\overline{Z_T} = 1\ k\Omega\underline{/0°}$, which is pure resistive.

$$PF = \cos(0°) = \textbf{1.0 or 100\%}$$

(e) $\overline{Z_T} = -j3000\ \Omega = 3\ k\Omega\underline{/-90°}$, which is pure capacitive.

$$PF = \cos(-90°) = \textbf{0.0 or 0\% leading}$$

EXAMPLE 24.8 A load with PF = 50% lagging draws 3.2 A of current when connected across the 110-V, 60-Hz line. (a) Determine P_A, P_T, P_{QL}. (b) Determine the impedance of the load.

Solution: (a) $P_A = E_S \times I_T = 110\ V \times 3.2\ A$
$$= 352\ VA$$

$$\frac{P_T}{P_A} = PF = 50\% = 0.5$$

therefore

$$P_T = 0.5\ P_A = \textbf{176 W}$$
$$P_{QL} = \sqrt{P_A^2 - P_T^2}$$
$$= \textbf{305 vars}$$

(b) The magnitude of Z_T equals $110\ V/3.2\ A = 34.4$ ohms. The angle of Z_T can be determined from the PF. Since $\cos\theta = PF = 0.5$, then $\theta = 60°$. Thus,

$$\overline{Z_T} = 34.4\ \Omega\underline{/60°} \text{ (inductive)}$$

EXAMPLE 24.9 An inductive load draws 20 A from a 110-V source. The load's reactive power is 500 vars. Determine the PF of the load.

Solution: The apparent power $P_A = 110\ V \times 20\ A = 2200\ VA$.

The reactive power P_{QL} is given as 500 vars.

The power triangle is drawn in Fig. 24–8. From the triangle we can see that

$$\sin(\theta) = \frac{P_{QL}}{P_A} = \frac{500}{2200} = 0.227$$

$$\therefore \theta = 13.1°$$

$$PF = \cos(\theta) = \cos(13.1°)$$

$$= 0.974 \text{ or } 97.4\% \text{ lagging}$$

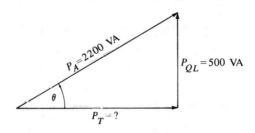

FIGURE 24–8

24.8 Determining Power Without Using Z_T

Many times the total load on a power source consists of many parallel branches with different impedances. In such cases it can be a very tedious task to calculate $\overline{Z_T}$ in order to determine the apparent power, true power, and reactive power of the entire system. It is *not* necessary to determine $\overline{Z_T}$. The total VA, watts, and vars of *any* system can be found by following this procedure:

(1) Find the true power (watts) and reactive power (vars) for each branch in the circuit.

(2) The total watts in the system equals the *sum* of the watts for each branch. This gives us P_T (total) for the entire system.

(3) The total vars is obtained by combining all the inductive vars (P_{QL}) in the system and combining all the capacitive vars (P_{QC}) in the system, and then *subtracting P_{QC} from P_{QL}*. That is,

$$P_Q(\text{total}) = P_{QL}(\text{total}) - P_{QC}(\text{total}) \tag{25-15}$$

(4) The total apparent power P_A(total) and the overall system power factor can then be found using the power triangle. That is,

$$P_A(\text{total})^2 = P_T(\text{total})^2 + P_Q(\text{total})^2$$

$$PF = \cos\theta = \frac{P_T(\text{total})}{P_A(\text{total})}$$

(5) If P_Q(total) is net capacitive, then the system PF will be leading; if P_Q(total) is net inductive, then the system PF will be lagging.

From the procedure outlined here, two significant points must be noticed. First, the total P_A must be found from the total P_T and the total P_Q, and is *not* found by adding the apparent powers of each branch. Second, the total reactive power P_Q is determined by the *difference* between the total inductive reactive power P_{QL} and the total capacitive reactive power P_{QC}. If $P_{QL} > P_{QC}$, the system power triangle will be similar to the one in Fig. 24–5(c); if $P_{QL} < P_{QC}$ the triangle will be like the one in Fig. 24–6(b).

EXAMPLE 24.10 (a) For the network in Fig. 24–9(a) determine the overall circuit P_A, P_T, P_Q, and power factor. (b) Find Z_T and I_T.

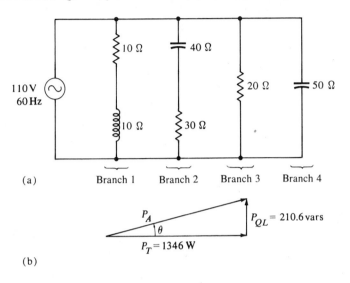

FIGURE 24-9

Solution: (a) We treat each branch separately. The impedance in branch 1 is $(10 + j10) = 14.1\ \Omega\underline{/45°}$. Thus, $I_1 = 110\text{ V}/14.1\ \Omega = 7.8$ A. This current through the 10-ohm resistor produces a true power $P_{T1} = I_1^2 R = (7.8)^2 \times 10 = 608.4$ W. This current through the 10-ohm X_L produces a reactive power $P_{QL} = (7.8)^2 \times 10 = 608.4$ vars.

This same procedure is repeated for each branch in the circuit. The results are tabulated:

Branch	I	Watts	Vars
1	7.8 A	608.4	608.4 (inductive)
2	2.2 A	132.6	176.8 (capacitive)
3	5.5 A	605.0	0
4	2.2 A	0	221 (capacitive)
		P_T(total) = 1346 W	P_Q(total) = 608.4 − 176.8 − 221 = 210.6 vars (inductive)

The power triangle is shown in Fig. 24–9(b), from which

$$P_A = \sqrt{1346^2 + 210.6^2} = 1362\text{ VA}$$

The overall circuit power factor is

$$\text{PF} = \frac{P_T}{P_A} = \frac{1346}{1362} = 0.988 = 98.8\%\text{ lagging}$$

Notice that the individual branch P_As were not found since they are not needed.

(b) The total apparent power supplied by the source is 1370 VA. Thus, we have

$$E_S \times I_T = 1370 \text{ VA}$$

$$I_T = \frac{1370 \text{ VA}}{110 \text{ V}} = 12.4 \text{ A}$$

Therefore,

$$Z_T = \frac{E_S}{I_T} = \frac{110 \text{ V}}{12.4 \text{ A}} = 8.9 \text{ ohms}$$

The angle of the impedance is the same as θ in the power triangle. Since

$$\cos \theta = \text{PF} = 0.992$$

$$\therefore \theta = 7.2°$$

Thus, $\overline{Z_T} = 8.9 \text{ } \Omega \underline{/+7.2°}$ (inductive)

The 110-V, 60-Hz line voltage supplies power to all types of loads such as motors, heaters, lamps, etc. Such loads are often specified in terms of their power ratings and PF rather than in terms of their actual impedance. For example, a motor load might be rated at 6 kVA at 80% lagging PF when driven from the 110-V source. This rating indicates the apparent power required by the motor (P_A = 6 kVA) and it tells that the motor is inductive with PF = 80%. From the definition of PF, the *true* power used by the motor is 80% × 6 kVA = 4.8 kW. The following example shows a typical power system employing loads with power rating specifications.

EXAMPLE 24.11 For the circuit in Fig. 24–10(a) determine the total P_A, P_Q, and P_T and the overall PF.

Solution: Load 1

$$P_{A1} = 1 \text{ kVA} = 1000 \text{ VA; PF} = 70\% \text{ inductive}$$

Thus,

$$P_{T1} = 70\% \times 1000 \text{ VA} = 700 \text{ W}$$

$$P_{Q1} = \sqrt{P_{A1}^2 - P_{T1}^2} = 714 \text{ vars (inductive)}$$

Load 2

$$P_{A2} = 5000 \text{ VA; PF} = 60\% \text{ capacitive}$$

Thus,

$$P_{T2} = 60\% \times 5000 = 3000 \text{ W}$$

$$P_{Q2} = \sqrt{5000^2 - 3000^2} = 4000 \text{ vars (capacitive)}$$

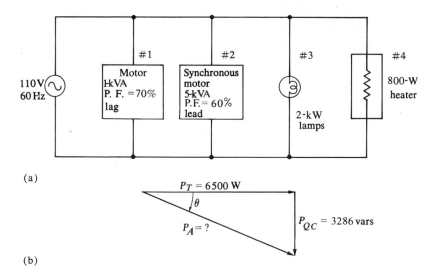

FIGURE 24-10

Load 3

$$P_{T3} = 2000 \text{ W} \quad \text{and} \quad P_{Q3} = 0 \text{ vars}$$

Load 4

$$P_{T4} = 800 \text{ W} \quad \text{and} \quad P_{Q4} = 0 \text{ vars}$$

These values are tabulated:

Load	Watts	Vars
1	700	714 (inductive)
2	3000	4000 (capacitive)
3	2000	0
4	800	0
	$P_{T(\text{total})} =$ **6500 W**	$P_{Q(\text{total})} = 4000-714$ = **3286 vars (capacitive)**

The power triangle is drawn in Fig. 24-10(b), from which

$$P_A(\text{total}) = \sqrt{P_T^2 + P_Q^2}$$
$$= 7283 \text{ VA}$$

$$\text{PF} = \frac{P_T}{P_A} = \frac{6500}{7283} = 0.892 = 89.2\% \text{ leading}$$

EXAMPLE 24.12 A certain motor is rated at 5 kVA, 110 V at 80% lagging. Express the motor's impedance in rectangular form.

Solution: $P_A = 5000 \text{ VA}$

Thus,
$$5000 = \frac{E_S^2}{Z_T} = \frac{(110\text{ V})^2}{Z_T}$$

or
$$Z_T = \frac{(110)^2}{5000} = 2.42 \text{ ohms}$$

The angle of the impedance can be found from
$$\cos(\theta) = PF = 0.8$$

therefore,
$$\theta = 37°$$

Thus,
$$\overline{Z_T} = 2.42 \text{ }\Omega\underline{/37°} = (1.94 + j1.45)\Omega$$

This represents a 1.94-Ω resistance in series with a 1.45-Ω inductive reactance.

24.9 Power Factor Correction

Power companies charge their customers only for the *true* power consumed. However, they must make provisions for the additional *reactive* power that is transferred back and forth because of the reactive portions of the loads. Because of this reactive power, the total *apparent* power delivered by the source is greater than the actual power consumed by the loads. This of course means that the total current supplied by the source will be larger than it would be if it were supplying only true power (watts). As far as the power companies are concerned, the cables supplying the various circuits must be heavy enough to carry the total current due to the true power and the reactive power combined.

From this standpoint, then, the power companies would like to see load power factors which are close to unity. If PF $\approx$ 1, this means that $P_A \approx P_T$ and $P_Q \approx 0$ (no net reactive power), so that no extra current is needed to supply P_Q. Keeping PF close to unity reduces excessive apparent power demands and avoids the use of heavier gauge wire. If a load has a low PF, some type of PF *correction* can be used to bring the PF close to one.

To illustrate power factor correction, consider the situation in Fig. 24–11(a). The load has a PF of 0.6 lagging and requires $P_A = 1500$ VA. Thus, $P_T = 0.6 \times 1500 = 900$ W. The power triangle for this load can be used to find $P_{QL} = 1200$ vars. The total current which must be supplied by the source is determined by E_S and P_A. That is,

$$I_T = \frac{P_A}{E_S} = \frac{1500 \text{ VA}}{110 \text{ V}} = 13.6 \text{ A}$$

A good portion of this current is needed to supply the 1200-vars inductive reactive power. This reactive power serves no useful purpose but is simply present due to the inductive reactance of the load (motor). If we could reduce the vars to zero then P_A would equal P_T, and PF = 1. This reduction in vars can be

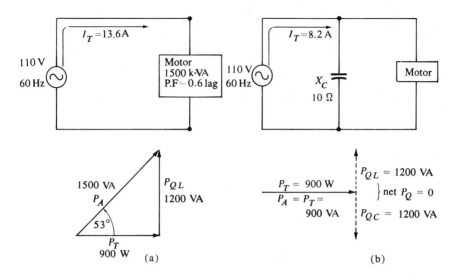

FIGURE 24-11 *Illustration of power factor correction to reduce current drawn from source.*

accomplished by connecting a *capacitive* reactance across the source such that the *capacitive* vars delivered to the capacitor completely cancel out the 1200 inductive vars delivered to the load. In other words, since $P_{QL} = 1200$ vars we need $P_{QC} = 1200$ vars to give a net $P_Q = P_{QL} - P_{QC} = 0$ vars.

The value of X_C which must be connected across the source [see part (b) of Fig. 24-11] is determined as

$$P_{QC} = \frac{E_S^2}{X_C} = 1200 \text{ vars}$$

or

$$X_C = \frac{E_S^2}{1200} = \frac{(110)^2}{1200} \approx 10 \text{ ohms}$$

The *total* load seen by the source now consists of the original load shunted by this *correction capacitor*. The new power triangle (shown in the figure) reduces to simply $P_A = P_T = 900$ W since the net $P_Q = 0$ vars. This of course is unity PF.

In the corrected circuit, P_A is only 900 VA, which exactly equals the 900-W true power needed by the motor. With this reduction in P_A from 1500 VA to 900 VA, the current drawn from the source is reduced from 13.6 A to

$$\frac{900 \text{ VA}}{110 \text{ V}} = 8.2 \text{ A}$$

This dramatic reduction in source current means that less heavy wire cables are needed between source and load.

The power factor correction method just shown can also be performed on systems with a *leading* PF. In such cases, the load is capacitive and requires capacitive vars. Thus, to reduce the vars to zero an appropriate inductive reactance can be placed in parallel with the load to produce unity PF.

24.10 Maximum Power Transfer Principle

In chap. 10 we considered the transfer of power from a practical voltage source that had an internal source resistance R_S. We discovered that when $R_L = R_S$, this resulted in the maximum possible power output from the source. We can extend this maximum power principle to alternating current circuits in which the practical ac source can have an internal source *impedance* that is partially resistive and partially reactive. Figure 24-12 represents a practical ac voltage source $\overline{E_S}$ with a source impedance $\overline{Z_S} = R_S \pm jX_S$. The series combination of $\overline{E_S}$ and $\overline{Z_S}$ can represent an actual voltage source *or* it could represent the Thevenin equivalent circuit for a complex network. In either case we are interested in the condition which will produce maximum *true* power output from the source.

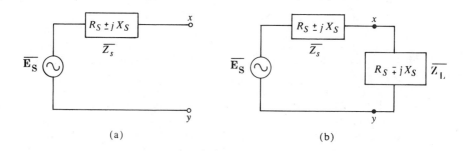

FIGURE 24-12 (a) *Practical ac source.* (b) *terminated with load Z_L for maximum true power transformer.*

The maximum power transfer principle for the ac situation can be stated as:

An ac source will supply maximum true power output to a load impedance when the resistive portion of $\overline{Z_L}$ equals the resistive portion of $\overline{Z_S}$ and when the reactive portion of $\overline{Z_L}$ cancels exactly (is equal and opposite to) the reactance portion of $\overline{Z_S}$.

For example, if a source has an internal impedance $\overline{Z_S} = (10 + j2)$ ohms the maximum *true* power will be delivered to a load $\overline{Z_L} = (10 - j2)$ ohms. In other words, $\overline{Z_L}$ has the same resistance as $\overline{Z_S}$ and has an *equal but opposite* reactance. This is illustrated in Fig. 24-12(b).

This maximum power transfer principle can be reasoned as follows. If the reactive portion of the load exactly cancels out the reactive portion of the source impedance, then the total circuit is pure resistive with R_S and R_L in series. But this is exactly the case considered in our discussion of resistive circuits where it was concluded that $R_L = R_S$ for maximum load power. In fact the maximum power delivered to the load is given by

$$P_{max} = \frac{E_S^2}{4R_S} \qquad (24\text{-}16)$$

which is the same as the purely resistive case.

24.11 Wattmeters

If a purely resistive load (PF = 1) is connected to a source, the total true power delivered from the source to the load can be calculated as the product of E_S and I_T. These values can be easily measured using a voltmeter and ammeter respectively. However, if the load is a complex impedance (PF $\neq$ 1) the product of the voltmeter and ammeter readings gives us only the *apparent* power supplied to the load. Thus, this method would not give us a measure of the *true* power consumed by the load.

A wattmeter is a type of meter that can be used to measure true power in any circuit. A wattmeter is connected between the source and the load as shown in Fig. 24–13(a). The wattmeter has *three* terminals. Terminal x is connected to the source, y is connected to the top end of the load, and z is connected to the common connection between the source and load.

The internal operation and construction of the wattmeter (as well as other measuring instruments) is discussed in Appendix D. For our purposes here we can treat the wattmeter as a combination of an ammeter and voltmeter as illustrated in part (b) of Fig. 24–13. This representation indicates that the internal meters measure I_L and V_L respectively. The wattmeter's scale indication always gives the value of true load power; that is, $V_L \times I_L \times$ power factor. In other words, a wattmeter operates so that its scale indication will be *true* power no matter what the PF of the load. This can be useful in measuring the watts in a large power system containing many parallel loads.

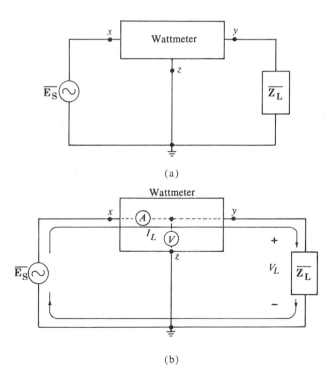

FIGURE 24–13 *Wattmeter used to measure true power to a load.*

EXAMPLE 24.13 A 100-V, 400-Hz source which is driving many parallel loads has its total current measured as 42 A. A wattmeter connected between the source and the loads indicates a reading of 3.5 kW. Determine the overall system power factor.

Solution: The total apparent power supplied by the source is

$$P_A = 100 \text{ V} \times 42 \text{ A} = 4.2 \text{ kVA}$$

The total true power supplied by the source is 3.5 kW. Thus,

$$\text{PF} = \frac{P_T}{P_A} = \frac{3.5 \text{ kW}}{4.2 \text{ kVA}} = 0.833 \text{ or } 83.3\%$$

EXAMPLE 24.14 In a previous chapter we discussed the *effective* resistance of a practical inductor. We saw then that this effective resistance R_e is not simply equal to the resistance of the wire as would be measured on a dc ohmmeter. In an ac circuit, the value of R_e will be greater than just the wire resistance and, in fact, R_e increases with frequency (due to skin effect, hysteresis, etc.).

Figure 24–14 shows a method for measuring the R_e of a coil by using an ammeter and a wattmeter. The ammeter is used to measure the coil current when a source E_S is applied to it. The wattmeter is used to measure the true power dissipated in the coil. These two readings can be used to determine R_e.

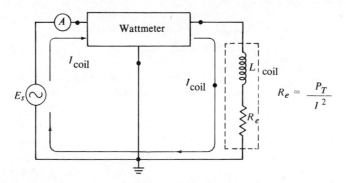

FIGURE 24–14 *Arrangement used to determine* R_e *of a coil.*

Determine R_e if the ammeter reads 60 mA and the wattmeter indicates $P_T = 200$ mW.

Solution: The true power dissipated in the coil is due to R_e and is equal to $I^2 R_e$. Thus,

$$P_T = I^2 R_e$$

or

$$R_e = \frac{P_T}{I^2} = \frac{200 \text{ mW}}{(60 \text{ mA})^2} = 55.5 \text{ ohms}$$

Note in the arrangement of Fig. 24–14 that the ammeter is connected on the source side of the wattmeter. This is done so that the ammeter's own power dissipation is not included in the wattmeter's reading.

Chapter Summary

1. A resistance dissipates true power ($P_T = V_R \times I_R$). The units for true power are watts.

2. A pure inductance does not dissipate true power ($P_T = 0$); it alternately accepts and returns power to the source. This transferred power is called reactive power ($P_{QL} = V_L \times I_L$). The units for reactive power are vars.

3. A pure capacitance does not dissipate true power ($P_T = 0$). It also alternately accepts and returns reactive power to the source ($P_{QC} = V_C \times I_C$).

4. The product of the source voltage and total source current is called apparent power P_A. For a pure resistive circuit, $P_A = P_T$; for a pure reactive circuit, $P_A = P_Q$.

5. For any ac circuit $P_A^2 = P_T^2 + P_Q^2$ where P_Q is the net reactive power ($P_{QL} - P_{QC}$).

6. Power factor is the ratio of true power P_T to apparent power P_A for a circuit. PF $= P_T/P_A = \cos\theta$ where θ is the circuit phase angle.

7. Power factor correction is used to increase PF of a system to unity in order to decrease the total apparent power and current drawn from the source.

8. An ac source will supply maximum true power output to a load impedance when the resistive portion of the load equals the resistive portion of the source impedance, and the reactance portion of the load cancels exactly the reactance portion of the source impedance.

9. A wattmeter is used to measure true power dissipation of a circuit.

Questions/Problems

Sections 24.2–24.4

24–1 Consider the circuit in Fig. 24–15. For each element in the circuit calculate its true power or reactive power.

24–2 If the frequency is increased in Fig. 24–15, how will that affect the answers to 24–1?

24–3 (a) In your own words, explain why a pure inductor does not consume *true* power. (b) Repeat for pure capacitance.

24–4 A 250-ohm resistor is connected in series with a 500-ohm inductive reactance. Calculate P_T for the resistor and P_{QL} for the coil if the applied voltage is 60 V.

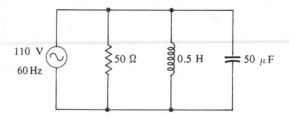

FIGURE 24–15

24–5 How will a decrease in frequency affect the amount of *true* power dissipation in the circuit of problem 24–4?

24–6 Answer *true* or *false:*
   a. A light bulb requires reactive power.
   b. True power in a resistor is not dependent on frequency.
   c. True power is returned to the source by a resistor.
   d. Reactive power is not consumed power.
   e. In a *pure* inductor no heat will be dissipated no matter how large the current becomes.
   f. Only reactances can dissipate reactive power.
   g. True power and reactive power are both measured in watts.

Sections 24.5–24.6

24–7 A 60-V source is applied to a network with $\overline{Z_T} = 50\ \Omega\underline{/15°}$. Determine the apparent power delivered by the source.

24–8 An 18-V source drives a network that has $\overline{Z_T} = 100\ \Omega\underline{/-45°}$. Calculate P_A, P_T, and P_Q. Draw the power triangle.

24–9 An inductive network draws 50 A of current from a 110-V source. The source supplies 1 kW of true power. Determine P_A, P_Q, and Z_T of the network.

Section 24.7

24–10 Determine the PF of the network of problem 24–8. Repeat for the network of problem 24–9.

24–11 A certain load impedance has PF = 0.9 lagging. The load requires 500 VA from a 50-V source. How much power is dissipated in this load?

24–12 What is the PF of a load with $\overline{Z} = (10 + j8)\Omega$?

24–13 What is the PF of a load with $\overline{Z} = 36\ \Omega\underline{/-30°}$?

24–14 A 1-k VA load with PF = 0.707 lagging is connected across a 110-V, 60-Hz source. Determine P_A, P_T, and P_Q. How much current is drawn from the source?

Section 24.8

24–15 Consider the circuit in Fig. 24–16. (a) Determine P_A, P_T, P_Q, and overall power factor. (b) How much current is drawn from the source? (c) If the input frequency is increased slightly, what would happen to the PF?

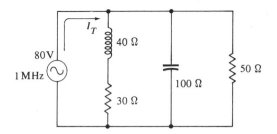

FIGURE 24-16

24-16 A 110-V, 60-Hz source supplies power to the following loads connected in parallel: (1) a motor rated at 600 VA, 80% PF lagging; (2) heater rated at 1 kW, 100% PF; (3) twenty 150-W light bulbs; (4) synchronous motor rated at 1.2 kVA, 60% PF leading. Determine P_A, P_Q, P_T, and the overall power factor. Draw the power triangle.

Section 24.9

24-17 For the circuit in Fig. 24-16 determine the value of the element which must be placed across the source in order to increase the PF to unity. Compare the values of I_T *before* and *after* PF correction.

24-18 Repeat problem 24-17 for the circuit described in problem 24-16.

24-19 Explain why power factor correction is desirable.

24-20 Answer true or false:
   a. Apparent power can never be less than true power consumption.
   b. For a pure resistive load, PF = 0.
   c. For a series *RL* circuit, PF decreases as frequency decreases.
   d. For a parallel *RC* circuit, PF increases as frequency decreases.
   e. PF is the ratio of true power to reactive power.
   f. PF correction increases the true power delivered by the source.
   g. When a source drives several parallel loads, the total P_A of the source is obtained by adding the apparent powers of each load.
   h. Reactive power represents energy lost in an inductor or a capacitor.

Sections 24.10–24.11

24-21 A 24-V source has an internal source impedance $\overline{Z_S} = 10\underline{/20°}$. Determine the load impedance which will draw maximum power from this source. Calculate the true power delivered to this load.

24-22 A television signal generator has an internal source impedance of 50 $\Omega\underline{/-37°}$ at 83.25 MHz. What should be the impedance of the transmitting antenna which serves as the load for the generator if maximum power transfer is desired? Express in polar form.

24-23 The effective resistance of an air-core coil is measured using the ammeter-wattmeter method of Fig. 24-14. When the coil is connected to a 110-V, 60-Hz source, the ammeter reads 1.5 A and the wattmeter reads 45 W. If an iron core is inserted into the coil, the readings are 0.2 A and 900 mW respectively.
   a. Determine R_e for the coil *without* the iron core.
   b. Determine R_e for the coil *with* the iron core.
   c. What causes R_e to increase when the iron core is used?

CHAPTER 25

Transformers

25.1 Introduction

A transformer is a useful electrical device that operates on the principle of *mutual electromagnetic induction* (chap. 14). Typically, a transformer consists of two coils so mounted that the magnetic *flux* produced by the current flowing in one coil will also link the second coil. As we shall see, this transformer action produces several useful effects.

The transformer is used today in many electrical engineering applications, most notable of which is its utilization in power distribution systems. The major applications of transformers utilize their ability to: (1) step up or step down ac voltage and current, (2) act as an impedance matching device, and (3) electrically isolate a signal source from a load.

Transformers can be used at power-line frequencies (50–60 Hz), audio frequencies (20 Hz–20 kHz), ultrasonic frequencies (20–100 kHz), and radio frequencies (above 100 kHz). The physical design and construction of a transformer depend on the frequency range for which it is intended. Power transformers and audio frequency transformers are usually built on some type of iron or steel core to achieve a high degree of magnetic coupling. At higher frequencies (*rf*), a transformer may be built on a nonmagnetic core (air core) or it may be wound on a core of ferrite material which is a high-frequency ferromagnetic material.

25.2 Mutual Induction and Transformer Action

In chap. 15 we encountered mutual induction as an illustration of Faraday's and Lenz's laws. We saw then that a *changing* current in one coil could induce an emf (electromotive force, voltage) in a second coil whose turns were linked by the flux lines produced by the current in the first coil. This action was termed *mutual induction*. The transformer is a device which depends on mutual induction for its operation.

Figure 25-1 shows a basic transformer arrangement with two separate coils wound on an iron core. The two coils are given special names. The coil to which a source of voltage is to be applied is called the *primary winding* and the other coil is called the *secondary winding*. Usually a load is connected across the terminals of the secondary winding.

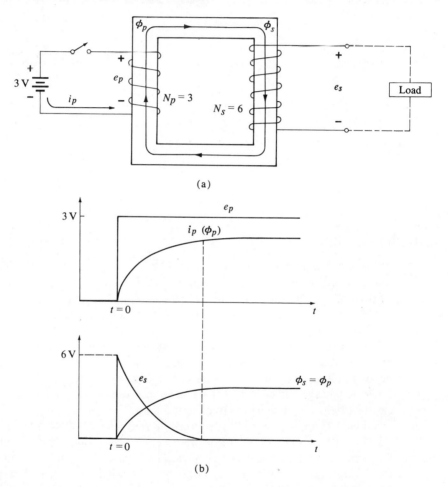

FIGURE 25-1 (a) *Basic transformer.* (b) *Circuit waveforms upon closing of primary switch.*

Upon closing the switch in the primary circuit, the following action takes place.

(1) The current through the primary coil i_p begins to increase, gradually building up to a steady dc value.

(2) This current buildup in the primary produces a flux ϕ_p through the primary coil, which also gradually increases to a steady value.

(3) This same flux will travel the iron-core path so that it also links the secondary coil. If we assume that none of the primary flux ϕ_p *leaks* outside the iron core, then we can say that $\phi_s = \phi_p$, where ϕ_s is the flux linking the secondary coil. This is a valid assumption since the iron core provides a low-reluctance magnetic path for ϕ_p. This would not be true for a nonmagnetic core material.

(4) The increasing flux in the core, then, will induce an emf across the secondary winding. Using Lenz's law and the left-hand rule, the polarity of secondary voltage will be as shown in the figure. (At this point, the student may wish to review this process in chap. 15.) A secondary voltage e_s will be present only as long as the flux in the core is changing. Thus, when i_p levels off, the flux remains constant and e_s becomes *zero*.

The waveforms for this situation are shown in part (b) of Fig. 25–1. Note that as i_p reaches its steady-state value, the secondary voltage drops to zero. This process is gradual as i_p begins to level off and the flux ϕ_p is increasing at a slower rate thereby inducing a smaller voltage in the secondary coil (recall Faraday's law says that the induced emf is proportional to *rate of change of flux*).

Symbols e_p and e_s In the diagram in Fig. 25–1(a) the symbols e_p and e_s are used to denote the voltages across the transformer primary and secondary windings respectively. Up to now we have used the symbols e and E strictly for *source* voltages. This practice will be deviated from in our discussion of transformers since usually the primary voltage e_p will equal the applied source voltage, and the secondary voltage e_s essentially *acts* as a source voltage for any load connected to its terminals.

Voltage Transformation It might have been noted in the waveform of Fig. 25–1 that the secondary voltage e_s jumps to a peak of 6 V when the primary voltage jumps to 3 V upon closing the switch. The primary voltage is, of course, equal to the 3-V source voltage. Why is e_s twice as large? To answer this question we must apply the formula of Faraday's law to our situation. Namely,

$$\text{emf} = N \frac{d\phi}{dt} \qquad (25\text{--}1)$$

which states that the emf induced by a change in the flux linking a coil equals the number of turns in the coil (N) multiplied by the rate of change of flux ($d\phi/dt$).

For the situation in Fig. 25–1, as soon as the switch is closed, 3 V appears across the primary coil. This means that the primary flux must be increasing at such a rate as to induce 3 V across the primary winding. Since the primary coil has 3 turns, then $N_p = 3$ and

$$e_p = N_p \frac{d\phi_p}{dt}$$

$$3 \text{ V} = 3 \times \frac{d\phi_p}{dt}$$

or

$$\frac{d\phi_p}{dt} = \frac{3}{3} = 1 \text{ Weber/second}$$

Thus, at the instant of switch closure the primary flux begins to increase at the rate of one weber per second.

Since $\phi_s = \phi_p$ then it follows that the secondary flux linking the secondary coil must also be initially increasing at this rate ($d\phi_s/dt = 1$ Wb/s). This along with the fact that the secondary winding has 6 turns ($N_S = 6$) allows us to calculate the secondary induced voltage as

$$e_s = N_s \frac{d\phi_s}{dt}$$
$$= 6 \times 1 \text{ Wb/s} = 6 \text{ V}$$

which verifies the waveform amplitude in Fig. 25–1(b). Thus, the secondary voltage amplitude is *twice* the primary amplitude because the secondary winding has *twice* as many turns as the primary winding. If the number of secondary turns and primary turns has been equal, then e_s would have been 3 V; if N_s had been *less* than N_p the value of e_s would have been *less* than 3 V. Stated generally,

The amplitude of secondary voltage in an iron-core transformer is related to the amplitude of the primary voltage by the same ratio as the number of secondary turns N_s to the number of primary turns N_p.

Stated mathematically,

$$\frac{e_p}{e_s} = \frac{N_p}{N_s} \qquad (25\text{–}2)$$

EXAMPLE 25.1 If N_p in Fig. 25–1 is increased to 12 turns, calculate the amplitude of e_s.

Solution: $\dfrac{e_p}{e_s} = \dfrac{N_p}{N_s} = \dfrac{12}{6} = 2$

so that $e_p = 2e_s$ or $e_s = 0.5\, e_p = $ **1.5 V**. Thus, $e_s < e_p$ since $N_s < N_p$.

25.3 AC Voltages Applied to the Primary

Transformers are most often used with ac voltages applied to the input. The situation in Fig. 25–1 used a dc input source, but the secondary output voltage was not a dc output, but rather a short duration pulse (spike). Transformers operate on the principle of mutual induction which requires a *changing* flux and therefore a changing primary current. Thus, if a continuous secondary output is required then the primary input cannot be *pure* dc but must instead be changing in some manner.

Figure 25–2(a) shows the same transformer with an ac sine wave now applied to the primary winding. Once again the primary winding has 3 turns and the secondary winding has 6 turns of wire. The primary applied voltage e_p will produce a current i_p in the primary coil which *lags* e_p by 90° [see waveforms in part (b) of figure]. This continuously changing primary current will produce a corresponding flux variation in the core. That is, the graph of ϕ_p (and ϕ_s) versus t follows the same waveform as i_p. This is shown in Fig. 25–2(c). Therefore, ϕ_p lags e_p by 90°.

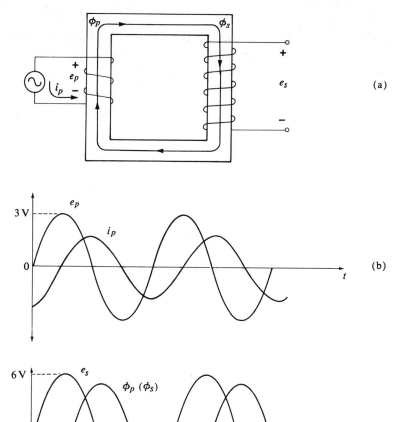

FIGURE 25-2 *Operation of iron-core transformer with ac input and open-circuited secondary.*

The flux waveform ϕ_p shows the manner in which the core's flux is changing so as to induce an emf in the primary coil that is equal to e_p. In our example, e_p is a sine wave with peak amplitude of 3 V. This same flux waveform links the secondary coil; thus, (assuming no flux leakage) we have $\phi_s = \phi_p$. Since the same continuously changing flux waveform is linking the secondary coil, we can expect that the emf induced across the secondary e_s will follow the same variation as e_p; however, since the secondary winding has *twice* as many turns we can expect e_s to have *twice* the amplitude of e_p (see the waveforms in the figure).

To summarize this process, a sine wave of voltage applied to the primary will induce a sine wave of voltage across the secondary such that

$$\frac{e_p}{e_s} = \frac{N_p}{N_s} \tag{25-2}$$

at any instant of time. The symbols e_p and e_s represent instantaneous values. The same relationship will hold true for the rms values E_p and E_s. Thus,

$$\frac{E_p}{E_s} = \frac{N_p}{N_s} \qquad (25\text{-}3)$$

Turns Ratio Relationship (25–3) can be rewritten as

$$E_p = \left(\frac{N_p}{N_s}\right) E_s \qquad (25\text{-}4a)$$

Here, the ratio N_p/N_s is called the *primary-to-secondary* turns ratio. It is often given the symbol a. That is,

$$a = \frac{N_p}{N_s}$$

so that

$$E_p = aE_s \text{ or } E_s = \frac{E_p}{a} \qquad (25\text{-}4b)$$

When the number of primary turns N_p is smaller than the number of secondary turns, N_s, the turns ratio a is less than one so that $E_p < E_s$. In such cases the induced secondary voltage will be greater than the applied primary voltage and we say that the transformer is a *step-up* transformer. On the other hand, when $N_p > N_s$, the turns ratio a is greater than one and $E_p > E_s$ so that we have a *step-down* transformer.

EXAMPLE 25.2 A 12-V, 1000-Hz source is applied to the primary of a transformer whose primary has 20 turns and whose secondary has 5 turns. Determine the secondary voltage.

Solution: This problem can be solved using Eq. (25–4b).

$$a = \frac{N_p}{N_s} = \frac{20}{5} = 4$$

$$E_s = \frac{E_p}{a}$$

$$\therefore E_s = \frac{12 \text{ V}}{4} = 3 \text{ V}$$

It could also have been solved using simple common sense. Since the primary has *four* times as many turns as the secondary, the transformer is a *step-down* transformer with a step-down ratio of four to one (4:1). Thus, $E_s = E_p/4 = 12/4 = 3$ V.

EXAMPLE 25.3 The voltage of the previous example is applied to a transformer with $N_p = 1000$ and $N_s = 250$. Determine the secondary voltage.

Solution: A little thought will bring us to the conclusion that this case is the same as that in the preceding example since we still have a *step-down* transformer ratio of 4:1. Thus, **E$_s$ = 3 V.**

EXAMPLE 25.4 A transformer has a 50-turn secondary. How many turns must the primary winding have if the transformer is to step up a 10-V input to produce a 25-V secondary output?

Solution: $\dfrac{E_p}{E_s} = \dfrac{N_p}{N_s} \rightarrow \dfrac{10 \text{ V}}{25 \text{ V}} = \dfrac{N_p}{50}$

$$\therefore N_p = 20 \text{ turns}$$

Transformer Circuit Symbol Figure 25–3 shows the standard circuit symbols for iron-core transformers. In (a) the transformer is a step-up transformer as indicated by the greater number of secondary turns. In (b) the greater number of primary turns indicates a step-down transformer.

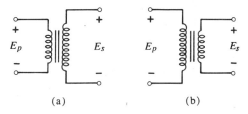

FIGURE 25–3 *Symbols for* (a) *step-up and* (b) *step-down iron-core transformers.*

Summary Thus Far Our analysis of the iron-core transformer thus far has been simplified by assuming some *ideal* conditions. The assumptions which have been implicit in our study of the *ideal* transformer are: (1) no leakage flux; all the primary flux ϕ_p links all of the secondary turns so that $\phi_s = \phi_p$; (2) no resistance in the primary and secondary windings; (3) a magnetic core material in which ϕ varies in direct proportion to current in the primary coil. (This implies a linear *B-H* curve.)

We will continue to analyze the *ideal* transformer because, in practice, a good iron-core transformer behaves very closely to the ideal transformer under many conditions. Later we will look at some of the effects produced in practical, non-ideal transformers.

Thus far for our ideal iron-core transformer we have seen that the ratio of the primary-induced emf to the secondary-induced emf is the same as the ratio of the number of primary turns to the number of secondary turns. A step-up or step-down transformer is obtained by choosing a proper N_p/N_s turns ratio.

25.4 Loaded Secondary

We have seen how a transformer can take an applied voltage and step it up or down according to its turns ratio. The voltage at the secondary winding terminals is present as a result of magnetic induction. In most transformer applications a load is connected across the secondary winding, thereby causing current to flow in the secondary circuit. It is useful when considering a load on the secondary to treat the induced secondary voltage e_s as a *voltage source*. In other words, as far as the transformer is concerned, the voltage source is applied to the primary; but as far as the load is concerned, the transformer's secondary voltage acts as the source.

Before considering the effects of a load on the transformer's secondary, let us review the no-load situation. With an open-circuited secondary as in Fig. 25–2, an ac voltage applied to the primary winding produces a current i_p which lags e_p by 90°. The value of i_p will depend on the frequency of e_p and on the self-inductance of the primary winding. This inductance is given the symbol L_p. Thus, using rms values we have

$$I_p = \frac{E_p}{X_p} = \frac{E_p}{2\pi f L_p}$$

The phasor diagram for the no-load situation is shown in Fig. 25–4. Note that I_p lags E_p by 90°. Also note that the secondary voltage E_s is *in phase* with E_p.

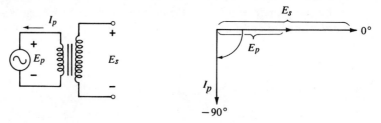

FIGURE 25–4 *Phasor diagram for no-load situation.*

The primary current which flows under no-load conditions is necessary to produce the magnetic flux in the core which induces both E_p and E_s. This current is therefore given the special name of *magnetizing current* since it magnetizes the core. We will give this primary magnetizing current the symbol I_{pm}. As we shall see, under loaded conditions this magnetizing current will be a very small portion of the total primary current which will flow.

EXAMPLE 25.5 A 5:1 step-down transformer has a primary inductance $L_p = 2$ H. A 24-V, 1000-Hz voltage is applied to the primary. Determine the primary magnetizing current I_{pm} and the secondary voltage E_s.

Solution: $E_s = \dfrac{E_p}{5} = \dfrac{24 \text{ V}}{5} = 4.8$ V (step-down)

$$I_{pm} = \frac{E_p}{X_p} = \frac{24 \text{ V}}{2\pi \times 10^3 \times 2} = \frac{24 \text{ V}}{12.6 \text{ k}\Omega}$$
$$= 1.9 \text{ mA}$$

In the no-load condition there is no *true* power supplied by the input source since the input current I_{pm} is 90° out of phase with the voltage. This of course assumes that any resistance in the primary winding is so small that it dissipates negligible power.

Resistive Load on Secondary Figure 25-5(a) shows an ideal iron-core transformer with a load R_L connected across the secondary terminals. From our previous discussions we know that the secondary voltage E_s will be *twice* the primary voltage E_p (since $N_s = 6$ and $N_p = 3$) and will be in phase with E_p. This secondary voltage acts as a voltage source for the load and establishes a current through the load and the secondary winding. This secondary current is given the symbol I_s. For a resistive load R_L, the secondary current is determined from

$$I_s = \frac{E_s}{R_L} \tag{25-5}$$

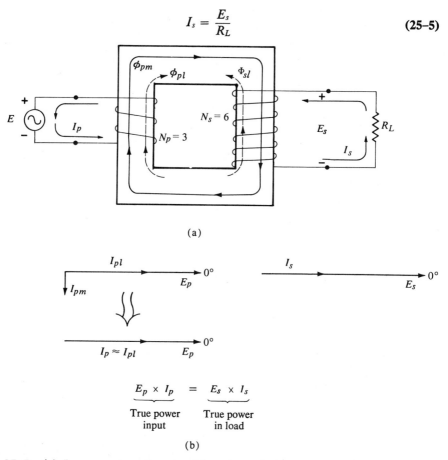

FIGURE 25-5 (a) *Iron-core transformer with resistive load.* (b) *Primary and secondary phasor diagrams.*

and is *in phase* with E_s. As a result of this flow of current in the secondary, there will be an accompanying *increase* in current in the primary winding. In other words,

> Going from a no-load condition to a loaded condition on the secondary of a transformer will cause the current in the primary winding to increase.

Why does the primary current increase when a load is connected to the secondary? This question can be answered if we invoke the principle of conservation of energy which demands that the energy output from any machine or device must equal the energy input. When the load R_L is connected to the transformer secondary, energy is dissipated in the load. The rate at which this energy is delivered to the load is the load power P_L. Since the load is resistive the load power is *true* power being consumed by the load. Thus,

$$P_L = E_s \times I_s \qquad (25\text{-}6)$$

This power being consumed by the load must come from someplace. In the transformer circuit, the only *source* of power is the voltage E_p applied to the primary. Any energy supplied to the load must come from this source. Thus, in order to supply a load power $E_s \times I_s$, the input source current must increase, thereby increasing the transformer primary current. This additional primary current which flows because of the load on the secondary is given the symbol I_{pl}. Since E_p has to supply *true* power to the transformer, then I_{pl} must be in phase with E_p.

Part (b) of Fig. 25-5 illustrates the phasor relationships for the primary and secondary voltages and currents. Note that there are *two* components of primary current: I_{pm} is the magnetizing current which lags E_p by 90° and I_{pl} is the in-phase primary current which delivers the power. The total primary current I_p is actually the vector sum of these two; that is,

$$\overline{I_p} = \overline{I_{pm}} + \overline{I_{pl}} \qquad (25\text{-}7)$$

However, in most transformers I_{pm} will be much smaller than I_{pl} under normal load conditions. Thus, to simplify our analysis we will neglect I_{pm} and assume that $I_p \approx I_{pl}$.

The true power input to the transformer is simply

$$P_{in} = E_p I_{pl} \approx E_p I_p$$

If we assume that no power is dissipated in the transformer windings or used up as heat in the core (hysteresis and eddy current losses), then P_{in} must equal P_L. Therefore,

$$E_p I_p = E_s I_s \qquad (25\text{-}8)$$

Using this relationship, we can determine I_p for a given load condition.

EXAMPLE 25.6 A 10-V source is applied to the transformer of Fig. 25-5. A load of 100 ohms is connected to the secondary. Determine E_s, I_s, and I_p.

Solution: The circuit schematic is drawn in Fig. 25-6. The transformer has a 1:2 step-up ratio. Therefore, with $E_p = 10$ V the secondary voltage is $E_s = 20$ V. With $R_L = 100$ ohms, the secondary load current is

$$I_s = \frac{20 \text{ V}}{100 \text{ }\Omega} = 0.2 \text{ A} = 200 \text{ mA}$$

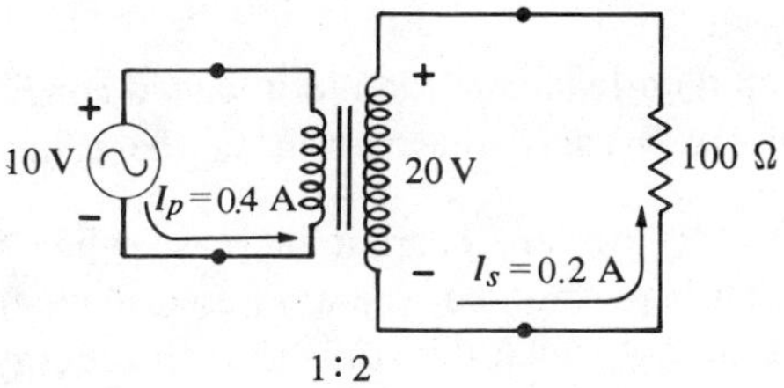

FIGURE 25-6

The load power is therefore
$$P_L = 20 \text{ V} \times 200 \text{ mA} = 4 \text{ W}$$
The input power is $E_p I_p$ and must also equal 4 W. Thus,
$$P_{\text{in}} = E_p I_p = 4 \text{ W}$$
The input power is $E_p I_p$ and must also equal 4 W. Thus,
$$P_{\text{in}} = E_p I_p$$
$$\therefore I_p = \frac{P_{\text{in}}}{E_p} = \frac{4 \text{ W}}{10 \text{ V}} = 0.4 \text{ A}$$

Note that I_p is *twice* as large as I_s while E_s is *twice* as large as I_s.

EXAMPLE 25.7 Repeat the last example for a load of 10 ohms.

Solution: E_s is still **20 V**. Therefore,
$$I_s = \frac{20 \text{ V}}{10 \text{ }\Omega} = 2 \text{ A}$$
and
$$P_L = 20 \text{ V} \times 2 \text{ A} = 40 \text{ W}$$
$$P_{\text{in}} = 10 \text{ V} \times I_p = P_{\text{load}} = 40 \text{ W}$$
Thus,
$$I_p = \frac{40 \text{ W}}{10 \text{ V}} = 4 \text{ A}$$

The primary current has increased due to the smaller R_L, which draws more current and consumes more power.

Transfer of Energy As we have seen, a load connected to the secondary will dissipate power. The energy consumed in the load must be supplied by the input source connected to the primary. A logical question at this point concerns the manner in which the primary source can supply energy (power) to the load when there is no electrical connection between the two. Although there is no electrical connection, there is a magnetic connection via the common flux lines which link both windings. It is these flux lines which transfer energy from the primary to the secondary load. The ac primary current produces a flux which alternately magnetizes the core in one direction, then in the other. This changing flux induces an ac voltage in the secondary which then produces current in the load.

When the load resistance value is changed, the secondary current I_s will change accordingly, thereby changing the amount of power required by the load. The primary current adjusts itself to the changes in the required load power. For example, if R_L is decreased, I_s will increase so that $P_L = E_s I_s$ also increases. Thus, I_p has to increase so that $P_{\text{in}} = E_p I_p$ can equal P_L. The opposite changes take place if R_L is increased.

Just how does the primary current know when to increase or decrease? This is a reasonable question to ask since there is no electrical connection between the source and the load. The answer again lies in the magnetic coupling between secondary and primary. The following is a step-by-step explanation (refer to Fig. 25-5).

(1) With no load on the secondary, the only current flowing in the primary winding will be the magnetizing current I_{pm}. The flux ϕ_{pm} produced in the core will have an amplitude which induces a voltage on the primary winding equal to the applied voltage E_p.

(2) When a load is connected to the secondary, a current flows through the secondary winding. This current in the secondary coil produces a *counterflux* ϕ_{sl} in the core which opposes (Lenz's law) the flux produced by the primary current, thereby tending to reduce the total flux in the core. [See dotted lines in Fig. 25-5(a)].

(3) The total flux in the core must maintain the amplitude needed to induce E_p in the primary winding (step 1). Thus, the primary current has to increase so as to produce an additional flux ϕ_{pl} needed to counteract the counterflux produced by the secondary current, thereby keeping the amplitude of ϕ constant at the original value (step 1).

(4) The same type of action occurs as the load is increased or decreased and the primary current changes accordingly.

25.5 Transformer Current Ratio

We saw in the previous section that for an ideal transformer, the input power to the primary exactly equals the power consumed by the secondary load. Accordingly, for a resistive load we saw that

$$E_p I_p = E_s I_s \tag{25-8}$$

This equality can be rearranged as

$$\frac{I_p}{I_s} = \frac{E_s}{E_p}$$

The ratio E_s/E_p is equal to N_s/N_p for the iron-core transformer. Thus,

$$\frac{I_p}{I_s} = \frac{N_s}{N_p} = \frac{1}{a} \tag{25-9}$$

This last relationship shows that the primary and secondary currents are related in just the opposite manner as the primary and secondary voltages. Table 25-1 summarizes these relationships.

As the table indicates, a transformer which is designed to step up the input voltage will step down the input current and vice versa. This is logical if the input and output powers are to be equal. The following examples illustrate.

Transformers

TABLE 25-1

Voltage relationship	$\dfrac{E_p}{E_s} = \dfrac{N_p}{N_s} = a$		step-up for $a < 1$ step-down for $a > 1$
Current relationship	$\dfrac{I_p}{I_s} = \dfrac{N_s}{N_p} = \dfrac{1}{a}$		step-up for $a > 1$ step-down for $a < 1$
Power relationship	$E_p I_p = E_s I_s$		$P_{in} = P_{load}$

EXAMPLE 25.8 A transformer with $N_p = 100$ and $N_s = 20$ has a 110-V ac source applied to its primary. A load of 10 ohms is connected to the secondary. Calculate E_s, I_s, and I_p.

Solution:
$$\frac{E_s}{E_p} = \frac{N_s}{N_p} = \frac{20}{100} = 0.2$$

$$E_s = 0.2\, E_p = 0.2 \times 110 \text{ V} = \mathbf{22 \text{ V}}$$

$$I_s = \frac{E_s}{R_L} = \frac{22 \text{ V}}{10\, \Omega} = \mathbf{2.2 \text{ A}}$$

There are two possible ways for calculating the primary current.

Method I:
$$E_p I_p = E_s I_s$$
$$110 \times I_p = 22 \times 2.2$$
$$\therefore I_p = \frac{48.4}{110} = \mathbf{0.44 \text{ A}}$$

Method II:
$$\frac{I_p}{I_s} = \frac{N_s}{N_p} = \frac{20}{100} = 0.2$$
$$I_p = 0.2\, I_s = 0.2 \times 2.2 \text{ A} = \mathbf{0.44 \text{ A}}$$

In this example the transformer acted as a *step-down* for voltage so that $E_s < E_p$. On the other hand, it acted as a *step-up* for current since $I_s > I_p$.

EXAMPLE 25.9 A certain transformer has a 1:10 step-up ratio. Determine how much input voltage and current are needed to supply 100 W of power to a 25-ohm load.

Solution: The secondary voltage E_s is across the 25-ohm load. Thus,

$$P_L = \frac{E_s^2}{R_L}$$

$$100 \text{ W} = \frac{E_s^2}{25}$$

Therefore, $E_s = 50$ V.

The secondary current is thus given as

$$I_s = \frac{50 \text{ V}}{25 \text{ } \Omega} = 2 \text{ A}$$

The transformer is a 1:10 step-up transformer. This means that it steps up the *voltage* by a factor of 10. Thus,

$$E_p = \frac{E_s}{10} = \frac{50 \text{ V}}{10} = 5 \text{ V}$$

We also know that the same transformer will *step down* the current by a factor of ten. Thus,

$$\frac{I_p}{I_s} = \frac{10}{1}$$

so that

$$I_p = 10 \, I_s = 20 \text{ A}$$

As a check, we can calculate input power

$$P_{\text{in}} = E_p I_p = 5 \text{ V} \times 20 \text{ A} = 100 \text{ W} = P_L$$

Impedance Load on Secondary For an ideal transformer the voltage and current relationships in Table 25–1 hold true for any type of load. The load can be capacitive, inductive, or a complex impedance. The following example illustrates.

EXAMPLE 25.10 For the transformer circuit in Fig. 25–7 determine (a) $\overline{E}_s$, $\overline{I}_s$, and $\overline{I}_p$; (b) the true power consumed by the load; (c) the true power delivered by the source.

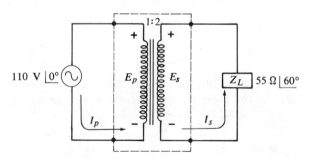

FIGURE 25–7 *Transformer driving an impedance load.*

Solution: (a) Since the transformer is a 1:2 step-up transformer, the secondary voltage will be

$$\overline{E}_s = 2 \, \overline{E}_p = 2 \times 110 \text{ V}\underline{/0°}$$
$$= 220 \text{ V}\underline{/0°}$$

$$\overline{I}_s = \frac{\overline{E}_s}{\overline{Z}_L} = \frac{220 \text{ V}\underline{/0°}}{55 \text{ }\Omega\underline{/60°}} = 4 \text{ A}\underline{/-60°}$$

The transformer acts as a 2:1 step-down for current. Therefore,

$$\overline{I_p} = 2 \times \overline{I_s} = 8 \text{ A}/\!-\!60°$$

(b) The load has a power factor equal to

$$PF = \cos(60°) = 0.5$$

The apparent load power is

$$P_A(\text{load}) = E_s I_s = 220 \text{ V} \times 4 \text{ A} = 880 \text{ VA}$$

Thus,

$$P_T(\text{load}) = PF \times P_A = 0.5 \times 880 = 440 \text{ W}$$

(c) The true power delivered by the input source has to equal the true power consumed by the load.

We can check this as follows:

$$P_T(\text{input}) = E_p \times I_p \times \cos(60°)$$
$$= 110 \text{ V} \times 8 \text{ A} \times 0.5$$
$$= 440 \text{ W}$$

25.6 Impedance Transformation

Consider the circuit in Fig. 25-8(a). The transformer has a 5:1 step-down ratio. Thus, with $E_p = 10$ V the secondary load voltage is $E_s = 2$ V. This voltage across the 2-ohm load will produce a secondary current $I_s = 1$ A. The current in the primary will therefore be 5 times *smaller;* that is

$$\frac{I_p}{I_s} = \frac{N_s}{N_p} = \frac{1}{5} = 0.2 \text{ A}$$

or $I_p = 0.2 I_s = 0.2$ A. This primary current represents the total current drawn from the source (neglecting the small magnetizing current I_{pm}).

Now consider the situation in Fig. 25-8(b) where the same transformer is used. However, this time the secondary is open circuited (no load) and a 50-ohm resistance is placed in parallel with the source and the primary winding. Clearly, since there is no secondary current then the only current through the primary winding will be the negligibly small magnetizing current I_{pm}. The 50-ohm resistance will draw 10 V/50 ohms = 0.2 A from the source. The voltage across the secondary winding will still be 2 V because of the 5:1 step-down ratio.

In comparing the two cases of Fig. 25-8, it is clear that in each case the 10-V input source has to deliver 0.2 A to the circuit and therefore 2 W of power. Thus, as far as the source is concerned, both circuits are equivalent. The source in Fig. 25-8(a) acts just as if it were driving a 50-ohm load even though the

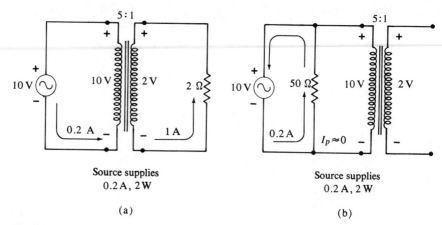

FIGURE 25-8 *A load placed across the secondary of a step-down transformer acts as a larger impedance as seen by the primary source.*

transformer secondary output is driving a 2-ohm load. In other words, the source acts as though it were supplying power to a 50-ohm load. The transformer *transforms* this power and supplies it to the 2-ohm load.

According to this analysis, it is seen that *for a step-down transformer a load connected to the secondary appears as a larger impedance to the primary source.* This seems reasonable in view of the fact that I_p is *less* than I_s in a step-down transformer. If we repeated the analysis for a *step-up* transformer the results would be just the opposite, as the following example illustrates.

EXAMPLE 25.11 Change the transformer in Fig. 25-8(a) so that it has a step-up ratio of 1:5. Determine the equivalent load seen by the primary source.

Solution: $E_s = 5E_p = 50$ V

$$I_s = \frac{E_s}{R_L} = \frac{50 \text{ V}}{2 \text{ }\Omega} = 25 \text{ A}$$

$$I_p = 5 I_s = 125 \text{ A}$$

Thus, the source sees an equivalent load of only

$$\frac{E_p}{I_p} = \frac{10 \text{ V}}{125 \text{ A}} = 0.08 \text{ }\Omega$$

Thus, *for a step-up transformer the secondary load appears as a smaller impedance to the primary source.* This seems reasonable since I_p is greater than I_s for a step-up transformer.

Impedance Transformation Formula The concepts discussed here can be expressed conveniently in one formula. If we connect a load Z_L to the secondary of a transformer we can express E_s and I_s as:

$$E_s = \frac{N_s}{N_p} = \frac{E_p}{a}$$

$$I_s = \frac{E_s}{Z_L} = \frac{E_p}{aZ_L}$$

The primary current I_p can be found using

$$I_p = \frac{N_s}{N_p} \times I_s = \frac{1}{a} \times I_s$$

$$= \frac{1}{a} \times \frac{E_p}{aZ_L}$$

Rearranging this last expression results in

$$\frac{E_p}{I_p} = a^2 Z_L$$

But, E_p/I_p represents the equivalent impedance Z_p seen by the primary source. Thus,

$$Z_p = \frac{E_p}{I_p} = a^2 Z_L \qquad (25\text{--}10)$$

where Z_p is the letter symbol for the equivalent primary impedance. This formula is valid for *any* complex load impedance, and to be technically correct should be written in vector form as

$$\overline{Z}_p = a^2 \overline{Z}_L \qquad (25\text{--}11)$$

This formula is called the *impedance transformation formula* and is used to determine the equivalent primary input impedance Z_p for a load Z_L on the secondary. This formula verifies our earlier quantitative analysis. In a *step-down* transformer the turns ratio $a = N_p/N_s$ is greater than one so that $Z_p = a^2 Z_L > Z_L$. For the step-up case, a is less than one so that $Z_p < Z_L$.

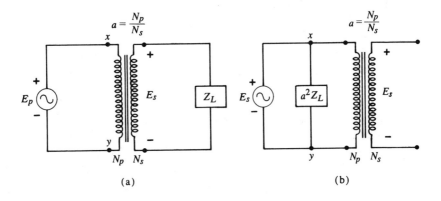

FIGURE 25-9 Secondary load Z_L reflected back to the primary as an equivalent impedance $a^2 Z_L$.

Formula (25–11) further tells us that a load Z_L connected to the secondary is *reflected* back to the primary to appear as if it has been multiplied by the *square* of the turns ratio (a^2). This Z_p is often referred to as the *primary reflected impedance*. Figure 25-9 illustrates the reflected load impedance concept. The secondary load Z_L [part (a) of the figure] affects the *source* the same as if it were multiplied by a^2 and connected across the primary [part (b) of figure].

EXAMPLE 25.12 What is the equivalent resistance seen by an ac source connected to the primary of a transformer which has $N_p = 120$ and $N_s = 30$ and which has a 125-ohm load connected to the secondary?

Solution: The transformer is a step-down with a turns ratio $a = 120/30 = 4$. Thus,

$$Z_p = a^2 Z_L = 4^2 \times 125$$
$$= 2000 \text{ ohms}$$

EXAMPLE 25.13 A certain transformer secondary drives a load $\overline{Z_L} = 2.5 \text{ k}\Omega/\underline{40°}$. The primary input impedance is $\overline{Z_P} = 100 \ \Omega/\underline{40°}$. What is the transformer's turns ratio?

Solution: $\overline{Z_P} = a^2 \overline{Z_L}$

$$a^2 = \frac{\overline{Z_P}}{\overline{Z_L}} = \frac{100 /\underline{40°}}{2.5 \times 10^3 /\underline{40°}} = \frac{1}{25}$$

therefore

$$a = 1/5$$

EXAMPLE 25.14 An audio power amplifier must often supply power to a speaker that has a very low resistance (8 Ω or 16 Ω, typically). Usually the power amplifier has an output impedance (source impedance) that is much larger than the speaker's resistance. This situation is highly undesirable if maximum power is to be delivered to the speaker. We know that the *maximum power principle* requires the load resistance to equal the source resistance for maximum load power. A transformer can be used as a means of connecting the speaker load to the amplifier output so that the load appears as a larger resistance to the amplifier. In this application the transformer is acting as an *impedance matching* device.

Figure 25–10 is an example of the situation described here. The amplifier's output resistance is 500 ohms. Determine the turns ratio which is required to make the 16-Ω speaker act as a 500-ohm load on the amplifier output.

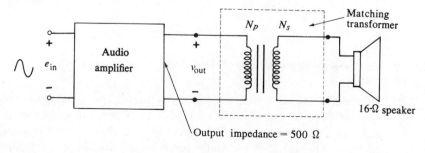

FIGURE 25–10 *Transformer used as an impedance matching device.*

Solution: $Z_p = a^2 Z_L$

$$\therefore a^2 = \frac{Z_p}{Z_L} = \frac{500}{16} = 31.3$$

$$a = 5.6 = \frac{N_p}{N_s}$$

The transformer has to be a *step-down* transformer since $a > 1$.

25.7 Multiple Secondaries and Tapped Secondary

Many iron-core transformers are constructed with more than one secondary winding; each secondary winding will have a voltage induced across its terminals when a changing voltage is applied to the primary. *Multiple secondaries are used so that one input source applied to the primary can produce several different output voltages on the various secondaries.*

Figure 25–11 shows the schematic for a transformer with two secondary windings. The primary winding has $N_p = 1000$ turns; the upper secondary winding has $N_{s1} = 3000$ turns and the lower secondary has $N_{s2} = 60$. The voltages across each secondary are determined in the same manner as for the conventional transformer. Thus,

$$E_{s1} = \frac{N_{s1}}{N_p} \times E_p$$

$$= \frac{3000}{1000} \times 110 \text{ V} = 330 \text{ V}$$

and

$$E_{s2} = \frac{N_{s2}}{N_p} \times E_p$$

$$= \frac{60}{1000} \times 110 \text{ V} = 6.6 \text{ V}$$

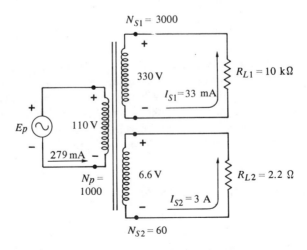

FIGURE 25–11 *With multiple secondaries, total primary current is determined by combining the effects of each secondary current.*

With the load values shown, the current in each secondary can be calculated as

$$I_{s1} = \frac{E_{s1}}{R_{L1}} = \frac{330 \text{ V}}{10 \text{ k}\Omega} = 33 \text{ mA}$$

$$I_{s2} = \frac{6.6 \text{ V}}{2.2} = 3 \text{ A}$$

The current in the primary winding can be calculated by equating *total* power input to *total* load power. The total power consumed by the loads is the sum of the individual load power dissipations.

$$\begin{aligned} P_{\text{loads}} &= E_{s1}I_{s1} + E_{s2}I_{s2} \\ &= (330 \times 0.033) + (6.6 \times 3) \\ &= 10.9 + 19.8 \\ &= 30.7 \text{ W} \end{aligned}$$

Thus,

$$P_{\text{input}} = E_p I_p = 30.7 \text{ W}$$

$$\therefore I_p = \frac{30.7 \text{ W}}{110 \text{ V}} = 279 \text{ mA}$$

The primary current could also be determined by using the current transformation ratio formula and calculating the component of primary current caused by each secondary current and then summing these components. Let's call I_{p1} the primary current component which is due to I_{s1}. Therefore,

$$I_{p1} = \frac{N_{s1}}{N_p} \times I_{s1} = 3 \times 33 \text{ mA} = 99 \text{ mA}$$

Similarly, I_{p2} is the primary current due to I_{s2}.

$$I_{p2} = \frac{N_{s2}}{N_p} \times I_{s2} = \frac{60}{1000} \times 3 \text{ A} = 180 \text{ mA}$$

Thus, the total primary current is

$$I_p = I_{p1} + I_{p2} = 279 \text{ mA}$$

which agrees with our previous result.

Tapped Secondary A transformer with a tapped secondary is shown in Fig. 25–12. A third connection (called a tap) is made to the secondary somewhere between the ends of the coil. When it is in the center, it is called a *center tap*. Two separate loads can be connected between the tap and either end of the coil as shown. This arrangement operates similar to the multiple arrangement transformer and can be analyzed in the same way.

25.8 Phase Inverting Transformer

In our earlier discussion of transformer action we saw that the secondary-induced voltage E_s is *in phase* with the applied primary voltage (see Fig. 25–2). This was simply a result of the way in which each winding was oriented on the core. If the secondary winding is wound in the opposite way to the primary winding, then the secondary voltage will be *180° out of phase* with the primary

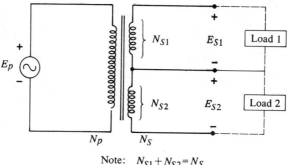

Note: $N_{S1} + N_{S2} = N_S$

FIGURE 25–12 *Transformer with tapped secondary.*

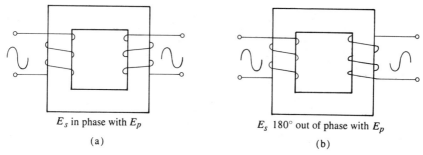

E_s in phase with E_p E_s 180° out of phase with E_p

(a)   (b)

FIGURE 25–13 *Relative orientation of windings determines phase shift between primary and secondary voltages.*

voltage. This can be verified using Lenz's law. Figure 25–13 shows the two possible arrangements. Note the difference in the secondary winding orientation in the two cases.

For the winding orientation in (b), the secondary voltage is 180° out of phase with the primary voltage. A transformer with such a winding arrangement is often called an inverting transformer because the output is the complete inverse of the input. The other arrangement is therefore a *noninverting* transformer.

Dot Convention The diagrams in Fig. 25–13 are too cumbersome to use in circuit schematics. For this reason, some method must be employed to distinguish between inverting and noninverting transformers while retaining the basic transformer circuit symbol. Figure 25–14 illustrates the common *dot convention* which is used for such a purpose. One end of each winding is marked with a dot. For a given transformer the dots indicate points which will be at the same polarity. In other words, the dotted end of the secondary winding will be positive whenever the dotted end of the primary winding is positive, and negative whenever the dotted end of the primary winding is negative. Clearly, then, the transformer in Fig. 25–14(a) is a noninverting transformer and the transformer in (b) is an inverting transformer.

In many applications the transformer's main function is to transfer power from the primary source to the secondary load, and the relative phase difference between the windings is not important. As such, the dot convention is omitted in those cases and is only used where the phase relationship is important.

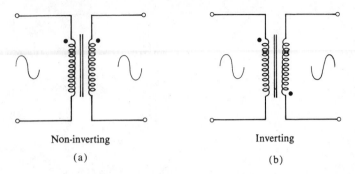

Non-inverting
(a)

Inverting
(b)

FIGURE 25-14 *Dot convention for indicating relative phase between windings.*

25.9 The Autotransformer

An autotransformer (Fig. 25–15) combines the primary and secondary into a single winding that is tapped somewhere along its length to provide three terminals (x,y,z). The total winding included between x-z is considered the *primary*, while the winding between y-z is the *secondary*. Although the primary and secondary are connected at point y, the autotransformer operates according to the same principles as the more conventional transformer with separate windings. That is, the primary and secondary voltages and currents are still related by the turns ratio.

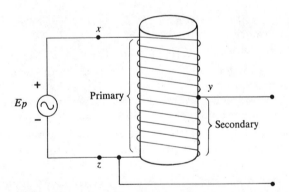

FIGURE 25–15 *Autotransformer.*

The circuit in Fig. 25–16(a) will be used to illustrate the operation of the autotransformer. In this circuit, the standard circuit symbol for the autotransformer is used. The primary has 300 turns and the secondary has 100 turns. Thus, with $E_p = 60$ V the secondary voltage will be 60 V/3 = 20 V. This 20 V produces 4 A through the 5-ohm load. Thus, the power delivered to the load is 4 A × 20 V = 80 W. The power input must therefore be 80 W so that

$$E_p I_p = 80 \text{ W}$$
$$\therefore I_p = \frac{80 \text{ W}}{60 \text{ V}} = 1.33 \text{ A}$$

This current is shown flowing through the source.

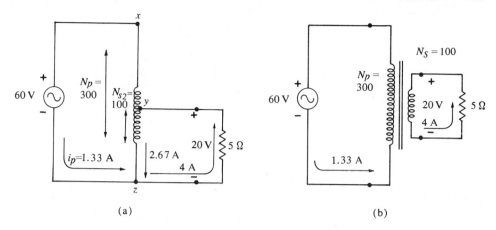

FIGURE 25-16 (a) *Current flowing through secondary winding y-z is less than load current in autotransformer.* (b) *Same case with conventional transformer.*

Let us momentarily focus our attention on point z. The 1.33-A source current flows *into z*. The 4-A load current flows *out of z*. To satisfy KCL, then, a current of 2.67 A must flow from *y into z* as shown. This result illustrates a significant fact: the secondary winding *y-z* does not conduct the full load current. In this autotransformer the 4-A load current is made of 1.33 A conducted through the primary winding and 2.67 A conducted through the secondary winding. Since the secondary winding only carries part of the load current, it can be made of less heavy wire than it would have to be in the equivalent conventional transformer [see part (b) of Fig. 25–16].

Besides this advantage, the autotransformer also offers a savings in the amount of wire used in the windings. This is due to the fact that part of the winding (*y-z*) is *shared* by both the secondary and primary. For example, only 300 total turns are needed in the autotransformer while 400 total turns are needed in the conventional transformer (Fig. 25–16b.) The principal disadvantage of the autotransformer is the lack of *electrical isolation* between the source and the load since they are connected at point z. As we shall see later, the electrical isolation offered by a separate-winding transformer is often necessary in many applications.

The tap on the winding of an autotransformer is often made movable so that the secondary output voltage can be varied as the tap is moved up or down. The *variable autotransformer* can then be used to control the output voltage over a continuous range.

Step-Up Autotransformer The autotransformer discussed in the preceding section was a step-down type since $N_s < N_p$. Figure 25–17 shows an arrangement used to produce a *step-up* autotransformer.

25.10 Transformer as an Isolation Device

We have seen how a transformer can be used to step up or step down voltage and current, and how it is used as an impedance matching device. A transformer is also often used to electrically isolate one portion of a circuit from another; that is, it is used to transfer electrical power without any electrical connection.

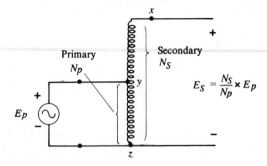

FIGURE 25-17 *Step-up autotransformer.*

Most sophisticated electrical equipment uses an input isolation transformer where the 60-Hz power line comes into the equipment chassis. The reason for this is the fact that one side of the ac line (so-called "neutral") is generally connected to earth ground (usually via a cold-water pipe). Thus, it is possible for a person to be connected to one side of the ac line by simply standing on a damp floor, or touching a water pipe, a faucet, etc. If the ac plug is inserted in such a direction so that the chassis of a transformerless piece of equipment is tied to the "hot" side of the line, then it is possible to receive a severe shock by touching the chassis [part (a) of Fig. 25–18]. By using an isolation transformer, then, the hot side of the ac line cannot be connected to the chassis [see part (b) of figure].

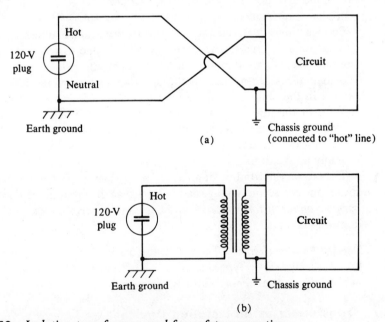

FIGURE 25-18 *Isolation transformer used for safety precaution.*

As another example of transformer isolation refer to Fig. 25–19. The circuit is an SCR phase-control circuit whose SCRs require positive gate-to-cathode pulses to turn ON. The source of these trigger pulses must supply pulses to both SCRs. However, neither the gates nor cathodes of the two SCRs are connected in common. Therefore, to supply both pulses from one source, a *pulse transformer* is used with one primary winding and two identical secondary windings. A pulse transformer is simply one which is designed to operate with narrow input pulses rather than pure sine waves; however, the principle of operation is the same.

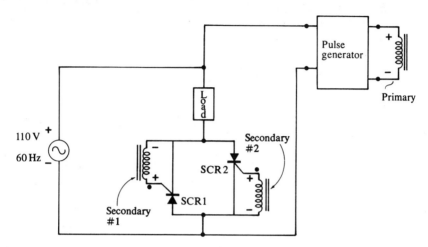

FIGURE 25–19 *Pulse transformer used to produce two electrically isolated output pulses.*

The dots on the three windings indicate the relative polarity for each. When the trigger source makes the dotted end of the primary positive relative to the undotted end, then the same thing happens at each secondary winding. Thus, each SCR simultaneously receives a positive pulse at its gate relative to its cathode. Of course, only the SCR with the positive anode-to-cathode voltage will be turned ON by its gate pulse.

The transformer in this example is used to electrically isolate the trigger source from the two loads because of the lack of any common connection point between the two. The three windings in the circuit diagram are shown separated for convenience. Of course, physically, they are in close proximity on the same iron core.

25.11 Transformer Power Losses

In our study of transformers thus far, we have assumed that in an ideal transformer there are zero power losses so that all of the primary source input power is transferred to the secondary load. In most practical situations, this assumption would rarely produce discrepancies of greater than ten percent. We will now briefly discuss the power losses that are present in practical iron-core transformers.

Copper Losses In practical transformers, both windings possess a certain amount of *resistance*. Since the primary and secondary currents must flow through these windings, a certain amount of power is dissipated as heat in these winding resistances. This power loss (I^2R) is often called the transformer's *copper loss* since it represents power lost as heat in the copper windings. The effective winding resistances of low-frequency (10-kHz) transformers can be easily determined using a dc ohmmeter across the windings. Generally speaking, a winding with more turns has a greater resistance than a winding with a smaller number of turns.

EXAMPLE 25.15 A certain transformer's secondary is connected to a 10-ohm load. The secondary current is 5 A and the primary current is 0.5 A. The transformer's winding resistances are $R_p = 1\ \Omega$, $R_s = 0.1\ \Omega$. Calculate the load power and the copper loss.

Solution: $P_L = I_s^2 \times R_L$
$= 5^2 \times 10$
$= 250\ \text{W}$

Copper losses $= I_p^2 R_p + I_s^2 R_s$
$= 0.5^2 \times 1 + 5^2 \times 0.1$
$= 0.25 + 2.5 = 2.75\ \text{W}$

In this typical case, the copper losses are only about 1 percent of the power consumed in the load.

Core Losses In addition to the power lost in the winding resistances, a certain amount of power is lost as heat within the iron core of the transformer. These *core losses* are caused mainly by two factors: *hysteresis losses* and *eddy current losses*. These effects were described thoroughly in our discussion of practical inductors in chap. 19 and are equally applicable to iron-core transformers.

Unlike the copper losses, the core losses do not depend on the size of the current flowing through the windings. As such, the amount of power lost due to core losses is usually measured with an open-circuited secondary ($I_s = 0$), as shown in Fig. 25-20. Since the only current flowing is I_{pm} (small magnetization current), the copper losses are negligible and are not included in the wattmeter reading. The wattmeter thus measures the true power dissipated in the core.

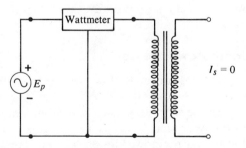

FIGURE 25-20 *Wattmeter measures transformer core losses with open-circuited secondary.*

Transformer Efficiency The actual performance of a practical transformer can be measured by its *efficiency*, which is defined as the ratio of load power output to source power input in percent. That is,

$$\text{efficiency } (\eta) = \frac{P_L}{P_{in}} \times 100\% \qquad (25\text{–}12a)$$

The source input power P_{in} has to equal the sum of all the power losses plus the power delivered to the load. Thus,

$$\eta = \frac{P_L}{(\text{copper loss} + \text{core loss} + P_L)} \times 100\% \qquad (25\text{–}12b)$$

For the ideal transformer, the copper and core losses were assumed to be zero so that its efficiency would be 100%. Typically, practical iron-core transformers operate with efficiencies from 90 to 98%.

EXAMPLE 25.16 The transformer of Ex. 25–15 has a core loss of 5 W. Determine the transformer's efficiency for the conditions of Ex. 25–15.

Solution: Using formula (25–12b)

$$\eta = \frac{250 \text{ W}}{2.75 \text{ W} + 5 \text{ W} + 250 \text{ W}} \times 100\%$$
$$= 97\%$$

Because of the power losses in a transformer, not all of the primary input power is transferred to the load. This factor can be taken into account in calculating the primary current if the transformer's efficiency is known. Recall that for the ideal transformer we assumed $P_{in} = P_L$ which led to the relationship $I_p/I_s = N_s/N_p$. This relationship is not quite true for an efficiency of less than 100%.

EXAMPLE 25.17 A transformer has $N_p = 1000$, $N_s = 200$, and an ac voltage of 110 V applied to the primary. If the transformer has a 90% efficiency, what will be the primary current when a 50-ohm load is connected to the secondary?

Solution: $E_s = \frac{N_s}{N_p} \times E_p = \frac{1}{5} \times 110 \text{ V} = 22 \text{ V}$

$$I_s = \frac{E_s}{R_L} = \frac{22 \text{ V}}{50 \text{ }\Omega} = 0.44 \text{ A}$$

Thus,

$$P_L = 22 \text{ V} \times 0.44 \text{ A} = 9.68 \text{ W}$$

Since $\eta = 90\%$, then

$$\frac{P_L}{P_{in}} = 0.9 \text{ or } P_{in} = \frac{9.68}{0.9} = 10.77 \text{ W}$$

Since $P_{in} = E_p I_p$, then

$$I_p = \frac{P_{in}}{E_p} = \frac{10.77 \text{ W}}{110 \text{ V}} = .098 \text{ A}$$

Note that if the *ideal* ratio $I_p/I_s = N_s/N_p$ is used, the value of I_p is calculated as .088 A, which is lower than the value calculated here. More I_p is required in order to supply transformer losses as well as the load power.

25.12 Transformer Leakage Flux

Another major assumption which was made in discussing the ideal transformer was that *all* of the flux established by the current in the primary winding must link the secondary winding. In practical iron-core transformers this assumption is a valid one because the iron core provides an "easy" path for the flux lines. However, there is always a certain small percentage of the flux lines produced by the primary magnetizing current which do not link the secondary coil even in the best transformers. This is illustrated in Fig. 25-21.

The flux lines that do not link the secondary winding are called the *primary leakage flux*, ϕ_l. The remainder of the flux lines do link the secondary and are called the *mutual flux*, ϕ_m. Typically, in iron-core transformers ϕ_l is less than one percent of ϕ_m.

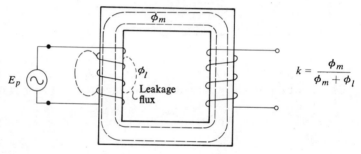

FIGURE 25-21 *Transformer leakage flux.*

Coefficient of Coupling The coefficient of coupling between the primary and secondary windings of a transformer is defined as the ratio of the number of flux lines linking the secondary to the number of flux lines originating in the primary. That is,

$$k \equiv \frac{\phi_m}{\phi_m + \phi_l} \tag{25-13}$$

where k is the *coefficient of coupling*. For ideal transformers $\phi_l = 0$ so that $k = 1$. We say that an ideal transformer has unity coupling.

In a good iron-core transformer k will be 0.98 or 0.99 and such transformers are said to be *closely* coupled. In air-core transformers, however, the leakage flux ϕ_l can be considerable, resulting in coefficients of coupling which are sometimes as low as 0.01. Such transformers are said to be *loosely* coupled.

Leakage Inductances The effect of the primary leakage flux is as if a small portion of the primary winding was not producing flux to link the secondary winding. In other words, a small portion of the primary coil does not take part in the transformer action but instead acts as a normal inductance in series with the primary winding. Thus, we can consider the primary coil to be divided into two parts: the *primary magnetizing inductance* L_{pm} and the *primary leakage* *inductance L_{pl}*. The primary magnetizing inductance L_{pm} is that portion of the total primary inductance L_p which produces flux in the iron core. The primary leakage inductance L_{pl} is that portion of L_p which does not produce flux in the core. For a coefficient of coupling k, we have

$$L_{pm} = kL_p$$
$$L_{pl} = (1 - k)L_p \qquad (25\text{--}14)$$

For values of k which are close to one (close coupling), it can be seen that $L_{pm} \simeq L_p$ and the leakage inductance L_{pl} is very small.

A similar analysis can be made concerning the secondary winding so that we can consider the secondary leakage flux and a corresponding *secondary leakage inductance L_{sl}*. Thus, we have

$$L_{sm} = kL_s$$
$$L_{sl} = (1 - k)L_s \qquad (25\text{--}15)$$

where L_s is the total secondary inductance and L_{sm} is that portion of L_s which takes part in magnetizing the core.

Practical Transformer Equivalent Circuit The effects of the primary and secondary leakage inductances can be taken into account in the *equivalent* circuit for a practical iron-core transformer shown in Fig. 25–22. On the primary side, the leakage inductance L_{pl} and the winding resistance R_p are shown in series with L_{pm}, the magnetizing inductance. Similarly, on the secondary side L_{sl} and R_s are in series with L_{sm}, the useful portion of the secondary winding inductance. The primary and secondary magnetizing inductances L_{pm} and L_{sm} (enclosed in dotted lines) still operate essentially as an ideal transformer. The nonideal effects are taken into account in the series resistances and inductances. Note that the actual transformer primary terminals are x-y and the secondary terminals are w-z so that the winding resistances and leakage inductances are not physically accessible. They are actually distributed throughout the windings.

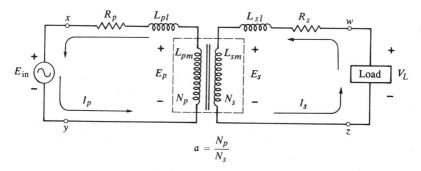

FIGURE 25–22 *Equivalent circuit for practical transformer including the effects of winding resistances and leakage inductances.*

25.13 Loading Effects in the Practical Transformer

If we examine the equivalent circuit in Fig. 25–22 it is possible to predict the effects the winding resistances and leakage inductances will have on the transformer operation. Consider the primary side first. The flow of primary current I_p will cause some of the input voltage E_{in} to appear across R_p and L_{pl}. This means that the full E_{in} will not appear across the primary of the ideal transformer portion (E_p) to be stepped up or down according to the turns ratio. Thus, the secondary induced voltage E_s will be less than it would be in the ideal transformer where R_p and L_{pl} are both zero.

On the secondary side a similar effect occurs because of the loss of voltage across R_s and L_{sl} as secondary current flows. This means that the full secondary induced voltage E_s does not reach the load. The overall effect of the winding resistances and leakage inductances, then, is to reduce the load voltage from what it would be in the ideal transformer case. This reduction in load voltage will, of course, be more significant as the load impedance is made lower. The lower load impedance causes I_s and I_p to be larger and thereby increases the voltage lost across the winding resistances and leakage inductances. Therefore, *the value of V_L will decrease as Z_L decreases.*

This reduction in transformer output voltage under load is similar to the behavior of a practical voltage source under load. The output voltage will be maximum in the no-load condition or when Z_L is very large since the currents flowing will be so small as to produce no voltage losses. As Z_L is made smaller, more current is drawn and the output voltage decreases.

EXAMPLE 25.18 An iron-core transformer has a primary inductance of 0.2 H, a secondary inductance of 50 mH, and a turns ratio of 2:1. The winding resistances are measured as $R_p = 2$ ohms and $R_s = 1$ ohm.

(a) Draw the transformer's equivalent circuit if its coefficient of coupling is $k = 0.995$.

(b) Determine the load voltage for a resistor load of 8 ohms and an input of 16 V at 1 kHz. Compare to the output at no-load.

(c) What will happen to V_L as the input frequency increases?

Solution: (a) The equivalent circuit is drawn in Fig. 25–23(a). The inductance values are found using Eqs. (25–14) and (25–15).

$$L_{pm} = kL_p \approx 0.2 \text{ H}; \quad L_{sm} = kL_s \approx 50 \text{ mH}$$

$$L_{pl} = (1-k)L_p = 1 \text{ mH}; \quad L_{sl} = (1-k)L_s = 0.25 \text{ mH}$$

(b) At $f = 1$ kHz the reactances of the leakage inductances can be found as:

$$X_{pl} = 2\pi f L_{pl} = 6.3 \text{ ohms}$$

$$X_{sl} = 2\pi f L_{sl} = 1.6 \text{ ohms}$$

The equivalent circuit is redrawn in Fig. 25–23(b) showing the circuit values.

In order to solve for V_L we must find the voltage across R_L. This can be accomplished using the impedance reflection concept discussed in an earlier section. The total impedance across the secondary winding includes the leakage inductive reactance, the winding resistance, and the load resistance. This total secondary impedance can be reflected to the primary side after multiplying it by the square of the turns ratio (a^2). In this case $a = 2$ so that $a^2 = 4$. Figure 25–23(c) shows the result of this operation. Note that the 8-ohm load appears as a 32-ohm *reflected* load.

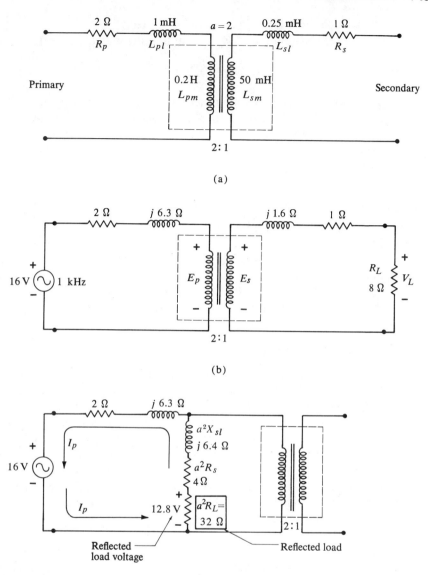

FIGURE 25–23

Using this circuit we can find the voltage across the *reflected* load by first finding I_p. The total impedance seen by the source can be obtained by combining all the series impedances. That is,

$$\overline{Z}_p = (2 + j6.3 + j6.4 + 4 + 32) \text{ ohms}$$
$$= (38 + j12.7) \text{ ohms}$$
$$= 40 \text{ }\Omega\underline{/18.4°}$$

Note that the 0.2-H primary magnetizing inductance has been neglected. This is valid since its reactance at 1 kHz is 1256 ohms so that the small current

(I_{pm}) which flows through it can be neglected.

Using the magnitude of $\overline{Z_p}$ we can find the magnitude of $\overline{I_p}$.

$$I_p = \frac{E_{\text{in}}}{Z_p} = \frac{16 \text{ V}}{40 \text{ }\Omega} = 0.4 \text{ A}$$

This current develops a voltage across the reflected load equal to 0.4 A × 32 Ω = 12.8 V. We are interested in the *actual* load voltage, not the *reflected* load voltage. Since the transformer is a 2:1 step-down transformer, the 12.8 V across the load on the *primary* side means that the actual load voltage on the *secondary* side is

$$V_L = \frac{12.8 \text{ V}}{2} = 6.4 \text{ V}$$

Under no-load conditions, the output voltage would simply be $E_{\text{in}}/2 = 8$ V. Thus, the 8-ohm load causes the output voltage to drop by approximately 20%.

(c) If the input source frequency is increased, the reactances of the leakage inductances will increase, thereby causing a greater loss of voltage across these components so that V_L **will decrease**. The student can verify this by repeating the preceding analysis at some higher frequency, say 5 kHz.

25.14 Transformer Frequency Response

An iron-core transformer will operate satisfactorily over a certain range of input frequencies. At very low frequencies and very high frequencies the transformer output will drop drastically. In order to examine these frequency effects we must include two additional elements in the equivalent circuit for the transformer: the equivalent primary and secondary shunt capacitances caused by the distributed capacitances between the turns of each winding. These capacitances C_p and C_s are included in the more complete transformer equivalent circuit shown in Fig. 25–24(a). Note that C_p and C_s are shown across the terminals of each winding. Also included in this circuit is the internal resistance of the primary source.

As we discuss the effects of frequency keep in mind that the induced secondary voltage E_s depends on the voltage E_p which actually appears across the useful portion of the primary coil L_{pm}. Thus, any portion of E_{in} which does not reach E_p cannot be transformed and applied to the load.

Low-Frequency Effects At low frequencies the effects of C_p and C_s can be neglected since they will have very high reactances. In addition, the leakage inductances will have very low reactances compared to R_p and R_s and can also be neglected. Thus, at low frequencies the equivalent circuit can be simplified as shown in part (b) of Fig. 25–24. If we examine this low-frequency equivalent circuit, it can be seen that the reactance of L_{pm} will get smaller and smaller as frequency decreases. As such, more of E_{in} will appear across R_{int} and R_p, and less will reach L_{pm}. This decrease in E_p will of course cause E_s and therefore V_L to decrease. The limiting case is $f = 0$ Hz (dc) where $X_{pm} = 0$ and therefore E_p, E_s, and V_L are all zero.

For a transformer to have a good low-frequency response, it should be designed with a large primary inductance so that X_{pm} will remain high at low fre-

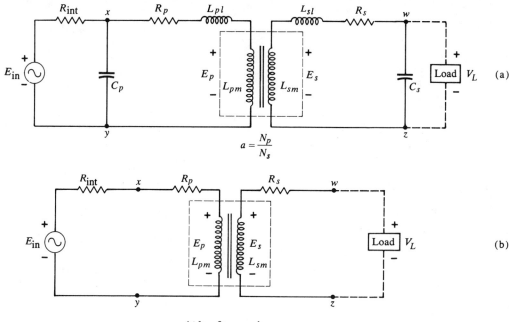

FIGURE 25-24 (a) *Complete equivalent circuit;* (b) *simplified at low frequencies.*

quencies. However, a larger primary inductance requires more turns of wire. This could have the additional effect of creating more winding capacitance which, as we shall see, affects the high-frequency response.

High-Frequency Effects To consider what happens at the higher frequencies, refer back to Fig. 25-24(a). Looking at the primary side first, we can see that the winding capacitance C_p will have a *shunting* effect as its reactance becomes smaller at higher frequencies. As such, it will "load down" the source so that the full E_{in} will not reach the primary terminals x-y. This of course means that E_p will be decreased.

The voltage which does reach E_p is transformed to E_s. The complete voltage E_s, however, is prevented from reaching the output terminals w-z because of the presence of L_{sl} and C_s. At high frequencies the reactance of L_{sl} becomes larger and the reactance of C_s becomes smaller. This combined action causes *less* voltage to reach the load. L_{sl} *blocks* and C_s *shunts* so that V_L drops rapidly at higher frequencies.

Figure 25-25 shows a typical frequency-response curve for an audio transformer. This response curve is similar to that of a band-pass filter. It is relatively flat over a range from about 100 Hz to 5 kHz. These frequencies represent the low- and high-frequency cutoffs at which the output has dropped by 3 dB (0.707 of mid-frequency value). The high-frequency cutoff can be extended if the transformer is designed so that the secondary capacitance C_s and the secondary leakage inductance L_{sl} are resonant at approximately 20 kHz. The resonant rise of voltage across this capacitance (recall series resonance effects) will offset the natural roll-off at the high frequency end (see dotted curve in figure).

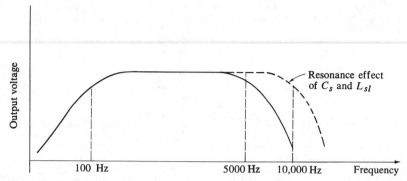

FIGURE 25-25 *Typical audio transformer response curve.*

25.15 Air-Core Transformers

Up to now we have discussed only closely coupled ($k \approx 1$) iron-core transformers. There are many applications for *air-core* transformers, especially in radio circuitry. In air-core transformers the coupling between the primary and the secondary can range from very close ($k \approx 1$) to very loose ($k \ll 1$) depending on the physical arrangement. Figure 25-26(a) shows an arrangement where two coils are wound on a nonmagnetic core so that almost all of the primary flux links the secondary coil. Figure 25-26(b) shows a loosely coupled arrangement where the coefficient of coupling might be typically $k = 0.01$.

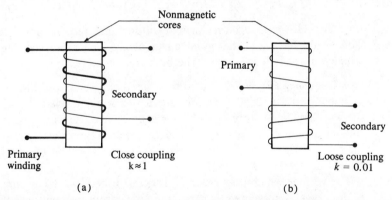

FIGURE 25-26 (a) *Closely coupled air-core transformer.* (b) *Loosely coupled version.*

If a closely coupled transformer is needed for a radio application, it is not practical to use an iron core because the core losses become prohibitively large at high frequencies. Therefore, in such applications an air-core arrangement like that in Fig. 25-26(a) would be used. In *rf* applications requiring a *tuned* transformer, very loose coupling is used and an arrangement like that in Fig. 25-26(b) would be used.

25.16 Mutual Inductance

In loosely coupled transformers a very small portion of the primary flux links the secondary. As such, we can consider that many of the primary turns are ineffective as far as mutual induction is concerned. Thus, it is not useful to talk about turns ratios for loosely coupled transformers. Instead, a parameter called *mutual inductance* is used to express the interaction between the primary and secondary windings. We know that every coil has *self-inductance* (*L*) which responds to any changing current by inducing a voltage across the coil ($V = L \times di/dt$). When two coils are magnetically coupled so that the current in one induces a voltage in the other, this interaction is called *mutual inductance* (*M*).

The mutual inductance between two coils has the same properties as the self-inductance of one coil except that it is the changing current in one coil which induces a voltage in the other coil, and vice versa. For example, Fig. 25–27 shows an air-core transformer with the primary self-inductance L_p and secondary self-inductance L_s. Note the absence of the double bar in the symbol for the air-core transformer. The double bar is used only to represent an iron-core transformer. If a current i_p is flowing through the primary coil, then a voltage e_s will be induced in the secondary according to

$$e_s = M \frac{di_p}{dt} \quad (25\text{--}16)$$

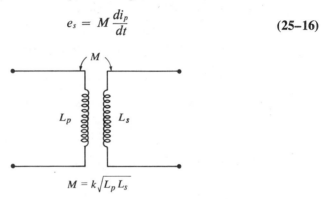

FIGURE 25–27 *Air-core transformer showing mutual inductance coupling.*

This is very similar to the current-voltage relationship for a self-inductance except that *M* is used. A corresponding relationship exists for a current i_s in the secondary; it will induce a voltage in the primary coil according to

$$e_p = M \frac{di_s}{dt} \quad (25\text{--}17)$$

The value of mutual inductance depends on the values of self-inductance of each coil and also on the coefficient of coupling between coils. Mathematically the relationship is

$$M = k\sqrt{L_p L_s} \quad (25\text{--}18)$$

For close coupling the value of *k* is larger and *M* will also be larger indicating that the coils will have a greater effect on each other. Loosely coupled coils have a low value of *k* and therefore less mutual inductance.

EXAMPLE 25.19 Two coils are wound on a cardboard spool so that $k = 0.04$. Determine the mutual inductance between the two coils if one has a 4-mH inductance and the other has 9-mH inductance.

Solution: $M = k\sqrt{L_pL_s} = 0.04\sqrt{4 \text{ mH} \times 9 \text{ mH}}$
$= 0.04 \times 6 \text{ mH}$
$= \textbf{0.24 mH}$

EXAMPLE 25.20 When the coils of the last example are moved closer together, k is increased to 0.1. What is the new value of M?

Solution: $M = 0.1\sqrt{4 \text{ mH} \times 9 \text{ mH}} = \textbf{0.6 mH}$

EXAMPLE 25.21 For the situation of the preceding example, determine the induced voltage in the 9-mH coil as the current in the 4-mH coil increases at the rate of 500 A/s.

Solution: Let's call the 4-mH coil the primary coil. Thus, using Eq. (25–16) we can find the voltage induced in the 9-mH secondary.

$$e_s = M\left(\frac{di_p}{dt}\right)$$
$$= 0.6 \text{ mH} \times 500 \text{ A/s}$$
$$= \textbf{0.3 volts}$$

25.17 Air-Core Transformer with AC Voltages

When an ac voltage is applied to the primary winding, an ac current will flow in the primary. This ac current will induce an ac voltage in the secondary coil. If I_p is the rms value of primary current, then the rms value of the secondary voltage E_s will be given by

$$E_s = X_M I_p \qquad (25\text{–}19)$$

where $X_M = 2\pi f M$ is the *mutual reactance*. We know that the ac voltage across a coil always leads the current by 90°. The same is true here; the secondary induced voltage leads I_p by 90°. Thus, we have

$$\overline{E_s} = X_M\underline{/90°} \times \overline{I_p} \qquad (25\text{–}20)$$

which shows the 90° phase angle of X_M which causes $\overline{E_s}$ to be 90° ahead of the $\overline{I_p}$ phasor.

EXAMPLE 25.22 Determine the secondary induced voltage for the situation in Fig. 25–28(a). Draw a phasor diagram showing $\overline{E_p}$, $\overline{I_p}$, and $\overline{E_s}$.

Solution: The self-inductance L_p has a reactance at 1 kHz

$$X_p = 2\pi f L_p$$
$$= 2\pi \times 10^3 \times 50 \times 10^{-3} = 314 \text{ }\Omega$$

The primary current I_p is found using Ohm's law

$$I_p = \frac{\overline{E_p}}{\overline{X_p}} = \frac{10 \text{ V}\underline{/0°}}{314 \text{ }\Omega\underline{/90°}} = \textbf{31.8 mA}\underline{\textbf{/—90°}}$$

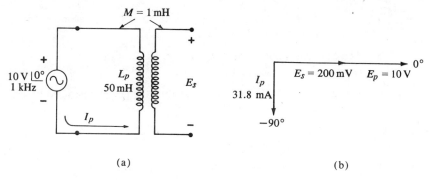

FIGURE 25-28

The secondary voltage can be calculated using Eq. (25-20) since M is shown as 1 mH. Thus,

$$\overline{E}_s = 2\pi f M \underline{/90°} \times \overline{I}_p$$
$$= 6.28 \, \Omega \underline{/90°} \times 31.8 \text{ mA} \underline{/-90°}$$
$$= 200 \text{ mV} \underline{/0°}$$

The phasor diagram is drawn in part (b) of the figure. Note that E_s is much smaller than E_p due to the loose coupling.

Chapter Summary

1. The basic transformer consists of two coils so arranged that a voltage applied to one coil (primary) produces a current and a resultant magnetic flux linking both the primary coil and the other coil (secondary). The flux linking the secondary coil induces a secondary voltage according to Faraday's law.

2. In an ideal iron-core transformer it is assumed that all the flux ϕ_p produced by the flow of current in the primary will also link the secondary winding. As a result, in an ideal iron-core transformer the secondary induced voltage e_s is related to the applied primary voltage e_p by the ratio of the number of turns in each winding: $e_p/e_s = N_p/N_s = a$. For a step-up transformer $a < 1$; for step-down $a > 1$.

3. The current which flows in the primary under no-load conditions is called the magnetization current I_{pm}.

4. When a load is placed across the secondary of an *ideal* transformer, current is drawn from the primary source such that the power input from this source equals the power consumed by the load ($E_p I_p = E_s I_s$).

5. In a transformer the electrical energy supplied by the primary input source is converted to magnetic energy in the core. This magnetic energy is reconverted to electrical energy in the secondary and delivered to the load.

6. In an ideal transformer the primary and secondary currents are related by: $i_p/i_s = N_s/N_p = 1/a$.

7. In an ideal transformer a load Z_L connected to the secondary appears as a reflected primary load $Z_p = a^2 Z_L$.

8. The chief causes of power losses in a practical transformer are copper losses (due to winding resistances) and core losses (due to hysteresis and eddy current effects).

9. A transformer's efficiency (η) is equal to the percentage ratio of secondary load power P_L to the primary input power P_{in}.

10. In a practical transformer, primary and secondary leakage fluxes give rise to leakage inductances that do not contribute to the transformer action.

11. In a practical transformer, the load voltage will drop as Z_L is decreased due to voltage losses across the winding resistances and leakage inductances.

12. In a practical transformer, the output voltage will drop at lower and higher frequencies.

13. Air-core transformers have coefficients of coupling which can vary from zero to almost one depending on the physical arrangement of the coils.

14. Mutual inductance (M) is a measure of the interaction of two magnetically coupled coils. M increases as coils are more closely coupled.

Questions/Problems

Sections 25.2–25.3

25-1 With the aid of the diagram in Fig. 25–1(a) briefly describe the process by which a voltage is induced in the secondary of the transformer.

25-2 Using Lenz's law, verify that the top end of the secondary coil in Fig. 25–1(a) will become positive when the switch is closed. What will happen to e_p and e_s when the switch is opened?

25-3 Why doesn't a dc input to a transformer produce a dc output on the secondary?

25-4 A 110-V, 60-Hz source is applied to the primary of a transformer that has $N_p = 120$ and $N_s = 15$. What is the voltage across the secondary?

25-5 A certain transformer is designed to step up a 12-V signal to 48 V. If $N_p = 25$ turns, how many turns should be in the secondary winding?

25-6 A certain transformer is a step-down transformer by a factor of 30. If $E_s = 2.5$ V and $N_s = 900$, find E_p and N_p.

Sections 25.4–25.5

25-7 The transformer of problem 25–6 has a primary winding inductance of 1.8 H. If $E_p = 120$ V, 400 Hz, determine the magnitude of the magnetizing current I_{pm}. Draw a phasor diagram showing E_p, E_s, and I_{pm}.

25-8 A transformer with $N_p/N_s = 0.1$ is used to supply power from a 12-V source to a 50-ohm load. Determine: (a) load voltage; (b) load current; (c) load power; (d) primary current; (e) input power from source.

25-9 For each of the situations shown in Fig. 25-29 determine the unknown quantities.

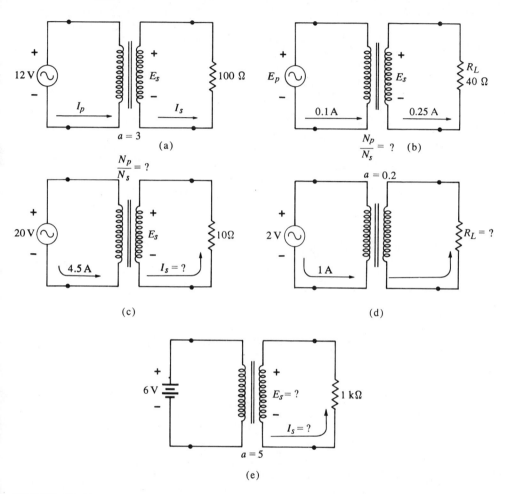

FIGURE 25-29

25-10 Transformers find extensive use in the transmission and distribution of power from utility company to the users (homes, factories, etc). The cables which are used to transmit the power must carry large amounts of current which can heat the cables and cause a loss of power. For this reason the cable currents must be kept as small as possible. The power companies accomplish this by transmitting very high voltages (>100 kV) over the cables and then using a transformer to step the voltages down before they reach the consumers. By transmitting high voltages, the current will be much lower for a given amount of power.

This can be illustrated by considering the two cases shown in Fig. 25-30. In both cases the consumer requires 240 V and 80 A. In (a) the transmitted voltage is 1200 V, which is stepped down to 240 V when it reaches the consumer.

In (b) the transmitted voltage is 120 kV.
a. Determine the current carried by the transmission cables in each case.
b. In which case will the cables have to be larger?
c. What is another advantage to transmitting high voltage–low current rather than low voltage–high current? (Hint: consider the resistance of the cable.)

25-11 A coil with $L = 10$ mH and $R = 150$ ohms is the load on the secondary of the transformer described in problem 25-8. If $\overline{E_p} = 18$ V$\underline{/0°}$ at 3600 Hz determine: (a) $\overline{E_s}, \overline{I_s}$; (b) $\overline{I_p}$; (c) P_T (load); and P_T (input)

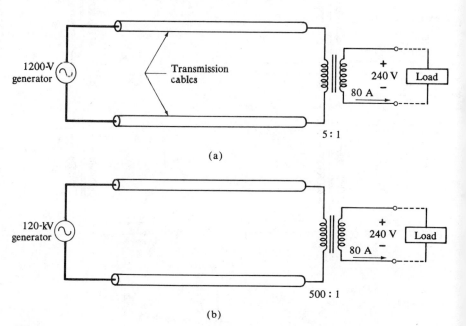

FIGURE 25-30

25-12 Indicate which of the following statements are *true* concerning an ideal transformer.
   a. Magnetizing current lags input voltage by 90°.
   b. Total primary current always lags input voltage by 90°.
   c. With capacitive load connected to secondary, I_p will lead E_p.
   d. With capacitive load connected to secondary, E_s lags E_p.
   e. Counterflux produced by I_s causes reduction in I_p.
   f. If $E_s > E_p$, then $I_s > I_p$.
   g. The amplitude of the flux waveform in the core decreases if secondary current is flowing.
   h. A pure dc voltage on the primary produces no secondary voltage (in steady state).

25-13 Explain how energy is transferred from the input source to the load when there is no electrical connection between primary and secondary of a transformer.

Section 25.6

25-14 A step-up transformer with $a = 0.125$ has a 152-ohm load connected to its secondary. What is the impedance seen by a source connected to the primary?

25–15 A certain transformer has an 800-turn primary winding. The load on its secondary consists of a 120-ohm resistor in series with a 1-μF capacitor. The input source frequency is 1 kHz. How many turns are required on the secondary winding if the primary reflected impedance is to have a magnitude of 5000 Ω?

25–16 In Fig. 25–31 the 24-V source and the 320-ohm resistance represent the Thevenin equivalent for the output of an audio amplifier which is to supply power to an 8-ohm speaker. Terminals x-y are the amplifier's output terminals.
   a. Connect the 8-ohm speaker directly to the amplifier output and calculate the power delivered to the speaker.
   b. Use an appropriate matching transformer to match the speaker load to the amplifier. Determine the required turns ratio and then determine the power delivered to the speaker.

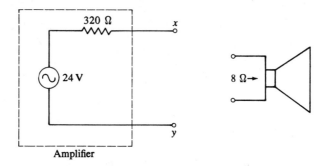

FIGURE 25–31

Section 25.7

25–17 Determine the primary current for the circuit in Fig. 25–32.

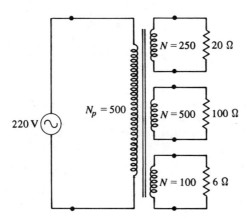

FIGURE 25–32

25–18 The transformer in Fig. 25–12 has N_p = 1000 and N_s = 300. The tap is placed one-quarter of the way from the top. The applied voltage is 200 V. If R_{L1} = 1 kΩ and R_{L2} = 2 kΩ, determine the current drawn from the source.

25–19 Is there any difference between a multiple secondary and a tapped secondary?

Is there a situation where a multiple secondary could be used but a tapped secondary could not?

Section 25.8

25-20 Sketch the waveforms of e_{in} and v_{xy} for the circuit of Fig. 25-33 showing the phase relationship between the two.

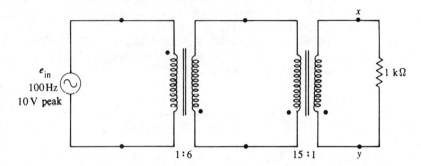

FIGURE 25-33

Sections 25.9–25.10

25-21 For the step-up autotransformer in Fig. 25-34, determine the load current and the currents through each portion of the winding.

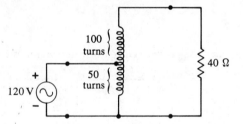

FIGURE 25-34

25-22 Figure 25-35 shows two ways of producing a variable ac voltage. One uses an autotransformer and the other uses a resistive potentiometer. Why is the autotransformer more desirable for this application?

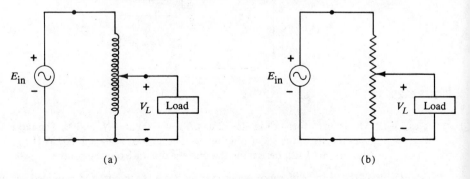

FIGURE 25-35

Section 25.11

25-23 A certain transformer has an efficiency of 92 percent. It has $N_p = 1200$, $N_s = 300$, and has 60 V applied to its primary. If the secondary load is 6 ohms, calculate I_p.

25-24 The transformer in problem 25-23 has winding resistances $R_p = 0.8$ ohm and $R_s = 0.2$ ohm. Determine the transformer copper losses and core losses.

Sections 25.12–25.13

25-25 Draw the equivalent circuit for a transformer that has the following characteristics:
$L_{pl} = 2$ mH; $L_{sl} = 120$ µH
$N_p/N_s = 4$; $L_p = 200$ mH; $L_s = 12$ mH
$R_p = 1.6$ ohms, $R_s = 0.5$ ohms

25-26 For the transformer of the last problem determine the voltage across a 4-ohm load when $E_{in} = 20$ V at 1 kHz. Compare to the no-load output voltage.

25-27 What value must E_{in} be increased to in the previous problem if V_L is to equal 5 V when $R_L = 4$ ohms? (Hint: remember the linearity principle.)

25-28 In Ex. 25.18 it was stated that V_L would decrease as the input frequency is increased. Verify this by repeating that example problem using $f = 5$ kHz.

25-29 For the situation in problem 25-27 indicate what will happen to V_L for each of the following circuit changes:
 a. increase in frequency
 b. increase in load resistance
 c. inclusion of E_{in}'s source resistance
 d. decrease in transformer's coefficient of coupling
 e. doubling the number of turns of both windings

Section 25.14

25-30 What causes the output of a transformer to drop at low frequencies?

25-31 What causes the transformer output to drop at high frequencies?

Sections 25.15–25.17

25-32 Consider the situations shown in Fig. 25-36. In which case is the coefficient of coupling greatest? In which case is it lowest?

25-33 Two coils are wound on a Bakelite core. The primary coil has $L_p = 60$ mH and the secondary coil has $L_s = 2.4$ mH. The coils are arranged so that $k = 0.08$.
 a. Determine M.
 b. Determine the *magnitude* of the induced secondary voltage if the primary current is decreasing at a rate of 10^4 A/s.
 c. Determine the rms value of E_s when $E_p = 15$ V at 1 MHz.
 d. Repeat (c) for a frequency of 1 kHz.

25-34 An air-core transformer with $M = 10$ mH has $E_s = 0.5$ V when $E_p = 18$ V at 100 kHz. Indicate the effect each of the following will have on the magnitude of E_s:

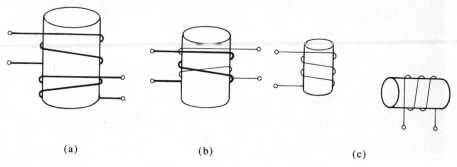

(a) (b) (c)

FIGURE 25-36

a. decrease in input frequency
b. doubling L_p and L_s
c. primary winding resistance R_p
d. increase in the coupling between coils

25-35. For the circuit in Fig. 25-37 determine I_p and E_s. Sketch the phasor diagram showing $\overline{E_p}$, $\overline{I_p}$, and $\overline{E_s}$. Note placement of dots on transformer.

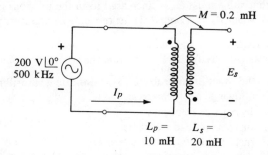

FIGURE 25-37

CHAPTER 26

Nonsinusoidal Waveforms and Harmonics

26.1 Introduction

There are a relatively small number of applications in electronics in which the waveforms of current and voltage are pure sine waves. Most electronic systems contain waveforms that are more complex nonsinusoidal waveforms. In our discussion of the sinusoidal waveform in chap. 16, one of the reasons given for its importance was that all nonsinusoidal waveforms can be thought of as being composed of combinations of various sinusoidal waveforms. This important concept will be elaborated on in this chapter.

One of the most familiar applications of nonsinusoidal waveforms is in the area of audio reproduction. The musical sound which occurs when a pure sine wave (such as 256 Hz) is fed into a loudspeaker is very uninteresting to the ear. This is a single-frequency tone. However, if the "middle C" key is struck on a piano, the vibrating string will send out not only the tone of the 256-Hz frequency (called the fundamental frequency) but also a double-frequency tone of 512 Hz (called the second harmonic frequency) and often a triple-frequency tone of 768 Hz (third harmonic). The presence of these "harmonics" gives the piano sound its richness of tone.

26.2 Harmonics

The term *harmonic* means a sine wave that has a frequency that is equal to an integral multiple of the frequency of a basic *fundamental* sine wave. To illustrate, refer to Fig. 26–1 where several sine waves are shown. The lowest frequency sine wave is called the *fundamental* and has a frequency $f_1 = 100$ Hz.

The second sine wave has a frequency which is twice that of the fundamental. This sine wave is called the *second harmonic*. The third sine wave has a frequency which is three times that of the fundamental. This sine wave is called the *third harmonic*. A sine wave with a frequency four times the fundamental would be the *fourth harmonic*, and so on. There is no first harmonic as such; the fundamental is actually the first harmonic.

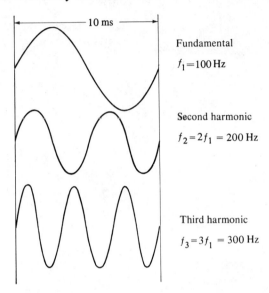

FIGURE 26–1 *Harmonics.*

EXAMPLE 26.1 A sine wave has a frequency of 1.3 kHz. (a) List the frequencies of the first five harmonics of this sine wave. (b) What is the frequency of the 1317th harmonic?

Solution: (a) 2nd harmonic = 2 × 1.3 kHz = **2.6 kHz**
3rd harmonic = 3 × 1.3 kHz = **3.9 kHz**
4th harmonic = 4 × 1.3 kHz = **5.2 kHz**
5th harmonic = 5 × 1.3 kHz = **6.5 kHz**
6th harmonic = 6 × 1.3 kHz = **7.8 kHz**

(b) 1317th harmonic = 1317 × 1.3 kHz = **1.712 MHz**

26.3 Combining Harmonics

It was mentioned earlier that the musical tone produced by a piano key could contain a fundamental, a second harmonic, and a third harmonic all combined to produce the resultant sound. If this sound were fed into a microphone and converted to an electrical signal, the result might appear as shown in Fig. 26–2. The dotted waveforms represent the various frequency components which go to make up the total signal. Notice that the fundamental sine wave has the largest amplitude, the 2nd harmonic is the next largest, and so on. This is usually the case with musical tone and with many electronic waveforms.

Nonsinusoidal Waveforms and Harmonics

The resultant waveform is obtained by adding together the instantaneous values of the three component waveforms at each point in time. The process is a tedious one and is not important enough to perform here. One important point to note here is that *the resultant waveform is not sinusoidal.* In previous chapters we combined sine waves of the same frequency and ended up with a sine wave. However, when sine waves of different frequencies are combined the result is generally a nonsinusoidal waveform.

Another important point illustrated in Fig. 26-2 is that the repetition frequency of the resultant waveform is the same as the fundamental frequency. This is true for all nonsinusoidal waveforms. That is,

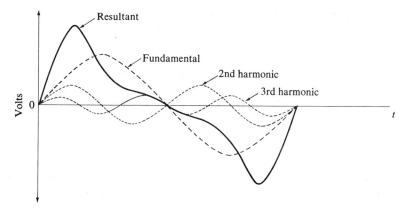

FIGURE 26-2 *A nonsinusoidal resultant waveform which possesses a fundamental, a second harmonic, and a third harmonic.*

Any periodic nonsinusoidal waveform has a fundamental frequency which is equal to the frequency of the waveform.

EXAMPLE 26.2 A certain square wave has a period of 4 ms. What is its fundamental frequency?

Solution: The square wave's repetition frequency is

$$f = \frac{1}{T} = \frac{1}{4 \text{ ms}} = 250 \text{ Hz}$$

This is also the frequency of the fundamental harmonic of the square wave.

26.4 Composition of Nonsinusoidal Waveforms

Any periodic nonsinusoidal waveform with a fundamental frequency $f_1 = 1/T$ is composed of: (1) Sine-wave harmonics at frequencies which are integral multiples of the fundamental frequency f_1. The second harmonic has a frequency of $2f_1$, the third harmonic has a frequency of $3f_1$, and so on. In general, the different harmonics have different amplitudes and they are not necessarily all in phase. (2) A dc component which is equal to the waveform's *average* value averaged over one complete cycle.

Stated another way, we can think of any periodic waveform as being formed by adding together a dc value and an infinite number of sine-wave harmonics with appropriate amplitudes and phase. This is shown graphically in Fig. 26–3 for a nonsinusoidal voltage source.

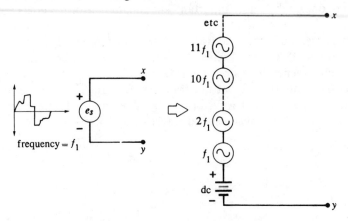

FIGURE 26–3 *A nonsinusoidal voltage and its equivalent harmonic representation.*

In practice, it is not necessary to consider all harmonics out to infinity since in most cases the amplitudes of the harmonics decrease rapidly for higher order harmonics. That is, the second harmonic amplitude is smaller than the fundamental amplitude; the third harmonic amplitude is smaller than the second harmonic amplitude; and so on. As such, many of the higher harmonic components can be neglected since their amplitudes are very small and contribute very little to the total waveform.

The dc component referred to and shown in Fig. 26–3 is equal to the average value of the waveform over one complete cycle. If a nonsinusoidal waveform has a *zero* average value, then its dc component will be zero and the waveform is pure ac. Such nonsinusoidal waveforms are made up solely of sine-wave harmonics.

EXAMPLE 26.3 Figure 26–4(a) shows a sweep waveform. List the frequencies of the components which make up this waveform.

Solution: The sweep waveform has a period of 40 μs. Thus, its fundamental frequency is

$$f_1 = \frac{1}{T} = \frac{1}{40 \ \mu s} = 25 \ \text{kHz}$$

Therefore the waveform has sine-wave harmonics at the following frequencies: **25 kHz (fundamental), 50 kHz, 75 kHz, 100 kHz** . . . etc.

In addition the waveform has a **positive dc** component since its average value is positive.

EXAMPLE 26.4 Repeat the process of Ex. 26.4 for the complex waveform in Fig. 26–4(b).

Solution: The waveform's fundamental frequency is $f_1 = 1/0.002 = 500$ Hz.

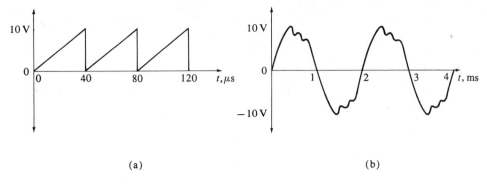

(a)   (b)

FIGURE 26-4 *Nonsinusoidal waveforms.*

Therefore the waveform has sine-wave harmonics at **500 Hz, 1 kHz, 1.5 kHz**, etc. Since the waveform is symmetrical above and below the zero axis, its average value is zero. Thus the waveform has **no dc component**.

26.5 Harmonic Analysis of a Square Wave

The square wave is one of the most important nonsinusoidal waveforms. Figure 26–5(a) shows an ac square wave with a peak-to-peak amplitude of 10 V. The square wave's period is shown as 1 s and its fundamental frequency is therefore 1 Hz. The square wave belongs to a special class of waveforms that possess only *odd* harmonics. In other words, its *even* harmonics have zero amplitude. Thus, this square wave has harmonics at 3 Hz, 5 Hz, 7 Hz, etc. It does not possess 2nd, 4th, 6th, etc. harmonics.

Part (b) of Fig. 26–5 shows the fundamental and the third and fifth harmonics of this square wave with their proper phase and amplitudes. We will not concern ourselves with how the amplitudes were obtained since the process (called *Fourier analysis*) involves higher mathematics. Since this square wave makes equal excursions above and below the zero axis, its average value is zero so that it has no dc component.

Examination of Fig. 26–5(b) reveals several interesting points. First, it can be seen that the amplitude of each successive harmonic gets smaller and smaller; the first harmonic has an amplitude of 6.37 V-peak, the third harmonic has an amplitude of 2.13 V-peak, and the fifth harmonic has an amplitude of 1.27 V-peak. These three harmonics are not the only harmonics. In fact, the square wave has all the odd harmonics out to infinity. However, the amplitudes of each successive harmonic get even smaller. For example, the 63rd harmonic of this square wave has an amplitude of only about 0.1 volt. Most common waveforms have harmonics whose amplitudes behave similarly to those of a square wave.

Second, it can be seen that all of the square-wave harmonics are in phase. That is, at the beginning of each fundamental period T all of the harmonics are at zero and are increasing in the positive direction. In general, the harmonics of any nonsinusoidal waveform are *not* all in phase. The phase relationship among the harmonics depends on the shape of the nonsinusoidal waveform.

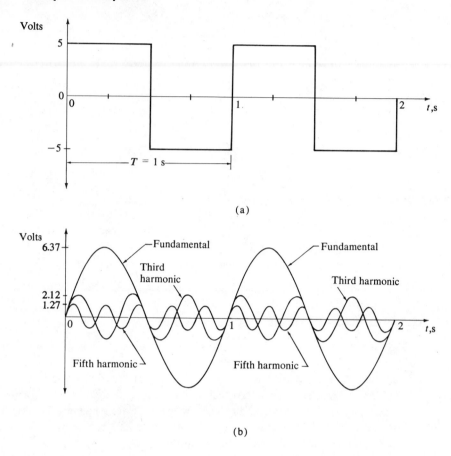

FIGURE 26–5 (a) *ac square wave.* (b) *First three harmonics of the square wave.*

Third, we might wonder why the amplitude of the fundamental is larger than the amplitude of the square wave itself. This apparent dilemma can be resolved if we realize that the square wave is made up of the *sum* of all its component harmonics. Although we don't have all of the harmonics, we can use the harmonics shown to get a feel for what happens. Figure 26–6(a) shows the result of adding the *first* and *third* harmonics point by point. As can be seen, even using only the first two harmonics produces a resultant which begins to resemble the square wave. Notice that the resultant has a dip which occurs at the point where the fundamental is at its positive peak and the third harmonic is at its negative peak. This indicates that even though the amplitude of the fundamental is 6.37 V, some of it is canceled out by the negative portions of the third harmonic.

By including more and more of the harmonics we can construct a resultant which more closely resembles the square wave. Part (b) of Fig. 26–6 shows the result of including the fifth harmonic. The resultant waveform is beginning to look like the original square wave. From a theoretical viewpoint, an infinite number of harmonics has to be included to completely construct the original square wave. However, in practice it is only necessary to consider a limited

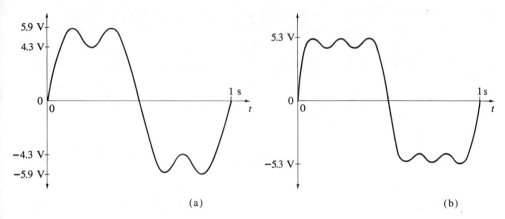

FIGURE 26-6 *Resultant produced by* (a) *sum of fundamental and third harmonics;* (b) *sum of fundamental, third, and fifth harmonics.*

number of harmonics. This is because beyond a certain point the higher harmonics contribute very little to the overall waveform.

EXAMPLE 26.5 Determine the harmonic content of the square wave in Fig. 26-7.

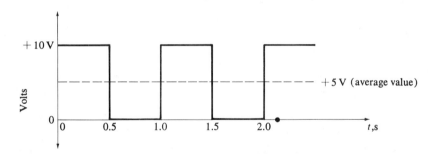

FIGURE 26-7 *Square wave with dc average value.*

Solution: The square wave in Fig. 26-7 is really the same as the one in Fig. 26-5 except that it is shifted upward by 5 V. In other words, it has a +5 V dc value added to it. Except for the dc component, both waveforms are 10-V peak-peak square waves. As such, the square wave of Fig. 26-7 has the same harmonics as the square wave of Fig. 26-5.

26.6 Waveforms With Only Odd Harmonics

Certain nonsinusoidal waveforms possess only odd-numbered harmonics. We saw one example of this in our discussion of the square wave in the last section. Figure 26-8 contains more examples of such waveforms. In viewing these waveforms it can be seen that each one of them possesses a certain type of *symmetry* which is a characteristic of waveforms which contain *only* odd harmonics. The second half-cycle of the waveform (between $t = T/2$) and

$t = T$) goes through the same variations as the first half-cycle (between $t = 0$ and $t = T/2$) except in the *opposite polarity*.

This type of symmetry is often called *mirror-image* symmetry and can be easily checked, as shown in Fig. 26–8(a), by sliding the second half-cycle underneath the first half-cycle (dotted lines) and noting that each half is the mirror image of the other. Using this technique, it is easy to recognize any nonsinusoidal waveform which possesses only *odd* harmonics. There is no similar test for *even* harmonics.

26.7 Finding Average Value of a Waveform

The average value of any sine wave is *zero* since its negative half-cycle is exactly the same as its positive half-cycle. Thus, the average value of any harmonic

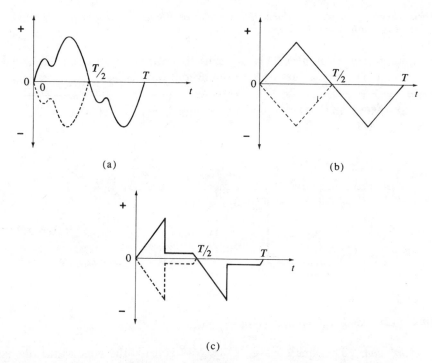

FIGURE 26–8 *Examples of waveforms which possess only odd haromonics and no even harmonics.*

must be zero. As such, the average value of a nonsinusoidal waveform is taken into account by adding a dc term along with the harmonics. Before addressing ourselves to the problem of finding the average value of a nonsinusoidal waveform, let us examine the sine wave a little more closely.

In Fig. 26–9(a) a sine wave is shown with the area between the waveform and the time axis shaded in. The area of the *positive* half-cycle is labeled $A+$, and the area of the *negative* half-cycle is labeled $A-$. For the sine wave it is obvious that $A+ = A-$, indicating that the waveform is equally positive and negative so that its average value is zero. The same reasoning can be used for the ac square wave in Fig. 26–9(b) and for all the waveforms in Fig. 26–8. Thus, each of these waveforms has a zero average value or dc component. They are *pure ac* waveforms and consist only of sine-wave harmonics.

Many nonsinusoidal waveforms will have nonzero average values. For such waveforms the positive area $A+$ will not equal the negative area $A-$. The average value of these waveforms can be determined by calculating the net area of the waveform over a single period, and then dividing by the length of the period. Stated mathematically,

$$AV = \frac{(A+) - (A-)}{T} \tag{26-1}$$

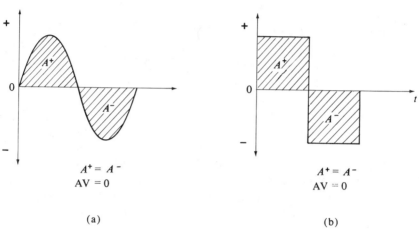

FIGURE 26-9 *For pure ac waveforms, the area above zero $(A+)$ equals the area below zero $(A-)$.*

In this formula AV represents the "average value" and T is the period of one cycle of the waveform.

EXAMPLE 26.6 Determine the average value of the waveforms in Fig. 26-10.

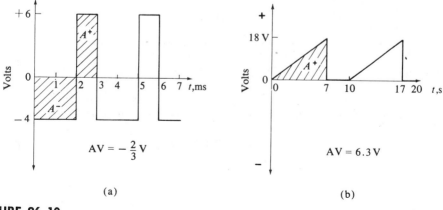

FIGURE 26-10

Solution: (a) The negative area is

$$A- = 4 \text{ V} \times 2 \text{ ms} = 8 \text{ V-ms}$$

The positive area is

$$A+ = 6 \text{ V} \times 1 \text{ ms} = 6 \text{ V-ms}$$

The waveform period is $T = 3$ ms. Thus,

$$AV = \frac{(6) \text{ V-ms} - (8) \text{ V-ms}}{3 \text{ ms}} = \frac{-2}{3} \text{ volt}$$

(b) This waveform has no negative area so that $A- = 0$. The positive area is found using the formula for the area of a triangle ($\frac{1}{2}$ base × height). Thus,

$$A+ = \tfrac{1}{2} \times 18 \text{ V} \times 7 \text{ s} = 63 \text{ V-s}$$

The waveform period is $T = 10$ s. Thus,

$$AV = \frac{63 \text{ V-s}}{10 \text{ s}} = 6.3 \text{ V}$$

The average values of the waveforms in Fig. 26–10 were easy to calculate using simple geometric formulas. For more complex waveshapes, the areas under the waveforms could be hard to calculate exactly. As such, some means of approximation has to be used. One useful method for approximating the area under a waveform is to use a number of rectangles, triangles, or other familiar shapes to reproduce the waveform. This method is illustrated in Fig. 26–11 for a *half-wave rectified sine wave*. One cycle of the waveform is shown.

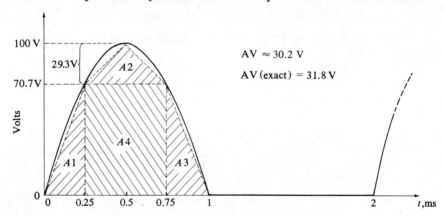

FIGURE 26–11

The area under the waveform is approximated by the three triangles $A1$, $A2$, and $A3$, and a rectangle $A4$. The area of the four shapes is

$$A1 = \frac{70.7 \times 0.25}{2} = 8.84 \text{ V-ms}$$

$$A2 = \frac{29.3 \times 0.5}{2} = 7.33 \text{ V-ms}$$

$$A3 = A1 = 8.84 \text{ V-ms}$$

$$A4 = 70.7 \times 0.5 = 35.35 \text{ V-ms}$$

Total area is the sum of these. Thus,

$$A+ = 60.36 \text{ V-ms}$$

Since the waveform does not go negative, there is no $A-$. The average value is therefore,

$$AV \approx \frac{60.36 \text{ V-ms}}{2 \text{ ms}} = 30.2 \text{ V}$$

The exact average value of this waveform is 31.8 V, so our approximate technique is fairly accurate (within 7%). A better approximation could be obtained by using more smaller pieces to reproduce the waveform area more exactly.

Measuring dc Average Value A nonsinusoidal waveform's average value is actually its dc component. As such, it can be measured using a dc current or voltmeter. For example, when the waveform in Fig. 26-11 is applied to a dc voltmeter, the voltmeter pointer will indicate 31.8 V. The pointer might vibrate somewhat around this reading due to the ac harmonics in the waveform.

The average value of any waveform can also be measured with an oscilloscope. Most oscilloscopes have a *dc-ac coupling* switch which can be used to select either dc or ac coupling (of the waveform to be displayed) into the scope. When dc coupling is used, the waveform is fed into the scope directly. When ac coupling is used, a series blocking capacitor is used to remove the dc component of the waveform before it is fed to the scope amplifiers so that the displayed waveform has *zero* average value. Thus, by noting the amount of vertical shifting of the *entire* waveform display as the scope is switched from dc to ac coupling, the average value can be determined.

To illustrate, Fig. 26-12 shows the scope display for the waveform of Fig. 26-11 for both dc and ac coupling. Note that using ac coupling causes the waveform to shift downward by 31.8 V (the waveform's average value) since this dc component was removed.

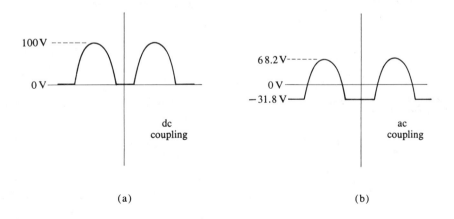

FIGURE 26-12 *Switching oscilloscope from dc to ac coupling causes waveform to shift by an amount equal to* AV.

EXAMPLE 26.7 A nonsinusoidal waveform with a period of 40 μs shifts *upward* by 30 V when the scope displaying it is switched from dc to ac coupling. (a) What is the waveform's dc component? (b) What are the frequencies of its harmonics?

Solution: (a) Since the waveform shifted *upward* when its dc component was removed, its average value must have been negative by 30 V. Thus, its dc component = -30 V.

(b) The fundamental frequency is $f_1 = 1/T = 25$ kHz. The waveform has harmonics at 50 kHz, 75 kHz, 100 kHz, etc. The harmonics and the -30-V dc component make up the complete nonsinusoidal waveform.

26.8 Nonsinusoidal Waveforms Applied to Linear Networks

Since a nonsinusoidal waveform consists of the sum of all its harmonics plus its dc component, then we can think of it as being the same as a series combination of generators as shown in Fig. 26-13(a). There is a dc source representing the dc component of the waveform and there are ac sinusoidal generators each representing one of the waveform harmonics. Each generator, in general, has a different amplitude and phase. The total series combination of the various sources is equivalent to the original waveform.

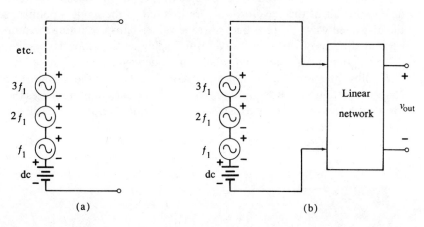

FIGURE 26–13 (a) *Equivalent representation of a nonsinusoidal waveform* (b) *driving a linear network.*

If the waveform is applied to a linear network as in part (b) of Fig. 26-13, then it is possible to use the principle of *superposition* to determine the circuit output. That is, we can determine the circuit output for each individual generator (replacing the others by a short circuit) and then sum up all of the various outputs to get the total resultant output. It's easy to see that such a process would be prohibitively tedious even if we included only a limited number of harmonics. Fortunately, such an analysis is rarely performed (except in textbooks). When it is, it is usually done with the aid of a computer.

It is not necessary to apply the superposition principle explicitly when dealing with nonsinusoidal waveforms. Rather, it is used implicitly in the thought processes involved in determining how various linear networks affect the harmonic content of a given waveform. These processes are not intended for coverage in this text; they are studied in a subsequent volume on pulse circuits. However, some examples will serve to provide an introduction to the reasoning involved.

Resistive Networks Any waveform of voltage applied to a *linear* resistive network will emerge at the output with the same shape. In general, it will be attenuated but the output will have exactly the same shape as the input. This can be easily reasoned since we know that a resistive network behaves the same for all frequencies. Thus, the network will treat each frequency component of the input in exactly the same manner. For example, suppose the network is a voltage divider with an attenuation ratio of 0.5. This means that the network will attenuate the dc component of the input, and each harmonic of the input by the *same* factor (0.5); as such, the output waveform's frequency components will maintain the same relative amplitudes, thereby resulting in the same nonsinusoidal waveshape as the input.

Networks Containing L or C A network which contains inductance, capacitance, or both will respond differently to different frequencies. We saw examples of *frequency-dependent* networks in our study of *RC* and *RL* filters and in our discussion of resonant circuits. When a *sinusoidal* input is applied to such a network, the output waveform will generally have a different amplitude and phase, but it will be a sine wave with the same frequency as the input sine wave. In fact, the sine wave is unique in this respect. When any other nonsinusoidal waveform is applied to a frequency dependent network, the output waveform will generally have a *different* shape than the input. This output *distortion* can be reasoned from our understanding of harmonics as follows.

(1) A nonsinusoidal waveform contains harmonics at many different frequencies.

(2) Each harmonic has an appropriate amplitude and phase such that the result of adding all the harmonics (plus the dc component) produces the given waveform.

(3) Since each harmonic is a sine wave of a different frequency, then it is apparent that the frequency-dependent network will affect each harmonic differently.

(4) Thus, as the various harmonics reach the output, they will no longer have the same relative amplitudes and phases; therefore, when they are combined to produce the output waveform, this waveform will not be an exact replica of the input waveform.

The amount of distortion that a frequency-dependent network produces on a nonsinusoidal input will depend on the frequency-response characteristics of the network and the harmonic content of the input. This is illustrated in Fig. 26–14(a) for a square wave applied to an *RC* low-pass filter.

The input square wave e_s [part (b) of figure] is a dc square wave since it is of only one polarity. Its peak value is 10 V and its average value is +5 V (student can verify this using area method). Its frequency is 1 kHz since its period is 1 ms. Thus, this square wave is made up of the following frequency components: dc, 1 kHz (fundamental), 3 kHz, 5 kHz, 7 kHz, 9 kHz, 11 kHz, etc. Figure 26–14(c) shows the output of the low-pass filter when the filter's high-frequency cutoff f_{hc} is 10 kHz. The waveform is somewhat distorted but maintains the same general shape as the input. The distortion is due to the fact that the circuit attenuates all of the input harmonics above 10 kHz. In other words, it attenuates the higher frequency harmonics of the input (11 kHz, 13 kHz, ...) while passing the fundamental and lower frequency harmonics. The low-pass filter

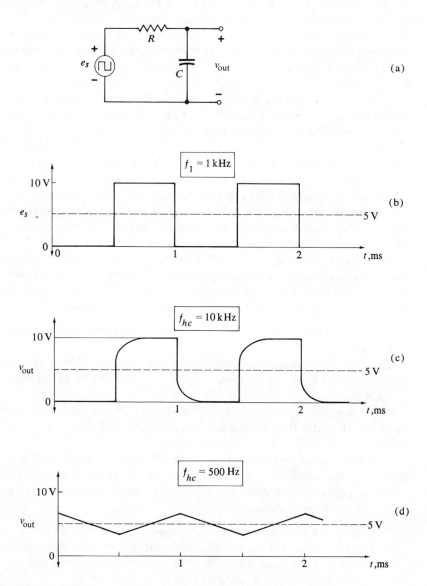

FIGURE 26–14 RC low-pass filter with square-wave input.

also passes the dc component of the input without attenuation. It can be seen that the v_{out} waveform has a +5-V average dc value, the same as the input.

If the low-pass filter's cutoff frequency is lowered to 500 Hz, the output will appear as shown in Fig. 26–14(d). It can be seen that the v_{out} waveform is now considerably different than the input. It still has a +5-V dc average value because a low-pass filter always passes dc. However, the filter now attenuates all of the square wave's harmonics to some degree, thereby producing significant distortion in the output. If f_{hc} is lowered even further, eventually the v_{out} waveform will simply contain the +5-V dc level since all the harmonics will be completely attenuated.

Another example is shown in Fig. 26–15, where a 1-kHz ac square wave is applied to a series resonant circuit. The output is taken across C. We know that the output will be maximum at input frequencies at or near f_R, the circuit's resonant frequency. If the resonant circuit is tuned so that f_R is equal to the fundamental frequency of the square wave, then the output will consist mainly of the sine-wave fundamental harmonic while rejecting the other harmonics, assuming that the circuit Q is high enough. In other words, it is possible to *extract* the fundamental harmonic of 1 kHz and reproduce it in the output. The same procedure could be used to extract the third harmonic by setting $f_R = 3$ kHz. For higher harmonics, the process becomes more difficult because the amplitudes of the square wave's harmonics become successively smaller and smaller. As such, sharper filtering is needed to be able to pass a higher harmonic while attenuating the lower harmonics.

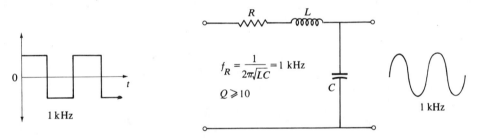

FIGURE 26–15 *Resonant circuit used to extract sine-wave fundamental from a square wave.*

The examples discussed here should begin to give the reader some insight into the methods used to qualitatively analyze linear circuits with nonsinusoidal inputs. These qualitative thought processes are the ones used most often in everyday work. Further work in this area, both qualitative and quantitative, will no doubt be presented in subsequent pulse circuit courses.

Chapter Summary

1. Any periodic waveform has a fundamental frequency which is equal to the frequency of the waveform, and harmonic frequencies which are integer multiples of the fundamental frequency.

2. Any periodic nonsinusoidal waveform is composed of a dc component and an infinite number of sine-wave harmonics.

3. Nonsinusoidal waveforms which possess mirror-image symmetry have only odd harmonics (3rd, 5th, 7th. . .).

4. The average value of a periodic waveform is obtained by finding the net area under one cycle of the waveform and dividing it by T, the waveform's period.

5. When a nonsinusoidal signal is applied to a network which contains inductance and/or capacitance, the output waveform will generally be a distorted version of the input.

Questions/Problems

Sections 26.2–26.5

26-1 A sine wave has a frequency of 200 Hz. (a) List the frequencies of the second through eleventh harmonics of this sine wave. (b) What is the frequency of the 3211th harmonic?

26-2 A certain nonsinusoidal waveform has a ninth harmonic of 567 kHz. What is the waveform's period?

26-3 Determine the fundamental and next ten harmonic frequencies of the waveform in Fig. 26–16(a). Does the waveform have a dc component?

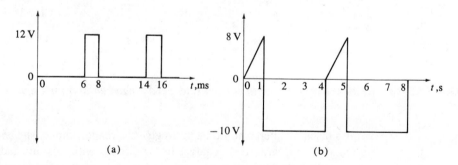

FIGURE 26-16

26-4 Repeat problem 26-3 for the waveform in Fig. 26–16(b).

26-5 Which of the following statements are *always* true:
   a. Every nonsinusoidal waveform has a dc component.
   b. Adding two sine waves produces another sine wave.
   c. A square wave has only even harmonics.
   d. The average value of a pure ac waveform is zero.
   e. Every nonsinusoidal waveform has an infinite number of harmonics.

Sections 26.6–26.7

26–6 Refer to the waveforms in Fig. 26–17. Which of these waveforms possesses only odd harmonics?

26–7 Determine the dc component (average value) of each waveform in Fig. 26–16.

26–8 The waveform in Fig. 26–16(a) is to be displayed on an oscillosocope. Show the display for both dc and ac coupling of the scope input.

26–9 A waveform displayed on a scope shifts *downward* by 5 V when the scope is switched from ac coupling to dc coupling. (a) What is the waveform's average

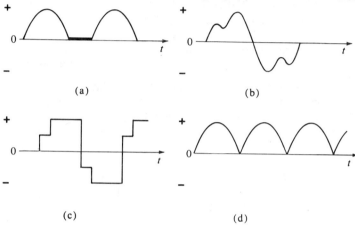

FIGURE 26–17

value? (b) If the period of the waveform is 10 μs, what are the harmonic frequencies?

Section 26.8

26–10 Explain why a network that contains L or C usually produces distortion of a nonsinusoidal input.

26–11 Figure 26–18 shows a square wave applied to an RC high-pass filter. Which of the following values of low-frequency cutoff f_{lc} will produce an output with the *most* distortion: (a) 100 Hz, (b) 2 Hz, (c) 10 kHz?

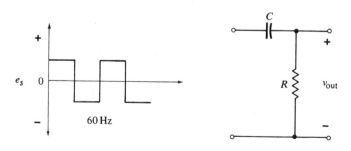

FIGURE 26–18

26–12 Repeat problem 26–11 for *least* distortion.

26–13 Refer to Fig. 26–15. Choose values for L and C which will successively extract the *third* harmonic sine wave. Assume R = 10 ohms. (Hint: circuit BW should be chosen so that all the other harmonics are outside the BW.)

26–14 Answer true or false to each of the following:
   a. The average value of the output of a high-pass filter is always zero.
   b. A low-pass filter will *always* produce significant distortion of a square-wave input.
   c. Any resistive network (linear or nonlinear) will pass a nonsinusoidal waveform with no distortion.
   d. If a 1-kHz square wave is applied to a resonant filter tuned to 2 kHz, the output will be zero.
   e. If an ac square wave with a frequency of 200 Hz is applied to a low-pass filter with $R = 5\,k\Omega$ and $C = 16\,\mu F$, the filter output will be essentially zero.

26–15 If a pure 1-kHz sine wave is applied to the circuit in Fig. 26–19, the output is not a sine wave. Assume the diode is silicon and sketch the output waveform. Determine the harmonic frequencies present in the output. This example shows how a *nonlinear* network distorts a sine wave and produces harmonics in the output.

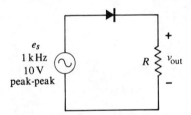

FIGURE 26–19

APPENDICES

A Table of $\log_{10} N$ — 777

B Trigonometric Functions: Decimal Degrees — 781

C Values of ϵ^x and ϵ^{-x} — 787

D Meters — 791

D.1 The Moving-Coil (d'Arsonval) Meter Movement, 791
D.2 Voltmeters, 794
D.3 The Ohmmeter, 797
D.4 Meter Loading, 797
D.5 AC Meters, 799

APPENDIX A

Tables of Log₁₀ N

N	0	1	2	3	4	5	6	7	8	9
10	0000	0043	0086	0128	0170	0212	0253	0294	0334	0374
11	0414	0453	0492	0531	0569	0607	0645	0682	0719	0755
12	0792	0828	0864	0899	0934	0969	1004	1038	1072	1106
13	1139	1173	1206	1239	1271	1303	1335	1367	1399	1430
14	1461	1492	1523	1553	1584	1614	1644	1673	1703	1732
15	1761	1790	1818	1847	1875	1903	1931	1959	1987	2014
16	2041	2068	2095	2122	2148	2175	2201	2227	2253	2279
17	2304	2330	2355	2380	2405	2430	2455	2480	2504	2529
18	2553	2577	2601	2625	2648	2672	2695	2718	2742	2765
19	2788	2810	2833	2856	2878	2900	2923	2945	2967	2989
20	3010	3032	3054	3075	3096	3118	3139	3160	3181	3201
21	3222	3243	3263	3284	3304	3324	3345	3365	3385	3404
22	3424	3444	3464	3483	3502	3522	3541	3560	3579	3598
23	3617	3636	3655	3674	3692	3711	3729	3747	3766	3784
24	3802	3820	3838	3856	3874	3892	3909	3927	3945	3962
25	3979	3997	4014	4031	4048	4065	4082	4099	4116	4133
26	4150	4166	4183	4200	4216	4232	4249	4265	4281	4298
27	4314	4330	4346	4362	4378	4393	4409	4425	4440	4456
28	4472	4487	4502	4518	4533	4548	4564	4579	4594	4609
29	4624	4639	4654	4669	4683	4698	4713	4728	4742	4757
30	4771	4786	4800	4814	4829	4843	4857	4871	4886	4900
31	4914	4928	4942	4955	4969	4983	4997	5011	5024	5038
32	5051	5065	5079	5092	5105	5119	5132	5145	5159	5172
33	5185	5198	5211	5224	5237	5250	5263	5276	5289	5302
34	5315	5328	5340	5353	5366	5378	5391	5403	5416	5428
35	5441	5453	5465	5478	5490	5502	5514	5527	5539	5551
36	5563	5575	5587	5599	5611	5623	5635	5647	5658	5670
37	5682	5694	5705	5717	5729	5740	5752	5763	5775	5786
38	5798	5809	5821	5832	5843	5855	5866	5877	5888	5899
39	5911	5922	5933	5944	5955	5966	5977	5988	5999	6010
40	6021	6031	6042	6053	6064	6075	6085	6096	6107	6117
N	0	1	2	3	4	5	6	7	8	9

N	0	1	2	3	4	5	6	7	8	9
40	6021	6031	6042	6053	6064	6075	6085	6096	6107	6117
41	6128	6138	6149	6160	6170	6180	6191	6201	6212	6222
42	6232	6243	6253	6263	6274	6284	6294	6304	6314	6325
43	6335	6345	6355	6365	6375	6385	6395	6405	6415	6425
44	6435	6444	6454	6464	6474	6484	6493	6503	6513	6522
45	6532	6542	6551	6561	6571	6580	6590	6599	6609	6618
46	6628	6637	6646	6656	6665	6675	6684	6693	6702	6712
47	6721	6730	6739	6749	6758	6767	6776	6785	6794	6803
48	6812	6821	6830	6839	6848	6857	6866	6875	6884	6893
49	6902	6911	6920	6928	6937	6946	6955	6964	6972	6981
50	6990	6998	7007	7016	7024	7033	7042	7050	7059	7067
51	7076	7084	7093	7101	7110	7118	7126	7135	7143	7152
52	7160	7168	7177	7185	7193	7202	7210	7218	7226	7235
53	7243	7251	7259	7267	7275	7284	7292	7300	7308	7316
54	7324	7332	7340	7348	7356	7364	7372	7380	7388	7396
55	7404	7412	7419	7427	7435	7443	7451	7459	7466	7474
56	7482	7490	7497	7505	7513	7520	7528	7536	7543	7551
57	7559	7566	7574	7582	7589	7597	7604	7612	7619	7627
58	7634	7642	7649	7657	7664	7672	7679	7686	7694	7701
59	7709	7716	7723	7731	7738	7745	7752	7760	7767	7774
60	7782	7789	7796	7803	7810	7818	7825	7832	7839	7846
61	7853	7860	7868	7875	7882	7889	7896	7903	7910	7917
62	7924	7931	7938	7945	7952	7959	7966	7973	7980	7987
63	7993	8000	8007	8014	8021	8028	8035	8041	8048	8055
64	8062	8069	8075	8082	8089	8096	8102	8109	8116	8122
65	8129	8136	8142	8149	8156	8162	8169	8176	8182	8189
66	8195	8202	8209	8215	8222	8228	8235	8241	8248	8254
67	8261	8267	8274	8280	8287	8293	8299	8306	8312	8319
68	8325	8331	8338	8344	8351	8357	8363	8370	8376	8382
69	8388	8395	8401	8407	8414	8420	8426	8432	8439	8445
70	8451	8457	8463	8470	8476	8482	8488	8494	8500	8506
N	0	1	2	3	4	5	6	7	8	9

N	0	1	2	3	4	5	6	7	8	9
70	8451	8457	8463	8470	8476	8482	8488	8494	8500	8506
71	8513	8519	8525	8531	8537	8543	8549	8555	8561	8567
72	8573	8579	8585	8591	8597	8603	8609	8615	8621	8627
73	8633	8639	8645	8651	8657	8663	8669	8675	8681	8686
74	8692	8698	8704	8710	8716	8722	8727	8733	8739	8745
75	8751	8756	8762	8768	8774	8779	8785	8791	8797	8802
76	8808	8814	8820	8825	8831	8837	8842	8848	8854	8859
77	8865	8871	8876	8882	8887	8893	8899	8904	8910	8915
78	8921	8927	8932	8938	8943	8949	8954	8960	8965	8971
79	8976	8982	8987	8993	8998	9004	9009	9015	9020	9025
80	9031	9036	9042	9047	9053	9058	9063	9069	9074	9079
81	9085	9090	9096	9101	9106	9112	9117	9122	9128	9133
82	9138	9143	9149	9154	9159	9165	9170	9175	9180	9186
83	9191	9196	9201	9206	9212	9217	9222	9227	9232	9238
84	9243	9248	9253	9258	9263	9269	9274	9279	9284	9289
85	9294	9299	9304	9309	9315	9320	9325	9330	9335	9340
86	9345	9350	9355	9360	9365	9370	9375	9380	9385	9390
87	9395	9400	9405	9410	9415	9420	9425	9430	9435	9440
88	9445	9450	9455	9460	9465	9469	9474	9479	9484	9489
89	9494	9499	9504	9509	9513	9518	9523	9528	9533	9538
90	9542	9547	9552	9557	9562	9566	9571	9576	9581	9586
91	9590	9595	9600	9605	9609	9614	9619	9624	9628	9633
92	9638	9643	9647	9652	9657	9661	9666	9671	9675	9680
93	9685	9689	9694	9699	9703	9708	9713	9717	9722	9727
94	9731	9736	9741	9745	9750	9754	9759	9763	9768	9773
95	9777	9782	9786	9791	9795	9800	9805	9809	9814	9818
96	9823	9827	9832	9836	9841	9845	9850	9854	9859	9863
97	9868	9872	9877	9881	9886	9890	9894	9899	9903	9908
98	9912	9917	9921	9926	9930	9934	9939	9943	9948	9952
99	9956	9961	9965	9969	9974	9978	9983	9987	9991	9996
N	0	1	2	3	4	5	6	7	8	9

APPENDIX B

Trigonometric Functions: Decimal Degrees

Deg.	Sin	Tan	Cot	Cos	Deg.
0.0	.00000	.00000	∞	1.0000	90.0
.1	.00175	.00175	573.0	1.0000	89.9
.2	.00349	.00349	286.5	1.0000	.8
.3	.00524	.00524	191.0	1.0000	.7
.4	.00698	.00698	143.24	1.0000	.6
.5	.00873	.00873	114.59	1.0000	.5
.6	.01047	.01047	95.49	0.9999	.4
.7	.01222	.01222	81.85	.9999	.3
.8	.01396	.01396	71.62	.9999	.2
.9	.01571	.01571	63.66	.9999	89.1
1.0	.01745	.01746	57.29	0.9998	89.0
.1	.01920	.01920	52.08	.9998	88.9
.2	.02094	.02095	47.74	.9998	.8
.3	.02269	.02269	44.07	.9997	.7
.4	.02443	.02444	40.92	.9997	.6
.5	.02618	.02619	38.19	.9997	.5
.6	.02792	.02793	35.80	.9996	.4
.7	.02967	.02968	33.69	.9996	.3
.8	.03141	.03143	31.82	.9995	.2
.9	.03316	.03317	30.14	.9995	88.1
2.0	.03490	.03492	28.64	0.9994	88.0
.1	.03664	.03667	27.27	.9993	87.9
.2	.03839	.03842	26.03	.9993	.8
.3	.04013	.04016	24.90	.9992	.7
.4	.04188	.04191	23.86	.9991	.6
.5	.04362	.04366	22.90	.9990	.5
.6	.04536	.04541	22.02	.9990	.4
.7	.04711	.04716	21.20	.9989	.3
.8	.04885	.04891	20.45	.9988	.2
.9	.05059	.05066	19.74	.9987	87.1
3.0	.05234	.05241	19.081	0.9986	87.0
.1	.05408	.05416	18.464	.9985	86.9
.2	.05582	.05591	17.886	.9984	.8
.3	.05756	.05766	17.343	.9983	.7
.4	.05931	.05941	16.832	.9982	.6
.5	.06105	.06116	16.350	.9981	.5
.6	.06279	.06291	15.895	.9980	.4
.7	.06453	.06467	15.464	.9979	.3
.8	.06627	.06642	15.056	.9978	.2
.9	.06802	.06817	14.669	.9977	86.1
4.0	.06976	.06993	14.301	0.9976	86.0
.1	.07150	.07168	13.951	.9974	85.9
.2	.07324	.07344	13.617	.9973	.8
.3	.07498	.07519	13.300	.9972	.7
.4	.07672	.07695	12.996	.9971	.6
.5	.07846	.07870	12.706	.9969	.5
.6	.08020	.08046	12.429	.9968	.4
.7	.08194	.08221	12.163	.9966	.3
.8	.08368	.08397	11.909	.9965	.2
.9	.08542	.08573	11.664	.9963	85.1
5.0	.08716	.08749	11.430	0.9962	85.0
Deg.	Cos	Cot	Tan	Sin	Deg.

Deg.	Sin	Tan	Cot	Cos	Deg.
5.0	.08716	.08749	11.430	0.9962	85.0
.1	.08889	.08925	11.205	.9960	84.9
.2	.09063	.09101	10.988	.9959	.8
.3	.09237	.09277	10.780	.9957	.7
.4	.09411	.09453	10.579	.9956	.6
.5	.09585	.09629	10.385	.9954	.5
.6	.09758	.09805	10.199	.9952	.4
.7	.09932	.09981	10.019	.9951	.3
.8	.10106	.10158	9.845	.9949	.2
.9	.10279	.10334	9.677	.9947	84.1
6.0	.10453	.10510	9.514	0.9945	84.0
.1	.10626	.10687	9.357	.9943	83.9
.2	.10800	.10863	9.205	.9942	.8
.3	.10973	.11040	9.058	.9940	.7
.4	.11147	.11217	8.915	.9938	.6
.5	.11320	.11394	8.777	.9936	.5
.6	.11494	.11570	8.643	.9934	.4
.7	.11667	.11747	8.513	.9932	.3
.8	.11840	.11924	8.386	.9930	.2
.9	.12014	.12101	8.264	.9928	83.1
7.0	.12187	.12278	8.144	0.9925	83.0
.1	.12360	.12456	8.028	.9923	82.9
.2	.12533	.12633	7.916	.9921	.8
.3	.12706	.12810	7.806	.9919	.7
.4	.12880	.12988	7.700	.9917	.6
.5	.13053	.13165	7.596	.9914	.5
.6	.13226	.13343	7.495	.9912	.4
.7	.13399	.13521	7.396	.9910	.3
.8	.13572	.13698	7.300	.9907	.2
.9	.13744	.13876	7.207	.9905	82.1
8.0	.13917	.14054	7.115	0.9903	82.0
.1	.14090	.14232	7.026	.9900	81.9
.2	.14263	.14410	6.940	.9898	.8
.3	.14436	.14588	6.855	.9895	.7
.4	.14608	.14767	6.772	.9893	.6
.5	.14781	.14945	6.691	.9890	.5
.6	.14954	.15124	6.612	.9888	4
.7	.15126	.15302	6.535	.9885	.3
.8	.15299	.15481	6.460	.9882	.2
.9	.15471	.15660	6.386	.9880	81.1
9.0	.15643	.15838	6.314	0.9877	81.0
.1	.15816	.16017	6.243	.9874	80.9
.2	.15988	.16196	6.174	.9871	.8
.3	.16160	.16376	6.107	.9869	.7
.4	.16333	.16555	6.041	.9866	.6
.5	.16505	.16734	5.976	.9863	.5
.6	.16677	.16914	5.912	.9860	.4
.7	.16849	.17093	5.850	.9857	.3
.8	.17021	.17273	5.789	.9854	.2
.9	.17193	.17453	5.730	.9851	80.1
10.0	.1736	.1763	5.671	0.9848	80.0
Deg.	Cos	Cot	Tan	Sin	Deg.

Deg.	Sin	Tan	Cot	Cos	Deg.
10.0	.1736	.1763	5.671	0.9848	**80.0**
.1	.1754	.1781	5.614	.9845	79.9
.2	.1771	.1799	5.558	.9842	.8
.3	.1788	.1817	5.503	.9839	.7
.4	.1805	.1835	5.449	.9836	.6
.5	.1822	.1853	5.396	.9833	.5
.6	.1840	.1871	5.343	.9829	.4
.7	.1857	.1890	5.292	.9826	.3
.8	.1874	.1908	5.242	.9823	.2
.9	.1891	.1926	5.193	.9820	79.1
11.0	.1908	.1944	5.145	0.9816	**79.0**
.1	.1925	.1962	5.097	.9813	78.9
.2	.1942	.1980	5.050	.9810	.8
.3	.1959	.1998	5.005	.9806	.7
.4	.1977	.2016	4.959	.9803	.6
.5	.1994	.2035	4.915	.9799	.5
.6	.2011	.2053	4.872	.9796	.4
.7	.2028	.2071	4.829	.9792	.3
.8	.2045	.2089	4.787	.9789	.2
.9	.2062	.2107	4.745	.9785	78.1
12.0	.2079	.2126	4.705	0.9781	**78.0**
.1	.2096	.2144	4.665	.9778	77.9
.2	.2113	.2162	4.625	.9774	.8
.3	.2130	.2180	4.586	.9770	.7
.4	.2147	.2199	4.548	.9767	.6
.5	.2164	.2217	4.511	.9763	.5
.6	.2181	.2235	4.474	.9759	.4
.7	.2198	.2254	4.437	.9755	.3
.8	.2215	.2272	4.402	.9751	.2
.9	.2233	.2290	4.366	.9748	77.1
13.0	.2250	.2309	4.331	0.9744	**77.0**
.1	.2267	.2327	4.297	.9740	76.9
.2	.2284	.2345	4.264	.9736	.8
.3	.2300	.2364	4.230	.9732	.7
.4	.2317	.2382	4.198	.9728	.6
.5	.2334	.2401	4.165	.9724	.5
.6	.2351	.2419	4.134	.9720	.4
.7	.2368	.2438	4.102	.9715	.3
.8	.2385	.2456	4.071	.9711	.2
.9	.2402	.2475	4.041	.9707	76.1
14.0	.2419	.2493	4.011	0.9703	**76.0**
.1	.2436	.2512	3.981	.9699	75.9
.2	.2453	.2530	3.952	.9694	.8
.3	.2470	.2549	3.923	.9690	.7
.4	.2487	.2568	3.895	.9686	.6
.5	.2504	.2586	3.867	.9681	.5
.6	.2521	.2605	3.839	.9677	.4
.7	.2538	.2623	3.812	.9673	.3
.8	.2554	.2642	3.785	.9668	.2
.9	.2571	.2661	3.758	.9664	75.1
15.0	0.2588	0.2679	3.732	0.9659	**75.0**
Deg.	Cos	Cot	Tan	Sin	Deg.

Deg.	Sin	Tan	Cot	Cos	Deg.
15.0	0.2588	0.2679	3.732	0.9659	**75.0**
.1	.2605	.2698	3.706	.9655	74.9
.2	.2622	.2717	3.681	.9650	.8
.3	.2639	.2736	3.655	.9646	.7
.4	.2656	.2754	3.630	.9641	.6
.5	.2672	.2773	3.606	.9636	.5
.6	.2689	.2792	3.582	.9632	.4
.7	.2706	.2811	3.558	.9627	.3
.8	.2723	.2830	3.534	.9622	.2
.9	.2740	.2849	3.511	.9617	74.1
16.0	0.2756	0.2867	3.487	0.9613	**74.0**
.1	.2773	.2886	3.465	.9608	73.9
.2	.2790	.2905	3.442	.9603	.8
.3	.2807	.2924	3.420	.9598	.7
.4	.2823	.2943	3.398	.9593	.6
.5	.2840	.2962	3.376	.9588	.5
.6	.2857	.2981	3.354	.9583	.4
.7	.2874	.3000	3.333	.9578	.3
.8	.2890	.3019	3.312	.9573	.2
.9	.2907	.3038	3.291	.9568	73.1
17.0	0.2924	0.3057	3.271	0.9563	**73.0**
.1	.2940	.3076	3.251	.9558	72.9
.2	.2957	.3096	3.230	.9553	.8
.3	.2974	.3115	3.211	.9548	.7
.4	.2990	.3134	3.191	.9542	.6
.5	.3007	.3153	3.172	.9537	.5
.6	.3024	.3172	3.152	.9532	.4
.7	.3040	.3191	3.133	.9527	.3
.8	.3057	.3211	3.115	.9521	.2
.9	.3074	.3230	3.096	.9516	72.1
18.0	0.3090	.03249	3.078	0.9511	**72.0**
.1	.3107	.3269	3.060	.9505	71.9
.2	.3123	.3288	3.042	.9500	.8
.3	.3140	.3307	3.024	.9494	.7
.4	.3156	.3327	3.006	.9489	.6
.5	.3173	.3346	2.989	.9483	.5
.6	.3190	.3365	2.971	.9478	.4
.7	.3206	.3385	2.954	.9472	.3
.8	.3223	.3404	2.937	.9466	.2
.9	.3239	.3424	2.921	.9461	71.1
19.0	0.3256	0.3443	2.904	0.9455	**71.0**
.1	.3272	.3463	2.888	.9449	70.9
.2	.3289	.3482	2.872	.9444	.8
.3	.3305	.3502	2.856	.9438	.7
.4	.3322	.3522	2.840	.9432	.6
.5	.3338	.3541	2.824	.9426	.5
.6	.3355	.3561	2.808	.9421	.4
.7	.3371	.3581	2.793	.9415	.3
.8	.3387	.3600	2.778	.9409	.2
.9	.3404	.3620	2.762	.9403	70.1
20.0	0.3420	0.3640	2.747	0.9397	**70.0**
Deg.	Cos	Cot	Tan	Sin	Deg.

Appendix B

Deg.	Sin	Tan	Cot	Cos	Deg.
20.0	0.3420	0.3640	2.747	0.9397	**70.0**
.1	.3437	.3659	2.733	.9391	69.9
.2	.3453	.3679	2.718	.9385	.8
.3	.3469	.3699	2.703	.9379	.7
.4	.3486	.3719	2.689	.9373	.6
.5	.3502	.3739	2.675	.9367	.5
.6	.3518	.3759	2.660	.9361	.4
.7	.3535	.3779	2.646	.9354	.3
.8	.3551	.3799	2.633	.9348	.2
.9	.3567	.3819	2.619	.9342	69.1
21.0	0.3584	0.3839	2.605	0.9336	**69.0**
.1	.3600	.3859	2.592	.9330	68.9
.2	.3616	.3879	2.578	.9323	.8
.3	.3633	.3899	2.565	.9317	.7
.4	.3649	.3919	2.552	.9311	.6
.5	.3665	.3939	2.539	.9304	.5
.6	.3681	.3959	2.526	.9298	.4
.7	.3697	.3979	2.513	.9291	.3
.8	.3714	.4000	2.500	.9285	.2
.9	.3730	.4020	2.488	.9278	68.1
22.0	0.3746	0.4040	2.475	0.9272	**68.0**
.1	.3762	.4061	2.463	.9265	67.9
.2	.3778	.4081	2.450	.9259	.8
.3	.3795	.4101	2.438	.9252	.7
.4	.3811	.4122	2.426	.9245	.6
.5	.3827	.4142	2.414	.9239	.5
.6	.3843	.4163	2.402	.9232	.4
.7	.3859	.4183	2.391	.9225	.3
.8	.3875	.4204	2.379	.9219	.2
.9	.3891	.4224	2.367	.9212	67.1
23.0	0.3907	0.4245	2.356	0.9205	**67.0**
.1	.3923	.4265	2.344	.9198	66.9
.2	.3939	.4286	2.333	.9191	.8
.3	.3955	.4307	2.322	.9184	.7
.4	.3971	.4327	2.311	.9178	.6
.5	.3987	.4348	2.300	.9171	.5
.6	.4003	.4369	2.289	.9164	.4
.7	.4019	.4390	2.278	.9157	.3
.8	.4035	.4411	2.267	.9150	.2
.9	.4051	.4431	2.257	.9143	66.1
24.0	0.4067	0.4452	2.246	0.9135	**66.0**
.1	.4083	.4473	2.236	.9128	65.9
.2	.4099	.4494	2.225	.9121	.8
.3	.4115	.4515	2.215	.9114	.7
.4	.4131	.4536	2.204	.9107	.6
.5	.4147	.4557	2.194	.9100	.5
.6	.4163	.4578	2.184	.9092	.4
.7	.4179	.4599	2.174	.9085	.3
.8	.4195	.4621	2.164	.9078	.2
.9	.4210	.4642	2.154	.9070	65.1
25.0	0.4226	0.4663	2.145	0.9063	**65.0**
Deg.	Cos	Cot	Tan	Sin	Deg.

Deg.	Sin	Tan	Cot	Cos	Deg.
25.0	0.4226	0.4663	2.145	0.9063	**65.0**
.1	.4242	.4684	2.135	.9056	64.9
.2	.4258	.4706	2.125	.9048	.8
.3	.4274	.4727	2.116	.9041	.7
.4	.4289	.4748	2.106	.9033	.6
.5	4305	4770	2.097	.9026	.5
.6	.4321	.4791	2.087	.9018	.4
.7	.4337	.4813	2.078	.9011	.3
.8	.4352	.4834	2.069	.9003	.2
.9	.4368	.4856	2.059	.8996	64.1
26.0	0.4384	0.4877	2.050	0.8988	**64.0**
.1	.4399	.4899	2.041	.8980	63.9
.2	.4415	.4921	2.032	.8973	.8
.3	.4431	.4942	2.023	.8965	.7
.4	.4446	.4964	2.014	.8957	.6
.5	.4462	.4986	2.006	.8949	.5
.6	.4478	.5008	1.997	.8942	.4
.7	.4493	.5029	1.988	.8934	.3
.8	.4509	.5051	1.980	.8926	.2
.9	.4524	.5073	1.971	.8918	63.1
27.0	0.4540	0.5095	1.963	0.8910	**63.0**
.1	.4555	.5117	1.954	.8902	62.9
.2	.4571	.5139	1.946	.8894	.8
.3	.4586	.5161	1.937	.8886	.7
.4	.4602	.5184	1.929	.8878	.6
.5	.4617	.5206	1.921	.8870	.5
.6	.4633	.5228	1.913	.8862	.4
.7	.4648	.5250	1.905	.8854	.3
.8	.4664	.5272	1.897	.8846	.2
.9	.4679	.5295	1.889	.8838	62.1
28.0	0.4695	0.5317	1.881	0.8829	**62.0**
.1	.4710	.5340	1.873	.8821	61.9
.2	.4726	.5362	1.865	.8813	.8
.3	.4741	.5384	1.857	.8805	.7
.4	.4756	.5407	1.849	.8796	.6
.5	.4772	.5430	1.842	.8788	.5
.6	4787	.5452	1.834	.8780	.4
.7	.4802	.5475	1.827	.8771	.3
.8	.4818	.5498	1.819	.8763	.2
.9	.4833	.5520	1.811	.8755	61.1
29.0	0.4848	0.5543	1.804	0.8746	**61.0**
.1	.4863	.5566	1.797	.8738	60.9
.2	.4879	.5589	1.789	.8729	.8
.3	.4894	.5612	1.782	.8721	.7
.4	.4909	.5635	1.775	.8712	.6
.5	.4924	.5658	1.767	.8704	.5
.6	.4939	.5681	1.760	.8695	.4
.7	.4955	.5704	1.753	.8686	.3
.8	.4970	.5727	1.746	.8678	.2
.9	.4985	.5750	1.739	.8669	60.1
30.0	0.5000	0.5774	1.732	0.8660	**60.0**
Deg.	Cos	Cot	Tan	Sin	Deg.

Deg.	Sin	Tan	Cot	Cos	Deg.
30.0	0.5000	0.5774	1.7321	0.8660	**60.0**
.1	.5015	.5797	1.7251	.8652	59.9
.2	.5030	.5820	1.7182	.8643	.8
.3	.5045	.5844	1.7113	.8634	.7
.4	.5060	.5867	1.7045	.8625	.6
.5	.5075	.5890	1.6977	.8616	.5
.6	.5090	.5914	1.6909	.8607	.4
.7	.5105	.5938	1.6842	.8599	.3
.8	.5120	.5961	1.6775	.8590	.2
.9	.5135	.5985	1.6709	.8581	59.1
31.0	0.5150	0.6009	1.6643	0.8572	**59.0**
.1	.5165	.6032	1.6577	.8563	58.9
.2	.5180	.6056	1.6512	.8554	.8
.3	.5195	.6080	1.6447	.8545	.7
.4	.5210	.6104	1.6383	.8536	.6
.5	.5225	.6128	1.6319	.8526	.5
.6	.5240	.6152	1.6255	.8517	.4
.7	.5255	.6176	1.6191	.8508	.3
.8	.5270	.6200	1.6128	.8499	.2
.9	.5284	.6224	1.6066	.8490	58.1
32.0	0.5299	0.6249	1.6003	0.8480	**58.0**
.1	.5314	.6273	1.5941	.8471	57.9
.2	.5329	.6297	1.5880	.8462	.8
.3	.5344	.6322	1.5818	.8453	.7
.4	.5358	.6346	1.5757	.8443	.6
.5	.5373	.6371	1.5697	.8434	.5
.6	.5388	.6395	1.5637	.8425	.4
.7	.5402	.6420	1.5577	.8415	.3
.8	.5417	.6445	1.5517	.8406	.2
.9	.5432	.6469	1.5458	.8396	57.1
33.0	0.5446	0.6494	1.5399	0.8387	**57.0**
.1	.5461	.6519	1.5340	.8377	56.9
.2	.5476	.6544	1.5282	.8368	.8
.3	.5490	.6569	1.5224	.8358	.7
.4	.5505	.6594	1.5166	.8348	.6
.5	.5519	.6619	1.5108	.8339	.5
.6	.5534	.6644	1.5051	.8329	.4
.7	.5548	.6669	1.4994	.8320	.3
.8	.5563	.6694	1.4938	.8310	.2
.9	.5577	.6720	1.4882	.8300	56.1
34.0	0.5592	0.6745	1.4826	0.8290	**56.0**
.1	.5606	.6771	1.4770	.8281	55.9
.2	.5621	.6796	1.4715	.8271	.8
.3	.5635	.6822	1.4659	.8261	.7
.4	.5650	.6847	1.4605	.8251	.6
.5	.5664	.6873	1.4550	.8241	.5
.6	.5678	.6899	1.4496	.8231	.4
.7	.5693	.6924	1.4442	.8221	.3
.8	.5707	.6950	1.4388	.8211	.2
.9	.5721	.6976	1.4335	.8202	55.1
35.0	0.5736	0.7002	1.4281	0.8192	**55.0**
Deg.	Cos	Cot	Tan	Sin	Deg.

Deg.	Sin	Tan	Cot	Cos	Deg.
35.0	0.5736	0.7002	1.4281	0.8192	**55.0**
.1	.5750	.7028	1.4229	.8181	54.9
.2	.5764	.7054	1.4176	.8171	.8
.3	.5779	.7080	1.4124	.8161	.7
.4	.5793	.7107	1.4071	.8151	.6
.5	.5807	.7133	1.4019	.8141	.5
.6	.5821	.7159	1.3968	.8131	.4
.7	.5835	.7186	1.3916	.8121	.3
.8	.5850	.7212	1.3865	.8111	.2
.9	.5864	.7239	1.3814	.8100	54.1
36.0	0.5878	0.7265	1.3764	0.8090	**54.0**
.1	.5892	.7292	1.3713	.8080	53.9
.2	.5906	.7319	1.3663	.8070	.8
.3	.5920	.7346	1.3613	.8059	.7
.4	.5934	.7373	1.3564	.8049	.6
.5	.5948	.7400	1.3514	.8039	.5
.6	.5962	.7427	1.3465	.8028	.4
.7	.5976	.7454	1.3416	.8018	.3
.8	.5990	.7481	1.3367	.8007	.2
.9	.6004	.7508	1.3319	.7997	53.1
37.0	0.6018	0.7536	1.3270	0.7986	**53.0**
.1	.6032	.7563	1.3222	.7976	52.9
.2	.6046	.7590	1.3175	.7965	.8
.3	.6060	.7618	1.3127	.7955	.7
.4	.6074	.7646	1.3079	.7944	.6
.5	.6088	.7673	1.3032	.7934	.5
.6	.6101	.7701	1.2985	.7923	.4
.7	.6115	.7729	1.2938	.7912	.3
.8	.6129	.7757	1.2892	.7902	.2
.9	.6143	.7785	1.2846	.7891	52.1
38.0	0.6157	0.7813	1.2799	0.7880	**52.0**
.1	.6170	.7841	1.2753	.7869	51.9
.2	.6184	.7869	1.2708	.7859	.8
.3	.6198	.7898	1.2662	.7848	.7
.4	.6211	.7926	1.2617	.7837	.6
.5	.6225	.7954	1.2572	.7826	.5
.6	.6239	.7983	1.2527	.7815	.4
.7	.6252	.8012	1.2482	.7804	.3
.8	.6266	.8040	1.2437	.7793	.2
.9	.6280	.8069	1.2393	.7782	51.1
39.0	0.6293	0.8098	1.2349	0.7771	**51.0**
.1	.6307	.8127	1.2305	.7760	50.9
.2	.6320	.8156	1.2261	.7749	.8
.3	.6334	.8185	1.2218	.7738	.7
.4	.6347	.8214	1.2174	.7727	.6
.5	.6361	.8243	1.2131	.7716	.5
.6	.6374	.8273	1.2088	.7705	.4
.7	.6388	.8302	1.2045	.7694	.3
.8	.6401	.8332	1.2002	.7683	.2
.9	.6414	.8361	1.1960	.7672	50.1
40.0	0.6428	0.8391	1.1918	0.7660	**50.0**
Deg.	Cos	Cot	Tan	Sin	Deg.

Deg.	Sin	Tan	Cot	Cos	Deg.
40.0	0.6428	0.8391	1.1918	0.7660	**50.0**
.1	.6441	.8421	1.1875	.7649	49.9
.2	.6455	.8451	1.1833	.7638	.8
.3	.6468	.8481	1.1792	.7627	.7
.4	.6481	.8511	1.1750	.7615	.6
40.5	0.6494	0.8541	1.1708	0.7604	**49.5**
.6	.6508	.8571	1.1667	.7593	.4
.7	.6521	.8601	1.1626	.7581	.3
.8	.6534	.8632	1.1585	.7570	.2
.9	.6547	.8662	1.1544	.7559	49.1
41.0	0.6561	0.8693	1.1504	0.7547	**49.0**
.1	.6574	.8724	1.1463	.7536	48.9
.2	.6587	.8754	1.1423	.7524	.8
.3	.6600	.8785	1.1383	.7513	.7
.4	.6613	.8816	1.1343	.7501	.6
.5	.6626	.8847	1.1303	.7490	.5
.6	.6639	.8878	1.1263	.7478	.4
.7	.6652	.8910	1.1224	.7466	.3
.8	.6665	.8941	1.1184	.7455	.2
.9	.6678	.8972	1.1145	.7443	48.1
42.0	0.6691	0.9004	1.1106	0.7431	**48.0**
.1	.6704	.9036	1.1067	.7420	47.9
.2	.6717	.9067	1.1028	.7408	.8
.3	.6730	.9099	1.0990	.7396	.7
.4	.6743	.9131	1.0951	.7385	.6
.5	.6756	.9163	1.0913	.7373	.5
.6	.6769	.9195	1.0875	.7361	.4
.7	.6782	.9228	1.0837	.7349	.3
.8	.6794	.9260	1 0799	.7337	.2
.9	.6807	.9293	1.0761	.7325	47.1
43.0	0.6820	0.9325	1.0724	0.7314	**47.0**
.1	.6833	.9358	1.0686	.7302	46.9
.2	.6845	.9391	1.0649	.7290	.8
.3	.6858	.9424	1.0612	.7278	.7
.4	.6871	.9457	1.0575	.7266	.6
.5	.6884	.9490	1.0538	.7254	.5
.6	.6896	.9523	1.0501	.7242	.4
.7	.6909	.9556	1.0464	.7230	.3
.8	.6921	.9590	1.0428	.7218	.2
.9	.6934	.9623	1.0392	.7206	46.1
44.0	0.6947	0.9657	1.0355	0.7193	**46.0**
.1	.6959	.9691	1.0319	.7181	45.9
.2	.6972	.9725	1.0283	.7169	.8
.3	.6984	.9759	1.0247	.7157	.7
.4	.6997	.9793	1.0212	.7145	.6
.5	.7009	.9827	1.0176	.7133	.5
.6	.7022	.9861	1.0141	.7120	.4
.7	.7034	.9896	1.0105	.7108	.3
.8	.7046	.9930	1.0070	.7096	.2
.9	.7059	.9965	1.0035	.7083	45.1
45.0	0.7071	1.0000	1.0000	0.7071	**45.0**
Deg.	Cos	Cot	Tan	Sin	Deg.

APPENDIX C

Values of ϵ^x and ϵ^{-x}

x	Function	0.00	0.01	0.02	0.03	0.04	0.05	0.06	0.07	0.08	0.09
0.0	ϵ^x	1.0000	1.0101	1.0202	1.0305	1.0408	1.0513	1.0618	1.0725	1.0833	1.0942
	ϵ^{-x}	1.0000	0.9900	0.9802	0.9704	0.9608	0.9512	0.9418	0.9324	0.9231	0.9139
0.1	ϵ^x	1.1052	1.1163	1.1275	1.1388	1.1503	1.1618	1.1735	1.1853	1.1972	1.2093
	ϵ^{-x}	0.9048	0.8958	0.8869	0.8781	0.8694	0.8607	0.8521	0.8437	0.8353	0.8270
0.2	ϵ^x	1.2214	1.2337	1.2461	1.2546	1.2712	1.2840	1.2969	1.3100	1.3231	1.3364
	ϵ^{-x}	0.8187	0.8106	0.8025	0.7945	0.7856	0.7788	0.7711	0.7634	0.7558	0.7483
0.3	ϵ^x	1.3499	1.3634	1.3771	1.3910	1.4049	1.4191	1.4333	1.4477	1.4623	1.4770
	ϵ^{-x}	0.7408	0.7334	0.7261	0.7189	0.7118	0.7047	0.6977	0.6907	0.6839	0.6771
0.4	ϵ^x	1.4918	1.5068	1.5220	1.5373	1.5527	1.5683	1.5841	1.6000	1.6161	1.6323
	ϵ^{-x}	0.6703	0.6637	0.6570	0.6505	0.6440	0.6376	0.6313	0.6250	0.6188	0.6126
0.5	ϵ^x	1.6487	1.6653	1.6820	1.6989	1.7160	1.7333	1.7507	1.7683	1.7860	1.8040
	ϵ^{-x}	0.6065	0.6005	0.5945	0.5886	0.5827	0.5769	0.5712	0.5655	0.5599	0.5543
0.6	ϵ^x	1.8221	1.8404	1.8589	1.8776	1.8965	1.9155	1.9348	1.9542	1.9739	1.9939
	ϵ^{-x}	0.5488	0.5434	0.5379	0.5326	0.5273	0.5220	0.5169	0.5117	0.5066	0.5017
0.7	ϵ^x	2.0138	2.0340	2.0544	2.0751	2.0959	2.1170	2.1383	2.1598	2.1815	2.2034
	ϵ^{-x}	0.4966	0.4916	0.4868	0.4819	0.4771	0.4724	0.4677	0.4630	0.4584	0.4538
0.8	ϵ^x	2.2255	2.2479	2.2705	2.2933	2.3164	2.3396	2.3632	2.3869	2.4109	2.4351
	ϵ^{-x}	0.4493	0.4449	0.4404	0.4360	0.4317	0.4274	0.4232	0.4190	0.4148	0.4107
0.9	ϵ^x	2.4596	2.4843	2.5093	2.5345	2.5600	2.5857	2.6117	2.6379	2.6645	2.6912
	ϵ^{-x}	0.4066	0.4025	0.3985	0.3946	0.3906	0.3867	0.3829	0.3791	0.3753	0.3716
1.0	ϵ^x	2.7183	2.7456	2.7732	2.8011	2.8292	2.8577	2.8864	2.9154	2.9447	2.9743
	ϵ^{-x}	0.3679	0.3642	0.3606	0.3570	0.3535	0.3499	0.3465	0.3430	0.3396	0.3362
1.1	ϵ^x	3.0042	3.0344	3.0649	3.0957	3.1268	3.1582	3.1899	3.2220	3.2544	3.2871
	ϵ^{-x}	0.3329	0.3296	0.3263	0.3230	0.3198	0.3166	0.3135	0.3104	0.3073	0.3042
1.2	ϵ^x	3.3201	3.3535	3.3872	3.4212	3.4556	3.4903	3.5254	3.5609	3.5966	3.6328
	ϵ^{-x}	0.3012	0.2982	0.2952	0.2923	0.2894	0.2865	0.2837	0.2808	0.2780	0.2753
1.3	ϵ^x	3.6693	3.7062	3.7434	3.7810	3.8190	3.8574	3.8962	3.9354	3.9749	4.0149
	ϵ^{-x}	0.2725	0.2698	0.2671	0.2645	0.2618	0.2592	0.2567	0.2541	0.2516	0.2491
1.4	ϵ^x	4.0552	4.0960	4.1371	4.1787	4.2207	4.2631	4.3060	4.3492	4.3929	4.4371
	ϵ^{-x}	0.2466	0.2441	0.2417	0.2393	0.2369	0.2346	0.2322	0.2299	0.2276	0.2254
1.5	ϵ^x	4.4817	4.5267	4.5722	4.6182	4.6646	4.7115	4.7588	4.8066	4.8550	4.9037
	ϵ^{-x}	0.2231	0.2209	0.2187	0.2165	0.2144	0.2122	0.2101	0.2080	0.2060	0.2039
1.6	ϵ^x	4.9530	5.0028	5.0531	5.1039	5.1552	5.2070	5.2593	5.3122	5.3656	5.4195
	ϵ^{-x}	0.2019	0.1999	0.1979	0.1959	0.1940	0.1920	0.1901	0.1882	0.1864	0.1845
1.7	ϵ^x	5.4739	5.5290	5.5845	5.6407	5.6973	5.7546	5.8124	5.8709	5.9299	5.9895
	ϵ^{-x}	0.1827	0.1809	0.1791	0.1773	0.1755	0.1738	0.1720	0.1703	0.1686	0.1670
1.8	ϵ^x	6.0496	6.1104	6.1719	6.2339	6.2965	6.3598	6.4237	6.4883	6.5535	6.6194
	ϵ^{-x}	0.1653	0.1637	0.1620	0.1604	0.1588	0.1572	0.1557	0.1541	0.1526	0.1511
1.9	ϵ^x	6.6859	6.7531	6.8210	6.8895	6.9588	7.0287	7.0993	7.1707	7.2427	7.3155
	ϵ^{-x}	0.1496	0.1481	0.1466	0.1451	0.1437	0.1423	0.1409	0.1395	0.1381	0.1367

x	Function	0.00	0.01	0.02	0.03	0.04	0.05	0.06	0.07	0.08	0.09
2.0	ϵ^x	7.3891	7.4633	7.5383	7.6141	7.6906	7.7679	7.8460	7.9248	8.0045	8.8049
	ϵ^{-x}	0.1353	0.1340	0.1327	0.1313	0.1300	0.1287	0.1275	0.1262	0.1249	0.1237
2.1	ϵ^x	8.1662	8.2482	8.3311	8.4149	8.4994	8.5849	8.6711	8.7583	8.8463	8.9352
	ϵ^{-x}	0.1225	0.1212	0.1200	0.1188	0.1177	0.1165	0.1153	0.1142	0.1130	0.1119
2.2	ϵ^x	9.0250	9.1157	9.2073	9.2999	9.3933	9.4877	9.5831	9.6794	9.7767	9.8749
	ϵ^{-x}	0.1108	0.1097	0.1086	0.1075	0.1065	0.1054	0.1044	0.1033	0.1023	0.1013
2.3	ϵ^x	9.9742	10.074	10.716	10.278	10.381	10.486	10.591	10.697	10.805	10.913
	ϵ^{-x}	0.1003	0.0993	0.0983	0.0973	0.0963	0.0954	0.0944	0.0935	0.0926	0.0916
2.4	ϵ^x	11.023	11.134	11.246	11.359	11.473	11.588	11.705	11.822	11.941	12.061
	ϵ^{-x}	0.0907	0.0898	0.0889	0.0880	0.0872	0.0863	0.0854	0.0846	0.0837	0.0829
2.5	ϵ^x	12.182	12.305	12.429	12.553	12.680	12.807	12.936	13.066	13.197	13.330
	ϵ^{-x}	0.0821	0.0813	0.0805	0.0797	0.0789	0.0781	0.0773	0.0765	0.0758	0.0750
2.6	ϵ^x	13.464	13.599	13.736	13.874	14.013	14.154	14.296	14.440	14.585	14.732
	ϵ^{-x}	0.0743	0.0735	0.0728	0.0721	0.0714	0.0707	0.0699	0.0693	0.0686	0.0679
2.7	ϵ^x	14.880	15.029	15.180	15.333	15.487	15.643	15.800	15.959	16.119	16.281
	ϵ^{-x}	0.0672	0.0665	0.0659	0.0652	0.0646	0.0639	0.0633	0.0627	0.0620	0.0614
2.8	ϵ^x	16.445	16.610	16.777	16.945	17.116	17.288	17.462	17.637	17.814	17.993
	ϵ^{-x}	0.0608	0.0602	0.0596	0.0590	0.0584	0.0578	0.0573	0.0567	0.0561	0.0556
2.9	ϵ^x	18.174	18.357	18.541	18.728	18.916	19.106	19.298	19.492	19.688	19.886
	ϵ^{-x}	0.0550	0.0545	0.0539	0.0534	0.0529	0.0523	0.0518	0.0513	0.0508	0.0503
3.0	ϵ^x	20.086	20.287	20.491	20.697	20.905	21.115	21.328	21.542	21.758	21.977
	ϵ^{-x}	0.0498	0.0493	0.0488	0.0483	0.0478	0.0474	0.0469	0.0464	0.0460	0.0455
3.1	ϵ^x	22.198	22.421	22.646	22.874	23.104	23.336	23.571	23.807	24.047	24.288
	ϵ^{-x}	0.0450	0.0446	0.0442	0.0437	0.0433	0.0429	0.0424	0.0420	0.0416	0.0412
3.2	ϵ^x	24.533	24.779	25.028	25.280	25.534	25.790	26.050	26.311	26.576	26.843
	ϵ^{-x}	0.0408	0.0404	0.0400	0.0396	0.0392	0.0388	0.0384	0.0380	0.0376	0.0373
3.3	ϵ^x	27.113	26.385	27.660	27.938	28.219	28.503	28.789	29.079	29.371	29.666
	ϵ^{-x}	0.0369	0.0365	0.0362	0.0358	0.0354	0.0351	0.0347	0.0344	0.0340	0.0337
3.4	ϵ^x	29.964	30.265	30.569	30.877	31.187	31.500	31.817	32.137	32.460	32.786
	ϵ^{-x}	0.0334	0.0330	0.0327	0.0324	0.0321	0.0317	0.0314	0.0311	0.0308	0.0305
3.5	ϵ^x	33.115	33.448	33.784	34.124	34.467	34.813	35.163	35.517	35.874	36.234
	ϵ^{-x}	0.0302	0.0299	0.0296	0.0293	0.0290	0.0287	0.0284	0.0282	0.0279	0.0276
3.6	ϵ^x	36.598	36.966	37.338	37.713	38.092	38.475	38.861	39.252	39.646	40.045
	ϵ^{-x}	0.0273	0.0271	0.0268	0.0265	0.0263	0.0260	0.0257	0.0255	0.0252	0.0250
3.7	ϵ^x	40.447	40.854	41.264	41.679	42.098	42.521	42.948	43.380	43.816	44.256
	ϵ^{-x}	0.0247	0.0245	0.0242	0.0240	0.0238	0.0235	0.0233	0.0231	0.0228	0.0226
3.8	ϵ^x	44.701	45.150	45.604	46.063	46.525	46.993	47.465	47.942	48.424	48.911
	ϵ^{-x}	0.0224	0.0221	0.0219	0.0217	0.0215	0.0213	0.0211	0.0209	0.0207	0.0204
3.9	ϵ^x	49.402	49.899	50.400	50.907	51.419	51.935	52.457	52.985	53.517	54.055
	ϵ^{-x}	0.0202	0.0200	0.0198	0.0196	0.0195	0.0193	0.0191	0.0189	0.0187	0.0185

x	Function	0.00	0.01	0.02	0.03	0.04	0.05	0.06	0.07	0.08	0.09
4.0	ϵ^x	54.598	55.147	55.701	56.261	56.826	57.397	57.974	58.557	59.145	59.740
	ϵ^{-x}	0.0183	0.0181	0.0180	0.0178	0.0176	0.0174	0.0172	0.0171	0.0169	0.0167
4.1	ϵ^x	60.340	60.947	61.559	62.178	62.803	63.434	64.072	64.714	65.366	66.023
	ϵ^{-x}	0.0166	0.0164	0.0162	0.0161	0.0159	0.0158	0.0156	0.0155	0.0153	0.0151
4.2	ϵ^x	66.686	67.357	68.033	68.717	69.408	70.105	70.810	71.522	72.240	72.966
	ϵ^{-x}	0.0150	0.0148	0.0147	0.0146	0.0144	0.0143	0.0141	0.0140	0.0138	0.0137
4.3	ϵ^x	73.700	74.440	75.189	75.944	76.708	77.478	78.257	79.044	79.838	80.640
	ϵ^{-x}	0.0136	0.0134	0.0133	0.0132	0.0130	0.0129	0.0128	0.0127	0.0125	0.0124
4.4	ϵ^x	81.451	82.269	83.096	83.931	84.775	85.627	86.488	87.357	88.235	89.121
	ϵ^{-x}	0.0123	0.0122	0.0120	0.0119	0.0118	0.0117	0.0116	0.0114	0.0113	0.0112
4.5	ϵ^x	90.017	90.922	91.836	92.759	93.691	94.632	95.583	96.544	97.514	98.494
	ϵ^{-x}	0.0111	0.0110	0.0109	0.0108	0.0107	0.0106	0.0105	0.0104	0.0103	0.0102
4.6	ϵ_x	99.484	100.48	101.49	102.51	103.54	104.58	105.64	106.70	107.77	108.85
	ϵ^{-x}	0.0101	0.0100	0.0099	0.0098	0.0097	0.0096	0.0095	0.0094	0.0093	0.0092
4.7	ϵ^x	109.95	111.05	112.17	113.30	114.43	115.58	116.75	117.92	119.10	120.30
	ϵ^{-x}	0.0091	0.0090	0.0089	0.0088	0.0087	0.0087	0.0086	0.0085	0.0084	0.0083
4.8	ϵ^x	121.51	122.73	123.97	125.21	126.47	127.74	129.02	130.32	131.63	132.95
	ϵ^{-x}	0.0082	0.0081	0.0081	0.0080	0.0079	0.0078	0.0078	0.0077	0.0076	0.0075
4.9	ϵ^x	134.29	135.64	137.00	138.38	139.77	141.17	142.59	144.03	145.47	146.94
	ϵ^{-x}	0.0074	0.0074	0.0073	0.0072	0.0072	0.0071	0.0070	0.0069	0.0069	0.0068
5.0	ϵ^x	148.41	149.90	151.41	152.93	154.47	156.02	157.59	159.17	160.77	162.39
	ϵ^{-x}	0.0067	0.0067	0.0066	0.0065	0.0065	0.0064	0.0063	0.0063	0.0062	0.0062
5.1	ϵ^x	164.02	165.67	167.34	169.02	170.72	172.43	174.16	175.91	177.68	179.47
	ϵ^{-x}	0.0061	0.0060	0.0060	0.0059	0.0059	0.0058	0.0057	0.0057	0.0056	0.0056
5.2	ϵ^x	181.27	183.09	184.93	186.79	188.67	190.57	192.48	194.42	196.37	198.34
	ϵ^{-x}	0.0055	0.0055	0.0054	0.0054	0.0053	0.0052	0.0052	0.0051	0.0051	0.0050
5.3	ϵ^x	200.34	202.35	204.38	206.44	208.51	210.61	212.72	214.86	217.02	219.20
	ϵ^{-x}	0.0050	0.0049	0.0049	0.0048	0.0048	0.0047	0.0047	0.0047	0.0046	0.0046
5.4	ϵ^x	221.41	223.63	225.88	228.15	230.44	232.76	235.10	237.46	239.85	242.26
	ϵ^{-x}	0.0045	0.0045	0.0044	0.0044	0.0043	0.0043	0.0043	0.0042	0.0042	0.0041
5.5	ϵ^x	244.69	247.15	249.64	252.14	254.68	257.24	259.82	262.43	265.07	267.74
	ϵ^{-x}	0.0041	0.0040	0.0040	0.0040	0.0039	0.0039	0.0038	0.0038	0.0038	0.0037
5.6	ϵ^x	270.43	273.14	275.89	278.66	281.46	284.29	287.15	290.03	292.95	295.89
	ϵ^{-x}	0.0037	0.0037	0.0036	0.0036	0.0036	0.0035	0.0035	0.0034	0.0034	0.0034
5.7	ϵ^x	298.87	301.87	304.90	307.97	311.06	314.19	317.35	320.54	323.76	327.01
	ϵ^{-x}	0.0033	0.0033	0.0033	0.0032	0.0032	0.0032	0.0032	0.0031	0.0031	0.0031
5.8	ϵ_x	330.30	333.62	336.97	340.36	343.78	347.23	350.72	354.25	357.81	361.41
	ϵ^{-x}	0.0030	0.0030	0.0030	0.0029	0.0029	0.0029	0.0029	0.0028	0.0028	0.0028
5.9	ϵ^x	365.04	368.71	372.41	376.15	379.93	383.75	387.61	391.51	395.44	399.41
	ϵ^{-x}	0.0027	0.0027	0.0027	0.0027	0.0026	0.0026	0.0026	0.0026	0.0025	0.0025

APPENDIX D

Meters

In working with electricity it is very important to be able to understand and properly use devices which can measure the electrical quantities current, voltage, resistance, and power. The devices used to measure these quantities are, respectively: ammeters (or milliammeters), voltmeters, ohmmeters, and wattmeters.

1. The Moving Coil (d'Arsonval) Meter Movement

The basis of most modern meters is the d'Arsonval moving-coil meter movement. The principle of its operation is illustrated in Fig. D-1. The moving coil is wound on a small magnetic core called the *armature* which is suspended by pivots so that it is free to rotate between two poles of a permanent magnet. A spiral spring is used to hold the armature and the pointer attached to it against a stop (on the left) when the movement is not in use. When a constant dc current is passed through the coil in the direction shown, the armature becomes an electromagnet whose magnetic field is established with a north pole on the left end and a south pole on the right end. This can be verified using the left-hand rule for solenoids (chap. 14).

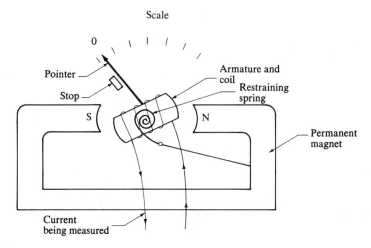

FIGURE D-1

The magnetic fields of the permanent magnet and the armature will interact such that the armature will be forced to rotate clockwise. This is easily seen from the principle that "unlike magnetic poles attract each other." Since the left end of the armature is made a north pole by the current in the coil, it will be attracted by the south pole of the permanent magnet. Similarly, the right end of the armature will be attracted toward the north pole of the permanent magnet. If there is sufficient current through the coil, this force of attraction will overcome the spring's restraining force and will cause the armature and the attached pointer to move in a clockwise (cw) direction. The greater the coil current, the farther the pointer will move across the scale. The pointer rotation is proportional to the current in the coil.

If a current is passed through the coil in the opposite direction, the armature becomes magnetized in the opposite polarity. In this situation it is obvious that the armature and pointer will be forced to move ccw due to the repulsion of the permanent magnet. However, they are kept from moving leftward by the stop. If too large of a current is passed in this direction, the large force could cause the pointer mechanism to become damaged.

From this discussion, it is apparent that the basic moving-coil meter movement cannot be used for alternating current since the pointer will read up-scale only for one direction of current. As we shall see later, though, a simple modification can make the movement useful for ac measurements.

Meter Movement Characteristics The meter movement has two important characteristics: its *full-scale current rating* (sensitivity) and its *internal resistance*. The full-scale current is the value of current through the coil which will make the pointer read at the highest mark on the scale. This maximum current is given the symbol I_M and depends on several factors including the design of the coil, the strength of the permanent magnet, and the tension in the spring. Meter movements are available with I_M values that are mostly standard decimal units (i.e., 100 A, 10 A, 1 A, 100 mA, 10 mA, 1 mA, 100 μA, etc.).

The value of I_M for a given meter movement represents the current needed for a full-scale deflection. For currents lower than I_M, the pointer deflection will be proportionately less. For example, if a 100-μA meter movement is passing a current of 25 μA, the pointer will be deflected one-quarter of full scale; if the meter is passing 75 μA, the pointer will be deflected three-quarters of full scale. In other words, the meter's deflection is proportional to the current so that the scale is a *linear scale*.

The current which is being measured by the moving-coil meter movement must pass through the coil on the armature. This coil possesses some resistance, which is called the movement's *internal resistance R_M*. The value of this resistance is typically small (0-500 ohms). A current meter is placed in series with the circuit in which it is to measure the current. As such, the meter's resistance R_M is *added* to the circuit's total resistance and can sometimes have a significant effect on the circuit current. This effect will be discussed in more detail later. In general, however, R_M is kept as *small* as possible to minimize its effect on the circuit being measured.

Increasing Current Ranges The maximum current which a meter movement can read is I_M. However, larger values of current can be measured if a *meter shunt* is used. Figure D-2 illustrates. In the figure, the meter movement shown has a full-scale current of 1 mA and $R_M = 18$ ohms. Thus, the meter's pointer will read full scale when the current through the *movement* is 1 mA. A resistor $R_S = 2$ ohms is placed in parallel (shunt) with the meter movement. R_S is called a meter shunt and its purpose is to bypass current around the meter movement. The parallel combination connected between points *a-b* forms a *new* meter whose maximum current will be 10 mA. This can be reasoned as follows:

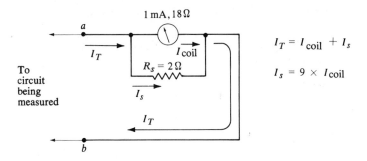

FIGURE D-2

(1) Since $R_S = 2$ ohms is *nine* times smaller than R_M, the current through the shunt (I_S) will be *nine* times greater than the current through the meter movement's coil (I_{coil}).

(2) Thus, when $I_{coil} = I_M = 1$ mA (full-scale deflection), the current through the shunt will be $I_S = 9$ mA.

(3) The total current I_T through the *new* meter formed by the parallel combination is equal to $I_{coil} + I_S = 10$ mA when the meter movement is at its full-scale deflection. Thus, with the shunt present the new meter behaves as if $I_M = 10$ mA and its scale can be calibrated accordingly.

As these steps explain, a meter shunt is used to extend the range of a meter movement beyond its own value of I_M. Most of use are familiar with VOMs which employ several different current ranges. In these instruments, each time a different current range is selected, a new value of R_S is shunted across the meter movement. The following example illustrates how the values of R_S are chosen.

EXAMPLE: A certain meter movement has $I_M = 50$ μA and $R_M = 900$ ohms. Determine the necessary values of R_S needed to extend the meter range to (a) 200 μA; (b) 1 mA; (c) 120 mA.

Solution: (a) $I_T = I_{coil} + I_S$

$$200 \text{ μA} = 50 \text{ μA} + I_S$$

Therefore,

$$I_S = 150 \text{ μA}$$

which is three times *larger* than I_{coil}. Therefore, R_S has to be three times *smaller* than R_M. **$R_S = 300$ ohms.**

(b) $I_S = I_T - I_{coil} = 1$ mA $- 50$ μA
$= 1000$ μA $- 50$ μA

$$I_S = 950 \text{ μA}$$

Since $I_S = 19 \times I_{coil}$, then $R_S = \dfrac{R_M}{19} \approx$ **47.3 ohms.**

(c) $I_S = I_T - I_{coil} = 120{,}000 \ \mu A - 50 \ \mu A$
$\approx 120{,}000 \ \mu A$

Since $I_S = 2400 \times I_{coil}$, then $R_S = \dfrac{R_M}{2400} = 0.375$ ohms.

Figure D-3 illustrates one method for switching the current meter between these three ranges.

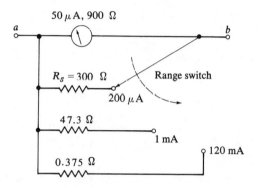

FIGURE D-3

2. Voltmeters

The basic meter movement that we have discussed responds to the flow of current and is therefore basically an ammeter. However, the addition of a high *series* resistance can convert it to a voltmeter. Recall that a voltmeter is always connected in *parallel* with the circuit whose voltage is being measured. Thus, the total internal resistance of a voltmeter should be kept as large as possible. Figure D-4 illustrates how a 1-mA, 100-ohm meter movement is converted to a voltmeter with a full-scale deflection of 10 V. A resistor $R_1 = 9900$ ohms has been connected in series with the meter movement. The terminals *a-b* of this series combination form the terminals of the voltmeter.

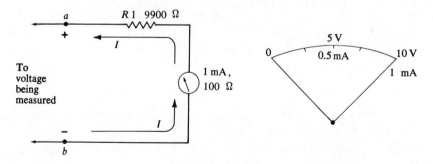

FIGURE D-4

Any voltage applied to terminals *a-b* will cause a current to flow through R_1 and the meter movement. For example, when $V_{ab} = 10$ V the current will be

$$I = \frac{10\text{ V}}{R_1 + R_M} = \frac{10\text{ V}}{10{,}000\text{ ohms}} = 1\text{ mA}$$

This 1 mA flows through the meter movement and produces a full-scale deflection. When, $V_{ab} = 5$ V the current will be

$$I = \frac{5\text{ V}}{10{,}000\text{ ohms}} = 0.5\text{ mA}$$

and the meter will deflect to half-scale. When $V_{ab} = 2.5$ V the current will be 0.25 mA and the meter will deflect to quarter-scale and so forth. Thus, the meter indication is proportional to the voltage V_{ab} being measured, so that the meter scale can be calibrated to read in volts rather than mA (see Fig. D–4).

The range of the voltmeter can be changed by changing the value of R_1. For this reason, R_1 is referred to as a *multiplier* resistor. The following examples will illustrate how the values of R_1 are chosen to produce different voltmeter ranges.

EXAMPLE: A 100-μA, 1000-ohm meter movement is to be used as a voltmeter. Choose proper values for multiplier resistors to provide the following full-scale voltage ranges: (a) 0.5 V; (b) 5 V; (c) 50 V.

Solution: (a) The full-scale voltage (0.5 V) has to produce the full-scale meter movement current (100 μA). Therefore,

$$\frac{0.5\text{ V}}{R_M + R_1} = 100\ \mu\text{A}$$

or

$$R_M + R_1 = \frac{0.5\text{ V}}{100\ \mu\text{A}} = 5000\text{ ohms}$$

Since $R_M = 1000$ ohms, then $R_1 = 5000 - 1000 = \mathbf{4000}$ **ohms**.
(b) Using similar reasoning for 5-V full scale,

$$R_M + R_1 = \frac{5\text{ V}}{100\ \mu\text{A}} = 50\text{ k}\Omega$$

$$R_1 = 49\text{ k}\Omega$$

(c) $\quad R_M + R_1 = \dfrac{50\text{ V}}{100\ \mu\text{A}} = 500\text{ k}\Omega$

Therefore,

$$R_1 = 499\text{ k}\Omega$$

Figure D–5 shows the switching arrangement which can be used to allow switching between these various ranges.

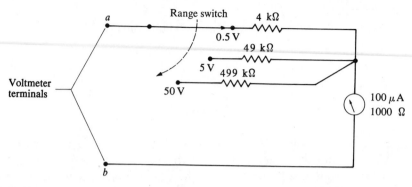

FIGURE D-5

Voltmeter Ohms-per-Volt Rating In the preceding example, a meter movement with $I_M = 100$ μA was used as part of the voltmeter. To convert this meter movement to a voltmeter with a full-scale deflection of V_M, it was necessary to provide the exact amount of total series resistance which would limit the current to I_M when V_M was applied. In equation form,

$$R_{total} = R_1 + R_M = V_M/I_M$$

This shows that R_{total} of a voltmeter depends on the meter movement's current sensitivity I_M. For each volt of the voltmeter's full-scale deflection V_M, a resistance equal to $1/I_M$ Ω is needed. This quantity $(1/I_M)$ is called the voltmeter's *ohms-per-volt* rating.

$$\Omega/V = 1/I_M$$

For the meter in the previous example (Fig. D-5)

$$\Omega/V = 1/100 \text{ μA} = 10{,}000 \text{ Ω}/V$$

If a meter movement with $I_M = 1$ mA had been used, then

$$\Omega/V = 1/1 \text{ mA} = 1000 \text{ Ω}/V$$

This shows that a meter movement with a lower I_M will produce a voltmeter with a larger Ω/V rating.

The Ω/V rating (also called the voltmeter's sensitivity) is useful in helping to determine the internal resistance of the voltmeter on a given voltage range. For example, the voltmeter in Fig. D-5 has a Ω/V rating of 10,000 Ω/V. Thus, when it is used on the 5-V full-scale range, its total internal resistance is obtained by multiplying 5 V times 10,000 Ω/V = 50,000 ohms. In general, then,

$$R_{total} = V_M \times (\Omega/V \text{ rating})$$

where V_M is *always the full-scale voltage.*

EXAMPLE: A certain voltmeter has a sensitivity of 25,000 Ω/V. The meter is being used on the 20-V scale and it reads 14 V. What is the meter's internal resistance?

Solution: Although the reading is only 14 V, the internal resistance is determined by V_M, the full-scale voltage. Thus,

$$R_{total} = 20 \text{ V} \times 25{,}000 \text{ }\Omega/\text{V} = 500{,}000 \text{ ohms}$$

The Ω/V rating of most voltmeters is usually specified on the face of the meter so that the user can compute total internal resistance for whichever range is being used. Some voltmeters (VTVMs, FETVMs, DVMs) have a fixed internal resistance which is the same for all ranges.

3. The Ohmmeter

An ohmmeter can be constructed from the basic meter movement by the addition of a series voltage source (usually a battery) and an adjustable resistor as illustrated in Fig. D–6. The ohmmeter terminals are *a-b*. When these terminals are applied to an unknown resistor R_X, a current I will flow through the meter. Maximum current flows when *a-b* are short-circuited ($R_X = 0$). The variable resistor R_Z is used to adjust this short-circuited current so that $I = I_M$ and the meter indicates full scale. Therefore, as R_X is increased from zero the current will decrease and the meter will read below full scale. Of course, as R_X gets very large the current will become essentially zero and the meter will be at zero. Thus, as R_X gets larger the meter deflection gets smaller. The meter scale can be calibrated in ohms (see Fig. D–6) so that the value of R_X can be read from the scale directly. Notice that the scale is a backwards scale (increasing from right to left) and is a nonlinear scale.

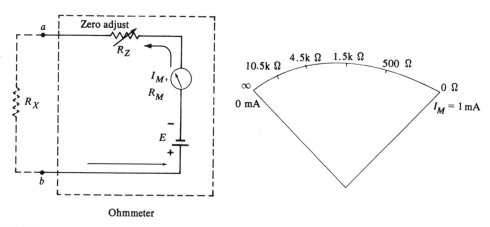

FIGURE D–6

An ohmmeter is never applied to a circuit which contains sources. The voltages in the circuit could produce excessive current through the meter movement and permanently damage it. The ohmmeter in Fig. D–6 can be used to measure resistances in the mid-range (100 ohms to 10-kilohms). For very large or very small resistances, a series-parallel arrangement is used.

4. Meter Loading

Perhaps the most important measurement aspect which a technician should understand is the loading effect caused by a meter's internal resistance. To illustrate the effect of loading caused by an *ammeter*, refer to Fig. D-7(a) where a 1-A current meter is being used to measure the current in a simple series circuit. Let's assume that this meter has an internal resistance of 2 ohms (including any meter shunts). We know from using Ohm's law that the current in the circuit (without the meter) should be 6 V/6 ohms = 1 A. The meter, however, adds its own 2 ohms to the circuit so that with the meter present the current will be 6 V/8 ohms = 0.75 A and this is what the meter will read. The reading is obviously not correct because of the *loading* effect caused by the meter's internal resistance. In Fig. D-7(b) the same circuit is shown using a meter with an internal resistance of only 0.1 ohm. Using Ohm's law, the current will be 6 V/6.1 ohms = 0.983 A, which is much closer to the correct value.

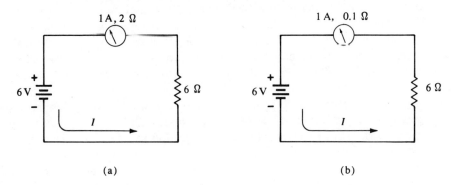

FIGURE D-7

From this discussion it is apparent that to minimize the effect of ammeter loading, the ammeter's internal resistance should be much less than the resistance in the circuit being measured. As a rule of thumb, if the ammeter's resistance, is at least *ten* times smaller than the resistance in the branch where it is inserted, then the error caused by meter loading will be less than 10%. If the ammeter's resistance is at least *twenty* times smaller, the error is only 5% or less.

EXAMPLE: Which of the ammeters in the circuit of Fig. D-8 is going to cause significant loading?

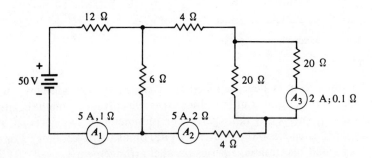

FIGURE D-8

Solution: Meter A_2 will cause loading since it has $R_M = 2$ ohms and is in series with a total of 4 ohms + 4 ohms + (20 || 20) ohms = 18 ohms.

Voltmeter Loading A voltmeter typically has a large internal resistance, but in some cases it may not be high enough and could cause a loading effect. Figure D–9(a) illustrates. A simple voltage divider circuit is shown and the voltage V_{xy} is to be measured. The circuit can be replaced by its Thevenin equivalent between points x-y as shown in part (b) of the figure (reader can verify Thevenin values). Thus, a voltmeter placed between points x-y sees the Thevenin equivalent circuit with $E_{Th} = 5$ V and $R_{Th} = 5$ kΩ. Ideally, the voltmeter should read a voltage equal to $E_{Th} = 5$ V. However, because of the meter's noninfinite resistance, a voltage divider is present between R_{Th} and R_M so that $V_{xy} < E_{Th}$. For example, with $R_M = 20$ kΩ the value of V_{xy} read by the meter will be

$$V_{xy} = \frac{20 \text{ k}\Omega}{5 \text{ k}\Omega + 20 \text{ k}\Omega} \times 5 \text{ V} = 4.0 \text{ V}$$

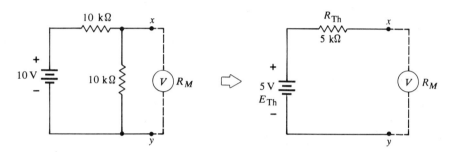

FIGURE D–9

which is much less than the accurate value of 5 V.

If R_M were increased to 200 kΩ, the measured V_{xy} would now be

$$V_{xy} = \frac{200}{5 + 200} \times 5 = 4.88 \text{ V}$$

which is much more accurate. Thus, the loading effect of a voltmeter becomes less as the meter's resistance is increased. For a voltmeter, a *large* internal resistance is desirable. Recall that for an ammeter a *small* internal resistance is desirable.

As a rule of thumb, a voltmeter's loading effect will be less than 10% if $R_M \geq 10\ R_{Th}$; it will be less than 5% if $R_M \geq 20\ R_{Th}$. The larger the R_M, the better.

5. AC Meters

Thus far we have discussed only the moving-coil type of meter movement which can operate only with direct current. We will now consider the measurement of alternating current and voltage. Meters which are used for ac measurements are basically of two types: dc meters which use a diode rectifier to change ac to pulsating dc; and meters which are fundamentally designed to operate with ac or with both dc and ac.

Rectified Meters A diode rectifier is a device which conducts current in only one direction. When an ac voltage is applied to the circuit in Fig. D–10, the diode allows only the positive portion of voltage to reach the voltmeter terminals. Thus, the dc voltmeter receives a dc voltage. The voltmeter deflection will be proportional to the average value of the pulsating dc waveform. The meter scale, however, can be calibrated to read rms or peak value as desired.

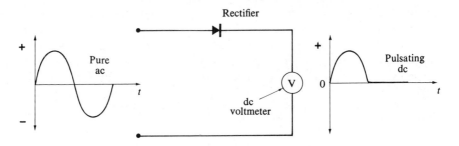

FIGURE D-10

A more complicated rectifier arrangement is needed to convert a dc ammeter to read ac. Essentially, the dc current meter is placed across a *bridge rectifier* circuit which is inserted in the circuit to be measured.

Electrodynamometer One of the most common meter movements which can be used directly with ac is the electrodynamometer. Its operating principle is illustrated in Fig. D-11. The construction is very similar to the moving-coil meter movement except that the horseshoe magnet is *not* a permanent magnet, but is also an electromagnet. The wire is coiled around both the fixed horseshoe and the movable armature. The coils are wound such that no matter what the direction of current is, the two magnets will repel each other so that the pointer rotates cw. This action is the same for dc or ac, and this meter movement can be used for both. Since the same current is used twice (once in the fixed coil and once in the moving coil), the force and meter deflection are proportional to the *square* of the current. The meter scale is accordingly a nonlinear one.

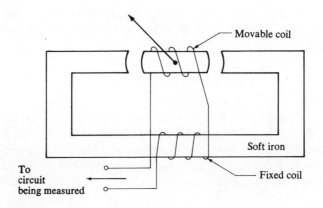

FIGURE D-11

The reader may well be wondering about the pointer deflection for ac. It might seem as though the pointer should move back and forth as the ac voltage changes. However, for relatively high frequencies (>10 Hz) the pointer cannot move fast enough to follow the ac variations. Instead, it responds to the average voltage and vibrates around it. At high enough frequencies the vibrations are not detectable.

Wattmeters A wattmeter measures *real* or *true* power dissipation. One type of wattmeter that is used for low-frequency power measurements uses the electrodynamometer principle. Figure D–12 shows such a wattmeter. The fixed coil and movable coil are now separated. The moving coil is used as an ammeter (in series with load) and the fixed coil is used as a voltmeter (across the load). Note the presence of a multiplier resistor R_1 for the voltmeter coil. In this arrangement, then, the magnetic field in the fixed coil is proportional to the voltage and the magnetic field in the moving coil is proportional to the current. The two magnetic fields interact to deflect the pointer by an amount which is proportional to the product $e \times i$. For high enough ac frequencies (>10 Hz) the pointer deflects proportional to the average of $e \times i$ product. This represents *average power* or *true power*. Thus, the wattmeter can be used to measure true power for any type of load regardless of the power factor. Of course, it can also measure dc power $E \times I$.

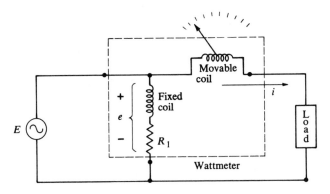

FIGURE D–12

ANSWERS TO SELECTED PROBLEMS

Chapter One

1–3 a
1–11 less angle

Chapter Two

2–3 (a) 0.5 N
(b) same
(c) 2 N, attraction
2–7 16 J
2–9 (a) 5 V
(b) 5 V
2–18 (a) 10 A
2–20 0.01 s
2–22 1 C
2–29 24 A
2–30 300 Ω
2–31 15 V
2–32 2880 W
2–33 0.27 W
2–34 0.625 A

Chapter Three

3–1 (a) 0.0000000001
(b) 100000
3–2 (a) 10^{-5}
(b) 10^{+5}
3–4 (a) 5.63×10^{-5}
(b) 1.294×10^7
(c) 3.77×10^1
3–5 (a) 735
(b) 0.0000815
(c) 66,530,000
3–6 (a) 1.2×10^4
(b) 2.4×10^{-3}
(c) 4.37×10^5
3–7 (a) 2.73×10^{-1}
(b) 2.61×10^{-8}
(c) 2.7×10^0
(e) 8×10^1
(f) -5×10^{-6}
3–8 (a) 1.8×10^{-3}
(d) 7×10^3
3–9 2×10^{-2} A
3–10 (a) 4.11×10^{-1}
(b) 3.77×10^2
3–13 (a) 257 mA
(b) 5.247 KV
(c) 0.037 μC
(d) 25 nA, .025 μA
(e) 4753 KΩ
(f) 950,000 pA
(g) 9.275 V
(h) 15 A
(i) 4.9×10^{-6} C
(j) 25×10^6 V
(k) 2.5×10^{-3} μC
(l) 95,000 MV
3–14 175 pC
3–15 94 V
3–16 4 μA
3–17 (a) 6 mA
(b) 20 μA
(c) 0.75 mA
(d) 0.1 mA

Chapter Four

4–1 (a) +3 V
(b) −3 V
(c) +3 V
4–2 (a) +8 V
(b) −5 V
(c) 0 V
4–3 $V_{FA} = -9$ V
$V_{AF} = +9$ V
$V_{AD} = +6$ V
$V_{CF} = +11$ V
4–4 +1 V
4–5 +6 V
4–6 (a) cw

Answers to Selected Problems

    (b) cw
    (c) no current
4–7 10 mA
4–9 +24 V, +9 V
4–10 −15 V, −24 V
4–11 −12 mA

Chapter Five

5–3 1.25 mS
5–4 h
5–5 800 ohms
5–6 a and c
5–10 (a) 6 µA
     (b) 0.4 A
5–11 230 V
5–12 2 MΩ
5–14 (a) 12 mA
     (b) zero
5–15 (a) 15.4 mS; 65 Ω
     (b) 56 mS; 17.9 Ω
     (c) 0 S; ∞ Ω
5–17 $r = 800\ \Omega;\ g = 1.25$ mS
5–18 $r = 3.125\ \Omega;\ g = 0.32$ S

Chapter Six

6–5 four
6–7 1.5 mA
6–8 2 V
6–10 (a) 40 mA
6–11 (a) 0.5 A
     (d) 22.2 µA
6–12 (a) 50 mV
     (d) 4.5 kV
6–13 (a) 12.5 V, 5 V, 7.5 V
6–15 $V_{R1} = 10$ V for $R1 = 200\ \Omega$
6–18 (a) 13.3 mA; 4 V, 2 V, 13.3 V, 2.67 V
     (b) 1 mA (ccw)
6–21 2.5 kΩ, 5 V
6–28 $V_{CB} = -11$ V
6–29 120 Ω, 90 Ω
6–31 (a) −40 V
     (b) −48 V
     (c) −24 V; −6 V
6–33 13.33 V
6–36 some possibilities:
     $R1 = 15$ kΩ
     $R2 = 2.5$ kΩ + 2.5 kΩ
     or
     $R1 = 3$ kΩ (three 1 kΩs)
     $R2 = 1$ kΩ
6–38 (a) 3 V
     (b) 4.8 V
     (c) 4.5 V
     (d) 0 V
6–40 $E_{in} = 58.5$ V
6–41 $V_x = 12$ V
6–42 6 V, 12 V, −18 V, −30 V
6–43 (a) 28.8 V
     (b) 0 V
6–44 9 V
6–47 110 Ω
6–49 15.3 Ω

Chapter Seven

7–3 3 mA, 5 mA, 2 mA, 10 mA; 3 kΩ
7–4 .33 mS, 3 kΩ
7–5 0.2 A
7–6 571 KΩ, 1.75 mA
7–7 (a) 25.5 Ω
     (b) 667 Ω
     (c) and (d) 1.88 kΩ
7–9 200 Ω
7–10 7 mA
7–12 16 kΩ
7–14 −1.25 mA
7–15 32 µA
7–17 (a) 2.5 mA
     (d) 16 mA
7–19 (a) 2.33 A, 2.33 A, 9.32 A
     (b) 10 mA, 30 mA
     (c) 1 mA, 9 mA
7–21 1 MΩ, 3 MΩ
7–22 7 A, 0, 7 A
7–23 0, 14 A, 0
7–24 8.3 A

Chapter Eight

8–2 $I_1 = 12$ mA, $I_2 = 4$ mA, $I_T = 16$ mA,
     $V_{wx} = 8$ V, $V_{xy} = 24$ V
8–3 $I_T = 0.214$ A
8–4 $I_T = 40$ mA, $V_{yz} = 4$ V
8–5 $I_T = 7.25$ mA, $V_{4k} = 16$ V
8–6 $V_{xy} = 10$ V
8–9 (a) 9 Ω
     (b) 1 kΩ
     (c) 120 Ω
8–11 120 V
8–14 4 kΩ resistor is faulty
8–15 1500 Ω
8–16 +2 V, −1.3 V
8–19 10 Ω, 43 Ω

Chapter Nine

9–3 50 W
9–4 180,000 J
9–5 3.6 W
9–7 58 mA
9–9 10.5 mW
9–10 47 mW
9–11 2.5 V
9–12 1.44 mW
9–13 48 mW
9–14 1.2 V
9–16 $R1$
9–17 7.5 Ω
9–22 23.3 %
9–23 9.3 A
9–24 25 W
9–25 720 mW
9–27 3.6×10^6 J
9–28 12

Chapter Ten

10–3 13 Ω

10–4 16.9 Ω
10–6 b
10–7 1.16 V @ 40°
10–8 (a) 8.8 A, 10.56 V
10–9 15 A, 48 Ω

Chapter Eleven

11–1 0.13 A, 5.2 V
11–2 0.155 A
11–3 2 V
11–4 24 V
11–5 $E_{th} = 15.6$ V, $R_{th} = 80$ Ω
11–6 270 mW
11–7 3 kΩ
11–8 0 V
11–9 $I_D = 0$
11–11 2.5 mA for $R_L = 10$ kΩ
11–13 1.8 V
11–15 $0 \leq R_L \leq 11$ kΩ
11–16 $I_N = 25$ mA
11–21 4.6 V, 1.48 kΩ
11–22 325 kΩ, 650 kΩ, 433 kΩ
11–23 6 kΩ, 10 kΩ, 15 kΩ
11–24 1 A
11–27 $I_1 = I_3 = 20$ mA, $I_2 = 0$
11–29 $V_x = 0$
11–30 $V_a = 4.3$ V, $V_c = 6.1$ V, $I_1 = 86$ μA, $I_2 = -68$ μA, $I_3 = -18$ μA, $I_4 = -59$ μA, $I_5 = 41$ μA

Chapter Twelve

12–3 (a) ≈ 80°
12–6 7 V, 0 V, −10 V
12–7 0.7 ms, 3.3 ms, 4.7 ms, 7.3 ms
12–10 12 V, 24 V
12–11 (a) 120 V
       (b) 33.8 V
12–12 5.5 mA, 20 μs
12–14 (a) 15 ms, 66.6 Hz
       (b) 8 μs, 125 kHz
       (c) 200 ns, 5 MHz
12–16 50 Hz
12–21 47 mA
12–23 2.5 mW
12–24 1.88 W
12–25 1.25 A

Chapter Thirteen

13–2 c
13–6 2.5 μF
13–7 5 mC
13–9 c
13–15 60 mA
13–16 8 V
13–17 (a) 12 kV /s
       (b) 4000 V /s, 2000 V /s
13–19 50 V
13–21 $\tau = 1$ ms
13–22 7.56 V, 22.2 mA
13–23 $v_{cap} = -22.8$ V at $t = 180$ ns
       $v_R = -3.4$ V at $t = 120$ ns
13–24 $v_{xy} = -23.1$ V, $v_{wx} = -0.9$ V at $t = 180$ ns
13–25 c
13–26 (b) −400 V /s, (c) −56 V /s, (d) 0.6 V
13–27 (a) 6.4 V, (b) 0.36 ms (c) 1.62 mA
13–28 10.6 V
13–29 (b) −0.3 V
       (c) 1.05 ms
13–31 (a) $50 (1 - \varepsilon^{-t/\tau})$
13–32 $50 - 40\varepsilon^{-t/\tau}$
13–33 (a) 6.29 V
       (b) 0.346 ms
       (c) 1.6 mA
13–34 10.34 V
13–35 (c) 160 μF
13–37 163.5 V, 81.75 V, 54.5 V
13–38 −2.7 V, 109 V
13–39 10.67 V
13–40 2.22 pF
13–42 0.04 mil
13–45 b, c, f
13–47 1.5 MΩ
13–48 1100 s
13–49 0.0144 J; 0.0144 J
13–51 (b) 0.5 J

Chapter Fourteen

14–17 flux is ccw
14–29 80 μWb
14–31 15 Wb/m²
14–32 2×10^{-4} Wb
14–35 5 A-t
14–37 C
14–38 a; $H = 600$ A-t/m
14–40 (a) 3.15×10^{-4} T
       (b) .378 T
14–41 $\mu = 2.2 \times 10^{-4}$
       $\mu_r = 180$
14–42 (a) 600 A-t
       (d) 5.7×10^{-4} Wb
14–50 (a) 1.05×10^6 A-t/Wb
       (b) 1.05×10^6 A-t/Wb
14–51 (d) $B = 3.3$ T
14–52 $\mathcal{R}$ (total) $= 6.66 \times 10^6$ A-t/Wb
       $\phi = 9 \times 10^6$ Wb
       $B = 0.03$ T

Chapter Fifteen

15–4 (a) 0.1 V
      (b) 0.2 V
      (c) 0.4 V
15–6 (a) 0.02 Wb
      (b) 500 V, 2000 V
15–9 Bar magnet has N-pole on left end
15–10 x positive relative to y
       a positive relative to b
15–12 d
15–14 3×10^2 mH
15–16 5 Ω
15–17 2 H
15–19 −24 V
15–20 −2.5 V
15–21 4000 A/s, decreasing

15-24 51.5 µH
15-25 (c) 1000 µs
       (d) 20 mA
15-27 (a) 12.6 mA
       (b) 6.9 V
       (c) 0.08 ms
15-28 1.25 A
15-30 0.22 ms
15-32 6000 V
15-33 Coil voltage equals 200 V
15-34 0.535 A
15-35 (a) 13.7 µs
       (b) 37.5 µJ
15-38 (a) $V_{xy} = 50$ V
       (b) $V_{xy} = 17.5$ V
15-39 80 V, 120 V
15-41 15 mH
15-42 1.25 mH

Chapter Sixteen

16-3 45°
16-4 30°
16-5 0.766; 0.94; −0.292; −0.73; −0.94
16-6 cos 36° = 0.809
      cos 226° = −0.695
      tan 152° = −0.53
      tan −65° = −2.15
16-7 (a) 0.68 rad
      (b) 2.04 rad
      (c) 86°
      (d) 36°
      (e) −540°
16-9 (a) 16 mA
      (b) −30 mA
16-11 72°, 76 V
16-12 500 Hz; 10 kHz
16-14 5 µs
16-15 −8.66 mA
16-16 118 V
16-17 637 Hz
16-18 (a) 25 V
      (b) 2.5 ms
      (c) 400 Hz
      (d) 2512 rad/s
16-20 216 mW
16-21 106 V
16-22 (a) 35.4 V
      (b) 7.07 mA
16-23 211 V; 17.6 mA
16-24 $i = 2.83 \sin(2\pi \times 60t)$
16-25 0.178 A$_{p-p}$
16-26 3.7 W
16-27 2.12 A
16-28 75.6 Ω
16-29 A leads B by 67.5°
16-32 (b) 20.5 mA, −206.6 mA

Chapter Seventeen

17-2 brightest—c
      dimmest—a
17-3 (a) 37.7 Ω
      (b) 942 Ω
17-4 (a) 0.0126 Ω
      (b) 0.314 Ω
17-5 (a) 6.6 mA
      (b) 0.33 A
      (c) 20 mA
17-6 0.8 H
17-7 190 µH
17-9 200 kHz
17-10 4.7 mV
17-11 (a) 0.08 A
      (b) 4 V, 8 V
17-12 (a) 14.3 Ω
      (b) 0.42 A
      (c) 0.3 A; 0.12 A
17-16 0.188 V [sin (377t + π/2)]
17-17 0.6 A [sin (25120t − π/2)]
17-21 brightest—a
      dimmest—c
17-22 (a) 26.5 kΩ
      (b) 1.06 kΩ
17-23 (a) 79.5 MΩ
      (b) 3.18 MΩ
17-24 (a) 315 mA
      (b) 6.3 mA
      (c) 126 mA
17-25 0.016 µF
17-26 0.022 µF
17-27 250 Ω
17-28 80 Hz
17-29 384 V
17-30 (a) 0.4 A
      (b) $V_1 = 8$ V, $V_2 = 4$ V
17-31 0.42 A, 0.12 A, 0.3 A
17-36 0.8 V [sin (377t − π/2)]
17-37 4.5 mA [sin (6280t + π/2)]
17-43 8000 pF
17-44 1.6 kΩ resistor

Chapter Eighteen

18-3 100 V peak, −30°
18-5 223 V, −63.5°
18-6 1118 Ω, −63.5°
18-7 $\overline{E_S} = 223$ V $\underline{/-63.5°}$
18-10 (a) 707 Ω
      (b) −54°
      (c) $R = 416, C = 2.8$ µF
18-12 (a) 300 V $\underline{/-15°}$
      (c) 25 mA $\underline{/50°}$
18-13 (a) $\overline{Z} = 318$ kΩ $\underline{/-86.4°}$
      at $f = 1$ Hz
      (c) $f_T = 15.9$ Hz; $Z = 28.2$ kΩ; θ = −45°
18-15 $R = 1$ kΩ, $C = 0.32$ µF
18-16 $R = 50$ kΩ, $C = 0.01$ µF
18-17 102.1 pF
18-18 11 µF
18-19 21.6 dB; −40 dB; 77 dB; −10.5 dB;
      100 dB; −60 dB
18-20 5; 0.001; 0.71; 10^6
18-21 100 V
18-22 1000 pF, 1590 Ω
18-25 0.2 µF, 500 Ω
18-28 (b) 1000 Hz
      (c) 1 kΩ, 0.16 µF

18–31 (a) 17.9 mA
(b) 63.5°
(c) 1.8 kΩ $/-63.5°$
18–33 7 kΩ, 0.057 µF
18–36 21.2 µF
18–37 (b) 2 kΩ, 0.08 µF

Chapter Nineteen

19–2 13 V, 22.6°
19–3 (a) $E_S = 223.6$ V
19–4 1116 Ω $/63.4°$
19–6 (a) 590 Ω
(b) 40.6 mA, 20.3 V, 12.8 V
(c) $/-32.1°$
19–7 27 µH, 212 Ω
19–10 $Z = 5$ kΩ $/0.6°$ at 80 Hz
19–12 0.92 H
19–13 60 Ω, 0.64 mH
19–16 4 kΩ, 0.318 H
19–19 0.318 H
19–21 (a) 17.9 mA
(b) 63.5°
(c) 1.8 kΩ $/63.5°$
19–24 5.66 kΩ, 2.25 H
19–25 7 kΩ, 11.1 H
19–26 2 kΩ, 32 mH

Chapter Twenty

20–1 -4 mA $/32°$, 4 mA $/-148°$ or 4 mA $/212°$
20–4 (a) $(20 + j27)$ Ω
(c) $(20 + j20)$ Ω
(e) $(90 + j0)$ Ω
20–5 2853 Ω $- j927$ Ω
20–6 244Ω $+ j548$Ω
20–8 (a) $(19.2 + j14.4)$ V
20–9 (a) 22.4 Ω $/-26.6°$
(b) 20 A $/+53°$
(c) 17 V $/135°$
(d) 6 mA $/-90°$
20–10 (a) 33.6 Ω $/53.5°$
(c) 28.2 Ω $/45°$
20–11 (a) 3000 $/0°$
(b) 149.6 $/18.6°$
(c) $-1 /-62°$
(d) 121.5 $/76°$
20–12 (a) 1700 Ω $/62°$, 1300 Ω $/-22.6°$
(b) 20.6 mA $/-62°$, 26.9 mA $/22.6°$
(c) 35.4 mA $/-12.8°$
(d) 989 Ω $/12.8°$
20–14 42 V $/29.5°$

Chapter Twenty-one

21–1 (a) 0.39 A $/-45.7°$
(b) $V_R = 11.7$ V, $V_L = 97.6$ V
$V_C = 61.9$ V
21–4 796 Hz; 90 Ω; 0.555 A; 111 V; 111 V
21–5 2.22
21–6 10 mV, 1.2 V
21–7 (a) 80
(b) 65 MHz
(c) 10.2 Ω

21–8 2.5 mH
21–10 BW = 795 Hz
21–13 75 Ω
21–14 4 Ω
21–16 22.5
21–17 40; 30 Ω
21–18 10.9 pF to 16.4 pF
21–21 (a) 1920 V
(b) 22 Ω
21–22 (a) $\overline{I_T} = 15.2$ mA $/-49°$
(b) $\overline{I_T} = 10$ mA $/0°$
(c) $\overline{I_T} = 11.1$ mA $/26.1°$
21–24 (a) 637 kHz
(b) 80 kΩ; 0.2 mA
(c) 40
21–26 63.7 mH; 250 Ω; 159 pF
21–27 50 kHz; 80; 250 Ω
21–28 BW = 6.4 kHz
21–30 12.5
21–32 12
21–34 600 kΩ

Chapter Twenty-two

22–3 c
22–4 c
22–7 $f_{hc} = 31.8$ kHz at -40 dB/decade
22–8 $C \geq 21$ µF
22–10 (a) 408 pF
(b) 105
22–12 1 kΩ, 1836 pF
22–16 a, b

Chapter Twenty-three

23–1 $\overline{Z_T} = 220$ Ω $/39.5°$
$\overline{I_T} = 0.227$ A $/-39.5°$
23–3 $\overline{I_T} = 0.82$ A $/21.4°$
23–4 $\overline{Z_T} = 63.1$ Ω $/14.6°$
23–6 0.603 mS (cap.)
23–7 180 mA $/56.3°$
23–8 192 Ω $+ j144$ Ω
23–9 $\overline{I_T} = 0.39$ A $/-23°$
$\overline{V_{XY}} = 14.5$ V $/6.6°$
23–10 0.4 A $/-41.3°$
23–13 14.5 V $/6.6°$, 0.42 A $/-38.4°$
23–18 (a) 100 Ω $/-90°$
(b) 1.6 µF
23–19 5 Ω
23–20 (a) 200 Ω
(b) -1.27 V
(c) $+1$ V
23–21 833 pF

Chapter Twenty-four

24–1 242 W, 64.2 vars, 228.3 vars
24–4 2.86 W, 5.76 vars
24–7 72 VA
24–8 2.3 W, 2.3 VA, 3.24 VA
24–9 $\overline{Z_T} = 2.2$ Ω $/79.5°$
24–10 0.707 leading; 0.182 lagging
24–11 450 W
24–12 0.78 lagging

24–13 0.866 leading
24–15 (a) 204.8 W; 38.1 vars; 208.3 VA
       (b) 2.6 A
24–16 PF = 99.4%
24–17 946.4 pF; 2.56 A
24–18 55.4 mH; 47.3 A
24–21 15.32 W
24–23 (a) 20 Ω
       (b) 22.5 Ω

Chapter Twenty-five

25–4   13.75 V
25–5   100 turns
25–6   27,000 turns; 75V
25–8   (a) 120 V
       (b) 2.4 A
       (c) 288 W
       (d) 24 A
       (e) 288 W
25–9   (a) $E_S = 4$ V; $I_S = 40$ mA
       $I_p = 13.3$ mA
       (c) $I_S = 3$ A; $E_S = 30$ V
25–11 (a) 180 V $\underline{/0°}$; 0.66 A $\underline{/-56.4°}$
       (b) 6.6 A $\underline{/-56.4°}$
       (c) 66.6 W each
25–14 2.4 Ω
25–15 160 turns
25–16 (a) 42.7 mW
       (b) 6.32
       (c) 451 mW
25–17 6.42 A
25–18 6.13 mA
25–21 9 A; 27 A
25–23 0.68 A
25–24 1.62 W; 1.64 W
25–25 3.45 V
25–27 4.12 V
25–28 24.3 V
25–33 (a) 0.96 mH
       (b) 9.6 V
       (c) 0.24 V $\underline{/0°}$

Chapter Twenty-six

26–6   c
26–7   (a) +3 V
       (b) −6.5 V
26–11 10 kHz
26–12 100 Hz
26–13 $Q = 10$; $L = 5.3$ mH
       $C = 0.53$ μF

INDEX

Admittance, 666
Alternating current, 66, 258
Ammeter, 98, 791
Ampere, 5, 25
Amplitude, 262
Angular frequency, 439
Apparent power, 693
Atom, 8
Attenuators, 117
Autotransformer, 734
Average value, 764

Band-pass, 641
Band-reject, 641
Bandwidth
   of parallel resonant circuit, 627
   of series resonant circuit, 609
B-H curves
   described, 358
   hysteresis, 360
Branch current method, 241
Bridge circuit, 182, 679

Capacitance
   defined, 286
   formula, 319
   parasitic, 327
   physical factors, 319
Capacitors
   charging, 288
   combinations of, 308
   defined, 285

   discharging, 301
   energy, 322
   leakage, 320
   phase angle, 479
   reactance, 473
   time constant, 297
Charges, 6-8
Coils (*See* Inductor)
Conductance
   defined, 73
   in parallel, 149
Conductors, 10
Coulomb, 16
Coulomb's law, 8, 16
Coupling coefficient, 740
Current
   ampere, 5, 25
   definition, 21
   direction, 23, 64
   divider, 156
Current sources
   ideal, 66
   practical, 223
Current-voltage graphs
   linear, 71
   nonlinear, 83
Cycle, 266

Damping, 631
Decibels, 508
Delta-wye transformations
   impedance circuits, 678
   resistive circuits, 236

Dielectric breakdown, 320
Direct current, 66, 258
Domain theory of magnetism, 342
Dynamic resistance and conductance, 85

Eddy currents, 546
Electric current (*See* Current)
Electric field intensity, 284
Electricity, 1
Electromagnet induction, 375
Electromagnetism, 345
Electromotive force (emf), 29
Electron
  charge, 8
  valence, 9
Electronics, 1
Electrostatic force, 283
Energy
  capacitor, 322
  definition, 3
  inductor, 407
  kinetic, 4
  potential, 4
Exponential form, 308, 400
Exponential function, 308
Exponential waveform, 260

Faraday's law, 377
Ferromagnetism, 342
Filters
  band-pass, 651
  general, 510
  RC high pass, 511
  RC low pass, 515
  resonant, 652
  RL high pass, 552
  RL low pass, 555
  types, 641
Flux (*See* Magnetic field lines)
Flux density, 352
Flux linkages, 376
Force, 2
Frequency, 266
Frequency response, 512
Functions, 253
Fundamental harmonic, 757
Fuses, 213

Graphs, 254
Ground, 62

Harmonics
  defined, 757
  fundamental, 757
Hysteresis
  described, 360
  effects in inductor, 546

Impedance
  defined, 494
  of parallel *RC*, 521
  of parallel *RL*, 560
  of series *RC*, 494
  of series *RL*, 542
Inductance
  factors determining, 391
  formula, 385
  phase angle, 467
  reactance, 459
  self, 384
Induction, 375
  mutual, 378, 713
  self, 384
Inductive kick, 405
Inductor
  combinations of, 409
  defined, 385
  discharging an, 403
  energy, 407
  phase angle, 467
  practical, 388
  rise of current in, 392
  time constant, 395
Insulator, 10
Insulator breakdown, 25
Internal resistance
  defined, 128
  measurement of, 132

J-operator, 576
Joules, 6

Kinetic energy, 4
Kirchhoff's current law (KCL), 153
Kirchhoff's voltage law (KVL)
  algebraic method, 114

Kirchhoff's voltage law (KVL) (*cont.*)
   general form of, 111
   stated, 106

Leakage flux
   defined, 351
   in transformers, 740
Left-hand rule, 345, 348
Lenz's law, 381
Linearity principle, 110, 178

Magnetic field intensity, H, 355
Magnetic field lines (flux)
   around wires, 345
   described, 340
   symbol, 351
Magnetic permeability (μ), 357
Magnetism, 337
Magnets, 338
Mass, 1
Maximum power transfer
   impedance circuit, 706
   resistive circuit, 199
Maxwell bridge, 683
Meters, 791
MMF, 354
Mutual inductance, 746
Mutual induction, 378, 713

Node voltage method, 243
Norton's theorem
   impedance circuits, 677
   resistive circuits, 232

Ohm, 30, 73
Ohmmeter, 797
Ohm's law, 30, 75
Ohm's law for magnetic circuits, 362
Open circuits, 83

Parallel circuits
   defined, 143
   resistance of, 147
Period, 266
Phase angle
   of capacitor, 479
   defined, 447
   of inductor, 467
Phase-shifting networks, 654

Phasor
   defined, 430
   notation, 497
Potential, 18
Potential difference, 19
Potential energy
   definition, 4
   electrostatic, 17
Power
   in ac circuits, 689
   apparent, 693
   conversion, 198
   defined, 31, 191
   dissipation, 194
   ratings, 197
   reactive, 691
   real (true), 690
   in a sine wave, 445
   watt, 32
Power factor, 698
Power triangle, 695
Pulses, 263

Quality factor, Q
   of coil, 613
   in parallel RLC, 626
   in series RLC, 606

Radian, 429
Reactance
   capacitive, 473
   inductive, 459
Reactive power, 691
Rectangular-to-polar, 584
Reluctance, 363
Resistance
   definition, 73
   dynamic, 85
   equation, 207
   internal, 128
   source, 128
   static, 85
   temperature effects, 212
Resistivity, 207
Resistors
   in parallel, 143
   in series, 98
Resonance
   parallel, 620

Resonance (*cont.*)
   series, 602
Rms amplitude, 440
Roll-off, 645

Scientific notation, 39
Semiconductors, 10
Series circuits
   defined, 95
   resistance of, 98
Series-parallel circuits
   defined, 169
   resistance of, 176
   troubleshooting, 180
Series resonance, 602
Short circuits, 82
Siemens, 75
Sine wave
   description, 260
   function, 430
Sinusoidal waveform (*See* Sine wave)
Skin effect, 545
Slope, 255
Solenoids, 348
Square wave
   described, 268
   harmonic analysis of, 761
Superposition method
   for impedance circuits, 672
   limitations of, 221
   for resistive circuits, 219
Susceptance, 666

Tank circuit, 621
Thevenin's theorem
   impedance circuits, 675
   limitations of, 227
   resistive circuits, 223
Time constants
   capacitor, 297
   graphs, 303, 396
   inductor, 395
Transformers, 713
   air-core, 746
   auto-transformer, 734
   basic principle, 714
   efficiency, 738
   frequency response, 744
   loaded, 719, 741
   multiple secondary, 741
   power losses, 737
   turns ratio, 716
Transition frequency, 504
Trig functions, 426
Tuning, 616

Units
   International System of, 5
   prefixed, 47

Vector algebra, 573
Volt, 20
Voltage
   conventions, 57
   definition, 20
   sources, 20
Voltage dividers
   impedances, 663
   loaded, 183
   unloaded, 117
Voltage sources, 20
   ideal, 66
   internal resistance, 127
   in parallel, 159
   practical, 127, 223
   in series, 62
Voltmeter, 146, 794

Wattmeter, 707, 801
Waveforms, 65, 253
   applied to resistors, 273
   basic types, 259
   periodic, 266
   sinusoidal, 260, 421
   sources of, 269
Weber, 351
Wheatstone bridge, 681
Wire tables, 209
Work, 3

```
621.319      T631I      87-01595
TOCCI  RONALD J
INTRODUCTION TO ELECTRIC C
COPY     0001
```